w should we model the opportunities created by access to data and to computing, how robust and resilient are these models? What information is needed for analysis, is it obtained and applied? If you are looking for answers to these questions, then e is no better place to start than this book."

Robert Calderbank, *Duke University*

s pioneering book elucidates the emerging cross-disciplinary area of information-retic data science. The book's contributors are leading information theorists who pellingly explain the close connections between the fields of information theory data science. Readers will learn about theoretical foundations for designing data ction and analysis systems, from data acquisition to data compression, computation, learning."

Al Hero, *University of Michigan*

Information-Theoretic Methods in Data Sc

Learn about the state-of-the-art at the interfac
science with this first unified treatment of th
in a clear, tutorial style, and using consistent
shows how information-theoretic methods ar
representation, data analysis, and statistics and

Coverage is broad, with chapters on sign
compressive sensing, data communication, rep
statistics, and much more. Each chapter includ
problems, emerging and open problems, and an
to develop in-depth knowledge and understand

Providing a thorough survey of the current r
is essential reading for graduate students and re
signal processing, machine learning, and statis

Miguel R. D. Rodrigues is Professor of Informat
ment of Electronic and Electrical Engineering,
Fellow at the Alan Turing Institute, London.

Yonina C. Eldar is a Professor in the Faculty
the Weizmann Institute of Science, a Fellow o
the Israel Academy of Sciences and Humanit
(Cambridge, 2015), and co-editor of *Compre*
(Cambridge, 2019), *Compressed Sensing* (Ca
in Signal Processing and Communications (C

Information-Theoretic Methods in Data Science

Edited by
MIGUEL R. D. RODRIGUES
University College London

YONINA C. ELDAR
Weizmann Institute of Science

CAMBRIDGE
UNIVERSITY PRESS

CAMBRIDGE
UNIVERSITY PRESS

University Printing House, Cambridge CB2 8BS, United Kingdom

One Liberty Plaza, 20th Floor, New York, NY 10006, USA

477 Williamstown Road, Port Melbourne, VIC 3207, Australia

314–321, 3rd Floor, Plot 3, Splendor Forum, Jasola District Centre, New Delhi – 110025, India

79 Anson Road, #06-04/06, Singapore 079906

Cambridge University Press is part of the University of Cambridge.

It furthers the University's mission by disseminating knowledge in the pursuit of education, learning, and research at the highest international levels of excellence.

www.cambridge.org
Information on this title: www.cambridge.org/9781108427135
DOI: 10.1017/9781108616799

First published 2021

Printed in the United Kingdom by TJ Books Ltd, Padstow Cornwall

A catalogue record for this publication is available from the British Library.

ISBN 978-1-108-42713-5 Hardback

To my husband Shalomi and children Yonatan, Moriah, Tal, Noa, and Roei for their boundless love and for filling my life with endless happiness

YE

To my wife Eduarda and children Isabel and Diana for their unconditional love, encouragement, and support

MR

Contents

Preface

Since its introduction in 1948, the field of information theory has proved instrumental in the analysis of problems pertaining to compressing, storing, and transmitting data. For example, information theory has allowed analysis of the fundamental limits of data communication and compression, and has shed light on practical communication system design for decades. Recent years have witnessed a renaissance in the use of information-theoretic methods to address problems beyond data compression, data communications, and networking, such as compressive sensing, data acquisition, data analysis, machine learning, graph mining, community detection, privacy, and fairness. In this book, we explore a broad set of problems on the interface of signal processing, machine learning, learning theory, and statistics where tools and methodologies originating from information theory can provide similar benefits. The role of information theory at this interface has indeed been recognized for decades. A prominent example is the use of information-theoretic quantities such as mutual information, metric entropy and capacity in establishing minimax rates of estimation back in the 1980s. Here we intend to explore modern applications at this interface that are shaping data science in the twenty-first century.

There are of course some notable differences between standard information-theoretic tools and signal-processing or data analysis methods. Globally speaking, information theory tends to focus on asymptotic limits, using large blocklengths, and assumes the data is represented by a finite number of bits and viewed through a noisy channel. The standard results are not concerned with complexity but focus more on fundamental limits characterized via achievability and converse results. On the other hand, some signal-processing techniques, such as sampling theory, are focused on discrete-time representations but do not necessarily assume the data is quantized or that there is noise in the system. Signal processing is often concerned with concrete methods that are optimal, namely, achieve the developed limits, and have bounded complexity. It is natural therefore to combine these tools to address a broader set of problems and analysis which allows for quantization, noise, finite samples, and complexity analysis.

This book is aimed at providing a survey of recent applications of information-theoretic methods to emerging data-science problems. The potential reader of this book could be a researcher in the areas of information theory, signal processing, machine learning, statistics, applied mathematics, computer science or a related research area, or a graduate student seeking to learn about information theory and data science and to scope out open problems at this interface. The particular design of this volume ensures that it can serve as both a state-of-the-art reference for researchers and a textbook for students.

The book contains 16 diverse chapters written by recognized leading experts worldwide, covering a large variety of topics that lie on the interface of signal processing, data science, and information theory. The book begins with an introduction to information theory which serves as a background for the remaining chapters, and also sets the notation to be used throughout the book. The following chapters are then organized into four categories: data acquisition (Chapters 2–4), data representation and analysis (Chapters 5–9), information theory and machine learning (Chapters 10 and 11), and information theory, statistics, and compression (Chapters 12–15). The last chapter, Chapter 16, connects several of the book's themes via a survey of Fano's inequality in a diverse range of data-science problems. The chapters are self-contained, covering the most recent research results in the respective topics, and can all be treated independently of each other. A brief summary of each chapter is given next.

Chapter 1 by Rodrigues, Draper, Bajwa, and Eldar provides an introduction to information theory concepts and serves two purposes: It provides background on classical information theory, and presents a taster of modern information theory applied to emerging data-science problems.

Chapter 2 by Kipnis, Eldar, and Goldsmith extends the notion of rate-distortion theory to continuous-time inputs deriving bounds that characterize the minimal distortion that can be achieved in representing a continuous-time signal by a series of bits when the sampler is constrained to a given sampling rate. For an arbitrary stochastic input and given a total bitrate budget, the authors consider the lowest sampling rate required to sample the signal such that reconstruction of the signal from a bit-constrained representation of its samples results in minimal distortion. It turns out that often the signal can be sampled at sub-Nyquist rates without increasing the distortion.

Chapter 3 by Jalali and Poor discusses the interplay between compressed sensing and compression codes. In particular, the authors consider the use of compression codes to design compressed sensing recovery algorithms. This allows the expansion of the class of structures used by compressed sensing algorithms to those used by data compression codes, which is a much richer class of inputs and relies on decades of developments in the field of compression.

Chapter 4 by Pilanci develops information-theoretical lower bounds on sketching for solving large statistical estimation and optimization problems. The term sketching is used for randomized methods that aim to reduce data dimensionality in computationally intensive tasks for gains in space, time, and communication complexity. These bounds allow one to obtain interesting trade-offs between computation and accuracy and shed light on a variety of existing methods.

Chapter 5 by Shakeri, Sarwate, and Bajwa treats the problem of dictionary learning, which is a powerful signal-processing approach for data-driven extraction of features from data. The chapter summarizes theoretical aspects of dictionary learning for vector- and tensor-valued data and explores lower and upper bounds on the sample complexity of dictionary learning which are derived using information-theoretic tools. The dependence of sample complexity on various parameters of the dictionary learning problem is highlighted along with the potential advantages of taking the structure of tensor data into consideration in representation learning.

Chapter 6 by Riegler and Bölcskei presents an overview of uncertainty relations for sparse signal recovery starting from the work of Donoho and Stark. These relations are then extended to richer data structures and bases, which leads to the recently discovered set-theoretic uncertainty relations in terms of Minkowski dimension. The chapter also explores the connection between uncertainty relations and the "large sieve," a family of inequalities developed in analytic number theory. It is finally shown how uncertainty relations allow one to establish fundamental limits of practical signal recovery problems such as inpainting, declipping, super-resolution, and denoising of signals.

Chapter 7 by Reeves and Pfister examines high-dimensional inference problems through the lens of information theory. The chapter focuses on the standard linear model for which the performance of optimal inference is studied using the replica method from statistical physics. The chapter presents a tutorial of these techniques and presents a new proof demonstrating their optimality in certain settings.

Chapter 8 by Shah discusses the question of learning distributions over permutations of a given set of choices based on partial observations. This is central to capturing choice in a variety of contexts such as understanding preferences of consumers over a collection of products based on purchasing and browsing data in the setting of retail and e-commerce. The chapter focuses on the learning task from marginal distributions of two types, namely, first-order marginals and pair-wise comparisons, and provides a comprehensive review of results in this area.

Chapter 9 by Raman and Varshney studies universal clustering, namely, clustering without prior access to the statistical properties of the data. The chapter formalizes the problem in information theory terms, focusing on two main subclasses of clustering that are based on distance and dependence. A review of well-established clustering algorithms, their statistical consistency, and their computational and sample complexities is provided using fundamental information-theoretic principles.

Chapter 10 by Raginsky, Rakhlin, and Xu introduces information-theoretic measures of algorithmic stability and uses them to upper-bound the generalization bias of learning algorithms. The notion of stability implies that its output does not depend too much on any individual training example and therefore these results shed light on the generalization ability of modern learning techniques.

Chapter 11 by Piantanida and Vega introduces the information bottleneck principle and explores its use in representation learning, namely, in the development of computational algorithms that learn the different explanatory factors of variation behind high-dimensional data. Using these tools, the authors obtain an upper bound on the generalization gap corresponding to the cross-entropy risk. This result provides an interesting connection between mutual information and generalization, and helps to explain why noise injection during training can improve the generalization ability of encoder models.

Chapter 12 by Ding, Yang, and Tarokh discusses fundamental limits of inference and prediction based on model selection principles from modern data analysis. Using information-theoretic tools the authors analyze several state-of-the-art model selection techniques and introduce two recent advances in model selection approaches, one concerning a new information criterion and the other concerning modeling-procedure selection.

Chapter 13 by Wu and Xu provides an exposition on some of the methods for determining the information-theoretical as well as computational limits for high-dimensional statistical problems with a planted structure. Planted structures refer to a ground truth structure (often of a combinatorial nature) which one is trying to discover in the presence of random noise. In particular, the authors discuss first- and second-moment methods for analyzing the maximum likelihood estimator, information-theoretic methods for proving impossibility results using mutual information and rate-distortion theory, and techniques originating from statistical physics. To investigate computational limits, they describe randomized polynomial-time reduction schemes that approximately map planted-clique problems to the problem of interest in total variation distance.

Chapter 14 by Zhao and Lai considers information-theoretic models for distributed statistical inference problems with compressed data. The authors review several research directions and challenges related to applying these models to various statistical learning problems. In these applications, data are distributed in multiple terminals, which can communicate with each other via limited-capacity channels. Information-theoretic tools are used to characterize the fundamental limits of the classical statistical inference problems using compressed data directly.

Chapter 15 by Feizi and Médard treats different aspects of the network functional compression problem. The goal is to compress a source of random variables for the purpose of computing a deterministic function at the receiver where the sources and receivers are nodes in a network. Traditional data compression schemes are special cases of functional compression, in which the desired function is the identity function. It is shown that for certain classes of functions considerable compression is possible in this setting.

Chapter 16 by Scarlett and Cevher provides a survey of Fano's inequality and its use in various statistical estimation problems. In particular, the chapter overviews the use of Fano's inequality for establishing impossibility results, namely, conditions under which a certain goal cannot be achieved by any estimation algorithm. The authors present several general-purpose tools and analysis techniques, and provide representative examples covering group testing, graphical model selection, sparse linear regression, density estimation, and convex optimization.

Within the chapters, the authors point to various open research directions at the interface of information theory, data acquisition, data analysis, machine learning, and statistics that will certainly see increasing attention in the years to come.

We would like to end by thanking all the authors for their contributions to this book and for their hard work in presenting the material in a unified and accessible fashion.

Notation

z	scalar (or value of random variable Z)
Z	random variable
$\mathbf{z}$	vector (or value of random vector $\mathbf{Z}$)
$\mathbf{Z}$	matrix (or random vector)
$\mathbf{z}_i$	ith entry of vector $\mathbf{z}$
$\mathbf{Z}_{i,j}$	(i, j)th entry of matrix $\mathbf{Z}$
$Z^n = (Z_1, \ldots, Z_n)$	sequence of n random variables
$z^n = (z_1, \ldots, z_n)$	value of sequence of n random variables Z^n
$Z_i^j = \left(Z_i, \ldots, Z_j\right)$	sequence of $j - i + 1$ random variables
$z_i^j = \left(z_i, \ldots, z_j\right)$	value of sequence of $j - i + 1$ random variables Z_i^j
$\lVert \cdot \rVert_p$	p-norm
$(\cdot)^{\mathrm{T}}$	transpose operator
$(\cdot)^*$	conjugate Hermitian operator
$(\cdot)^\dagger$	pseudo-inverse of the matrix argument
$\mathrm{tr}(\cdot)$	trace of the square matrix argument
$\det(\cdot)$	determinant of the square matrix argument
$\mathrm{rank}(\cdot)$	rank of the matrix argument
$\mathrm{range}(\cdot)$	range span of the column vectors of the matrix argument
$\lambda_{\max}(\cdot)$	maximum eigenvalue of the square matrix argument
$\lambda_{\min}(\cdot)$	minimum eigenvalue of the square matrix argument
$\lambda_i(\cdot)$	ith largest eigenvalue of the square matrix argument
$\mathbf{I}$	identity matrix (its size is determined from the context)
$\mathbf{0}$	matrix with zero entries (its size is determined from the context)
$\mathcal{T}$	standard notation for sets
$\lvert\mathcal{T}\rvert$	cardinality of set $\mathcal{T}$
$\mathbb{R}$	set of real numbers
$\mathbb{C}$	set of complex numbers
$\mathbb{R}^n$	set of n-dimensional vectors of real numbers
$\mathbb{C}^n$	set of n-dimensional vectors of complex numbers
j	imaginary unit
$\mathrm{Re}(x)$	real part of the complex number x
$\mathrm{Im}(x)$	imaginary part of the complex number x
$\lvert x\rvert$	modulus of the complex number x
$\arg(x)$	argument of the complex number x
$\mathbb{E}[\cdot]$	statistical expectation
$\mathbb{P}[\cdot]$	probability measure

$H(\cdot)$	entropy
$H(\cdot\|\cdot)$	conditional entropy
$h(\cdot)$	differential entropy
$h(\cdot\|\cdot)$	conditional differential entropy
$D(\cdot\\|\cdot)$	relative entropy
$I(\cdot;\cdot)$	mutual information
$I(\cdot;\cdot\|\cdot)$	conditional mutual information
$\mathcal{N}(\mu,\sigma^2)$	scalar Gaussian distribution with mean μ and variance σ^2
$\mathcal{N}(\mu,\mathbf{\Sigma})$	multivariate Gaussian distribution with mean μ and covariance matrix $\mathbf{\Sigma}$

Contributors

Waheed U. Bajwa
Department of Electrical and Computer Engineering
Rutgers University

Helmut Bölcskei
Department of Information Technology and Electrical Engineering
Department of Mathematics
ETH Zürich

Volkan Cevher
Laboratory for Information and Inference Systems
Institute of Electrical Engineering
School of Engineering, EPFL

Jie Ding
School of Statistics
University of Minnesota

Stark C. Draper
Department of Electrical and Computer Engineering
University of Toronto

Yonina C. Eldar
Faculty of Mathematics and Computer Science
Weizmann Institute of Science

Soheil Feizi
Department of Computer Science
University of Maryland, College Park

Andrea J. Goldsmith
Department of Electrical Engineering
Stanford University

Shirin Jalali
Nokia Bell Labs

Alon Kipnis
Department of Statistics
Stanford University

Lifeng Lai
Department of Electrical and Computer Engineering
University of California, Davis

Muriel Médard
Department of Electrical Engineering and Computer Science
Massachussets Institute of Technology

Henry D. Pfister
Department of Electrical and Computer Engineering
Duke University

Pablo Piantanida
Laboratoire des Signaux et Systèmes
Université Paris Saclay
CNRS-CentraleSupélec

Montreal Institute for Learning Algorithms (Mila), Université de Montréal

Mert Pilanci
Department of Electrical Engineering
Stanford University

H. Vincent Poor
Department of Electrical Engineering
Princeton University

Maxim Raginsky
Department of Electrical and Computer Engineering
University of Illinois at Urbana-Champaign

Alexander Rakhlin
Department of Brain & Cognitive Sciences
Massachusetts Institute of Technology

Ravi Kiran Raman
Department of Electrical and Computer Engineering
University of Illinois at Urbana-Champaign

Galen Reeves
Department of Electrical and Computer Engineering
Department of Statistical Science
Duke University

Leonardo Rey Vega
Department of Electrical Engineering
School of Engineering
Universidad de Buenos Aires
CSC-CONICET

Erwin Riegler
Department of Information Technology and Electrical Engineering
ETH Zürich

Miguel R. D. Rodrigues
Department of Electronic and Electrical Engineering
University College London

Anand D. Sarwate
Department of Electrical and Computer Engineering
Rutgers University

Jonathan Scarlett
Department of Computer Science and Department of Mathematics
National University of Singapore

Devavrat Shah
Department of Electrical Engineering and Computer Science
Massachusetts Institute of Technology

Zahra Shakeri
Electronic Arts

Vahid Tarokh
Department of Electrical and Computer Engineering
Duke University

Lav R. Varshney
Department of Electrical and Computer Engineering
University of Illinois at Urbana-Champaign

Yihong Wu
Department of Statistics and Data Science
Yale University

Aolin Xu
Department of Electrical and Computer Engineering
University of Illinois at Urbana-Champaign

Jiaming Xu
Fuqua School of Business
Duke University

Yuhong Yang
School of Statistics
University of Minnesota

Wenwen Zhao
Department of Electrical and Computer Engineering
University of California, Davis

1 Introduction to Information Theory and Data Science

Miguel R. D. Rodrigues, Stark C. Draper, Waheed U. Bajwa, and Yonina C. Eldar

Summary

The field of information theory – dating back to 1948 – is one of the landmark intellectual achievements of the twentieth century. It provides the philosophical and mathematical underpinnings of the technologies that allow accurate representation, efficient compression, and reliable communication of sources of data. A wide range of storage and transmission infrastructure technologies, including optical and wireless communication networks, the internet, and audio and video compression, have been enabled by principles illuminated by information theory. Technological breakthroughs based on information-theoretic concepts have driven the "information revolution" characterized by the anywhere and anytime availability of massive amounts of data and fueled by the ubiquitous presence of devices that can capture, store, and communicate data.

The existence and accessibility of such massive amounts of data promise immense opportunities, but also pose new challenges in terms of how to extract useful and actionable knowledge from such data streams. Emerging data-science problems are different from classical ones associated with the transmission or compression of information in which the semantics of the data was unimportant. That said, we are starting to see that information-theoretic methods and perspectives can, in a new guise, play important roles in understanding emerging data-science problems. The goal of this book is to explore such new roles for information theory and to understand better the modern interaction of information theory with other data-oriented fields such as statistics and machine learning.

The purpose of this chapter is to set the stage for the book and for the upcoming chapters. We first overview classical information-theoretic problems and solutions. We then discuss emerging applications of information-theoretic methods in various data-science problems and, where applicable, refer the reader to related chapters in the book. Throughout this chapter, we highlight the perspectives, tools, and methods that play important roles in classic information-theoretic paradigms and in emerging areas of data science. Table 1.1 provides a summary of the different topics covered in this chapter and highlights the different chapters that can be read as a follow-up to these topics.

Table 1.1. Major topics covered in this chapter and their connections to other chapters

Section(s)	Topic	Related chapter(s)
1.1–1.4	An introduction to information theory	15
1.6	Information theory and data acquisition	2–4, 6, 16
1.7	Information theory and data representation	5, 11
1.8	Information theory and data analysis/processing	6–16

1.1 Classical Information Theory: A Primer

Claude Shannon's 1948 paper "A mathematical theory of communications," *Bell Systems Technical Journal*, July/Oct. 1948, laid out a complete architecture for digital communication systems [1]. In addition, it articulated the philosophical decisions for the design choices made. *Information theory*, as Shannon's framework has come to be known, is a beautiful and elegant example of engineering science. It is all the more impressive as Shannon presented his framework decades before the first digital communication system was implemented, and at a time when digital computers were in their infancy.

Figure 1.1 presents a general schematic of a digital communication system. This figure is a reproduction of Shannon's "Figure 1" from his seminal paper. Before 1948 no one had conceived of a communication system in this way. Today nearly all digital communication systems obey this structure.

The flow of information through the system is as follows. An *information source* first produces a random message that a transmitter wants to convey to a destination. The message could be a word, a sentence, or a picture. In information theory, all information sources are modeled as being sampled from a set of possibilities according to some probability distribution. Modeling information sources as stochastic is a key aspect of Shannon's approach. It allowed him to quantify uncertainty as the lack of knowledge and reduction in uncertainty as the gaining of knowledge or "information."

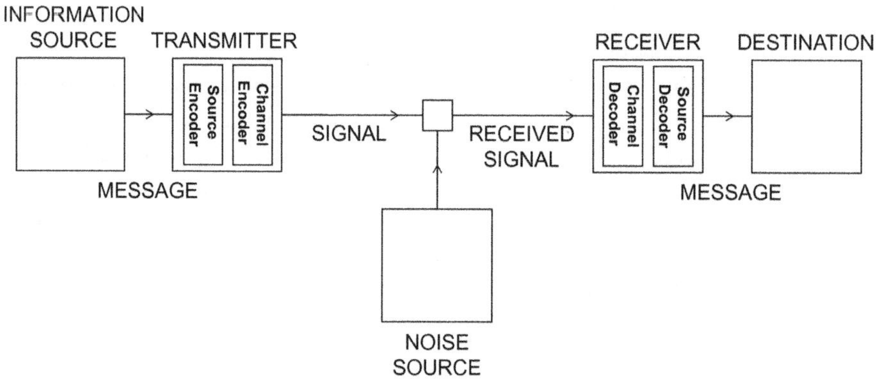

Figure 1.1 Reproduction of Shannon's Figure 1 in [1] with the addition of the source and channel encoding/decoding blocks. In Shannon's words, this is a "schematic diagram of a general communication system."

The message is then fed into a transmission system. The transmitter itself has two main sub-components: the *source encoder* and the *channel encoder*. The source encoder converts the message into a sequence of 0s and 1s, i.e., a bit sequence. There are two classes of source encoders. *Lossless source coding* removes predictable redundancy that can later be recreated. In contrast, *lossy* source coding is an irreversible process wherein some distortion is incurred in the compression process. Lossless source coding is often referred to as *data compression* while lossy coding is often referred to as *rate-distortion* coding. Naturally, the higher the distortion the fewer the number of bits required.

The bit sequence forms the data payload that is fed into a channel encoder. The output of the channel encoder is a signal that is transmitted over a noisy communication medium. The purpose of the *channel code* is to convert the bits into a set of possible signals or *codewords* that can be reliably recovered from the noisy received signal.

The communication medium itself is referred to as the *channel*. The channel can model the physical separation of the transmitter and receiver. It can also, as in data storage, model separation in time.

The destination observes a signal that is the output of the communication channel. Similar to the transmitter, the receiver has two main components: a *channel decoder* and a *source decoder*. The former maps the received signal into a bit sequence that is, one hopes, the same as the bit sequence produced by the transmitter. The latter then maps the estimated bit sequence to an estimate of the original message.

If lossless compression is used, then an apt *measure of performance* is the probability that the message estimate at the destination is not equal to the original message at the transmitter. If lossy compression (rate distortion) is used, then other measures of goodness, such as mean-squared error, are more appropriate.

Interesting questions addressed by information theory include the following.

1. Architectures
   - What trade-offs in performance are incurred by the use of the architecture detailed in Figure 1.1?
   - When can this architecture be improved upon; when can it not?
2. Source coding: lossless data compression
   - How should the information source be modeled; as stochastic, as arbitrary but unknown, or in some other way?
   - What is the shortest bit sequence into which a given information source can be compressed?
   - What assumptions does the compressor work under?
   - What are basic compression techniques?
3. Source coding: rate-distortion theory
   - How do you convert an analog source into a digital bitstream?
   - How do you reconstruct/estimate the original source from the bitstream?
   - What is the trade-off involved between the number of bits used to describe a source and the distortion incurred in reconstruction of the source?

4. Channel coding
   - How should communication channels be modeled?
   - What throughput, measured in bits per second, at what reliability, measured in terms of probability of error, can be achieved?
   - Can we quantify fundamental limits on the realizable trade-offs between throughput and reliability for a given channel model?
   - How does one build computationally tractable channel coding systems that "saturate" the fundamental limits?
5. Multi-user information theory
   - How do we design systems that involve multiple transmitters and receivers?
   - How do many (perhaps correlated) information sources and transmission channels interact?

The decades since Shannon's first paper have seen fundamental advances in each of these areas. They have also witnessed information-theoretic perspectives and thinking impacting a number of other fields including security, quantum computing and communications, and cryptography. The basic theory and many of these developments are documented in a body of excellent texts, including [2–9]. Some recent advances in network information theory, which involves multiple sources and/or multiple destinations, are also surveyed in Chapter 15. In the next three sections, we illustrate the basics of information-theoretic thinking by focusing on simple (point-to-point) binary sources and channels. In Section 1.2, we discuss the compression of binary sources. In Section 1.3, we discuss channel coding over binary channels. Finally, in Section 1.4, we discuss computational issues, focusing on linear codes.

1.2 Source Coding: Near-Lossless Compression of Binary Sources

To gain a feel for the tools and results of classical information theory consider the following lossless source coding problem. One observes a length-n string of random coin flips, $X_1, X_2, \ldots, X_n$, each $X_i \in \{\text{heads}, \text{tails}\}$. The flips are independent and identically distributed with $\mathbb{P}(X_i = \text{heads}) = p$, where $0 \leq p \leq 1$ is a known parameter. Suppose we want to map this string into a bit sequence to store on a computer for later retrieval. Say we are going to assign a *fixed* amount of memory to store the sequence. How much memory must we allocate?

Since there are 2^n possible sequences, all of which could occur if p is not equal to 0 or 1, if we use n bits we can be 100% certain we could index any heads/tails sequence that we might observe. However, certain sequences, while possible, are much less likely than others. Information theory exploits such non-uniformity to develop systems that can trade off between efficiency (the storage of fewer bits) and reliability (the greater certainty that one will later be able to reconstruct the observed sequence). In the following, we accept some (arbitrarily) small probability $\epsilon > 0$ of observing a sequence that we choose not to be able to store a description of.[1] One can think of ϵ as the probability of the system failing. Under this assumption we derive bounds on the number of bits that need to be stored.

1.2.1 Achievability: An Upper Bound on the Rate Required for Reliable Data Storage

To figure out which sequences we may choose not to store, let us think about the statistics. In expectation, we observe np heads. Of the 2^n possible heads/tails sequences there are $\binom{n}{np}$ sequences with np heads. (For the moment we ignore non-integer effects and deal with them later.) There will be some variability about this mean but, at a minimum, we must be able to store all these expected realizations since these realizations all have the same probability. While $\binom{n}{np}$ is the cardinality of the set, we prefer to develop a good approximation that is more amenable to manipulation. Further, rather than counting cardinality, we will count the log-cardinality. This is because given k bits we can index 2^k heads/tails source sequences. Hence, it is the exponent in which we are interested.

Using Stirling's approximation to the factorial, $\log_2 n! = n\log_2 n-(\log_2 e)n+O(\log_2 n)$, and ignoring the order term, we have

$$\log\binom{n}{np} \simeq n\log_2 n - n(1-p)\log_2(n(1-p)) - np\log_2(np) \tag{1.1}$$

$$= n\log_2\left(\frac{1}{1-p}\right) + np\log_2\left(\frac{1-p}{p}\right)$$

$$= n\left[(1-p)\log_2\left(\frac{1}{1-p}\right) + p\log_2\left(\frac{1}{p}\right)\right]. \tag{1.2}$$

In (1.1), the $(\log_2 e)n$ terms have canceled and the term in square brackets in (1.2) is called the *(binary) entropy*, which we denote as $H_B(p)$, so

$$H_B(p) = -p\log_2 p - (1-p)\log_2(1-p), \tag{1.3}$$

where $0 \le p \le 1$ and $0\log 0 = 0$. The binary entropy function is plotted in Fig. 1.2 within Section 1.3. One can compute that when $p = 0$ or $p = 1$ then $H_B(0) = H_B(1) = 0$. The interpretation is that, since there is only one all-tails and one all-heads sequence, and we are quantifying log-cardinality, there is only one sequence to index in each case so $\log_2(1) = 0$. In these cases, we *a priori* know the outcome (respectively, all the heads or all tails) and so do not need to store any bits to describe the realization. On the other hand, if the coin is fair then $p = 0.5$, $H_B(0.5) = 1$, $\binom{n}{n/2} \sim 2^n$, and we must use n bits of storage. In other words, on an exponential scale almost all binary sequences are 50% heads and 50% tails. As an intermediate value, if $p = 0.11$ then $H_B(0.11) \simeq 0.5$.

[1] In source coding, this is termed *near-lossless* source coding as the arbitrarily small ϵ bounds the probability of system failure and thus loss of the original data. In the *variable-length* source coding paradigm, one stores a variable amount of bits per sequence, and minimizes the expected number of bits stored. We focus on the near-lossless paradigm as the concepts involved more closely parallel those in channel coding.

The operational upshot of (1.2) is that if one allocates $nH_B(p)$ bits then basically all expected sequences can be indexed. Of course, there are caveats. First, np need not be integer. Second, there will be variability about the mean. To deal with both, we allocate a few more bits, $n(H_B(p)+\delta)$ in total. We use these bits not just to index the expected sequences, but also the *typical sequences*, those sequences with empirical entropy close to the entropy of the source.[2] In the case of coin flips, if a particular sequence consists of n_H heads (and $n - n_H$ tails) then we say that the sequence is "typical" if

$$H_B(p)-\delta \leq \left[\frac{n_H}{n}\log_2\left(\frac{1}{p}\right)+\frac{n-n_H}{n}\log_2\left(\frac{1}{1-p}\right)\right] \leq H_B(p)+\delta. \tag{1.4}$$

It can be shown that the cardinality of the set of sequences that satisfies condition (1.4) is upper-bounded by $2^{n(H_B(p)+\delta)}$. Therefore if, for instance, one lists the typical sequences lexicographically, then any typical sequence can be described using $n(H_B(p)+\delta)$ bits. One can also show that for any $\delta > 0$ the probability of the source *not* producing a typical sequence can be upper-bounded by any $\epsilon > 0$ as n grows large. This follows from the law of large numbers. As n grows the distribution of the fraction of heads in the realized source sequence concentrates about its expectation. Therefore, as long as n is sufficiently large, and as long as $\delta > 0$, any $\epsilon > 0$ will do. The quantity $H_B(p)+\delta$ is termed the *storage "rate"* R. For this example $R = H_B(p)+\delta$. The rate is the amount of memory that must be made available per source symbol. In this case, there were n symbols (n coin tosses), so one normalizes $n(H_B(p)+\delta)$ by n to get the rate $H_B(p)+\delta$.

The above idea can immediately be extended to independent and identically distributed (i.i.d.) finite-alphabet (and more general) sources as well. The general definition of the *entropy of a finite-alphabet random variable* X with probability mass function (p.m.f.) p_X is

$$H(X) = -\sum_{x\in\mathcal{X}} p_X(x)\log_2 p_X(x), \tag{1.5}$$

where "finite-alphabet" means the sample space $\mathcal{X}$ is finite.

Regardless of the distribution (binary, non-binary, even non-i.i.d.), the simple coin-flipping example illustrates one of the central tenets of information theory. That is, to focus one's design on what is likely to happen, i.e., the typical events, rather than on worst-case events. The partition of events into typical and atypical is, in information theory, known as the *asymptotic equipartition property* (AEP). In a nutshell, the simplest form of the AEP says that for long i.i.d. sequences one can, up to some arbitrarily small probability ϵ, partition all possible outcomes into two sets: the typical set and the atypical set. The probability of observing an event in the typical set is at least $1-\epsilon$. Furthermore, on an exponential scale all typical sequences are of equal probability. Designing for typical events is a hallmark of information theory.

[2] In the literature, these are termed the "weakly" typical sequences. There are other definitions of typicality that differ in terms of their mathematical use. The overarching concept is the same.

1.2.2 Converse: A Lower Bound on the Rate Required for Reliable Data Storage

A second hallmark of information theory is the emphasis on developing bounds. The source coding scheme described above is known as an *achievability result*. Achievability results involve describing an operational system that can, in principle, be realized in practice. Such results provide *(inner) bounds* on what is possible. The performance of the best system is at least this good. In the above example, we developed a source coding technique that delivers high-reliability storage and requires a rate of $H(X)+\delta$, where both the error ϵ and the slack δ can be arbitrarily small if n is sufficiently large.

An important coupled question is how much (or whether) we can reduce the rate further, thereby improving the efficiency of the scheme. In information theory, *outer bounds* on what is possible – e.g., showing that if the encoding rate is too small one cannot guarantee a target level of reliability – are termed *converse results*.

One of the key lemmas used in converse results is *Fano's inequality* [7], named for Robert Fano. The statement of the inequality is as follows: For any pair of random variables $(U,V)\in\mathcal{U}\times\mathcal{V}$ jointly distributed according to $p_{U,V}(\cdot,\cdot)$ and for any estimator $G:\mathcal{U}\to\mathcal{V}$ with probability of error $P_e=\Pr[G(U)\neq V]$,

$$H(V|U)\leq H_B(P_e)+P_e\log_2(|\mathcal{V}|-1). \tag{1.6}$$

On the left-hand side of (1.6) we encounter the *conditional entropy* $H(V|U)$ of the joint p.m.f. $p_{U,V}(\cdot,\cdot)$. We use the notation $H(V|U=u)$ to denote the entropy in V when the realization of the random variable U is set to $U=u$. Let us name this the "pointwise" conditional entropy, the value of which can be computed by applying our formula for entropy (1.5) to the p.m.f. $p_{V|U}(\cdot|u)$. The conditional entropy is the expected pointwise conditional entropy:

$$H(V|U)=\sum_{u\in\mathcal{U}}p_U(u)H(V|U=u)=\sum_{u\in\mathcal{U}}p_U(u)\left[\sum_{v\in\mathcal{V}}p_{V|U}(v|u)\log_2\left(\frac{1}{p_{V|U}(v|u)}\right)\right]. \tag{1.7}$$

Fano's inequality (1.6) can be interpreted as a bound on the ability of any hypothesis test function G to make a (single) correct guess of the realization of V on the basis of its observation of U. As the desired error probability $P_e\to 0$, both terms on the right-hand side go to zero, implying that the conditional entropy must be small. Conversely, if the left-hand side is not too small, that asserts a non-zero lower bound on P_e. A simple explicit bound is achieved by upper-bounding $H_B(P_e)$ as $H_B(P_e)\leq 1$ and rearranging to find that $P_e\geq(H(V|U)-1)/\log_2(|\mathcal{V}|-1)$.

The usefulness of Fano's inequality stems, in part, from the weak assumptions it makes. One can apply Fano's inequality to any joint distribution. Often identification of an applicable joint distribution is part of the creativity in the use of Fano's inequality. For instance in the source coding example above, one takes V to be the stored data sequence, so $|\mathcal{V}|=2^{n(H_B(p)+\delta)}$, and U to be the original source sequence, i.e., $U=X^n$. While we do not provide the derivation herein, the result is that to achieve an error probability of at most P_e the storage rate R is lower-bounded by $R\geq H(X)-P_e\log_2|\mathcal{X}|-H_B(P_e)/n$, where $|\mathcal{X}|$ is the source alphabet size; for the binary example $|\mathcal{X}| = 2$. As we let $P_e \to 0$ we see that the lower bound on the achievable rate is $H(X)$ which, letting $\delta \to 0$, is also our upper bound. Hence we have developed an operational approach to data compression where the rate we achieve matches the converse bound.

We now discuss the interaction between achievability and converse results. As long as the compression rate $R > H(X)$ then, due to concentration in measure, in the achievability case the failure probability $\epsilon > 0$ and rate slack $\delta > 0$ can both be chosen to be arbitrarily small. Concentration of measure occurs as the *blocklength* n becomes large. In parallel with n getting large, the total number of bits stored nR also grows.

The entropy $H(X)$ thus specifies a boundary between two regimes of operation. When the rate R is larger than $H(X)$, achievability results tell us that arbitrarily reliable storage is possible. When R is smaller than $H(X)$, converse results imply that reliable storage is not possible. In particular, rearranging the converse expression and once again noting that $H_B(P_e) \leq 1$, the error probability can be lower-bounded as

$$P_e \geq \frac{H(X) - R - 1/n}{\log_2 |\mathcal{X}|}. \tag{1.8}$$

If $R < H(X)$, then for n sufficiently large P_e is bounded away from zero.

The entropy $H(X)$ thus characterizes a *phase transition* between one state, the possibility of reliable data storage, and another, the impossibility. Such sharp information-theoretic phase transitions also characterize classical information-theoretic results on data transmission which we discuss in the next section, and applications of information-theoretic tools in the data sciences which we turn to later in the chapter.

1.3 Channel Coding: Transmission over the Binary Symmetric Channel

Shannon applied the same mix of ideas (typicality, entropy, conditional entropy) to solve the, perhaps at first seemingly quite distinct, problem of reliable and efficient digital communications. This is typically referred to as Shannon's "channel coding" problem in contrast to the "source coding" problem already discussed.

To gain a sense of the problem we return to the simple binary setting. Suppose our source coding system has yielded a length-k string of "information bits." For simplicity we assume these bits are randomly distributed as before, i.i.d. along the sequence, but are now fair; i.e., each is equally likely to be "0" or a "1." The objective is to convey this sequence over a communications channel to a friend. Importantly we note that, since the bits are uniformly distributed, our result on source coding tells us that no further compression is possible. Thus, uniformity of message bits is a worst-case assumption.

The channel we consider is the *binary symmetric channel* (BSC). We can transmit binary symbols over a BSC. Each input symbol is conveyed to the destination, but not entirely accurately. The binary symmetric channel "flips" each channel input symbol ($0 \to 1$ or $1 \to 0$) with probability p, $0 \leq p \leq 1$. Flips occur independently. The challenge is for the destination to deduce, one hopes with high accuracy, the k information bits

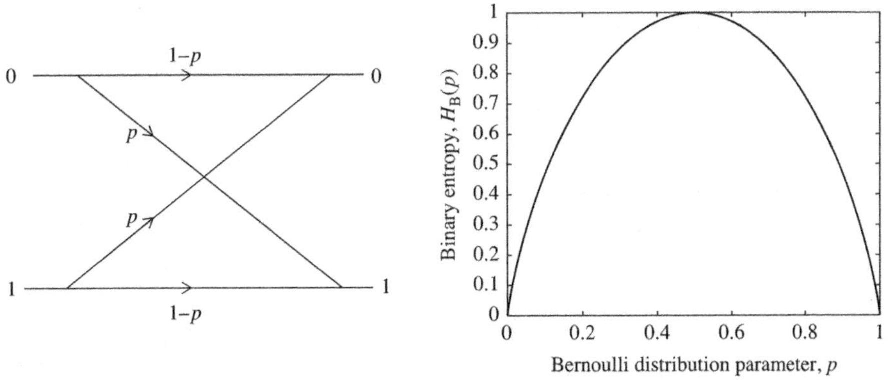

Figure 1.2 On the left we present a graphical description of the binary symmetric channel (BSC). Each transmitted binary symbol is represented as a 0 or 1 input on the left. Each received binary observation is represented by a 0 or 1 output on the right. The stochastic relationship between inputs and outputs is represented by the connectivity of the graph where the probability of transitioning each edge is represented by the edge label p or $1-p$. The channel is "symmetric" due to the symmetries in these transition probabilities. On the right we plot the binary entropy function $H_B(p)$ as a function of p, $0 \leq p \leq 1$. The capacity of the BSC is $C_{BSC} = 1 - H_B(p)$.

transmitted. Owing to the symbol flipping noise, we get some slack; we transmit $n \geq k$ binary channel symbols. For efficiency's sake, we want n to be as close to k as possible, while meeting the requirement of high reliability. The ratio k/n is termed the "rate" of communication. The length-n binary sequence transmitted is termed the "codeword." This "bit flipping" channel can be used, e.g., to model data storage errors in a computer memory. A graphical representation of the BSC is depicted in Fig. 1.2.

1.3.1 Achievability: A Lower Bound on the Rate of Reliable Data Communication

The idea of channel coding is analogous to human-evolved language. The length-k string of information bits is analogous to what we think, i.e., the concept we want to impart to the destination. The length-n codeword string of binary channel symbols is analogous to what we say (the sentence). There is redundancy in spoken language that makes it possible for spoken language to be understood in noisy (albeit not too noisy) situations. We analogously engineer redundancy into what a computer transmits in order to be able to combat the expected (the typical!) noise events. For the BSC those would be the expected bit-flip sequences.

We now consider the noise process. For any chosen length-n codeword there are about $\binom{n}{np}$ typical noise patterns which, using the same logic as in our discussion of source compression, is a set of roughly $2^{nH_B(p)}$ patterns. If we call X^n the codeword and E^n the noise sequence, then what the receiver measures is $Y^n = X^n + E^n$. Here addition is vector addition over $\mathbb{F}_2$, i.e., coordinate-wise, where the addition of two binary symbols is implemented using the *XOR* operator. The problem faced by the receiver is to identify the transmitted codeword. One can imagine that if the possible codewords are far apart in the sense that they differ in many entries (i.e., their *Hamming distance* is large) then the receiver will be less likely to make an error when deciding on the transmitted codeword. Once such a codeword estimate has been made it can then be mapped back to the length-k information bit sequence. A natural decoding rule, in fact the *maximum-likelihood* rule, is for the decoder to pick the codeword closest to Y^n in terms of Hamming distance.

The design of the codebook (analogous to the choice of grammatically correct – and thus allowable – sentences in a spoken language) is a type of probabilistic packing problem. The question is, how do we select the set of codewords so that the probability of a decoding error is small? We can develop a simple upper bound on how large the set of reliably decodable codewords can be. There are 2^n possible binary output sequences. For any codeword selected there are roughly $2^{nH_{\mathrm{B}}(p)}$ typical output sequences, each associated with a typical noise sequence, that form a *noise ball* centered on the codeword. If we were simply able to divide up the output space into disjoint sets of cardinality $2^{nH_{\mathrm{B}}(p)}$, we would end up with $2^n/2^{nH_{\mathrm{B}}(p)} = 2^{n(1-H_{\mathrm{B}}(p))}$ distinct sets. This *sphere-packing* argument tells us that the best we could hope to do would be to transmit this number of distinct codewords reliably. Thus, the number of information bits k would equal $n(1-H_{\mathrm{B}}(p))$. Once we normalize by the number n of channels uses we get a transmission rate of $1-H_{\mathrm{B}}(p)$.

Perhaps quite surprisingly, as n gets large, $1-H_{\mathrm{B}}(p)$ is the supremum of achievable rates at (arbitrarily) high reliability. This is the Shannon capacity $C_{\mathrm{BSC}} = 1-H_{\mathrm{B}}(p)$. The result follows from the law of large numbers, which can be used to show that the typical noise balls concentrate. Shannon's proof that one can actually find a configuration of codewords while keeping the probability of decoding error small was an early use of the *probabilistic method*. For any rate $R = C_{\mathrm{BSC}} - \delta$, where $\delta > 0$ is arbitrarily small, a randomized choice of the positioning of each codeword will with high probability, yield a code with a small probability of decoding error. To see the plausibility of this statement we revisit the sphere-packing argument. At rate $R = C_{\mathrm{BSC}} - \delta$ the 2^{nR} codewords are each associated with a typical noise ball of $2^{nH_{\mathrm{B}}(p)}$ sequences. If the noise balls were all (in the worst case) disjointed, this would be a total of $2^{nR}2^{nH_{\mathrm{B}}(p)} = 2^{n(1-H_{\mathrm{B}}(p)-\delta)+nH_{\mathrm{B}}(p)} = 2^{n(1-\delta)}$ sequences. As there are 2^n binary sequences, the fraction of the output space taken up by the union of typical noise spheres associated with the codewords is $2^{n(1-\delta)}/2^n = 2^{-n\delta}$. So, for any $\delta > 0$ fixed, as the blocklength $n \to \infty$, only an exponentially disappearing fraction of the output space is taken up by the noise balls. By choosing the codewords independently at random, each uniformly chosen over all length-n binary sequences, one can show that the expected (over the choice of codewords and channel noise realization) average probability of error is small. Hence, at least one codebook exists that performs at least as well as this expectation.

While Shannon showed the existence of such a code (actually a sequence of codes as $n \to \infty$), it took another half-century for researchers in error-correction coding to find asymptotically optimal code designs and associated decoding algorithms that were computationally tractable and therefore implementable in practice. We discuss this computational problem and some of these recent code designs in Section 1.4.

While the above example is set in the context of a binary-input and binary-output channel model, the result is a prototype of the result that holds for *discrete memoryless channels*. A discrete memoryless channel is described by the conditional distribution

$p_{Y|X} : \mathcal{X} \to \mathcal{Y}$. Memoryless means that output symbols are conditionally independent given the input codeword, i.e., $p_{Y^n|X^n}(y^n|x^n) = \prod_{i=1}^{n} p_{Y|X}(y_i|x_i)$. The supremum of achievable rates is the *Shannon capacity* C, where

$$C = \sup_{p_X}[H(Y) - H(Y|X)] = \sup_{p_X} I(X;Y). \tag{1.9}$$

In (1.9), $H(Y)$ is the entropy of the output space, induced by the choice of *input distribution* p_X via $p_Y(y) = \sum_{x\in\mathcal{X}} p_X(x)p_{Y|X}(y|x)$, and $H(Y|X)$ is the conditional entropy of $p_X(\cdot)p_{Y|X}(\cdot|\cdot)$. For the BSC the optimal choice of $p_X(\cdot)$ is uniform. We shortly develop an operational intuition for this choice by connecting it to hypothesis testing. We note that this choice induces the uniform distribution on Y. Since $|\mathcal{Y}| = 2$, this means that $H(Y) = 1$. Further, plugging the channel law of the BSC into (1.7) yields $H(Y|X) = H_B(p)$. Putting the pieces together recovers the Shannon capacity result for the BSC, $C_{BSC} = 1 - H_B(p)$.

In (1.9), we introduce the equality $H(Y) - H(Y|X) = I(X;Y)$, where $I(X;Y)$, denotes the *mutual information* of the joint distribution $p_X(\cdot)p_{Y|X}(\cdot|\cdot)$. The mutual information is another name for the *Kullback–Leibler (KL) divergence* between the joint distribution $p_X(\cdot)p_{Y|X}(\cdot|\cdot)$ and the product of the joint distribution's marginals, $p_X(\cdot)p_Y(\cdot)$. The general formula for the KL divergence between a pair of distributions p_U and p_V defined over a common alphabet $\mathcal{A}$ is

$$D(p_U\|p_V) = \sum_{a\in\mathcal{A}} p_U(a)\log_2\left(\frac{p_U(a)}{p_V(a)}\right). \tag{1.10}$$

In the definition of mutual information over $\mathcal{A} = \mathcal{X}\times\mathcal{Y}$, $p_{X,Y}(\cdot,\cdot)$ plays the role of $p_U(\cdot)$ and $p_X(\cdot)p_Y(\cdot)$ plays the role of $p_V(\cdot)$.

The KL divergence arises in hypothesis testing, where it is used to quantify the error exponent of a binary hypothesis test. Conceiving of channel decoding as a hypothesis-testing problem – which one of the codewords was transmitted? – helps us understand why (1.9) is the formula for the Shannon capacity. One way the decoder can make its decision regarding the identity of the true codeword is to test each codeword against independence. In other words, does the empirical joint distribution of any particular codeword X^n and the received data sequence Y^n look jointly distributed according to the channel law or does it look independent? That is, does (X^n, Y^n) look like it is distributed i.i.d. according to $p_{XY}(\cdot,\cdot) = p_X(\cdot)p_{Y|X}(\cdot|\cdot)$ or i.i.d. according to $p_X(\cdot)p_Y(\cdot)$? The exponent of the error in this test is $-D(p_{XY}\|p_Xp_Y) = -I(X;Y)$. Picking the input distribution p_X to maximize (1.9) maximizes this exponent. Finally, via an application of the union bound we can assert that, roughly, $2^{nI(X;Y)}$ codewords can be allowed before more than one codeword in the codebook appear to be jointly distributed with the observation vector Y^n according to p_{XY}.

1.3.2 Converse: An Upper Bound on the Rate of Reliable Data Communication

An application of Fano's inequality (1.6) shows that C is also an upper bound on the achievable communication rate. This application of Fano's inequality is similar to that used in source coding. In this application of (1.6), we set $V = X^n$ and $U = Y^n$. The greatest additional subtlety is that we must leverage the memoryless property of the channel to *single-letterize* the bound. To single-letterize means to express the final bound in terms of only the $p_X(\cdot)p_{Y|X}(\cdot|\cdot)$ distribution, rather than in terms of the joint distribution of the length-n input and output sequences. This is an important step because n is allowed to grow without bound. By single-letterizing we express the bound in terms of a fixed distribution, thereby making the bound computable.

As at the end of the discussion of source coding, in channel coding we find a boundary between two regimes of operation: the regime of efficient and reliable data transmission, and the regime wherein such reliable transmission is impossible. In this instance, it is the Shannon capacity C that demarks the phase-transition between these two regimes of operation.

1.4 Linear Channel Coding

In the previous sections, we discussed the sharp phase transitions in both source and channel coding discovered by Shannon. These phase transitions delineate fundamental boundaries between what is possible and what is not. In practice, one desires schemes that "saturate" these bounds. In the case of source coding, we can saturate the bound if we can design source coding techniques with rates that can be made arbitrarily close to $H(X)$ (from above). For channel coding we desire coding methods with rates that can be made arbitrarily close to C (from below). While Shannon discovered and quantified the bounds, he did not specify realizable schemes that attained them.

Decades of effort have gone into developing methods of source and channel coding. For lossless compression of memoryless sources, as in our motivating examples, good approaches such as Huffman and arithmetic coding were found rather quickly. On the other hand, finding computationally tractable and therefore implementable schemes of error-correction coding that got close to capacity took much longer. For a long time it was not even clear that computationally tractable techniques of error correction that saturated Shannon's bounds were even possible. For many years researchers thought that there might be a second phase transition at the *cutoff rate*, only below which computationally tractable methods of reliable data transmission existed. (See [10] for a nice discussion.) Indeed, only with the emergence of modern coding theory in the 1990s and 2000s that studies turbo, low-density parity-check (LDPC), spatially coupled LDPC, and Polar codes has the research community, even for the BSC, developed computationally tractable methods of error correction that closed the gap to Shannon's bound.

In this section, we introduce the reader to linear codes. Almost all codes in use have linear structure, structure that can be exploited to reduce the complexity of the decoding process. As in the previous sections we only scratch the surface of the discipline of error-correction coding. We point the reader to the many excellent texts on the subject, e.g., [6, 11–15].

1.4.1 Linear Codes and Syndrome Decoding

Linear codes are defined over finite fields. As we have been discussing the BSC, the field we will focus on is $\mathbb{F}_2$. The set of codewords of a length-n binary linear code corresponds to a subspace of the vector space $\mathbb{F}_2^n$. To encode we use a matrix–vector multiplication defined over $\mathbb{F}_2$ to map a length-k column vector $\mathbf{b} \in \mathbb{F}_2^k$ of "information bits" into a length-n column vector $\mathbf{x} \in \mathbb{F}_2^n$ of binary "channel symbols" as

$$\mathbf{x} = \mathbf{G}^{\mathrm{T}}\mathbf{b}, \tag{1.11}$$

where $\mathbf{G} \in \mathbb{F}_2^{k \times n}$ is a $k \times n$ binary "generator" matrix and $\mathbf{G}^{\mathrm{T}}$ denotes the transpose of $\mathbf{G}$. Assuming that $\mathbf{G}$ is full rank, all 2^k possible binary vectors $\mathbf{b}$ are mapped by $\mathbf{G}$ into 2^k distinct codewords $\mathbf{x}$, so the set of possible codewords (the "*codebook*") is the row-space of $\mathbf{G}$. We compute the rate of the code as $R = k/n$.

Per our earlier discussion, the channel adds the length-n noise sequence $\mathbf{e}$ to $\mathbf{x}$, yielding the channel output $\mathbf{y} = \mathbf{x} + \mathbf{e}$. To decode, the receiver pre-multiplies $\mathbf{y}$ by the *parity-check matrix* $\mathbf{H} \in \mathbb{F}_2^{m \times n}$ to produce the length-m *syndrome* $\mathbf{s}$ as

$$\mathbf{s} = \mathbf{H}\mathbf{y}. \tag{1.12}$$

We caution the reader not to confuse the parity-check matrix $\mathbf{H}$ with the entropy function $H(\cdot)$. By design, the rows of $\mathbf{H}$ are all orthogonal to the rows of $\mathbf{G}$ and thus span the null-space of $\mathbf{G}$.[3] When the columns of $\mathbf{G}$ are linearly independent, the dimension of the null-space of $\mathbf{G}$ is $n - k$ and the relation $m = n - k$ holds.

Substituting in the definition for $\mathbf{x}$ into the expression for $\mathbf{y}$ and thence into (1.12), we compute

$$\mathbf{s} = \mathbf{H}(\mathbf{G}^{\mathrm{T}}\mathbf{b} + \mathbf{e}) = \mathbf{H}\mathbf{G}^{\mathrm{T}}\mathbf{b} + \mathbf{H}\mathbf{e} = \mathbf{H}\mathbf{e}, \tag{1.13}$$

where the last step follows because the rows of $\mathbf{G}$ and $\mathbf{H}$ are orthogonal by design so that $\mathbf{H}\mathbf{G}^{\mathrm{T}} = \mathbf{0}$, the $m \times k$ all-zeros matrix. Inspecting (1.13), we observe that the computation of the syndrome $\mathbf{s}$ yields m linear constraints on the noise vector $\mathbf{e}$.

Since $\mathbf{e}$ is of length n and $m = n - k$, (1.13) specifies an under-determined set of linear equations in $\mathbb{F}_2$. However, as already discussed, while $\mathbf{e}$ *could* be any vector, when the blocklength n becomes large, concentration of measure comes into play. With high probability the realization of $\mathbf{e}$ will concentrate around those sequences that contain only np non-zero elements. We recall that $p \in [0, 1]$ is the bit-flip probability and note that in $\mathbb{F}_2$ any non-zero element must be a one. In coding theory, we are therefore faced with the problem of solving an under-determined set of linear equations subject to a *sparsity constraint*: There are only about np non-zero elements in the solution vector. Consider $p \leq 0.5$. Then, as error vectors $\mathbf{e}$ containing fewer bit flips are more likely, the maximum-likelihood solution for the noise vector $\mathbf{e}$ is to find the maximally sparse vector that satisfies the syndrome constraints, i.e.,

$$\widehat{\mathbf{e}} = \arg\min_{\mathbf{e} \in \mathbb{F}_2^n} d_{\mathrm{H}}(\mathbf{e}) \quad \text{such that} \quad \mathbf{s} = \mathbf{H}\mathbf{e}, \tag{1.14}$$

where $d_{\mathrm{H}}(\cdot)$ is the Hamming weight (or distance from 0^n) of the argument. As mentioned before, the Hamming weight is the number of non-zero entries of $\mathbf{e}$. It plays a role

[3] Note that in finite fields vectors can be self-orthogonal; e.g., in $\mathbb{F}_2$ any even-weight vector is orthogonal to itself.

analogous to the cardinality function in $\mathbb{R}^n$ (sometimes denoted $||\cdot||_0$), which is often used to enforce sparsity in the solution to optimization problems.

We observe that there are roughly $2^{nH_B(p)}$ typical binary bit-flip sequences, each with roughly np non-zero elements. The syndrome $\mathbf{s}$ provides m linear constraints on the noise sequence. Each constraint is binary so that, if all constraints are linearly independent, each constraint reduces by 50% the set of possible noise sequences. Thus, if the number m of constraints exceeds $\log_2(2^{nH_B(p)}) = nH_B(p)$ we should be able to decode correctly.[4]

Decoders can thus be thought of as solving a binary search problem where the measurements/queries are fixed ahead of time, and the decoder uses the results of the queries, often in an iterative fashion, to determine $\mathbf{e}$. Once $\widehat{\mathbf{e}}$ has been calculated, the codeword estimate $\widehat{\mathbf{x}} = \mathbf{y} + \widehat{\mathbf{e}} = \mathbf{G}^T\mathbf{b} + (\mathbf{e} - \widehat{\mathbf{e}})$. If $\widehat{\mathbf{e}} = \mathbf{e}$, then the term in brackets cancels and $\mathbf{b}$ can uniquely and correctly be recovered from $\mathbf{G}^T\mathbf{b}$. This last point follows since the codebook is the row-space of $\mathbf{G}$ and $\mathbf{G}$ is full rank.

Noting from the previous section that the capacity of the BSC channel is $C = 1 - H_B(p)$ and the rate of the code is $R = k/n$, we would achieve capacity if $1 - H_B(p) = k/n$ or, equivalently, if the syndrome length $m = n - k = n(1 - k/n) = nH_B(p)$. This is the objective of coding theory: to find "good" codes (specified by their generator matrix $\mathbf{G}$ or, alternately, by their parity-check matrix $\mathbf{H}$) and associated decoding algorithms (that attempt to solve (1.14) in a computationally efficient manner) so as to be able to keep $R = k/n$ as close as possible to $C_{BSC} = 1 - H_B(p)$.

1.4.2 From Linear to Computationally Tractable: Polar Codes

To understand the challenge of designing computationally tractable codes say that, in the previous discussion, one picked $\mathbf{G}$ (or $\mathbf{H}$) according to the Bernoulli-0.5 random i.i.d. measure. Then for any fixed rate R if one sets the blocklength n to be sufficiently large the generator (or parity-check) matrix produced will, with arbitrarily high probability, specify a code that is capacity-achieving. Such a selection of $\mathbf{G}$ or $\mathbf{H}$ is respectively referred to as the Elias or the Gallager ensemble.

However, attempting to use the above capacity-achieving scheme can be problematic from a computational viewpoint. To see the issue consider blocklength $n = 4000$ and rate $R = 0.5$, which are well within normal operating ranges for these parameters. For these choices there are $2^{nR} = 2^{2000}$ codewords. Such an astronomical number of codewords makes the brute-force solution of (1.14) impossible. Hence, the selection of $\mathbf{G}$ or $\mathbf{H}$ according to the Elias or Gallager ensembles, while yielding linear structure, does not by itself guarantee that a computationally efficient decoder will exist. Modern methods of coding – LDPC codes, spatially coupled codes, and Polar codes – while being linear codes also design additional structure into the code ensemble with the express intent of

[4] We comment that this same syndrome decoding can also be used to provide a solution to the near-lossless source coding problem of Section 1.2. One pre-multiplies the source sequence by the parity-check matrix $\mathbf{H}$, and stores the syndrome of the source sequence. For a biased binary source, one can solve (1.14) to recover the source sequence with high probability. This approach does not feature prominently in source coding, with the exception of *distributed source coding*, where it plays a prominent role. See [7, 9] for further discussion.

making it compatible with computationally tractable decoding algorithms. To summarize, in coding theory the design of a channel coding scheme involves the joint design of the codebook and the decoding algorithm.

Regarding the phase transitions discovered by Shannon for source and channel coding, a very interesting code construction is Erdal Arikan's Polar codes [16]. Another tractable code construction that connects to phase transitions is the spatial-coupling concept used in convolutionally structured LDPC codes [17–19]. In [16], Arikan considers symmetric channels and introduces a symmetry-breaking transformation. This transformation is a type of pre-coding that combines pairs of symmetric channels to produce a pair of virtual channels. One virtual channel is "less noisy" than the original channel and one is more noisy. Arikan then applies this transformation recursively. In the limit, the virtual channels polarize. They either become noiseless and so have capacity one, or become useless and have capacity zero. Arikan shows that, in the limit, the fraction of virtual channels that become noiseless is equal to the capacity of the original symmetric channel; e.g., $1 - H_B(p)$, if the original channel were the BSC. One transmits bits uncoded over the noiseless virtual channels, and does not use the useless channels. The recursive construction yields log-linear complexity in encoding and decoding, $O(n \log n)$, making Polar codes computationally attractive. In many ways, the construction is information-theoretic in nature, focusing on mutual information rather than Hamming distance as the quantity of importance in the design of capacity-achieving codes.

To conclude, we note that many important concepts, such as fundamental limits, achievability results, converse results, and computational limitations, that arise in classical information theory also arise in modern data-science problems. In classical information theory, as we have seen, such notions have traditionally been considered in the context of data compression and transmission. In data science similar notions are being studied in the realms of acquisition, data representation, analysis, and processing. There are some instances where one can directly borrow classical information-theoretic tools used to determine limits in, e.g., the channel coding problem to compute limits in data-science tasks. For example, in compressive sensing [20] and group testing [21] achievability results have been derived using the probabilistic method and converse results have been developed using Fano's inequality [22]. However, there are various other data-science problems where information-theoretic methods have not yet been directly applied. We elaborate further in the following sections how information-theoretic ideas, tools, and methods are also gradually shaping data science.

1.5 Connecting Information Theory to Data Science

Data science – a loosely defined concept meant to bring together various problems studied in statistics, machine learning, signal processing, harmonic analysis, and computer science under a unified umbrella – involves numerous other challenges that go beyond the traditional source coding and channel coding problems arising in communication or storage systems. These challenges are associated with the need to acquire, represent, and analyze information buried in data in a reliable and computationally efficient manner in the presence of a variety of constraints such as security, privacy, fairness, hardware resources, power, noise, and many more.

Figure 1.3 presents a typical data-science pipeline, encompassing functions such as data acquisition, data representation, and data analysis, whose overarching purpose is to turn data captured from the physical world into insights for decision-making. It is also common to consider various other functions within a data-science "system" such as data preparation, data exploration, and more. We restrict ourselves to this simplified version because it serves to illustrate how information theory is helping shape data science. The goals of the different blocks of the data-science pipeline in Fig. 1.3 are as follows.

- The data-acquisition block is often concerned with the act of turning physical-world continuous-time analog signals into discrete-time digital signals for further digital processing.
- The data-representation block concentrates on the extraction of relevant attributes from the acquired data for further analysis.
- The data-analysis block concentrates on the extraction of meaningful actionable information from the data features for decision-making.

From the description of these goals, one might think that information theory – a field that arose out of the need to study communication systems in a principled manner – has little to offer to the principles of data acquisition, representation, analysis, or processing. But it turns out that information theory has been advancing our understanding of data science in three major ways.

- First, information theory has been leading to new system architectures for the different elements of the data-science pipeline. Representative examples associated with new architectures for data acquisition are overviewed in Section 1.6.
- Second, information-theoretic methods can be used to unveil fundamental operational limits in various data-science tasks, including in data acquisition, representation, analysis, and processing. Examples are overviewed in Sections 1.6–1.8.
- Third, information-theoretic measures can be used as the basis for developing algorithms for various data-science tasks. We allude to some examples in Sections 1.7 and 1.8.

In fact, the questions one can potentially ask about the data-science pipeline depicted in Fig. 1.3 exhibit many parallels to the questions one asks about the communications

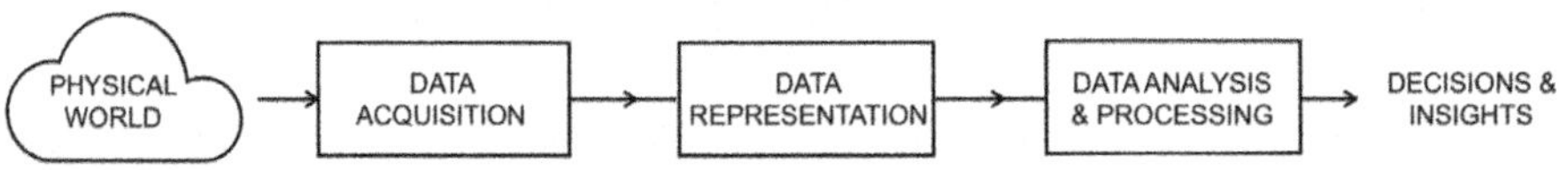

Figure 1.3 A simplified data-science pipeline encompassing functions such as data acquisition, data representation, and data analysis and processing.

system architecture shown in Fig. 1.1. Specifically, what are the trade-offs in performance incurred by adopting this data-science architecture? In particular, are there other systems that do not involve the separation of the different data-science elements and exhibit better performance? Are there fundamental limits on what is possible in data acquisition, representation, analysis, and processing? Are there computationally feasible algorithms for data acquisition, representation, analysis, and processing that attain such limits?

There has been progress in data science in all three of these directions. As a concrete example that showcases many similarities between the data-compression and – communication problems and data-science problems, information-theoretic methods have been providing insight into the various operational regimes associated with the following different data-science tasks: (1) the regime where there is no algorithm – regardless of its complexity – that can perform the desired task subject to some accuracy; this "regime of impossibility" in data science has the flavor of converse results in source coding and channel coding in information theory; (2) the regime where there are algorithms, potentially very complex and computationally infeasible, that can perform the desired task subject to some accuracy; this "regime of possibility" is akin to the initial discussion of linear codes and the Elias and Gallager ensembles in channel coding; and (3) the regime where there are computationally feasible algorithms to perform the desired task subject to some accuracy; this "regime of computational feasibility" in data science has many characteristics that parallel those in design of computationally tractable source and channel coding schemes in information theory.

Interestingly, in the same way that the classical information-theoretic problems of source coding and channel coding exhibit phase transitions, many data-science problems have also been shown to exhibit sharp phase transitions in the *high-dimensional setting*, where the number of data samples and the dimension of the data approach infinity. Such phase transitions are typically expressed as a function of various parameters associated with the data-science problem. The resulting *information-theoretic* limit/threshold/barrier (a.k.a. *statistical phase transition*) partitions the problem parameter space into two regions [23–25]: one defining problem instances that are impossible to solve and another defining problem instances that can be solved (perhaps only with a brute-force algorithm). In turn, the *computational* limit/threshold/barrier (a.k.a. *computational phase transition*) partitions the problem parameter space into a region associated with problem instances that are easy to solve and another region associated with instances that are hard to solve [26, 27].

There can, however, be differences in how one establishes converse and achievability results – and therefore phase transitions – in classical information-theoretic problems and data-science ones. Converse results in data science can often be established using Fano's inequality or variations on it (see also Chapter 16). In contrast, achievability results often cannot rely on classical techniques, such as the probabilistic method, necessitating instead the direct analysis of the algorithms. Chapter 13 elaborates on some emerging tools that may be used to establish statistical and computational limits in data-science problems.

In summary, numerous information-theoretic tools, methods, and quantities are increasingly becoming essential to cast insight into data science. It is impossible to capture all the recent developments in a single chapter, but the following sections sample a number of recent results under three broad themes: data acquisition, data representation, and data analysis and processing.

1.6 Information Theory and Data Acquisition

Data acquisition is a critical element of the data-science architecture shown in Fig. 1.3. It often involves the conversion of a *continuous-time analog* signal into a *discrete-time digital* signal that can be further processed in digital signal-processing pipelines.[5]

Conversion of a continuous-time analog signal $x(t)$ into a discrete-time digital representation typically entails two operations. The first operation – known as sampling – involves recording the values of the original signal $x(t)$ at particular instants of time. The simplest form of sampling is direct uniform sampling in which the signal is recorded at uniform sampling times $x(kT_S) = x(k/F_S)$, where T_S denotes the *sampling period* (in seconds), F_S denotes the *sampling frequency* (in Hertz), and k is an integer. Another popular form of sampling is generalized shift-invariant sampling in which $x(t)$ is first filtered by a linear time-invariant (LTI) filter, or a bank of LTI filters, and only then sampled uniformly [28]. Other forms of generalized and non-uniform sampling have also been studied. Surprisingly, under certain conditions, the sampling process can be shown to be lossless: for example, the classical *sampling theorem* for bandlimited processes asserts that it is possible to perfectly recover the original signal from its uniform samples provided that the sampling frequency F_S is at least twice the signal bandwidth B. This minimal sampling frequency $F_{NQ} = 2B$ is referred to as the *Nyquist rate* [28].

The second operation – known as *quantization* – involves mapping the continuous-valued signal samples onto discrete-valued ones. The levels are taken from a finite set of levels that can be represented using a finite sequence of bits. In (optimal) *vector quantization* approaches, a series of signal samples are converted simultaneously to a bit sequence, whereas in (sub-optimal) *scalar quantization*, each individual sample is mapped to bits. The quantization process is inherently lossy since it is impossible to accurately represent real-valued samples using a finite set of bits. *Rate-distortion theory* establishes a trade-off between the average number of bits used to encode each signal sample – referred to as the *rate* – and the average distortion incurred in the reconstruction of each signal sample – referred to simply as the *distortion* – via two functions. The *rate-distortion function* $R(D)$ specifies the smallest number of bits required on average per sample when one wishes to represent each sample with average distortion less than D, whereas the *distortion-rate function* $D(R)$ specifies the lowest average distortion achieved per sample when one wishes to represent on average each sample with R bits [7]. A popular measure of distortion in the recovery of the original signal

[5] Note that it is also possible that the data are already presented in an inherently digital format; Chapters 3, 4, and 6 deal with such scenarios.

samples from the quantized ones is the mean-squared error (MSE). Note that this class of problems – known as lossy source coding – is the counterpart of the lossless source coding problems discussed earlier.

The motivation for this widely used data-acquisition architecture, involving (1) a sampling operation at or just above the Nyquist rate and (2) scalar or vector quantization operations, is its simplicity that leads to a practical implementation. However, it is well known that the separation of the sampling and quantization operations is not necessarily optimal. Indeed, the optimal strategy that attains Shannon's distortion-rate function associated with arbitrary continuous-time random signals with known statistics involves a general mapping from continuous-time signal space to a sequence of bits that does not consider any practical constraints in its implementation [1, 3, 29]. Therefore, recent years have witnessed various generalizations of this data-acquisition paradigm informed by the principles of information theory, on the one hand, and guided by practical implementations, on the other.

One recent extension considers a data-acquisition paradigm that illuminates the dependence between these two operations [30–32]. In particular, given a total rate budget, Kipnis *et al.* [30–32] draw on information-theoretic methods to study the lowest sampling rate required to sample a signal such that the reconstruction of the signal from the bit-constrained representation of its samples results in minimal distortion. The sampling operation consists of an LTI filter, or bank of filters, followed by pointwise sampling of the outputs of the filters. The authors also show that, without assuming any particular structure on the input analog signal, this sampling rate is often below the signal's Nyquist rate. That is, due to the fact that there is loss encountered by the quantization operation, there is no longer in general a requirement to sample the signal at the Nyquist rate.

As an example, consider the case where $x(t)$ is a stationary random process bandlimited to B with a triangular power spectral density (PSD) given formally by

$$S(f) = \frac{\sigma_x^2}{B}[1 - |f/B|]_+ \tag{1.15}$$

with $[a]_+ = \max(a,0)$. In this case, the Nyquist sampling rate is $2B$. However, when quantization is taken into account, the sampling rate can be lowered without introducing further distortion. Specifically, assuming a bitrate leading to distortion D, the minimal sampling rate is shown to be equal to [32]

$$f_{\mathrm{R}} = 2B\sqrt{1 - D/\sigma_x^2}. \tag{1.16}$$

Thus, as the distortion grows, the minimal sampling rate is reduced. When we do not allow any distortion, namely, no quantization takes place, $D = 0$ and $f_{\mathrm{R}} = 2B$ so that Nyquist rate sampling is required.

Such results show how information-theoretic methods are leading to new insights about the interplay between sampling and quantization. In particular, these new results can be seen as an extension of the classical sampling theorem applicable to bandlimited random processes in the sense that they describe the minimal amount of excess distortion in the reconstruction of a signal due to lossy compression of its samples, leading to the minimal sampling frequency required to achieve this distortion.[6] In general, this sampling frequency is below the Nyquist rate. Chapter 2 surveys some of these recent results in data acquisition.

Another generalization of the classical data-acquisition paradigm considers scenarios where the end goal is not to reconstruct the original analog signal $x(t)$ but rather to perform some other operation on it [33]. For example, in the context of parameter estimation, Rodrigues *et al.* [33] show that the number of bits per sample required to achieve a certain distortion in such *task-oriented* data acquisition can be much lower than that required for *task-ignorant* data acquisition. More recently, Shlezinger *et al.* [34, 35] study task-oriented *hardware-efficient* data-acquisition systems, where optimal vector quantizers are replaced by practical ones. Even though the optimal rate-distortion curve cannot be achieved by replacing optimal vector quantizers by simple serial scalar ones, it is shown in [34, 35] that one can get close to the minimal distortion in settings where the information of interest is not the signal itself, but rather a low-dimensional parameter vector embedded in the signal. A practical application of this setting is in massive multiple-input multiple-output (MIMO) systems where there is a strong need to utilize simple low-resolution quantizers due to power and memory constraints. In this context, it is possible to design a simple quantization system, consisting of scalar uniform quantizers and linear pre- and post-processing, leading to minimal channel estimation distortion.

These recent results also showcase how information-theoretic methods can provide insight into the interplay between data acquisition, representation, and analysis, in the sense that knowledge of the data-analysis goal can influence the data-acquisition process. These results therefore also suggest new architectures for the conventional data-science pipeline that do not involve a strict separation between the data-acquisition, data-representation, and data-analysis and processing blocks.

Beyond this data-acquisition paradigm involving the conversion of continuous-time signals to digital ones, recent years have also witnessed the emergence of various other data-acquisition approaches. Chapters 3, 4, and 6 cover further data-acquisition strategies that are also benefiting from information-theoretic methods.

1.7 Information Theory and Data Representation

The outputs of the data-acquisition block – often known as *"raw" data* – typically need to be turned into "meaningful" representations – known as *features* – for further data analysis. Note that the act of transforming raw data into features, where the number of dimensions in the features is lower than that in the raw data, is also referred to as *dimensionality reduction*.

Recent years have witnessed a shift from *model-based data representations*, relying on predetermined transforms – such as wavelets, curvelets, and shearlets – to compute the features from raw data, to *data-driven representations* that leverage a

[6] In fact, this theory can be used even when the input signal is not bandlimited.

number of (raw) data "examples"/"instances" to first learn a (linear or nonlinear) data representation transform conforming to some postulated data generative model [36–39]. Mathematically, given N (raw) data examples $\{\mathbf{y}^i \in \mathcal{Y}\}_{i=1}^N$ (referred to as *training samples*), data-driven representation learning often assumes generative models of the form

$$\mathbf{y}^i = F(\mathbf{x}^i) + \mathbf{w}^i, \quad i = 1, \ldots, N, \tag{1.17}$$

where $\mathbf{x}^i \in \mathcal{X}$ denote feature vectors that are assumed to be realizations of a random vector $\mathbf{X}$ distributed as p_X, $\mathbf{w}^i$ denote acquisition noise and/or modeling errors that are realizations of random noise $\mathbf{W}$ distributed as p_W, and $F : \mathcal{X} \to \mathcal{Y}$ denotes the true (linear or nonlinear) representation transform that belongs to some postulated class of transforms $\mathcal{F}$. The operational challenge in representation learning is to estimate $F(\cdot)$ using the training samples, after which the features can be obtained using the inverse images of data samples returned by

$$G \stackrel{\text{def}}{=} F^{-1} : \mathcal{Y} \to \mathcal{X}. \tag{1.18}$$

Note that if $F(\cdot)$ is not a bijection then $G(\cdot)$ will not be the inverse operator.[7]

In the literature, $F(\cdot)$ and $G(\cdot)$ are sometimes also referred to as the *synthesis operator* and the *analysis operator*, respectively. In addition, the representation-learning problem as stated is referred to as *unsupervised representation learning* [37, 39]. Another category of representation learning is *supervised representation learning* [40, 41], in which training data correspond to tuples $(\mathbf{y}^i, \ell^i)$ with ℓ^i termed the *label* associated with training sample $\mathbf{y}^i$. Representation learning in this case involves obtaining an analysis/synthesis operator that results in the best task-specific (e.g., classification and regression) performance. Another major categorization of representation learning is in terms of the linearity of $G(\cdot)$, with the resulting classes referred to as *linear representation learning* and *nonlinear representation learning*, respectively.

The problem of learning (estimating) the true transformation from a given postulated generative model poses various challenges. One relates to the design of appropriate algorithms for estimating $F(\cdot)$ and computing inverse images $G(\cdot)$. Another challenge involves understanding information-theoretic and computational limitations in representation learning in order to identify regimes where existing algorithms are nearly optimal, regimes where existing algorithms are clearly sub-optimal, and to guide the development of new algorithms. These challenges are also being addressed using information-theoretic tools. For example, researchers often map the representation learning problem onto a channel coding problem, where the transformation $F(\cdot)$ represents the message that needs to be decoded at the output of a channel that maps $F(\cdot)$ onto $F(\mathbf{X}) + \mathbf{W}$. This allows leveraging of information-theoretic tools such as Fano's inequality for derivation of fundamental limits on the estimation error of $F(\cdot)$ as a function of the number of training samples [25, 42–45]. We next provide a small sampling of representation learning results that involve the use of information-theoretic tools and methods.

[7] In some representation learning problems, instead of using $F(\cdot)$ to obtain the inverse images of data samples, $G(\cdot)$ is learned directly from training samples.

1.7.1 Linear Representation Learning

Linear representation learning constitutes one of the oldest and, to date, the most prevalent data-representation technique in data science. While there are several different variants of linear representation learning both in unsupervised and in supervised settings, all these variants are based on the assumption that the raw data samples lie near a *low-dimensional (affine) subspace*. Representation learning in this case is therefore equivalent to learning the subspace(s) underlying raw data. This will be a single subspace in the unsupervised setting, as in *principal component analysis (PCA)* [46–48] and *independent component analysis (ICA)* [49, 50], and multiple subspaces in the supervised setting, as in *linear discriminant analysis (LDA)* [51–53] and *quadratic discriminant analysis (QDA)* [53].

Mathematically, linear representation learning operates under the assumption of the raw data space being $\mathcal{Y} = \mathbb{R}^m$, the feature space being $\mathcal{X} = \mathbb{R}^k$ with $k \ll m$, the raw data samples being given by

$$\mathbf{Y} = \mathbf{A}\mathbf{X} + \mathbf{W} \tag{1.19}$$

with $\mathbf{A} \in \mathcal{F} \subset \mathbb{R}^{m\times k}$, and the feature estimates being given by $\widehat{\mathbf{X}} = \mathbf{B}\mathbf{Y}$ with $\mathbf{B} \in \mathbb{R}^{k\times m}$. In this setting, $(F,G) = (\mathbf{A},\mathbf{B})$ and representation learning reduces to estimating the linear operators $\mathbf{A}$ and/or $\mathbf{B}$ under various assumptions on $\mathcal{F}$ and the generative model.[8] In the case of PCA, for example, it is assumed that $\mathcal{F}$ is the Stiefel manifold in $\mathbb{R}^m$ and the feature vector $\mathbf{X}$ is a random vector that has zero mean and uncorrelated entries. On the other hand, ICA assumes $\mathbf{X}$ to have zero mean and independent entries. (The zero-mean assumption in both PCA and ICA is for ease of analysis and can be easily removed at the expense of extra notation.)

Information-theoretic frameworks have long been used to develop computational approaches for estimating $(\mathbf{A},\mathbf{B})$ in ICA and its variants; see, e.g., [49, 50, 54, 55]. Recent years have also seen the use of information-theoretic tools such as Fano's inequality to derive sharp bounds on the feasibility of linear representation learning. One such result that pertains to PCA under the so-called *spiked covariance model* is described next.

Suppose the training data samples are N independent realizations according to (1.19), i.e.,

$$\mathbf{y}^i = \mathbf{A}\mathbf{x}^i + \mathbf{w}^i, \quad i = 1,\ldots,N, \tag{1.20}$$

where $\mathbf{A}^{\mathrm{T}}\mathbf{A} = \mathbf{I}$ and both $\mathbf{x}^i$ and $\mathbf{w}^i$ are independent realizations of $\mathbf{X}$ and $\mathbf{W}$ that have zero mean and diagonal covariance matrices given by

$$\mathbb{E}[\mathbf{X}\mathbf{X}^{\mathrm{T}}] = \mathrm{diag}(\lambda_1,\ldots,\lambda_k), \quad \lambda_1 \geq \lambda_2 \geq \cdots \geq \lambda_k > 0, \tag{1.21}$$

and $\mathbb{E}[\mathbf{W}\mathbf{W}^{\mathrm{T}}] = \sigma^2\mathbf{I}$, respectively. Note that the ideal $\mathbf{B}$ in this PCA example is given by $\mathbf{B} = \mathbf{A}^{\mathrm{T}}$. It is then shown in [43, Theorem 5] using various analytical

[8] Supervised learning typically involves estimation of multiple $\mathbf{A}$s and/or $\mathbf{B}$s.

tools, which include Fano's inequality, that **A** can be reliably estimated from N training samples only if[9]

$$\frac{N\lambda_k^2}{k(m-k)(1+\lambda_k)} \to \infty. \tag{1.22}$$

This is the "converse" for the spiked covariance estimation problem.

The "achievability" result for this problem is also provided in [43]. Specifically, when the condition given in (1.22) is satisfied, a practical algorithm exists that allows reliable estimation of **A** [43]. This algorithm involves taking $\widehat{\mathbf{A}}$ to be the k eigenvectors corresponding to the k largest eigenvalues of the sample covariance $(1/N)\sum_{i=1}^{N}\mathbf{y}^i\mathbf{y}^{i^{\mathrm{T}}}$ of the training data. We therefore have a sharp information-theoretic phase transition in this problem, which is characterized by (1.22). Notice here, however, that while the converse makes use of information-theoretic tools, the achievability result does not involve the use of the probabilistic method; rather, it requires analysis of an explicit (deterministic) algorithm.

The sharp transition highlighted by the aforementioned result can be interpreted in various ways. One of the implications of this result is that it is impossible to reliably estimate the PCA features when $m > N$ and $m, N \to \infty$. In such *high-dimensional PCA* settings, it is now well understood that *sparse PCA*, in which the columns of **A** are approximately "sparse," is more appropriate for linear representation learning. We refer the reader to works such as [43, 56, 57] that provide various information-theoretic limits for the sparse PCA problem.

We conclude by noting that there has been some recent progress regarding bounds on the computational feasibility of linear representation learning. For example, the fact that there is a practical algorithm to learn a linear data representation in some high-dimensional settings implies that computational barriers can almost coincide with information-theoretic ones. It is important to emphasize, though, that recent work – applicable to the detection of a subspace structure within a data matrix [25, 58–62] – has revealed that classical computationally feasible algorithms such as PCA cannot always approach the information-theoretic detection threshold [25, 61].

1.7.2 Nonlinear Representation Learning

While linear representation learning techniques tend to have low computational complexity, they often fail to capture relevant information within complex physical phenomena. This, coupled with a meteoric rise in computing power, has led to widespread adoption of nonlinear representation learning in data science.

There is a very wide portfolio of nonlinear representation techniques, but one of the most well-known classes, which has been the subject of much research during the last two decades, postulates that (raw) data lie near a *low-dimensional manifold* embedded in a higher-dimensional space. Representation learning techniques belonging to this class

[9] Reliable estimation here means that the error between $\widehat{\mathbf{A}}$ and **A** converges to 0 with increasing number of samples.

include local linear embedding [63], Isomap [64], kernel entropy component analysis (ECA) [65], and nonlinear generalizations of linear techniques using the *kernel trick* (e.g., kernel PCA [66], kernel ICA [67], and kernel LDA [68]). The use of information-theoretic machinery in these methods has mostly been limited to formulations of the algorithmic problems, as in kernel ECA and kernel ICA. While there exist some results that characterize the regime in which manifold learning is impossible, such results leverage the probabilistic method rather than more fundamental information-theoretic tools [69].

Recent years have seen the data-science community widely embrace another non-linear representation learning approach that assumes data lie near a *union of subspaces (UoS)*. This approach tends to have several advantages over manifold learning because of the linearity of individual components (subspaces) in the representation learning model. While there exist methods that learn the subspaces explicitly, one of the most popular classes of representation learning under the UoS model in which the subspaces are implicitly learned is referred to as *dictionary learning* [38]. Formally, dictionary learning assumes the data space to be $\mathcal{Y} = \mathbb{R}^m$, the feature space to be

$$\mathcal{X} = \{\mathbf{x} \in \mathbb{R}^p : \|\mathbf{x}\|_0 \leq k\} \tag{1.23}$$

with $k \ll m \leq p$, and the random generative model to be

$$\mathbf{Y} = \mathbf{D}\mathbf{X} + \mathbf{W} \tag{1.24}$$

with $\mathbf{D} \in \mathcal{F} = \{\mathbf{D} \in \mathbb{R}^{m \times p} : \|\mathbf{D}_i\|_2 = 1, i = 1, \ldots, p\}$ representing a dictionary and $\mathbf{W} \in \mathbb{R}^m$ representing the random noise vector. This corresponds to the random data vector $\mathbf{Y}$ lying near a union of $\binom{p}{k}$ k-dimensional subspaces. Notice that, while $F(\cdot) = \mathbf{D}$ is a linear operator,[10] its inverse image $G(\cdot)$ is highly nonlinear and typically computed as

$$\widehat{\mathbf{X}} = G(\mathbf{Y}) = \arg \min_{\mathbf{X}:\|\mathbf{X}\|_0 \leq k} \|\mathbf{Y} - \mathbf{D}\mathbf{X}\|_2^2, \tag{1.25}$$

with (1.25) referred to as *sparse coding*.

The last 15 years have seen the development of a number of algorithms that enable learning of the dictionary $\mathbf{D}$ both in unsupervised and in supervised settings [41, 70–72]. Sample complexity of these algorithms in terms of both infeasibility (converse) and achievability has been a more recent effort [44, 45, 73–76]. In particular, it is established in [44] using Fano's inequality that the number of samples N, which are independent realizations of the generative model (1.24), i.e.,

$$\mathbf{y}^i = \mathbf{D}\mathbf{x}^i + \mathbf{w}^i, \quad i = 1, \ldots, N, \tag{1.26}$$

must scale at least as fast as $N = O(mp^2\epsilon^{-2})$ in order to ensure recovery of an estimate $\widehat{\mathbf{D}}$ such that $\|\widehat{\mathbf{D}} - \mathbf{D}\|_F \leq \epsilon$. This lower bound on sample complexity, which is derived in the minimax sense, is akin to the converse bounds in source and channel coding in classical information theory. However, general tightness of this lower bound, which requires analyzing explicit (deterministic) dictionary-learning algorithms and deriving

[10] Strictly speaking, $\mathbf{D}$ restricted to $\mathcal{X}$ is also nonlinear.

matching achievability results, remains an open problem. Computational limits are also in general open for dictionary learning.

Recent years have also seen extension of these results to the case of data that have a multidimensional (tensor) structure [45]. We refer the reader to Chapter 5 in the book for a more comprehensive review of dictionary-learning results pertaining to both vector and tensor data.

Linear representation learning, manifold learning, and dictionary learning are all based on a geometric viewpoint of data. It is also possible to view these representation-learning techniques from a purely numerical linear algebra perspective. Data representations in this case are referred to as *matrix factorization-based representations*. The matrix factorization perspective of representation learning allows one to expand the classes of learning techniques by borrowing from the rich literature on linear algebra. Non-negative matrix factorization [77], for instance, allows one to represent data that are inherently non-negative in terms of non-negative features that can be assigned physical meanings. We refer the reader to [78] for a more comprehensive overview of matrix factorizations in data science; [79] also provides a recent information-theoretic analysis of non-negative matrix factorization.

1.7.3 Recent Trends in Representation Learning

Beyond the subspace and UoS models described above, another emerging approach to learning data representations relates to the use of *deep neural networks* [80]. In particular, this involves designing a nonlinear transformation $G : \mathcal{Y} \to \mathcal{X}$ consisting of a series of stages, with each stage encompassing a linear and a nonlinear operation, that can be used to produce a data representation $\mathbf{x} \in \mathcal{X}$ given a data instance $\mathbf{y} \in \mathcal{Y}$ as follows:

$$\mathbf{x} = G(\mathbf{y}) = f_L(\mathbf{W}^L \cdot f_{L-1}(\mathbf{W}^{L-1} \cdot (\cdots f_1(\mathbf{W}^1 \mathbf{y} + \mathbf{b}^1) \cdots) + \mathbf{b}^{L-1}) + \mathbf{b}^L), \tag{1.27}$$

where $\mathbf{W}^i \in \mathbb{R}^{n_i \times n_{i-1}}$ is a weight matrix, $\mathbf{b}^i \in \mathbb{R}^{n_i}$ is a bias vector, $f_i : \mathbb{R}^{n_i} \to \mathbb{R}^{n_i}$ is a nonlinear operator such as a *rectified linear unit (ReLU)*, and L corresponds to the number of layers in the deep neural network. The challenge then relates to how to learn the set of weight matrices and bias vectors associated with the deep neural network. For example, in classification problems where each data instance $\mathbf{x}$ is associated with a discrete label ℓ, one typically relies on a training set $(\mathbf{y}^i, \ell^i), i = 1, \ldots, N$, to define a loss function that can be used to tune the various parameters of the network using algorithms such as gradient descent or stochastic gradient descent [81].

This approach to data representation underlies some of the most spectacular advances in areas such as computer vision, speech recognition, speech translation, natural language processing, and many more, but this approach is also not fully understood. However, information-theoretically oriented studies have also been recently conducted to gain insight into the performance of deep neural networks by enabling the analysis of the learning process or the design of new learning algorithms. For example, Tishby *et al.* [82] propose an information-theoretic analysis of deep neural networks based on the *information bottleneck* principle. They view the neural network learning process as a trade-off between compression and prediction that leads up to the extraction of a set of minimal sufficient statistics from the data in relation to the target task. Shwartz-Ziv and Tishby [83] – building upon the work in [82] – also propose an information-bottleneck-based analysis of deep neural networks. In particular, they study *information paths* in the so-called *information plane* capturing the evolution of a pair of items of mutual information over the network during the training process: one relates to the mutual information between the ith layer output and the target data label, and the other corresponds to the mutual information between the ith layer output and the data itself. They also demonstrate empirically that the widely used stochastic gradient descent algorithm undergoes a "fitting" phase – where the mutual information between the data representations and the target data label increases – and a "compression" phase – where the mutual information between the data representations and the data decreases. See also related works investigating the flow of information in deep networks [84–87].

Achille and Soatto [88] also use an information-theoretic approach to understand deep-neural-networks-based data representations. In particular, they show how deep neural networks can lead to minimal sufficient representations with properties such as invariance to nuisances, and provide bounds that connect the amount of information in the weights and the amount of information in the activations to certain properties of the activations such as invariance. They also show that a new information-bottleneck Lagrangian involving the information between the weights of a network and the training data can overcome various overfitting issues.

More recently, information-theoretic metrics have been used as a proxy to learn data representations. In particular, Hjelm *et al.* [89] propose unsupervised learning of representations by maximizing the mutual information between an input and the output of a deep neural network.

In summary, this body of work suggests that information-theoretic quantities such as mutual information can inform the analysis, design, and optimization of state-of-the-art representation learning approaches. Chapter 11 covers some of these recent trends in representation learning.

1.8 Information Theory and Data Analysis and Processing

The outputs of the data-representation block – the features – are often the basis for further data analysis or processing, encompassing both *statistical inference* and *statistical learning* tasks such as estimation, regression, classification, clustering, and many more.

Statistical inference forms the core of classical statistical signal processing and statistics. Broadly speaking, it involves use of explicit stochastic data models to understand various aspects of data samples (features). These models can be *parametric*, defined as those characterized by a finite number of parameters, or *non-parametric*, in which the number of parameters continuously increases with the number of data samples. There is a large portfolio of statistical inference tasks, but we limit our discussion to the problems of *model selection*, *hypothesis testing*, *estimation*, and *regression*.

Briefly, model selection involves the use of data features/samples to select a stochastic data model from a set of candidate models. Hypothesis testing, on the other hand, is the task of determining whether a certain postulated hypothesis (stochastic model) underlying the data is true or false. This is referred to as *binary* hypothesis testing, as opposed to *multiple* hypothesis testing in which the data are tested against several hypotheses. Statistical estimation, often studied under the umbrella of *inverse problems* in many disciplines, is the task of inferring some parameters underlying the stochastic data model. In contrast, regression involves estimating the relationships between different data features that are divided into the categories of *response variable(s)* (also known as *dependent variables*) and *predictors* (also known as *independent variables*).

Statistical learning, along with *machine learning*, primarily concentrates on approaches to find structure in data. In particular, while the boundary between statistical inference and statistical learning is not a hard one, statistical learning tends not to focus on explicit stochastic models of data generation; rather, it often treats the data generation mechanism as a black box, and primarily concentrates on learning a "model" with good prediction accuracy [90]. There are two major paradigms in statistical learning: *supervised learning* and *unsupervised learning*.

In supervised learning, one wishes to determine predictive relationships between and/or across data features. Representative supervised learning approaches include classification problems where the data features are mapped to a discrete set of values (a.k.a. *labels*) and regression problems where the data features are mapped instead to a continuous set of values. Supervised learning often involves two distinct phases of *training* and *testing*. The training phase involves use of a dataset, referred to as *training data*, to learn a model that finds the desired predictive relationship(s). These predictive relationships are often implicitly known in the case of training data, and the goal is to leverage this knowledge for learning a model during training that *generalizes* these relationships to as-yet unseen data. Often, one also employs a *validation dataset* in concert with training data to tune possible *hyperparameters* associated with a statistical-learning model. The testing phase involves use of another dataset with known characteristics, termed *test data*, to estimate the learned model's generalization capabilities. The error incurred by the learned model on training and test data is referred to as *training error* and *testing error*, respectively, while the error that the model will incur on future unseen data can be captured by the so-called *generalization error*. One of the biggest challenges in supervised learning is understanding the generalization error of a statistical learning model as a function of the number of data samples in training data.

In unsupervised learning, one wishes instead to determine the underlying structure within the data. Representative unsupervised learning approaches include *density estimation*, where the objective is to determine the underlying data distribution given a set of data samples, and *clustering*, where the aim is to organize the data points onto different groups so that points belonging to the same group exhibit some degree of similarity and points belonging to different groups are distinct.

Challenges arising in statistical inference and learning also involve analyzing, designing and optimizing inference and learning algorithms, and understanding statistical and computational limits in inference and learning tasks. We next provide a small sampling of statistical inference and statistical learning results that involve the use of information-theoretic tools and methods.

1.8.1 Statistical Inference

We now survey some representative results arising in model selection, estimation, regression, and hypothesis-testing problems that benefit from information-theoretic methods. We also offer an example associated with community detection and recovery on graphs where information-theoretic and related tools can be used to determine statistical and computational limits.

Model Selection

On the algorithmic front, the problem of model selection has been largely impacted by information-theoretic tools. *Given a data set, which statistical model "best" describes the data?* A huge array of work, dating back to the 1970s, has tackled this question using various information-theoretic principles. The *Akaike information criterion (AIC)* for model selection [91], for instance, uses the KL divergence as the main tool for derivation of the final criterion. The *minimum description length (MDL)* principle for model selection [92], on the other hand, makes a connection between source coding and model selection and seeks a model that best compresses the data. The AIC and MDL principles are just two of a number of information-theoretically inspired model-selection approaches; we refer the interested reader to Chapter 12 for further discussion.

Estimation and Regression

Over the years, various information-theoretic tools have significantly advanced our understanding of the interconnected problems of estimation and regression. In statistical inference, a typical estimation/regression problem involving a scalar random variable $Y \in \mathbb{R}$ takes the form

$$Y = f(\mathbf{X};\boldsymbol{\beta}) + W, \tag{1.28}$$

where the random vector $\mathbf{X} \in \mathbb{R}^p$ is referred to as a *covariate* in statistics and *measurement vector* in signal processing, $\boldsymbol{\beta} \in \mathbb{R}^p$ denotes the unknown parameter vector, termed *regression parameters* and *signal* in statistics and signal processing, respectively, and W represents observation noise/modeling error.[11] Both estimation and regression problems in statistical inference concern themselves with recovering $\boldsymbol{\beta}$ from N realizations $\{(y^i,\mathbf{x}^i)\}_{i=1}^N$ from the model (1.28) under an assumed $f(\cdot;\cdot)$. In estimation, one is interested in recovering a $\widehat{\boldsymbol{\beta}}$ that is as close to the true $\boldsymbol{\beta}$ as possible; in regression, on the other hand, one is concerned with prediction, i.e., how close $f(\mathbf{X};\widehat{\boldsymbol{\beta}})$ is to $f(\mathbf{X};\boldsymbol{\beta})$ for the random vector $\mathbf{X}$. Many modern setups in estimation/regression problems correspond to the high-dimensional setting in which $N \ll p$. Such setups often lead to seemingly ill-posed mathematical problems, resulting in the following important question: *How small can the estimation and/or regression errors be as a function of N, p, and properties of the covariates and parameters?*

Information-theoretic methods have been used in a variety of ways to address this question for a number of estimation/regression problems. The most well known of these

[11] The assumption is that raw data have been transformed into its features, which correspond to $\mathbf{X}$.

results are for the *generalized linear model (GLM)*, where the realizations of $(Y, \mathbf{X}, W)$ are given by

$$y^i = \mathbf{x}^{i^{\mathrm{T}}}\boldsymbol{\beta} + w^i \Longrightarrow \mathbf{y} = \widetilde{\mathbf{X}}\boldsymbol{\beta} + \mathbf{w}, \tag{1.29}$$

with $\mathbf{y} \in \mathbb{R}^N, \widetilde{\mathbf{X}} \in \mathbb{R}^{N \times p}$, and $\mathbf{w} \in \mathbb{R}^N$ denoting concatenations of y^i, $\mathbf{x}^{i^{\mathrm{T}}}$, and w^i, respectively. Fano's inequality has been used to derive lower bounds on the errors in GLMs under various assumptions on the matrix $\widetilde{\mathbf{X}}$ and $\boldsymbol{\beta}$ [93–95]. Much of this work has been limited to the case of *sparse* $\boldsymbol{\beta}$, in which it is assumed that no more than a few (say, $s \ll N$) regression parameters are non-zero [93, 94]. The work by Raskutti *et al.* [95] extends many of these results to $\boldsymbol{\beta}$ that is not strictly sparse. This work focuses on *approximately* sparse regression parameters, defined as lying within an ℓ_q ball, $q \in [0,1]$ of radius R_q as follows:

$$\mathcal{B}_q(R_q) = \left\{\boldsymbol{\beta} : \sum_{i=1}^{p} |\beta_i|^q \leq R_q\right\}, \tag{1.30}$$

and provides matching minimax lower and upper bounds (i.e., the optimal minimax rate) both for the estimation error, $\|\widehat{\boldsymbol{\beta}} - \boldsymbol{\beta}\|_2^2$, and for the prediction error, $(1/n)\|\widetilde{\mathbf{X}}(\widehat{\boldsymbol{\beta}} - \boldsymbol{\beta})\|_2^2$. In particular, it is established that, under suitable assumptions on $\widetilde{\mathbf{X}}$, it is possible to achieve estimation and prediction errors in GLMs that scale as $R_q(\log p/N)^{1-q/2}$. The corresponding result for exact sparsity can be derived by setting $q = 0$ and $R_q = s$. Further, there exist no algorithms, regardless of their computational complexity, that can achieve errors smaller than this rate for every $\boldsymbol{\beta}$ in an ℓ_q ball. As one might expect, Fano's inequality is the central tool used by Raskutti *et al.* [95] to derive this lower bound (the "converse"). The achievability result requires direct analysis of algorithms, as opposed to use of the probabilistic method in classical information theory. Since both the converse and the achievability bounds coincide in regression and estimation under the GLM, we end up with a sharp statistical phase transition. Chapters 6, 7, 8, and 16 elaborate further on various other recovery and estimation problems arising in data science, along with key tools that can be used to gain insight into such problems.

Additional information-theoretic results are known for the *standard linear model* – where $\mathbf{Y} = \sqrt{s}\mathbf{X}\boldsymbol{\beta} + \mathbf{W}$, with $\mathbf{Y} \in \mathbb{R}^n, \mathbf{X} \in \mathbb{R}^{n \times p}, \boldsymbol{\beta} \in \mathbb{R}^p, \mathbf{W} \in \mathbb{R}^n \sim \mathcal{N}(\mathbf{0}, \mathbf{I})$, and s a scaling factor representing a signal-to-noise ratio. In particular, subject to mild conditions on the distribution of the parameter vector, it has been established that the mutual information and the minimum mean-squared error obey the so-called *I-MMSE* relationship given by [96]:

$$\frac{dI\left(\boldsymbol{\beta}; \sqrt{s}\mathbf{X}\boldsymbol{\beta} + \mathbf{W}\right)}{ds} = \frac{1}{2} \cdot \mathrm{mmse}\left(\mathbf{X}\boldsymbol{\beta} | \sqrt{s}\mathbf{X}\boldsymbol{\beta} + \mathbf{W}\right), \tag{1.31}$$

where $I\left(\boldsymbol{\beta}; \sqrt{s}\mathbf{X}\boldsymbol{\beta} + \mathrm{W}\right)$ corresponds to the mutual information between the standard linear model input and output and

$$\mathrm{mmse}\left(\mathbf{X}\boldsymbol{\beta} | \sqrt{s}\mathbf{X}\boldsymbol{\beta} + \mathbf{W}\right) = \mathbb{E}\left\{\left\|\mathbf{X}\boldsymbol{\beta} - \mathbb{E}\left\{\mathbf{X}\boldsymbol{\beta} | \sqrt{s}\mathbf{X}\boldsymbol{\beta} + \mathbf{W}\right\}\right\|_2^2\right\} \tag{1.32}$$

is the minimum mean-squared error associated with the estimation of $\mathbf{X}\beta$ given $\sqrt{s}\mathbf{X}\beta + \mathbf{W}$. Other relations involving information-theoretic quantities, such as mutual information, and estimation-theoretic ones have also been established in a wide variety of settings in recent years, such as Poisson models [97]. These relations have been shown to have important implications in classical information-theoretic problems – notably in the analysis and design of communications systems (e.g., [98–101]) – and, more recently, in data-science ones. In particular, Chapter 7 elaborates further on how the I-MMSE relationship can be used to gain insight into modern high-dimensional inference problems.

Hypothesis Testing

Information-theoretic tools have also been advancing our understanding of hypothesis-testing problems (one of the most widely used statistical inference techniques). In general, we can distinguish between *binary hypothesis-testing* problems, where the data are tested against two hypotheses often known as the *null* and the *alternate* hypotheses, and *multiple-hypothesis-testing* problems in which the data are tested against multiple hypotheses. We can also distinguish between *Bayesian* approaches to hypothesis testing, where one specifies a prior probability associated with each of the hypotheses, and *non-Bayesian* ones, in which one does not specify *a priori* any prior probability.

Formally, a classical formulation of the binary hypothesis-testing problem involves testing whether a number of i.i.d. data samples (features) $\mathbf{x}^1, \mathbf{x}^2, \ldots, \mathbf{x}^N$ of a random variable $\mathbf{X} \in \mathcal{X} \sim p_X$ conform to one or other of the hypotheses $\mathcal{H}_0 : p_X = p_0$ and $\mathcal{H}_1 : p_X = p_1$, where under the first hypothesis one postulates that the data are generated i.i.d. according to model (distribution) p_0 and under the second hypothesis one assumes the data are generated i.i.d. according to model (distribution) p_1. A binary hypothesis test $T : \mathcal{X} \times \cdots \times \mathcal{X} \to \{\mathcal{H}_0, \mathcal{H}_1\}$ is a mapping that outputs an estimate of the hypothesis given the data samples.

In non-Bayesian settings, the performance of such a binary hypothesis test can be described by two *error probabilities*. The *type-I error probability*, which relates to the rejection of a true null hypothesis, is given by

$$P_{e|0}(T) = \mathbb{P}\left(T\left(\mathbf{X}^1, \mathbf{X}^2, \ldots, \mathbf{X}^N\right) = \mathcal{H}_1 | \mathcal{H}_0\right) \tag{1.33}$$

and the *type-II error probability*, which relates to the failure to reject a false null hypothesis, is given by

$$P_{e|1}(T) = \mathbb{P}\left(T\left(\mathbf{X}^1, \mathbf{X}^2, \ldots, \mathbf{X}^N\right) = \mathcal{H}_0 | \mathcal{H}_1\right). \tag{1.34}$$

In this class of problems, one is typically interested in minimizing one of the error probabilities subject to a constraint on the other error probability as follows:

$$P_e(\alpha) = \min_{T : P_{e|0}(T) \le \alpha} P_{e|1}(T), \tag{1.35}$$

where the minimum can be achieved using the well-known *Neymann–Pearson test* [102].

Information-theoretic tools – such as typicality [7] – have long been used to analyze the performance of this class of problems. For example, the classical Stein lemma asserts that asymptotically with the number of data samples approaching infinity [7]

$$\lim_{\alpha\to 0}\lim_{N\to\infty}\frac{1}{N}\cdot\log P_e(\alpha) = -D(p_0\|p_1), \tag{1.36}$$

where $D(\cdot\|\cdot)$ is the Kullback–Leibler distance between two different distributions.

In Bayesian settings, the performance of a hypothesis-testing problem can be described by the *average error probability* given by

$$\begin{aligned} P_e(T) = &\mathbb{P}(\mathcal{H}_0)\cdot\mathbb{P}\left(T\left(\mathbf{X}^1,\mathbf{X}^2,\ldots,\mathbf{X}^N\right)=\mathcal{H}_1|\mathcal{H}_0\right)\\ &+\mathbb{P}(\mathcal{H}_1)\cdot\mathbb{P}\left(T\left(\mathbf{X}^1,\mathbf{X}^2,\ldots,\mathbf{X}^N\right)=\mathcal{H}_0|\mathcal{H}_1\right), \end{aligned} \tag{1.37}$$

where $\mathbb{P}(\mathcal{H}_0)$ and $\mathbb{P}(\mathcal{H}_1)$ relate to the prior probabilities ascribed to each of the hypothesis. It is well known that the *maximum a posteriori test* (or *maximum a posteriori decision rule*) minimizes this average error probability [102].

Information-theoretic tools have been used to analyze the performance of Bayesian hypothesis-testing problems too. For example, consider a simple M-ary Bayesian hypothesis-testing problem involving M possible hypotheses, which are modeled by a random variable C drawn according to some prior distribution p_C, and the data are modeled by a random variable X drawn according to the distribution $p_{X|C}$. In particular, since it is often difficult to characterize in closed form the minimum average error probability associated with the optimal maximum *a posteriori* test, information-theoretic measures can be used to upper- or lower-bound this quantity. A lower bound on the minimum average error probability – derived from Fano's inequality – is given by

$$P_{e,\min} = \min_T P_e(T) \geq 1 - \frac{H(C|X)}{\log_2(M-1)}. \tag{1.38}$$

An upper bound on the minimum average error probability is [103]

$$P_{e,\min} = \min_T P_e(T) \leq 1 - \exp(-H(C|X)). \tag{1.39}$$

A number of other bounds on the minimum average error probability involving *Shannon information measures*, *Rényi information measures*, or other generalizations have also been devised over the years [104–106] and have led to stronger converse results not only in classical information-theory problems but also in data-science ones [107].

Example: Community Detection and Estimation on Graphs

We now briefly offer examples of hypothesis testing and estimation problems arising in modern data analysis that exhibit sharp statistical and computational phase transitions which can be revealed using emerging information-theoretic methods.

To add some context, in modern data analysis it is increasingly common for datasets to consist of various items exhibiting complex relationships among them, such as pair-wise or multi-way interactions between items. Such datasets can therefore be represented by a *graph* or a *network* of interacting items where the network *vertices* denote different items and the network *edges* denote pair-wise interactions between the items.[12] Our example – involving a concrete challenge arising in the analysis of such networks of interacting items – relates to the *detection* and *recovery* of *community* structures within the graph. A community consists of a subset of vertices within the graph that are densely connected to one another but sparsely connected to other vertices within the graph [108].

Concretely, consider a simple instance of such problems where one wishes to discern whether the underlying graph is random or whether it contains a dense subgraph (a community). Mathematically, we can proceed by considering two objects: (1) an Erdős–Rényi random graph model $\mathcal{G}(N,q)$ consisting of N vertices where each pair of vertices is connected independently with probability q; and (2) a planted dense subgraph model $\mathcal{G}(N,K,p,q)$ with N vertices where each vertex is assigned to a random set $\mathcal{S}$ with probability K/N ($K \leq N$) and each pair of vertices are connected with probability p if both of them are in the set $\mathcal{S}$ and with probability q otherwise ($p > q$). We can then proceed by constructing a hypothesis-testing problem where under one hypothesis one postulates that the observed graph is drawn from $\mathcal{G}(N,q)$ and under the other hypothesis one postulates instead that the observed graph is drawn from $\mathcal{G}(N,K,p,q)$. It can then be established in the asymptotic regime $p = cq = O(N^{-\alpha})$, $K = O\left(N^{-\beta}\right)$, $N \to \infty$, that (a) one can detect the community with arbitrarily low error probability with simple linear-time algorithms when $\beta > \frac{1}{2} + \alpha/4$; (b) one can detect the community with arbitrarily low error probability only with no-polynomial-time algorithms when $\alpha < \beta < \frac{1}{2} + \alpha/4$; and (c) there is no test – irrespective of its complexity – that can detect the community with arbitrarily low error probability when $\beta < \min\left(\alpha, \frac{1}{2} + \alpha/4\right)$ [109]. It has also been established that the recovery of the community exhibits identical statistical and computational limits.

This problem in fact falls under a much wider problem class arising in modern data analysis, involving the detection or recovery of structures planted in random objects such as graphs, matrices, or tensors. The characterization of statistical limits in detection or recovery of such structures can typically be done by leveraging various tools: (1) statistical tools such as the first- and the second-moment methods; (2) information-theoretic methods such as mutual information and rate distortion; and (3) statistical physics-based tools such as the interpolation method. In contrast, the characterization of computational limits associated with these statistical problems often involves finding an approximate randomized polynomial-time reduction, mapping certain graph-theoretic problems such as the planted-clique problem approximately to the statistical problem under consideration, in order to show that the statistical problem is at least as hard as the planted-clique problem. Chapter 13 provides a comprehensive overview of emerging methods – including information-theoretic ones – used to establish both statistical and computational limits in modern data analysis.

[12] Some datasets can also be represented by *hyper-graphs* of interacting items, where *vertices* denote the different objects and *hyper-edges* denotes *multi-way* interactions between the different objects.

1.8.2 Statistical Learning

We now survey emerging results in statistical learning that are benefiting from information-theoretic methods.

Supervised Learning

In the supervised learning setup, one desires to learn a hypothesis based on a set of data examples that can be used to make predictions given new data [90]. In particular, in order to formalize the problem, let $\mathcal{X}$ be the *domain set*, $\mathcal{Y}$ be the *label set*, $\mathcal{Z} = \mathcal{X} \times \mathcal{Y}$ be the *examples domain*, μ be a distribution on $\mathcal{Z}$, and $\mathcal{W}$ a *hypothesis class* (i.e., $\mathcal{W} = \{W\}$ is a set of *hypotheses* $W : \mathcal{X} \to \mathcal{Y}$). Let also $\mathcal{S} = \{\mathbf{z}^1, \dots, \mathbf{z}^N\} = \{(\mathbf{x}^1, y^1), \dots, (\mathbf{x}^N, y^N)\} \in \mathcal{Z}^N$ be the *training set* – consisting of a number of data points and their associated labels – drawn i.i.d. from $\mathcal{Z}$ according to μ. A *learning algorithm* is a Markov kernel that maps the *training set* $\mathcal{S}$ to an element W of the hypothesis class $\mathcal{W}$ according to the probability law $p_{W|\mathcal{S}}$.

A key challenge relates to understanding the generalization ability of the learning algorithm, where the *generalization error* corresponds to the difference between the *expected (or true) error* and the *training (or empirical) error*. In particular, by considering a non-negative *loss function* $L : \mathcal{W} \times \mathcal{Z} \to \mathbb{R}^+$, one can define the expected error and the training error associated with a hypothesis W as follows:

$$\mathsf{loss}_\mu(W) = \mathbb{E}\{L(W, \mathbf{Z})\} \qquad \text{and} \qquad \mathsf{loss}_\mathcal{S}(W) = \frac{1}{N}\sum_{i=1}^{N} L(W, \mathbf{z}^i),$$

respectively. The generalization error is given by

$$\mathsf{gen}(\mu, W) = \mathsf{loss}_\mu(W) - \mathsf{loss}_\mathcal{S}(W)$$

and its expected value is given by

$$\mathsf{gen}(\mu, p_{W|\mathcal{S}}) = \mathbb{E}\left\{\mathsf{loss}_\mu(W) - \mathsf{loss}_\mathcal{S}(W)\right\},$$

where the expectation is with respect to the joint distribution of the algorithm input (the training set) and the algorithm output (the hypothesis).

A number of approaches have been developed throughout the years to characterize the generalization error of a learning algorithm, relying on either certain complexity measures of the hypothesis space or certain properties of the learning algorithm. These include VC-based bounds [110], algorithmic stability-based bounds [111], algorithmic robustness-based bounds [112], PAC-Bayesian bounds [113], and many more. However, many of these generalization error bounds cannot explain the generalization abilities of a variety of machine-learning methods for various reasons: (1) some of the bounds depend only on the hypothesis class and not on the learning algorithm, (2) existing bounds do not easily exploit dependences between different hypotheses, and (3) existing bounds also do not exploit dependences between the learning algorithm input and output.

More recently, approaches leveraging information-theoretic tools have been emerging to characterize the generalization ability of various learning methods. Such approaches often express the generalization error in terms of certain information measures between the algorithm input (the training dataset) and the algorithm output (the hypothesis), thereby incorporating the various ingredients associated with the learning problem, including the dataset distribution, the hypothesis space, and the learning algorithm itself. In particular, inspired by [114], Xu and Raginsky [115] derive an upper bound on the generalization error, applicable to σ-sub-Gaussian loss functions, given by

$$\left|\mathsf{gen}(\mu, p_{W|S})\right| \leq \sqrt{\frac{2\sigma^2}{N} \cdot I(S; W)},$$

where $I(S; W)$ corresponds to the mutual information between the input – the dataset – and the output – the hypothesis – of the algorithm. This bound supports the intuition that the less information the output of the algorithm contains about the input to the algorithm the less it will overfit, providing a means to strike a balance between the ability to fit data and the ability to generalize to new data by controlling the algorithm's input–output mutual information. Raginsky *et al.* [116] also propose similar upper bounds on the generalization error based on several information-theoretic measures of algorithmic stability, capturing the idea that the output of a stable learning algorithm cannot depend "too much" on any particular training example. Other generalization error bounds involving information-theoretic quantities appear in [117, 118]. In particular, Asadi *et al.* [118] combine chaining and mutual information methods to derive generalization error bounds that significantly outperform existing ones.

Of particular relevance, these information-theoretically based generalization error bounds have also been used to gain further insight into machine-learning models and algorithms. For example, Pensia *et al.* [119] build upon the work by Xu and Raginsky [115] to derive very general generalization error bounds for a broad class of iterative algorithms that are characterized by bounded, noisy updates with Markovian structure, including stochastic gradient Langevin dynamics (SGLD) and variants of the stochastic gradient Hamiltonian Monte Carlo (SGHMC) algorithm. This work demonstrates that mutual information is a very effective tool for bounding the generalization error of a large class of iterative *empirical risk minimization* (ERM) algorithms. Zhang *et al.* [120], on the other hand, build upon the work by Xu and Raginsky [115] to study the expected generalization error of deep neural networks, and offer a bound that shows that the error decreases exponentially to zero with the increase of convolutional and pooling layers in the network. Other works that study the generalization ability of deep networks based on information-theoretic considerations and measures include [121, 122]. Chapters 10 and 11 scope these directions in supervised learning problems.

Unsupervised Learning

In unsupervised learning setups, one desires instead to understand the structure associated with a set of data examples. In particular, multivariate information-theoretic functionals such as partition information, minimum partition information, and multi-information have been recently used in the formulation of unsupervised clustering problems [123, 124]. Chapter 9 elaborates further on such approaches to unsupervised learning problems.

1.8.3 Distributed Inference and Learning

Finally, we add that there has also been considerable interest in the generalization of the classical statistical inference and learning problems overviewed here to the distributed setting, where a statistician/learner has access only to data distributed across various terminals via a series of limited-capacity channels. In particular, much progress has been made in distributed-estimation [125–127], hypothesis-testing [128–138], learning [139, 140], and function-computation [141] problems in recent years. Chapters 14 and 15 elaborate further on how information theory is advancing the state-of-the-art for this class of problems.

1.9 Discussion and Conclusion

This chapter overviewed the classical information-theoretic problems of data compression and communication, questions arising within the context of these problems, and classical information-theoretic tools used to illuminate fundamental architectures, schemes, and limits for data compression and communication.

We then discussed how information-theoretic methods are currently advancing the frontier of data science by unveiling new data-processing architectures, data-processing limits, and algorithms. In particular, we scoped out how information theory is leading to a new understanding of data-acquisition architectures, and provided an overview of how information-theoretic methods have been uncovering limits and algorithms for linear and nonlinear representation learning problems, including deep learning. Finally, we also overviewed how information-theoretic tools have been contributing to our understanding of limits and algorithms for statistical inference and learning problems.

Beyond the typical data-acquisition, data-representation, and data-analysis tasks covered throughout this introduction, there are also various other emerging challenges in data science that are benefiting from information-theoretic techniques. For example, *privacy* is becoming a very relevant research area in data science in view of the fact that data analysis can not only reveal useful insights but can also potentially disclose sensitive information about individuals. Differential privacy is an inherently information-theoretic framework that can be used as the basis for the development of data-release mechanisms that control the amount of private information leaked to a data analyst while retaining some degree of utility [142]. Other information-theoretic frameworks have also been used to develop data pre-processing mechanisms that strike a balance between the amount of useful information and private information leaked to a data analyst (e.g., [143]).

Fairness is likewise becoming a very relevant area in data science because data analysis can also potentially exacerbate biases in decision-making, such as discriminatory treatments of individuals according to their membership of a legally protected group such as race or gender. Such biases may arise when protected variables (or correlated ones) are used explicitly in the decision-making. Biases also arise when learning algorithms that inherit biases present in training sets are then used in decision-making.

Recent works have concentrated on the development of information-theoretically based data pre-processing schemes that aim simultaneously to control discrimination and preserve utility (e.g., [144]).

Overall, we anticipate that information-theoretic methods will play an increasingly important role in our understanding of data science in upcoming years, including in shaping data-processing architectures, in revealing fundamental data-processing limits, and in the analysis, design, and optimization of new data-processing algorithms.

Acknowledgments

The work of Miguel R. D. Rodrigues and Yonina C. Eldar was supported in part by the Royal Society under award IE160348. The work of Stark C. Draper was supported in part by a Discovery Research Grant from the Natural Sciences and Engineering Research Council of Canada (NSERC). The work of Waheed U. Bajwa was supported in part by the National Science Foundation under award CCF-1453073 and by the Army Research Office under award W911NF-17-1-0546.

References

[1] C. E. Shannon, "A mathematical theory of communications," *Bell System Technical J.*, vol. 27, nos. 3–4, pp. 379–423, 623–656, 1948.

[2] R. G. Gallager, *Information theory and reliable communications*. Wiley, 1968.

[3] T. Berger, *Rate distortion theory: A mathematical basis for data compression*. Prentice-Hall, 1971.

[4] I. Csiszár and J. Körner, *Information theory: Coding theorems for discrete memoryless systems*. Cambridge University Press, 2011.

[5] A. Gersho and R. M. Gray, *Vector quantization and signal compression*. Kluwer Academic Publishers, 1991.

[6] D. J. C. MacKay, *Information theory, inference and learning algorithms*. Cambridge University Press, 2003.

[7] T. M. Cover and J. A. Thomas, *Elements of information theory*. John Wiley & Sons, 2006.

[8] R. W. Yeung, *Information theory and network coding*. Springer, 2008.

[9] A. El Gamal and Y.-H. Kim, *Network information theory*. Cambridge University Press, 2011.

[10] E. Arikan, "Some remarks on the nature of the cutoff rate," in *Proc. Workshop Information Theory and Applications (ITA '06)*, 2006.

[11] R. E. Blahut, *Theory and practice of error control codes*. Addison-Wesley Publishing Company, 1983.

[12] S. Lin and D. J. Costello, *Error control coding*. Pearson, 2005.

[13] R. M. Roth, *Introduction to coding theory*. Cambridge University Press, 2006.

[14] T. Richardson and R. Urbanke, *Modern coding theory*. Cambridge University Press, 2008.

[15] W. E. Ryan and S. Lin, *Channel codes: Classical and modern*. Cambridge University Press, 2009.

[16] E. Arikan, "Channel polarization: A method for constructing capacity-achieving codes for symmetric binary-input memoryless channels," *IEEE Trans. Information Theory*, vol. 55, no. 7, pp. 3051–3073, 2009.

[17] A. Jiménez-Feltström and K. S. Zigangirov, "Time-varying periodic convolutional codes with low-density parity-check matrix," *IEEE Trans. Information Theory*, vol. 45, no. 2, pp. 2181–2191, 1999.

[18] M. Lentmaier, A. Sridharan, D. J. J. Costello, and K. S. Zigangirov, "Iterative decoding threshold analysis for LDPC convolutional codes," *IEEE Trans. Information Theory*, vol. 56, no. 10, pp. 5274–5289, 2010.

[19] S. Kudekar, T. J. Richardson, and R. L. Urbanke, "Threshold saturation via spatial coupling: Why convolutional LDPC ensembles perform so well over the BEC," *IEEE Trans. Information Theory*, vol. 57, no. 2, pp. 803–834, 2011.

[20] E. J. Candès and M. B. Wakin, "An introduction to compressive sampling," *IEEE Signal Processing Mag.*, vol. 25, no. 2, pp. 21–30, 2008.

[21] H. Q. Ngo and D.-Z. Du, "A survey on combinatorial group testing algorithms with applications to DNA library screening," *Discrete Math. Problems with Medical Appl.*, vol. 55, pp. 171–182, 2000.

[22] G. K. Atia and V. Saligrama, "Boolean compressed sensing and noisy group testing," *IEEE Trans. Information Theory*, vol. 58, no. 3, pp. 1880–1901, 2012.

[23] D. Donoho and J. Tanner, "Observed universality of phase transitions in high-dimensional geometry, with implications for modern data analysis and signal processing," *Phil. Trans. Roy. Soc. A: Math., Phys. Engineering Sci.*, pp. 4273–4293, 2009.

[24] D. Amelunxen, M. Lotz, M. B. McCoy, and J. A. Tropp, "Living on the edge: Phase transitions in convex programs with random data," *Information and Inference*, vol. 3, no. 3, pp. 224–294, 2014.

[25] J. Banks, C. Moore, R. Vershynin, N. Verzelen, and J. Xu, "Information-theoretic bounds and phase transitions in clustering, sparse PCA, and submatrix localization," *IEEE Trans. Information Theory*, vol. 64, no. 7, pp. 4872–4894, 2018.

[26] R. Monasson, R. Zecchina, S. Kirkpatrick, B. Selman, and L. Troyansky, "Determining computational complexity from characteristic 'phase transitions,'" *Nature*, vol. 400, no. 6740, pp. 133–137, 1999.

[27] G. Zeng and Y. Lu, "Survey on computational complexity with phase transitions and extremal optimization," in *Proc. 48th IEEE Conf. Decision and Control (CDC '09)*, 2009, pp. 4352–4359.

[28] Y. C. Eldar, *Sampling theory: Beyond bandlimited systems*. Cambridge University Press, 2014.

[29] C. E. Shannon, "Coding theorems for a discrete source with a fidelity criterion," *IRE National Convention Record*, vol. 4, no. 1, pp. 142–163, 1959.

[30] A. Kipnis, A. J. Goldsmith, Y. C. Eldar, and T. Weissman, "Distortion-rate function of sub-Nyquist sampled Gaussian sources," *IEEE Trans. Information Theory*, vol. 62, no. 1, pp. 401–429, 2016.

[31] A. Kipnis, Y. C. Eldar, and A. J. Goldsmith, "Analog-to-digital compression: A new paradigm for converting signals to bits," *IEEE Signal Processing Mag.*, vol. 35, no. 3, pp. 16–39, 2018.

[32] A. Kipnis, Y. C. Eldar, and A. J. Goldsmith, "Fundamental distortion limits of analog-to-digital compression," *IEEE Trans. Information Theory*, vol. 64, no. 9, pp. 6013–6033, 2018.

[33] M. R. D. Rodrigues, N. Deligiannis, L. Lai, and Y. C. Eldar, "Rate-distortion trade-offs in acquisition of signal parameters," in *Proc. IEEE International Conference or Acoustics, Speech, and Signal Processing (ICASSP '17)*, 2017.

[34] N. Shlezinger, Y. C. Eldar, and M. R. D. Rodrigues, "Hardware-limited task-based quantization," submitted to *IEEE Trans. Signal Processing*, accepted 2019.

[35] N. Shlezinger, Y. C. Eldar, and M. R. D. Rodrigues, "Asymptotic task-based quantization with application to massive MIMO," submitted to *IEEE Trans. Signal Processing*, accepted 2019.

[36] A. Argyriou, T. Evgeniou, and M. Pontil, "Convex multi-task feature learning," *Machine Learning*, vol. 73, no. 3, pp. 243–272, 2008.

[37] A. Coates, A. Ng, and H. Lee, "An analysis of single-layer networks in unsupervised feature learning," in *Proc. 14th International Conference on Artificial Intelligence and Statistics (AISTATS '11)*, 2011, pp. 215–223.

[38] I. Tosic and P. Frossard, "Dictionary learning," *IEEE Signal Processing Mag.*, vol. 28, no. 2, pp. 27–38, 2011.

[39] Y. Bengio, A. Courville, and P. Vincent, "Representation learning: A review and new perspectives," *IEEE Trans. Pattern Analysis Machine Intelligence*, vol. 35, no. 8, pp. 1798–1828, 2013.

[40] S. Yu, K. Yu, V. Tresp, H.-P. Kriegel, and M. Wu, "Supervised probabilistic principal component analysis," in *Proc. 12th ACM SIGKDD International Conference on Knowledge Discovery and Data Mining (KDD '06)*, 2006, pp. 464–473.

[41] J. Mairal, F. Bach, J. Ponce, G. Sapiro, and A. Zisserman, "Supervised dictionary learning," in *Proc. Advances in Neural Information Processing Systems (NeurIPS '09)*, 2009, pp. 1033–1040.

[42] V. Vu and J. Lei, "Minimax rates of estimation for sparse PCA in high dimensions," in *Proc. 15th International Conference on Artificial Intelligence and Statistics (AISTATS '12)*, 2012, pp. 1278–1286.

[43] T. T. Cai, Z. Ma, and Y. Wu, "Sparse PCA: Optimal rates and adaptive estimation," *Annals Statist.*, vol. 41, no. 6, pp. 3074–3110, 2013.

[44] A. Jung, Y. C. Eldar, and N. Görtz, "On the minimax risk of dictionary learning," *IEEE Trans. Information Theory*, vol. 62, no. 3, pp. 1501–1515, 2016.

[45] Z. Shakeri, W. U. Bajwa, and A. D. Sarwate, "Minimax lower bounds on dictionary learning for tensor data," *IEEE Trans. Information Theory*, vol. 64, no. 4, 2018.

[46] H. Hotelling, "Analysis of a complex of statistical variables into principal components," *J. Educ. Psychol.*, vol. 6, no. 24, pp. 417–441, 1933.

[47] M. E. Tipping and C. M. Bishop, "Probabilistic principal component analysis," *J. Roy. Statist. Soc. Ser. B*, vol. 61, no. 3, pp. 611–622, 1999.

[48] I. T. Jolliffe, *Principal component analysis*, 2nd edn. Springer-Verlag, 2002.

[49] P. Comon, "Independent component analysis: A new concept?" *Signal Processing*, vol. 36, no. 3, pp. 287–314, 1994.

[50] A. Hyvärinen, J. Karhunen, and E. Oja, *Independent component analysis*. John Wiley & Sons, 2004.

[51] P. Belhumeur, J. Hespanha, and D. Kriegman, "Eigenfaces vs. Fisherfaces: Recognition using class specific linear projection," *IEEE Trans. Pattern Analysis Machine Intelligence*, vol. 19, no. 7, pp. 711–720, 1997.

[52] J. Ye, R. Janardan, and Q. Li, "Two-dimensional linear discriminant analysis," in *Proc. Advances in Neural Information Processing Systems (NeurIPS '04)*, 2004, pp. 1569–1576.

[53] T. Hastie, R. Tibshirani, and J. Friedman, *The elements of statistical learning: Data mining, inference, and prediction*, 2nd edn. Springer, 2016.

[54] A. Hyvärinen, "Fast and robust fixed-point algorithms for independent component analysis," *IEEE Trans. Neural Networks*, vol. 10, no. 3, pp. 626–634, 1999.

[55] D. Erdogmus, K. E. Hild, Y. N. Rao, and J. C. Príncipe, "Minimax mutual information approach for independent component analysis," *Neural Comput.*, vol. 16, no. 6, pp. 1235–1252, 2004.

[56] A. Birnbaum, I. M. Johnstone, B. Nadler, and D. Paul, "Minimax bounds for sparse PCA with noisy high-dimensional data," *Annals Statist.*, vol. 41, no. 3, pp. 1055–1084, 2013.

[57] R. Krauthgamer, B. Nadler, and D. Vilenchik, "Do semidefinite relaxations solve sparse PCA up to the information limit?" *Annals Statist.*, vol. 43, no. 3, pp. 1300–1322, 2015.

[58] Q. Berthet and P. Rigollet, "Representation learning: A review and new perspectives," *Annals Statist.*, vol. 41, no. 4, pp. 1780–1815, 2013.

[59] T. Cai, Z. Ma, and Y. Wu, "Optimal estimation and rank detection for sparse spiked covariance matrices," *Probability Theory Related Fields*, vol. 161, nos. 3–4, pp. 781–815, 2015.

[60] A. Onatski, M. Moreira, and M. Hallin, "Asymptotic power of sphericity tests for high-dimensional data," *Annals Statist.*, vol. 41, no. 3, pp. 1204–1231, 2013.

[61] A. Perry, A. Wein, A. Bandeira, and A. Moitra, "Optimality and sub-optimality of PCA for spiked random matrices and synchronization," *arXiv:1609.05573*, 2016.

[62] Z. Ke, "Detecting rare and weak spikes in large covariance matrices," *arXiv:1609.00883*, 2018.

[63] D. L. Donoho and C. Grimes, "Hessian eigenmaps: Locally linear embedding techniques for high-dimensional data," *Proc. Natl. Acad. Sci. USA*, vol. 100, no. 10, pp. 5591–5596, 2003.

[64] J. B. Tenenbaum, V. de Silva, and J. C. Langford, "A global geometric framework for nonlinear dimensionality reduction," *Science*, vol. 290, no. 5500, pp. 2319–2323, 2000.

[65] R. Jenssen, "Kernel entropy component analysis," *IEEE Trans. Pattern Analysis Machine Intelligence*, vol. 32, no. 5, pp. 847–860, 2010.

[66] B. Schölkopf, A. Smola, and K.-R. Müller, "Kernel principal component analysis," in *Proc. Intl. Conf. Artificial Neural Networks (ICANN '97)*, 1997, pp. 583–588.

[67] J. Yang, X. Gao, D. Zhang, and J.-Y. Yang, "Kernel ICA: An alternative formulation and its application to face recognition," *Pattern Recognition*, vol. 38, no. 10, pp. 1784–1787, 2005.

[68] S. Mika, G. Ratsch, J. Weston, B. Schölkopf, and K. R. Mullers, "Fisher discriminant analysis with kernels," in *Proc. IEEE Workshop Neural Networks for Signal Processing IX*, 1999, pp. 41–48.

[69] H. Narayanan and S. Mitter, "Sample complexity of testing the manifold hypothesis," in *Proc. Advances in Neural Information Processing Systems (NeurIPS '10)*, 2010, pp. 1786–1794.

[70] K. Kreutz-Delgado, J. F. Murray, B. D. Rao, K. Engan, T.-W. Lee, and T. J. Sejnowski, "Dictionary learning algorithms for sparse representation," *Neural Comput.*, vol. 15, no. 2, pp. 349–396, 2003.

[71] M. Aharon, M. Elad, and A. Bruckstein, "K-SVD: An algorithm for designing overcomplete dictionaries for sparse representation," *IEEE Trans. Signal Processing*, vol. 54, no. 11, pp. 4311–4322, 2006.

[72] Q. Zhang and B. Li, "Discriminative K-SVD for dictionary learning in face recognition," in *Proc. IEEE Conference on Computer Vision and Pattern Recognition (CVPR '10)*, 2010, pp. 2691–2698.

[73] Q. Geng and J. Wright, "On the local correctness of ℓ^1-minimization for dictionary learning," in *Proc. IEEE International Symposium on Information Theory (ISIT '14)*, 2014, pp. 3180–3184.

[74] A. Agarwal, A. Anandkumar, P. Jain, P. Netrapalli, and R. Tandon, "Learning sparsely used overcomplete dictionaries," in *Proc. 27th Conference on Learning Theory (COLT '14)*, 2014, pp. 123–137.

[75] S. Arora, R. Ge, and A. Moitra, "New algorithms for learning incoherent and overcomplete dictionaries," in *Proc. 27th Conference on Learning Theory (COLT '14)*, 2014, pp. 779–806.

[76] R. Gribonval, R. Jenatton, and F. Bach, "Sparse and spurious: Dictionary learning with noise and outliers," *IEEE Trans. Information Theory*, vol. 61, no. 11, pp. 6298–6319, 2015.

[77] D. D. Lee and H. S. Seung, "Algorithms for non-negative matrix factorization," in *Proc. Advances in Neural Information Processing Systems 13 (NeurIPS '01)*, 2001, pp. 556–562.

[78] A. Cichocki, R. Zdunek, A. H. Phan, and S.-I. Amari, *Nonnegative matrix and tensor factorizations: Applications to exploratory multi-way data analysis and blind source separation*. John Wiley & Sons, 2009.

[79] M. Alsan, Z. Liu, and V. Y. F. Tan, "Minimax lower bounds for nonnegative matrix factorization," in *Proc. IEEE Statistical Signal Processing Workshop (SSP '18)*, 2018, pp. 363–367.

[80] Y. LeCun, Y. Bengio, and G. Hinton, "Deep learning," *Nature*, vol. 521, pp. 436–444, 2015.

[81] I. Goodfellow, Y. Bengio, and A. Courville, *Deep learning*. MIT Press, 2016, www.deeplearningbook.org.

[82] N. Tishby and N. Zaslavsky, "Deep learning and the information bottleneck principle," in *Proc. IEEE Information Theory Workshop (ITW '15)*, 2015.

[83] R. Shwartz-Ziv and N. Tishby, "Opening the black box of deep neural networks via information," *arXiv:1703.00810*, 2017.

[84] C. W. Huang and S. S. Narayanan, "Flow of Rényi information in deep neural networks," in *Proc. IEEE International Workshop Machine Learning for Signal Processing (MLSP '16)*, 2016.

[85] P. Khadivi, R. Tandon, and N. Ramakrishnan, "Flow of information in feed-forward deep neural networks," *arXiv:1603.06220*, 2016.

[86] S. Yu, R. Jenssen, and J. Príncipe, "Understanding convolutional neural network training with information theory," *arXiv:1804.09060*, 2018.

[87] S. Yu and J. Príncipe, "Understanding autoencoders with information theoretic concepts," *arXiv:1804.00057*, 2018.

[88] A. Achille and S. Soatto, "Emergence of invariance and disentangling in deep representations," *arXiv:1706.01350*, 2017.

[89] R. D. Hjelm, A. Fedorov, S. Lavoie-Marchildon, K. Grewal, P. Bachman, A. Trischler, and Y. Bengio, "Learning deep representations by mutual information estimation and maximization," in *International Conference on Learning Representations (ICLR '19)*, 2019.

[90] S. Shalev-Shwartz and S. Ben-David, *Understanding machine learning: From theory to algorithms*. Cambridge University Press, 2014.

[91] H. Akaike, "A new look at the statistical model identification," *IEEE Trans. Automation Control*, vol. 19, no. 6, pp. 716–723, 1974.

[92] A. Barron, J. Rissanen, and B. Yu, "The minimum description length principle in coding and modeling," *IEEE Trans. Information Theory*, vol. 44, no. 6, pp. 2743–2760, 1998.

[93] M. J. Wainwright, "Information-theoretic limits on sparsity recovery in the high-dimensional and noisy setting," *IEEE Trans. Information Theory*, vol. 55, no. 12, pp. 5728–5741, 2009.

[94] M. J. Wainwright, "Sharp thresholds for high-dimensional and noisy sparsity recovery using ℓ_1-constrained quadratic programming (lasso)," *IEEE Trans. Information Theory*, vol. 55, no. 5, pp. 2183–2202, 2009.

[95] G. Raskutti, M. J. Wainwright, and B. Yu, "Minimax rates of estimation for high-dimensional linear regression over ℓ_q-balls," *IEEE Trans. Information Theory*, vol. 57, no. 10, pp. 6976–6994, 2011.

[96] D. Guo, S. Shamai, and S. Verdú, "Mutual information and minimum mean-square error in Gaussian channels," *IEEE Trans. Information Theory*, vol. 51, no. 4, pp. 1261–1282, 2005.

[97] D. Guo, S. Shamai, and S. Verdú, "Mutual information and conditional mean estimation in Poisson channels," *IEEE Trans. Information Theory*, vol. 54, no. 5, pp. 1837–1849, 2008.

[98] A. Lozano, A. M. Tulino, and S. Verdú, "Optimum power allocation for parallel Gaussian channels with arbitrary input distributions," *IEEE Trans. Information Theory*, vol. 52, no. 7, pp. 3033–3051, 2006.

[99] F. Pérez-Cruz, M. R. D. Rodrigues, and S. Verdú, "Multiple-antenna fading channels with arbitrary inputs: Characterization and optimization of the information rate," *IEEE Trans. Information Theory*, vol. 56, no. 3, pp. 1070–1084, 2010.

[100] M. R. D. Rodrigues, "Multiple-antenna fading channels with arbitrary inputs: Characterization and optimization of the information rate," *IEEE Trans. Information Theory*, vol. 60, no. 1, pp. 569–585, 2014.

[101] A. G. C. P. Ramos and M. R. D. Rodrigues, "Fading channels with arbitrary inputs: Asymptotics of the constrained capacity and information and estimation measures," *IEEE Trans. Information Theory*, vol. 60, no. 9, pp. 5653–5672, 2014.

[102] S. M. Kay, *Fundamentals of statistical signal processing: Detection theory*. Prentice Hall, 1998.

[103] M. Feder and N. Merhav, "Relations between entropy and error probability," *IEEE Trans. Information Theory*, vol. 40, no. 1, pp. 259–266, 1994.

[104] I. Sason and S. Verdú, "Arimoto–Rényi conditional entropy and Bayesian M-ary hypothesis testing," *IEEE Trans. Information Theory*, vol. 64, no. 1, pp. 4–25, 2018.

[105] Y. Polyanskiy, H. V. Poor, and S. Verdú, "Channel coding rate in the finite blocklength regime," *IEEE Trans. Information Theory*, vol. 56, no. 5, pp. 2307–2359, 2010.

[106] G. Vazquez-Vilar, A. T. Campo, A. Guillén i Fàbregas, and A. Martinez, "Bayesian M-ary hypothesis testing: The meta-converse and Verdú–Han bounds are tight," *IEEE Trans. Information Theory*, vol. 62, no. 5, pp. 2324–2333, 2016.

[107] R. Venkataramanan and O. Johnson, "A strong converse bound for multiple hypothesis testing, with applications to high-dimensional estimation," *Electron. J. Statist*, vol. 12, no. 1, pp. 1126–1149, 2018.

[108] E. Abbe, "Community detection and stochastic block models: Recent developments," *J. Machine Learning Res.*, vol. 18, pp. 1–86, 2018.

[109] B. Hajek, Y. Wu, and J. Xu, "Computational lower bounds for community detection on random graphs," in *Proc. 28th Conference on Learning Theory (COLT '15)*, Paris, 2015, pp. 1–30.

[110] V. N. Vapnik, "An overview of statistical learning theory," *IEEE Trans. Neural Networks*, vol. 10, no. 5, pp. 988–999, 1999.

[111] O. Bousquet and A. Elisseeff, "Stability and generalization," *J. Machine Learning Res.*, vol. 2, pp. 499–526, 2002.

[112] H. Xu and S. Mannor, "Robustness and generalization," *Machine Learning*, vol. 86, no. 3, pp. 391–423, 2012.

[113] D. A. McAllester, "PAC-Bayesian stochastic model selection," *Machine Learning*, vol. 51, pp. 5–21, 2003.

[114] D. Russo and J. Zou, "How much does your data exploration overfit? Controlling bias via information usage," *arXiv:1511.05219*, 2016.

[115] A. Xu and M. Raginsky, "Information-theoretic analysis of generalization capability of learning algorithms," in *Proc. Advances in Neural Information Processing Systems (NeurIPS '17)*, 2017.

[116] M. Raginsky, A. Rakhlin, M. Tsao, Y. Wu, and A. Xu, "Information-theoretic analysis of stability and bias of learning algorithms," in *Proc. IEEE Information Theory Workshop (ITW '16)*, 2016.

[117] R. Bassily, S. Moran, I. Nachum, J. Shafer, and A. Yehudayof, "Learners that use little information," *arXiv:1710.05233*, 2018.

[118] A. R. Asadi, E. Abbe, and S. Verdú, "Chaining mutual information and tightening generalization bounds," *arXiv:1806.03803*, 2018.

[119] A. Pensia, V. Jog, and P. L. Loh, "Generalization error bounds for noisy, iterative algorithms," *arXiv:1801.04295v1*, 2018.

[120] J. Zhang, T. Liu, and D. Tao, "An information-theoretic view for deep learning," *arXiv:1804.09060*, 2018.

[121] M. Vera, P. Piantanida, and L. R. Vega, "The role of information complexity and randomization in representation learning," *arXiv:1802.05355*, 2018.

[122] M. Vera, L. R. Vega, and P. Piantanida, "Compression-based regularization with an application to multi-task learning," *arXiv:1711.07099*, 2018.

[123] C. Chan, A. Al-Bashadsheh, and Q. Zhou, "Info-clustering: A mathematical theory of data clustering," *IEEE Trans. Mol. Biol. Multi-Scale Commun.*, vol. 2, no. 1, pp. 64–91, 2016.

[124] R. K. Raman and L. R. Varshney, "Universal joint image clustering and registration using multivariate information measures," *IEEE J. Selected Topics Signal Processing*, vol. 12, no. 5, pp. 928–943, 2018.

[125] Z. Zhang and T. Berger, "Estimation via compressed information," *IEEE Trans. Information Theory*, vol. 34, no. 2, pp. 198–211, 1988.

[126] T. S. Han and S. Amari, "Parameter estimation with multiterminal data compression," *IEEE Trans. Information Theory*, vol. 41, no. 6, pp. 1802–1833, 1995.

[127] Y. Zhang, J. C. Duchi, M. I. Jordan, and M. J. Wainwright, "Information-theoretic lower bounds for distributed statistical estimation with communication constraints," in *Proc. Advances in Neural Information Processing Systems (NeurIPS '13)*, 2013.

[128] R. Ahlswede and I. Csiszár, "Hypothesis testing with communication constraints," *IEEE Trans. Information Theory*, vol. 32, no. 4, pp. 533–542, 1986.

[129] T. S. Han, "Hypothesis testing with multiterminal data compression," *IEEE Trans. Information Theory*, vol. 33, no. 6, pp. 759–772, 1987.

[130] T. S. Han and K. Kobayashi, "Exponential-type error probabilities for multiterminal hypothesis testing," *IEEE Trans. Information Theory*, vol. 35, no. 1, pp. 2–14, 1989.

[131] T. S. Han and S. Amari, "Statistical inference under multiterminal data compression," *IEEE Trans. Information Theory*, vol. 44, no. 6, pp. 2300–2324, 1998.

[132] H. M. H. Shalaby and A. Papamarcou, "Multiterminal detection with zero-rate data compression," *IEEE Trans. Information Theory*, vol. 38, no. 2, pp. 254–267, 1992.

[133] G. Katz, P. Piantanida, R. Couillet, and M. Debbah, "On the necessity of binning for the distributed hypothesis testing problem," in *Proc. IEEE International Symposium on Information Theory (ISIT '15)*, 2015.

[134] Y. Xiang and Y. Kim, "Interactive hypothesis testing against independence," in *Proc. IEEE International Symposium on Information Theory (ISIT '13)*, 2013.

[135] W. Zhao and L. Lai, "Distributed testing against independence with conferencing encoders," in *Proc. IEEE Information Theory Workshop (ITW '15)*, 2015.

[136] W. Zhao and L. Lai, "Distributed testing with zero-rate compression," in *Proc. IEEE International Symposium on Information Theory (ISIT '15)*, 2015.

[137] W. Zhao and L. Lai, "Distributed detection with vector quantizer," *IEEE Trans. Signal Information Processing Networks*, vol. 2, no. 2, pp. 105–119, 2016.

[138] W. Zhao and L. Lai, "Distributed testing with cascaded encoders," *IEEE Trans. Information Theory*, vol. 64, no. 11, pp. 7339–7348, 2018.

[139] M. Raginsky, "Learning from compressed observations," in *Proc. IEEE Information Theory Workshop (ITW '07)*, 2007.

[140] M. Raginsky, "Achievability results for statistical learning under communication constraints," in *Proc. IEEE International Symposium on Information Theory (ISIT '09)*, 2009.

[141] A. Xu and M. Raginsky, "Information-theoretic lower bounds for distributed function computation," *IEEE Trans. Information Theory*, vol. 63, no. 4, pp. 2314–2337, 2017.

[142] C. Dwork and A. Roth, "The algorithmic foundations of differential privacy," *Foundations and Trends Theoretical Computer Sci.*, vol. 9, no. 3–4, pp. 211–407, 2014.

[143] J. Liao, L. Sankar, V. Y. F. Tan, and F. P. Calmon, "Hypothesis testing under mutual information privacy constraints in the high privacy regime," *IEEE Trans. Information Forensics Security*, vol. 13, no. 4, pp. 1058–1071, 2018.

[144] F. P. Calmon, D. Wei, B. Vinzamuri, K. N. Ramamurthy, and K. R. Varshney, "Data pre-processing for discrimination prevention: Information-theoretic optimization and analysis," *IEEE J. Selected Topics Signal Processing*, vol. 12, no. 5, pp. 1106–1119, 2018.

2 An Information-Theoretic Approach to Analog-to-Digital Compression

Alon Kipnis, Yonina C. Eldar, and Andrea J. Goldsmith

Summary

Processing, storing, and communicating information that originates as an analog phenomenon involve conversion of the information to bits. This conversion can be described by the combined effect of sampling and quantization, as illustrated in Fig. 2.1. The digital representation in this procedure is achieved by first sampling the analog signal so as to represent it by a set of discrete-time samples and then quantizing these samples to a finite number of bits. Traditionally, these two operations are considered separately. The sampler is designed to minimize information loss due to sampling on the basis of prior assumptions about the continuous-time input [1]. The quantizer is designed to represent the samples as accurately as possible, subject to the constraint on the number of bits that can be used in the representation [2]. The goal of this chapter is to revisit this paradigm by considering the joint effect of these two operations and to illuminate the dependence between them.

2.1 Introduction

Consider the minimal sampling rate that arises in classical sampling theory due to Whittaker, Kotelnikov, Shannon, and Landau [1, 3, 4]. These works establish the Nyquist rate or the spectral occupancy of the signal as the critical sampling rate above which the signal can be perfectly reconstructed from its samples. This statement, however, focuses only on the condition for perfectly reconstructing a bandlimited signal from its infinite-precision samples; it does not incorporate the quantization precision of the samples and does not apply to signals that are not bandlimited. In fact, as follows from lossy source coding theory, it is impossible to obtain an exact representation of any continuous-amplitude sequence of samples by a digital sequence of numbers due to finite quantization precision, and therefore any digital representation of an analog signal is prone to error. That is, no continuous amplitude signal can be reconstructed from its quantized samples with zero distortion regardless of the sampling rate, even when the signal is bandlimited. This limitation raises the following question. In converting a signal to bits via sampling and quantization at a given bit precision, can the signal be

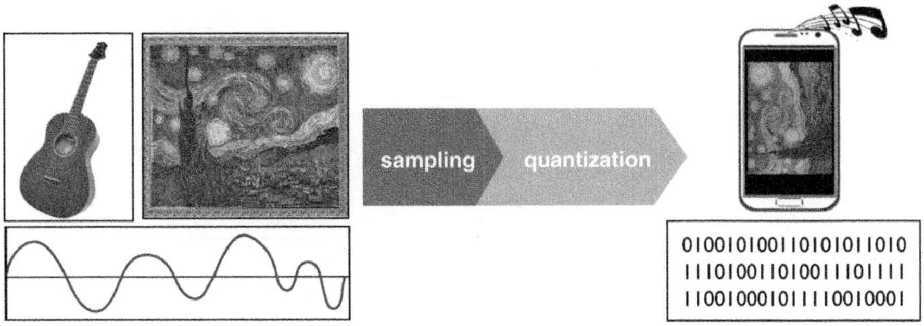

Figure 2.1 Analog-to-digital conversion is achieved by combining sampling and quantization.

reconstructed from these samples with minimal distortion based on sub-Nyquist sampling? One of the goals of this chapter is to discuss this question by extending classical sampling theory to account for quantization and for non-bandlimited inputs. Namely, for an arbitrary stochastic input and given a total bitrate budget, we consider the lowest sampling rate required to sample the signal such that reconstruction of the signal from a bit-constrained representation of its samples results in minimal distortion. As we shall see, without assuming any particular structure for the input analog signal, this sampling rate is often below the signal's Nyquist rate.

The minimal distortion achievable in recovering a signal from its representation by a finite number of bits per unit time depends on the particular way the signal is quantized or, more generally, encoded, into a sequence of bits. Since we are interested in the fundamental distortion limit in recovering an analog signal from its digital representation, we consider all possible encoding and reconstruction (decoding) techniques. As an example, in Fig. 2.1 the smartphone display may be viewed as a reconstruction of the real-world painting *The Starry Night* from its digital representation. No matter how fine the smartphone screen, this recovery is not perfect since the digital representation of the analog image is not accurate. That is, loss of information occurs during the transformation from analog to digital. Our goal is to analyze this loss as a function of hardware limitations on the sampling mechanism and the number of bits used in the encoding. It is convenient to normalize this number of bits by the signal's free dimensions, that is, the dimensions along which new information is generated. For example, the free dimensions of a visual signal are usually the horizontal and vertical axes of the frame, and the free dimension of an audio wave is time. For simplicity, we consider analog signals with a single free dimension, and we denote this dimension as *time*. Therefore, our restriction on the digital representation is given in terms of its *bitrate* – the number of bits per unit time.

For an arbitrary continuous-time random signal with a known distribution, the fundamental distortion limit due to the encoding of the signal using a limited bitrate is given by Shannon's distortion-rate function (DRF) [5–7]. This function provides the optimal trade-off between the bitrate of the signal's digital representation and the distortion in recovering the original signal from this representation. Shannon's DRF is described only in terms of the distortion criterion, the probability distribution of the continuous-time signal, and the maximal bitrate allowed in the digital representation. Consequently, the

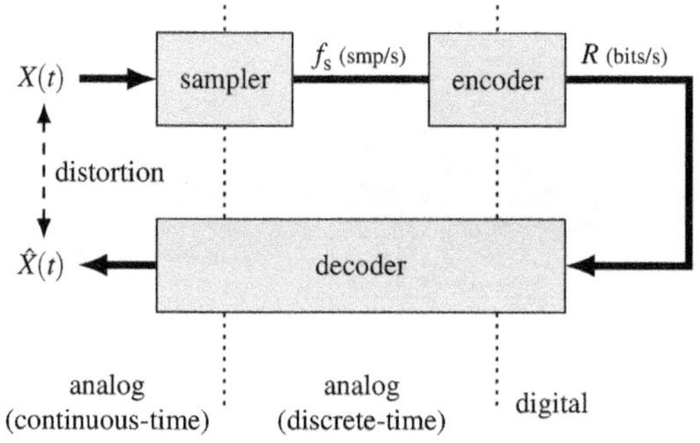

Figure 2.2 Analog-to-digital compression (ADX) and reconstruction setting. Our goal is to derive the minimal distortion between the signal and its reconstruction from any encoding at bitrate R of the samples of the signal taken at sampling rate f_s.

optimal encoding scheme that attains Shannon's DRF is a general mapping from the space of continuous-time signals to bits that does not consider any practical constraints in implementing such a mapping. In practice, the minimal distortion in recovering analog signals from their mapping to bits considers the digital encoding of the signal *samples*, with a constraint on both the *sampling rate* and the *bitrate* of the system [8–10]. Here the sampling rate f_s is defined as the number of samples per unit time of the continuous-time source signal; the bitrate R is the number of bits per unit time used in the representation of these samples. The resulting system describing our problem is illustrated in Fig. 2.2, and is denoted as the *analog-to-digital compression* (ADX) setting.

The digital representation in this setting is obtained by transforming a continuous-time continuous-amplitude random source signal $X(\cdot)$ through the concatenated operation of a *sampler* and an *encoder*, resulting in a bit sequence. The *decoder* estimates the original analog signal from this bit sequence. The *distortion* is defined to be the mean-squared error (MSE) between the input signal $X(\cdot)$ and its reconstruction $\widehat{X}(\cdot)$. Since we are interested in the fundamental distortion limit subject to a sampling constraint, we allow optimization over the encoder and the decoder as the time interval over which $X(\cdot)$ is sampled goes to infinity. When $X(\cdot)$ is bandlimited and the sampling rate f_s exceeds its Nyquist rate f_{Nyq}, the encoder can recover the signal using standard interpolation and use the optimal source code at bitrate R to attain distortion equal to Shannon's DRF of the signal [11]. Therefore, for bandlimited signals, a non-trivial interplay between the sampling rate and the bitrate arises only when f_s is below a signal's Nyquist rate. In addition to the optimal encoder and decoder, we also explore the optimal sampling mechanism, but limit ourselves to the class of linear and continuous deterministic samplers. Namely, each sample is defined by a bounded linear functional over a class of signals. Finally, in order to account for system imperfections or those due to external interferences, we assume that the signal $X(\cdot)$ is corrupted by additive noise prior to sampling. The noise-free version is obtained from our results by setting the intensity of this noise to zero.

The minimal distortion in the ADX system of Fig. 2.2 is bounded from below by two extreme cases of the sampling rate and the bitrate, as illustrated in Fig. 2.3: (1) when the bitrate R is unlimited, the minimal ADX distortion reduces to the minimal

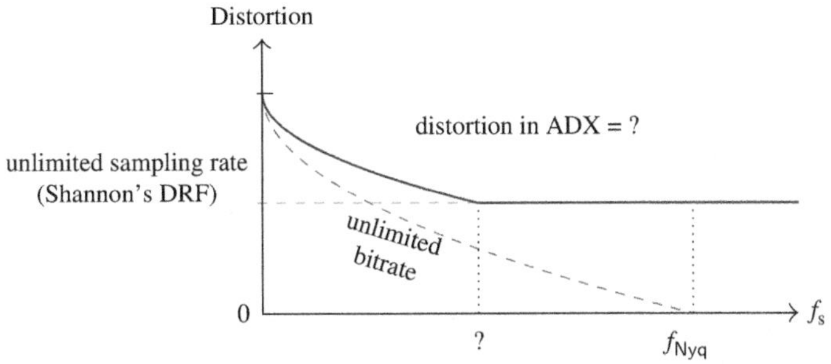

Figure 2.3 The minimal sampling rate for attaining the minimal distortion achievable under a bitrate-limited representation is usually below the Nyquist rate f_{Nyq}. In this figure, the noise is assumed to be zero.

MSE (MMSE) in interpolating a signal from its noisy samples at rate f_s [12, 13]. (2) When the sampling rate f_s is unlimited or above the Nyquist rate of the signal and when the noise is zero, the ADX distortion reduces to Shannon's DRF of the signal. Indeed, in this case, the optimal encoder can recover the original continuous-time source without distortion and then encode this recovery in an optimal manner according to the optimal lossy compression scheme attaining Shannon's DRF. When f_s is unlimited or above the Nyquist rate and the noise is not zero, the minimal distortion is the *indirect* (or *remote*) DRF of the signal given its noise-corrupted version, see Section 3.5 of [7] and [14]. Our goal is therefore to characterize the MSE due to the joint effect of a finite bitrate constraint and sampling at a sub-Nyquist sampling rate. In particular, we are interested in the minimal sampling rate for which Shannon's DRF, or the indirect DRF, describing the minimal distortion subject to a bitrate constraint, is attained. As illustrated in Fig. 2.3, and as will be explained in more detail below, this sampling rate is usually below the Nyquist rate of $X(\cdot)$, or, more generally, the spectral occupancy of $X(\cdot)$ when non-uniform or generalized sampling techniques are allowed. We denote this minimal sampling rate as the *critical sampling rate* subject to a bitrate constraint, since it describes the minimal sampling rate required to attain the optimal performance in systems operating under quantization or bitrate restrictions. The critical sampling rate extends the minimal-distortion sampling rate considered by Shannon, Nyquist, and Landau. It is only as the bitrate goes to infinity that sampling at the Nyquist rate is necessary to attain minimal (namely zero) distortion.

In order to gain intuition as to why the minimal distortion under a bitrate constraint may be attained by sampling below the Nyquist rate, we first consider in Section 2.2 a simpler version of the ADX setup involving the lossy compression of linear projections of signals represented as finite-dimensional random real vectors. Next, in Section 2.3 we formalize the combined sampling and source coding problem arising from Fig. 2.2 and provide basic properties of the minimal distortion in this setting. In Section 2.4, we fully characterize the minimal distortion in ADX as a function of the bitrate and sampling rate and derive the critical sampling rate that leads to optimal performance. We conclude this chapter in Section 2.5, where we consider uniform samplers, in particular single-branch and more general multi-branch uniform samplers, and show that these samplers attain the fundamental distortion limit.

2.2 Lossy Compression of Finite-Dimensional Signals

Let X^n be an n-dimensional Gaussian random vector with covariance matrix Σ_{X^n}, and let Y^m be a projected version of X^n defined by

$$Y^m = \mathbf{H}X^n, \tag{2.1}$$

where $\mathbf{H} \in \mathbb{R}^{m\times n}$ is a deterministic matrix and $m < n$. This projection of X^n into a lower-dimensional space is the counterpart of sampling the continuous-time analog signal $X(\cdot)$ in the ADX setting. We consider the normalized MMSE estimate of X^n from a representation of Y^m using a limited number of bits.

Without constraining the number of bits, the distortion in this estimation is given by

$$\mathsf{mmse}(X^n|Y^m) \triangleq \frac{1}{n}\,\mathrm{tr}\left(\Sigma_{X^n} - \Sigma_{X^n|Y^m}\right), \tag{2.2}$$

where $\Sigma_{X^n|Y^m}$ is the conditional covariance matrix. However, when Y^m is to be encoded using a code of no more than nR bits, the minimal distortion cannot be smaller than the indirect DRF of X^n given Y^m, denoted by $D_{X^n|Y^m}(R)$. This function is given by the following parametric expression [14]:

$$\begin{aligned} D(R_\theta) &= \mathrm{tr}\left(\Sigma_{X^n}\right) - \sum_{i=1}^{m}\left[\lambda_i(\Sigma_{X^n|Y^m}) - \theta\right]^+, \\ R_\theta &= \frac{1}{2}\sum_{i=1}^{m}\log^+\left[\lambda_i(\Sigma_{X^n|Y^m})/\theta\right], \end{aligned} \tag{2.3}$$

where $x^+ = \max\{x,0\}$ and $\lambda_i(\Sigma_{X^n|Y^m})$ is the ith eigenvalue of $\Sigma_{X^n|Y^m}$.

It follows from (2.2) that X^n can be recovered from Y^m with zero MMSE if and only if

$$\lambda_i(\Sigma_{X^n}) = \lambda_i(\Sigma_{X^n|Y^m}), \tag{2.4}$$

for all $i = 1,\ldots,n$. When this condition is satisfied, (2.3) takes the form

$$\begin{aligned} D(R_\theta) &= \sum_{i=1}^{n}\min\{\lambda_i(\Sigma_{X^n}),\theta\}, \\ R_\theta &= \frac{1}{2}\sum_{i=1}^{n}\log^+\left[\lambda_i(\Sigma_{X^n})/\theta\right], \end{aligned} \tag{2.5}$$

which is Kolmogorov's reverse water-filling expression for the DRF of the vector Gaussian source X^n [15], i.e., the minimal distortion in encoding X^n using codes of rate R bits per source realization. The key insight is that the requirements for equality between (2.3) and (2.5) are not as strict as (2.4): all that is needed is equality among those eigenvalues that affect the value of (2.5). In particular, assume that for a point (R,D) on $D_{X^n}(R)$, only $\lambda_n(\Sigma_{X^n}),\ldots,\lambda_{n-m+1}(\Sigma_{X^n})$ are larger than θ, where the eigenvalues are organized in ascending order. Then we can choose the rows of $\mathbf{H}$ to be the m left eigenvectors corresponding to $\lambda_n(\Sigma_{X^n}),\ldots,\lambda_{n-m+1}(\Sigma_{X^n})$. With this choice of $\mathbf{H}$, the m largest eigenvalues of $\Sigma_{X^n|Y^m}$ are identical to the m largest eigenvalues of Σ_{X^n}, and (2.5) is equal to (2.3).

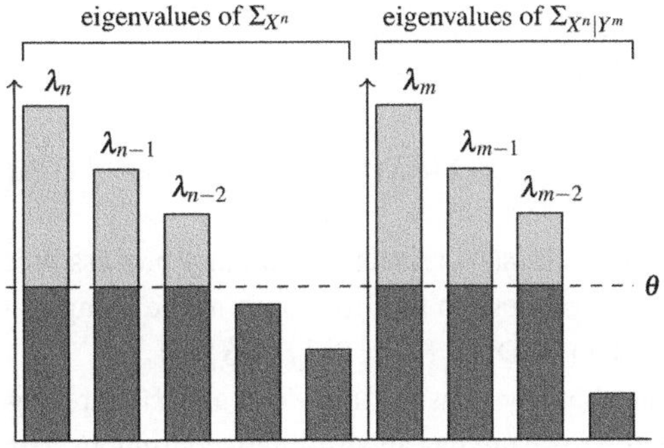

Figure 2.4 Optimal sampling occurs whenever $D_{X^n}(R) = D_{X^n|Y^m}(R)$. This condition is satisfied even for $m < n$, as long as there is equality among the eigenvalues of Σ_{X^n} and $\Sigma_{X^n|Y^m}$ which are larger than the water-level parameter θ.

Since the rank of the sampling matrix is now $m < n$, we effectively performed sampling below the "Nyquist rate" of X^n without degrading the performance dictated by its DRF. One way to understand this phenomenon is as an alignment between the range of the sampling matrix $\mathbf{H}$ and the subspace over which X^n is represented, according to Kolmogorov's expression (2.5). That is, when Kolmogorov's expression implies that not all degrees of freedom are utilized by the optimal distortion-rate code, sub-sampling does not incur further performance loss, provided that the sampling matrix is aligned with the optimal code. This situation is illustrated in Fig. 2.4. Sampling with an $\mathbf{H}$ that has fewer rows than the rank of Σ_{X^2} is the finite-dimensional analog of sub-Nyquist sampling in the infinite-dimensional setting of continuous-time signals.

In the rest of this chapter, we explore the counterpart of the phenomena described above in the richer setting of continuous-time stationary processes that may or may not be bandlimited, and whose samples may be corrupted by additive noise.

2.3 ADX for Continuous-Time Analog Signals

We now explore the fundamental ADX distortion in the setting of continuous-time stationary processes that may be corrupted by noise prior to sampling. We consider the system of Fig. 2.5 in which $X(\cdot) \triangleq \{X(t), t \in \mathbb{R}\}$ is a stationary process with power

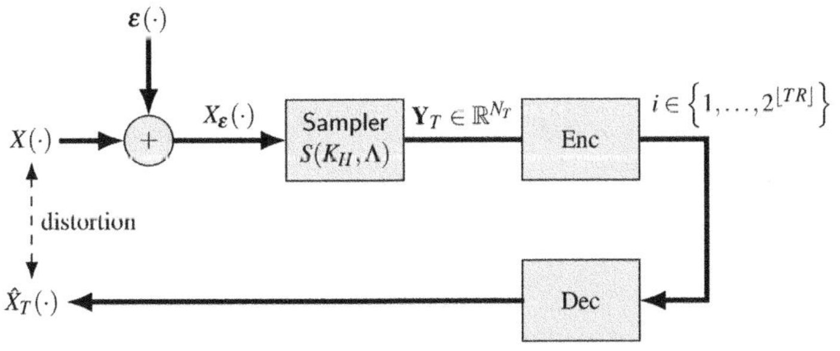

Figure 2.5 Combined sampling and source coding setting.

spectral density (PSD) $S_X(f)$. This PSD is assumed to be a real, symmetric, and absolute integrable function that satisfies

$$\mathbb{E}[X(t)X(s)] = \int_{-\infty}^{\infty} S_X(f) e^{2\pi j(t-s)f} df, \quad t,s \in \mathbb{R}. \tag{2.6}$$

The noise process $\varepsilon(\cdot)$ is another stationary process independent of $X(\cdot)$ with PSD $S_\varepsilon(f)$ of similar properties, so that the input to the sampler is the stationary process $X_\varepsilon(\cdot) \triangleq X(\cdot)+\varepsilon(\cdot)$ with PSD $S_{X_\varepsilon}(f) = S_X(f)+S_\varepsilon(f)$.

We note that, by construction, $X(\cdot)$ and $\varepsilon(\cdot)$ are regular processes in the sense that their spectral measure has an absolutely continuous density with respect to the Lebesgue measure. If in addition the support of $S_X(f)$, denoted by supp S_X, is contained[1] within a bounded interval, then we say that $X(\cdot)$ is bandlimited and denote by f_{Nyq} its Nyquist rate, defined as twice the maximal element in suppS_X. The *spectral occupancy* of $X(\cdot)$ is defined to be the Lebesgue measure of supp S_X.

Although this is not necessary for all parts of our discussion, we assume that the processes $X(\cdot)$ and $\varepsilon(\cdot)$ are Gaussian. This assumption leads to closed-form characterizations for many of the expressions we consider. In addition, it follows from [16, 17] that a lossy compression policy that is optimal under a Gaussian distribution can be used to encode non-Gaussian signals with matching second-order statistics, while attaining the same distortion as if the signals were Gaussian. Hence, the optimal sampler and encoding system we use to obtain the fundamental distortion limit for Gaussian signals attains the same distortion limit for non-Gaussian signals as long as the second-order statistics of the two signals are the same.

2.3.1 Bounded Linear Sampling

The sampler in Fig. 2.5 outputs a finite-dimensional vector of samples where, most generally, each sample is defined by a linear and bounded (hence, continuous) functional of the process $X_\varepsilon(\cdot)$. For this reason, we denote a sampler of this type as a *bounded linear sampler*. In order to consider this sampler in applications, it is most convenient to define it in terms of a bilinear kernel $K_H(t,s)$ on $\mathbb{R}\times\mathbb{R}$ and a discrete *sampling set* $\Lambda \subset \mathbb{R}$, as illustrated in Fig. 2.6. The kernel $K_H(t,s)$ defines a time-varying linear system on a suitable class of signals [18], and hence each element $t_n \in \Lambda$ defines a linear bounded functional $K(t_n,s)$ on this class by

$$Y_n \triangleq \int_{-\infty}^{\infty} X_\varepsilon(s) K_H(t_n,s) ds.$$

For a time horizon T, we denote by $\mathbf{Y}_T$ the finite-dimensional vector obtained by sampling at times $t_1,\ldots,t_n \in \Lambda_T$, where

$$\Lambda_T \triangleq \Lambda \cap [-T/2, T/2].$$

[1] Since the PSD is associated with an absolutely continuous spectral measure, sets defined in term of the PSD, e.g., supp S_X, are understood to be unique up to symmetric difference of Lebesgue measure zero.

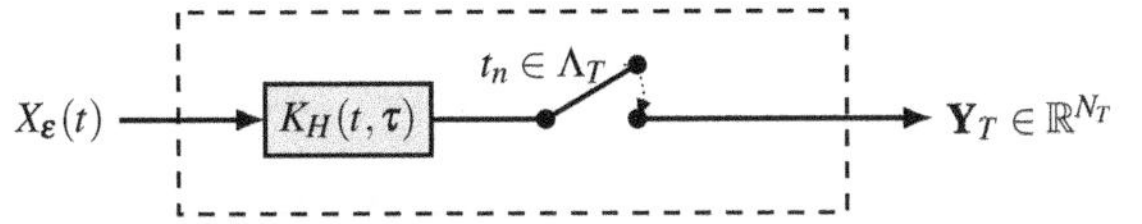

Figure 2.6 Bounded linear sampler ($N_T \triangleq |\Lambda_T|$).

We assume in addition that Λ is uniformly discrete in the sense that there exists $\varepsilon > 0$ such that $|t - s| > \varepsilon$ for every non-identical $t, s \in \Lambda$. The density of Λ_T is defined as the number of points in Λ_T divided by T and denoted here by $d(\Lambda_T)$. Whenever it exists, we define the limit

$$d(\Lambda) = \lim_{T\to\infty} d(\Lambda_T) = \lim_{T\to\infty} \frac{|\Lambda \cap [-T/2, T/2]|}{T}$$

as the *symmetric density* of Λ, or simply it's density.

Linear Time-Invariant Uniform Sampling

An important special case of the bounded linear sampler is that of a linear time-invariant (LTI) uniform sampler [1], illustrated in Fig. 2.7. For this sampler, the sampling set is a uniform grid $\mathbb{Z}T_s = \{nT_s, n \in \mathbb{Z}\}$, where $T_s = f_s^{-1} > 0$. The kernel is of the form $K_H(t,s) = h(t-s)$ where $h(t)$ is the impulse response of an LTI system with frequency response $H(f)$. Therefore, the entries of $\mathbf{Y}_T$ corresponding to sampling at times $nT_s \in \Lambda$ are given by

$$Y_n \triangleq \int_{-\infty}^{\infty} h(nT_s - s) X_\varepsilon(s) ds.$$

It is easy to check that $d(T_s\mathbb{Z}) = f_s$ and hence, in this case, the density of the sampling set has the usual interpretation of sampling rate.

Multi-Branch Linear Time-Invariant Uniform Sampling

A generalization of the uniform LTI sampler incorporates several of these samplers in parallel, as illustrated in Fig. 2.8. Each of the L branches in Fig. 2.8 consists of a LTI system with frequency response $H_l(f)$ followed by a uniform pointwise sampler with sampling rate f_s/L, so that the overall sampling rate is f_s. The vector $\mathbf{Y}_T$ consists of the concatenation of the vectors $Y_{1,T}, \ldots, Y_{L,T}$ obtained from each of the sampling branches.

2.3.2 Encoding and Reconstruction

For a time horizon T, the encoder in Fig. 2.5 can be any function of the form

$$f : \mathbb{R}^{N_T} \to \left\{1, \ldots, 2^{\lfloor TR \rfloor}\right\}, \tag{2.7}$$

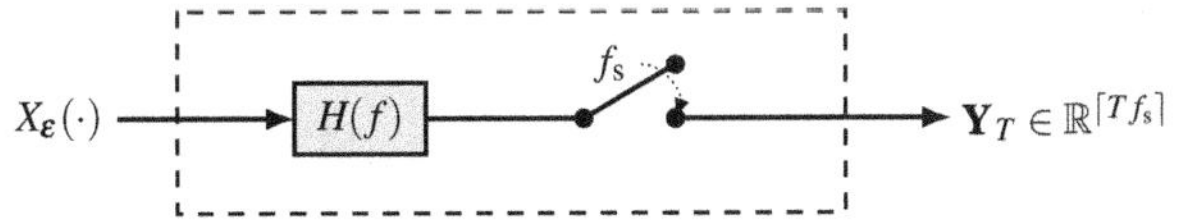

Figure 2.7 Uniform linear time-invariant sampler.

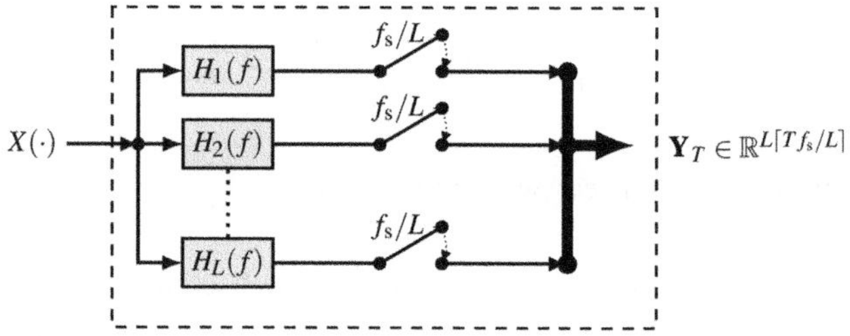

Figure 2.8 Multi-branch linear time-invariant uniform sampler.

where $N_T = \dim(\mathbf{Y}_T) = |\Lambda_T|$. That is, the encoder receives the vector of samples $\mathbf{Y}_T$ and outputs an index out of $2^{\lfloor TR \rfloor}$ possible indices. The decoder receives this index, and produces an estimate $\widehat{X}(\cdot)$ for the signal $X(\cdot)$ over the interval $[-T/2, T/2]$. Thus, it is a mapping

$$g : \left\{1, \ldots, 2^{\lfloor TR \rfloor}\right\} \to \mathbb{R}^{[-T/2, T/2]}. \tag{2.8}$$

The goal of the joint operation of the encoder and the decoder is to minimize the expected mean-squared error (MSE)

$$\frac{1}{T} \int_{-T/2}^{T/2} \mathbb{E}\left(X(t) - \widehat{X}(t)\right)^2 dt.$$

In practice, an encoder may output a finite number of samples that are then interpolated to the continuous-time estimate $\widehat{X}(\cdot)$. Since our goal is to understand the limits in converting signals to bits, this separation between decoding and interpolation, as well as the possible restrictions each of these steps encounters in practice, are not explored within the context of ADX.

Given a particular bounded linear sampler $S = (\Lambda, K_H)$ and a bitrate R, we are interested in characterizing the function

$$D_T(S,R) \triangleq \inf_{f,g} \frac{1}{T} \int_{-T/2}^{T/2} \mathbb{E}\left(X(t) - \widehat{X}(t)\right)^2 dt, \tag{2.9}$$

or its limit as $T \to \infty$, where the infimum is over all encoders and decoders of the form (2.7) and (2.8). The function $D_T(S,R)$ is defined only in terms of the sampler S and the bitrate R, and in this sense measures the minimal distortion that can be attained using the sampler S subject to a bitrate constraint R on the representation of the samples.

2.3.3 Optimal Distortion in ADX

From the definition of $D_T(S,R)$ and the ADX setting, it immediately follows that $D_T(S,R)$ is non-increasing in R. Indeed, any encoding into a set of $2^{\lfloor TR \rfloor}$ elements can be obtained as a special case of encoding to a set of $2^{\lfloor T(R+r) \rfloor}$ elements, for $r > 0$. In addition, by using the trivial encoder $g \equiv 0$ we see that $D_T(S,R)$ is bounded from above by the variance σ_X^2 of $X(\cdot)$, which is given by

$$\sigma_X^2 \triangleq \int_{-\infty}^{\infty} S_X(f) df.$$

In what follows, we explore additional important properties of $D_T(S,R)$.

Optimal Encoding

Denote by $\widetilde{X}_T(\cdot)$ the process that is obtained by estimating $X(\cdot)$ from the output of the sampler according to an MSE criterion. That is

$$\widetilde{X}_T(t) \triangleq \mathbb{E}[X(t)|\mathbf{Y}_T], \quad t \in \mathbb{R}. \tag{2.10}$$

From properties of the conditional expectation and MSE, under any encoder f we may write

$$\frac{1}{T}\int_{-T/2}^{T/2} \mathbb{E}\left(X(t)-\widehat{X}(t)\right)^2 dt = \mathsf{mmse}_T(S) + \mathsf{mmse}\left(\widetilde{X}_T|f(\mathbf{Y}_T)\right), \tag{2.11}$$

where

$$\mathsf{mmse}_T(S) \triangleq \frac{1}{T}\int_{-T/2}^{T/2} \mathbb{E}\left(X(t)-\widetilde{X}_T(t)\right)^2 dt \tag{2.12}$$

is the distortion due to sampling and

$$\mathsf{mmse}\left(\widetilde{X}_T|f(\mathbf{Y}_T)\right) \triangleq \frac{1}{T}\int_{-T/2}^{T/2} \mathbb{E}\left(\widetilde{X}_T(t)-g(f(\mathbf{Y}_T))\right)^2 dt$$

is the distortion associated with the lossy compression procedure, and depends on the sampler only through $\widetilde{X}_T(\cdot)$.

The decomposition (2.11) already provides important clues on an optimal encoder and decoder pair that attains $D_T(S,R)$. Specifically, it follows from (2.11) that there is no loss in performance if the encoder tries to describe the process $\widetilde{X}_T(\cdot)$ subject to the bitrate constraint, rather than the process $X(\cdot)$. Consequently, the optimal decoder outputs the conditional expectation of $\widetilde{X}_T(\cdot)$ given $f(\mathbf{Y}_T)$. The decomposition (2.11) was first used in [14] to derive the indirect DRF of a pair of stationary Gaussian processes, and later in [19] to derive indirect DRF expressions in other settings. An extension of the principle presented in this decomposition to arbitrary distortion measures is discussed in [20].

The decomposition (2.11) also sheds light on the behavior of the optimal distortion $D_T(S,R)$ under the two extreme cases of unlimited bitrate and unrestricted sampling rate, each of which is illustrated in Fig. 2.3. We discuss these two cases next.

Unlimited Bitrate

If we remove the bitrate constraint in the ADX setting (formally, letting $R \to \infty$), loss of information is only due to noise and sampling. In this case, the second term in the RHS of (2.11) disappears, and the distortion in ADX is given by $\mathsf{mmse}_T(S)$. Namely, we have

$$\lim_{R\to\infty} D_T(S,R) = \mathsf{mmse}_T(S).$$

The unlimited bitrate setting reduces the ADX problem to a classical problem in sampling theory: the MSE under a given sampling system. Of particular interest is the case of optimal sampling, i.e., when this MSE vanishes as $T \to \infty$. For example, by considering the noiseless case and assuming that $K_H(t,s) = \delta(t-s)$ is the identity operator, the sampler is defined solely in terms of Λ. The condition on $\mathsf{mmse}_T(S)$ to converge to zero is related to the conditions for stable sampling in Paley–Wiener spaces studied by Landau and Beurling [21, 22]. In order to see this connection more precisely, note that (2.6) defines an isomorphism between the Hilbert spaces of finite-variance random variables measurable with respect to the sigma algebra generated by $X(\cdot)$ and the Hilbert space generated by the inverse Fourier transform of $\left\{e^{2\pi jtf}\sqrt{S_X(f)}, t\in\mathbb{R}\right\}$ [23]. Specifically, this isomorphism is obtained by extending the map

$$X(t) \longleftrightarrow \mathcal{F}^{-1}\left\{e^{2\pi jtf}\sqrt{S_X(f)}\right\}(s)$$

to the two aforementioned spaces. It follows that sampling and reconstructing $X(\cdot)$ with vanishing MSE is equivalent to the same operation in the Paley–Wiener space of analytic functions whose Fourier transform vanishes outside supp S_X. In particular, the condition $\mathsf{mmse}_T(S) \overset{T\to\infty}{\longrightarrow} 0$ holds whenever Λ is a set of stable sampling in this Paley–Wiener space, i.e., there exists a universal constant $A > 0$ such that the L_2 norm of each function in this space is bounded by A times the energy of the samples of this function. Landau [21] showed that a necessary condition for this property is that the number of points in Λ that fall within the interval $[-T/2, T/2]$ is at least the spectral occupancy of $X(\cdot)$ times T, minus a constant that is logarithmic in T. For this reason, this spectral occupancy is termed the *Landau rate* of $X(\cdot)$, and we denote it here by f_{Lnd}. In the special case where supp S_X is an interval (symmetric around the origin since $X(\cdot)$ is real), the Landau and Nyquist rates coincide.

Optimal Sampling

The other special case of the ADX setting is obtained when there is no loss of information due to sampling. For example, this is the case when $\mathsf{mmse}_T(S)$ goes to zero under the conditions mentioned above of zero noise, identity kernel, and sampling density exceeding the spectral occupancy. More generally, this situation occurs whenever $\widetilde{X}_T(\cdot)$ converges (in expected norm) to the MMSE estimator of $X(\cdot)$ from $X_\varepsilon(\cdot)$. This MMSE estimator is a stationary process obtained by non-causal Wiener filtering, and its PSD is

$$S_{X|X_\varepsilon}(f) \triangleq \frac{S_X^2(f)}{S_X(f)+S_\varepsilon(f)}, \tag{2.13}$$

where in (2.13) and in similar expressions henceforth we interpret the expression to be zero whenever both the numerator and denominator are zero. The resulting MMSE is given by

$$\mathsf{mmse}(X|X_\varepsilon) \triangleq \int_{-\infty}^{\infty} [S_X(f) - S_{X|X_\varepsilon}(f)]df. \tag{2.14}$$

Since our setting does not limit the encoder from computing $\widetilde{X}_T(\cdot)$, the ADX problem reduces in this case to the indirect source coding problem of recovering $X(\cdot)$ from a bitrate R representation of its corrupted version $X_\varepsilon(\cdot)$. This problem was considered and

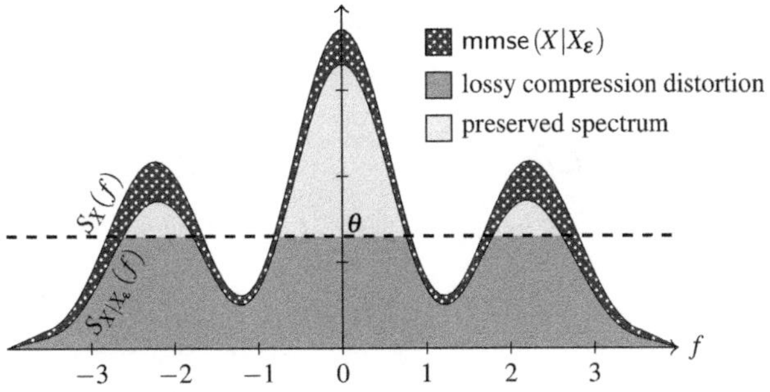

Figure 2.9 Water-filling interpretation of (2.15). The distortion is the sum of $\mathsf{mmse}(X|X_\varepsilon)$ and the lossy compression distortion.

solved by Dobrushin and Tsybakov in [14], where the following expression was given for the optimal trade-off between bitrate and distortion:

$$D_{X|X_\varepsilon}(R_\theta) \triangleq \mathsf{mmse}(X|X_\varepsilon) + \int_{-\infty}^{\infty} \min\{S_{X|X_\varepsilon}(f), \theta\} df, \tag{2.15a}$$

$$R_\theta = \frac{1}{2}\int_{-\infty}^{\infty} \log^+[S_{X|X_\varepsilon}(f)/\theta] df. \tag{2.15b}$$

A graphical water-filling interpretation of (2.15) is given in Fig. 2.9. When the noise $\varepsilon(\cdot)$ is zero, $S_{X|X_\varepsilon}(f) = S_X(f)$, and hence (2.15) reduces to

$$D_X(R_\theta) \triangleq \int_{-\infty}^{\infty} \min\{S_X(f), \theta\} df, \tag{2.16a}$$

$$R_\theta = \frac{1}{2}\int_{-\infty}^{\infty} \log^+[S_X(f)/\theta] df, \tag{2.16b}$$

which is Pinsker's expression [15] for the DRF of the process $X(\cdot)$, denoted here by $D_X(R)$. Note that (2.16) is the continuous-time counterpart of (2.3).

From the discussion above, we conclude that

$$D_T(S,R) \geq D_{X|X_\varepsilon}(R) \geq \max\{D_X(R), \mathsf{mmse}(X|X_\varepsilon)\}. \tag{2.17}$$

Furthermore, when the estimator $\mathbb{E}[X(t)|X_\varepsilon]$ can be obtained from $\mathbf{Y}_T$ as $T \to \infty$, we have that $D_T(S,R) \xrightarrow{T\to\infty} D_{X|X_\varepsilon}(R)$. In this situation, we say that the conditions for optimal sampling are met, since the only distortion is due to the noise and the bitrate constraint.

The two lower bounds in Fig. 2.3 describe the behavior of $D_T(S,R)$ in the two special cases of unrestricted bitrate and optimal sampling. Our goal in the next section is to characterize the intermediate case of non optimal sampling and a finite bitrate constraint.

2.4 The Fundamental Distortion Limit

Given a particular bounded linear sampler $S = (\Lambda, K_H)$ and a bitrate R, we defined the function $D_T(S,R)$ as the minimal distortion that can be attained in the combined sampling and lossy compression setup of Fig. 2.5. Our goal in this section is to derive and analyze a function $D^\star(f_s,R)$ that bounds from below $D_T(S,R)$ for any such bounded linear sampler with symmetric density of Λ not exceeding f_s. The achievability of this lower bound is addressed in the next section.

2.4.1 Definition of $D^\star(f_s,R)$

In order to define $D^\star(f_s,R)$, we let $F^\star(f_s) \subset \mathbb{R}$ be any set that maximizes

$$\int_F S_{X|X_\varepsilon}(f)df = \int_F \frac{S_X^2(f)}{S_X(f)+S_\varepsilon(f)}df \tag{2.18}$$

over all Lebesgue measurable sets F whose Lebesgue measure does not exceed f_s. In other words, $F^\star(f_s)$ consists of the f_s spectral bands with the highest energy in the spectrum of the process $\{\mathbb{E}[X(t)|X_\varepsilon(\cdot)],\ t\in\mathbb{R}\}$. Define

$$D^\star(f_s,R_\theta) = \mathsf{mmse}^\star(f_s) + \int_{F^\star(f_s)} \min\{S_{X|X_\varepsilon}(f),\theta\}df, \tag{2.19a}$$

$$R_\theta = \frac{1}{2}\int_{F^\star(f_s)} \log^+[S_{X|X_\varepsilon}(f)/\theta]df, \tag{2.19b}$$

where

$$\mathsf{mmse}^\star(f_s) \triangleq \sigma_X^2 - \int_{F^\star(f_s)} S_{X|X_\varepsilon}(f)df = \int_{-\infty}^{\infty}\left[S_X(f) - S_{X|X_\varepsilon}(f)\mathbf{1}_{F^\star(f_s)}\right]df.$$

Graphical interpretations of $D^\star(f_s,R)$ and $\mathsf{mmse}^\star(f_s)$ are provided in Fig. 2.10.

The importance of the function $D^\star(f_s,R)$ can be deduced from the following two theorems:

THEOREM 2.1 (converse) *Let $X(\cdot)$ be a Gaussian stationary process corrupted by a Gaussian stationary noise $\varepsilon(\cdot)$, and sampled using a bounded linear sampler $S = (K_H,\Lambda)$.*

(i) *Assume that for any $T>0$, $d(\Lambda_T)\le f_s$. Then, for any bitrate R,*

$$D_T(S,R) \ge D^\star(f_s,R).$$

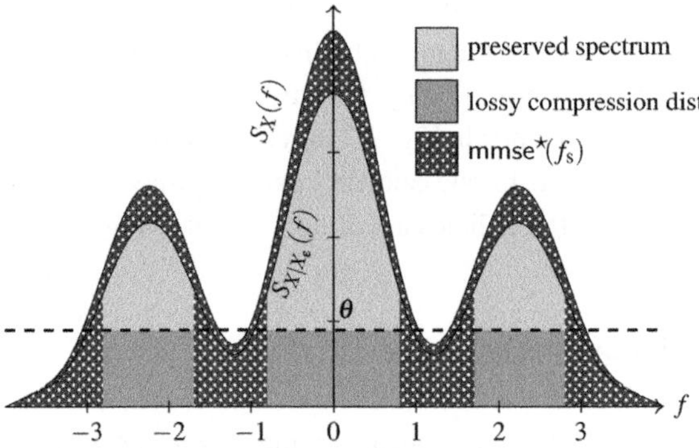

Figure 2.10 Water-filling interpretation of $D^\star(f_s,R_\theta)$: the fundamental distortion limit under any bounded linear sampling. This distortion is the sum of the fundamental estimation error $\mathsf{mmse}^\star(f_s)$ and the lossy compression distortion.

(ii) *Assume that the symmetric density of* Λ *exists and satisfies* $d(\Lambda) \leq f_s$. *Then, for any bitrate R,*

$$\liminf_{T \to \infty} D_T(S,R) \geq D^\star(f_s,R).$$

In addition to the negative statement of Theorem 2.1, we show in the next section the following positive coding result.

THEOREM 2.2 (achievability) *Let* $X(\cdot)$ *be a Gaussian stationary process corrupted by a Gaussian stationary noise* $\varepsilon(\cdot)$. *Then, for any* f_s *and* $\varepsilon > 0$, *there exists a bounded linear sampler S with a sampling set of symmetric density not exceeding* f_s *such that, for any R, the distortion in ADX attained by sampling* $X_\varepsilon(\cdot)$ *using S over a large enough time interval T, and encoding these samples using* $\lfloor TR \rfloor$ *bits, does not exceed* $D^\star(f_s,R)+\varepsilon$.

A full proof of Theorem 2.1 can be found in [24]. Intuition for Theorem 2.1 may be obtained by representing $X(\cdot)$ according to its Karhunen–Loève (KL) expansion over $[-T/2,T/2]$, and then using a sampling matrix that keeps only $N_T \triangleq \lfloor Tf_s \rfloor$ of these coefficients. The function $D^\star(f_s,R)$ arises as the limiting expression in the noisy version of (2.5), when the sampling matrix is tuned to keep those KL coefficients corresponding to the N_T largest eigenvalues in the expansion.

In Section 2.5, we provide a constructive proof of Theorem 2.2 that also shows that $D^\star(f_s,R)$ is attained using a multi-branch LTI uniform sampler with an appropriate choice of pre-sampling filters. The rest of the current section is devoted to studying properties of the minimal ADX distortion $D^\star(f_s,R)$.

2.4.2 Properties of $D^\star(f_s,R)$

In view of Theorems 2.1 and 2.2, the function $D^\star(f_s,R)$ trivially satisfies the properties mentioned in Section 2.3.3 for the optimal distortion in ADX. It is instructive to observe how these properties can be deduced directly from the definition of $D^\star(f_s,R)$ in (2.19).

Unlimited Bitrate

As $R \to \infty$, the parameter θ goes to zero and (2.19a) reduces to $\mathsf{mmse}^\star(f_s)$. This function describes the MMSE that can be attained by any bounded linear sampler with symmetric density at most f_s. In particular, in the non-noisy case, $\mathsf{mmse}^\star(f_s) = 0$ if and only if f_s exceeds the Landau rate of $X(\cdot)$. Therefore, in view of the explanation in Section 2.3.3 and under unlimited bitrate, zero noise, and the identity pre-sampling operation, Theorem 2.1 agrees with the necessary condition derived by Landau for stable sampling in the Paley–Wiener space [21].

Optimal Sampling

The other extreme in the expression for $D^\star(f_s,R)$ is when f_s is large enough that it does not impose any constraint on sampling. In this case, we expect the ADX distortion to coincide with the function $D_{X|X_\varepsilon}(R)$ of (2.15), since the latter is the minimal distortion only due to noise and lossy compression at bitrate R. From the definition of $F^\star(f_s)$, we observe that $F^\star(f_s) = \text{supp}\, S_X$ (almost everywhere) whenever f_s is equal to or greater than the Landau rate of $X(\cdot)$. By examining (2.19), we see that this equality implies that

$$D^\star(f_s, R) = D_{X|X_\varepsilon}(R). \tag{2.20}$$

In other words, the condition $f_s \geq f_{\mathsf{Lnd}}$ means that there is no loss due to sampling in the ADX system. This property of the minimal distortion is not surprising. It merely expresses the fact anticipated in Section 2.3.3 that, when (2.10) vanishes as T goes to infinity, the estimator $\mathbb{E}[X(t)|X_\varepsilon]$ is obtained from the samples in this limit and thus the only loss of information after sampling is due to the noise.

In the next section we will see that, under some conditions, the equality (2.20) is extended to sampling rates smaller than the Landau rate of the signal.

2.4.3 Optimal Sampling Subject to a Bitrate Constraint

We now explore the minimal sampling rate f_s required in order to attain equality in (2.20), that is, the minimal sampling rate at which the minimal distortion in ADX equals the indirect DRF of $X(\cdot)$ given $X_\varepsilon(\cdot)$, describing the minimal distortion subject only to a bitrate R constraint and additive noise. Intuition for this sampling rate is obtained by exploring the behavior of $D^\star(f_s, R)$ as a function of f_s for a specific PSD and a fixed bitrate R. For simplicity, we explore this behavior under the assumption of zero noise ($\varepsilon \equiv 0$) and signal $X(\cdot)$ with a unimodal PSD as in Fig. 2.11. Note that in this case we

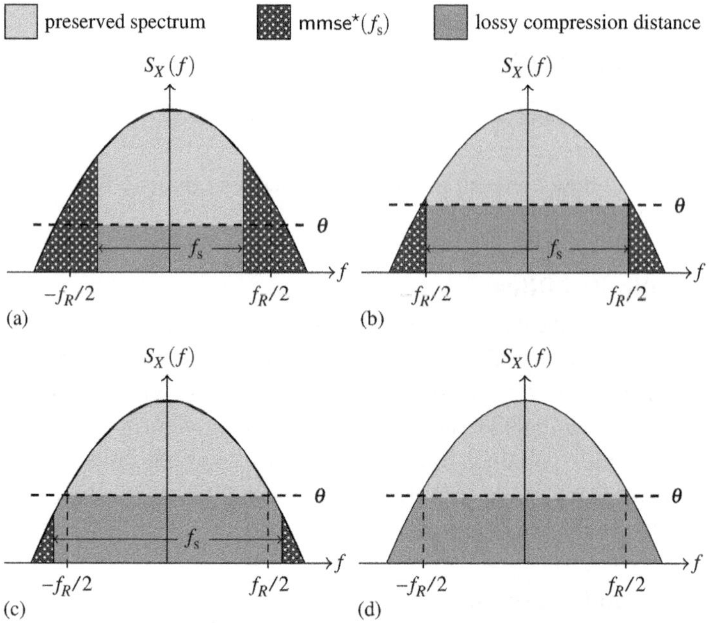

Figure 2.11 Water-filling interpretation for the function $D^\star(f_s, R)$ under zero noise, a fixed bitrate R, and three sampling rates: (a) $f_s < f_R$, (b) $f_s = f_R$, and (c) $f_s > f_R$. (d) corresponds to the DRF of $X(\cdot)$ at bitrate R. This DRF is attained whenever $f_s \geq f_R$, where f_R is smaller than the Nyquist rate.

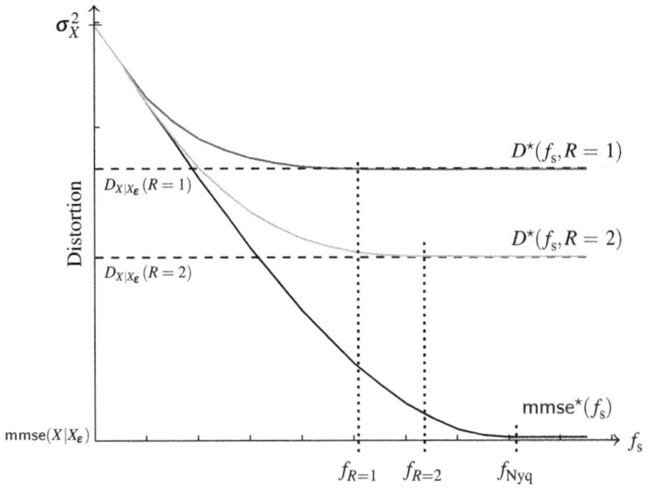

Figure 2.12 The function $D^\star(f_s, R)$ for the PSD of Fig. 2.11 and two values of the bitrate R. Also shown is the DRF of $X(\cdot)$ at these values that is attained at the sub-Nyquist sampling rates marked by f_R.

have $S_{X|X_\varepsilon}(f) = S_X(f)$ since the noise is zero, $f_{\mathsf{Lnd}} = f_{\mathsf{Nyq}}$ since $S_X(f)$ has a connected support, and $F^\star(f_s)$ is the interval of length f_s centered around the origin since $S_X(f)$ is unimodal. In all cases in Fig. 2.11 the bitrate R is fixed and corresponds to the preserved part of the spectrum through (2.19b). The distortion $D^\star(f_s, R)$ changes with f_s, and is given by the sum of two terms in (2.19a): $\mathsf{mmse}^\star(f_s)$ and the lossy compression distortion. For example, the increment in f_s from (a) to (b) reduces $\mathsf{mmse}^\star(f_s)$ and increases the lossy compression distortion, although the overall distortion decreases due to this increment. However, the increase in f_s leading from (b) to (c) is different: while (c) shows an additional reduction in $\mathsf{mmse}^\star(f_s)$ compared with (b), the sum of the two distortion terms is identical in both cases and, as illustrated in (d), equals the DRF of $X(\cdot)$ from (2.16). It follows that, in the case of Fig. 2.11, the optimal ADX performance is attained at some sampling rate f_R that is smaller than the Nyquist rate, and depends on the bitrate R through expression (2.16). The full behavior of $D^\star(f_s, R)$ as a function of f_s is illustrated in Fig. 2.12 for two values of R.

The phenomenon described above and in Figs. 2.11 and 2.12 can be generalized to any Gaussian stationary process with arbitrary PSD and noise in the ADX setting, according to the following theorem.

THEOREM 2.3 (optimal sampling rate [24]) *Let $X(\cdot)$ be a Gaussian stationary process with PSD $S_X(f)$ corrupted by a Gaussian noise $\varepsilon(\cdot)$. For each point (R, D) on the graph of $D_{X|X_\varepsilon}(R)$ associated with a water-level θ via* (2.15), *let f_R be the Lebesgue measure of the set*

$$F_\theta \triangleq \{f : S_{X|X_\varepsilon}(f) \geq \theta\}.$$

Then, for all $f_s \geq f_R$,

$$D^\star(f_s, R) = D_{X|X_\varepsilon}(R).$$

The proof of Theorem 2.3 is relatively straightforward and follows from the definition of F_θ and $D^\star(f_s, R)$.

We emphasize that the critical frequency f_R depends only on the PSDs $S_X(f)$ and $S_\varepsilon(f)$, and on the operating point on the graph of $D^\star(f_s, R)$. This point may be parametrized by D, R, or the water-level θ using (2.15). Furthermore, we can consider a version of Theorem 2.3 in which the bitrate is a function of the distortion and the sampling rate, by inverting $D^\star(f_s, R)$ with respect to R. This inverse function, $R^\star(f_s, D)$, is the minimal number of bits per unit time one must provide on the samples of $X_\varepsilon(\cdot)$, obtained by any bounded linear sampler with sampling density not exceeding f_s, in order to attain distortion not exceeding D. The following representation of $R^\star(f_s, D)$ in terms of f_R is equivalent to Theorem 2.3.

THEOREM 2.4 (rate-distortion lower bound) *Consider the samples of a Gaussian stationary process $X(\cdot)$ corrupted by a Gaussian noise $\varepsilon(\cdot)$ obtained by a bounded linear sampler of maximal sampling density f_s. The bitrate required to recover $X(\cdot)$ with MSE at most $D >$ mmse$^\star(f_s)$ is at least*

$$R^\star(f_s, D) = \begin{cases} \frac{1}{2}\int_{F^\star(f_s)} \log^+\left(\frac{f_s S_{X|X_\varepsilon}(f)}{D - \mathsf{mmse}^\star(f_s)}\right) df, & f_s < f_R, \\ R_{X|X_\epsilon}(D), & f_s \geq f_R, \end{cases} \tag{2.21}$$

where

$$R_{X|X_\varepsilon}(D_\theta) = \frac{1}{2}\int_{-\infty}^{\infty} \log^+[S_{X|X_\varepsilon}(f)/\theta]\, df$$

is the indirect rate-distortion function of $X(\cdot)$ given $X_\varepsilon(\cdot)$, and θ is determined by

$$D_\theta = \mathsf{mmse}^\star(f_s) + \int_{-\infty}^{\infty} \min\{S_{X|X_\varepsilon}(f), \theta\}\, df.$$

Theorems 2.3 and 2.4 imply that the equality in (2.20), which was previously shown to hold for $f_s \geq f_{\mathsf{Lnd}}$, is extended to all sampling rates above $f_R \leq f_{\mathsf{Lnd}}$. As R goes to infinity, $D^\star(f_s, R)$ converges to mmse$^\star(f_s)$, the water-level θ goes to zero, the set F_θ coincides with the support of $S_X(f)$, and f_R converges to f_{Lnd}. Theorem 2.3 then implies that mmse$^\star(f_s) = 0$ for all $f_s \geq f_{\mathsf{Lnd}}$, a fact that agrees with Landau's characterization of sets of sampling for perfect recovery of signals in the Paley–Wiener space, as explained in Section 2.3.3.

An intriguing way to explain the critical sampling rate subject to a bitrate constraint arising from Theorem 2.3 follows by considering the degrees of freedom in the representation of the analog signal pre- and post-sampling and with lossy compression of the samples. For stationary Gaussian signals with zero sampling noise, the degrees of freedom in the signal representation are those spectral bands in which the PSD is non-zero. When the signal energy is not uniformly distributed over these bands, the optimal lossy compression scheme described by (2.16) calls for discarding those bands with the lowest energy, i.e., the parts of the signal with the lowest uncertainty.

The degree to which the new critical rate f_R is smaller than the Nyquist rate depends on the energy distribution of $X(\cdot)$ across its spectral occupancy. The more uniform this

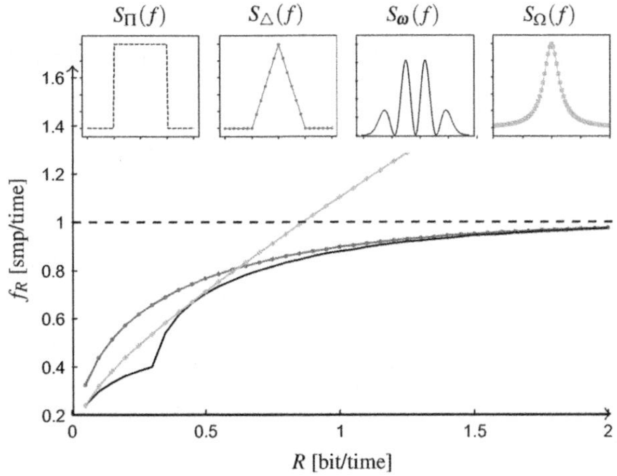

Figure 2.13 The critical sampling rate f_R as a function of the bitrate R for the PSDs given in the small frames at the top of the figure. For the bandlimited PSDs $S_\Pi(f)$, $S_\Delta(f)$, and $S_\omega(f)$, the critical sampling rate is always at or below the Nyquist rate. The critical sampling rate is finite for any R, even for the non-bandlimited PSD $S_\Omega(f)$.

distribution, the more degrees of freedom are required to represent the lossy compressed signal and therefore the closer f_R is to the Nyquist rate. In the examples below we derive the precise relation between f_R and R for various PSDs. These relations are illustrated in Fig. 2.13 for the signals $S_\Pi(f)$, $S_\Delta(f)$, $S_\omega(f)$, and $S_\Omega(f)$ defined below.

2.4.4 Examples

Example 2.1 Consider the Gaussian stationary process $X_\Delta(\cdot)$ with PSD

$$S_\Delta(f) \triangleq \sigma^2 \frac{[1-|f/W|]^+}{W},$$

for some $W > 0$. Assuming that the noise is zero,

$$F_\theta = [W(W\theta - 1), W(1 - W\theta)]$$

and thus $f_R = 2W(1 - W\theta)$. The exact relation between f_R and R is obtained from (2.19b) and found to be

$$R = \frac{1}{2}\int_{-f_R/2}^{f_R/2} \log\left(\frac{1-|f/W|}{1-f_R/2W}\right) df = W \log\left(\frac{2W}{2W - f_R}\right) - \frac{f_R}{2\ln 2}.$$

In particular, note that $R \to \infty$ leads to $f_R \to f_{\mathsf{Nyq}} = 2W$, as anticipated.

Example 2.2 Let $X_\Pi(\cdot)$ be the process with PSD

$$S_\Pi(f) = \frac{\mathbf{1}_{|f|<W}(f)}{2W}. \tag{2.22}$$

Assume that $\varepsilon(\cdot)$ is noise with a flat spectrum within the band $(-W, W)$ such that $\gamma \triangleq S_\Pi(f)/S_\varepsilon(f)$ is the SNR at the spectral component f. Under these conditions, the water-level θ in (2.15) satisfies

$$\theta = \sigma^2 \frac{\gamma}{1+\gamma} 2^{-R/W},$$

and hence

$$\frac{D_{X|X_\varepsilon} R)}{\sigma^2} = \frac{1}{1+\gamma} + \frac{\gamma}{1+\gamma} 2^{-R/W}. \tag{2.23}$$

In particular, $F_\theta = [-W, W]$, so that $f_R = 2W = f_{\mathsf{Nyq}}$ for any bitrate R and $D^\star(f_s, R) = D_{X|X_\varepsilon}(R)$ only for $f_s \geq f_{\mathsf{Nyq}}$. That is, for the process $X_\Pi(\cdot)$, optimal sampling under a bitrate constraint occurs only at or above its Nyquist rate.

Example 2.3 Consider the PSD

$$S_\Omega(f) \triangleq \frac{\sigma^2/f_0}{(\pi f/f_0)^2 + 1},$$

for some $f_0 > 0$. The Gaussian stationary process $X_\Omega(\cdot)$ corresponding to this PSD is a Markov process, and it is in fact the unique Gaussian stationary process that is also Markovian (a.k.a. the Ornstein–Uhlenbeck process). Since the spectral occupancy of $X_\Omega(\cdot)$ is the entire real line, its Nyquist and Landau rates are infinite and it is impossible to recover it with zero MSE using any bounded linear sampler. Assuming $\varepsilon(\cdot) \equiv 0$ and noting that $S_\Omega(f)$ is unimodal, $F^\star(f_s) = (-f_2/2, f_s/2)$, and thus

$$\mathsf{mmse}^\star(f_s) = 2\int_{f_s/2}^{\infty} S_\Omega(f)df = \sigma^2\left[1 - \frac{2}{\pi}\arctan\left(\frac{\pi f_s}{2f_0}\right)\right].$$

Consider now a point (R, D) on the DRF of $X_\Omega(\cdot)$ and its corresponding water-level θ determined from (2.16). It follows that $f_R = (2f_s/\pi)\sqrt{1/\theta f_s - 1}$, so that Theorem 2.3 implies that the distortion cannot be reduced below D by sampling above this rate. The exact relation between R and f_R is found to be

$$R = \frac{1}{\ln 2}\left(f_R - \frac{2f_0}{\pi}\arctan\left(\frac{\pi f_R}{2f_0}\right)\right). \tag{2.24}$$

Note that, although the Nyquist rate of $X_\Omega(\cdot)$ is infinite, for any finite R there exists a critical sampling frequency f_R satisfying (2.24) such that $D_{X_\Omega}(R)$ is attained by sampling at or above f_R.

The asymptotic behavior of (2.24) as R goes to infinity is given by $R \sim f_R/\ln 2$. Thus, for R sufficiently large, the optimal sampling rate is linearly proportional to R. The ratio R/f_s is the average number of bits per sample used in the resulting digital representation. It follows from (2.24) that, asymptotically, the "right" number of bits per sample converges to $1/\ln 2 \approx 1.45$. If the number of bits per sample is below this value, then the distortion in ADX is dominated by the DRF $D_{X_\Omega}(\cdot)$, as there are not enough bits to represent the information acquired by the sampler. If the number of bits per sample is greater than this value, then the distortion in ADX is dominated by the sampling distortion, as there are not enough samples for describing the signal up to a distortion equal to its DRF.

As a numerical example, assume that we encode $X_\Omega(t)$ using two bits per sample, i.e., $f_s = 2R$. As $R \to \infty$, the ratio between the minimal distortion $D^\star(f_s, R)$ and $D_{X_\Omega}(R)$ converges to approximately 1.08, whereas the ratio between $D^\star(f_s, R)$ and $\mathsf{mmse}^\star(f_s)$ converges to approximately 1.48. In other words, it is possible to attain the optimal encoding performance within an approximately 8% gap by providing one sample per each two bits per unit time used in this encoding. On the other hand, it is possible to attain the optimal sampling performance within an approximately 48% gap by providing two bits per each sample taken.

2.5 ADX under Uniform Sampling

We now analyze the distortion in the ADX setting of Fig. 2.5 under the important class of single- and multi-branch LTI uniform samplers. Our goal in this section is to show that for any source and noise PSDs, $S_X(f)$ and $S_\varepsilon(f)$, respectively, the function $D^\star(f_s, R)$ describing the fundamental distortion limit in ADX is attainable using a multi-branch LTI uniform sampler. By doing so, we also provide a proof of Theorem 2.2.

We begin by analyzing the ADX system of Fig. 2.5 under an LTI uniform sampler. As we show, the asymptotic distortion in this case can be obtained in a closed form that depends only on the signal and noise PSDs, the sampling rate, the bitrate, and the pre-sampling filter $H(f)$. We then show that, by taking $H(f)$ to be a low-pass filter with cutoff frequency $f_s/2$, we can attain the fundamental distortion limit $D^\star(f_s, R)$ whenever the function $S_{X|X_\varepsilon}(f)$ of (2.13) attains its maximum at the origin. In the more general case of an arbitrarily shaped $S_{X|X_\varepsilon}(f)$, we use multi-branch sampling in order to achieve $D^\star(f_s, R)$.

2.5.1 Single-Branch LTI Uniform Sampling

Assume that the sampler S is the LTI uniform sampler defined in Section 2.3.1 and illustrated in Fig. 2.7. This sampler is characterized by its sampling rate f_s and the frequency response $H(f)$ of the pre-sampling filter.

In Section 2.3.3, we saw that, for any bounded linear sampler, optimal encoding in ADX is obtained by first forming the estimator $\widetilde{X}_T(\cdot)$ from $\mathbf{Y}_T$, and then encoding $\widetilde{X}_T(\cdot)$ in an optimal manner subject to the bitrate constraint. That is, the encoder performs estimation under an MSE criterion followed by optimal source coding for this estimate. Under the LTI uniform sampler, the process $\widetilde{X}_T(\cdot)$ has an asymptotic distribution described by the conditional expectation of $X(\cdot)$ given the sigma algebra generated by $\{X_\varepsilon(n/f_s), n \text{ in } \mathbb{Z}\}$. Using standard linear estimation techniques, this conditional expectation has a representation similar to that of a Wiener filter given by [12]:

$$\widetilde{X}(t) \triangleq \mathbb{E}[X(t)|\{X(n/f_s), n \in \mathbb{Z}\}] = \sum_{n\in\mathbb{Z}} X_\varepsilon(n/f_s) w(t - n/f_s), \quad t \in \mathbb{R}, \tag{2.25}$$

where the Fourier transform of $w(t)$ is

$$W(f) = \frac{S_X(f)|H(f)|^2}{\sum_{k\in\mathbb{Z}} S_{X_\varepsilon}(f - kf_s)|H(f - kf_s)|^2}.$$

Moreover, the resulting MMSE, which is the asymptotic value of $\mathsf{mmse}_T(S)$, is given by

$$\mathsf{mmse}_H(f_s) \triangleq \sum_{n\in\mathbb{Z}} \int_{-\frac{f_s}{2}}^{\frac{f_s}{2}} \left[S_X(f - nf_s) - \widetilde{S}_X(f)\right] df, \tag{2.26}$$

where

$$\widetilde{S}_X(f) \triangleq \frac{S_X^2(f - f_s n)|H(f - f_s n)|^2}{\sum_{n\in\mathbb{Z}} S_{X_\varepsilon}(f - f_s n)|H(f - f_s n)|^2}. \tag{2.27}$$

From the decomposition (2.11), it follows that, when S is an LTI uniform sampler, the distortion can be expressed as

$$D_H(f_s, R) \triangleq \liminf_{T\to\infty} D_T(S, R) = \mathsf{mmse}_H(f_s) + D_{\widetilde{X}}(R),$$

where $D_{\widetilde{X}}(R)$ is the DRF of the Gaussian process $\widetilde{X}(\cdot)$ defined by (2.25), satisfying the law of the process $\widetilde{X}_T(\cdot)$ in the limit as T goes to infinity.

Note that, whenever $f_s \geq f_{\mathsf{Nyq}}$ and supp S_X is included within the passband of $H(f)$, we have that $\widetilde{S}_X(f) = S_{X|X_\varepsilon}(f)$ and thus $\mathsf{mmse}_H(f_s) = \mathsf{mmse}(X|X_\varepsilon)$, i.e., no distortion due to sampling. Moreover, in this situation, $\widetilde{X}(t) = \mathbb{E}[X(t)|X_\varepsilon(\cdot)]$ and

$$D_H(f_s, R) = D_{X|X_\varepsilon}(R). \tag{2.28}$$

The equality (2.28) is a special case of (2.20) for LTI uniform sampling, and says that there is no loss due to sampling in ADX whenever the sampling rate exceeds the Nyquist rate of $X(\cdot)$.

When the sampling rate is below f_{Nyq}, (2.25) implies that the estimator $\widetilde{X}(\cdot)$ has the form of a stationary process modulated by a deterministic pulse, and is therefore a *block-stationary* process, also called a *cyclostationary* process [25]. The DRF for this class of processes can be described by a generalization of the orthogonal transformation and rate allocation that leads to the water-filling expression (2.16) [26]. Evaluating the resulting expression for the DRF of the cyclostationary process $\widetilde{X}(\cdot)$ leads to a closed-form expression for $D_H(f_s, R)$, which was initially derived in [27].

THEOREM 2.5 (achievability for LTI uniform sampling) *Let $X(\cdot)$ be a Gaussian stationary process corrupted by a Gaussian stationary noise $\varepsilon(\cdot)$. The minimal distortion in*

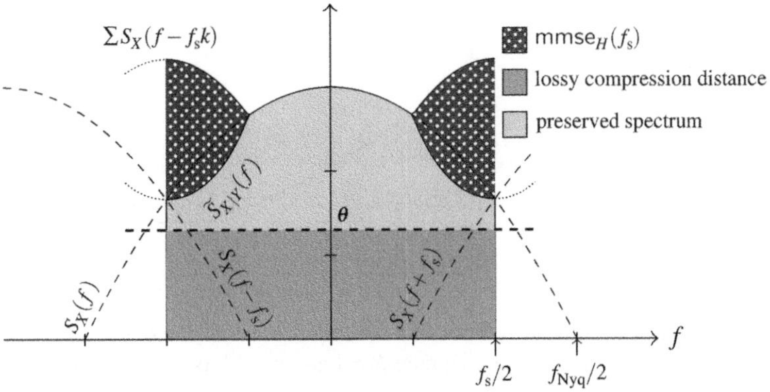

Figure 2.14 Water-filling interpretation of (2.29) with an all-pass filter $H(f)$. The function $\sum_{k\in\mathbb{Z}} S_X(f-f_s k)$ is the aliased PSD that represents the full energy of the original signal within the discrete-time spectrum interval $(-f_s/2, f_s/2)$. The part of the energy recovered by the $\widetilde{X}(\cdot)$ is $\widetilde{S}_X(f)$. The distortion due to lossy compression is obtained by water-filling over the recovered energy according to (2.29a). The overall distortion $D_H(f_s,R)$ is the sum of the sampling distortion and the distortion due to lossy compression.

ADX at bitrate R with an LTI uniform sampler with sampling rate f_s and pre-sampling filter $H(f)$ is given by

$$D_H(f_s,R_\theta) = \mathsf{mmse}_H(f_s) + \int_{-\frac{f_s}{2}}^{\frac{f_s}{2}} \min\left\{\widetilde{S}_X(f),\theta\right\} df, \tag{2.29a}$$

$$R_\theta = \frac{1}{2}\int_{-\frac{f_s}{2}}^{\frac{f_s}{2}} \log_2^+\left[\widetilde{S}_X(f)/\theta\right] df, \tag{2.29b}$$

where $\mathsf{mmse}_H(f_s)$ *and* $\widetilde{S}_X(f)$ *are given by* (2.26) *and* (2.27), *respectively.*

A graphical water-filling interpretation of (2.29) is provided in Fig. 2.14. This expression combines the MMSE (2.26), which depends only on f_s and $H(f)$, with the expression for the indirect DRF of (2.15), which also depends on the bitrate R. The function $\widetilde{S}_X(f)$ arises in the MMSE estimation of $X(\cdot)$ from its samples and can be interpreted as an average over the PSD of polyphase components of the cyclostationary process $\widetilde{X}(\cdot)$ [26].

As is implicit in the analysis above, the coding scheme that attains $D_H(f_s,R)$ is described by the decomposition of the non-causal MSE estimate of $X(\cdot)$ from its samples $\mathbf{Y}_T$. This estimate is encoded using a codebook with $2^{\lfloor TR \rfloor}$ elements that attains the DRF of the Gaussian process $\widetilde{X}(\cdot)$ at bitrate R, and the decoded codewords (which is a waveform over $[-T/2,T/2]$) are used as the reconstruction of $X(\cdot)$. For any $\rho > 0$, the MSE resulting from this process can be made smaller than $D_H(f_s,R-\rho)$ by taking T to be sufficiently large.

Example 2.4 (continuation of Example 2.2) As a simple example for using formula (2.29), consider the process $X_\Pi(\cdot)$ of Example 2.2. Assuming that the noise $\varepsilon(\cdot) \equiv 0$ (equivalently, $\gamma \to \infty$) and that $H(f)$ passes all frequencies $f \in [-W,W]$, the relation between the distortion in (2.29a) and the bitrate in (2.29b) is given by

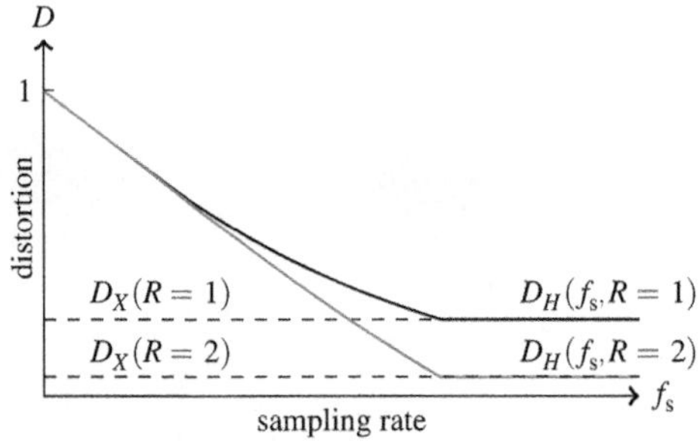

Figure 2.15 Distortion as a function of sampling rate for the source with PSD $S_\Pi(f)$ of (2.22), zero noise, and source coding rates $R = 1$ and $R = 2$ bits per time unit.

$$D_H(f_s,R) = \begin{cases} \mathsf{mmse}_H(f_s) + \sigma^2 \frac{f_s}{2W} 2^{-\frac{2R}{f_s}}, & f_s < 2W, \\ \sigma^2 2^{-\frac{R}{W}}, & f_s \geq 2W, \end{cases} \tag{2.30}$$

where $\mathsf{mmse}_H(f_s) = \sigma^2[1 - f_s/2W]^+$. Expression (2.30) is shown in Fig. 2.15 for two fixed values of the bitrate R. It has a very intuitive structure: for frequencies below $f_{\mathsf{Nyq}} = 2W$, the distortion as a function of the bitrate increases by a constant factor due to the error as a result of non-optimal sampling. This factor completely vanishes once the sampling rate exceeds the Nyquist frequency, in which case $D_H(f_s,R)$ coincides with the DRF of $X(\cdot)$.

In the noisy case when $\gamma = S_\Pi(f)/S_\varepsilon(f)$, we have $\mathsf{mmse}_H(f_s) = \sigma^2(1 - f_s/(2W(1+\gamma)))$ and the distortion takes the form

$$D^\star(f_s,R) = \sigma^2 \begin{cases} \mathsf{mmse}_H(f_s) + (f_s/2W)(\gamma/(1+\gamma))2^{-2R/f_s}, & f_s < 2W, \\ \mathsf{mmse}(X|X_\varepsilon) + (\gamma/(1+\gamma))2^{-R/W}, & f_s \geq 2W, \end{cases} \tag{2.31}$$

where $\mathsf{mmse}(X|X_\varepsilon) = 1/(1+\gamma)$.

Next, we show that, when $S_{X|X_\varepsilon}(f)$ is unimodal, an LTI uniform sampler can be used to attain $D^\star(f_s,R)$.

2.5.2 Unimodal PSD and Low-Pass Filtering

Under the assumption that $H(f)$ is an ideal low-pass filter with cutoff frequency $f_s/2$, (2.29) becomes

$$D_{\mathrm{LPF}}(f_s,R_\theta) = \int_{-\frac{f_s}{2}}^{\frac{f_s}{2}} [S_X(f) - S_{X|X_\varepsilon}(f)]df + \int_{-\frac{f_s}{2}}^{\frac{f_s}{2}} \min\{S_{X|X_\varepsilon}(f),\theta\}df, \tag{2.32a}$$

$$R_\theta = \frac{1}{2}\int_{-\frac{f_s}{2}}^{\frac{f_s}{2}} \log_2^+[S_{X|X_\varepsilon}(f)/\theta]df. \tag{2.32b}$$

Comparing (2.32) with (2.19), we see that the two expressions coincide whenever the interval $[-f_s/2, f_s/2]$ minimizes (2.18). Therefore, we conclude that, when the function $S_{X|X_\varepsilon}(f)$ is unimodal in the sense that it attains its maximal value at the origin, the fundamental distortion in ADX is attained using an LTI uniform sampler with a low-pass filter of cutoff frequency $f_s/2$ as its pre-sampling operation.

An example for a PSD for which (2.32) describes its fundamental distortion limit is the one in Fig. 2.11. Note the LPF with cutoff frequency $f_s/2$ in cases (a)–(c) there. Another example of this scenario for a unimodal PSD is given in Example 2.4 above.

Example 2.5 (continuation of Examples 2.2 and 2.4) In the case of the process $X_\Pi(\cdot)$ with a flat spectrum noise as in Examples 2.2 and 2.4, (2.32) leads to (2.31). It follows that the fundamental distortion limit in ADX with respect to $X_\Pi(\cdot)$ and a flat spectrum noise is given by (2.31), which was obtained from (2.29). Namely, the fundamental distortion limit in this case is obtained using any pre-sampling filter whose passband contains $[-W, W]$, and using an LPF is unnecessary.

In particular, the distortion in (2.31) corresponding to $f_s \geq f_{\text{Nyq}}$ equals the indirect DRF of $X(\cdot)$ given $X_\varepsilon(\cdot)$, which can be found directly from (2.15). Therefore, (2.31) implies that optimal sampling for $X_\Pi(\cdot)$ under LTI uniform sampling occurs only at or above its Nyquist rate. This conclusion is not surprising since, according to Example 2.2, super-Nyquist sampling of $X_\Pi(\cdot)$ is necessary for (2.20) to hold under any bounded linear sampler.

The analysis above implies in particular that a distortion of $D^\star(f_s, R)$ is achievable using LTI uniform sampling for any signal $X(\cdot)$ with a unimodal PSD in the noiseless case or unimodal $S_{X|X_\varepsilon}(f)$ in the noisy case. Therefore, Theorem 2.5 implies Theorem 2.2 in these special cases. In what follows, we use multi-branch sampling in order to show that Theorem 2.2 holds for an arbitrary $S_{X|X_\varepsilon}(f)$.

2.5.3 Multi-Branch LTI Uniform Sampling

We now consider the ADX system of Fig. 2.5 where the sampler is the multi-branch sampler defined in Section 2.3.1 and illustrated in Fig. 2.8. This sampler is characterized by L filters $H_1(f), \ldots, H_L(f)$ and a sampling rate f_s.

The generalization of Theorem 2.5 under this sampler is as follows [27].

THEOREM 2.6 *Let $X(\cdot)$ be a Gaussian stationary process corrupted by a Gaussian stationary noise $\varepsilon(\cdot)$. The minimal distortion in ADX at bitrate R with a multi-branch LTI uniform sampler is given by*

$$D_{H_1,\ldots,H_L}(f_s, R) = \mathsf{mmse}_{H_1,\ldots,H_L}(f_s) + \sum_{l=1}^{L} \int_{-\frac{f_s}{2}}^{\frac{f_s}{2}} \min\{\lambda_l(f)\} df, \tag{2.33a}$$

$$R_\theta = \frac{1}{2} \sum_{l=1}^{L} \int_{-\frac{f_s}{2}}^{\frac{f_s}{2}} \log_2^+ [\lambda_l(f)/\theta] df, \tag{2.33b}$$

where $\lambda_1,\ldots,\lambda_L$ are the eigenvalues of the matrix

$$\widetilde{\mathbf{S}}_X(f) = \left(\mathbf{S}_Y^{-\frac{1}{2}}(f)\right)^H \mathbf{K}(f)\mathbf{S}_Y^{-\frac{1}{2}}(f),$$

with

$$(\mathbf{S}_Y(f))_{i,j} = \sum_{n\in\mathbb{Z}} S_{X_\varepsilon}(f-f_s n)H_i(f-f_s n)H_j^*(f-f_s n), \quad i,j=1,\ldots,L,$$
$$(\mathbf{K}(f))_{i,j} = \sum_{n\in\mathbb{Z}} S_X^2(f-f_s n)H_i(f-f_s n)H_j^*(f-f_s n), \quad i,j=1,\ldots,L.$$

In addition,

$$\mathsf{mmse}_{H_1,\ldots,H_L}(f_s) \triangleq \sigma_X^2 - \int_{-\frac{f_s}{2}}^{\frac{f_s}{2}} tr\left(\widetilde{\mathbf{S}}_X(f)\right)df$$

is the minimal MSE in estimating $X(\cdot)$ from the combined output of the L sampling branches as T approaches infinity.

The most interesting feature in the extension of (2.29) provided by (2.33) is the dependences between samples obtained over different branches, expressed in the definition of the matrix $\widetilde{\mathbf{S}}_X(f)$. In particular, if $f_s \geq f_{\mathsf{Nyq}}$, then we may choose the bandpasses of the L filters to be a set of L disjoint intervals, each of length at most f_s/L, such that the union of their supports contains the support of this choice, the matrix $\widetilde{\mathbf{S}}_X(f)$ is diagonal and its eigenvalues are

$$\lambda_l = \widetilde{S}_l(f) \triangleq \frac{\sum_{n\in\mathbb{Z}} S_X^2(f-f_s n)}{\sum_{n\in\mathbb{Z}} S_{X+\varepsilon}(f-f_s n)} \mathbf{1}_{\mathrm{supp} H_l}(f).$$

Since the union of the filters' support contains the support of $S_{X|X_\varepsilon}$, we have

$$D_{H_1,\ldots,H_L}(f_s,R) = D_{X|X_\varepsilon}(R).$$

While it is not surprising that a multi-branch sampler attains the optimal sampling distortion when f_s is above the Nyquist rate, we note that at each branch the sampling rate can be as small as f_{Nyq}/L. This last remark suggests that a similar principle may be used under sub-Nyquist sampling to sample those particular parts of the spectrum of maximal energy whenever $S_{X|X_\varepsilon}(f)$ is not unimodal.

Our goal now is to prove Theorem 2.2 by showing that, for any PSDs $S_X(f)$ and $S_\varepsilon(f)$, the distortion in (2.33) can be made arbitrarily close to the fundamental distortion limit $D^\star(f_s,R)$ with an appropriate choice of the number of sampling branches and their filters. Using the intuition gained above, given a sampling rate f_s we cover the set of maximal energy $F^\star(f_s)$ of (2.18) using L disjoint intervals, such that the length of each interval does not exceed f_s/L. For any $\varepsilon > 0$, it can be shown that there exists L large enough such $\int_\Delta S_{X|X_\varepsilon}(f)df < \varepsilon$, where Δ is the part that is not covered by the L intervals [28].

From this explanation, we conclude that, for any PSD $S_{X|X_\varepsilon}(f)$, $f_s > 0$, and $\varepsilon > 0$, there exists an integer L and a set of L pre-sampling filters $H_1(f),\ldots,H_L(f)$ such that, for every bitrate R,

$$D_{H_1,\ldots,H_L}(f_s,R) \leq D^\star(f_s,R)+\varepsilon. \tag{2.34}$$

Since $D_{H_1,\ldots,H_L}(f_s,R)$ is obtained in the limit as T approaches infinity of the minimal distortion in ADX under the aforementioned multi-branch uniform sampler, the fundamental distortion limit in ADX is achieved up to an arbitrarily small constant.

The description starting from Theorem 2.6 and ending in (2.34) sketches the proof of the achievability side of the fundamental ADX distortion (Theorem 2.2). Below we summarize the main points in the procedure described in this section.

(i) Given a sampling rate f_s, use a multi-branch LTI uniform sampler with a sufficient number of sampling branches L that the effective passband of all branches is close enough to $F^\star$, which is a set of Lebesgue measure f_s that maximizes (2.18).

(ii) Estimate the signal $X(\cdot)$ under an MSE criterion, leading to $\widetilde{X}_T(\cdot)$ defined in (2.10). As $T \to \infty$ this process converges in $\mathbf{L}_2$ norm to $\widetilde{X}(\cdot)$ defined in (2.25).

(iii) Given a bitrate constraint R, encode a realization of $\widetilde{X}_T(\cdot)$ in an optimal manner subject to an MSE constraint as in standard source coding [7]. For example, for $\rho > 0$ arbitrarily small, we may use a codebook consisting of $2^{\lfloor T(R+\rho)\rfloor}$ waveforms of duration T generated by independent draws from the distribution defined by the preserved part of the spectrum in Fig. 2.10. We then use minimum distance encoding with respect to this codebook.

2.6 Conclusion

The processing, communication, and digital storage of an analog signal requires first representing it as a bit sequence. Hardware and modeling constraints in processing analog information imply that the digital representation is obtained by first sampling the analog waveform and then quantizing or encoding its samples. That is, the transformation from analog signals to bits involves the composition of sampling and quantization or, more generally, lossy compression operations.

In this chapter we explored the minimal sampling rate required to attain the fundamental distortion limit in reconstructing a signal from its quantized samples subject to a strict constraint on the bitrate of the system. We concluded that, when the energy of the signal is not uniformly distributed over its spectral occupancy, the optimal signal representation can be attained by sampling at some critical rate that is lower than the Nyquist rate or, more generally, the Landau rate, in bounded linear sampling. This critical sampling rate depends on the bitrate constraint, and converges to the Nyquist or Landau rates in the limit of infinite bitrate. This reduction in the optimal sampling rate under finite bit precision is made possible by designing the sampling mechanism to sample only those parts of the signals that are not discarded due to optimal lossy compression.

The information-theoretic approach to analog-to-digital compression explored in this chapter can be extended in various directions. First, while we considered the minimal sampling rate and resulting distortion under an ideal encoding of the samples, such an encoding is rarely possible in practice. Indeed, in most cases, the encoding of the samples is subject to additional constraints in addition to the bit resolution, such as complexity, time delay, or limited information on the distribution of the signal and the noise. It is therefore important to characterize the optimal sampling rate and resulting distortion under these limitations. In addition, the reduction in the optimal sampling rate under the bitrate constraint from the Nyquist rate to f_R can be understood as the result of a reduction in degrees of freedom in the compressed signal representation compared with the original source. It is interesting to consider whether a similar principle holds for non-stationary [29] or non-Gaussian [30, 31] signal models (e.g., sparse signals).

References

[1] Y. C. Eldar, *Sampling theory: Beyond bandlimited systems*. Cambridge University Press, 2015.

[2] R. M. Gray and D. L. Neuhoff, "Quantization," *IEEE Trans. Information Theory*, vol. 44, no. 6, pp. 2325–2383, 1998.

[3] C. E. Shannon, "Communication in the presence of noise," *IRE Trans. Information Theory*, vol. 37, pp. 10–21, 1949.

[4] H. Landau, "Sampling, data transmission, and the Nyquist rate," *Proc. IEEE*, vol. 55, no. 10, pp. 1701–1706, 1967.

[5] C. E. Shannon, "A mathematical theory of communication," *Bell System Technical J.*, vol. 27, pp. 379–423, 623–656, 1948.

[6] C. E. Shannon, "Coding theorems for a discrete source with a fidelity criterion," *IRE National Convention Record*, vol. 4, no. 1, pp. 142–163, 1959.

[7] T. Berger, *Rate-distortion theory: A mathematical basis for data compression*. Prentice-Hall, 1971.

[8] R. Walden, "Analog-to-digital converter survey and analysis," *IEEE J. Selected Areas in Communications*, vol. 17, no. 4, pp. 539–550, 1999.

[9] J. Candy, "A use of limit cycle oscillations to obtain robust analog-to-digital converters," *IEEE Trans. Communications*, vol. 22, no. 3, pp. 298–305, 1974.

[10] B. Oliver, J. Pierce, and C. Shannon, "The philosophy of PCM," *IRE Trans. Information Theory*, vol. 36, no. 11, pp. 1324–1331, 1948.

[11] D. L. Neuhoff and S. S. Pradhan, "Information rates of densely sampled data: Distributed vector quantization and scalar quantization with transforms for Gaussian sources," *IEEE Trans. Information Theory*, vol. 59, no. 9, pp. 5641–5664, 2013.

[12] M. Matthews, "On the linear minimum-mean-squared-error estimation of an undersampled wide-sense stationary random process," *IEEE Trans. Signal Processing*, vol. 48, no. 1, pp. 272–275, 2000.

[13] D. Chan and R. Donaldson, "Optimum pre- and postfiltering of sampled signals with application to pulse modulation and data compression systems," *IEEE Trans. Communication Technol.*, vol. 19, no. 2, pp. 141–157, 1971.

[14] R. Dobrushin and B. Tsybakov, "Information transmission with additional noise," *IRE Trans. Information Theory*, vol. 8, no. 5, pp. 293–304, 1962.

[15] A. Kolmogorov, "On the Shannon theory of information transmission in the case of continuous signals," *IRE Trans. Information Theory*, vol. 2, no. 4, pp. 102–108, 1956.

[16] A. Lapidoth, "On the role of mismatch in rate distortion theory," *IEEE Trans. Information Theory*, vol. 43, no. 1, pp. 38–47, 1997.

[17] I. Kontoyiannis and R. Zamir, "Mismatched codebooks and the role of entropy coding in lossy data compression," *IEEE Trans. Information Theory*, vol. 52, no. 5, pp. 1922–1938, 2006.

[18] A. H. Zemanian, *Distribution theory and transform analysis: An introduction to generalized functions, with applications*. Courier Corporation, 1965.

[19] J. Wolf and J. Ziv, "Transmission of noisy information to a noisy receiver with minimum distortion," *IEEE Trans. Information Theory*, vol. 16, no. 4, pp. 406–411, 1970.

[20] H. Witsenhausen, "Indirect rate distortion problems," *IEEE Trans. Information Theory*, vol. 26, no. 5, pp. 518–521, 1980.

[21] H. Landau, "Necessary density conditions for sampling and interpolation of certain entire functions," *Acta Mathematica*, vol. 117, no. 1, pp. 37–52, 1967.

[22] A. Beurling and L. Carleson, *The collected works of Arne Beurling: Complex analysis*. Birkhäuser, 1989, vol. 1.

[23] F. J. Beutler, "Sampling theorems and bases in a Hilbert space," *Information and Control*, vol. 4, nos. 2–3, pp. 97–117, 1961.

[24] A. Kipnis, Y. C. Eldar, and A. J. Goldsmith, "Fundamental distortion limits of analog-to-digital compression," *IEEE Trans. Information Theory*, vol. 64, no. 9, pp. 6013–6033, 2018.

[25] W. Bennett, "Statistics of regenerative digital transmission," *Bell Labs Technical J.*, vol. 37, no. 6, pp. 1501–1542, 1958.

[26] A. Kipnis, A. J. Goldsmith, and Y. C. Eldar, "The distortion rate function of cyclostationary Gaussian processes," *IEEE Trans. Information Theory*, vol. 64, no. 5, pp. 3810–3824, 2018.

[27] A. Kipnis, A. J. Goldsmith, Y. C. Eldar, and T. Weissman, "Distortion rate function of sub-Nyquist sampled Gaussian sources," *IEEE Trans. Information Theory*, vol. 62, no. 1, pp. 401–429, 2016.

[28] A. Kipnis, "Fundamental distortion limits of analog-to-digital compression," Ph.D. dissertation, Stanford University, 2018.

[29] A. Kipnis, A. J. Goldsmith, and Y. C. Eldar, "The distortion-rate function of sampled Wiener processes," in *IEEE Transactions on Information Theory*, vol. 65, no. 1, pp. 482–499, Jan. 2019. doi: 10.1109/TIT.2018.2878446

[30] A. Kipnis, G. Reeves, and Y. C. Eldar, "Single letter formulas for quantized compressed sensing with Gaussian codebooks," in *2018 IEEE International Symposium on Information Theory (ISIT)*, 2018, pp. 71–75.

[31] A. Kipnis, G. Reeves, Y. C. Eldar, and A. J. Goldsmith, "Compressed sensing under optimal quantization," in *2017 IEEE International Symposium on Information Theory (ISIT)*, 2017, pp. 2148–2152.

3 Compressed Sensing via Compression Codes

Shirin Jalali and H. Vincent Poor

Summary

Compressed sensing (CS) refers to the following fundamental data-acquisition problem. A signal $\mathbf{x} \in \mathbb{R}^n$ is measured as $\mathbf{y} = A\mathbf{x} + \mathbf{z}$, where $A \in \mathbb{R}^{m \times n}$ $(m < n)$ and $\mathbf{z} \in \mathbb{R}^m$ denote the sensing matrix and the measurement noise, respectively. The goal of a CS recovery algorithm is to estimate $\mathbf{x}$ from under-determined measurements $\mathbf{y}$. CS is possible because the signals we are interested in capturing are typically highly structured. Recovery algorithms take advantage of such structures to solve the described under-determined system of linear equations. On the other hand, data compression, i.e., efficiently encoding of an acquired signal, is one of the pillars of information theory. Like CS, data-compression codes are designed to take advantage of a signal's structure to encode it efficiently. Studying modern image and video compression codes on one hand and CS recovery algorithms on the other hand reveals that, to a great extent, structures used by compression codes are much more elaborate than those used by CS algorithms. Using more complex structures in CS, similar to those employed by data-compression codes, potentially leads to more efficient recovery methods that require fewer linear measurements or have a better reconstruction quality. In this chapter, we establish fundamental connections between the two seemingly independent problems of data compression and CS. This connection leads to CS recovery methods that are based on compression codes. Such compression-based CS recovery methods, indirectly, take advantage of all structures used by compression codes. This process, with minimal effort, elevates the class of structures used by CS algorithms to those used by compression codes and hence leads to more efficient CS recovery methods.

3.1 Compressed Sensing

Data acquisition refers to capturing signals that lie in the physical world around us and converting them to processable digital signals. It is a basic operation that takes place before any other data processing can begin to happen. Audio recorders, cameras, X-ray computed tomography (CT) scanners, and magnetic resonance imaging (MRI) machines are some examples of data-acquisition devices that employ different mechanisms to perform this crucial step.

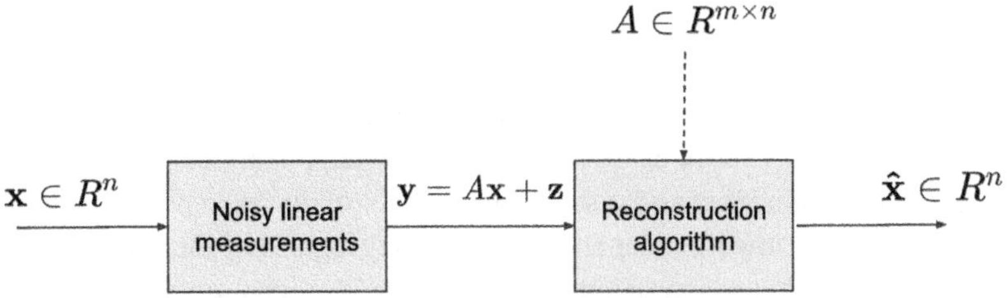

Figure 3.1 Block diagram of a CS measurement system. Signal $\mathbf{x}$ is measured as $\mathbf{y} \in \mathbb{R}^m$ and later recovered as $\widehat{\mathbf{x}}$.

For decades the Nyquist–Shannon sampling theorem served as the theoretical foundation of most data-acquisition systems. The sampling theorem states that for a band-limited signal with a maximum frequency f_c, a sampling rate of $2f_c$ is enough for perfect reconstruction. CS is arguably one of the most disruptive ideas conceived after this fundamental theorem [1, 2]. CS proves that, for sparse signals, the sampling rate can in fact be substantially below the rate required by the Nyquist–Shannon sampling theorem while perfect reconstruction still being possible.

CS is a special data-acquisition technique that is applicable to various measurement devices that can be described as a noisy linear measurement system. Let $\mathbf{x} \in \mathbb{R}^n$ denote the desired signal that is to be measured. Then, the measured signal $\mathbf{y} \in \mathbb{R}^m$ in such systems can be written as

$$\mathbf{y} = A\mathbf{x} + \mathbf{z}, \tag{3.1}$$

where $A \in \mathbb{R}^{m \times n}$ and $\mathbf{z} \in \mathbb{R}^m$ denote the sensing matrix and the measurement noise, respectively. A reconstruction algorithm then recovers $\mathbf{x}$ from measurements $\mathbf{y}$, while having access to sensing matrix A. Fig. 3.1 shows a block diagram representation of the described measurement system.

For $\mathbf{x}$ to be recoverable from $\mathbf{y} = A\mathbf{x} + \mathbf{z}$, for years, the conventional wisdom has been that, since there are n unknown variables, the number of measurements m should be equal to or larger than n. However, in recent years, researchers have observed that, since most signals we are interested in acquiring do not behave like random noise, and are typically "structured," therefore, employing our knowledge about their structure might enable us to recover them, even if the number of measurements (m) is smaller than the ambient dimension of the signal (n). CS theory proves that this is in fact the case, at least for signals that have sparse representations in some transform domain.

More formally, the problem of CS can be stated as follows. Consider a class of signals denoted by a set Q, which is a compact subset of $\mathbb{R}^n$. A signal $\mathbf{x} \in Q$ is measured by m noisy linear projections as $\mathbf{y} = A\mathbf{x} + \mathbf{z}$. A CS recovery (reconstruction) algorithm estimates the signal $\mathbf{x}$ from measurements $\mathbf{y}$, while having access to the sensing matrix A and knowing the set Q.

While the original idea of CS was developed for sparse or approximately sparse signals, the results were soon extended to other types of structure, such as block-sparsity and low-rankness as well. (See [3–26] for some examples of this line of work.)

Despite such extensions, for one-dimensional signals, and to a great extent for higher-dimensional signals such as images and video files, the main focus has still remained on sparsity and its extensions. For two-dimensional signals, in addition to sparse signals, there has also been extensive study of low-rank matrices.

The main reasons for such a focus on sparsity have been two-fold. First, sparsity is a relevant structure that shows up in many signals of interest. For instance, images are known to have (approximately) sparse wavelet representations. In most applications of CS in wireless communications, the desired coefficients that are to be estimated are sparse. The second reason for focusing on sparsity has been theoretical results that show that the ℓ_0-"norm" can be replaced with the ℓ_1-norm. The ℓ_0-"norm" of a signal $\mathbf{x} \in \mathbb{R}^n$ counts the number of non-zero elements of $\mathbf{x}$, and hence serves as a measure of its sparsity. For solving $\mathbf{y} = A\mathbf{x}$, when $\mathbf{x}$ is known to be sparse, a natural optimization is the following:

$$\min \|\mathbf{u}\|_0 \text{ s.t. } A\mathbf{u} = \mathbf{y}, \tag{3.2}$$

where $\mathbf{u} \in \mathbb{R}^n$. This is an NP-hard combinatorial optimization. However, replacing the ℓ_0-"norm" with the ℓ_1-norm leads to a tractable convex optimization problem.

Since the inception of the idea of CS there has been a large body of work on developing practical, efficient, and robust CS algorithms that are able to recover sparse signals from their under-sampled measurements. Such algorithms have found compelling applications for instance in MRI machines and other computational imaging systems. Since most focus in CS has been on sparsity and its extensions, naturally, most reconstruction algorithms are also designed for recovering sparse signals from their noisy linear projections. For instance, widely used optimizations and algorithms such as basis pursuit [27], the least absolute shrinkage and selection operator (LASSO) [28], fast iterative shrinkage-thresholding algorithm (FISTA) [29], orthogonal matching pursuit (OMP) [30], the iterative thresholding method of [31], least-angle regression (LARS) [32], iterative hard thresholding (IHT) [33], CoSaMP [34], approximate message passing (AMP) [35], and subspace pursuit [36] have been developed for recovering sparse signals.

3.2 Compression-Based Compressed Sensing

While sparsity is an important and attractive structure that is present in many signals of interest, most such signals, including images and video files, in addition to exhibiting sparsity, follow structures that are beyond sparsity. A CS recovery algorithm that takes advantage of the full structure that is present in a signal, potentially, outperforms those that merely rely on a simple structure. In other words, such an algorithm would potentially require fewer measurements, or, for an equal number of measurements, present a better reconstruction quality.

To have high-performance CS recovery methods, it is essential to design algorithms that take advantage of the full structure that is present in a signal. To address this issue, researchers have proposed different potential solutions. One line of work has been based on using already-existing denoising algorithms to develop denoising-based CS recovery methods. Denoising, especially image denoising, compared with CS is a very well-established research area. In fact, denoising-based recovery methods [37] show very good performance in practice, as they take advantage of more complex structures. Most other directions, such as those based on universal CS [38–40], or those based on learning general structure [41], while interesting from a theoretical viewpoint, have yet to yield efficient practical algorithms.

In this chapter, we focus on another approach for addressing this issue, which we refer to as compression-based CS. As we discuss later in this chapter, compression-based CS leads to theoretically tractable efficient algorithms that achieve state-of-the-art performance in some applications.

Data compression is a crucial data-processing task, which is important from the perspectives of both data communication and data storage. For instance, modern image and video compression codes play a key role in current wired and wireless communication systems, as they substantially reduce the size of such files. A data-compression code provides an encoding and a decoding mechanism than enables storage of data using as few bits as possible. For image, audio, and video files, compared with CS, data compression is a very well-established research area with, in some cases, more than 50 years of active research devoted to developing efficient codes.

As with CS, the success of data-compression codes hinges upon their ability to take advantage of structures that are present in signals. Comparing the set of structures used by data-compression codes and those employed by CS algorithms reveals that, especially for images and video files, the class of structures used by compression codes is much richer and much more complex. Given this discrepancy in the scope of structures used in the two cases, and the shortcomings of existing recovery methods highlighted earlier, the following question arises: is it possible to use existing compression codes as part of a CS recovery algorithm to have "compression-based CS" recovery algorithms? The motivation for such a method is to provide a shortcut to the usual path of designing recovery methods, which consists of the following two rather challenging steps: (i) (explicitly) specifying the structure that is present in a class of signals, and (ii) finding an efficient and robust algorithm that, given $\mathbf{y} = A\mathbf{x} + \mathbf{z}$, finds $\widehat{\mathbf{x}}$ that is consistent with both the measurements and the discovered structure. Given this motivation, in the rest of this chapter, we try to address the following questions.

QUESTION 1 *Can compression codes be used as a tool to design CS recovery algorithms?*

QUESTION 2 *If compression-based CS is possible, how well do such recovery algorithms perform, in terms of their required sampling rate, robustness to noise, and reconstruction quality?*

QUESTION 3 *Are there efficient compression-based CS algorithms with low computational complexity that employ off-the-shelf compression codes, such as JPEP or JPEG2000 for images and MPEG for videos?*

Throughout this chapter, we mainly focus on deterministic signal models. However, at the end, we describe how most of the results can be extended to stochastic processes as well. We also discuss the implications of such generalizations. Before proceeding to the main argument, in the following section we review some basic definitions and notation used throughout this chapter.

3.3 Definitions

3.3.1 Notation

For $x \in \mathbb{R}$, δ_x denotes the Dirac measure with an atom at x. Consider $x \in \mathbb{R}$ and $b \in \mathbb{N}^+$. Every real number can be written as $x = \lfloor x \rfloor + x_q$, where $\lfloor x \rfloor$ denotes the largest integer smaller than or equal to x and $x_q = x - \lfloor x \rfloor$. Since $x - 1 < \lfloor x \rfloor \leq x$, $x_q \in [0, 1)$. Let $0.a_1a_2\ldots$ denote the binary expansion of x_q. That is,

$$x_q = \sum_{i=1}^{\infty} a_i 2^{-i}.$$

Then, define the b-bit quantized version of x as

$$[x]_b = \lfloor x \rfloor + \sum_{i=1}^{b} a_i 2^{-i}.$$

Using this definition,

$$x - 2^{-b} < [x]_b \leq x.$$

For a vector $x^n \in \mathbb{R}^n$, let $[x^n]_b = ([x_1]_b, \ldots, [x_n]_b)$.

Given two vectors $\mathbf{x}$ and $\mathbf{y}$ both in $\mathbb{R}^n$, $\langle \mathbf{x}, \mathbf{y} \rangle = \sum_{i=1}^{n} x_i y_i$ denotes their inner product. Throughout this chapter, ln and log denote the natural logarithm and logarithm to base 2, respectively.

Sets are denoted by calligraphic letters and the size of set $\mathcal{A}$ is denoted by $|\mathcal{A}|$. The ℓ_0-"norm" of $\mathbf{x} \in \mathbb{R}^n$ is defined as $\|\mathbf{x}\|_0 = |\{i : x_i \neq 0\}|$.

3.3.2 Compression

Data compression is about efficiently storing an acquired signal, and hence is a step that happens after the data have been collected. The main goal of data-compression codes is to take advantage of the structure of the data and represent it as efficiently as possible, by minimizing the required number of bits. Data-compression algorithms are either lossy or lossless. In this part, we briefly review the definition of a lossy compression scheme for real-valued signals in $\mathbb{R}^n$. (Note that lossless compression of real-valued signals is not feasible.)

Consider Q, a compact subset of $\mathbb{R}^n$. A fixed-rate compression code for set Q is defined via encoding and decoding mappings $(\mathcal{E}, \mathcal{D})$, where

$$\mathcal{E} : \mathbb{R}^n \to \{1, \ldots, 2^r\}$$

and

$$\mathcal{D} : \{1, \ldots, 2^r\} \to \mathbb{R}^n.$$

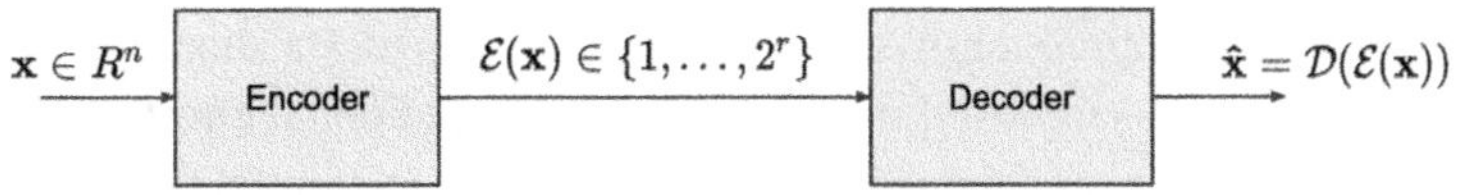

Figure 3.2 Encoder maps signal $\mathbf{x}$ to r bits, $\mathcal{E}(\mathbf{x})$, and, later, the decoder maps the coded bits to $\widetilde{\mathbf{x}}$.

Here r denotes the rate of the code. The codebook of such code is defined as

$$C = \{\mathcal{D}(\mathcal{E}(\mathbf{x})) : \mathbf{x} \in Q\}.$$

Note that C is always a finite set with at most 2^r distinct members. Fig. 3.2 shows a block diagram presentation of the described compression code. The compression code defined by $(\mathcal{E}, \mathcal{D})$ encodes signal $\mathbf{x} \in Q$ into r bits as $\mathcal{E}(\mathbf{x})$, and decodes the encoded bits to $\widetilde{\mathbf{x}} = \mathcal{D}(\mathcal{E}(\mathbf{x}))$. The performance of this code is measured in terms of its rate r and its induced distortion δ defined as

$$\delta = \sup_{\mathbf{x} \in Q} \|\mathbf{x} - \widetilde{\mathbf{x}}\|_2.$$

Consider a family of compression codes $\{(\mathcal{E}_r, \mathcal{D}_r) : r > 0\}$ for set Q indexed by their rate r. The (deterministic) distortion-rate function of this family of compression codes is defined as

$$\delta(r) = \sup_{\mathbf{x} \in Q} \|\mathbf{x} - \mathcal{D}_r(\mathcal{E}_r(\mathbf{x}))\|_2.$$

In other words, $\delta(r)$ denotes the distortion of the code operating at rate r. The corresponding rate-distortion function of this family of compression codes is defined as

$$r(\delta) = \inf\{r : \delta(r) \leq \delta\}.$$

Finally, the α-dimension of a family of compression codes $\{(\mathcal{E}_r, \mathcal{D}_r) : r > 0\}$ is defined as [42]

$$\alpha = \limsup_{\delta \to 0} \frac{r(\delta)}{\log(1/\delta)}. \tag{3.3}$$

This dimension, as shown later, serves as a measure of structuredness and plays a key role in understanding the performance of compression-based CS schemes.

3.4 Compressible Signal Pursuit

Consider the standard problem of CS, which is recovering a "structured" signal $\mathbf{x} \in \mathbb{R}^n$ from an under-sampled set of linear measurements $\mathbf{y} = A\mathbf{x} + \mathbf{z}$, where $\mathbf{y} \in \mathbb{R}^m$, and m is (typically) much smaller than n. The CS theory shows that for various classes of structured signals solving this problem is possible. For different types of structure such as sparsity and low-rankness, it also provides various theoretically analyzable efficient and robust recovery algorithms. However, there is a considerable gap between structures covered by CS theory and those existing in actual signals of interest such as images and videos. Compression-based CS potentially provides a shortcut to generalizing the class of structures employed in CS without much additional effort. In this section, we review compressible signal pursuit (CSP), as the first attempt at designing compression-based CS recovery schemes [42].

Consider a class of signals represented by a compact set $Q \subset \mathbb{R}^n$. Further, assume that there exists a family of compression algorithms for set Q, denoted by $\{(\mathcal{E}_r, \mathcal{D}_r) : r > 0\}$. Let $r(\delta)$ denote the deterministic rate-distortion function of this family of codes, as defined in Section 3.3.2. As an example, consider the class of natural images and the JPEG compression code [43] used at different rates. In general, a compression code might take advantage of signals' sparsity in a certain transform domain such as the wavelet domain, or a very different type of structure. Instead of being interested in structures and mechanisms used by the compression code, we are interested in recovering $\mathbf{x} \in Q$ from the under-sampled measurements $\mathbf{y} = A\mathbf{x} + \mathbf{z}$, using compression algorithms $\{(\mathcal{E}_r, \mathcal{D}_r) : r > 0\}$.

CSP is a compression-based CS recovery approach that recovers $\mathbf{x}$ from $\mathbf{y} = A\mathbf{x} + \mathbf{z}$, by having access to the sensing matrix A and a rate-r compression code characterized by mappings $(\mathcal{E}_r, \mathcal{D}_r)$ and a corresponding codebook C_r. CSP operates according to Occam's principle and outputs a *compressible* signal (a signal in the codebook of the code) that best matches the measurements. More precisely, CSP estimates $\mathbf{x}$ by solving the following optimization:

$$\widehat{\mathbf{x}} = \arg\min_{\mathbf{c} \in C_r} \|\mathbf{y} - A\mathbf{c}\|_2^2. \tag{3.4}$$

The following theorem considers the noise-free regime where $\mathbf{z} = 0$, and connects the number of measurements m and the reconstruction quality of CSP $\|\widehat{\mathbf{x}} - \mathbf{x}\|_2$ to the properties of the compression code, i.e., its rate and its distortion.

THEOREM 3.1 *Consider compact set $Q \subset \mathbb{R}^n$ with a rate-r compression code $(\mathcal{E}_r, \mathcal{D}_r)$ operating at distortion δ. Let $A \in \mathbb{R}^{m \times n}$, where $A_{i,j}$ are independently and identically distributed (i.i.d.) as $\mathcal{N}(0,1)$. For $\mathbf{x} \in Q$, let $\widehat{\mathbf{x}}$ denote the reconstruction of $\mathbf{x}$ from $\mathbf{y} = A\mathbf{x}$, generated by the CSP optimization employing code $(\mathcal{E}_r, \mathcal{D}_r)$. Then,*

$$\|\widehat{\mathbf{x}} - \mathbf{x}\|_2 \le \delta \sqrt{\frac{1+\tau_1}{1-\tau_2}},$$

with probability at least

$$1 - 2^r e^{\frac{m}{2}(\tau_2 + \log(1-\tau_2))} - e^{-\frac{m}{2}(\tau_1 - \log(1+\tau_1))},$$

where $\tau_1 > 0$ and $\tau_2 \in (0,1)$ are arbitrary.

Proof Let $\widetilde{\mathbf{x}} = \mathcal{D}_r(\mathcal{E}_r(\mathbf{x}))$ and $\widehat{\mathbf{x}} = \arg\min_{\mathbf{c} \in C_r} \|\mathbf{y} - A\mathbf{c}\|_2^2$. Since $\widehat{\mathbf{x}}$ is the minimizer of $\|\mathbf{y} - A\mathbf{c}\|_2^2$ over all $\mathbf{c} \in C_r$, and since $\widetilde{\mathbf{x}} \in C_r$, it follows that $\|\mathbf{y} - A\widehat{\mathbf{x}}\|_2 \le \|\mathbf{y} - A\widetilde{\mathbf{x}}\|_2$. That is,

$$\|A\mathbf{x} - A\widehat{\mathbf{x}}\|_2 \le \|A\mathbf{x} - A\widetilde{\mathbf{x}}\|_2. \tag{3.5}$$

Given $\mathbf{x}$, define set $\mathcal{U}$ to denote the set of all possible normalized error vectors as

$$\mathcal{U} = \left\{ \frac{\mathbf{x}-\mathbf{c}}{\|\mathbf{x}-\mathbf{c}\|} : \mathbf{c} \in C_r \right\}.$$

Note that since $|C_r| \leq 2^r$ and $\mathbf{x}$ is fixed, $|\mathcal{U}| \leq 2^r$ as well.

For $\tau_1 > 0$ and $\tau_2 \in (0,1)$, define events $\mathcal{E}_1$ and $\mathcal{E}_2$ as

$$\mathcal{E}_1 \triangleq \{\|A(\mathbf{x}-\widetilde{\mathbf{x}})\|_2^2 \leq m(1+\tau_1)\|\mathbf{x}-\widetilde{\mathbf{x}}\|^2\}$$

and

$$\mathcal{E}_2 \triangleq \{\|A\mathbf{u}\|^2 \geq m(1-\tau_2) : \boldsymbol{u} \in \mathcal{U}\},$$

respectively.

For a fixed $\boldsymbol{u} \in \mathcal{U}$, since the entries of A are i.i.d. $\mathcal{N}(0,1)$, $A\mathbf{u}$ is a vector of m i.i.d. standard normal random variables. Hence, by Lemma 2 in [39],

$$\mathrm{P}(\|A\mathbf{u}\|_2^2 \geq m(1+\tau_1)) \leq e^{-\frac{m}{2}(\tau_1 - \log(1+\tau_1))} \tag{3.6}$$

and

$$\mathrm{P}(\{\|A\mathbf{u}\|_2^2 \leq m(1-\tau_2)\}) \leq e^{\frac{m}{2}(\tau_2 + \log(1-\tau_2))}. \tag{3.7}$$

Since $\widetilde{\mathbf{x}} \in C_r$,

$$\frac{\mathbf{x}-\widetilde{\mathbf{x}}}{\|\mathbf{x}-\widetilde{\mathbf{x}}\|_2} \in \mathcal{U}$$

and, from (3.6),

$$\mathrm{P}(\mathcal{E}_1^c) \leq e^{-\frac{m}{2}(\tau_1 - \log(1+\tau_1))}.$$

Moreover, since $|\mathcal{U}| \leq 2^r$, by the union bound, we have

$$\mathrm{P}(\mathcal{E}_2^c) \leq 2^r e^{\frac{m}{2}(\tau_2 + \log(1-\tau_2))}. \tag{3.8}$$

Hence, by the union bound,

$$\begin{aligned}\mathrm{P}(\mathcal{E}_1 \cap \mathcal{E}_2) &= 1 - \mathrm{P}(\mathcal{E}_1^c \cup \mathcal{E}_2^c) \\ &\geq 1 - 2^r e^{\frac{m}{2}(\tau_2 + \log(1-\tau_2))} - e^{-\frac{m}{2}(\tau_1 - \log(1+\tau_1))}.\end{aligned}$$

By definition, $\widetilde{\mathbf{x}}$ is the reconstruction of $\mathbf{x}$ using code $(\mathcal{E}_r, \mathcal{D}_r)$. Hence, $\|\mathbf{x}-\widetilde{\mathbf{x}}\|_2 \leq \delta$, and, conditioned on $\mathcal{E}_1$,

$$\|\mathbf{y} - A\widetilde{\mathbf{x}}\|_2 \leq \delta\sqrt{m(1+\tau_1)}. \tag{3.9}$$

On the other hand, conditioned on $\mathcal{E}_2$,

$$\|\mathbf{y} - A\widehat{\mathbf{x}}\|_2 \geq \|\mathbf{x}-\widehat{\mathbf{x}}\|\sqrt{m(1-\tau_2)}. \tag{3.10}$$

Combining (3.9) and (3.10), conditioned on $\mathcal{E}_1 \cap \mathcal{E}_2$, we have

$$\|\mathbf{x}-\widehat{\mathbf{x}}\| \leq \delta\sqrt{\frac{1+\tau_1}{1-\tau_2}}.$$

To understand the implications of Theorem 3.1, the following corollary considers a specific choice of parameters τ_1 and τ_2 and relates the required number of measurements by the CSP algorithm to the α-dimension of the compression code.

COROLLARY 3.1 (Corollary 1 in [42]) *Consider a family of compression codes $(\mathcal{E}_r, \mathcal{D}_r)$ for set Q with corresponding codebook C_r and rate-distortion function $r(\delta)$. Let $A \in \mathbb{R}^{m\times n}$, where $A_{i,j}$ are i.i.d. $\mathcal{N}(0,1)$. For $\mathbf{x} \in Q$ and $\mathbf{y} = A\mathbf{x}$, let $\widehat{\mathbf{x}}$ denote the solution of (3.4). Given $\nu > 0$ and $\eta > 1$, such that $\eta/\log(1/e\delta) < \nu$, let*

$$m = \frac{\eta r}{\log(1/e\delta)}.$$

Then,

$$P\Big(\|\mathbf{x} - \widehat{\mathbf{x}}\|_2 \geq \theta\delta^{1-\frac{1+\nu}{\eta}}\Big) \leq e^{-0.8m} + e^{-0.3\nu r},$$

where $\theta = 2e^{-(1+\nu)/\eta}$.

Proof In Theorem 3.1, let $\tau_1 = 3$, $\tau_2 = 1 - (e\delta)^{2(1+\epsilon)/\eta}$. For $\tau_1 = 3$, $0.5(\tau_1 - \log(1+\tau_1)) > 0.8$. For $\tau_2 = 1 - (e\delta)^{2(1+\epsilon)/\eta}$,

$$\begin{aligned}
&r\ln 2 + \frac{m}{2}(\tau_2 + \ln(1-\tau_2)) \\
&= r\ln 2 + \frac{m}{2}\left(1 - (e\delta)^{2(1+\epsilon)/\eta} + \frac{2(1+\epsilon)}{\eta}\ln(e\delta)\right) \\
&\leq r\left(\ln 2 + \frac{\eta}{2\log(1/e\delta)} - (1+\epsilon)\ln 2\right) \\
&\overset{(a)}{\leq} r\ln 2\left(1 + \frac{\epsilon}{2} - (1+\epsilon)\right) \\
&\leq -0.3\epsilon r, \qquad (3.11)
\end{aligned}$$

where (a) is due to the fact that

$$\frac{\eta}{\log(1/e\delta)} = \frac{\eta\ln 2}{\ln(1/e\delta)} < \frac{\epsilon}{2}.$$

Finally,

$$\begin{aligned}
\delta\sqrt{\frac{1+\tau_1}{1-\tau_2}} &= \delta\sqrt{\frac{4}{(e\delta)^{2(1+\epsilon)/\eta}}} \\
&= \theta\delta^{1-(1+\epsilon)/\eta}, \qquad (3.12)
\end{aligned}$$

where $\theta = 2e^{-(1+\epsilon)/\eta}$.

Corollary 3.1 implies that, using a family of compression codes $\{(\mathcal{E}_r, \mathcal{D}_r) : r > 0\}$, as $r \to \infty$ and $\delta \to 0$, the achieved reconstruction error converges to zero (as $\lim_{\delta\to 0}\theta\delta^{1-(1+\epsilon)/\eta} = 0$), while the number of measurements converges to $\eta\alpha$, where α is the α dimension of the compression algorithms and $\eta > 1$ is a free parameter. In other words, as long as the number of measurements m is larger than α, using an appropriate compression code, CSP recovers $\mathbf{x}$.

Theorem 3.1 and its corollary both ignore the effect of noise and consider noise-free measurements. In practice, noise is always present. So it is important to understand the effect of noise on the performance. For instance, the following theorem characterizes the effect of a deterministic noise with a bounded ℓ_2-norm on the performance of CSP.

THEOREM 3.2 (Theorem 2 in [42]) *Consider compression code $(\mathcal{E},\mathcal{D})$ operating at rate r and distortion δ on set $Q \subset \mathbb{R}^n$. For $\mathbf{x} \in Q$, and $\mathbf{y} = A\mathbf{x}+\mathbf{z}$ with $\|\mathbf{z}\|_2 \le \zeta$, let $\widehat{\mathbf{x}}$ denote the reconstruction of $\mathbf{x}$ from $\mathbf{y}$ offered by the CSP optimization employing code $(\mathcal{E},\mathcal{D})$. Then,*

$$\|\widehat{\mathbf{x}}-\mathbf{x}\|_2 \le \delta\sqrt{\frac{1+\tau_1}{1-\tau_2}} + \frac{2\zeta}{\sqrt{(1-\tau_2)d}},$$

with probability exceeding

$$1-2^r e^{\frac{d}{2}(\tau_2+\log(1-\tau_2))} - e^{-\frac{d}{2}(\tau_1-\log(1+\tau_1))},$$

where $\tau_1 > 0$ and $\tau_2 \in (0,1)$ are free parameters.

CSP optimization is a discrete optimization that minimizes a convex cost function over a discrete set of exponential size. Hence, solving CSP in its original form is computationally prohibitive. In the next section, we study this critical issue and review an efficient algorithm with low computational complexity that is designed to approximate the solution of the CSP optimization.

3.5 Compression-Based Gradient Descent (C-GD)

As discussed in the previous section, CSP is based on an exhaustive search over exponentially many codewords and as a result is computationally infeasible. Compression-based gradient descent (C-GD) is a computationally efficient and theoretically analyzable approach to approximating the solution of CSP. The C-GD algorithm [44, 45], inspired by the *projected gradient descent* (PGD) algorithm [46], works as follows. Start from some $\mathbf{x}^0 \in \mathbb{R}^n$. For $t = 1,2,\ldots$, proceed as follows:

$$\mathbf{s}^{t+1} = \mathbf{x}^t + \eta A^{\mathrm{T}}(\mathbf{y} - A\mathbf{x}^t), \tag{3.13}$$

and

$$\mathbf{x}^{t+1} = \mathrm{P}_{C_r}(\mathbf{s}^{t+1}), \tag{3.14}$$

where $\mathrm{P}_{C_r}(\cdot)$ denotes projection into the set of codewords. In other words, for $\mathbf{x} \in \mathbb{R}^n$,

$$\mathrm{P}_{C_r}(\cdot) = \arg\min_{\mathbf{c}\in C_r} \|\mathbf{x}-\mathbf{c}\|_2. \tag{3.15}$$

Here index t denotes the iteration number and $\eta \in \mathbb{R}$ denotes the step size. Each iteration of this algorithm involves performing two operations. The first step is moving in the direction of the negative of the gradient of $\|\mathbf{y} - A\mathbf{x}\|_2^2$ with respect to $\mathbf{x}$ to find solutions that are closer to the $\mathbf{y} = A\mathbf{x}$ hyperplane. The second step, i.e., the projection step, ensures that the estimate C-GD obtains belongs to the codebook and hence conforms with the source structure. The following theorem characterizes the convergence performance of the described C-GD algorithm.

THEOREM 3.3 (Theorem 2 in [45]) *Consider* $\mathbf{x} \in \mathbb{R}^n$. *Let* $\mathbf{y} = A\mathbf{x} + \mathbf{z}$ *and assume that the entries of the sensing matrix A are i.i.d.* $\mathcal{N}(0,1)$ *and that* z_i, $i = 1,\ldots,m$, *are i.i.d.* $\mathcal{N}(0,\sigma_z^2)$. *Let*

$$\eta = \frac{1}{m}$$

and define $\widetilde{\mathbf{x}} = \mathrm{P}_{\mathcal{C}_r}(\mathbf{x})$, *where* $\mathrm{P}_{\mathcal{C}_r}(\cdot)$ *is defined in* (3.15). *Given* $\epsilon > 0$, *for* $m \geq 80r(1+\epsilon)$, *with a probability larger than* $1 - e^{-\frac{m}{2}} - 2^{-40r\epsilon} - 2^{-2\epsilon r + 0.5} - e^{-0.15m}$, *we have*

$$\|\mathbf{x}^{t+1} - \widetilde{\mathbf{x}}\|_2 \leq 0.9\|\mathbf{x}^t - \widetilde{\mathbf{x}}\|_2 + 2\left(2 + \sqrt{\frac{n}{m}}\right)^2 \delta + \sigma_z \sqrt{\frac{32(1+\epsilon)r}{m}}, \tag{3.16}$$

for $k = 0,1,2,\ldots$.

As will be clear in the proof, the choice of 0.9 in Theorem 3.3 is arbitrary, and the result could be derived for any positive value strictly smaller than one. We present the result for this choice as it clarifies the statement of the result and its proof.

As stated earlier, each iteration of the C-GD involves two steps: (i) moving in the direction of the gradient and (ii) projection of the result onto the set of codewords of the compression code. The first step is straightforward and requires two matrix–vector multiplications. For the second step, optimally solving (3.15) might be challenging. However, for any "good" compression code, it is reasonable to assume that employing the code's encoder and decoder consecutively well approximates this operation. In fact, it can be proved that, if $\|\mathrm{P}_{\mathcal{C}_r}(\mathbf{x}) - \mathcal{D}_r(\mathcal{E}_r(\mathbf{x}))\|$ is smaller than ϵ for all $\mathbf{x}$, then replacing $\mathrm{P}_{\mathcal{C}_r}(\cdot)$ with $\mathcal{D}_r(\mathcal{E}_r(\cdot))$ only results in an additive error of ϵ in (3.16). (Refer to Theorem 3 in [45].) Under this simplification, the algorithm's two steps can be summarized as

$$\mathbf{x}^{t+1} = \mathcal{D}_r(\mathcal{E}_r(\mathbf{x}^t + \eta A^{\mathrm{T}}(\mathbf{y} - A\mathbf{x}^t))).$$

Proof By definition, $\mathbf{x}^{t+1} = \mathrm{P}_{\mathcal{C}_r}(\mathbf{s}^{t+1})$ and $\widetilde{\mathbf{x}} = \mathrm{P}_{\mathcal{C}_r}(\mathbf{x})$. Hence, since $\mathbf{x}^{t+1}$ is the closest vector in $\mathcal{C}_r$ to $\mathbf{s}^{t+1}$ and $\widetilde{\mathbf{x}}$ is also in $\mathcal{C}_r$, we have $\|\mathbf{x}^{t+1} - \mathbf{s}^{t+1}\|_2^2 \leq \|\widetilde{\mathbf{x}} - \mathbf{s}^{t+1}\|_2^2$. Equivalently, $\|(\mathbf{x}^{t+1} - \widetilde{\mathbf{x}}) - (\mathbf{s}^{t+1} - \widetilde{\mathbf{x}})\|_2^2 \leq \|\widetilde{\mathbf{x}} - \mathbf{s}^{t+1}\|_2^2$, or

$$\|\mathbf{x}^{t+1} - \widetilde{\mathbf{x}}\|_2^2 \leq 2\langle \mathbf{x}^{t+1} - \widetilde{\mathbf{x}}, \mathbf{s}^{t+1} - \widetilde{\mathbf{x}} \rangle. \tag{3.17}$$

For $k = 0,1,\ldots$, define the error vector and its normalized version as

$$\boldsymbol{\theta}^t \triangleq \mathbf{x}^t - \widetilde{\mathbf{x}}$$

and

$$\underline{\boldsymbol{\theta}}^t \triangleq \frac{\boldsymbol{\theta}^t}{\|\boldsymbol{\theta}^t\|},$$

respectively.

Also, given $\theta^t \in \mathbb{R}^n$, $\theta^{t+1} \in \mathbb{R}^n$, $\eta \in \mathbb{R}^+$, and $A \in \mathbb{R}^{m\times n}$, define the coefficient μ as

$$\mu(\theta^{t+1}, \theta^t, \eta) \triangleq \langle \underline{\theta}^{t+1}, \underline{\theta}^t \rangle - \eta \langle A\underline{\theta}^{t+1}, A\underline{\theta}^t \rangle.$$

Using these definitions and substituting for s^{t+1} and $\mathbf{y} = A\mathbf{x} + \mathbf{z}$, it follows from (3.17) that

$$\begin{aligned} \|\theta^{t+1}\|_2^2 &\leq 2\langle \mathbf{x}^{t+1} - \widetilde{\mathbf{x}}, \mathbf{x}^t + \eta A^{\mathrm{T}}(A\mathbf{x} + \mathbf{z} - A\mathbf{x}^t) - \widetilde{\mathbf{x}} \rangle \\ &= 2\langle \mathbf{x}^{t+1} - \widetilde{\mathbf{x}}, \mathbf{x}^t - \widetilde{\mathbf{x}} \rangle + 2\eta \langle \mathbf{x}^{t+1} - \widetilde{\mathbf{x}}, A^{\mathrm{T}}A(\mathbf{x} - \mathbf{x}^t) \rangle + 2\eta \langle \mathbf{x}^{t+1} - \widetilde{\mathbf{x}}, A^{\mathrm{T}}\mathbf{z} \rangle \\ &= 2\langle \theta^{t+1}, \theta^t \rangle - 2\eta \langle A\theta^{t+1}, A\theta^t \rangle + 2\eta \langle A\theta^{t+1}, A(\mathbf{x} - \widetilde{\mathbf{x}}) \rangle + 2\eta \langle \theta^{t+1}, A^{\mathrm{T}}\mathbf{z} \rangle. \end{aligned} \quad (3.18)$$

Hence, on dividing both sides by (3.18), we have

$$\begin{aligned} \|\theta^{t+1}\|_2 &\leq 2\left(\langle \underline{\theta}^{t+1}, \underline{\theta}^t \rangle - \eta \langle A\underline{\theta}^{t+1}, A\underline{\theta}^t \rangle\right)\|\theta^t\|_2 + 2\eta(\sigma_{\max}(A))^2 \|\mathbf{x} - \widetilde{\mathbf{x}}\|_2 \\ &\quad + 2\eta \langle \underline{\theta}^{t+1}, A^{\mathrm{T}}\mathbf{z} \rangle. \end{aligned} \quad (3.19)$$

In the following we bound the three main random terms on the right-hand side of (3.19), namely, $\langle \underline{\theta}^{t+1}, \underline{\theta}^t \rangle - \eta \langle A\underline{\theta}^{t+1}, A\underline{\theta}^t \rangle$, $(\sigma_{\max}(A))^2$, and $\langle \underline{\theta}^{t+1}, A^{\mathrm{T}}\mathbf{z} \rangle$.

(i) Bounding $\langle \underline{\theta}^{t+1}, \underline{\theta}^t \rangle - \eta \langle A\underline{\theta}^{t+1}, A\underline{\theta}^t \rangle$. We prove that with high probability this term is smaller than 0.45. Let $\mathcal{F}$ define the set of all possible normalized error vectors defined as

$$\mathcal{F} \triangleq \left\{ \frac{\widehat{\mathbf{x}}_1 - \widehat{\mathbf{x}}_2}{\|\widehat{\mathbf{x}}_1 - \widehat{\mathbf{x}}_2\|_2} : \forall\, \widehat{\mathbf{x}}_1, \widehat{\mathbf{x}}_2 \in C_r, \widehat{\mathbf{x}}_1 \neq \widehat{\mathbf{x}}_2 \right\}. \quad (3.20)$$

Now define event $\mathcal{E}_1$ as

$$\mathcal{E}_1 \triangleq \{ \langle \underline{\theta}^{t+1}, \underline{\theta}^t \rangle - \eta \langle A\underline{\theta}^{t+1}, A\underline{\theta}^t \rangle < 0.45 : \forall\, \boldsymbol{u}, \boldsymbol{v} \in \mathcal{F} \}. \quad (3.21)$$

From Lemma 6 in [41], given $\boldsymbol{u}, \boldsymbol{v} \in \mathcal{F}$, we have

$$P\left(\langle \underline{\theta}^{t+1}, \underline{\theta}^t \rangle - \eta \langle A\underline{\theta}^{t+1}, A\underline{\theta}^t \rangle \geq 0.45 \right) \leq 2^{-\frac{m}{20}}. \quad (3.22)$$

Hence, by the union bound,

$$P(\mathcal{E}_1^c) \leq |\mathcal{F}|^2 2^{-\frac{m}{20}}. \quad (3.23)$$

Note that by construction $|\mathcal{F}| \leq |C_r|^2 \leq 2^{2r}$. Therefore,

$$P(\mathcal{E}_1) \geq 1 - 2^{(4r - 0.05m)}.$$

By the theorem assumption, $m \geq 80r(1+\epsilon)$, where $\epsilon > 0$. Therefore, with probability at least $1 - 2^{-40r\epsilon}$, event $\mathcal{E}_1$ happens.

(ii) Bounding $(\sigma_{\max}(A))^2$. Define event $\mathcal{E}_2$ as

$$\mathcal{E}_2 \triangleq \{ \sigma_{\max}(A) \leq 2\sqrt{m} + \sqrt{n} \}.$$

As proved in [47], we have

$$P(\mathcal{E}_2^c) \leq e^{-\frac{m}{2}}.$$

(iii) Bounding $\langle \underline{\theta}^{t+1}, A^{\mathrm{T}}\mathbf{z} \rangle$. Note that

$$\langle \underline{\theta}^{t+1}, A^{\mathrm{T}}\mathbf{z} \rangle = \langle A\underline{\theta}^{t+1}, \mathbf{z} \rangle.$$

Let $\boldsymbol{A}_i^{\mathrm{T}}$ denote the ith row of random matrix A. Then,

$$A\underline{\theta}^{t+1} = [\langle \boldsymbol{A}_1, \underline{\theta}^{t+1}\rangle, \ldots, \langle \boldsymbol{A}_m, \underline{\theta}^{t+1}\rangle]^{\mathrm{T}}.$$

But, for fixed $\underline{\theta}^{t+1}$, $\left\{\langle \boldsymbol{A}_i, \underline{\theta}^{t+1}\rangle\right\}_{i=1}^{n}$ are i.i.d. $\mathcal{N}(0,1)$ random variables. Hence, from Lemma 3 in [39], $\langle \underline{\theta}^{t+1}, A^{\mathrm{T}}\mathbf{z}\rangle$ has the same as $\|\mathbf{z}\|_2 \langle \underline{\theta}^{t+1}, \boldsymbol{g}\rangle$, where $\boldsymbol{g} = [g_1, \ldots, g_n]^{\mathrm{T}}$ is independent of $\|\mathbf{z}\|_2$ and $g_i \overset{\text{i.i.d.}}{\sim} \mathcal{N}(0,1)$. Given $\tau_1 > 0$ and $\tau_2 > 0$, define events $\mathcal{E}_3$ and $\mathcal{E}_4$ as follows:

$$\mathcal{E}_3 \triangleq \{\|\mathbf{z}\|_2^2 \le (1+\tau_1)\sigma_z^2\sigma_z^2\}$$

and

$$\mathcal{E}_4 \triangleq \{\left|\langle \underline{\theta}, \boldsymbol{g}\rangle\right|^2 \le 1+\tau_2, \forall\, \underline{\theta} \in \mathcal{F}\}.$$

From Lemma 2 in [39], we have

$$P(\mathcal{E}_3^{\mathrm{c}}) \le e^{-\frac{m}{2}(\tau_1 - \ln(1+\tau_1))}, \tag{3.24}$$

and by the same lemma, for fixed $\underline{\theta}^{t+1}$, we have

$$P\left(\left|\langle \underline{\theta}^{t+1}, \boldsymbol{g}\rangle\right|^2 \ge 1+\tau_2\right) \le e^{-\frac{1}{2}(\tau_2 - \ln(1+\tau_2))}. \tag{3.25}$$

Hence, by the union bound,

$$\begin{aligned} P(\mathcal{E}_4^{\mathrm{c}}) &\le |\mathcal{F}| e^{-\frac{\tau_2}{2}} \\ &\le 2^{2r} e^{-\frac{1}{2}(\tau_2 - \ln(1+\tau_2))} \\ &\le 2^{2r - \frac{\tau_2}{2}}, \end{aligned} \tag{3.26}$$

where the last inequality holds for all $\tau_2 > 7$. Setting $\tau_2 = 4(1+\epsilon)r - 1$, where $\epsilon > 0$, ensures that $P(\mathcal{E}_4^{\mathrm{c}}) \le 2^{-2\epsilon r + 0.5}$. Setting $\tau_1 = 1$,

$$P(\mathcal{E}_3^{\mathrm{c}}) \le e^{-0.15m}. \tag{3.27}$$

Now using the derived bounds and noting that $\|\mathbf{x} - \widetilde{\mathbf{x}}\|_2 \le \delta$, conditioned on $\mathcal{E}_1 \cap \mathcal{E}_2 \cap \mathcal{E}_3 \cap \mathcal{E}_4$, we have

$$2\left(\langle \underline{\theta}^{t+1}, \underline{\theta}^{t}\rangle - \eta\langle A\underline{\theta}^{t+1}, A\underline{\theta}^{t}\rangle\right) \le 0.9, \tag{3.28}$$

$$\frac{2}{m}(\sigma_{\max}(A))^2 \|\mathbf{x} - \widetilde{\mathbf{x}}\|_2 \le \frac{2}{m}\left(2\sqrt{m} + \sqrt{n}\right)^2 \delta = 2\left(2 + \sqrt{\frac{n}{m}}\right)^2 \delta, \tag{3.29}$$

and

$$\begin{aligned} 2\eta\langle \underline{\theta}^{t+1}, A^{\mathrm{T}}\mathbf{z}\rangle &= \frac{2}{m}\langle \underline{\theta}^{t+1}, A^{\mathrm{T}}\mathbf{z}\rangle \\ &\le \frac{2}{m}\sqrt{\sigma_z^2(1+\tau_1)m(1+\tau_2)} \\ &= \frac{2\sigma_z}{m}\sqrt{8m(1+\epsilon)r} \\ &= \sigma_z\sqrt{\frac{32(1+\epsilon)r}{m}}. \end{aligned} \tag{3.30}$$

Hence, combining (3.28), (3.29), and (3.30) yields the desired error bound. Finally, note that, by the union bound,

$$P(\mathcal{E}_1 \cap \mathcal{E}_2 \cap \mathcal{E}_3 \cap \mathcal{E}_4) \geq 1 - \sum_{i=1}^{4} P(\mathcal{E}_i)$$
$$\geq 1 - e^{-\frac{m}{2}} - 2^{-40r\epsilon} - 2^{-2\epsilon r+0.5} - e^{-0.15m}.$$

3.6 Stylized Applications

In this section, we consider three well-studied classes of signals, namely, (i) sparse signals, (ii) piecewise polynomials, and (iii) natural images. For each class of signal, we explore the implications of the main results discussed so far for these specific classes of signals. For the first two classes, we consider simple families of compression codes to study the performance of the compression-based CS methods. For images, on the other hand, we use standard compression codes such as JPEG and JPEG2000. These examples enable us to shed light on different aspects of the CSP optimization and the C-GD algorithm, such as (i) their required number of measurements, (ii) the reconstruction error in a noiseless setting, (iii) the reconstruction error in the presence of noise, and (iv) the convergence rate of the C-GD algorithm.

3.6.1 Sparse Signals and Connections to IHT

The focus of this chapter has been on moving beyond simple structures such as sparsity in the context of CS. However, to better understand the performance of compression-based CS algorithms and to compare them with other results in the literature, as our first application, we consider the standard class of sparse signals and study the performance of CSP and C-GD for this special class of signals.

Let $\mathcal{B}_2^n(\rho)$ and $\Gamma_k^n(\rho)$ denote the ball of radius ρ in $\mathbb{R}^n$ and the set of all k-sparse signals in $\mathcal{B}_2^n(\rho)$, respectively. That is,

$$\mathcal{B}_2^n(\rho) = \{\mathbf{x} \in \mathbb{R}^n : \|\mathbf{x}\|_2 \leq \rho\}, \tag{3.31}$$

and

$$\Gamma_k^n(\rho) = \{\mathbf{x} \in \mathcal{B}_2^n(\rho) : \|\mathbf{x}\|_0 \leq k\}. \tag{3.32}$$

Consider the following compression code for signals in $\Gamma_k^n(\rho)$. Consider $\mathbf{x} \in \Gamma_k^n(\rho)$. By definition, $\mathbf{x}$ contains at most k non-zero entries. The encoder encodes $\mathbf{x}$ by first describing the locations of the k non-zero entries and then the values of those non-zero entries, each b-bit quantized. To encode $\mathbf{x} \in \Gamma_k^n(\rho)$, the described code spends at most r bits, where

$$\begin{aligned} r &\le k\lceil \log n\rceil + k(b+\lceil \log \rho\rceil) \\ &\le k(b+\log\rho+\log n+2). \end{aligned} \tag{3.33}$$

The resulting distortion δ of this compression code can can be bounded as

$$\begin{aligned} \delta &= \sup_{\mathbf{x}\in\Gamma_k^n(\rho)} \|\mathbf{x}-\widehat{\mathbf{x}}\|_2 \\ &= \sup_{\mathbf{x}\in\Gamma_k^n(\rho)} \sqrt{\sum_{i:x_i\neq 0}(x_i-\widehat{x}_i)^2} \\ &= \sqrt{\sum_{i:x_i\neq 0} 2^{-2b}} \\ &= 2^{-b}\sqrt{k}. \end{aligned} \tag{3.34}$$

The α-dimension of this compression code can be bounded as

$$\begin{aligned} \alpha &= \lim_{\delta\to 0}\frac{r(\delta)}{\log(1/\delta)} \\ &\le \lim_{b\to\infty}\frac{k(b+\log\rho+\log n+2)}{b-\log k} = k. \end{aligned} \tag{3.35}$$

It can in fact be shown that the α-dimension is equal to k.

Consider using this specific compression algorithm in the C-GD framework. The resulting algorithm is very similar to the well-known iterative hard thresholding (IHT) algorithm [33]. The IHT algorithm, a CS recovery algorithm for sparse signals, is an iterative algorithm. Its first step, as with C-GD, is moving in the opposite direction of the gradient of the cost function. At the projection step, IHT keeps the k largest elements and sets the rest to zero. The C-GD algorithm, on the other hand, after moving in the opposite direction to the gradient, finds the codeword in the described code that is closest to the result. For the special code described earlier, this is equivalent to first finding the k largest entries and setting the rest to zero. Then, each remaining non-zero entry x_i is first clipped between $[-1,1]$ as

$$x_i \mathbf{1}_{x_i\in(-1,1)} + \mathrm{sign}(x_i)\mathbf{1}_{|x_i|\ge 1},$$

where $\mathbf{1}_{\mathcal{A}}$ is an indicator of event $\mathcal{A}$, and then quantized to $b+1$ bits.

Consider $\mathbf{x}\in\Gamma_k^n(1)$ and let $\mathbf{y}=A\mathbf{x}+\mathbf{z}$, where $A_{i,j}\overset{\text{i.i.d.}}{\sim}\mathcal{N}(0,1/n)$ and $z_i\overset{\text{i.i.d.}}{\sim}\mathcal{N}(0,\sigma_z^2)$. Let $\widetilde{\mathbf{x}}$ denote the projection of $\mathbf{x}$, as described earlier. The following corollary of Theorem 3.3 characterizes the convergence performance of C-GD applied to $\mathbf{y}$ when using this code.

COROLLARY 3.2 (Corollary 2 of [45]) *Given $\gamma>0$, set $b=\lceil\gamma\log n+\frac{1}{2}\log k\rceil$ bits. Also, set $\eta=n/m$. Then, given $\epsilon>0$, for $m\ge 80\widetilde{r}(1+\epsilon)$, where $\widetilde{r}=(1+\gamma)k\log n+(k/2)\log k+2k$,*

$$\frac{1}{\sqrt{n}}\|\mathbf{x}^{t+1}-\widetilde{\mathbf{x}}\|_2 \le \frac{0.9}{\sqrt{n}}\|\mathbf{x}^t-\widetilde{\mathbf{x}}\|_2 + 2\left(2+\sqrt{\frac{n}{m}}\right)^2 n^{-1/2-\gamma} + \sigma_z\sqrt{\frac{8(1+\epsilon)\widetilde{r}}{m}}, \tag{3.36}$$

for $t=0,1,\ldots,$ with probability larger than $1-2^{-2\epsilon\widetilde{r}}$.

Proof Consider $u \in [-1,1]$. Quantizing u by a uniform quantizer that uses $b+1$ bits yields $\widehat{u}$, which satisfies $|u-\widehat{u}| < 2^{-b}$. Therefore, using $b+1$ bits to quantize each non-zero element of $\mathbf{x} \in \Gamma_k^n$ yields a code which achieves distortion $\delta \leq 2^{-b}\sqrt{k}$. Hence, for $b+1 = \lceil \gamma \log n + \frac{1}{2}\log k \rceil + 1$,

$$\delta \leq n^{-\gamma}.$$

On the other hand, the code rate r can be upper-bounded as

$$r \leq \sum_{i=0}^{k} \log \binom{n}{i} + k(b+1) \leq \log n^{t+1} + k(b+1) = (k+1)\log n + k(b+1),$$

where the last inequality holds for all n large enough. The rest of the proof follows directly from inserting these numbers in to the statement of Theorem 3.3.

In the noiseless setting, according to this corollary, if the number of measurements m satisfies $m = \Omega(k \log n)$, using (3.36), the final reconstruction error can be bounded as

$$\lim_{t\to\infty} \frac{1}{\sqrt{n}} \|\mathbf{x}^{t+1} - \widetilde{\mathbf{x}}\|_2 = O\left(\left(2 + \sqrt{\frac{n}{m}}\right)^2 n^{-1/2-\gamma}\right),$$

or

$$\lim_{t\to\infty} \frac{1}{\sqrt{n}} \|\mathbf{x}^{t+1} - \widetilde{\mathbf{x}}\|_2 = O\left(\frac{n^{\frac{1}{2}-\gamma}}{m}\right).$$

Hence, if $\gamma > 0.5$, the error vanishes as the ambient dimension of the signal space grows to infinity. There are two key observations regarding the number of measurements used by C-GD and CSP that one can make.

1. The α-dimension of the described compression code is equal to k. This implies that, using slightly more than k measurements, CSP is able to almost accurately recover the signal from its under-determined linear measurements. However, solving CSP involves an exhaustive search over all of the exponentially many codebooks. On the other hand, the computationally efficient C-GD method employs $k \log n$ measurements, which is more than what is required by CSP by a factor $\log n$.
2. The results derived for C-GD are slightly weaker than those derived for IHT, as (i) the C-GD algorithm's reconstruction is not exact, even in the noiseless setting, and (ii) C-GD requires $O(k \log n)$ measurements, compared with $O(k \log(n/k))$ required by IHT.

In the case where the measurements are distorted by additive white Gaussian noise, Corollary 3.2 states that there will be an additive

$$O\left(\sigma_z \sqrt{\frac{k \log n}{m}}\right)$$

distortion term. While no similar result on the performance of the IHT algorithm in the presence of stochastic noise exists, the noise sensitivity of C-GD is comparable to the noise-sensitivity performance of other algorithms based on convex optimization, such as LASSO and the Dantzig selector [48, 49].

3.6.2 Piecewise-Polynomial Functions

Consider the class of piecewise-polynomial functions $p : [0,1] \to [0,1]$, with at most Q singularities,[1] and maximum degree of N. Let Poly_N^Q denote the defined class of signals. For $p \in \text{Poly}_N^Q$, let $(x_1, x_2, \ldots, x_n)$ be the samples of p at

$$0, \frac{1}{n}, \ldots, \frac{n-1}{n}.$$

Let N_ℓ and $\{a_i^\ell\}_{i=0}^{N_\ell}$, $\ell = 1, \ldots, Q$, denote the degree and the set of coefficients of the ℓth polynomial in p. Furthermore, assume the following.

1. The coefficients of each polynomial belong to the interval $[0,1]$.
2. For every ℓ, $\sum_{i=0}^{N_\ell} a_i^\ell < 1$.

Define $\mathcal{P}$, the class of signals derived from sampling piecewise-polynomial functions, as follows:

$$\mathcal{P} \triangleq \left\{ \mathbf{x} \in \mathbb{R}^n \mid x_i = p\left(\frac{i}{n}\right),\ p \in \text{Poly}_N^Q \right\}. \tag{3.37}$$

The described class of signals can be considered as a generalization of the class of piecewise-constant functions, which is a popular signal model in imaging applications. To apply C-GD to this class of signals, consider the following compression code for signals in $\mathcal{P}$. For $\mathbf{x} \in \mathcal{P}$, the encoder first describes the locations of the discontinuities of its corresponding piecewise-polynomial function, and then, using a uniform quantizer that spends b bits per coefficient, describes the quantized coefficients of the polynomials. Using b bits per coefficient, the described code, in total, spends at most r bits, where

$$r \leq (N+1)(Q+1)b + Q(\log n + 1). \tag{3.38}$$

Moreover, using b bits per coefficient, the distortion in approximating each point can be bounded as

$$\begin{aligned} \left| \sum_{i=0}^{N_\ell} a_i^\ell t^i - \sum_{i=0}^{N_\ell} [a_i^\ell]_b t^i \right| &\leq \sum_{i=0}^{N_\ell} |a_i^\ell - [a_i^\ell]_b| \\ &\leq (N_\ell + 1) 2^{-b} \\ &\leq (N+1) 2^{-b}. \end{aligned} \tag{3.39}$$

The C-GD algorithm combined with the described compression code provides a natural extension of IHT to piecewise-polynomial functions. At every iteration, the resulting C-GD algorithm projects its current estimate of the signal to the space of quantized piecewise-polynomial functions. As shown in Appendix B of [45], this projection step can be done efficiently using dynamic programming.

Consider $\mathbf{x} \in \mathcal{P}$ and $\mathbf{y} = A\mathbf{x} + \mathbf{z}$, where $A_{i,j} \overset{\text{i.i.d.}}{\sim} \mathcal{N}(0, 1/n)$ and $z_i \overset{\text{i.i.d.}}{\sim} \mathcal{N}(0, \sigma_z^2)$. As with Corollary 3.2, the following corollary characterizes the convergence performance of C-GD combined with the described compression code.

[1] A singularity is a point at which the function is not infinitely differentiable.

COROLLARY 3.3 (Corollary 3 in [45]) *Given $\gamma > 0$, let $\eta = n/m$ and*

$$b = \lceil (\gamma + 0.5) \log n + \log(N+1) \rceil.$$

Set

$$\tilde{r} = ((\gamma + 0.5)(N+1)(Q+1) + Q) \log n + (N+1)(Q+1)(\log(N+1) + 1) + 1.$$

Then, given $\epsilon > 0$, for $m \geq 80\tilde{r}(1+\epsilon)$, for $t = 0, 1, 2, \ldots$, we have

$$\frac{1}{\sqrt{n}} \|\mathbf{x}^{t+1} - \tilde{\mathbf{x}}\|_2 \leq \frac{0.9}{\sqrt{n}} \|\mathbf{x}^t - \tilde{\mathbf{x}}\|_2 + 2\left(2 + \sqrt{\frac{n}{m}}\right)^2 n^{-0.5-\gamma} + \sigma_z \sqrt{\frac{8(1+\epsilon)\tilde{r}}{m}}, \tag{3.40}$$

with a probability larger than $1 - 2^{-2\epsilon\tilde{r}+1}$.

Proof For $b = \lceil (\gamma + 0.5) \log n + \log(N+1) \rceil$, from (3.38),

$$r \leq ((\gamma + 0.5)(N+1)(Q+1) + Q) \log n + (N+1)(Q+1)(\log(N+1) + 1) + 1 = \tilde{r}.$$

Also, from (3.39), the overall error can be bounded as

$$\delta \leq \sqrt{n}(N+1)2^{-b},$$

and for $b = \lceil (\gamma + 0.5) \log n + \log(N+1) \rceil$,

$$\delta \leq n^{-\gamma}.$$

Inserting these numbers in Theorem 3.3 yields the desired result.

Here are the takeaways from this corollary.

1. If we assume that n is much larger than N and Q, then the required number of measurements is $\Omega((N+1)(Q+1)\log n)$. Given that there are $(N+1)(Q+1)$ free parameters that describe $\mathbf{x} \in \mathcal{P}$, no algorithm is expected to require less than $(N+1)(Q+1)$ observations.
2. Using an argument similar to the one used in the previous example, in the absence of measurement noise, the reconstruction error can be bounded as follows

$$\lim_{k \to \infty} \frac{1}{\sqrt{n}} \|\mathbf{x}^{t+1} - \tilde{\mathbf{x}}\|_2 = O\left(\frac{n^{\frac{1}{2}-\gamma}}{m}\right).$$

   Therefore, for $\gamma > 0.5$, the reconstruction distortion converges to zero, as n grows to infinity.
3. In the case of noisy measurements, the effect of the noise in the reconstruction error behaves as

$$O\left(\sigma_z \sqrt{(Q+1)(N+1)\frac{\log n}{m}}\right).$$

3.6.3 Practical Image Compressed Sensing

As the final application, we consider the case of natural images. In this case JPEG and JPEG2000 are well-known efficient compression codes that exploit complex structures that are present in images. Hence, C-GD combined by these compression codes yields a recovery algorithm that also takes advantage of the same structures.

To actually run the C-GD algorithm, there are three parameters that need to be specified:

1. step-size η,
2. compression rate r,
3. number of iterations.

Proper setting of the first two parameters is crucial to the success of the algorithm. Note that the theoretical results of Section 3.5 were derived by fixing η as

$$\eta = \frac{1}{m}.$$

However, choosing η in this way can lead to a very slow convergence rate of the algorithm in practice. Hence, this parameter can also be considered as a free parameter that needs to be specified. One approach to set the step size is to set it adaptively. Let η_t denote the step size in iteration t. Then, η_t can be set such that after projection the result yields the smallest measurement error. That is,

$$\eta_t = \arg\min_{\eta} \left\| \mathbf{y} - A\mathrm{P}_{\mathcal{C}_r}(\mathbf{x}^t + \eta A^{\mathrm{T}}(\mathbf{y} - A\mathbf{x}^t)) \right\|_2^2, \tag{3.41}$$

where the projection operation $\mathrm{P}_{\mathcal{C}_r}$ is defined in (3.15). The optimization described in (3.41) is a scalar optimization problem. Derivative-free methods, such as the Nelder–Mead method also known as the downhill simplex method [50] can be employed to solve it.

Tuning the parameter r is a more challenging task, which can be described as a model-selection problem. (See Chapter 7 of [51].) There are some standard techniques such as multi-fold cross validation that address this issue. However, finding a method with low computational complexity that works well for the C-GD algorithm is an open problem that has yet to be addressed.

Finally, controlling the number of iterations in the C-GD method can be done using standard stopping rules, such as limiting the maximum number of iterations, or lower-bounding the reduction of the squared error per iteration, i.e., $\|\mathbf{x}^{t+1} - \mathbf{x}^t\|_2$.

Using these parameter-selection methods results in an algorithm described below as Algorithm 3.1. Here, as usual, $\mathcal{E}$ and $\mathcal{D}$ denote the encoder and decoder of the compression code, respectively. $K_{1,\max}$ denotes the maximum number of iterations and ϵ_{T} denotes the lower bound in the reduction mentioned earlier. In all the simulation results cited later, $K_{1,\max} = 50$ and $\epsilon_{\mathrm{T}} = 0.001$.

Tables 3.1 and 3.2 show some simulation results reported in [45], for noiseless and noisy measurements, respectively. In both cases the measurement matrix is a random partial-Fourier matrix, generated by randomly choosing rows from a Fourier matrix.

Table 3.1. PSNR of 512×512 reconstructions with no noise – sampled by a random partial-Fourier measurement matrix

Method	m/n	Boat	House	Barbara
NLR-CS	10%	**23.06**	**27.26**	**20.34**
	30%	26.38	30.74	23.67
JPEG-CG	10%	18.38	24.11	16.36
	30%	24.70	30.51	20.37
JP2K-CG	10%	20.75	26.30	18.64
	30%	**27.73**	**38.07**	**24.89**

Note: The bold numbers highlight the best performance achieved in each case.

Algorithm 3.1 C-GD: Compression-based (projected) gradient descent

1: **Inputs**: compression code $(\mathcal{E}, \mathcal{D})$, measurements $\mathbf{y} \in \mathbb{R}^m$, sensing matrix $A \in \mathbb{R}^{m \times n}$
2: **Initialize:** $\mathbf{x}^0$, η_0, $K_{1,\max}$, $K_{2,\max}$, ϵ_{T}
3: **for** $1 \leq t \leq K_{1,\max}$ **do**
4: $\quad \eta_t \leftarrow$ Apply Nelder–Mead method with $K_{2,\max}$ iterations to solve (3.41).
5: $\quad \mathbf{x}^{t+1} \leftarrow \mathcal{D}\big(\mathcal{E}(\mathbf{x}^t + \eta_t A^{\mathrm{T}}(\mathbf{y} - A\mathbf{x}^t))\big)$
$\quad$ **if** $(1/\sqrt{n})\|\mathbf{x}^{t+1} - \mathbf{x}^t\|_2 < \epsilon_{\mathrm{T}}$ **then** return $\mathbf{x}^{t+1}$
$\quad$ **end if**
6: **end for**
7: **Output:** $\mathbf{x}^{t+1}$

The derived results are compared with a state-of-the-art recovery algorithm for partial-Fourier matrices, non-local low-rank regularization (NLR-CS) [52]. In both tables, when compression algorithm a is employed in the platform of C-GD, the resulting C-GD algorithm is referred to as a-GD. For instance, C-GD that uses JPEG compression code is referred to as JPEG-GD. The results are derived using the C-GD algorithm applied to a number of standard test images. In these results, standard JPEG and JPEG2000 codes available in the Matlab-R2016b image- and video-processing package are used as compression codes for images. Boat, House, and Barbara test images are the standard images available in the Matlab image toolbox.

The mean square error (MSE) between $\mathbf{x} \in \mathbb{R}^n$ and $\widehat{\mathbf{x}} \in \mathbb{R}^n$ is measured as follows:

$$\mathrm{MSE} = \frac{1}{n}\|\mathbf{x} - \widehat{\mathbf{x}}\|_2^2. \tag{3.42}$$

Then, the corresponding peak signal-to-noise (PSNR) is defined as

$$\mathrm{PSNR} = 20\log\left(\frac{255}{\sqrt{\mathrm{MSE}}}\right). \tag{3.43}$$

In the case of noisy measurements, $\mathbf{y} = A\mathbf{x} + \mathbf{z}$, the signal-to-noise ratio (SNR) is defined as

$$\mathrm{SNR} = 20\log\left(\frac{\|A\mathbf{x}\|_2}{\|\mathbf{z}\|_2}\right), \tag{3.44}$$

Table 3.2. PSNR of 512×512 reconstructions with Gaussian measurement noise with various SNR values – sampled by a random partial-Fourier measurement matrix

		MRI		Barbara		Snake	
Method	m/n	SNR=10	SNR=30	SNR=10	SNR=30	SNR=10	SNR=30
NLR-CS	10%	11.66	24.14	12.10	19.83	10.50	18.75
	30%	12.60	26.84	13.32	24.05	11.98	24.82
JPEG-GD	10%	14.34	20.50	15.60	18.60	12.33	15.67
	30%	19.20	24.70	18.17	22.89	14.40	22.37
JP2K-GD	10%	**17.33**	**25.40**	**16.53**	**21.65**	**18.00**	**23.12**
	30%	**21.56**	**35.38**	**21.82**	**28.19**	**21.06**	**29.30**

Note: The bold numbers highlight the best performance achieved in each case.

In both tables, JP2K-CG consistently outperforms JPEG-CG. This confirms the premise of this chapter that taking advantage of the full structure of a signal improves the performance of a recovery algorithm. As mentioned earlier, JP2K-CG and JPEG-CG employ JPEG2000 and JPEG codes, respectively. These tables suggest that the advantage of JPEG2000 compression over JPEG compression, in terms of their compression performance, directly translates to advantage of CS recovery algorithms that use them. In the case of noiseless measurements, when the sampling rate is small, $m/n = 0.1$, as reported in [45], for the majority of images (Boat, Barbara, House, and Panda), NLR-CS outperforms both JPEG-CG and JP2K-CG. For some images, though (Dog, Snake), JP2K presents the best result. As the sampling rate increases to $m/n = 0.3$, JP2K-CG consistently outperforms other algorithms. Finally, it can be observed that, in the case of noisy measurements too, JP2K-CG consistently outperforms NLR-CS and JPEG-CG both for $m/n = 0.1$ and for $m/n = 0.3$.

3.7 Extensions

In this section, we explore a couple of other areas in which extensions of compression-based compression sensing ideas can have meaningful impact.

3.7.1 Phase Retrieval

Compression algorithms take advantage of complex structures that are present in signals such as images and videos. Therefore, compression-based CS algorithms automatically take advantage of such structures without much extra effort. The same reasoning can be employed in other areas as well. In this section we briefly review some recent results on the application of compression codes to the problem of phase retrieval.

Phase retrieval refers to the problem of recovering a signal from its phaseless measurements. It is a well-known fundamental problem with applications such as X-ray crystallography [53], astronomical imaging [54], and X-ray tomography [55]. Briefly, the problem can be described as follows. Consider an n-dimensional complex-valued signal $\mathbf{x} \in \mathbb{C}^n$. The signal is captured by some phaseless measurements $\mathbf{y} \in \mathbb{R}^n$, where

$$\mathbf{y} = |A\mathbf{x}|^2 + \mathbf{z}.$$

Here, $A \in \mathbb{C}^{m \times n}$ and $\mathbf{z} \in \mathbb{R}^m$ denote the sensing matrix and the measurement noise, respectively. Also, $|\cdot|$ denotes the element-wise absolute value operation. That is, for $\mathbf{x} = (x_1, \ldots, x_n)$, $|\mathbf{x}| = (|x_1|, \ldots, |x_n|)$. The goal of a phase-retrieval recovery algorithm is to recover $\mathbf{x}$, up to a constant phase shift, from measurements $\mathbf{y}$. This is arguably a more complicated problem than the problem of CS as the measurement process itself is also nonlinear. Hence, solving this problem in the case where $\mathbf{x}$ is an arbitrary signal in $\mathbb{C}^n$ and m is larger than n is still challenging.

In recent years, there has been a fresh interest in the problem of phase retrieval and developing more efficient recovery algorithms. In one of the influential results in this area, Candès *et al.* propose lifting a phase-retrieval problem into a higher-dimensional problem, in which, as with CS, the measurement constraints can be stated as linear constraints [56, 57]. After this transform, ideas from convex optimization and matrix completion can be employed to efficiently and robustly solve this problem. (Refer to [58, 59] for a few other examples of works in this area.) One common aspect of these works is that, given the challenging structure of the problem, to a great extent, they all either ignore the structure of the signal or only consider sparsity. In general, however, signals follow complex patterns that if used efficiently can yield considerable improvement in the performance of the systems in terms of the required number of measurements or reconstruction quality.

Given our discussions so far on compression codes and compression-based compressed sensing, it is not surprising that using compression codes for the problem of phase retrieval provides a platform for addressing this problem. Consider, $\mathcal{Q}$, a compact subset of $\mathbb{C}^n$. Assume that $(\mathcal{E}_r, \mathcal{D}_r)$ denotes a rate-r compression code with maximum reconstruction distortion δ for signals in $\mathcal{Q}$. In other words, for all $\mathbf{x} \in \mathcal{Q}$,

$$\|\mathbf{x} - \mathcal{D}_r(\mathcal{E}_r(\mathbf{x}))\|_2 \leq \delta. \tag{3.45}$$

Consider $\mathbf{x} \in \mathcal{Q}$ and noise-free measurements $\mathbf{y} = |A\mathbf{x}|^2$. COmpressible PhasE Retrieval (COPER), a compression-based phase-retrieval solution proposed in [60], recovers $\mathbf{x}$, by solving the following optimization:

$$\widehat{\mathbf{x}} = \arg\min_{\mathbf{c} \in \mathcal{C}_r} \left\| \mathbf{y} - |A\mathbf{c}|^2 \right\|, \tag{3.46}$$

where as usual $\mathcal{C}_r$ denotes the codebook of the code defined by $(\mathcal{E}_r, \mathcal{D}_r)$. In other words, similarly to CSP, COPER also seeks the codeword that minimizes the measurement error. The following theorem characterizes the performance of COPER and connects the number of measurements m with the rate r, distortion δ, and reconstruction quality.

Before stating the theorem, note that, since the measurements are phaseless, for all values of $\theta \in [0, 2\pi]$, $e^{j\theta}\mathbf{x}$ and $\mathbf{x}$ will generate the same measurements. That is, $|A\mathbf{x}| = |A(e^{j\theta}\mathbf{x})|$. Hence, without any additional information about the phase of the signal, any recovery algorithm is expected only to recover $\mathbf{x}$, up to an unknown phase shift.

THEOREM 3.4 (Theorem 1 in [60]) *Consider* $\mathbf{x} \in Q$ *and measurements* $\mathbf{y} = |A\mathbf{x}|$. *Let* $\widehat{\mathbf{x}}$ *denote the solution from the COPER optimization, described in* (3.46). *Assume that the elements of matrix* $A \in \mathbb{C}^{m \times n}$ *are i.i.d.* $\mathcal{N}(0,1) + i\mathcal{N}(0,1)$. *Then,*

$$\inf_{\theta} \|e^{i\theta}\mathbf{x} - \widehat{\mathbf{x}}\|_2^2 \leq 16\sqrt{3}\frac{1+\tau_2}{\sqrt{\tau_1}} m\delta, \tag{3.47}$$

with a probability larger than

$$1 - 2^r e^{\frac{m}{2}(K + \ln \tau_1 - \ln m)} - e^{-2m(\tau_2 - \ln(1+\tau_2))},$$

where $K = \ln(2\pi e)$, *and* τ_1 and τ_2 *are arbitrary positive real numbers.*

To understand the implications of the above theorem, consider the following corollary.

COROLLARY 3.4 (Corollary 1 in [60]) *Consider the same setup as that in Theorem 3.4. For* $\eta > 1$, *let*

$$m = \eta \frac{r}{\log(1/\delta)}.$$

Then, given $\epsilon > 0$, *for large enough* r, *we have*

$$\mathrm{P}\left(\inf_{\theta} \|e^{i\theta}\mathbf{x} - \widehat{\mathbf{x}}\|^2 \leq C\delta^{\epsilon}\right) \geq 1 - 2^{-c_\eta r} - e^{-0.6m}, \tag{3.48}$$

where $C = 32\sqrt{3}$. *If* $\eta > 1/(1-\epsilon)$, c_η *is a positive number greater than* $\eta(1-\epsilon) - 1$.

Given a family of compression codes $\{(\mathcal{E}_r, \mathcal{D}_r) : r > 0\}$, indexed by their rates, as δ approaches zero,

$$\frac{r(\delta)}{\log(1/\delta)}$$

approaches the α dimension of this family of codes defined in (3.3). Hence, Corollary 3.4 states that, as δ approaches zero, COPER generates, by employing high-rate codes, an almost zero-distortion reconstruction of the input vector, using slightly more than α measurements. For instance, for the class of complex-valued bounded k-sparse signals, one could use the same strategy as used earlier to build a code for the class of real-valued k-sparse signals to find a code of this class of signals. It is straightforward to confirm that the α dimension of the resulting class of codes is equal to $2k$. This shows that COPER combined with an appropriate compression code is able to recover signals in this class from slightly more than $2k$ measurements.

3.7.2 Stochastic Processes

In this chapter, we mainly focus on deterministic signal classes that are defined as compact subsets of $\mathbb{R}^n$. In this section, we shift our focus to structured stochastic processes and show how most of the results derived so far carry over to such sources as well. First, in the next section, we start by reviewing information-theoretic lossy compression codes defined for stationary processes. As highlighted below, these definitions are slightly different than how we have defined compression codes earlier in this chapter. We also explain why we have adopted these non-mainstream definitions. In the following, we highlight the differences between the two sets of definitions, explain why we have adopted the alternative definitions as the main setup in this chapter, and finally specify how compression-based CS of stochastic processes raises new questions about fundamental limits of noiseless CS.

Lossy Compression

Consider a stationary stochastic process $\mathbf{X} = \{X_i\}_{i\in\mathbb{Z}}$, with alphabet $\mathcal{X}$, where $\mathcal{X} \subset \mathbb{R}$. A (fixed-length) lossy compression code for process $\mathbf{X}$ is defined by its blocklength n and encoding and decoding mappings $(\mathcal{E}, \mathcal{D})$, where

$$\mathcal{E} : \mathcal{X}^n \to \{1, \ldots, 2^{nR}\}$$

and

$$\mathcal{D} : \{1, \ldots, 2^{nR}\} \to \widehat{\mathcal{X}}^n.$$

Here, $\mathcal{X}$ and $\widehat{\mathcal{X}}$ denote the source and the reconstruction alphabet, respectively. Unlike the earlier definitions, where the rate r represented the *total* number of bits used to encode each source vector, here the rate R denotes the number of bits per source symbol. The distortion performance of the described code is measured in terms of its induced *expected* average distortion defined as

$$D = \frac{1}{n}\sum_{i=1}^{n} \mathrm{E}\left[d(X_i, \widehat{X}_i)\right],$$

where $\widehat{X}^n = \mathcal{D}(\mathcal{E}(X^n))$ and $d : \mathcal{X} \times \widehat{\mathcal{X}} \to \mathbb{R}^+$ denotes a per-letter distortion measure. Distortion D is said to be achievable at rate R if, for any $\epsilon > 0$, there exists a family of compression codes $(n, \mathcal{E}_n, \mathcal{D}_n)$ such that, for all large enough n,

$$\frac{1}{n}\sum_{i=1}^{n} \mathrm{E}\left[d(X_i, \widehat{X}_i)\right] \le D + \epsilon,$$

where $\widehat{X}^n = \mathcal{D}_n(\mathcal{E}_n(X^n))$. The distortion-rate function $D(\mathbf{X}, R)$ of process $\mathbf{X}$ is defined as the infimum of all achievable distortions at rate R. The rate-distortion function of process $\mathbf{X}$ is defined as $R(\mathbf{X}, D) = \inf\{R : D(\mathbf{X}, R) \le D\}$.

The distortion performance of such codes can alternatively be measured through their excess distortion probability, which is the probability that the distortion between the source output and its reconstruction offered by the code exceeds some fixed level, i.e., $\mathrm{P}(d_n(X^n, \widehat{X}^n) > D)$. The distortion D is said to be achievable at rate R under vanishing excess distortion probability if, for any $\epsilon > 0$, there exists a family of rate-R lossy compression codes $(n, \mathcal{E}_n, \mathcal{D}_n)$ such that, for all large enough n,

$$\mathrm{P}(d_n(X^n, \widehat{X}^n) > D) \le \epsilon,$$

where $\widehat{X}^n = \mathcal{D}_n(\mathcal{E}_n(X^n))$.

While in general these two definitions lead to two different distortion-rate functions, for stationary ergodic processes the distortion-rate functions under these definitions coincide and are identical [61–63].

Structured Processes

Consider a real-valued stationary stochastic process $\mathbf{X} = \{X_i\}_{i\in\mathbb{Z}}$. CS of process $\mathbf{X}$, i.e., recovering X^n from measurements $Y^m = AX^n + Z^m$, where $m < n$, is possible only if $\mathbf{X}$ is "structured." This raises the following important questions.

QUESTION 4 *What does it mean for a real-valued stationary stochastic process to be structured?*

QUESTION 5 *How can we measure the level of structuredness of a process?*

For stationary processes with a discrete alphabet, information theory provides us with a powerful tool, namely, the entropy rate function, that measures the complexity or compressibility of such sources [24, 64]. However, the entropy rate of all real-valued processes is infinite and hence this measure by itself is not useful to measure the complexity of real-valued sources.

In this section, we briefly address the above questions in the context of CS. Before proceeding, to better understand what it means for a process to be structured, let us review a classic example, which has been studied extensively in CS. Consider process $\mathbf{X}$ which is i.i.d., and for which X_i, $i \in \mathbb{N}$, is distributed as

$$(1-p)\delta_0 + p\nu_c. \tag{3.49}$$

Here, ν_c denotes the probability density function (p.d.f.) of an absolutely continuous distribution. For $p < 1$, the output vectors of this source are, with high probability, sparse, and therefore compressed sensing of them is feasible. Therefore, for $p < 1$, process $\mathbf{X}$ can be considered as a structured process. As the value of p increases, the percentage of non-zero entries in X^n increases as well. Hence, as the signal becomes less sparse, a CS recovery algorithm would require more measurements. This implies that, at least for such sparse processes, p can serve as a reasonable measure of structuredness. In general, a plausible measure of structuredness is expected to conform with such an intuition. The higher the level of structuredness, the larger the required sampling rate. In the following example we derive one such measure that satisfies this property.

Example 3.1 Consider random variable X distributed such that with probability $1-p$ it is equal to 0, and with probability p it is uniformly distributed between 0 and 1. That is,

$$X \sim (1-p)\delta_0 + p\nu,$$

where ν denotes the p.d.f. of Unif[0,1]. Consider quantizing X by b bits. It is straightforward to confirm that

$$\mathrm{P}([X]_b = 0) = 1 - p + p2^{-b}$$

and

$$\mathrm{P}([X]_b = i) = p2^{-b},$$

for $i \in \{2^{-b}, 2^{-b+1}, \ldots, 1-2^{-b}\}$. Hence,

$$\begin{aligned} H([X]_b) &= -(1-p+p2^{-b})\log(1-p+p2^{-b}) - (2^b-1)p2^{-b}\log(p2^{-b}) \\ &= pb - (1-p+p2^{-b})\log(1-p+p2^{-b}) - (1-2^{-b})p\log(p) - 2^{-b}p. \end{aligned} \tag{3.50}$$

As expected, $H([X]_b)$ grows to infinity, as b grows without bound. Dividing both sides by b, it follows that, for a fixed p,

$$\frac{H([X]_b)}{b} = p + \delta_b, \tag{3.51}$$

where $\delta_b = o(1)$. This suggests that $H([X]_b)$ grows almost linearly with b and the asymptotic slope of its growth is equal to

$$\lim_{b\to\infty} \frac{H([X]_b)}{b} = p.$$

The result derived in Example 3.1 on the slope of the growth of the quantized entropy function holds for any absolutely continuous distribution satisfying $H(\lfloor X \rfloor) < \infty$ [66]. In fact this slope is a well-known quantity referred to as the Rényi information dimension of X [66].

DEFINITION 3.1 (Rényi information dimension) *Given random variable X, the upper Rényi information dimension of X is defined as*

$$\bar{d}(X) = \limsup_{b\to\infty} \frac{H([X]_b)}{b}. \tag{3.52}$$

The lower Rényi information dimension of X, $\underline{d}(X)$ is defined similarly by replacing lim sup *with* lim inf. *In the case where $\bar{d}(X) = \underline{d}(X)$, the Rényi information dimension of X is defined as $d(X) = \bar{d}(X) = \underline{d}(X)$.*

For a stationary memoryless process $\mathbf{X} = \{X_i\}_{i\in\mathbb{Z}}$, under some regularity conditions on the distribution of $\mathbf{X}$, as n and m grow to infinity, while m/n stays constant, the Rényi information dimension of X_1 characterizes the minimum required sampling rate (m/n) for almost lossless recovery of X^n from measurements $Y^m = AX^n$ [67]. In other words, for such sources, $d(X_1)$ is the infimum of all sampling rates for which there exists a proper decoding function with $\mathrm{P}(X^n \neq \widehat{X}^n) \leq \epsilon$, where $\epsilon \in (0,1)$. This result provides an operational meaning for the Rényi information dimension and partially addresses the questions we raised earlier about measuring the structuredness of processes, at least for i.i.d. sources.

To measure the structuredness of general non-i.i.d. stationary processes, a generalized version of this definition is needed. Such a generalized definition would also take structures that manifest themselves as specific correlations between output symbols into account. In fact such a generalization has been proposed in [40], and more recently in [68]. In the following we briefly review the information dimension of stationary processes, as proposed in [40].

DEFINITION 3.2 (kth-order information dimension) *The upper kth-order information dimension of stationary process* $\mathbf{X}$ *is defined as*

$$\bar{d}_k(\mathbf{X}) = \limsup_{b\to\infty} \frac{H([X_{k+1}]_b | [X^k]_b)}{b}. \tag{3.53}$$

The lower kth-order information dimension of $\mathbf{X}$, $\underline{d}_k(\mathbf{X})$ *is defined similarly by replacing* $\limsup$ *with* $\liminf$. *In the case where* $\bar{d}_k(\mathbf{X}) = \underline{d}_k(\mathbf{X})$, *the kth-order information dimension of* $\mathbf{X}$ *is defined as* $d_k(\mathbf{X}) = \bar{d}(\mathbf{X}) = \underline{d}(\mathbf{X})$.

It can be proved that the kth-order information dimension is always a positive number between zero and one. Also, both $\bar{d}_k(\mathbf{X})$ and $\underline{d}_k(\mathbf{X})$ are non-increasing functions of k.

DEFINITION 3.3 (Information dimension) *The upper information dimension of stationary process* $\mathbf{X}$ *is defined as*

$$\bar{d}_o(\mathbf{X}) = \lim_{k\to\infty} \bar{d}_k(\mathbf{X}). \tag{3.54}$$

The lower information dimension of process $\mathbf{X}$ *is defined as* $\underline{d}_o(\mathbf{X}) = \lim_{k\to\infty} \underline{d}_k(\mathbf{X})$. *When* $\bar{d}_o(\mathbf{X}) = \underline{d}_o(\mathbf{X})$, *the information dimension of process* $\mathbf{X}$ *is defined as* $d_o(\mathbf{X}) = \underline{d}_o(\mathbf{X}) = \bar{d}_o(\mathbf{X})$.

Just like the Rényi information dimension, the information dimension of general stationary processes has an operational meaning in the context of *universal* CS of such sources.[2]

Another possible measure of structuredness for stochastic processes, which is more closely connected to compression-based CS, is the rate-distortion dimension defined as follows.

DEFINITION 3.4 (Rate-distortion dimension) *The upper rate-distortion dimension of stationary process* $\mathbf{X}$ *is defined as*

$$\overline{\dim}_R(\mathbf{X}) = \limsup_{D\to 0} \frac{2R(\mathbf{X}, D)}{\log(1/D)}. \tag{3.55}$$

The lower rate-distortion dimension of process $\mathbf{X}$, $\underline{\dim}_R(\mathbf{X})$, *is defined similarly by replacing* $\limsup$ *with* $\liminf$. *Finally, if*

$$\overline{\dim}_R(\mathbf{X}) = \underline{\dim}_R(\mathbf{X}),$$

the rate-distortion dimension of process $\mathbf{X}$ *is defined as* $\dim_R(\mathbf{X}) = \overline{\dim}_R(\mathbf{X}) = \underline{\dim}_R(\mathbf{X})$.

For memoryless stationary sources, the information dimension simplifies to the Rényi information dimension of the marginal distribution, which is known to be equal to the rate-distortion dimension [69]. For more general sources, under some technical conditions, the rate-distortion dimension and the information dimension can be proved to be equal [70].

[2] In information theory, an algorithm is called universal if it does not need to know the source distribution and yet, asymptotically, achieves the optimal performance.

CSP Applied to Stochastic Processes

In Section 3.4, we discussed the performance of the CSP optimization, a compression-based CS scheme, in the deterministic setting, where signal $\mathbf{x}$ is from compact set Q. The results can be extended to the case of stochastic processes as well. The following theorem characterizes the performance of the CSP in the stochastic setting in terms of the rate, distortion, and excess probability of the compression code.

THEOREM 3.5 (Theorem 4 in [70]) *Consider process $\mathbf{X}$ and a rate-R blocklength-n lossy compression code with distortion exceeding D, with a probability smaller than ϵ. Without any loss of generality assume that the source is normalized such that $D < 1$, and let C_n denote the codebook of this code. Let $Y^m = AX^n$, and assume that $A_{i,j}$ are i.i.d. as $\mathcal{N}(0,1)$. For arbitrary $\alpha > 0$ and $\eta > 1$, let $\delta = \eta / \ln(1/D) + \alpha$, and*

$$m = \left(\frac{2\eta R}{\log(1/D)}\right) n. \tag{3.56}$$

Further, let $\widetilde{X}^n$ denote the solution of the CSP optimization. That is, $\widetilde{X}^n = \arg\min_{\mathbf{c}\in C_n} \|A\mathbf{c} - Y^m\|_2^2$. Then,

$$\mathrm{P}\left(\frac{1}{\sqrt{n}}\| X^n - \widetilde{X}^n\|_2 \geq \left(2 + \sqrt{\frac{n}{m}}\right) D^{\frac{1}{2}\left(1-\frac{1+\delta}{\eta}\right)}\right) \leq \epsilon + 2^{-\frac{1}{2}nR\alpha} + e^{-\frac{m}{2}}.$$

COROLLARY 3.5 (Corollary 3 in [70]) *Consider a stationary process $\mathbf{X}$ with upper rate-distortion dimension $\overline{\dim}_R(\mathbf{X})$. Let $Y^m = AX^n$, where $A_{i,j}$ are i.i.d. $\mathcal{N}(0,1)$. For any $\Delta > 0$, if the number of measurements $m = m_n$ satisfies*

$$\liminf_{n\to\infty} \frac{m_n}{n} > \overline{\dim}_R(\mathbf{X}),$$

then there exists a family of compression codes which used by the CSP optimization yields

$$\lim_{n\to\infty} \mathrm{P}\left(\frac{1}{\sqrt{n}}\| X^n - \widetilde{X}^n\|_2 \geq \Delta\right) = 0,$$

where $\widetilde{X}^n$ denotes the solution of the CSP optimization.

This corollary has the following important implication for the problem of Bayesian CS in the absence of measurement noise. *Asymptotically, zero-distortion recovery is achievable, as long as the sampling rate m/n is larger than the upper rate-distortion dimension of the source.* This observation raises the following fundamental and open questions about noiseless CS of stationary stochastic processes.

QUESTION 6 *Does the rate-distortion dimension of a stationary process characterize the fundamental limit of the required sampling rate for zero-distortion recovery?*

QUESTION 7 *Are the fundamental limits of the required rates for zero-distortion recovery and for almost-lossless recovery the same?*

3.8 Conclusions and Discussion

Data-compression algorithms and CS algorithms are fundamentally different as they are designed to address different problems. Data-compression codes are designed to represent (encode) signals by as few bits as possible. The goal of CS recovery algorithms, on the other hand, is to recover a structured signal from its under-determined (typically linear) measurements. Both types of algorithm succeed by taking advantage of signals, structures and patterns. While the field of CS is only about a decade old, there has been more than 50 years of research on developing efficient image and video compression algorithms. Therefore, structures and patterns that are used by state-of-the-art compression codes are typically much more complex than those used by CS recovery algorithms. In this chapter, we discussed how using compression codes to solve the CS problem potentially enables us to expand the class of structures used by CS codes to those already used by data-compression codes.

This approach of using compression codes to address a different problem, namely, CS, can be used to address other emerging data-acquisition problems as well. In this chapter, we briefly reviewed the problem of phase retrieval, and how, theoretically, compression-based phase-retrieval recovery methods are able to recover a signal from its phase-less measurements.

References

[1] D. L. Donoho, "Compressed sensing," *IEEE Trans. Information Theory*, vol. 52, no. 4, pp. 1289–1306, 2006.

[2] E. J. Candès and T. Tao, "Near-optimal signal recovery from random projections: Universal encoding strategies?" *IEEE Trans. Information Theory*, vol. 52, no. 12, pp. 5406–5425, 2006.

[3] S. Bakin, "Adaptive regression and model selection in data mining problems," Ph.D. Thesis, Australian National University, 1999.

[4] Y. C. Eldar and M. Mishali, "Robust recovery of signals from a structured union of subspaces," *IEEE Trans. Information. Theory*, vol. 55, no. 11, pp. 5302–5316, 2009.

[5] Y. C. Eldar, P. Kuppinger, and H. Bölcskei, "Block-sparse signals: Uncertainty relations and efficient recovery," *IEEE Trans. Signal Processing*, vol. 58, no. 6, pp. 3042–3054, 2010.

[6] M. Yuan and Y. Lin, "Model selection and estimation in regression with grouped variables," *J. Roy. Statist. Soc. Ser. B*, vol. 68, no. 1, pp. 49–67, 2006.

[7] S. Ji, D. Dunson, and L. Carin, "Multi-task compressive sensing," *IEEE Trans. Signal Processing*, vol. 57, no. 1, pp. 92–106, 2009.

[8] A. Maleki, L. Anitori, Z. Yang, and R. G. Baraniuk, "Asymptotic analysis of complex lasso via complex approximate message passing (CAMP)," *IEEE Trans. Information Theory*, vol. 59, no. 7, pp. 4290–4308, 2013.

[9] M. Stojnic, "Block-length dependent thresholds in block-sparse compressed sensing," *arXiv:0907.3679*, 2009.

[10] M. Stojnic, F. Parvaresh, and B. Hassibi, "On the reconstruction of block-sparse signals with an optimal number of measurements," *IEEE Trans. Signal Processing*, vol. 57, no. 8, pp. 3075–3085, 2009.

[11] M. Stojnic, "ℓ_2/ℓ_1-optimization in block-sparse compressed sensing and its strong thresholds," *IEEE J. Selected Topics Signal Processing*, vol. 4, no. 2, pp. 350–357, 2010.

[12] L. Meier, S. Van De Geer, and P. Buhlmann, "The group Lasso for logistic regression," *J. Roy. Statist. Soc. Ser. B*, vol. 70, no. 1, pp. 53–71, 2008.

[13] V. Chandrasekaran, B. Recht, P. A. Parrilo, and A. S. Willsky, "The convex geometry of linear inverse problems," *Found. Comput. Math.*, vol. 12, no. 6, pp. 805–849, 2012.

[14] R. G. Baraniuk, V. Cevher, M. F. Duarte, and C. Hegde, "Model-based compressive sensing," *IEEE Trans. Information Theory*, vol. 56, no. 4, pp. 1982–2001, 2010.

[15] B. Recht, M. Fazel, and P. A. Parrilo, "Guaranteed minimum rank solutions to linear matrix equations via nuclear norm minimization," *SIAM Rev.*, vol. 52, no. 3, pp. 471–501, 2010.

[16] M. Vetterli, P. Marziliano, and T. Blu, "Sampling signals with finite rate of innovation," *IEEE Trans. Signal Processing*, vol. 50, no. 6, pp. 1417–1428, 2002.

[17] S. Som and P. Schniter, "Compressive imaging using approximate message passing and a Markov-tree prior," *IEEE Trans. Signal Processing*, vol. 60, no. 7, pp. 3439–3448, 2012.

[18] D. Donoho and G. Kutyniok, "Microlocal analysis of the geometric separation problem," *Comments Pure Appl. Math.*, vol. 66, no. 1, pp. 1–47, 2013.

[19] E. J. Candentss, X. Li, Y. Ma, and J. Wright, "Robust principal component analysis?" *J. ACM*, vol. 58, no. 3, pp. 1–37, 2011.

[20] A. E. Waters, A. C. Sankaranarayanan, and R. Baraniuk, "Sparcs: Recovering low-rank and sparse matrices from compressive measurements," in *Proc. Advances in Neural Information Processing Systems*, 2011, pp. 1089–1097.

[21] V. Chandrasekaran, S. Sanghavi, P. A. Parrilo, and A. Willsky, "Rank-sparsity incoherence for matrix decomposition," *SIAM J. Optimization*, vol. 21, no. 2, pp. 572–596, 2011.

[22] M. F. Duarte, W. U. Bajwa, and R. Calderbank, "The performance of group Lasso for linear regression of grouped variables," Technical Report TR-2010-10, Duke University, Department of Computer Science, Durham, NC, 2011.

[23] T. Blumensath and M. E. Davies, "Sampling theorems for signals from the union of finite-dimensional linear subspaces," *IEEE Trans. Information Theory*, vol. 55, no. 4, pp. 1872–1882, 2009.

[24] M. B. McCoy and J. A. Tropp, "Sharp recovery bounds for convex deconvolution, with applications," *arXiv:1205.1580*, 2012.

[25] C. Studer and R. G. Baraniuk, "Stable restoration and separation of approximately sparse signals," *Appl. Comp. Harmonic Analysis (ACHA)*, vol. 37, no. 1, pp. 12–35, 2014.

[26] G. Peyré and J. Fadili, "Group sparsity with overlapping partition functions," in *Proc. EUSIPCO Rev*, 2011, pp. 303–307.

[27] S. S. Chen, D. L. Donoho, and M. A. Saunders, "Atomic decomposition by basis pursuit," *SIAM Rev.*, vol. 43, no. 1, pp. 129–159, 2001.

[28] R. Tibshirani, "Regression shrinkage and selection via the Lasso," *J. Roy. Statist. Soc. Ser. B* vol. 58, no. 1, pp. 267–288, 1996.

[29] A. Beck and M. Teboulle, "A fast iterative shrinkage-thresholding algorithm for linear inverse problems," *SIAM J. Imaging Sci.*, vol. 2, no. 1, pp. 183–202, 2009.

[30] J. A. Tropp and A. C. Gilbert, "Signal recovery from random measurements via orthogonal matching pursuit," *IEEE Trans. Information Theory*, vol. 53, no. 12, pp. 4655–4666, 2007.

[31] I. Daubechies, M. Defrise, and C. De Mol, "An iterative thresholding algorithm for linear inverse problems with a sparsity constraint," *Commun. Pure Appl. Math.*, vol. 57, no. 11, pp. 1413–1457, 2004.

[32] B. Efron, T. Hastie, I. Johnstone, and R. Tibshirani, "Least angle regression," *Annals Statist.*, vol. 32, no. 2, pp. 407–499, 2004.

[33] T. Blumensath and M. E. Davies, "Iterative hard thresholding for compressed sensing," *Appl. Comp. Harmonic Analysis (ACHA)*, vol. 27, no. 3, pp. 265–274, 2009.

[34] D. Needell and J. A. Tropp, "CoSaMP: Iterative signal recovery from incomplete and inaccurate samples," *Appl. Comp. Harmonic Analysis (ACHA)*, vol. 26, no. 3, pp. 301–321, 2009.

[35] D. L. Donoho, A. Maleki, and A. Montanari, "Message passing algorithms for compressed sensing," *Proc. Natl. Acad. Sci. USA*, vol. 106, no. 45, pp. 18 914–18 919, 2009.

[36] W. Dai and O. Milenkovic, "Subspace pursuit for compressive sensing signal reconstruction," *IEEE Trans. Information Theory*, vol. 55, no. 5, pp. 2230–2249, 2009.

[37] C. A. Metzler, A. Maleki, and R. G. Baraniuk, "From denoising to compressed sensing," *IEEE Trans. Information Theory*, vol. 62, no. 9, pp. 5117–5144, 2016.

[38] J. Zhu, D. Baron, and M. F. Duarte, "Recovery from linear measurements with complexity-matching universal signal estimation," *IEEE Trans. Signal Processing*, vol. 63, no. 6, pp. 1512–1527, 2015.

[39] S. Jalali, A. Maleki, and R. G. Baraniuk, "Minimum complexity pursuit for universal compressed sensing," *IEEE Trans. Information Theory*, vol. 60, no. 4, pp. 2253–2268, 2014.

[40] S. Jalali and H. V. Poor, "Universal compressed sensing for almost lossless recovery," *IEEE Trans. Information Theory*, vol. 63, no. 5, pp. 2933–2953, 2017.

[41] S. Jalali and A. Maleki, "New approach to Bayesian high-dimensional linear regression," *Information and Inference*, vol. 7, no. 4, pp. 605–655, 2018.

[42] S. Jalali and A. Maleki, "From compression to compressed sensing," *Appl. Comput Harmonic Analysis*, vol. 40, no. 2, pp. 352–385, 2016.

[43] D. S. Taubman and M. W. Marcellin, *JPEG2000: Image compression fundamentals, standards and practice*. Kluwer Academic Publishers, 2002.

[44] S. Beygi, S. Jalali, A. Maleki, and U. Mitra, "Compressed sensing of compressible signals," in *Proc. IEEE International Symposium on Information Theory*, 2017, pp. 2158–2162.

[45] S. Beygi, S. Jalali, A. Maleki, and U. Mitra, "An efficient algorithm for compression-based compressed sensing," vol. 8, no. 2, pp. 343–375, June 2019.

[46] R. T. Rockafellar, "Monotone operators and the proximal point algorithm," *SIAM J. Cont. Opt.*, vol. 14, no. 5, pp. 877–898, 1976.

[47] E. J. Candès, J. Romberg, and T. Tao, "Decoding by linear programming," *IEEE Trans. Information Theory*, vol. 51, no. 12, pp. 4203–4215, 2005.

[48] E. Candès and T. Tao, "The Dantzig selector: Statistical estimation when p is much larger than n," *Annals Statist*, vol. 35, no. 6, pp. 2313–2351, 2007.

[49] P. J. Bickel, Y. Ritov, and A. B. Tsybakov, "Simultaneous analysis of Lasso and Dantzig selector," *Annals Statist*, vol. 37, no. 4, pp. 1705–1732, 2009.

[50] J. A. Nelder and R. Mead, "A simplex method for function minimization," *Comp. J*, vol. 7, no. 4, pp. 308–313, 1965.

[51] J. Friedman, T. Hastie, and R. Tibshirani, *The elements of statistical learning: Data mining, inference, and prediction*, 2nd edn. Springer, 2009.

[52] W. Dong, G. Shi, X. Li, Y. Ma, and F. Huang, "Compressive sensing via nonlocal low-rank regularization," *IEEE Trans. Image Processing*, 2014.

[53] R. W. Harrison, "Phase problem in crystallography," *J. Opt. Soc. America A*, vol. 10, no. 5, pp. 1046–1055, 1993.

[54] C. Fienup and J. Dainty, "Phase retrieval and image reconstruction for astronomy," *Image Rec.: Theory and Appl.*, pp. 231–275, 1987.

[55] F. Pfeiffer, T. Weitkamp, O. Bunk, and C. David, "Phase retrieval and differential phase-contrast imaging with low-brilliance X-ray sources," *Nature Physics*, vol. 2, no. 4, pp. 258–261, 2006.

[56] E. J. Candès, T. Strohmer, and V. Voroninski, "Phaselift: Exact and stable signal recovery from magnitude measurements via convex programming," *Commun. Pure Appl. Math.*, vol. 66, no. 8, pp. 1241–1274, 2013.

[57] E. J. Candès, Y. C. Eldar, T. Strohmer, and V. Voroninski, "Phase retrieval via matrix completion," *SIAM Rev.*, vol. 57, no. 2, pp. 225–251, 2015.

[58] H. Ohlsson, A. Yang, R. Dong, and S. Sastry, "CPRL – an extension of compressive sensing to the phase retrieval problem," in *Proc. Advances in Neural Information Processing Systems 25*, 2012, pp. 1367–1375.

[59] P. Schniter and S. Rangan, "Compressive phase retrieval via generalized approximate message passing," *IEEE Trans. Information Theory*, vol. 63, no. 4, pp. 1043–1055, 2015.

[60] M. Bakhshizadeh, A. Maleki, and S. Jalali, "Compressive phase retrieval of structured signals," in *Proc. IEEE International Symposium on Information Theory*, 2018, pp. 2291–2295.

[61] Y. Steinberg and S. Verdu, "Simulation of random processes and rate-distortion theory," *IEEE Trans. Information Theory*, vol. 42, no. 1, pp. 63–86, 1996.

[62] S. Ihara and M. Kubo, "Error exponent of coding for stationary memoryless sources with a fidelity criterion," *IEICE Trans. Fund. Elec., Comm. and Comp. Sciences*, vol. 88, no. 5, pp. 1339–1345, 2005.

[63] K. Iriyama, "Probability of error for the fixed-length lossy coding of general sources," *IEEE Trans. Information Theory*, vol. 51, no. 4, pp. 1498–1507, 2005.

[64] C. E. Shannon, "A mathematical theory of communication: Parts I and II," *Bell Systems Technical J.*, vol. 27, pp. 379–423 and 623–656, 1948.

[65] T. Cover and J. Thomas, *Elements of information theory*, 2nd edn. Wiley, 2006.

[66] A. Rényi, "On the dimension and entropy of probability distributions," *Acta Math. Acad. Sci. Hungarica*, vol. 10, no. 5, 1–2, pp. 193–215, 1959.

[67] Y. Wu and S. Verdú, "Rényi information dimension: Fundamental limits of almost lossless analog compression," *IEEE Trans. Information Theory*, vol. 56, no. 8, pp. 3721–3748, 2010.

[68] B. C. Geiger and T. Koch, "On the information dimension rate of stochastic processes," in *Proc. IEEE International Symposium on Information Theory*, 2017, pp. 888–892.

[69] T. Kawabata and A. Dembo, "The rate-distortion dimension of sets and measures," *IEEE Trans. Information Theory*, vol. 40, no. 5, pp. 1564–1572, 1994.

[70] F. E. Rezagah, S. Jalali, E. Erkip, and H. V. Poor, "Compression-based compressed sensing," *IEEE Trans. Information Theory*, vol. 63, no. 10, pp. 6735–6752, 2017.

4 Information-Theoretic Bounds on Sketching

Mert Pilanci

Summary

Approximate computation methods with provable performance guarantees are becoming important and relevant tools in practice. In this chapter, we focus on sketching methods designed to reduce data dimensionality in computationally intensive tasks. Sketching can often provide better space, time, and communication complexity trade-offs by sacrificing minimal accuracy. This chapter discusses the role of information theory in sketching methods for solving large-scale statistical estimation and optimization problems. We investigate fundamental lower bounds on the performance of sketching. By exploring these lower bounds, we obtain interesting trade-offs in computation and accuracy. We employ Fano's inequality and metric entropy to understand fundamental lower bounds on the accuracy of sketching, which is parallel to the information-theoretic techniques used in statistical minimax theory.

4.1 Introduction

In recent years we have witnessed an unprecedented increase in the amount of available data in a wide variety of fields. Approximate computation methods with provable performance guarantees are becoming important and relevant tools in practice to attack larger-scale problems. The term *sketching* is used for randomized algorithms designed to reduce data dimensionality in computationally intensive tasks. In large-scale problems, sketching allows us to leverage limited computational resources such as memory, time, and bandwidth, and also explore favorable trade-offs between accuracy and computational complexity.

Random projections are widely used instances of sketching, and have attracted substantial attention in the literature, especially very recently in the machine-learning, signal processing, and theoretical computer science communities [1–6]. Other popular sketching techniques include leverage score sampling, graph sparsification, core sets, and randomized matrix factorizations. In this chapter we overview sketching methods, develop lower bounds using information-theoretic techniques, and present upper bounds on their performance. In the next section we begin by introducing commonly used sketching methods.

This chapter focuses on the role of information theory in sketching methods for solving large-scale statistical estimation and optimization problems, and investigates fundamental lower bounds on their performance. By exploring these lower bounds, we obtain interesting trade-offs in computation and accuracy. Moreover, we may hope to obtain improved sketching constructions by understanding their information-theoretic properties. The lower-bounding techniques employed here parallel the information-theoretic techniques used in statistical minimax theory [7, 8]. We apply Fano's inequality and packing constructions to understand fundamental lower bounds on the accuracy of sketching.

Randomness and sketching also have applications in privacy-preserving queries [9, 10]. Privacy has become an important concern in the age of information where breaches of sensitive data are frequent. We will illustrate that randomized sketching offers a computationally simple and effective mechanism to preserve privacy in optimization and machine learning.

We start with an overview of different constructions of sketching matrices in Section 4.2. In Section 4.3, we briefly review some background on convex analysis and optimization. Then we present upper bounds on the performance of sketching from an optimization viewpoint in Section 4.4. To be able to analyze upper bounds, we introduce the notion of *localized Gaussian complexity*, which also plays an important role in the characterization of minimax statistical bounds. In Section 4.5, we discuss information-theoretic lower bounds on the statistical performance of sketching. In Section 4.6, we turn to non-parametric problems and information-theoretic lower bounds. Finally, in Section 4.7 we discuss privacy-preserving properties of sketching using a mutual information characterization, and communication-complexity lower bounds.

4.2 Types of Randomized Sketches

In this section we describe popular constructions of sketching matrices. Given a sketching matrix $\mathbf{S}$, we use $\{\mathbf{s}_i\}_{i=1}^m$ to denote the collection of its n-dimensional rows. Here we consider sketches which are zero mean, and are normalized, i.e., they satisfy the following two conditions:

$$\text{(a)} \quad \mathbb{E}\mathbf{S}^{\mathrm{T}}\mathbf{S} = \mathbf{I}_{d\times d}, \tag{4.1}$$

$$\text{(b)} \quad \mathbb{E}\mathbf{S} = \mathbf{0}_{n\times d}. \tag{4.2}$$

The reasoning behind the above conditions will become clearer when they are applied to sketching optimization problems involving data matrices.

A very typical use of sketching is to obtain compressed versions of a large data matrix $\mathbf{A}$. We obtain the matrix $\mathbf{SA} \in \mathbb{R}^{m\times d}$ using simple matrix multiplication. See Fig. 4.1 for an illustration. As we will see in a variety of examples, random matrices preserve most of the information in the matrix $\mathbf{A}$.

4.2.1 Gaussian Sketches

The most classical sketch is based on a random matrix $\mathbf{S} \in \mathbb{R}^{m\times n}$ with i.i.d. standard Gaussian entries. Suppose that we generate a random matrix $\mathbf{S} \in \mathbb{R}^{m\times n}$ with entries drawn

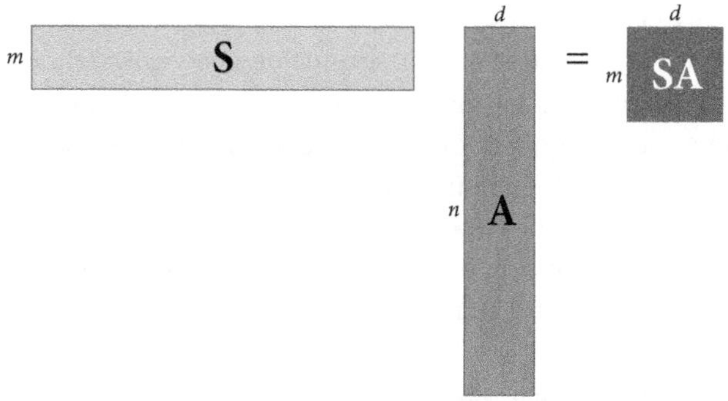

Figure 4.1 Sketching a tall matrix **A**. The smaller matrix $\mathbf{SA} \in \mathbb{R}^{m\times d}$ is a compressed version of the original data $\mathbf{A} \in \mathbb{R}^{n\times d}$.

from i.i.d. zero-mean Gaussian random variables with variance $1/m$. Note that we have $\mathbb{E}\mathbf{S} = \mathbf{0}_{m\times d}$ and also $\mathbb{E}\mathbf{S}^{\mathrm{T}}\mathbf{S} = \sum_{i=1}^{m} \mathbb{E}\mathbf{s}_i\mathbf{s}_i^{\mathrm{T}} = \sum_{i=1}^{m} \mathbf{I}_d(1/m) = \mathbf{I}_d$. Analyzing the Gaussian sketches is considerably easier than analyzing sketches of other types, because of the special properties of the Gaussian distribution such as rotation invariance. However, Gaussian sketches may not be the most computationally efficient choice for many data matrices, as we will discuss in the following sections.

4.2.2 Sub-Gaussian Sketches

A generalization of the previous construction is a random sketch with rows, drawn from i.i.d. sub-Gaussian random variables. In particular, a zero-mean random vector $\mathbf{s} \in \mathbb{R}^n$ is 1-sub-Gaussian if, for any $\mathbf{u} \in \mathbb{R}^n$, we have

$$\mathbb{P}[\langle \mathbf{s}, \mathbf{u}\rangle \geq \varepsilon\|\mathbf{u}\|_2] \leq e^{-\varepsilon^2/2} \qquad \text{for all } \varepsilon \geq 0. \tag{4.3}$$

For instance, a vector with i.i.d. $\mathcal{N}(0,1)$ entries is 1-sub-Gaussian, as is a vector with i.i.d. Rademacher entries (uniformly distributed over $\{-1,+1\}$). In many models of computation, multiplying numbers by random signs is simpler than multiplying by Gaussian variables, and only costs an addition operation. Note that multiplying by -1 only amounts to flipping the sign bit in the signed number representation of the number in the binary system. In modern computers, the difference between addition and multiplication is often not appreciable. However, the real disadvantage of sub-Gaussian and Gaussian sketches is that they require matrix–vector multiplications with unstructured and dense random matrices. In particular, given a data matrix $\mathbf{A} \in \mathbb{R}^{n\times d}$, computing its sketched version $\mathbf{SA}$ requires $O(mnd)$ basic operations using classical matrix multiplication algorithms, in general.

4.2.3 Randomized Orthonormal Systems

The second type of randomized sketch we consider is the *randomized orthonormal system* (ROS), for which matrix multiplication can be performed much more efficiently.

In order to define an ROS sketch, we first let $\mathbf{H} \in \mathbb{R}^{n\times n}$ be an orthonormal matrix with entries $\mathbf{H}_{ij} \in [(-1/\sqrt{n}), (1/\sqrt{n})]$. Standard classes of such matrices are the Hadamard or Fourier bases, for which matrix–vector multiplication can be performed in $O(n \log n)$ time via the fast Hadamard or Fourier transforms, respectively. For example, an $n \times n$ Hadamard matrix $\mathbf{H} = \mathbf{H}_n$ can be recursively constructed as follows:

$$\mathbf{H}_2 = \frac{1}{\sqrt{2}}\begin{bmatrix} 1 & 1 \\ 1 & -1 \end{bmatrix}, \quad \mathbf{H}_4 = \frac{1}{\sqrt{2}}\begin{bmatrix} \mathbf{H}_2 & \mathbf{H}_2 \\ \mathbf{H}_2 & -\mathbf{H}_2 \end{bmatrix}, \quad \mathbf{H}_{2^t} = \underbrace{\mathbf{H}_2 \otimes \mathbf{H}_2 \otimes \cdots \otimes \mathbf{H}_2}_{\text{Kronecker product } t \text{ times}}.$$

From any such matrix, a sketching matrix $\mathbf{S} \in \mathbb{R}^{m \times n}$ from a ROS ensemble can be obtained by sampling i.i.d. rows of the form

$$\mathbf{s}^{\mathrm{T}} = \sqrt{n}\mathbf{e}_j^{\mathrm{T}}\mathbf{H}\mathbf{D} \qquad \text{with probability } 1/n \text{ for } j = 1, \ldots, n,$$

where the random vector $\mathbf{e}_j \in \mathbb{R}^n$ is chosen uniformly at random from the set of all n canonical basis vectors, and $\mathbf{D} = \mathrm{diag}(\mathbf{r})$ is a diagonal matrix of i.i.d. Rademacher variables $\mathbf{r} \in \{-1, +1\}^n$, where $\mathbb{P}[\mathbf{r}_i = +1] = \mathbb{P}[\mathbf{r}_i = -1] = \frac{1}{2}\,\forall i$. Alternatively, the rows of the ROS sketch can be sampled without replacement and one can obtain similar guarantees to sampling with replacement. Given a fast routine for matrix–vector multiplication, ROS sketch $\mathbf{SA}$ of the data $\mathbf{A} \in \mathbb{R}^{n \times d}$ can be formed in $O(nd \log m)$ time (for instance, see [11]).

4.2.4 Sketches Based on Random Row Sampling

Given a probability distribution $\{\mathbf{p}_j\}_{j=1}^n$ over $[n] = \{1, \ldots, n\}$, another choice of sketch is to randomly sample the rows of the extended data matrix $\mathbf{A}$ a total of m times with replacement from the given probability distribution. Thus, the rows of S are independent and take on the values

$$\mathbf{s}^{\mathrm{T}} = \frac{\mathbf{e}_j^{\mathrm{T}}}{\sqrt{\mathbf{p}_j}} \qquad \text{with probability } \mathbf{p}_j \text{ for } j = 1, \ldots, n,$$

where $\mathbf{e}_j \in \mathbb{R}^n$ is the jth canonical basis vector. Different choices of the weights $\{\mathbf{p}_j\}_{j=1}^n$ are possible, including those based on the *leverage scores* of $\mathbf{A}$. Leverage scores are defined as

$$\mathbf{p}_j := \frac{\|\mathbf{u}_j\|_2^2}{\sum_{i=1}^n \|\mathbf{u}_i\|_2^2},$$

where $u_1, u_2, \ldots, u_n$ are the rows of $\mathbf{U} \in \mathbb{R}^{n \times d}$, which is the matrix of left singular vectors of $\mathbf{A}$. Leverage scores can be obtained using a singular value decomposition $\mathbf{A} = \mathbf{U}\mathbf{\Sigma}\mathbf{V}^{\mathrm{T}}$. Moreover, there also exist faster randomized algorithms to approximate the leverage scores (e.g., see [12]). In our analysis of lower bounds to follow, we assume that the weights are α-balanced, meaning that

$$\max_{j=1,\ldots,n} \mathbf{p}_j \leq \frac{\alpha}{n} \tag{4.4}$$

for some constant α that is independent of n.

4.2.5 Graph Sparsification via Sub-Sampling

Let $G = (V, E)$ be a weighted, undirected graph with d nodes and n edges, where V and E are the set of nodes and the set of edges, respectively. Let $\mathbf{A} \in \mathbb{R}^{n\times d}$ be the node–edge incidence matrix of the graph G. Suppose we randomly sample the edges in the graph a total of m times with replacement from a given probability distribution over the edges. The obtained graph is a weighted subgraph of the original, whose incidence matrix is $\mathbf{SA}$. Similarly to the row sampling sketch, the sketch can be written as

$$\mathbf{s}^{\mathrm{T}} = \frac{\mathbf{e}_j^{\mathrm{T}}}{\sqrt{\mathbf{p}_j}} \qquad \text{with probability } \mathbf{p}_j \text{ for } j = 1, \ldots, n.$$

We note that row and graph sub-sampling sketches satisfy the condition (4.1). However, they do not satisfy the condition (4.2). In many computational problems on graphs, sparsifying the graph has computational advantages. Notable examples of such problems are solving Laplacian linear systems and graph partitioning, where sparsification can be used. We refer the reader to Spielman and Srivastava [13] for details.

4.2.6 Sparse Sketches Based on Hashing

In many applications, the data matrices contain very few non-zero entries. For sparse data matrices, special constructions of the sketching matrices yield greatly improved performance. Here we describe the count-sketch construction from [14, 15]. Let $h : [n] \to [m]$ be a hash functions from a pair-wise independent family.[1] The entry $\mathbf{S}_{ij}$ of the sketch matrix is given by σ_j if $i = h(j)$, and otherwise it is zero, where $\sigma \in \{-1, +1\}^n$ is a random vector containing 4-wise independent variables. Therefore, the jth column of $\mathbf{S}$ is non-zero only in the row indexed by $h(j)$. We refer the reader to [14, 15] for the details. An example realization of the sparse sketch is given below, where each column contains a single non-zero entry which is uniformly random sign ± 1 at a uniformly random index:

$$\begin{bmatrix} 0 & 1 & 0 & 1 & 0 & 0 & 0 \\ -1 & 0 & 0 & 0 & 0 & 1 & 0 \\ 0 & 0 & 0 & 0 & -1 & 0 & 0 \\ 0 & 0 & 1 & 0 & 0 & 0 & -1 \end{bmatrix}.$$

Figure 4.2 shows examples of different randomized sketching matrices $S \in \mathbb{R}^{m\times n}$, where $m = 64, n = 1024$, which are drawn randomly. We refer readers to Nelson and Nguyên [16] for details on sparse sketches.

4.3 Background on Convex Analysis and Optimization

In this section, we first briefly review relevant concepts from convex analysis and optimization. A set $C \subseteq \mathbb{R}^d$ is convex if, for any $\mathbf{x}, \mathbf{y} \in C$,

$$t\mathbf{x} + (1-t)\mathbf{y} \in C \text{ for all } t \in [0, 1].$$

[1] A hash function is from a pair-wise independent family if $\mathbb{P}[h(j) = i, h(k) = l] = 1/m^2$ and $\mathbb{P}[h(j) = i] = 1/m$ for all i, j, k, l.

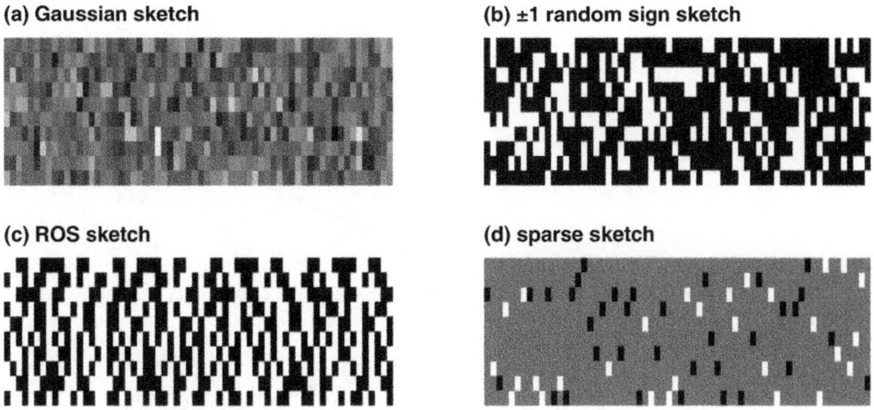

Figure 4.2 Different types of sketching matrices: (a) Gaussian sketch, (b) ±1 random sign sketch, (c) randomized orthogonal system sketch, and (d) sparse sketch.

Let $\mathcal{X}$ be a convex set. A function $f : \mathcal{X} \to \mathbb{R}$ is convex if, for any $\mathbf{x}, \mathbf{y} \in \mathcal{X}$,

$$f(t\mathbf{x} + (1-t)\mathbf{y}) \le t f(\mathbf{x}) + (1-t) f(\mathbf{y}) \text{ for all } t \in [0,1].$$

Given a matrix $\mathbf{A} \in \mathbb{R}^{n \times d}$, we define the linear transform of the convex set C as $\mathbf{A}C = \{\mathbf{A}\mathbf{x} \,|\, \mathbf{x} \in C\}$. It can be shown that $\mathbf{A}C$ is convex if C is convex.

A convex optimization problem is a minimization problem of the form

$$\min_{\mathbf{x} \in C} f(\mathbf{x}), \tag{4.5}$$

where $f(\mathbf{x})$ is a convex function and C is a convex set. In order to characterize optimality of solutions, we will define the tangent cone of C at a fixed vector $\mathbf{x}^*$ as follows:

$$\mathcal{T}_C(\mathbf{x}^*) = \{t(\mathbf{x} - \mathbf{x}^*) \quad | \quad t \ge 0 \text{ and } \mathbf{x} \in C\}. \tag{4.6}$$

Figures 4.3 and 4.4 illustrate[2] examples of tangent cones of a polyhedral convex set in $\mathbb{R}^2$. A first-order characterization of optimality in the convex optimization problem (4.5) is given by the tangent cone. If a vector $\mathbf{x}^*$ is optimal in (4.5), it holds that

$$\mathbf{z}^{\mathrm{T}} \nabla f(\mathbf{x}^*) \ge 0, \forall \mathbf{z} \in \mathcal{T}_C(\mathbf{x}^*). \tag{4.7}$$

We refer the reader to Hiriart-Urruty and Lemaréchal [17] for details on convex analysis, and Boyd and Vandenberghe [18] for an in-depth discussion of convex optimization problems and applications.

[2] Note that the tangent cones extend toward infinity in certain directions, whereas the shaded regions in Figs. 4.3 and 4.4 are compact for illustration.

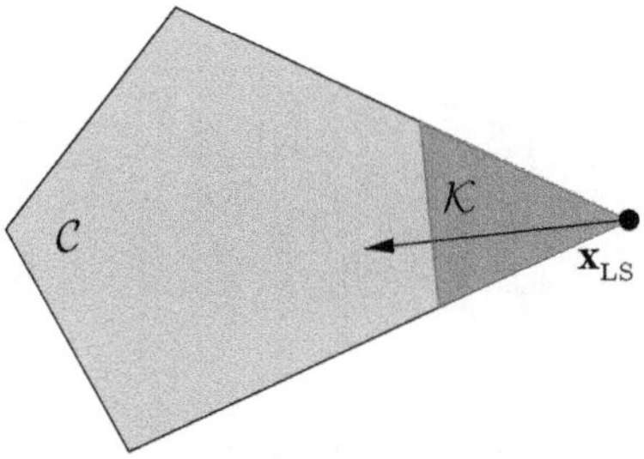

Figure 4.3 A narrow tangent cone where the Gaussian complexity is small.

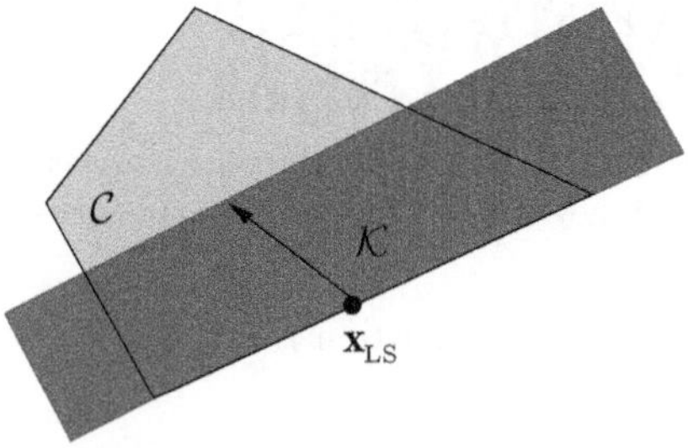

Figure 4.4 A wide tangent cone where the Gaussian complexity is large.

4.4 Sketching Upper Bounds for Regression Problems

Now we consider an instance of a convex optimization problem. Consider the least-squares optimization

$$\mathbf{x}^* = \arg\min_{\mathbf{x}\in C} \underbrace{\|\mathbf{A}\mathbf{x} - \mathbf{b}\|_2}_{f(\mathbf{x})}, \tag{4.8}$$

where $\mathbf{A} \in \mathbb{R}^{n\times d}$ and $\mathbf{b} \in \mathbb{R}^n$ are the input data and $C \subseteq \mathbb{R}^d$ is a closed and convex constraint set. In statistical and signal-processing applications, it is typical to use the constraint set to impose structure on the obtained solution $\mathbf{x}$. Important examples of the convex constraint C include the non-negative orthant, ℓ_1-ball for promoting sparsity, and the ℓ_∞-ball as a relaxation to the combinatorial set $\{0, 1\}^d$.

In the unconstrained case when $C = \mathbb{R}^d$, a closed-form solution exists for the solution of (4.8), which is given by $\mathbf{x}^* = (\mathbf{A}^T\mathbf{A})^{-1}\mathbf{A}^T\mathbf{b}$. However, forming the Gram matrix $\mathbf{A}^T\mathbf{A}$ and inverting using direct methods such as QR decomposition, or the singular value decomposition, typically requires $O(nd^2) + O(nd \min(n, d))$ operations. Faster iterative algorithms such as the conjugate gradient (CG) method can be used to obtain an approximate solution in $O(nd\kappa(\mathbf{A}))$ time, where $\kappa(\mathbf{A})$ is the condition number of the data matrix $\mathbf{A}$. Using sketching methods, it is possible to obtain even faster approximate solutions, as we will discuss in what follows.

In the constrained case, a variety of efficient iterative algorithms have been developed in the last couple of decades to obtain the solution, such as proximal and projected gradient methods, their accelerated variants, and barrier-based second-order methods. Sketching can also be used to improve the run-time of these methods.

4.4.1 Over-Determined Case ($n > d$)

In many applications, the number of observations, n, exceeds the number of unknowns, d, which gives rise to the tall $n \times d$ matrix A. In machine learning, it is very common to encounter datasets where n is very large and d is of moderate size. Suppose that we first compute the sketched data matrices **SA** and **Sb** from the original data **A** and **b**, then consider the following approximation to the above optimization problem:

$$\widehat{\mathbf{x}} = \arg\min_{\mathbf{x} \in C} \|\mathbf{SAx} - \mathbf{Sb}\|_2^2. \tag{4.9}$$

After applying the sketch to the data matrices, the sketched problem has dimensions $m \times d$, which is lower than the original dimensions when $m < n$. Note that the objective in the above problem (4.9) can be seen as an unbiased approximation of the original objective function (4.8), since it holds that

$$\mathbb{E}\|\mathbf{SAx} - \mathbf{Sb}\|_2^2 = \|\mathbf{Ax} - \mathbf{b}\|_2^2$$

for any fixed choice of **A**, **x**, and **b**. This is a consequence of the condition (4.1), which is satisfied by all of the sketching matrices considered in Section 4.2.

4.4.2 Gaussian Complexity

Gaussian complexity plays an important role in statistics, empirical process theory, compressed sensing, and the theory of Banach spaces [19–21]. Here we consider a localized version of the Gaussian complexity, which is defined as follows:

$$\mathcal{W}_t(C) := \mathbb{E}_g\Big[\sup_{\substack{\mathbf{z} \in C \\ \|\mathbf{z}\|_2 \le t}} |\langle \mathbf{g}, \mathbf{z} \rangle|\Big], \tag{4.10}$$

where $\mathbf{g}$ is a random vector with i.i.d. standard Gaussian entries, i.e., $\mathbf{g} \sim \mathcal{N}(\mathbf{0}, \mathbf{I}_n)$. The parameter $t > 0$ controls the radius at which the random deviations are localized. For a finite value of t, the supremum in (4.10) is always achieved since the constraint set is compact.

Analyzing the sketched optimization problem requires us to control the random deviations constrained to the set of possible descent directions $\{\mathbf{x} - \mathbf{x}^* \mid \mathbf{x} \in C\}$. We now define a transformed tangent cone at x^* as follows:

$$\mathcal{K} = \{t\mathbf{A}(\mathbf{x} - \mathbf{x}^*) \quad | \quad t \ge 0 \text{ and } \mathbf{x} \in C\},$$

which can be alternatively defined as $A\mathcal{T}_C(\mathbf{x}^*)$ using the definition given in (4.6). The next theorem provides an upper bound on the performance of the sketching method for constrained optimization based on localized Gaussian complexity.

THEOREM 4.1 *Let S be a Gaussian sketch, and let $\widehat{\mathbf{x}}$ be the solution of (4.9). Suppose that $m \ge c_0 \mathcal{W}_1(\mathcal{K})^2/\epsilon^2$, where c_0 is a universal constant, then it holds that*

$$\frac{\|A(\widehat{\mathbf{x}} - \mathbf{x}^*)\|_2}{f(\mathbf{x}^*)} \le \epsilon,$$

and consequently we have

$$f(\mathbf{x}^*) \le f(\widehat{\mathbf{x}}) \le f(\mathbf{x}^*)(1+\epsilon). \tag{4.11}$$

As predicted by the theorem, the approximation ratio improves as the sketch dimension m increases, and converges to one as $m \to \infty$. However, we are often interested in the rate of convergence of the approximation ratio. Theorem 4.1 characterizes this rate by relating the geometry of the constraint set to the accuracy of the sketching method (4.9). As an illustration, Figs. 4.3 and 4.4 show narrow and wide tangent cones in $\mathbb{R}^2$, respectively. The proof of Theorem 4.1 combines the convex optimality condition involving the tangent cone in (4.7) with results on empirical processes, and can be found in Pilanci and Wainwright [22]. An important feature of Theorem 4.1 is that the approximation quality is relative to the optimal value $f(\mathbf{x}^*)$. This is advantageous when $f(\mathbf{x}^*)$ is small, e.g., the optimal value can be zero in noiseless signal recovery problems. However, in problems where the signal-to-noise ratio is low, $f(\mathbf{x}^*)$ can be large, and hence negatively affects the approximation quality. We illustrate the implications of Theorem 4.1 on some concrete examples in what follows.

Example 4.1 Unconstrained Least Squares For unconstrained problems, we have $\mathcal{K} = \text{range}(\mathbf{A})$, i.e., the tangent cone is equal to the range space of the data matrix $\mathbf{A}$. In order to apply Theorem 4.1, we need the following lemma about the Gaussian complexity of a subspace.

LEMMA 4.1 *Let Q be a subspace of dimension q. The Gaussian complexity of Q satisfies*

$$\mathcal{W}_t(Q) \le t\sqrt{q}.$$

Proof Let $\mathbf{U}$ be an orthonormal basis for the subspace Q. We have the following representation: $\mathbf{L} = \{\mathbf{U}\mathbf{x} \mid \mathbf{x} \in \mathbb{R}^q\}$. Consequently the Gaussian complexity $\mathcal{W}_1(Q)$ can be written as

$$\begin{aligned}\mathbb{E}_{\mathbf{g}}\Big[\sup_{\substack{\mathbf{x}\\ \|\mathbf{U}\mathbf{x}\|_2 \le t}} \langle \mathbf{g}, \mathbf{U}\mathbf{x}\rangle\Big] = \mathbb{E}_{\mathbf{g}}\Big[\sup_{\substack{\mathbf{x}\\ \|\mathbf{x}\|_2 \le t}} \langle \mathbf{U}^{\mathrm{T}}\mathbf{g}, \mathbf{x}\rangle\Big] = t\,\mathbb{E}_{\mathbf{g}}\|\mathbf{U}^{\mathrm{T}}\mathbf{g}\|_2 &\le t\sqrt{\mathbb{E}\ \mathrm{tr}\ \mathbf{U}\mathbf{U}^{\mathrm{T}}\mathbf{g}\mathbf{g}^{\mathrm{T}}}\\ &= t\sqrt{\mathrm{tr}\ \mathbf{U}^{\mathrm{T}}\mathbf{U}}\\ &= t\sqrt{q}.\end{aligned}$$

Where the inequality follows from Jensen's inequality and concavity of the square root, and first and fifth equality follow since $\mathbf{U}^{\mathrm{T}}\mathbf{U} = \mathbf{I}_q$. Therefore, the Gaussian complexity of the range of $\mathbf{A}$ for $t = 1$ satisfies

$$\mathcal{W}_1(\text{range}(\mathbf{A})) \le \sqrt{\text{rank}(\mathbf{A})}.$$

Setting the dimension of the sketch $m \ge c_0\,\text{rank}(\mathbf{A})/\epsilon^2$ suffices to obtain an ϵ approximate solution in the sense of (4.11). We note that rank($\mathbf{A}$) might not be known *a priori*, but the upper bound $\text{rank}(\mathbf{A}) \le d$ may be useful when $n \gg d$.

Example 4.2 ℓ_1 Constrained Least Squares For ℓ_1-norm-constrained problems we have $C = \{\mathbf{x} \mid \|\mathbf{x}\|_1 \leq r\}$ for some radius parameter r. The tangent cone $\mathcal{K}$ at the optimal point $\mathbf{x}^*$ depends on the support[3] of $\mathbf{x}^*$, and hence on its cardinality $\|\mathbf{x}^*\|_0$. In [23], it is shown that the localized Gaussian complexity satisfies

$$\mathcal{W}_1(\mathcal{K}) \leq c_1 \frac{\gamma_{k_+}(\mathbf{A})}{\beta_{k_-}(\mathbf{A})} \sqrt{\|\mathbf{x}^*\|_0 \log d},$$

where c_1 is a universal constant and γ_k are β_k are ℓ_1-restricted maximum and minimum eigenvalues defined as follows:

$$\gamma_k := \max_{\substack{\|\mathbf{z}\|=1 \\ \|\mathbf{z}\|_1 \leq \sqrt{k}}} \|\mathbf{A}\mathbf{z}\|_2^2 \quad \text{and} \quad \beta_k := \min_{\substack{\|\mathbf{z}\|=1 \\ \|\mathbf{z}\|_1 \leq \sqrt{k}}} \|\mathbf{A}\mathbf{z}\|_2^2.$$

As a result, we conclude that for ℓ_1-constrained problems, the sketch dimension can be substantially smaller when ℓ_1-constrained eigenvalues are well behaved.

4.4.3 Under-Determined Case ($n \leq d$)

In many applications the dimension of the data vectors may be larger than the sample size. In these situations, it makes sense to reduce the dimensionality by applying the sketch on the right, i.e., $\mathbf{A}\mathbf{S}^{\mathrm{T}}$, and solve

$$\arg \min_{\substack{\mathbf{z} \in \mathbb{R}^m \\ \mathbf{S}^{\mathrm{T}}\mathbf{z} \in C}} \|(\mathbf{A}\mathbf{S}^{\mathrm{T}}\mathbf{z} - \mathbf{b})\|_2. \tag{4.12}$$

Note that the vector $\mathbf{z} \in \mathbb{R}^m$ is of smaller dimension than the original variable $\mathbf{x} \in \mathbb{R}^d$. After solving the reduced-dimensional problem and obtaining its optimal solution $\mathbf{z}^*$, the final estimate for the original variable $\mathbf{x}$ can be taken as $\widehat{\mathbf{x}} = \mathbf{S}^{\mathrm{T}}\mathbf{z}^*$. We will investigate this approach in Section 4.5 in non-parametric statistical estimation problems and present concrete theoretical guarantees.

It is instructive to note that, in the special case where we have ℓ_2 regularization and $C = \mathbb{R}^d$, we can easily transform the under-determined least-squares problem into an over-determined one using convex duality, or the matrix-inversion lemma. We first write the sketched problem (4.12) as the constrained convex program

$$\min_{\substack{\mathbf{z} \in \mathbb{R}^m, \mathbf{y} \in \mathbb{R}^n \\ \mathbf{y} = \mathbf{A}\mathbf{S}^{\mathrm{T}}\mathbf{z}}} \frac{1}{2}\|\mathbf{y} - \mathbf{b}\|_2^2 + \rho\|\mathbf{z}\|_2^2,$$

and form the convex dual. It can be shown that strong duality holds, and consequently primal and dual programs can be stated as follows:

$$\min_{\mathbf{z} \in \mathbb{R}^m} \frac{1}{2}\|\mathbf{A}\mathbf{S}^{\mathrm{T}}z - \mathbf{b}\|_2^2 + \rho\|\mathbf{z}\|_2^2 = \max_{\mathbf{x} \in \mathbb{R}^d} -\frac{1}{4\rho}\|\mathbf{S}\mathbf{A}^{\mathrm{T}}\mathbf{x}\|_2^2 - \frac{1}{2}\|\mathbf{x}\|_2^2 + \mathbf{x}^{\mathrm{T}}\mathbf{b},$$

[3] The term support refers to the set of indices where the solution has a non-zero value.

where the primal and dual solutions satisfy $\mathbf{z}^* = (1/2\rho)\mathbf{S}\mathbf{A}^{\mathrm{T}}\mathbf{x}^*$ at the optimum [18]. Therefore the sketching matrix applied from the right, $\mathbf{A}\mathbf{S}^{\mathrm{T}}$, corresponds to a sketch applied on the left, $\mathbf{S}\mathbf{A}^{\mathrm{T}}$, in the dual problem which parallels (4.9). This observation can be used to derive approximation results on the dual program. We refer the reader to [22] for an application in support vector machine classification where $\mathbf{b} = \mathbf{0}_n$.

4.5 Information-Theoretic Lower Bounds

4.5.1 Statistical Upper and Lower Bounds

In order to develop information-theoretic lower bounds, we consider a statistical observation model for the constrained regression problem. Consider the following model:

$$\mathbf{b} = \mathbf{A}\mathbf{x}^{\dagger} + \mathbf{w}, \quad \text{where } \mathbf{w} \sim \mathcal{N}(0, \sigma^2 \mathbf{I}_n) \text{ and } \mathbf{x}^{\dagger} \in \mathcal{C}_0, \tag{4.13}$$

where $\mathbf{x}^{\dagger}$ is the unknown vector to be estimated and $\mathbf{w}$ is an i.i.d. noise vector whose entries are distributed as $\mathcal{N}(0, \sigma^2)$. In this section we will focus on the observation model (4.13) and present a lower bound on all estimators which use the sketched data $(\mathbf{SA}, \mathbf{Sb})$ to form an estimate $\widehat{\mathbf{x}}$.

We assume that the unknown vector $\mathbf{x}^{\dagger}$ belongs to some set $\mathcal{C}_0 \subseteq \mathcal{C}$ that is star-shaped around zero.[4] In many cases of interest we have $\mathcal{C} = \mathcal{C}_0$, i.e., when the set $\mathcal{C}$ is convex and simple to describe. In this case, the constrained least-squares estimate $\mathbf{x}^*$ from equation (4.8) corresponds to the constrained maximum-likelihood estimator for estimating the unknown regression vector $\mathbf{x}^{\dagger}$ under the Gaussian observation model (4.13). However $\mathcal{C}_0$ may not be computationally tractable as an optimization constraint set, such as a non-convex set, and we can consider a set $\mathcal{C}$ which is a convex relaxation[5] of this set, such that $\mathcal{C} \subset \mathcal{C}_0$. An important example is the set of s sparse and bounded vectors given by $\mathcal{C}_0 = \{\mathbf{x} : \|\mathbf{x}\|_0 \le s, \|\mathbf{x}\|_\infty \le 1\}$, which has combinatorially many elements. The well-known ℓ_1 relaxation given by $\mathcal{C} = \{\mathbf{x} : \|\mathbf{x}\|_1 \le \sqrt{s}, \|\mathbf{x}\|_\infty \le 1\}$ satisfies $\mathcal{C} \subset \mathcal{C}_0$, which follows from the Cauchy–Schwartz inequality, and is widely used [24, 25] to find sparse solutions.

We now present a theoretical result on the statistical performance of the original constrained least-squares estimator in (4.8)

THEOREM 4.2 *Let $\mathcal{C}$ be any set that contains the true parameter $\mathbf{x}^{\dagger}$. Then the constrained estimator $\mathbf{x}^*$ in* (4.8) *under the observation model* (4.13) *has mean-squared error upper-bounded as*

$$\mathbb{E}_w\left[\frac{1}{n}\|\mathbf{A}(\mathbf{x}^* - \mathbf{x}^{\dagger})\|_2^2\right] \le c_2\Big(\delta^*(n)^2 + \frac{\sigma^2}{n}\Big),$$

[4] This assumption means that, for any $x \in \mathcal{C}_0$ and scalar $t \in [0,1]$, the point tx also belongs to $\mathcal{C}_0$.

[5] We may also consider an approximation of $\mathcal{C}_0$ which doesn't necessarily satisfy $\mathcal{C} \subset \mathcal{C}_0$, for example, the ℓ_1 and ℓ_0 unit balls.

where $\delta^*(n)$ is the critical radius, equal to the smallest positive solution $\delta > 0$ to the inequality

$$\frac{\mathcal{W}_\delta(C)}{\delta\sqrt{n}} \le \frac{\delta}{\sigma}. \tag{4.14}$$

We refer the reader to [20, 23] for a proof of this theorem. This result provides a baseline against which to compare the statistical recovery performance of the randomized sketching method. In particular, an important goal is characterizing the minimal projection dimension m that will enable us to find an estimate $\widehat{\mathbf{x}}$ with the error guarantee

$$(1/n)\|\mathbf{A}(\widehat{\mathbf{x}} - \mathbf{x}^\dagger)\|_2^2 \approx (1/n)\|\mathbf{A}(\mathbf{x}^* - \mathbf{x}^\dagger)\|_2^2,$$

in a computationally simpler manner using the compressed data $\mathbf{SA}, \mathbf{Sb}$.

An application of Theorem 4.1 will yield that the sketched solution $\widehat{\mathbf{x}}$ in (4.9), using the choice of sketch dimension $m = c_0 \mathcal{W}_1(\mathcal{K})^2/\epsilon^2$, satisfies the bound

$$\|\mathbf{A}(\widehat{\mathbf{x}} - \mathbf{x}^*)\| \le \epsilon \|\mathbf{A}\mathbf{x}^* - \mathbf{b}\|_2,$$

where $\|\mathbf{A}\mathbf{x}^* - \mathbf{b}\|_2 = f(\mathbf{x}^*)$ is the optimal value of the optimization problem (4.8). However, under the model (4.13) we have

$$\|\mathbf{A}\mathbf{x}^* - \mathbf{b}\|_2 = \|\mathbf{A}(\mathbf{x}^* - \mathbf{x}^\dagger) - \mathbf{w}\|_2 \le \|\mathbf{A}(\mathbf{x}^* - \mathbf{x}^\dagger)\|_2 + \|\mathbf{w}\|_2,$$

which is at least $O(\sigma\sqrt{n})$ because of the term $\|\mathbf{w}\|_2$. This upper bound suggests that $(1/n)\|\mathbf{A}(\widehat{\mathbf{x}} - \mathbf{x}^\dagger)\|_2^2$ is bounded by $O(\epsilon^2\sigma^2) = O(\sigma^2\mathcal{W}_1(\mathcal{K})^2/m)$. This can be considered as a negative result for the sketching method, since the error scales as $O(1/m)$ instead of $O(1/n)$. We will show that this upper bound is tight, and the $O(1/m)$ scaling is unavoidable for all methods that sketch the data once. In contrast, as we will discuss in Section 4.5.7, an iterative sketching method can achieve optimal prediction error using sketches of comparable dimension.

We will in fact show that, unless $m \ge n$, *any method* based on observing *only* the pair $(\mathbf{SA}, \mathbf{Sb})$ necessarily has a substantially larger error than the least-squares estimate. In particular, our result applies to an arbitrary measurable function $(\mathbf{SA}, \mathbf{Sb}) \mapsto \widehat{\mathbf{x}}$, which we refer to as an *estimator*.

More precisely, our lower bound applies to any random matrix $\mathbf{S} \in \mathbb{R}^{m \times n}$ for which

$$\|\mathbb{E}[\mathbf{S}^T(\mathbf{S}\mathbf{S}^T)^{-1}\mathbf{S}]\|_{\text{op}} \le \eta\,\frac{m}{n}, \tag{4.15}$$

where η is a constant that is independent of n and m, and $\|\mathbf{A}\|_{\text{op}}$ denotes the ℓ_2-operator norm, which reduces to the maximum eigenvalue for a symmetric matrix. These conditions hold for various standard choices of the sketching matrix, including most of those discussed in the Section 4.2: the Gaussian sketch, the ROS sketch,[6] the sparse sketch, and the α-balanced leverage sampling sketch. The following lemma shows that the condition (4.15) is satisfied for Gaussian sketches with equality and $\eta = 1$.

[6] See [23] for a proof of this fact for Gaussian and ROS sketches. To be more precise, for ROS sketches, the condition (4.15) holds when rows are sampled without replacement.

LEMMA 4.2 *Let $\mathbf{S} \in \mathbb{R}^{m \times n}$ be a random matrix with i.i.d. Gaussian entries. We have*

$$\left\| \mathbb{E}\left[\mathbf{S}^{\mathrm{T}}(\mathbf{S}\mathbf{S}^{\mathrm{T}})^{-1}\mathbf{S}\right] \right\|_{\mathrm{op}} = \frac{m}{n}.$$

Proof Let $\mathbf{S} = \mathbf{U}\mathbf{\Sigma}\mathbf{V}^{\mathrm{T}}$ denote the singular value decomposition of the random matrix $\mathbf{S}$. Note that we have $\mathbf{S}^{\mathrm{T}}(\mathbf{S}\mathbf{S}^{\mathrm{T}})^{-1}\mathbf{S} = \mathbf{V}\mathbf{V}^{\mathrm{T}}$. By virtue of the rotation invariance of the Gaussian distribution, columns of $\mathbf{V}$ denoted by $\{\mathbf{v}_i\}_{i=1}^m$ are uniformly distributed over the n-dimensional unit sphere, and it holds that $\mathbb{E}\mathbf{v}_i\mathbf{v}_i^{\mathrm{T}} = (1/n)\mathbf{I}_n$ for $i = 1, \ldots, m$. Consequently, we obtain

$$\mathbb{E}\left[\mathbf{S}^{\mathrm{T}}(\mathbf{S}\mathbf{S}^{\mathrm{T}})^{-1}\mathbf{S}\right] = \mathbb{E}\sum_{i=1}^{m} \mathbf{v}_i\mathbf{v}_i^{\mathrm{T}} = m\mathbb{E}\mathbf{v}_1\mathbf{v}_1^{\mathrm{T}} = \frac{m}{n}\mathbf{I}_n,$$

and the bound on the operator norm follows.

4.5.2 Fano's Inequality

Let X and Y represent two random variables with a joint probability distribution $P_{x,y}$, where X is discrete and takes values from a finite set $\mathcal{X}$. Let $\widehat{X} = g(Y)$ be the predicted value of X for some deterministic function g which also takes values in $\mathcal{X}$. Then Fano's inequality states that

$$P\left[X \neq \widehat{X}\right] \geq \frac{H(X|Y) - 1}{\log_2(|\mathcal{X}| - 1)}.$$

Fano's inequality follows as a simple consequence of the chain rule for entropy. However, it is very powerful for deriving lower bounds on the error probabilities in coding theory, statistics, and machine learning [7, 26–30].

4.5.3 Metric Entropy

For a given positive tolerance value $\delta > 0$, we define the δ-packing number $M_{\delta,\|\cdot\|}$ of a set $C \subseteq \mathbb{R}^d$ with respect to a norm $\|\cdot\|$ as the largest number of vectors $\{\mathbf{x}^j\}_{j=1}^M \subseteq C$ which are elements of C and satisfy

$$\|\mathbf{x}^k - \mathbf{x}^l\| > \delta \quad \forall k \neq l.$$

We define the *metric entropy* of the set C with respect to a norm $\|\cdot\|$ as the logarithm of the corresponding packing number

$$N_{\delta,\|\cdot\|}(C) = \log_2 M_{\delta,\|\cdot\|}.$$

The concept of metric entropy provides a way to measure the complexity, or effective size, of a set with infinitely many elements and dates back to the seminal work of Kolmogorov and Tikhomirov [31].

4.5.4 Minimax Risk

In this chapter, we will take a frequentist approach in modeling the unknown vector $\mathbf{x}^\dagger$ we are trying to estimate from the data. In order to assess the quality of estimation, we will consider a risk function associated with our estimation method. Note that, for a fixed value of the unknown vector $\mathbf{x}^\dagger$, there exist estimators which make no error for that particular vector $\mathbf{x}^\dagger$, such as the estimator which always returns $\mathbf{x}^\dagger$ regardless of the observation. We will take the worst-case risk approach considered in the statistical estimation literature, which focuses on the *minimax* risk. More precisely, we define the minimax risk as follows:

$$\mathcal{M}(Q) = \inf_{\widehat{\mathbf{x}} \in Q} \sup_{\mathbf{x}^\dagger \in \mathcal{X}} \mathbb{E}\left[\frac{1}{n} \|\mathbf{A}(\widehat{\mathbf{x}} - \mathbf{x}^\dagger)\|_2^2 \right], \tag{4.16}$$

where the infimum ranges over all estimators that use the input data $\mathbf{A}$ and $\mathbf{b}$ to estimate $\mathbf{x}^\dagger$.

4.5.5 Reduction to Hypothesis Testing

In this section we present a reduction of the minimax estimation risk to hypothesis testing. Suppose that we have a packing of the constraint set C given by the collection $\mathbf{z}^{(1)}, ..., \mathbf{z}^{(M)}$ with radius 2δ. More precisely, we have

$$\|A(\mathbf{z}^{(i)} - \mathbf{z}^{(j)})\|_2 \geq 2\delta \quad \forall i \neq j,$$

where $\mathbf{z}^{(i)} \in C$ for all $i = 1, ..., M$. Next, consider a set of probability distributions $\{P_{\mathbf{z}^{(j)}}\}_{j=1}^{M}$ corresponding to the distribution of the observation when the unknown vector is $\mathbf{x}^\dagger = \mathbf{z}^j$. Suppose that we have an M-ary hypothesis-testing problem constructed as follows. Let J_δ denote a random variable with uniform distribution over the index set $\{1, \ldots, M\}$ that allows us to pick an element of the packing set at random. Note that M is a function of δ, hence we keep the dependence of J_δ on δ explicit in our notation. Let us set the random variable Z according to the probability distribution $P_{\mathbf{z}^{(j)}}$ in the event that $J_\delta = j$, i.e.,

$$Z \sim \mathbb{P}_{\mathbf{x}^{(j)}} \quad \text{whenever} \quad J_\delta = j.$$

Now we will consider the problem of detecting the index set given the value of Z. The next lemma is a standard reduction in minimax theory, and relates the minimax estimation risk to the M-ary hypothesis-testing error (see Birgé [30] and Yu [7]).

LEMMA 4.3 *The minimax risk Q is lower-bounded by*

$$\mathcal{M}(Q) \geq \delta^2 \inf_{\psi} \mathbb{P}[\psi(Z) \neq J_\delta]. \tag{4.17}$$

A proof of this lemma can be found in Section A4.2. Lemma 4.3 allows us to apply Fano's method after transforming the estimation problem into a hypothesis-testing problem based on sketched data. Let us recall the condition on sketching matrices stated earlier,

$$\|\mathbb{E}[\mathbf{S}^{\mathrm{T}}(\mathbf{S}\mathbf{S}^{\mathrm{T}})^{-1}\mathbf{S}]\|_{\mathrm{op}} \leq \eta \frac{m}{n}, \tag{4.18}$$

where η is a constant that is independent of n and m. Now we are ready to present the lower bound on the statistical performance of sketching.

THEOREM 4.3 *For any random sketching matrix* $\mathbf{S} \in \mathbb{R}^{m \times n}$ *satisfying condition* (4.18), *any estimator* $(\mathbf{SA}, \mathbf{Sb}) \mapsto \mathbf{x}^\dagger$ *has MSE lower-bounded as*

$$\sup_{\mathbf{x}^\dagger \in C_0} \mathbb{E}_{\mathbf{S},\mathbf{w}}\left[\frac{1}{n}\|\mathbf{A}(\mathbf{x}^\dagger - \mathbf{x}^*)\|_2^2\right] \geq \frac{\sigma^2}{128\eta}\frac{\log_2(\frac{1}{2}M_{1/2})}{\min\{m,n\}}, \tag{4.19}$$

where $M_{1/2}$ *is the* $1/2$*-packing number of* $C_0 \cap \mathbb{B}_A(1)$ *in the semi-norm* $(1/\sqrt{n})\|\mathbf{A}(\cdot)\|_2$.

We defer the proof to Section 4.8, and investigate the implications of the lower bound in the next section. It can be shown that Theorem 4.3 is tight, since Theorem 4.1 provides a matching upper bound.

4.5.6 Implications of the Information-Theoretic Lower Bound

We now investigate some consequences of the lower bound given in Theorem 4.3. We will focus on concrete examples of popular statistical estimation and optimization problems to illustrate its applicability.

Example 4.3 Unconstrained Least Squares We first consider the simple unconstrained case, where the constraint is the entire d-dimensional space, i.e., $C = \mathbb{R}^d$. With this choice, it is well known that, under the observation model (4.13), the least-squares solution $\mathbf{x}^*$ has prediction mean-squared error upper-bounded as follows:[7]

$$\mathbb{E}\left[\frac{1}{n}\|\mathbf{A}(\mathbf{x}^* - \mathbf{x}^\dagger)\|_2^2\right] \lesssim \frac{\sigma^2 \text{rank}(\mathbf{A})}{n} \tag{4.20a}$$

$$\leq \frac{\sigma^2 d}{n}, \tag{4.20b}$$

where the expectation is over the noise variable w in (4.13). On the other hand, with the choice $C_0 = \mathbb{B}_2(1)$, it is well known that we can construct a $1/2$-packing with $M = 2^d$ elements, so that Theorem 4.3 implies that any estimator $\mathbf{x}^\dagger$ based on $(\mathbf{SA}, \mathbf{Sb})$ has prediction MSE lower-bounded as

$$\mathbb{E}_{\mathbf{S},\mathbf{w}}\left[\frac{1}{n}\|\mathbf{A}(\widehat{\mathbf{x}} - \mathbf{x}^\dagger)\|_2^2\right] \gtrsim \frac{\sigma^2 d}{\min\{m,n\}}. \tag{4.20c}$$

Consequently, the sketch dimension m must grow proportionally to n in order for the sketched solution to have a mean-squared error comparable to the original least-squares estimate. This may not be desirable for least-squares problems in which $n \gg d$, since it should be possible to sketch down to a dimension proportional to rank($\mathbf{A}$) which is always upper-bounded by d. Thus, Theorem 4.3 reveals a surprising gap between the classical least-squares sketch (4.9) and the accuracy of the original least-squares estimate. In the regime $n \gg m$, the prediction MSE of the sketched solution is $O(\sigma^2(d/m))$

[7] In fact, a closed-form solution exists for the prediction error, which it is straightforward to obtain from the closed-form solution of the least-squares estimator. However, this simple form is sufficient to illustrate information-theoretic lower bounds.

which is a factor of n/m larger than the optimal prediction MSE in (4.20b). In Section 4.5.7, we will see that this gap can be removed by iterative sketching algorithms which don't obey the information-theoretic lower bound (4.20c).

Example 4.4 ℓ_1 Constrained Least Squares We can consider other forms of constrained least-squares estimates as well, such as those involving an ℓ_1-norm constraint to encourage sparsity in the solution. We now consider the sparse variant of the linear regression problem, which involves the ℓ_0 "ball"

$$\mathbb{B}_0(s) := \left\{ \mathbf{x} \in \mathbb{R}^d \mid \sum_{j=1}^{d} \mathbb{I}[\mathbf{x}_j \neq 0] \leq s \right\},$$

corresponding to the set of all vectors with at most s non-zero entries. Fixing some radius $R \geq \sqrt{s}$, consider a vector $\mathbf{x}^\dagger \in C_0 := \mathbb{B}_0(s) \cap \{\|\mathbf{x}\|_1 = R\}$, and suppose that we have noisy observations of the form $\mathbf{b} = \mathbf{A}\mathbf{x}^\dagger + \mathbf{w}$.

Given this setup, one way in which to estimate $\mathbf{x}^\dagger$ is by computing the least-squares estimate $\mathbf{x}^*$ constrained to the ℓ_1-ball $C = \{\mathbf{x} \in \mathbb{R}^n \mid \|\mathbf{x}\|_1 \leq R\}$.[8] This estimator is a form of the Lasso [2, 32] which has been studied extensively in the context of statistical estimation and signal reconstruction.

On the other hand, the $1/2$-packing number M of the set C_0 can be lower-bounded as $\log_2 M \gtrsim s \log_2(ed/s)$. We refer the reader to [33] for a proof. Consequently, in application to this particular problem, Theorem 4.3 implies that any estimator $\widehat{\mathbf{x}}$ based on the pair $(\mathbf{SA}, \mathbf{Sb})$ has mean-squared error lower-bounded as

$$\mathbb{E}_{\mathbf{w},\mathbf{S}}\left[\frac{1}{n}\|\mathbf{A}(\widehat{\mathbf{x}} - \mathbf{x}^\dagger)\|_2^2\right] \gtrsim \frac{\sigma^2 s \log_2(ed/s)}{\min\{m, n\}}. \tag{4.21}$$

Again, we see that the projection dimension m must be of the order of n in order to match the mean-squared error of the constrained least-squares estimate $\mathbf{x}^*$ up to constant factors.

Example 4.5 Low-Rank Matrix Estimation In the problem of multivariate regression, the goal is to estimate a matrix $\mathbf{X}^\dagger \in \mathbb{R}^{d_1 \times d_2}$ model based on observations of the form

$$\mathbf{Y} = \mathbf{A}\mathbf{X}^\dagger + \mathbf{W}, \tag{4.22}$$

where $\mathbf{Y} \in \mathbb{R}^{n \times d_2}$ is a matrix of observed responses, $\mathbf{A} \in \mathbb{R}^{n \times d_1}$ is a data matrix, and $\mathbf{W} \in \mathbb{R}^{n \times d_2}$ is a matrix of noise variables. A typical interpretation of this model is a collection of d_2 regression problems, where each one involves a d_1-dimensional regression vector, namely a particular column of the matrix $\mathbf{X}^\dagger$. In many applications, including reduced-rank regression, multi-task learning, and recommender systems

[8] This setup is slightly unrealistic, since the estimator is assumed to know the radius $R = \|\mathbf{x}^\dagger\|_1$. In practice, one solves the least-squares problem with a Lagrangian constraint, but the underlying arguments are essentially the same.

(e.g., [34–37]), it is reasonable to model the matrix $\mathbf{X}^\dagger$ as being a low-rank matrix. Note that a rank constraint on the matrix $\mathbf{X}$ can be written as an ℓ_0-"norm" sparsity constraint on its singular values. In particular, we have

$$\text{rank}(\mathbf{X}) \le r \quad \text{if and only if} \quad \sum_{j=1}^{\min\{d_1,d_2\}} \mathbb{I}[\gamma_j(\mathbf{X}) > 0] \le r,$$

where $\gamma_j(\mathbf{X})$ denotes the jth singular value of $\mathbf{X}$. This observation motivates a standard relaxation of the rank constraint using the nuclear norm $|||\mathbf{X}|||_{\text{nuc}} := \sum_{j=1}^{\min\{d_1,d_2\}} \gamma_j(\mathbf{X})$.

Accordingly, let us consider the constrained least-squares problem

$$\mathbf{X}^* = \arg \min_{\mathbf{X}\in\mathbb{R}^{d_1\times d_2}} \left\{ \frac{1}{2} |||\mathbf{Y} - \mathbf{A}\mathbf{X}|||_{\text{fro}}^2 \right\} \qquad \text{such that } |||\mathbf{X}|||_{\text{nuc}} \le R, \tag{4.23}$$

where $|||\cdot|||_{\text{fro}}$ denotes the Frobenius norm on matrices, or equivalently the Euclidean norm on its vectorized version. Let C_0 denote the set of matrices with rank $r < \frac{1}{2}\min\{d_1, d_2\}$, and Frobenius norm at most one. In this case the constrained least-squares solution $\mathbf{X}^*$ satisfies the bound

$$\mathbb{E}\left[\frac{1}{n}\|\mathbf{A}(\mathbf{X}^* - \mathbf{X}^\dagger)\|_2^2\right] \lesssim \frac{\sigma^2 r(d_1 + d_2)}{n}. \tag{4.24a}$$

On the other hand, the 1/2-packing number of the set C_0 is lower-bounded as $\log_2 M \gtrsim r(d_1 + d_2)$ (see [36] for a proof), so that Theorem 4.3 implies that any estimator $\widehat{\mathbf{X}}$ based on the pair $(\mathbf{SA}, \mathbf{SY})$ has MSE lower-bounded as

$$\mathbb{E}_{\mathbf{w},\mathbf{S}}\left[\frac{1}{n}\|\mathbf{A}(\widehat{\mathbf{X}} - \mathbf{X}^\dagger)\|_2^2\right] \gtrsim \frac{\sigma^2 r(d_1 + d_2)}{\min\{m,n\}}. \tag{4.24b}$$

As with the previous examples, we see the sub-optimality of the sketched approach in the regime $m < n$.

4.5.7 Iterative Sketching

It is possible to improve the basic sketching estimator using adaptive measurements. Consider the constrained least-squares problem in (4.8):

$$x^* = \arg\min_{\mathbf{x}\in C} \frac{1}{2}\|\mathbf{Ax} - \mathbf{b}\|_2^2 \tag{4.25}$$

$$= \arg\min_{\mathbf{x}\in C} \underbrace{\frac{1}{2}\|\mathbf{Ax}\|_2^2 - \mathbf{b}^{\mathrm{T}}\mathbf{Ax} + \frac{1}{2}\|\mathbf{b}\|_2^2}_{f(\mathbf{x})}. \tag{4.26}$$

We may use an iterative method to obtain $\mathbf{x}^*$ which uses the gradient $\nabla f(\mathbf{x}) = \mathbf{A}^{\mathrm{T}}(\mathbf{Ax} - \mathbf{b})$ and Hessian $\nabla^2 f(\mathbf{x}) = \mathbf{A}^{\mathrm{T}}\mathbf{A}$ to minimize the second-order Taylor expansion of $f(\mathbf{x})$ at a current iterate $\mathbf{x}_t$ using $\nabla f(\mathbf{x}_t)$ and $\nabla^2 f(\mathbf{x}_t)$ as follows:

$$\mathbf{x}_{t+1} = \mathbf{x}_t + \arg\min_{\mathbf{x}\in C} \left\|\left[\nabla^2 f(\mathbf{x})\right]^{1/2}\mathbf{x}\right\|_2^2 + \mathbf{x}^{\mathrm{T}}\, \nabla f(\mathbf{x}_t) \tag{4.27}$$

$$= \mathbf{x}_t + \arg\min_{\mathbf{x}\in C} \|\mathbf{Ax}\|_2^2 - \mathbf{x}^{\mathrm{T}}\mathbf{A}^{\mathrm{T}}(\mathbf{b} - \mathbf{Ax}_t). \tag{4.28}$$

We apply a sketching matrix $\mathbf{S}$ to the data $\mathbf{A}$ on the formulation (4.28) and define this procedure as an *iterative sketch*

$$\mathbf{x}_{t+1} = \mathbf{x}_t + \arg\min_{\mathbf{x}\in C} \|\mathbf{S}\mathbf{A}\mathbf{x}\|_2^2 - 2\mathbf{x}^{\mathrm{T}}\mathbf{A}^{\mathrm{T}}(\mathbf{b}-\mathbf{A}\mathbf{x}_t). \tag{4.29}$$

Note that this procedure uses more information then the classical sketch (4.9), in particular it calculates the left matrix–vector multiplications with the data A in the following order:

$$\begin{gathered} \mathbf{s}_1^{\mathrm{T}}\mathbf{A} \\ \mathbf{s}_2^{\mathrm{T}}\mathbf{A} \\ \vdots \\ \mathbf{s}_m^{\mathrm{T}}\mathbf{A} \\ \vdots \\ (\mathbf{b}-\mathbf{A}\mathbf{x}_1)^{\mathrm{T}}\mathbf{A} \\ \vdots \\ (\mathbf{b}-\mathbf{A}\mathbf{x}_t)^{\mathrm{T}}\mathbf{A}, \end{gathered}$$

where $\mathbf{s}_1^{\mathrm{T}},\ldots,\mathbf{s}_m^{\mathrm{T}}$ are the rows of the sketching matrix $\mathbf{S}$. This can be considered as an adaptive form of sketching where the residual directions $(\mathbf{b}-\mathbf{A}\mathbf{x}_t)$ are used after the random directions $\mathbf{s}_1,\ldots,\mathbf{s}_m$. As a consequence, the information-theoretic bounds we considered in Section 4.4.6 do not apply to iterative sketching. In Pilanci and Wainwright [23], it is shown that this algorithm achieves the minimax statistical risk given in (4.16) using at most $O(\log_2 n)$ iterations while obtaining equivalent speedups from sketching. We also note that the iterative sketching method can also be applied to more general convex optimization problems other than the least-squares objective. We refer the reader to Pilanci and Wainwright [38] for the application of sketching in solving general convex optimization problems.

4.6 Non-Parametric Problems

4.6.1 Non-Parametric Regression

In this section we discuss an extension of the sketching method to non-parametric regression problems over Hilbert spaces. The goal of non-parametric regression is making predictions of a continuous response after observing a covariate, where they are related via

$$y_i = f^*(x_i) + v_i, \tag{4.30}$$

where $\mathbf{v} \sim \mathcal{N}(0,\sigma^2\mathbf{I}_n)$, and the function $f^*(\mathbf{x})$ needs to be estimated from $\{x_i,y_i\}_{i=1}^n$. We will consider the well-studied case where the function f^* is assumed to belong to a reproducing kernel Hilbert space (RKHS) $\mathcal{H}$, and has a bounded Hilbert norm $\|f\|_{\mathcal{H}}$

[39, 40]. For these regression problems it is customary to consider the kernel ridge regression (KRR) problem based on convex optimization

$$\widehat{f} = \arg\min_{f \in \mathcal{H}} \left\{ \frac{1}{2n} \sum_{i=1}^{n} (y_i - f(x_i))^2 + \lambda \|f\|_{\mathcal{H}}^2 \right\}. \tag{4.31}$$

An RKHS is generated by a kernel function which is positive semidefinite (PSD). A PSD kernel is a symmetric function $K : \mathcal{X} \times \mathcal{X} \to \mathbb{R}$ that satisfies

$$\sum_{i,j=1}^{r} y_i y_j K(x_i, x_j) \geq 0$$

for all collections of points $\{x_1, ..., x_n\}$, $\{y_1, ..., y_n\}$ and $\forall r \in \mathbb{Z}_+$. The vector space of all functions of the form

$$f(\cdot) = \sum_{i}^{r} y_i K(\cdot, x_i)$$

generates an RKHS by taking closure of all such linear combinations. It can be shown that this RKHS is uniquely associated with the kernel function K (see Aronszajn [41] for details). Let us define a finite-dimensional kernel matrix $\mathbf{K}$ using n covariates as follows

$$\mathbf{K}_{ij} = \frac{1}{n} K(x_i, x_j),$$

which is a positive semidefinite matrix. In the linear least-squares regression the kernel matrix reduces to the Gram matrix given by $\mathbf{K} = \mathbf{A}^{\mathrm{T}}\mathbf{A}$. It is also known that the above infinite-dimensional program can be recast as a finite-dimensional quadratic optimization problem involving the kernel matrix

$$\widehat{w} = \arg\min_{\mathbf{w} \in \mathbb{R}^n} \frac{1}{2} \|\mathbf{K}\mathbf{w} - (1/\sqrt{n})\mathbf{y}\|_2^2 + \lambda \mathbf{w}^{\mathrm{T}}\mathbf{K}\mathbf{w} \tag{4.32}$$

$$= \arg\min_{\mathbf{w} \in \mathbb{R}^n} \frac{1}{2} \mathbf{w}^{\mathrm{T}}\mathbf{K}^2\mathbf{w} - \mathbf{w}^{\mathrm{T}} \frac{\mathbf{K}\mathbf{y}}{\sqrt{n}} + \lambda \mathbf{w}^{\mathrm{T}}\mathbf{K}\mathbf{w}, \tag{4.33}$$

and we can find the optimal solution to the infinite-dimensional problem (4.31) via the following relation:[9]

$$\widehat{f}(\cdot) = \frac{1}{n} \sum_{i=1}^{n} \widehat{w}_i K(\cdot, x_i). \tag{4.34}$$

We now define a kernel complexity measure that is based on the eigenvalues of the kernel matrix $\mathbf{K}$. Let $\lambda_1 \geq \lambda_2 \geq \cdots \geq \lambda_n$ correspond to the real eigenvalues of the symmetric positive-definite kernel matrix $\mathbf{K}$. The kernel complexity is defined as follows.

[9] Our definition of the kernel optimization problem slightly differs from the literature. The classical kernel problem can be recovered by a variable change $\mathbf{w}' = \mathbf{K}^{1/2} w$, where $\mathbf{K}^{1/2}$ is the matrix square root. We refer the reader to [40] for more details on kernel-based methods.

DEFINITION 4.1 (Kernel complexity)

$$\mathcal{R}(\delta) = \sqrt{\sum_{i=1}^{n} \min\{\delta^2, \lambda_i\}},$$

which is the sum of eigenvalues truncated at level δ. As in (4.14), *we define a critical radius $\delta^*(n)$ as the smallest positive solution $\delta^*(n) > 0$ to the following inequality:*

$$\frac{R(\delta)}{\delta\sqrt{n}} \leq \frac{\delta}{\sigma}, \tag{4.35}$$

where σ is the noise standard deviation in the statistical model (4.30). *The existence of a unique solution is guaranteed for all kernel classes (see Bartlett et al. [20]). The critical radius plays an important role in the minimax risk through an information-theoretic argument. The next theorem provides a lower bound on the statistical risk of any estimator applied to the observation model* (4.30).

THEOREM 4.4 *Given n i.i.d. samples from the model* (4.30), *any estimator $\widehat{f}$ has prediction error lower-bounded as*

$$\sup_{\|f^*\|_{\mathcal{H}} \leq 1} \mathbb{E}\frac{1}{n}\sum_{i=1}^{n}(\widehat{f}(x_i) - f^*(x_i))^2 \geq c_0\delta^*(n)^2, \tag{4.36}$$

where c_0 is a numerical constant and $\delta^*(n)$ is the critical radius defined in (4.35).

The lower bound given by Theorem 4.4 can be shown to be tight, and is achieved by the kernel-based optimization procedure (4.33) and (4.34) (see Bartlett *et al.* [20]). The proof of Theorem 4.4 can be found in Yang *et al.* [42]. We may define the *effective dimension $d^*(n)$* of the kernel via the relation

$$d^*(n) := n\delta^*(n)^2.$$

This definition allows us to interpret the convergence rate in (4.36) as $d^*(n)/n$, which resembles the classical parametric convergence rate where the number of variables is $d^*(n)$.

4.6.2 Sketching Kernels

Solving the optimization problem (4.33) becomes a computational challenge when the sample size n is large, since it involves linear algebraic operations on an $n \times n$ matrix $\mathbf{K}$. There is a large body of literature on approximating kernel matrices using randomized methods [43–46]. Here we assume that the matrix $\mathbf{K}$ is available, and a sketching matrix $\mathbf{S} \in \mathbb{R}^{m \times n}$ can be applied to form a randomized approximation of the kernel matrix. We will present an extension of (4.9), which achieves optimal statistical accuracy. Specifically, the sketching method we consider solves

$$\widehat{\mathbf{v}} = \arg\min_{\mathbf{v}\in\mathbb{R}^m} \frac{1}{2}\mathbf{v}^{\mathrm{T}}(\mathbf{SK})(\mathbf{KS}^{\mathrm{T}})\mathbf{v} - \mathbf{v}^{\mathrm{T}}\frac{\mathbf{SKy}}{\sqrt{n}} + \lambda\mathbf{v}^{\mathrm{T}}\mathbf{SKS}^{\mathrm{T}}\mathbf{v}, \tag{4.37}$$

which involves smaller-dimensional sketched kernel matrices $\mathbf{SK}$, $\mathbf{SKS}^{\mathrm{T}}$ and a lower-dimensional decision variable $\mathbf{v} \in \mathbb{R}^m$. Then we can recover the original variable via

$\mathbf{w} = \mathbf{S}^{\mathrm{T}}\mathbf{v}$. The next theorem shows that the sketched kernel-based optimization method achieves the optimal prediction error.

THEOREM 4.5 *Let $\mathbf{S} \in \mathbb{R}^{m \times n}$ be a Gaussian sketching matrix where $m \geq c_3 d_n$, and choose $\lambda = 3\delta^*(n)$. Given n i.i.d. samples from the model* (4.30), *the sketching procedure* (4.42) *produces a regression estimate $\widehat{f}$ which satisfies the bound*

$$\frac{1}{n}\sum_{i=1}^{n}(\widehat{f}(x_i) - f^*(x_i))^2 \leq c_2\delta^*(n)^2,$$

where $\delta^*(n)$ is the critical radius defined in (4.35).

A proof of this theorem can be found in Yang *et al.* [42]. We note that a similar result holds for the ROS sketch matrices with extra logarithmic terms in the dimension of the sketch, i.e., when $m \geq c_4 d_n \log^4(n)$ holds. Notably, Theorem 4.5 guarantees that the sketched estimator achieves the optimal error. This is in contrast to the lower-bound case in Section 4.4.6, where the sketching method does not achieve a minimax optimal error. This is due to the fact that the sketched problem in (4.37) is using the observation $\mathbf{SKy}$ instead of $\mathbf{Sy}$. Therefore, the lower bound in Section 4.4.6 does not apply for this construction. It is worth noting that one can formulate the ordinary least-squares case as a kernel regression problem with kernel $K = \mathbf{A}\mathbf{A}^{\mathrm{T}}$, and then apply the sketching method (4.37), which is guaranteed to achieve the minimax optimal risk. However, computing the kernel matrix $\mathbf{A}\mathbf{A}^{\mathrm{T}}$ would cost $O(nd^2)$ operations, which is more than would be required for solving the original least-squares problem.

We note that some kernel approximation methods avoid computing the kernel matrix $\mathbf{K}$ and directly form low-rank approximations. We refer the reader to [43] for an example, which also provides an error guarantee for the approximate kernel.

4.7 Extensions: Privacy and Communication Complexity

4.7.1 Privacy and Information-Theoretic Bounds

Another interesting property of randomized sketching is privacy preservation in the context of optimization and learning. Privacy properties of random projections for various statistical tasks have been studied in the recent literature [10, 11, 47]. It is of great theoretical and practical interest to characterize fundamental privacy and optimization trade-offs of randomized algorithms. We first show the relation between sketching and a mutual information-based privacy measure.

4.7.2 Mutual Information Privacy

Suppose we model the data matrix $\mathbf{A} \in \mathbb{R}^{n \times d}$ as stochastic, where each entry is drawn randomly. One way we can assess the information revealed to the server is considering the mutual information per symbol, which is given by the formula

$$\frac{I(\mathbf{SA};\mathbf{A})}{nd} = \frac{1}{nd}\{H(\mathbf{A}) - H(\mathbf{A}|\mathbf{SA})\}$$
$$= \frac{1}{nd}D(\mathbb{P}_{\mathbf{SA},\mathbf{A}}||\mathbb{P}_{\mathbf{SA}}\mathbb{P}_{\mathbf{A}}),$$

where we normalize by nd since the data matrix $\mathbf{A}$ has nd entries in total. The following corollary is a direct application of Theorem 4.1.

COROLLARY 4.1 *Let the entries of the matrix* $\mathbf{A}$ *be i.i.d from an arbitrary distribution with finite variance* σ^2. *Using sketched data, we can obtain an* ϵ*-approximate*[10] *solution to the optimization problem while ensuring that the revealed mutual information satisfies*

$$\frac{I(\mathbf{SA};\mathbf{A})}{nd} \leq \frac{c_0}{\epsilon^2}\frac{\mathcal{W}^2(\mathbf{A}\mathcal{K})}{n}\log_2(2\pi e\sigma^2).$$

Therefore, we can guarantee the mutual information privacy of the sketching-based methods, whenever the term $\mathcal{W}(\mathbf{A}\mathcal{K})$ is small.

An alternative and popular characterization of privacy is referred to as the differential privacy (see Dwork *et al.* [9]), where other randomized methods, such as additive noise for preserving privacy, were studied. It is also possible to directly analyze differential privacy-preserving aspects of random projections as considered in Blocki *et al.* [10].

4.7.3 Optimization-Based Privacy Attacks

We briefly discuss a possible approach an adversary might take to circumvent the privacy provided by sketching. If the data matrix is sparse, then one might consider optimization-based recovery techniques borrowed from compressed sensing to recover the data $\mathbf{A}$ given the sketched data $\widetilde{\mathbf{A}} = \mathbf{SA}$,

$$\min_{\mathbf{A}} ||\mathbf{A}||_1$$
$$\text{s.t. } \mathbf{SA} = \widetilde{\mathbf{A}},$$

where we have used the matrix ℓ_1 norm $||\mathbf{A}||_1 := \sum_{i=1}^{n}\sum_{j=1}^{d}|\mathbf{A}_{ij}|$. The success of the above optimization method will critically depend on the sparsity level of the original data $\mathbf{A}$. Most of the randomized sketching constructions shown in Section 4.2 can be shown to be susceptible to data recovery via optimization (see Candès and Tao [25] and Candès *et al.* [48]). However, this method assumes that the sketching matrix $\mathbf{S}$ is available to the attacker. If $\mathbf{S}$ is not available to the adversary, then the above method cannot be used and the recovery is not straightforward.

4.7.4 Communication Complexity-Space Lower Bounds

In this section we consider a streaming model of computation, where the algorithm is allowed to make only one pass over the data. In this model, an algorithm receives updates to the entries of the data matrix $\mathbf{A}$ in the form "add a to $\mathbf{A}_{ij}$." An entry can

[10] Here ϵ-approximate solution refers to the approximation defined in Theorem 4.1, relative to the optimal value.

be updated more than once, and the value a is any arbitrary real number. The sketches introduced in this chapter provide a valuable data structure when the matrix is very large in size, and storing and updating the matrix directly can be impractical. Owing to the linearity of sketches, we can update the sketch $\mathbf{SA}$ by adding $a\mathbf{S}\mathbf{e}_i\mathbf{e}_j^{\mathrm{T}}$ to $\mathbf{SA}$, and maintain an approximation with limited memory.

The following theorem due to Clarkson and Woodruff [49] provides a lower bound of the space used by any algorithm for least-squares regression which performs a single pass over the data.

THEOREM 4.6 *Any randomized 1-pass algorithm which returns an ϵ-approximate solution to the unconstrained least-squares problem with probability at least $7/9$ needs $\Omega(d^2(1/\epsilon + \log(nd)))$ bits of space.*

This theorem confirms that the space complexity of sketching for unconstrained least-squares regression is near optimal. Because of the choice of the sketching dimension $m = O(d)$, the space used by the sketch $\mathbf{SA}$ is $O(d^2)$, which is optimal up to constants according to the theorem.

4.8 Numerical Experiments

In this section, we illustrate the sketching method numerically and confirm the theoretical predictions of Theorems 4.1 and 4.3. We consider both the classical low-dimensional statistical regime where $n > d$, and the ℓ_1-constrained least-squares minimization known as LASSO (see Tibshirani [51]):

$$\mathbf{x}^* = \arg \min_{\mathbf{x}\, s.t.\, \|\mathbf{x}\|_1 \le \lambda} \|\mathbf{Ax} - \mathbf{b}\|_2 .$$

We generate a random i.i.d. data matrix $\mathbf{A} \in \mathbb{R}^{n \times d}$, where $n = 10\,000$ and $d = 1000$, and set the observation vector $\mathbf{b} = \mathbf{Ax}^\dagger + \sigma\mathbf{w}$, where $\mathbf{x}^\dagger \in \{-1, 0, 1\}^d$ is a random s-sparse vector and $\mathbf{w}$ has i.i.d. $\mathcal{N}(0, 10^{-4})$ components. For the sketching matrix $\mathbf{S} \in \mathbb{R}^{m \times n}$, we consider the Gaussian and Rademacher (± 1 i.i.d.-valued) random matrices, where m ranges between 10 and 400. Consequently, we solve the sketched program

$$\widehat{\mathbf{x}} = \arg \min_{\mathbf{x}\, s.t.\, \|\mathbf{x}\|_1 \le \lambda} \|\mathbf{SAx} - \mathbf{Sb}\|_2 .$$

Figures 4.5 and 4.6 show the relative prediction mean-squared error given by the ratio

$$\frac{(1/n)\|\mathbf{A}(\widehat{\mathbf{x}} - \mathbf{x}^\dagger)\|_2^2}{(1/n)\|\mathbf{A}(\mathbf{x}^* - \mathbf{x}^\dagger)\|_2^2},$$

where it is averaged over 20 realizations of the sketching matrix, and $\widehat{\mathbf{x}}$ and $\mathbf{x}^*$ are the sketched and the original solutions, respectively. As predicted by the upper and lower bounds given in Theorems 4.1 and 4.3, the prediction mean-squared error of the sketched estimator scales as $O((s \log d)/m)$, since the corresponding Gaussian complexity $\mathcal{W}_1(\mathcal{K})^2$ is $O(s \log d)$. These plots reveal that the prediction mean-squared error of the sketched estimators for both Gaussian and Rademacher sketches are in agreement with the theory.

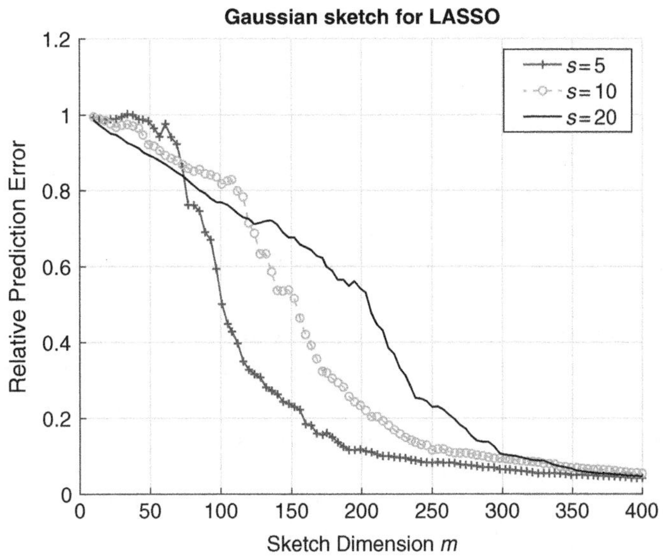

Figure 4.5 Sketching LASSO using Gaussian random projections.

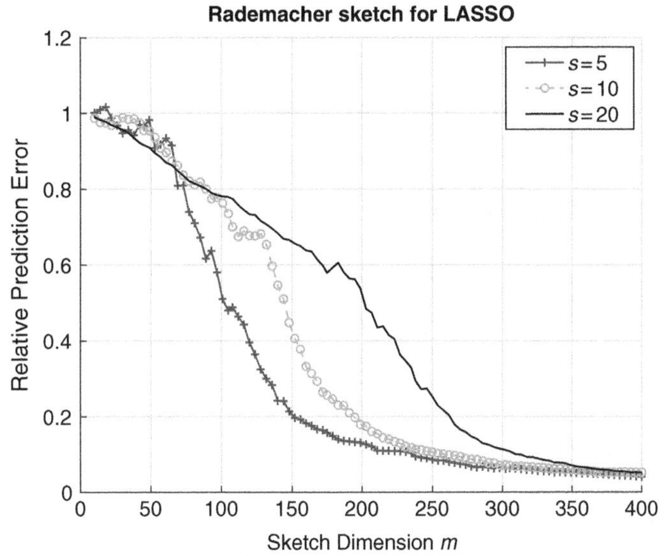

Figure 4.6 Sketching LASSO using Rademacher random projections.

4.9 Conclusion

This chapter presented an overview of random projection based methods for solving large-scale statistical estimation and constrained optimization problems. We investigated fundamental lower bounds on the performance of sketching using information-theoretic tools. Randomized sketching has interesting theoretical properties, and also has numerous practical advantages in machine-learning and optimization problems. Sketching yields faster algorithms with lower space complexity while maintaining strong approximation guarantees.

For the upper bound on the approximation accuracy in Theorem 4.2, Gaussian complexity plays an important role, and also provides a geometric characterization of the dimension of the sketch. The lower bounds given in Theorem 4.3 are statistical in nature, and involve packing numbers, and consequently metric entropy, which measures the complexity of the sets. The upper bounds on the Gaussian sketch can be extended to Rademacher sketches, sub-Gaussian sketches, and randomized orthogonal system sketches (see Pilanci and Wainwright [22] and also Yun *et al.* [42] for the proofs). However, the results for non-Gaussian sketches often involve superfluous logarithmic factors and large constants as artifacts of the analysis. As can be observed in Figs. 4.5 and 4.6, the mean-squared error curves for Gaussian and Rademacher sketches are in agreement with each other. It can be conjectured that the approximation ratio of sketching is universal for random matrices with entries sampled from well-behaved distributions. This is an important theoretical question for future research. We refer the reader to the work of Donoho and Tanner [51] for observations of the universality in compressed sensing.

Finally, a number of important limitations of the analysis techniques need to be considered. The minimax criterion (4.16) is a worst-case criterion in nature by virtue of its definition, and may not correctly reflect the average error of sketching when the unknown vector $\mathbf{x}^\dagger$ is randomly distributed. Furthermore, in some applications, it might be suitable to consider prior information on the unknown vector. As an interesting direction of future research, it would be interesting to study lower bounds for sketching in a Bayesian setting.

A4.1 Proof of Theorem 4.3

Let us define the shorthand notation $\|\cdot\|_{\mathbf{A}} := (1/\sqrt{n})\|\mathbf{A}(\cdot)\|_2$. Let $\{\mathbf{z}^j\}_{j=1}^M$ be a 1/2-packing of $C_0 \cap \mathbb{B}_A(1)$ in the semi-norm defined by $\|\cdot\|_{\mathbf{A}}$, and, for a fixed $\delta \in (0, 1/4)$, define $\mathbf{x}^j = 4\delta\mathbf{z}^j$. Since $4\delta \in (0,1)$, the star-shaped assumption guarantees that each $\mathbf{x}^j$ belongs to C_0. We thus obtain a collection of M vectors in C_0 such that

$$2\delta \leq \|\mathbf{x}^j - \mathbf{x}^k\|_{\mathbf{A}} \leq 8\delta \qquad \text{for all } j \neq k.$$

Letting J be a random index uniformly distributed over $\{1, \ldots, M\}$, suppose that, conditionally on $J = j$, we observe the sketched observation vector $\mathbf{Sb} = \mathbf{SAx}^j + \mathbf{Sw}$, as well as the sketched matrix $\mathbf{SA}$. Conditioned on $J = j$, the random vector $\mathbf{Sb}$ follows an $\mathcal{N}(\mathbf{SAx}^j, \sigma^2\mathbf{SS}^T)$ distribution, denoted by $\mathbb{P}_{\mathbf{x}^j}$. We let $\bar{Y}$ denote the resulting mixture variable, with distribution $(1/M)\sum_{j=1}^M \mathbb{P}_{\mathbf{x}^j}$.

Consider the multi-way testing problem of determining the index J by observing $\bar{Y}$. With this setup, we may apply Lemma 4.3 (see, e.g., [30, 46]), which implies that, for any estimator $\mathbf{x}^\dagger$, the worst-case mean-squared error is lower-bounded as

$$\sup_{\mathbf{x}^* \in C} \mathbb{E}_{\mathbf{S},\mathbf{w}} \|\mathbf{x}^\dagger - \mathbf{x}^*\|_{\mathbf{A}}^2 \geq \delta^2 \inf_{\psi} \mathbb{P}[\psi(\bar{Y}) \neq J], \tag{4.38}$$

where the infimum ranges over all testing functions ψ. Consequently, it suffices to show that the testing error is lower-bounded by 1/2.

In order to do so, we first apply Fano's inequality [27] conditionally on the sketching matrix $\mathbf{S}$ and get

$$\mathbb{P}[\psi(\bar{Y}) \neq J] = \mathbb{E}_{\mathbf{S}}\left\{\mathbb{P}[\psi(\bar{Y}) \neq J \mid \mathbf{S}]\right\} \geq 1 - \frac{\mathbb{E}_{\mathbf{S}}[I_{\mathbf{S}}(\bar{Y};J)] + 1}{\log_2 M}, \tag{4.39}$$

where $I_{\mathbf{S}}(\bar{Y};J)$ denotes the mutual information between $\bar{Y}$ and J with $\mathbf{S}$ fixed. Our next step is to upper-bound the expectation $\mathbb{E}_{\mathbf{S}}[I(\bar{Y};J)]$.

Letting $D(\mathbb{P}_{\mathbf{x}^j} \,\|\, \mathbb{P}_{\mathbf{x}^k})$ denote the Kullback–Leibler (KL) divergence between the distributions $\mathbb{P}_{\mathbf{x}^j}$ and $\mathbb{P}_{\mathbf{x}^k}$, the convexity of KL divergence implies that

$$\begin{aligned} I_{\mathbf{S}}(\bar{Y};J) &= \frac{1}{M}\sum_{j=1}^{M} D\left(\mathbb{P}_{\mathbf{x}^j} \,\|\, \frac{1}{M}\sum_{k=1}^{M}\mathbb{P}_{\mathbf{x}^k}\right) \\ &\leq \frac{1}{M^2}\sum_{j,k=1}^{M} D(\mathbb{P}_{\mathbf{x}^j} \,\|\, \mathbb{P}_{\mathbf{x}^k}). \end{aligned}$$

Computing the KL divergence for Gaussian vectors yields

$$I_{\mathbf{S}}(\bar{Y};J) \leq \frac{1}{M^2}\sum_{j,k=1}^{M} \frac{1}{2\sigma^2}(\mathbf{x}^j - \mathbf{x}^k)^{\mathrm{T}}\mathbf{A}^{\mathrm{T}}\left[\mathbf{S}^{\mathrm{T}}(\mathbf{S}\mathbf{S}^{\mathrm{T}})^{-1}\mathbf{S}\right]\mathbf{A}(\mathbf{x}^j - \mathbf{x}^k).$$

Thus, using condition (4.15), we have

$$\mathbb{E}_{\mathbf{S}}[I(\bar{Y};J)] \leq \frac{1}{M^2}\sum_{j,k=1}^{M}\frac{m\eta}{2n\sigma^2}\|\mathbf{A}(\mathbf{x}^j - \mathbf{x}^k)\|_2^2 \leq \frac{32m\eta}{\sigma^2}\delta^2,$$

where the final inequality uses the fact that $\|\mathbf{x}^j - \mathbf{x}^k\|_{\mathbf{A}} = 1/\sqrt{n}\|\mathbf{A}(\mathbf{x}^j - \mathbf{x}^k)\|_2 \leq 8\delta$ for all pairs.

Combined with our previous bounds (4.38) and (4.39), we find that

$$\sup_{\mathbf{x}^* \in C}\mathbb{E}\|\widehat{\mathbf{x}} - \mathbf{x}^*\|_2^2 \geq \delta^2\left\{1 - \frac{32(m\eta\delta^2/\sigma^2) + 1}{\log_2 M}\right\}.$$

Setting $\delta = \sigma^2 \log_2(M/2)/64\eta m$ yields the lower bound (4.19).

A4.2 Proof of Lemma 4.3

By Markov's inequality applied on the random variable $\|\widehat{\mathbf{x}} - \mathbf{x}^\dagger\|_{\mathbf{A}}^2$ we have

$$\mathbb{E}\|\widehat{\mathbf{x}} - \mathbf{x}^\dagger\|_{\mathbf{A}}^2 \geq \delta^2\,\mathbb{P}[\|\widehat{\mathbf{x}} - \mathbf{x}^\dagger\|_{\mathbf{A}}^2 \geq \delta^2]. \tag{4.40}$$

Now note that

$$\begin{aligned} \sup_{\mathbf{x}^* \in C}\mathbb{P}[\|\widehat{\mathbf{x}} - \mathbf{x}^\dagger\|_{\mathbf{A}} \geq \delta] &\geq \max_{j \in \{1,\ldots,M\}} \mathbb{P}[\,\|\widehat{\mathbf{x}} - \mathbf{x}^{(j)}\|_{\mathbf{A}} \geq \delta \mid J_\delta = j] \\ &\geq \frac{1}{M}\sum_{j=1}^{M}\mathbb{P}[\|\widehat{\mathbf{x}} - \mathbf{x}^{(j)}\|_{\mathbf{A}} \geq \delta \mid J_\delta = j], \end{aligned} \tag{4.41}$$

since every element of the packing set satisfies $\mathbf{x}^{(j)} \in C$ and the discrete maximum is upper-bounded by the average over $\{1,...,M\}$. Since we have $\mathbb{P}[J_\delta = j] = 1/M$, we equivalently have

$$\frac{1}{M}\sum_{j=1}^{M}\mathbb{P}[\|\widehat{\mathbf{x}}-\mathbf{x}^{(j)}\|_{\mathbf{A}}] \geq \delta \,|\, J_\delta = j] = \sum_{j=1}^{M}\mathbb{P}\Big[\|\widehat{\mathbf{x}}-\mathbf{x}^{(j)}\|_{\mathbf{A}} \geq \delta \,\Big|\, J_\delta = j\Big]\mathbb{P}[J_\delta = j]$$
$$= \mathbb{P}\Big[\|\widehat{\mathbf{x}}-\mathbf{x}^{(J_\delta)}\|_{\mathbf{A}} \geq \delta\Big]. \tag{4.42}$$

Now we will argue that, whenever the true index is $J_\delta = j$ and if $\|\widehat{x}-x^{(j)}\|_{\mathbf{A}} < \delta$, then we can form a hypothesis tester $\psi(Z)$ identifying the true index j. Consider the test

$$\psi(Z) := \arg\min_{j\in[M]} \|\mathbf{x}^{(j)} - \widehat{\mathbf{x}}\|_{\mathbf{A}}.$$

Now note that $\|\mathbf{x}^j - \widehat{\mathbf{x}}\|_{\mathbf{A}} < \delta$ ensures that

$$\|\mathbf{x}^{(i)} - \widehat{\mathbf{x}}\|_{\mathbf{A}} \geq \|\mathbf{x}^{(i)} - \mathbf{x}^{(j)}\|_{\mathbf{A}} - \|\mathbf{x}^{(j)} - \widehat{\mathbf{x}}\|_{\mathbf{A}} \geq 2\delta - \delta = \delta,$$

where the second inequality follows from the 2δ-packing construction of our collection $\mathbf{x}^{(1)},...,\mathbf{x}^{(M)}$. Consequently $\|\mathbf{x}^{(i)} - \widehat{\mathbf{x}}\|_{\mathbf{A}} > \delta$ for all $i \in \{1,...,N\} - \{j\}$, and the test $\psi(Z)$ identifies the true index $J = j$. Therefore we obtain

$$\Big\{\|\mathbf{x}^{(j)} - \widehat{\mathbf{x}}\|_{\mathbf{A}} < \delta\Big\} \quad \Rightarrow \quad \{\phi(Z) = j\},$$

and conclude that the complements of these events obey

$$\mathbb{P}\Big[\|\mathbf{x}^{(j)} - \widehat{\mathbf{x}}\|_{\mathbf{A}} \geq \delta \,|\, J_\delta = j\Big] \geq \mathbb{P}[\phi(Z) \neq j \,|\, J_\delta = j].$$

Taking averages over the indices $1,...,M$, we obtain

$$\mathbb{P}\Big[\|\mathbf{x}^{(J_\delta)} - \widehat{\mathbf{x}}\|_{\mathbf{A}} \geq \delta\Big] = \frac{1}{M}\sum_{j=1}^{M}\mathbb{P}\Big[\|\mathbf{x}^{(j)} - \widehat{\mathbf{x}}\|_{\mathbf{A}} \geq \delta \,|\, J_\delta = j\Big] \geq \mathbb{P}[\phi(Z) \neq J_\delta].$$

Combining the above with the earlier lower bound (4.41) and the identity (4.42), we obtain

$$\sup_{\mathbf{x}^*\in C}\mathbb{P}[\|\widehat{\mathbf{x}} - \mathbf{x}^*\|_{\mathbf{A}} \geq \delta] \geq \mathbb{P}[\phi(Z) \neq J_\delta] \geq \inf_{\phi}\mathbb{P}[\phi(Z) \neq J_\delta],$$

where the second inequality follows by taking the infimum over all tests, which can only make the probability smaller. Plugging in the above lower bound in (4.40) completes the proof of the lemma.

References

[1] S. Vempala, *The random projection method.* American Mathematical Society, 2004.

[2] E. J. Candès and T. Tao, "Near-optimal signal recovery from random projections: Universal encoding strategies?" *IEEE Trans. Information Theory*, vol. 52, no. 12, pp. 5406–5425, 2006.

[3] N. Halko, P. Martinsson, and J. A. Tropp, "Finding structure with randomness: Probabilistic algorithms for constructing approximate matrix decompositions," *SIAM Rev.*, vol. 53, no. 2, pp. 217–288, 2011.
[4] M. W. Mahoney, *Randomized algorithms for matrices and data.* Now Publishers, 2011.
[5] D. P. Woodruff, "Sketching as a tool for numerical linear algebra," *Foundations and Trends Theoretical Computer Sci.*, vol. 10, nos. 1–2, pp. 1–157, 2014.
[6] S. Muthukrishnan, "Data streams: Algorithms and applications," *Foundations and Trends Theoretical Computer Sci.*, vol. 1, no. 2, pp. 117–236, 2005.
[7] B. Yu, "Assouad, Fano, and Le Cam," in *Festschrift in Honor of Lucien Le Cam.* Springer, 1997, pp. 423–435.
[8] Y. Yang and A. Barron, "Information-theoretic determination of minimax rates of convergence," *Annals Statist.*, vol. 27, no. 5, pp. 1564–1599, 1999.
[9] C. Dwork, F. McSherry, K. Nissim, and A. Smith, "Calibrating noise to sensitivity in private data analysis," in *Proc. Theory of Cryptography Conference*, 2006, pp. 265–284.
[10] J. Blocki, A. Blum, A. Datta, and O. Sheffet, "The Johnson–Lindenstrauss transform itself preserves differential privacy," in *Proc. 2012 IEEE 53rd Annual Symposium on Foundations of Computer Science*, 2012, pp. 410–419.
[11] N. Ailon and B. Chazelle, "Approximate nearest neighbors and the fast Johnson-Lindenstrauss transform," in *Proc. 38th Annual ACM Symposium on Theory of Computing*, 2006, pp. 557–563.
[12] P. Drineas and M. W. Mahoney, "Effective resistances, statistical leverage, and applications to linear equation solving," *arXiv:1005.3097*, 2010.
[13] D. A. Spielman and N. Srivastava, "Graph sparsification by effective resistances," *SIAM J. Computing*, vol. 40, no. 6, pp. 1913–1926, 2011.
[14] M. Charikar, K. Chen, and M. Farach-Colton, "Finding frequent items in data streams," in *International Colloquium on Automata, Languages, and Programming*, 2002, pp. 693–703.
[15] D. M. Kane and J. Nelson, "Sparser Johnson–Lindenstrauss transforms," *J. ACM*, vol. 61, no. 1, article no. 4, 2014.
[16] J. Nelson and H. L. Nguyên, "Osnap: Faster numerical linear algebra algorithms via sparser subspace embeddings," in *Proc. 2013 IEEE 54th Annual Symposium on Foundations of Computer Science (FOCS)*, 2013, pp. 117–126.
[17] J. Hiriart-Urruty and C. Lemaréchal, *Convex analysis and minimization algorithms.* Springer, 1993, vol. 1.
[18] S. Boyd and L. Vandenberghe, *Convex optimization.* Cambridge University Press, 2004.
[19] M. Ledoux and M. Talagrand, *Probability in Banach spaces: Isoperimetry and processes.* Springer, 1991.
[20] P. L. Bartlett, O. Bousquet, and S. Mendelson, "Local Rademacher complexities," *Annals Statist.*, vol. 33, no. 4, pp. 1497–1537, 2005.
[21] V. Chandrasekaran, B. Recht, P. A. Parrilo, and A. S. Willsky, "The convex geometry of linear inverse problems," *Foundations Computational Math.*, vol. 12, no. 6, pp. 805–849, 2012.
[22] M. Pilanci and M. J. Wainwright, "Randomized sketches of convex programs with sharp guarantees," UC Berkeley, Technical Report, 2014, full-length version at *arXiv:1404.7203*; Presented in part at ISIT 2014.
[23] M. Pilanci and M. J. Wainwright, "Iterative Hessian sketch: Fast and accurate solution approximation for constrained least-squares," *J. Machine Learning Res.*, vol. 17, no. 1, pp. 1842–1879, 2016.

[24] S. Chen, D. L. Donoho, and M. A. Saunders, "Atomic decomposition by basis pursuit," *SIAM J. Sci. Computing*, vol. 20, no. 1, pp. 33–61, 1998.

[25] E. J. Candès and T. Tao, "Decoding by linear programming," *IEEE Trans. Information Theory*, vol. 51, no. 12, pp. 4203–4215, 2005.

[26] R. M. Fano and W. Wintringham, "Transmission of information," *Phys. Today*, vol. 14, p. 56, 1961.

[27] T. Cover and J. Thomas, *Elements of information theory*. John Wiley & Sons, 1991.

[28] P. Assouad, "Deux remarques sur l'estimation," *Comptes Rendus Acad. Sci. Paris*, vol. 296, pp. 1021–1024, 1983.

[29] I. A. Ibragimov and R. Z. Has'minskii, *Statistical estimation: Asymptotic theory*. Springer, 1981.

[30] L. Birgé, "Estimating a density under order restrictions: Non-asymptotic minimax risk," *Annals Statist.*, vol. 15, no. 3, pp. 995–1012, 1987.

[31] A. Kolmogorov and B. Tikhomirov, "ϵ-entropy and ϵ-capacity of sets in functional spaces," *Uspekhi Mat. Nauk*, vol. 86, pp. 3–86, 1959, English transl. *Amer. Math. Soc. Translations*, vol. 17, pp. 277–364, 1961.

[32] R. Tibshirani, "Regression shrinkage and selection via the Lasso," *J. Roy. Statist. Soc. Ser. B*, vol. 58, no. 1, pp. 267–288, 1996.

[33] G. Raskutti, M. J. Wainwright, and B. Yu, "Minimax rates of estimation for high-dimensional linear regression over ℓ_q-balls," *IEEE Trans. Information Theory*, vol. 57, no. 10, pp. 6976–6994, 2011.

[34] N. Srebro, N. Alon, and T. S. Jaakkola, "Generalization error bounds for collaborative prediction with low-rank matrices," in *Proc. Advances in Neural Information Processing Systems*, 2005, pp. 1321–1328.

[35] M. Yuan and Y. Lin, "Model selection and estimation in regression with grouped variables," *J. Roy. Statist. Soc. B*, vol. 1, no. 68, p. 49, 2006.

[36] S. Negahban and M. J. Wainwright, "Estimation of (near) low-rank matrices with noise and high-dimensional scaling," *Annals Statist.*, vol. 39, no. 2, pp. 1069–1097, 2011.

[37] F. Bunea, Y. She, and M. Wegkamp, "Optimal selection of reduced rank estimators of high-dimensional matrices," *Annals Statist.*, vol. 39, no. 2, pp. 1282–1309, 2011.

[38] M. Pilanci and M. J. Wainwright, "Newton sketch: A near linear-time optimization algorithm with linear-quadratic convergence," *SIAM J. Optimization*, vol. 27, no. 1, pp. 205–245, 2017.

[39] H. L. Weinert, (ed.), *Reproducing kernel hilbert spaces: Applications in statistical signal processing*. Hutchinson Ross Publishing Co., 1982.

[40] B. Schölkopf and A. Smola, *Learning with kernels*. MIT Press, 2002.

[41] N. Aronszajn, "Theory of reproducing kernels," *Trans. Amer. Math. Soc.*, vol. 68, pp. 337–404, 1950.

[42] Y. Yang, M. Pilanci, and M. J. Wainwright, "Randomized sketches for kernels: Fast and optimal nonparametric regression," *Annals Statist.*, vol. 45, no. 3, pp. 991–1023, 2017.

[43] A. Rahimi and B. Recht, "Random features for large-scale kernel machines," in *Proc. Advances in Neural Information Processing Systems*, 2008, pp. 1177–1184.

[44] A. Rahimi and B. Recht, "Weighted sums of random kitchen sinks: Replacing minimization with randomization in learning," in *Proc. Advances in Neural Information Processing Systems*, 2009, pp. 1313–1320.

[45] P. Drineas and M. W. Mahoney, "On the Nyström method for approximating a Gram matrix for improved kernel-based learning," *J. Machine Learning Res.*, vol. 6, no. 12, pp. 2153–2175, 2005.

[46] Q. Le, T. Sarlós, and A. Smola, "Fastfood – approximating kernel expansions in loglinear time," in *Proc. 30th International Conference on Machine Learning*, 2013, 9 unnumbered pages.

[47] S. Zhou, J. Lafferty, and L. Wasserman, "Compressed regression," *IEEE Trans. Information Theory*, vol. 55, no. 2, pp. 846–866, 2009.

[48] E. J. Candès, J. Romberg, and T. Tao, "Robust uncertainty principles: Exact signal reconstruction from highly incomplete frequency information," *IEEE Trans. Information Theory*, vol. 52, no. 2, pp. 489–509, 2004.

[49] K. L. Clarkson and D. P. Woodruff, "Numerical linear algebra in the streaming model," in *Proceedings of the Forty-First Annual ACM Symposium on Theory of Computing*, 2009, pp. 205–214.

[50] R. Tibshirani, "Regression shrinkage and selection via the lasso," *J. Roy. Statist. Soc. Ser. B (Methodological)*, vol. 58, no. 1, pp. 267–288, 1996.

[51] D. Donoho and J. Tanner, "Observed universality of phase transitions in high-dimensional geometry, with implications for modern data analysis and signal processing," *Phil. Trans. Roy. Soc. London A: Math., Phys. Engineering Sci.*, vol. 367, no. 1906, pp. 4273–4293, 2009.

5 Sample Complexity Bounds for Dictionary Learning from Vector- and Tensor-Valued Data

Zahra Shakeri, Anand D. Sarwate, and Waheed U. Bajwa

Summary

During the last decade, dictionary learning has emerged as one of the most powerful methods for data-driven extraction of features from data. While the initial focus on dictionary learning had been from an algorithmic perspective, recent years have seen an increasing interest in understanding the theoretical underpinnings of dictionary learning. Many such results rely on the use of information-theoretic analytic tools and help us to understand the fundamental limitations of different dictionary-learning algorithms. This chapter focuses on the theoretical aspects of dictionary learning and summarizes existing results that deal with dictionary learning both from vector-valued data and from tensor-valued (i.e., multi-way) data, which are defined as data having multiple modes. These results are primarily stated in terms of lower and upper bounds on the sample complexity of dictionary learning, defined as the number of samples needed to identify or reconstruct the true dictionary underlying data from noiseless or noisy samples, respectively. Many of the analytic tools that help yield these results come from the information-theory literature; these include restating the dictionary-learning problem as a channel coding problem and connecting the analysis of minimax risk in statistical estimation Fano's inequality. In addition to highlighting the effects of different parameters on the sample complexity of dictionary learning, this chapter also brings out the potential advantages of dictionary learning from tensor data and concludes with a set of open problems that remain unaddressed for dictionary learning.

5.1 Introduction

Modern machine learning and signal processing relies on finding meaningful and succinct representations of data. Roughly speaking, data representation entails transforming "raw" data from its original domain to another domain in which it can be processed more effectively and efficiently. In particular, the performance of any information-processing algorithm is dependent on the representation it is built on [1]. There are two major approaches to data representation. In *model-based approaches*, a *predetermined* basis is used to transform data. Such a basis can be formed using predefined transforms such as the Fourier transform [2], wavelets [3], and curvelets [4]. The *data-driven approach* infers transforms from the data to yield efficient representations. Prior works on data representation show that data-driven techniques generally outperform model-based techniques as the learned transformations are tuned to the input signals [5, 6].

Since contemporary data are often high-dimensional and high-volume, we need efficient algorithms to manage them. In addition, rapid advances in sensing and data-acquisition technologies in recent years have resulted in individual data samples or signals with *multimodal* structures. For example, a single observation may contain measurements from a two-dimensional array over time, leading to a data sample with three modes. Such data are often termed *tensors* or multi-way arrays [7]. Specialized algorithms can take advantage of this tensor structure to handle multimodal data more efficiently. These algorithms represent tensor data using fewer parameters than vector-valued data-representation methods by means of tensor decomposition techniques [8–10], resulting in reduced computational complexity and storage costs [11–15].

In this chapter, we focus on data-driven representations. As data-collection systems grow and proliferate, we will need efficient data representations for processing, storage, and retrieval. Data-driven representations have successfully been used for signal processing and machine-learning tasks such as data compression, recognition, and classification [5, 16, 17]. From a theoretical standpoint, there are several interesting questions surrounding data-driven representations. Assuming there is an unknown generative model forming a "true" representation of data, these questions include the following. (1) What algorithms can be used to learn the representation effectively? (2) How many data samples are needed in order for us to learn the representation? (3) What are the fundamental limits on the number of data samples needed in order for us to learn the representation? (4) How robust are the solutions addressing these questions to parameters such as noise and outliers? In particular, state-of-the-art data-representation algorithms have excellent empirical performance but their non-convex geometry makes analyzing them challenging.

The goal of this chapter is to provide a brief overview of some of the aforementioned questions for a class of data-driven-representation methods known as *dictionary learning* (DL). Our focus here will be both on the vector-valued and on the tensor-valued (i.e., multidimensional/multimodal) data cases.

5.1.1 Dictionary Learning: A Data-Driven Approach to Sparse Representations

Data-driven methods have a long history in representation learning and can be divided into two classes. The first class includes linear methods, which involve transforming (typically vector-valued) data using linear functions to exploit the latent structure in data [5, 18, 19]. From a geometric point of view, these methods effectively learn a low-dimensional subspace and projection of data onto that subspace, given some constraints. Examples of classical linear approaches for vector-valued data include principal component analysis (PCA) [5], linear discriminant analysis (LDA) [18], and independent component analysis (ICA) [19].

The second class consists of nonlinear methods. Despite the fact that historically linear representations have been preferred over nonlinear methods because of their lesser computational complexity, recent advances in available analytic tools and computational power have resulted in an increased interest in nonlinear representation learning. These techniques have enhanced performance and interpretability compared with linear techniques. In nonlinear methods, data are transformed into a higher-dimensional space, in which they lie on a low-dimensional manifold [6, 20–22]. In the world of nonlinear transformations, nonlinearity can take different forms. In manifold-based methods such as diffusion maps, data are projected onto a nonlinear manifold [20]. In kernel (nonlinear) PCA, data are projected onto a subspace in a higher-dimensional space [21]. Auto-encoders encode data according to the desired task [22]. DL uses a union of subspaces as the underlying geometric structure and projects input data onto one of the learned subspaces in the union. This leads to sparse representations of the data, which can be represented in the form of an overdetermined matrix multiplied by a sparse vector [6]. Although nonlinear representation methods result in non-convex formulations, we can often take advantage of the problem structure to guarantee the existence of a unique solution and hence an optimal representation.

Focusing specifically on DL, it is known to have slightly higher computational complexity than linear methods, but it surpasses their performance in applications such as image denoising and inpainting [6], audio processing [23], compressed sensing [24], and data classification [17, 25]. More specifically, given input training signals $\mathbf{y} \in \mathbb{R}^m$, the goal in DL is to construct a basis such that $\mathbf{y} \approx \mathbf{D}\mathbf{x}$. Here, $\mathbf{D} \in \mathbb{R}^{m \times p}$ is denoted as the dictionary that has unit-norm columns and $\mathbf{x} \in \mathbb{R}^p$ is the dictionary coefficient vector that has a few non-zero entries. While the initial focus in DL had been on algorithmic development for various problem setups, works in recent years have also provided fundamental analytic results that help us understand the fundamental limits and performance of DL algorithms for both vector-valued [26–33] and tensor-valued [12, 13, 15] data.

There are two paradigms in the DL literature: the dictionary can be assumed to be a complete or an overcomplete basis (effectively, a frame [34]). In both cases, columns of the dictionary span the entire space [27]; in complete dictionaries, the dictionary matrix is square ($m = p$), whereas in overcomplete dictionaries the matrix has more columns than rows ($m < p$). In general, overcomplete representations result in more flexibility to allow both sparse and accurate representations [6].

5.1.2 Chapter Outline

In this chapter, we are interested in summarizing key results in learning of overcomplete dictionaries. We group works according to whether the data are vector-valued (one-dimensional) or tensor-valued (multidimensional). For both these cases, we focus on works that provide fundamental limits on the sample complexity for reliable dictionary

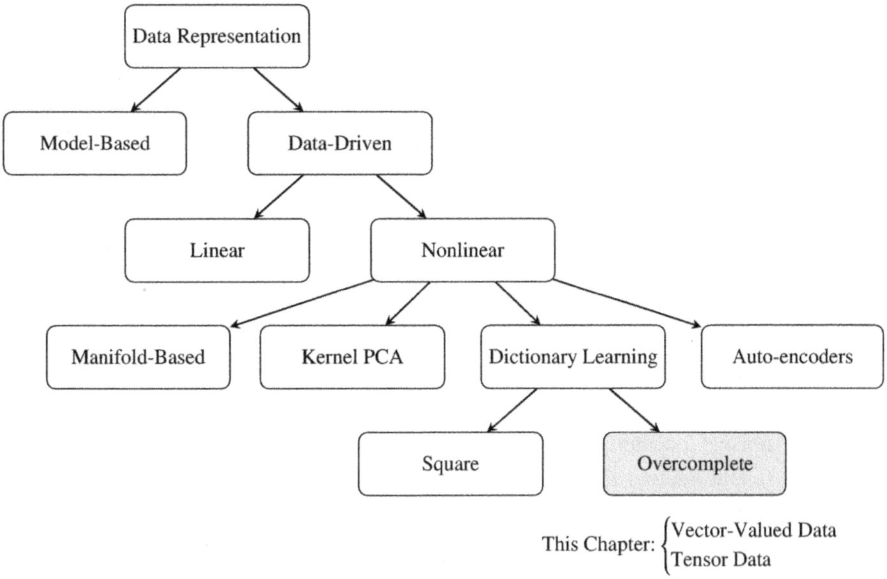

Figure 5.1 A graphical representation of the scope of this chapter in relation to the literature on representation learning.

estimation, i.e., the number of observations that are necessary to recover the true dictionary that generates the data up to some predefined error. The main information-theoretic tools that are used to derive these results range from reformulating the dictionary-learning problem as a channel coding problem to connecting the minimax risk analysis to Fano's inequality. We refer the reader to Fig. 5.1 for a graphical overview of the relationship of this chapter to other themes in representation learning.

We address the DL problem for vector-valued data in Section 5.2, and that for tensor data in Section 5.3. Finally, we talk about extensions of these works and some open problems in DL in Section 5.4. We focus here only on the problems of *identifiability* and *fundamental limits*; in particular, we do not survey DL algorithms in depth apart from some brief discussion in Sections 5.2 and 5.3. The monograph of Okoudjou [35] discusses algorithms for vector-valued data. Algorithms for tensor-valued data are relatively more recent and are described in our recent paper [13].

5.2 Dictionary Learning for Vector-Valued Data

We first address the problem of reliable estimation of dictionaries underlying data that have a single mode, i.e., are vector-valued. In particular, we focus on the subject of the sample complexity of the DL problem from two prospectives: (1) fundamental limits on the sample complexity of DL using *any* DL algorithm, and (2) the numbers of samples that are needed for different DL algorithms to reliably estimate a true underlying dictionary that generates the data.

5.2.1 Mathematical Setup

In the conventional vector-valued dictionary learning setup, we are given a total number N of vector-valued samples, $\{\mathbf{y}^n \in \mathbb{R}^m\}_{n=1}^N$, that are assumed to be generated from a fixed dictionary, $\mathbf{D}^0$, according to the following model:

$$\mathbf{y}^n = \mathbf{D}^0\mathbf{x}^n + \mathbf{w}^n, \quad n = 1,\ldots,N. \tag{5.1}$$

Here, $\mathbf{D}^0 \in \mathbb{R}^{m\times p}$ is a (deterministic) unit-norm frame ($m < p$) that belongs to the following compact set:[1]

$$\mathbf{D}^0 \in \mathcal{D} \triangleq \left\{\mathbf{D} \in \mathbb{R}^{m\times p}, \|\mathbf{D}_j\|_2 = 1 \ \forall j \in \{1,\ldots,p\}\right\}, \tag{5.2}$$

and is referred to as the *generating*, *true*, or *underlying* dictionary. The vector $\mathbf{x}^n \in \mathbb{R}^p$ is the *coefficient vector* that lies in some set $\mathcal{X} \subseteq \mathbb{R}^p$, and $\mathbf{w}^n \in \mathbb{R}^m$ denotes the observation noise. Concatenating the observations into a matrix $\mathbf{Y} \in \mathbb{R}^{m\times N}$, their corresponding coefficient vectors into $\mathbf{X} \in \mathbb{R}^{p\times N}$, and noise vectors into $\mathbf{W} \in \mathbb{R}^{m\times N}$, we get the following generative model:

$$\mathbf{Y} = \mathbf{D}^0\mathbf{X} + \mathbf{W}. \tag{5.3}$$

Various works in the DL literature impose different conditions on the coefficient vectors $\{\mathbf{x}^n\}$ to define the set $\mathcal{X}$. The most common assumption is that $\mathbf{x}^n$ is sparse with one of several probabilistic models for generating sparse $\mathbf{x}^n$. In contrast to exact sparsity, some works consider approximate sparsity and assume that $\mathbf{x}^n$ satisfies some decay profile [38], while others assume *group sparsity* conditions for $\mathbf{x}^n$ [39]. The latter condition comes up implicitly in DL for tensor data, as we discuss in Section 5.3. Similarly, existing works consider a variety of noise models, the most common being Gaussian white noise. Regardless of the assumptions on coefficient and noise vectors, all of these works assume that the observations are independent for $n = 1,2,\ldots,N$.

We are interested here in characterizing when it is possible to recover the true dictionary $\mathbf{D}^0$ from observations $\mathbf{Y}$. There is an inherent ambiguity in dictionary recovery: reordering the columns of $\mathbf{D}^0$ or multiplying any column by -1 yields a dictionary which can generate the same $\mathbf{Y}$ (with appropriately modified $\mathbf{X}$). Thus, each dictionary is equivalent to $2^p p!$ other dictionaries. To measure the distance between dictionaries, we can either define the distance between equivalence classes of dictionaries or consider errors within a local neighborhood of a fixed $\mathbf{D}^0$, where the ambiguity can potentially disappear.

The specific criterion that we focus on is sample complexity, defined as the number of observations necessary to recover the true dictionary up to some predefined error. The measure of closeness of the recovered dictionary and the true dictionary can be defined in several ways. One approach is to compare the *representation error* of these dictionaries. Another measure is the mean-squared error (MSE) between the estimated and generating dictionary, defined as

[1] A frame $\mathbf{F} \in \mathbb{R}^{m\times p}$, $m \leq p$, is defined as a collection of vectors $\{\mathbf{F}_i \in \mathbb{R}^m\}_{i=1}^p$ in some separable Hilbert space $\mathcal{H}$ that satisfy $c_1\|\mathbf{v}\|_2^2 \leq \sum_{i=1}^p |\langle \mathbf{F}_i, \mathbf{v}\rangle|^2 \leq c_2\|\mathbf{v}\|_2^2$ for all $\mathbf{v} \in \mathcal{H}$ and for some constants c_1 and c_2 such that $0 < c_1 \leq c_2 < \infty$. If $c_1 = c_2$, then $\mathbf{F}$ is a tight frame [36, 37].

$$\mathbb{E}_{\mathbf{Y}}\left\{d\left(\widehat{\mathbf{D}}(\mathbf{Y}),\mathbf{D}^0\right)^2\right\}, \tag{5.4}$$

where $d(\cdot,\cdot)$ is some distance metric and $\widehat{\mathbf{D}}(\mathbf{Y})$ is the recovered dictionary according to observations $\mathbf{Y}$. For example, if we restrict the analysis to a local neighborhood of the generating dictionary, then we can use the Frobenius norm as the distance metric.

We now discuss an optimization approach to solving the dictionary recovery problem. Understanding the objective function within this approach is the key to understanding the sample complexity of DL. Recall that solving the DL problem involves using the observations to estimate a dictionary $\widehat{\mathbf{D}}$ such that $\widehat{\mathbf{D}}$ is close to $\mathbf{D}^0$. In the ideal case, the objective function involves solving the *statistical risk minimization* problem as follows:

$$\widehat{\mathbf{D}} \in \arg\min_{\mathbf{D}\in\mathcal{D}}\left\{\inf_{\mathbf{x}\in\mathcal{X}}\mathbb{E}\left\{\frac{1}{2}\|\mathbf{y}-\mathbf{D}\mathbf{x}\|_2^2+\mathcal{R}(\mathbf{x})\right\}\right\}. \tag{5.5}$$

Here, $\mathcal{R}(\cdot)$ is a regularization operator that enforces the pre-specified structure, such as sparsity, on the coefficient vectors. Typical choices for this parameter include functions of $\|\mathbf{x}\|_0$ or its convex relaxation, $\|\mathbf{x}\|_1$.[2] However, solving (5.5) requires knowledge of exact distributions of the problem parameters as well as high computational power. Hence, works in the literature resort to algorithms that solve the *empirical risk minimization* (ERM) problem [40]:

$$\widehat{\mathbf{D}} \in \arg\min_{\mathbf{D}\in\mathcal{D}}\left\{\sum_{n=1}^{N}\inf_{\mathbf{x}^n\in\mathcal{X}}\left\{\frac{1}{2}\left\|\mathbf{y}^n-\mathbf{D}\mathbf{x}^n\right\|_2^2+\mathcal{R}(\mathbf{x}^n)\right\}\right\}. \tag{5.6}$$

In particular, to provide analytic results, many estimators solve this problem in lieu of (5.5) and then show that the solution of (5.6) is close to (5.5).

There are a number of computational algorithms that have been proposed to solve (5.6) directly for various regularizers, or indirectly using heuristic approaches. One of the most popular heuristic approaches is the K-SVD algorithm, which can be thought of as solving (5.6) with ℓ_0-norm regularization [6]. There are also other methods such as the *method of optimal directions* (MOD) [41] and online DL [25] that solve (5.6) with convex regularizers. While these algorithms have been known to perform well in practice, attention has shifted in recent years to theoretical studies to (1) find the fundamental limits of solving the statistical risk minimization problem in (5.5), (2) determine conditions on objective functions like (5.6) to ensure recovery of the true dictionary, and (3) characterize the number of samples needed for recovery using either (5.5) or (5.6). In this chapter, we are also interested in understanding the sample complexity for the DL statistical risk minimization and ERM problems. We summarize such results in the existing literature for the statistical risk minimization of DL in Section 5.2.2 and for the ERM problem in Section 5.2.3. Because the measure of closeness or error differs between these theoretical results, the corresponding sample complexity bounds are different.

[2] The so-called ℓ_0-norm counts the number of non-zero entries of a vector; it is not a norm.

REMARK 5.1 In this section, we assume that the data are available in a batch, centralized setting and the dictionary is deterministic. In the literature, DL algorithms have been proposed for other settings such as streaming data, distributed data, and Bayesian dictionaries [42–45]. Discussion of these scenarios is beyond the scope of this chapter. In addition, some works have looked at ERM problems that are different from (5.6). We briefly discuss these works in Section 5.4.

5.2.2 Minimax Lower Bounds on the Sample Complexity of DL

In this section, we study the fundamental limits on the accuracy of the dictionary recovery problem that is achievable by *any* DL method in the minimax setting. Specifically, we wish to understand the behavior of the *best estimator* that achieves the lowest *worst-case MSE* among all possible estimators. We define the error of such an estimator as the *minimax risk*, which is formally defined as

$$\varepsilon^* = \inf_{\widehat{\mathbf{D}}(\mathbf{Y})} \sup_{\mathbf{D}\in\widetilde{\mathcal{D}}} \mathbb{E}_{\mathbf{Y}}\left\{d\left(\widehat{\mathbf{D}}(\mathbf{Y}),\mathbf{D}\right)^2\right\}. \tag{5.7}$$

Note that the minimax risk does not depend on any specific DL method and provides a lower bound for the error achieved by any estimator.

The first result we present pertains to lower bounds on the minimax risk, i.e., minimax lower bounds, for the DL problem using the Frobenius norm as the distance metric between dictionaries. The result is based on the following assumption.

A1.1 (Local recovery) The true dictionary lies in the neighborhood of a fixed, known reference dictionary,[3] $\mathbf{D}^* \in \mathcal{D}$, i.e., $\mathbf{D}^0 \in \widetilde{\mathcal{D}}$, where

$$\widetilde{\mathcal{D}} = \left\{\mathbf{D}|\mathbf{D}\in\mathcal{D}, \left\|\mathbf{D}-\mathbf{D}^*\right\|_{\mathrm{F}} \le r\right\}. \tag{5.8}$$

The range for the neighborhood radius r in (5.8) is $(0,2\sqrt{p}]$. This conditioning comes from the fact that, for any $\mathbf{D},\mathbf{D}' \in \mathcal{D}$, $\|\mathbf{D}-\mathbf{D}'\|_{\mathrm{F}} \le \|\mathbf{D}\|_{\mathrm{F}} + \|\mathbf{D}'\|_{\mathrm{F}} = 2\sqrt{p}$. By restricting dictionaries to this class, for small enough r, ambiguities that are a consequence of using the Frobenius norm can be prevented. We also point out that any lower bound on ε^* is also a lower bound on the global DL problem.

THEOREM 5.1 (Minimax lower bounds [33]) *Consider a DL problem for vector-valued data with N i.i.d. observations and true dictionary $\mathbf{D}$ satisfying assumption A1.1 for some $r \in (0,2\sqrt{p}]$. Then, for any coefficient distribution with mean zero and covariance matrix $\mathbf{\Sigma}_x$, and white Gaussian noise with mean zero and variance σ^2, the minimax risk ε^* is lower-bounded as*

$$\varepsilon^* \ge c_1 \min\left\{r^2, \frac{\sigma^2}{N\|\mathbf{\Sigma}_x\|_2}(c_2 p(m-1)-1)\right\}, \tag{5.9}$$

for some positive constants c_1 and c_2.

[3] The use of a reference dictionary is an artifact of the proof technique and, for sufficiently large r, $\mathcal{D} \approx \widetilde{\mathcal{D}}$.

Theorem 5.1 holds both for square and for overcomplete dictionaries. To obtain this lower bound on the minimax risk, a standard information-theoretic approach is taken in [33] to reduce the dictionary-estimation problem to a multiple-hypothesis-testing problem. In this technique, given fixed $\mathbf{D}^*$ and r, and $L \in \mathbb{N}$, a packing $\mathcal{D}_L = \{\mathbf{D}^1, \mathbf{D}^2, \ldots, \mathbf{D}^L\} \subseteq \widetilde{\mathcal{D}}$ of $\widetilde{\mathcal{D}}$ is constructed. The distance of the packing is chosen to ensure a tight lower bound on the minimax risk. Given observations $\mathbf{Y} = \mathbf{D}^l\mathbf{X} + \mathbf{W}$, where $\mathbf{D}^l \in \mathcal{D}_L$ and the index l is chosen uniformly at random from $\{1, \ldots, L\}$, and any estimation algorithm that recovers a dictionary $\widehat{\mathbf{D}}(\mathbf{Y})$, a minimum-distance detector can be used to find the recovered dictionary index $\widehat{l} \in \{1, \ldots, L\}$. Then, Fano's inequality can be used to relate the probability of error, i.e., $\mathbb{P}(\widehat{l}(\mathbf{Y}) \neq l)$, to the mutual information between observations and the dictionary (equivalently, the dictionary index l), i.e., $I(\mathbf{Y}; l)$ [46].

Let us assume that r is sufficiently large that the minimizer of the left-hand side of (5.9) is the second term. In this case, Theorem 5.1 states that, to achieve any error $\varepsilon \geq \varepsilon^*$, we need the number of samples to be on the order of

$$N = \Omega\left(\frac{\sigma^2 mp}{\|\boldsymbol{\Sigma}_x\|_2 \varepsilon}\right).^4$$

Hence, the lower bound on the minimax risk of DL can be translated to a lower bound on the number of necessary samples, as a function of the desired dictionary error. This can further be interpreted as a lower bound on the sample complexity of the dictionary recovery problem.

We can also specialize this result to sparse coefficient vectors. Assume $\mathbf{x}^n$ has up to s non-zero elements, and the random support of the non-zero elements of $\mathbf{x}^n$ is assumed to be uniformly distributed over the set $\{\mathcal{S} \subseteq \{1, \ldots, p\} : |\mathcal{S}| = s\}$, for $n = \{1, \ldots, N\}$. Assuming that the non-zero entries of $\mathbf{x}^n$ are i.i.d. with variance σ_x^2, we get $\boldsymbol{\Sigma}_x = (s/p)\sigma_x^2\mathbf{I}_p$. Therefore, for sufficiently large r, the sample complexity scaling to achieve any error ε becomes $\Omega((\sigma^2 mp^2)/(\sigma_x^2 s\varepsilon))$. In this special case, it can be seen that, in order to achieve a fixed error ε, the sample complexity scales with the number of degrees of freedom of the dictionary multiplied by the number of dictionary columns, i.e., $N = \Omega(mp^2)$. There is also an inverse dependence on the sparsity level s. Defining the signal-to-noise ratio of the observations as $\text{SNR} = (s\sigma_x^2)/(m\sigma^2)$, this can be interpreted as an inverse relationship with the SNR. Moreover, if all parameters except the data dimension, m, are fixed, increasing m requires a linear increase in N. Evidently, this linear relation is limited by the fact that $m \leq p$ has to hold in order to maintain completeness or overcompleteness of the dictionary: increasing m by a large amount requires increasing p also.

While the tightness of this result remains an open problem, Jung *et al.* [33] have shown that for a special class of square dictionaries that are perturbations of the identity matrix, and for sparse coefficients following a specific distribution, this result is order-wise tight. In other words, a square dictionary that is perturbed from the identity matrix can be recovered from this sample size order. Although this result does not extend to overcomplete dictionaries, it suggests that the lower bounds may be tight.

[4] We use $f(n) = \Omega(g(n))$ and $f(n) = O(g(n))$ if, for sufficiently large $n \in \mathbb{N}$, $f(n) > c_1 g(n)$ and $f(n) < c_2 g(n)$, respectively, for some positive constants c_1 and c_2.

Finally, while distance metrics that are invariant to dictionary ambiguities have been used for achievable overcomplete dictionary recovery results [30, 31], obtaining minimax lower bounds for DL using these distance metrics remains an open problem.

In this subsection, we discussed the number of *necessary* samples for reliable dictionary recovery (the sample complexity lower bound). In the next subsection, we focus on achievability results, i.e., the number of *sufficient* samples for reliable dictionary recovery (the sample complexity upper bound).

5.2.3 Achievability Results

The preceding lower bounds on minimax risk hold for any estimator or computational algorithm. However, the proofs do not provide an understanding of how to construct effective estimators and provide little intuition about the potential performance of practical estimation techniques. In this section, we direct our attention to explicit reconstruction methods and their sample complexities that ensure reliable recovery of the underlying dictionary. Since these *achievability* results are tied to specific algorithms that are guaranteed to recover the true dictionary, the sample complexity bounds from these results can also be used to derive upper bounds on the minimax risk. As we will see later, there remains a gap between the lower bound and the upper bound on the minimax risk. Alternatively, one can interpret the sample complexity lower bound and upper bound as the necessary number of samples and sufficient number of samples for reliable dictionary recovery, respectively. In the following, we focus on *identifiability* results: the estimation procedures are not required to be computationally efficient.

One of the first achievability results for DL was derived in [27, 28] for square matrices. Since then, a number of works have been carried out for overcomplete DL involving vector-valued data [26, 29–32, 38]. These works differ from each other in terms of their assumptions on the true underlying dictionary, the dictionary coefficients, the presence or absence of noise and outliers, the reconstruction objective function, the distance metric used to measure the accuracy of the solution, and the local or global analysis of the solution. In this section, we summarize a few of these results in terms of various assumptions on the noise and outliers and provide a brief overview of the landscape of these results in Table 5.1. We begin our discussion with achievability results for DL for the case where $\mathbf{Y}$ is exactly given by $\mathbf{Y} = \mathbf{D}^0\mathbf{X}$, i.e., the noiseless setting.

Before proceeding, we provide a definition and an assumption that will be used for the rest of this section. We note that the constants that are used in the presented theorems change from one result to another.

(Worst-case coherence) For any dictionary $\mathbf{D} \in \mathcal{D}$, its worst-case coherence is defined as $\mu(\mathbf{D}) = \max_{i \neq j} \left|\langle \mathbf{D}_i, \mathbf{D}_j \rangle\right|$, where $\mu(\mathbf{D}) \in (0, 1)$ [36].

(Random support of sparse coefficient vectors) For any $\mathbf{x}^n$ that has up to s non-zero elements, the support of the non-zero elements of $\mathbf{x}^n$ is assumed to be distributed uniformly at random over the set $\{\mathcal{S} \subseteq \{1, \ldots, p\} : |\mathcal{S}| = s\}$, for $n = \{1, \ldots, N\}$.

Noiseless Recovery

We begin by discussing the first work that proves local identifiability of the overcomplete DL problem. The objective function that is considered in that work is

$$\left(\widehat{\mathbf{X}},\widehat{\mathbf{D}}\right) = \arg\min_{\mathbf{D}\in\mathcal{D},\mathbf{X}} \|\mathbf{X}\|_1 \text{ subject to } \mathbf{Y} = \mathbf{D}\mathbf{X}, \tag{5.10}$$

where $\|\mathbf{X}\|_1 \triangleq \sum_{i,j} |\mathbf{X}_{i,j}|$ denotes the sum of absolute values of $\mathbf{X}$.

This result is based on the following set of assumptions.

A2.1 (Gaussian random coefficients). The values of the non-zero entries of $\mathbf{x}^n$s are independent Gaussian random variables with zero mean and common standard deviation $\sigma_x = \sqrt{p/sN}$.

A2.2 (Sparsity level). The sparsity level satisfies $s \leq \min\left\{c_1/\mu(\mathbf{D}^0), c_2 p\right\}$ for some constants c_1 and c_2.

THEOREM 5.2 (Noiseless, local recovery [29]) *There exist positive constants* c_1, c_2 *such that if assumptions A2.1 and A2.2 are satisfied for true* $(\mathbf{X}, \mathbf{D}^0)$, *then* $(\mathbf{X}, \mathbf{D}^0)$ *is a local minimum of* (5.10) *with high probability.*

The probability in this theorem depends on various problem parameters and implies that $N = \Omega\left(sp^3\right)$ samples are sufficient for the desired solution, i.e., true dictionary and coefficient matrix, to be locally recoverable. The proof of this theorem relies on studying the local properties of (5.10) around its optimal solution and does not require defining a distance metric.

We now present a result that is based on the use of a combinatorial algorithm, which can provably and exactly recover the true dictionary. The proposed algorithm solves the objective function (5.6) with $\mathcal{R}(\mathbf{x}) = \lambda\|\mathbf{x}\|_1$, where λ is the regularization parameter and the distance metric that is used is the column-wise distance. Specifically, for two dictionaries $\mathbf{D}^1$ and $\mathbf{D}^2$, their column-wise distance is defined as

$$d(\mathbf{D}_j^1, \mathbf{D}_j^2) = \min_{l\in\{-1,1\}} \left\|\mathbf{D}_j^1 - l\mathbf{D}_j^2\right\|_2, \quad j \in \{1,\ldots,p\}, \tag{5.11}$$

where $\mathbf{D}_j^1$ and $\mathbf{D}_j^2$ are the jth column of $\mathbf{D}^1$ and $\mathbf{D}^2$, respectively. This distance metric avoids the sign ambiguity among dictionaries belonging to the same equivalence class. To solve (5.6), Agarwal *et al.* [30] provide a novel DL algorithm that consists of an initial dictionary-estimation stage and an alternating minimization stage to update the dictionary and coefficient vectors. The provided guarantees are based on using this algorithm to update the dictionary and coefficients. The forthcoming result is based on the following set of assumptions.

A3.1 (Bounded random coefficients) The non-zero entries of $\mathbf{x}^n$s are drawn from a zero-mean unit-variance distribution and their magnitude satisfies $x_{\min} \leq |\mathbf{x}_i^n| \leq x_{\max}$.

A3.2 (Sparsity level) The sparsity level satisfies $s \leq \min\{c_1/\sqrt{\mu(\mathbf{D}^0)}, c_2 m^{1/9}, c_3 p^{1/8}\}$ for some positive constants c_1, c_2, c_3 that depend on $x_{\min}$, $x_{\max}$, and the spectral norm of $\mathbf{D}^0$.

A3.3 (Dictionary assumptions) The true dictionary has bounded spectral norm, i.e., $\|\mathbf{D}^0\|_2 \leq c_4\sqrt{p/m}$, for some positive c_4.

THEOREM 5.3 (Noiseless, exact recovery [30]) *Consider a DL problem with N i.i.d. observations and assume that assumptions A3.1–A3.3 are satisfied. Then, there exists a universal constant c such that for given $\eta > 0$, if*

$$N \geq c\left(\frac{x_{\max}}{x_{\min}}\right)^2 p^2 \log\left(\frac{2p}{\eta}\right), \tag{5.12}$$

there exists a procedure consisting of an initial dictionary estimation stage and an alternating minimization stage such that after $T = O(\log(1/\varepsilon))$ iterations of the second stage, with probability at least $1 - 2\eta - 2\eta N^2$, $d(\widehat{\mathbf{D}}, \mathbf{D}^0) \leq \varepsilon, \forall \varepsilon > 0$.

This theorem guarantees that the true dictionary can be recovered to an arbitrary precision given $N = \Omega(p^2 \log p)$ samples. This result is based on two steps. The first step is guaranteeing an error bound for the initial dictionary-estimation step. This step involves using a clustering-style algorithm to approximate the dictionary columns. The second step is proving a local convergence result for the alternating minimization stage. This step involves improving estimates of the coefficient vectors and the dictionary through Lasso [47] and least-square steps, respectively. More details for this work can be found in [30].

While the works in [29, 30] study the sample complexity of the overcomplete DL problem, they do not take noise into account. Next, we present works that obtain sample complexity results for noisy reconstruction of dictionaries.

Noisy Reconstruction

The next result we discuss is based on the following objective function:

$$\max_{\mathbf{D} \in \mathcal{D}} \frac{1}{N} \sum_{n=1}^{N} \max_{|\mathcal{S}|=s} \|\mathbf{P}_{\mathcal{S}}(\mathbf{D})\mathbf{y}^n\|_2^2, \tag{5.13}$$

where $\mathbf{P}_{\mathcal{S}}(\mathbf{D})$ denotes projection of $\mathbf{D}$ onto the span of $\mathbf{D}_{\mathcal{S}} = \{\mathbf{D}_j\}_{j \in \mathcal{S}}$.[5] Here, the distance metric that is used is $d(\mathbf{D}^1, \mathbf{D}^2) = \max_{j \in \{1,\ldots,p\}} \|\mathbf{D}_j^1 - \mathbf{D}_j^2\|_2$. In addition, the results are based on the following set of assumptions.

A4.1 (Unit-norm tight frame) The true dictionary is a unit-norm tight frame, i.e., for all $\mathbf{v} \in \mathbb{R}^m$ we have $\sum_{j=1}^{p} |\langle \mathbf{D}_j^0, \mathbf{v}\rangle|^2 = p\|\mathbf{v}\|_2^2/m$.

[5] This objective function can be thought of as a manipulation of (5.6) with the ℓ_0-norm regularizer for the coefficient vectors. See equation 2 of [38] for more details.

A4.2 (Lower isometry constant) The lower isometry constant of $\mathbf{D}^0$, defined as $\delta_s(\mathbf{D}^0) \triangleq \max_{|\mathcal{S}|\leq s} \delta_{\mathcal{S}}(\mathbf{D}^0)$ with $1-\delta_{\mathcal{S}}(\mathbf{D}^0)$ denoting the minimal eigenvalue of $\mathbf{D}^{0\,*}_{\mathcal{S}}\mathbf{D}^0_{\mathcal{S}}$, satisfies $\delta_s(\mathbf{D}^0) \leq 1-(s/m)$.

A4.3 (Decaying random coefficients) The coefficient vector $\mathbf{x}^n$ is drawn from a symmetric decaying probability distribution ν on the unit sphere S^{p-1}.[6]

A4.4 (Bounded random noise) The vector $\mathbf{w}^n$ is a bounded random white-noise vector satisfying $\|\mathbf{w}^n\|_2 \leq M_w$ almost surely, $\mathbb{E}\{\mathbf{w}^n\} = \mathbf{0}$, and $\mathbb{E}\{\mathbf{w}^n\mathbf{w}^{n*}\} = \rho^2\mathbf{I}_m$.

A4.5 (Maximal projection constraint) Define $\mathbf{c}(\mathbf{x}^n)$ to be the non-increasing rearrangement of the absolute values of $\mathbf{x}^n$. Given a sign sequence $\mathbf{l} \in \{-1,1\}^p$ and a permutation operator $\pi : \{1,\dots,p\} \to \{1,\dots,p\}$, define $\mathbf{c}_{\pi,\mathbf{l}}(\mathbf{x}^n)$ whose ith element is equal to $\mathbf{l}_i\mathbf{c}(\mathbf{x}^n)_{\pi(i)}$ for $i \in \{1,\dots,p\}$. There exists $\kappa > 0$ such that, for $\mathbf{c}(\mathbf{x}^n)$ and $\mathcal{S}_\pi \triangleq \pi^{-1}(\{1,\dots,s\})$, we have

$$\nu\left(\min_{\pi,\mathbf{l}}\left(\left\|\mathbf{P}_{\mathcal{S}_\pi}(\mathbf{D}^0)\mathbf{D}^0\mathbf{c}_{\pi,\mathbf{l}}(\mathbf{x}^n)\right\|_2 - \max_{|\mathcal{S}|=s,\mathcal{S}\neq\mathcal{S}_\pi}\left\|\mathbf{P}_{\mathcal{S}}(\mathbf{D}^0)\mathbf{D}^0\mathbf{c}_{\pi,\mathbf{l}}(\mathbf{x}^n)\right\|_2\right) \geq 2\kappa + 2M_w\right) = 1. \tag{5.15}$$

THEOREM 5.4 (Noisy, local recovery [38]) *Consider a DL problem with N i.i.d. observations and assume that assumptions* ***A4.1–A4.5*** *are satisfied. If, for some $0 < q < 1/4$, the number of samples satisfies*

$$2N^{-q} + N^{-2q} \leq \frac{c_1\sqrt{1-\delta_s(\mathbf{D}^0)}}{\sqrt{s}\left(1 + c_2\sqrt{\log\left(c_3 p\binom{p}{s}\right)/(c_4 s(1-\delta_s(\mathbf{D}^0)))}\right)}, \tag{5.16}$$

then, with high probability, there is a local maximum of (5.13) *within distance at most $2N^{-q}$ of $\mathbf{D}^0$.*

The constants c_1, c_2, c_3, and c_4 in Theorem 5.4 depend on the underlying dictionary, coefficient vectors, and the underlying noise. The proof of this theorem relies on the fact that, for the true dictionary and its perturbations, the maximal response, i.e., $\left\|\mathbf{P}_{\mathcal{S}}(\widetilde{\mathbf{D}})\mathbf{D}^0\mathbf{x}^n\right\|_2$,[7] is attained for the set $\mathcal{S} = \mathcal{S}_\pi$ for most signals. A detailed explanation of the theorem and its proof can be found in the paper of Schnass [38].

In order to understand Theorem 5.4, let us set $q \approx \frac{1}{4} - ((\log p)/(\log N))$. We can then understand this theorem as follows. Given $N/\log N = \Omega(mp^3)$, except with probability $O(N^{-mp})$, there is a local minimum of (5.13) within distance $O(pN^{-1/4})$ of the true dictionary. Moreover, since the objective function that is considered in this work is also solved for the K-SVD algorithm, this result gives an understanding of the performance of the K-SVD algorithm. Compared with results with $\mathcal{R}(\mathbf{x})$ being a function of the ℓ_1-norm [29, 30], this result requires the true dictionary to be a tight frame. On the flip side,

[6] A probability measure ν on the unit sphere S^{p-1} is called symmetric if, for all measurable sets $\mathcal{X} \subseteq S^{p-1}$, for all sign sequences $\mathbf{l} \in \{-1,1\}^p$ and all permutations $\pi : \{1,\dots,p\} \to \{1,\dots,p\}$, we have

$$\nu(\mathbf{l}\mathcal{X}) = \nu(\mathcal{X}), \text{ where } \mathbf{l}\mathcal{X} = \left\{\left(\mathbf{l}_1\mathbf{x}_1,\dots,\mathbf{l}_p\mathbf{x}_p\right) : \mathbf{x} \in \mathcal{X}\right\},$$
$$\nu(\pi(\mathcal{X})) = \nu(\mathcal{X}), \text{ where } \pi(\mathcal{X}) = \left\{\left(\mathbf{x}_{\pi(1)},\dots,\mathbf{x}_{\pi(p)}\right) : \mathbf{x} \in \mathcal{X}\right\}. \tag{5.14}$$

[7] $\widetilde{\mathbf{D}}$ can be $\mathbf{D}^0$ itself or some perturbation of $\mathbf{D}^0$.

the coefficient vector in Theorem 5.4 is not necessarily sparse; instead, it only has to satisfy a decaying condition.

Next, we present a result obtained by Arora *et al.* [31] that is similar to that of Theorem 5.3 in the sense that it uses a combinatorial algorithm that can provably recover the true dictionary given noiseless observations. It further obtains dictionary reconstruction results for the case of noisy observations. The objective function considered in this work is similar to that of the K-SVD algorithm and can be thought of as (5.6) with $\mathcal{R}(\mathbf{x}) = \lambda\|\mathbf{x}\|_0$, where λ is the regularization parameter.

Arora *et al.* [31] define two dictionaries $\mathbf{D}^1$ and $\mathbf{D}^2$ to be *column-wise ε close* if there exists a permutation π and $l \in \{-1, 1\}$ such that $\left\|\mathbf{D}_j^1 - l\mathbf{D}_{\pi(j)}^2\right\|_2 \leq \varepsilon$. This distance metric captures the distance between equivalent classes of dictionaries and avoids the sign-permutation ambiguity. They propose a DL algorithm that first uses combinatorial techniques to recover the support of coefficient vectors, by clustering observations into overlapping clusters that use the same dictionary columns. To find these large clusters, a clustering algorithm is provided. Then, the dictionary is roughly estimated given the clusters, and the solution is further refined. The provided guarantees are based on using the proposed DL algorithm. In addition, the results are based on the following set of assumptions.

A5.1 (Bounded coefficient distribution) Non-zero entries of $\mathbf{x}^n$ are drawn from a zero-mean distribution and lie in $[-x_{\max}, -1] \cup [1, x_{\max}]$, where $x_{\max} = O(1)$. Moreover, conditioned on any subset of coordinates in $\mathbf{x}^n$ being non-zero, non-zero values of $\mathbf{x}_i^n$ are independent from each other. Finally, the distribution has bounded 3-wise moments, i.e., the probability that $\mathbf{x}^n$ is non-zero in any subset S of three coordinates is at most c^3 times $\prod_{i \in S} \mathbb{P}\left\{\mathbf{x}_i^n \neq 0\right\}$, where $c = O(1)$.[8]

A5.2 (Gaussian noise) The $\mathbf{w}^n$s are independent and follow a spherical Gaussian distribution with standard deviation $\sigma = o(\sqrt{m})$.

A5.3 (Dictionary coherence) The true dictionary is $\widetilde{\mu}$-incoherent, that is, for all $i \neq j$, $\langle \mathbf{D}_i^0, \mathbf{D}_j^0 \rangle \leq \widetilde{\mu}(\mathbf{D}^0)/\sqrt{m}$ and $\widetilde{\mu}(\mathbf{D}^0) = O(\log(m))$.

A5.4 (Sparsity level) The sparsity level satisfies $s \leq c_1 \min\{p^{2/5}, \sqrt{m}/\widetilde{\mu}(\mathbf{D}^0)\log m\}$, for some positive constant c_1.

THEOREM 5.5 (Noisy, exact recovery [31]) *Consider a DL problem with N i.i.d. observations and assume that assumptions **A5.1–A5.4** are satisfied. Provided that*

$$N = \Omega\left(\sigma^2\varepsilon^{-2}p\log p\left(\frac{p}{s^2} + s^2 + \log\left(\frac{1}{\varepsilon}\right)\right)\right), \tag{5.17}$$

there is a universal constant c_1 and a polynomial-time algorithm that learns the underlying dictionary. With high probability, this algorithm returns $\widehat{\mathbf{D}}$ that is column-wise ε close to $\mathbf{D}^0$.

[8] This condition is trivially satisfied if the set of the locations of non-zero entries of $\mathbf{x}^n$ is a random subset of size s.

For desired error ε, the run-time of the algorithm and the sample complexity depend on $\log(1/\varepsilon)$. With the addition of noise, there is also a dependence on ε^{-2} for N, which is inevitable for noisy reconstruction of the true dictionary [31, 38]. In the noiseless setting, this result translates into $N = \Omega\big(p \log p\big((p/s^2) + s^2 + \log(1/\varepsilon)\big)\big)$.

Noisy Reconstruction with Outliers

In some scenarios, in addition to observations $\mathbf{Y}$ drawn from $\mathbf{D}^0$, we encounter observations $\mathbf{Y}_{\text{out}}$ that are not generated according to $\mathbf{D}^0$. We call such observations outliers (as opposed to inliers). In this case, the observation matrix is $\mathbf{Y}_{\text{obs}} = [\mathbf{Y}, \mathbf{Y}_{\text{out}}]$, where $\mathbf{Y}$ is the inlier matrix and $\mathbf{Y}_{\text{out}}$ is the outlier matrix. In this part, we study the robustness of dictionary identification in the presence of noise and outliers. The following result studies (5.6) with $\mathcal{R}(\mathbf{x}) = \lambda\|\mathbf{x}\|_1$, where λ is the regularization parameter. Here, the Frobenius norm is considered as the distance metric. In addition, the result is based on the following set of assumptions.

A6.1 (Cumulative coherence) The cumulative coherence of the true dictionary $\mathbf{D}^0$, which is defined as

$$\mu_s(\mathbf{D}^0) \triangleq \sup_{|\mathcal{S}|\leq s} \sup_{j\notin\mathcal{S}} \left\|\mathbf{D}_{\mathcal{S}}^{0\mathrm{T}} \mathbf{D}_j^0\right\|_1, \tag{5.18}$$

satisfies $\mu_s(\mathbf{D}^0) \leq 1/4$.

A6.2 (Bounded random coefficients) Assume non-zero entries of $\mathbf{x}^n$ are drawn i.i.d. from a distribution with absolute mean $\mathbb{E}\{|x|\}$ and variance $\mathbb{E}\{x^2\}$. We denote $\mathbf{l}^n = \text{sign}(\mathbf{x}^n)$.[9] Dropping the index of $\mathbf{x}^n$ and $\mathbf{l}^n$ for simplicity of notation, the following assumptions are satisfied for the coefficient vector: $\mathbb{E}\{\mathbf{x}_{\mathcal{S}}\mathbf{x}_{\mathcal{S}}^{\mathrm{T}}|\mathcal{S}\} = \mathbb{E}\{x^2\}\mathbf{I}_s$, $\mathbb{E}\{\mathbf{x}_{\mathcal{S}}\mathbf{l}_{\mathcal{S}}^{\mathrm{T}}|\mathcal{S}\} = \mathbb{E}\{|x|\}\mathbf{I}_s$, $\mathbb{E}\{\mathbf{l}_{\mathcal{S}}\mathbf{l}_{\mathcal{S}}^{\mathrm{T}}|\mathcal{S}\} = \mathbf{I}_s$, $\|\mathbf{x}\|_2 \leq M_x$, and $\min_{i\in\mathcal{S}} |\mathbf{x}_i| \geq x_{\min}$. We define $\kappa_x \triangleq \mathbb{E}\{|x|\}/\sqrt{\mathbb{E}\{x^2\}}$ as a measure of the flatness of $\mathbf{x}$. Moreover, the following inequality is satisfied:

$$\frac{\mathbb{E}\{x^2\}}{M_x\mathbb{E}\{|x|\}} > \frac{cs}{(1-2\mu_s(\mathbf{D}^0))p}\left(\|\mathbf{D}^0\|_2 + 1\right)\left\|\mathbf{D}^{0\mathrm{T}}\mathbf{D}^0 - \mathbf{I}\right\|_{\mathrm{F}}, \tag{5.19}$$

where c is a positive constant.

A6.3 (Regularization parameter) The regularization parameter satisfies $\lambda \leq x_{\min}/4$.

A6.4 (Bounded random noise) Assume non-zero entries of $\mathbf{w}^n$ are drawn i.i.d. from a distribution with mean 0 and variance $\mathbb{E}\{w^2\}$. Dropping the index of vectors for simplicity, $\mathbf{w}$ is a bounded random white-noise vector satisfying $\mathbb{E}\{\mathbf{w}\mathbf{w}^{\mathrm{T}}|\mathcal{S}\} = \mathbb{E}\{w^2\}\mathbf{I}_m$, $\mathbb{E}\{\mathbf{w}\mathbf{x}^{\mathrm{T}}|\mathcal{S}\} = \mathbb{E}\{\mathbf{w}\mathbf{l}^{\mathrm{T}}|\mathcal{S}\} = \mathbf{0}$, and $\|\mathbf{w}\|_2 \leq M_w$. Furthermore, denoting $\bar{\lambda} \triangleq \lambda/\mathbb{E}\{|x|\}$,

$$\frac{M_w}{M_x} \leq \frac{7}{2}(c_{\max} - c_{\min})\bar{\lambda}, \tag{5.20}$$

[9] The sign of the vector $\mathbf{v}$ is defined as $\mathbf{l} = \text{sign}(\mathbf{v})$, whose elements are $\mathbf{l}_i = \mathbf{v}_i/|\mathbf{v}_i|$ for $\mathbf{v}_i \neq 0$ and $\mathbf{l}_i = 0$ for $\mathbf{v}_i = 0$, where i denotes any index of the elements of $\mathbf{v}$.

where $c_{\min}$ and $c_{\max}$ depend on problem parameters such as s, the coefficient distribution, and $\mathbf{D}^0$.

A6.5 (Sparsity level) The sparsity level satisfies $s \le p \big/ \left(16\left(\|\mathbf{D}^0\|_2 + 1\right)^2\right)$.

A6.6 (Radius range) The error radius $\varepsilon > 0$ satisfies $\varepsilon \in \left(\bar{\lambda} c_{\min}, \bar{\lambda} c_{\max}\right)$.

A6.7 (Outlier energy) Given inlier matrix $\mathbf{Y} = \{\mathbf{y}^n\}_{n=1}^N$ and outlier matrix $\mathbf{Y}_{\text{out}} = \{\mathbf{y}'^n\}_{n=1}^{N_{\text{out}}}$, the energy of $\mathbf{Y}_{\text{out}}$ satisfies

$$\frac{\|\mathbf{Y}_{\text{out}}\|_{1,2}}{N} \le \frac{c_1 \varepsilon \sqrt{s} \mathbb{E}\left\{\|\mathbf{x}\|_2^2\right\}}{\bar{\lambda} \mathbb{E}\{|x|\}} \left(\frac{A^0}{p}\right)^{3/2} \left[\frac{1}{p}\left(1 - \frac{c_{\min}\bar{\lambda}}{\varepsilon}\right) - c_2 \sqrt{\frac{mp+\eta}{N}}\right], \tag{5.21}$$

where $\|\mathbf{Y}_{\text{out}}\|_{1,2}$ denotes the sum of the ℓ_2-norms of the columns of $\mathbf{Y}_{\text{out}}$, c_1 and c_2 are positive constants, independent of parameters, and A^0 is the lower frame bound of $\mathbf{D}^0$, i.e., $A^0\|\mathbf{v}\|_2^2 \le \left\|\mathbf{D}^{0\text{T}}\mathbf{v}\right\|_2^2$ for any $\mathbf{v} \in \mathbb{R}^m$.

THEOREM 5.6 (Noisy with outliers, local recovery [32]) *Consider a DL problem with N i.i.d. observations and assume that assumptions A6.1–A6.6 are satisfied. Suppose*

$$N > c_0(mp+\eta)p^2 \left(\frac{M_x^2}{\mathbb{E}\left\{\|\mathbf{x}\|_2^2\right\}}\right)^2 \left(\frac{\varepsilon + \left((M_w/M_x) + \bar{\lambda}\right) + \left((M_w/M_x) + \bar{\lambda}\right)^2}{\varepsilon - c_{\min}\bar{\lambda}}\right), \tag{5.22}$$

then, with probability at least $1 - 2^{-\eta}$, (5.6) *admits a local minimum within distance ε of $\mathbf{D}^0$. In addition, this result is robust to the addition of outlier matrix $\mathbf{Y}_{\text{out}}$, provided that the assumption in **A6.7** is satisfied.*

The proof of this theorem relies on using the Lipschitz continuity property of the objective function in (5.6) with respect to the dictionary and sample complexity analysis using Rademacher averages and Slepian's Lemma [48]. Theorem 5.6 implies that

$$N = \Omega\left(\left(mp^3 + \eta p^2\right)\left(\frac{M_w}{M_x \varepsilon}\right)^2\right) \tag{5.23}$$

samples are sufficient for the existence of a local minimum within distance ε of true dictionary $\mathbf{D}^0$, with high probability. In the noiseless setting, this result translates into $N = \Omega\left(mp^3\right)$, and sample complexity becomes independent of the radius ε. Furthermore, this result applies to overcomplete dictionaries with dimensions $p = O(m^2)$.

5.2.4 Summary of Results

In this section, we have discussed DL minimax risk lower bounds [33] and achievability results [29–32, 38]. These results differ in terms of the distance metric they use. An interesting question arises here. Can these results be unified so that the bounds can be directly compared with one another? Unfortunately, the answer to this question is not as straightforward as it seems, and the inability to unify them is a limitation that we discuss in Section 5.4. A summary of the general scaling of the discussed results for sample complexity of (overcomplete) dictionary learning is provided in Table 5.1. We note that these are general scalings that ignore other technicalities. Here, the provided sample complexity results depend on the presence or absence of noise and outliers. All the presented results require that the underlying dictionary satisfies incoherence conditions in some way. For a one-to-one comparison of these results, the bounds for the case of absence of noise and outliers can be compared. A detailed comparison of the noiseless recovery for square and overcomplete dictionaries can be found in Table I of [32].

5.3 Dictionary Learning for Tensors

Many of today's data are collected using various sensors and tend to have a multidimensional/tensor structure (Fig. 5.2). Examples of tensor data include (1) hyperspectral images that have three modes, two spatial and one spectral; (2) colored videos that have four modes, two spatial, one depth, and one temporal; and (3) dynamic magnetic resonance imaging in a clinical trial that has five modes, three spatial, one temporal, and one subject. To find representations of tensor data using DL, one can follow two paths. A naive approach is to vectorize tensor data and use traditional vectorized representation learning techniques. A better approach is to take advantage of the multidimensional structure of data to learn representations that are specific to tensor data. While the main focus of the literature on representation learning has been on the former approach, recent works have shifted focus to the latter approaches [8–11]. These works use various tensor decompositions to decompose tensor data into smaller components. The representation learning problem can then be reduced to learning the components that represent different modes of the tensor. This results in a reduction in the number of degrees of freedom in the learning problem, due to the fact that the dimensions of the representations learned for each mode are significantly smaller than the dimensions of the representation learned for the vectorized tensor. Consequently, this approach gives rise to compact and efficient representation of tensors.

To understand the fundamental limits of dictionary learning for tensor data, one can use the sample complexity results in Section 5.2, which are a function of the underlying dictionary dimensions. However, considering the reduced number of degrees of freedom in the tensor DL problem compared with vectorized DL, this problem should be solvable with a smaller number of samples. In this section, we formalize this intuition and address the problem of reliable estimation of dictionaries underlying tensor data. As in the previous section, we will focus on the subject of sample complexity of the DL problem from two prospectives: (1) fundamental limits on the sample complexity of DL for tensor data using any DL algorithm, and (2) the numbers of samples that are needed for different DL algorithms in order to reliably estimate the true dictionary from which the tensor data are generated.

5.3.1 Tensor Terminology

A tensor is defined as a multi-way array, and the tensor order is defined as the number of components of the array. For instance, $\underline{\mathbf{X}} \in \mathbb{R}^{p_1 \times \cdots \times p_K}$ is a Kth-order tensor. For $K = 1$ and $K = 2$, the tensor is effectively a vector and a matrix, respectively. In order to

Table 5.1. Summary of the sample complexity results of various works

Reference	Jung *et al.* [33]	Geng *et al.* [29]	Agarwal *et al.* [30]	Schnass *et al.* [38]	Arora *et al.* [31]	Gribonval *et al.* [32]
Distance metric	$\left\|\mathbf{D}^1 - \mathbf{D}^2\right\|_F$	–	$\min_{l\in\{\pm1\}} \left\|\mathbf{D}_j^1 - l\mathbf{D}_{\pi(j)}^2\right\|_2$	$\max_j \left\|\mathbf{D}_j^1 - \mathbf{D}_j^2\right\|_2$	$\min_{l\in\{\pm1\}} \left\|\mathbf{D}_j^1 - l\mathbf{D}_j^2\right\|_2$	$\left\|\mathbf{D}^1 - \mathbf{D}^2\right\|_F$
Regularizer	ℓ_0	ℓ_1	ℓ_1	ℓ_1	ℓ_0	ℓ_1
Sparse coefficient distribution	Non-zero i.i.d. zero-mean, variance σ_x^2	Non-zero i.i.d. $\sim \mathcal{N}(0,\sigma_x)$	Non-zero zero-mean unit-variance $x_{\min} \le \|\mathbf{x}_i\| \le x_{\max}$	symmetric decaying (non-sparse)	Non-zero zero-mean $\mathbf{x}_i \in \pm[1, x_{\max}]$	Non-zero $\|\mathbf{x}_i\| > x_{\min}$, $\|\mathbf{x}\|_2 \le M_x$
Sparsity level	–	$O(\min\{1/\mu, p\})$	$O(\min\{1/\sqrt{\mu}, m^{1/9}, p^{1/8}\})$	$O(1/\mu)$	$O(\min\{1/(\mu \log m), p^{2/5}\})$	$O(m)$
Noise distribution	i.i.d. $\sim \mathcal{N}(0,\sigma)$	–	–	$\mathbb{E}\{\mathbf{w}\} = \mathbf{0}$, $\mathbb{E}\{\mathbf{w}\mathbf{w}^*\} = \rho^2\mathbf{I}_m$, $\|\mathbf{w}\|_2 \le M_w$	i.i.d. $\sim \mathcal{N}(0,\sigma)$	$\mathbb{E}\{\mathbf{w}\mathbf{w}^T \mid S\} = \mathbb{E}\{w^2\}\mathbf{I}_m$, $\mathbb{E}\{\mathbf{w}\mathbf{x}^T \mid S\} = \mathbf{0}$, $\|\mathbf{w}\|_2 \le M_w$
Outlier	–	–	–	–	–	Robust
Local/global	Local	Local	Global	Local	Global	Local
Sample complexity	$\frac{mp^2}{\varepsilon^2}$	sp^3	$p^2 \log p$	mp^3	$\frac{p}{\varepsilon^2} \log p(p/s^2 + s^2 + \log(1/\varepsilon))$	$\frac{mp^3}{\varepsilon^2}$

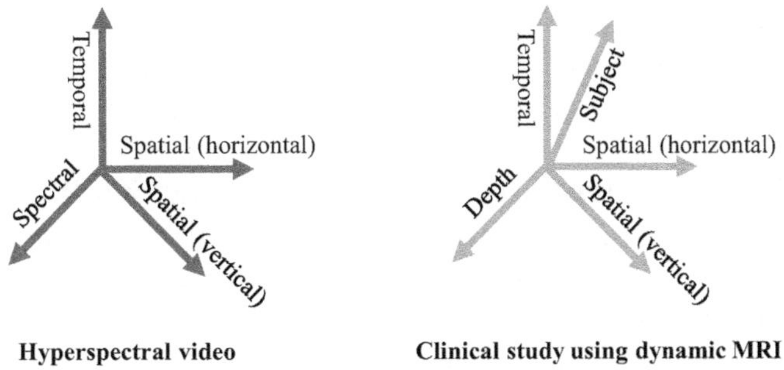

Figure 5.2 Two of countless examples of tensor data in today's sensor-rich world.

better understand the results reported in this section, we first need to define some tensor notation that will be useful throughout this section.

Tensor Unfolding: Elements of tensors can be rearranged to form matrices. Given a Kth-order tensor, $\underline{\mathbf{X}} \in \mathbb{R}^{p_1 \times \cdots \times p_K}$, its mode-$k$ unfolding is denoted as $\mathbf{X}_{(k)} \in \mathbb{R}^{p_k \times \prod_{i \neq k} p_i}$. The columns of $\mathbf{X}_{(k)}$ are formed by fixing all the indices, except one in the kth mode.

Tensor Multiplication: The mode-k product between the Kth-order tensor, $\underline{\mathbf{X}} \in \mathbb{R}^{p_1 \times \cdots \times p_K}$, and a matrix, $\mathbf{A} \in \mathbb{R}^{m_k \times p_k}$, is defined as

$$(\underline{\mathbf{X}} \times_k \mathbf{A})_{i_1,\ldots,i_{k-1},j,i_{k+1},\ldots,i_K} = \sum_{i_k=1}^{p_k} \underline{\mathbf{X}}_{i_1,\ldots,i_{k-1},i_k,i_{k+1},\ldots,i_K} \mathbf{A}_{j,i_k}. \tag{5.24}$$

Tucker Decomposition [49]: Given a Kth-order tensor $\underline{\mathbf{Y}} \in \mathbb{R}^{m_1 \times \cdots \times m_K}$ satisfying $\operatorname{rank}(\underline{\mathbf{Y}}_{(k)}) \leq p_k$, the Tucker decomposition decomposes $\underline{\mathbf{Y}}$ into a *core* tensor $\underline{\mathbf{X}} \in \mathbb{R}^{p_1 \times \cdots \times p_K}$ multiplied by *factor matrices* $\mathbf{D}_k \in \mathbb{R}^{m_k \times p_k}$ along each mode, i.e.,

$$\underline{\mathbf{Y}} = \underline{\mathbf{X}} \times_1 \mathbf{D}_1 \times_2 \mathbf{D}_2 \times_3 \cdots \times_K \mathbf{D}_K. \tag{5.25}$$

This can be restated as

$$\operatorname{vec}\left(\underline{\mathbf{Y}}_{(1)}\right) = (\mathbf{D}_K \otimes \mathbf{D}_{K-1} \otimes \cdots \otimes \mathbf{D}_1) \operatorname{vec}\left(\underline{\mathbf{X}}_{(1)}\right), \tag{5.26}$$

where $\otimes$ denotes the matrix Kronecker product [50] and $\operatorname{vec}(\cdot)$ denotes stacking of the columns of a matrix into one column. We will use the shorthand notation $\operatorname{vec}(\underline{\mathbf{Y}})$ to denote $\operatorname{vec}\left(\underline{\mathbf{Y}}_{(1)}\right)$ and $\bigotimes_k \mathbf{D}_k$ to denote $\mathbf{D}_1 \otimes \cdots \otimes \mathbf{D}_K$.

5.3.2 Mathematical Setup

To exploit the structure of tensors in DL, we can model tensors using various tensor decomposition techniques. These include Tucker decomposition, CANDECOMP/PARAFAC (CP) decomposition [51], and the t-product tensor factorization [52]. While the Tucker decomposition can be restated as the Kronecker product of matrices multiplied by a vector, other decompositions result in different formulations. In this chapter, we consider the Tucker decomposition for the following reasons: (1) it represents a sequence of independent transformations, i.e., factor matrices, for different data modes, and (2) Kronecker-structured matrices have successfully been used for data representation in applications such as magnetic resonance imaging, hyperspectral imaging, video acquisition, and distributed sensing [8, 9].

Kronecker-Structured Dictionary Learning (KS-DL)

In order to state the main results of this section, we begin with a generative model for tensor data that is based on Tucker decomposition. Specifically, we assume we have access to a total number of N tensor observations, $\underline{\mathbf{Y}}^n \in \mathbb{R}^{m_1 \times \cdots \times m_K}$, that are generated according to the following model:[10]

$$\mathrm{vec}(\underline{\mathbf{Y}}^n) = \left(\mathbf{D}_1^0 \otimes \mathbf{D}_2^0 \otimes \cdots \otimes \mathbf{D}_K^0\right) \mathrm{vec}(\underline{\mathbf{X}}^n) + \mathrm{vec}(\underline{\mathbf{W}}^n), \quad n = 1, \ldots, N. \tag{5.27}$$

Here, $\{\mathbf{D}_k^0 \in \mathbb{R}^{m_k \times p_k}\}_{k=1}^K$ are the true fixed *coordinate dictionaries*, $\underline{\mathbf{X}}^n \in \mathbb{R}^{p_1 \times \cdots \times p_K}$ is the coefficient tensor, and $\underline{\mathbf{W}}^n \in \mathbb{R}^{m_1 \times \cdots \times m_K}$ is the underlying noise tensor. In this case, the true dictionary $\mathbf{D}^0 \in \mathbb{R}^{m \times p}$ is Kronecker-structured (KS) and has the form

$$\mathbf{D}^0 = \bigotimes_k \mathbf{D}_k^0, \quad m = \prod_{k=1}^K m_k, \quad \text{and} \quad p = \prod_{k=1}^K p_k,$$

where

$$\mathbf{D}_k^0 \in \mathcal{D}_k = \left\{\mathbf{D}_k \in \mathbb{R}^{m_k \times p_k}, \left\|\mathbf{D}_{k,j}\right\|_2 = 1 \ \forall j \in \{1, \ldots, p_k\}\right\}. \tag{5.28}$$

We define the set of KS dictionaries as

$$\mathcal{D}_{\mathrm{KS}} = \left\{\mathbf{D} \in \mathbb{R}^{m \times p} : \mathbf{D} = \bigotimes_k \mathbf{D}_k, \mathbf{D}_k \in \mathcal{D}_k \ \forall k \in \{1, \ldots, K\}\right\}. \tag{5.29}$$

Comparing (5.27) with the traditional formulation in (5.1), it can be seen that KS-DL also involves vectorizing the observation tensor, but it has the main difference that the structure of the tensor is captured in the underlying KS dictionary. An illustration of this for a second-order tensor is shown in Fig. 5.3. As with (5.3), we can stack the vectorized observations, $\mathbf{y}^n = \mathrm{vec}(\underline{\mathbf{Y}}^n)$, vectorized coefficient tensors, $\mathbf{x}^n = \mathrm{vec}(\underline{\mathbf{X}}^n)$, and vectorized noise tensors, $\mathbf{w}^n = \mathrm{vec}(\underline{\mathbf{W}}^n)$, in columns of $\mathbf{Y}$, $\mathbf{X}$, and $\mathbf{W}$, respectively. We now discuss the role of sparsity in coefficient tensors for dictionary learning. While in vectorized DL it is usually assumed that the random support of non-zero entries of $\mathbf{x}^n$ is uniformly distributed, there are two different definitions of the random support of $\underline{\mathbf{X}}^n$ for tensor data.

(1) Random sparsity. The random support of $\mathbf{x}^n$ is uniformly distributed over the set $\{\mathcal{S} \subseteq \{1, \ldots, p\} : |\mathcal{S}| = s\}$.
(2) Separable sparsity. The random support of $\mathbf{x}^n$ is uniformly distributed over the set $\mathcal{S}$ that is related to $\{\mathcal{S}_1 \times \ldots \mathcal{S}_K : \mathcal{S}_k \subseteq \{1, \ldots, p_k\}, |\mathcal{S}_k| = s_k\}$ via lexicographic indexing. Here, $s = \prod_k s_k$.

Separable sparsity requires non-zero entries of the coefficient tensor to be grouped in blocks. This model also implies that the columns of $\mathbf{Y}_{(k)}$ have s_k-sparse representations with respect to $\mathbf{D}_k^0$ [53].

[10] We have reindexed the $\mathbf{D}_k$s here for simplicity of notation.

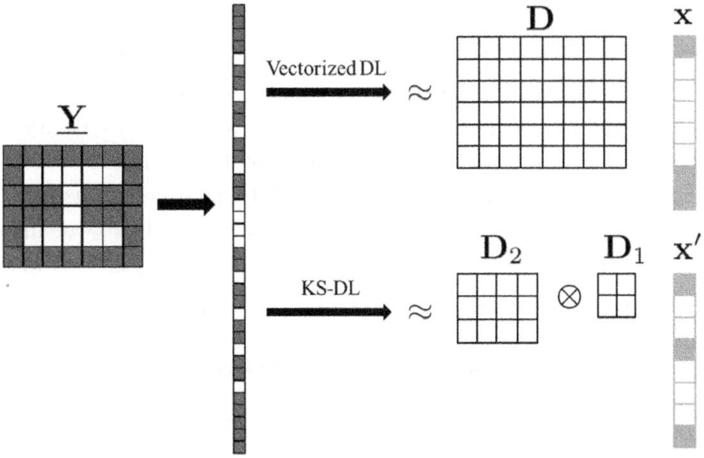

Figure 5.3 Illustration of the distinctions of KS-DL versus vectorized DL for a second-order tensor: both vectorize the observation tensor, but the structure of the tensor is exploited in the KS dictionary, leading to the learning of two coordinate dictionaries with fewer parameters than the dictionary learned in vectorized DL.

The aim in KS-DL is to estimate coordinate dictionaries, $\widehat{\mathbf{D}}_k$s, such that they are close to $\mathbf{D}_k^0$s. In this scenario, the statistical risk minimization problem has the form

$$\left(\widehat{\mathbf{D}}_1,\ldots,\widehat{\mathbf{D}}_K\right) \in \underset{\{\mathbf{D}_k\in\mathcal{D}_k\}_{k=1}^K}{\arg\min} \left\{ \inf_{\mathbf{x}\in\mathcal{X}} \mathbb{E}\left\{ \frac{1}{2}\left\| \mathbf{y} - \left(\bigotimes_k \mathbf{D}_k\right)\mathbf{x}\right\|_2^2 + \mathcal{R}(\mathbf{x})\right\}\right\}, \tag{5.30}$$

and the ERM problem is formulated as

$$\left(\widehat{\mathbf{D}}_1,\ldots,\widehat{\mathbf{D}}_K\right) \in \underset{\{\mathbf{D}_k\in\mathcal{D}_k\}_{k=1}^K}{\arg\min} \left\{ \sum_{n=1}^N \inf_{\mathbf{x}^n\in\mathcal{X}} \left\{ \frac{1}{2}\left\| \mathbf{y}^n - \left(\bigotimes_k \mathbf{D}_k\right)\mathbf{x}^n\right\|_2^2 + \mathcal{R}(\mathbf{x}^n)\right\}\right\}, \tag{5.31}$$

where $\mathcal{R}(\cdot)$ is a regularization operator on the coefficient vectors. Various KS-DL algorithms have been proposed that solve (5.31) heuristically by means of optimization tools such as alternative minimization [9] and tensor rank minimization [54], and by taking advantage of techniques in tensor algebra such as the higher-order SVD for tensors [55]. In particular, an algorithm is proposed in [11], which shows that the Kronecker product of any number of matrices can be rearranged to form a rank-1 tensor. In order to solve (5.31), therefore, in [11] a regularizer is added to the objective function that enforces this low-rankness on the rearrangement tensor. The dictionary update stage of this algorithm involves learning the rank-1 tensor and rearranging it to form the KS dictionary. This is in contrast to learning the individual coordinate dictionaries by means of alternating minimization [9].

In the case of theory for KS-DL, the notion of closeness can have two interpretations. One is the distance between the true KS dictionary and the recovered KS dictionary, i.e., $d\left(\widehat{\mathbf{D}}(\mathbf{Y}),\mathbf{D}^0\right)$. The other is the distance between each true coordinate dictionary and the corresponding recovered coordinate dictionary, i.e., $d\left(\widehat{\mathbf{D}}_k(\mathbf{Y}),\mathbf{D}_k^0\right)$. While small recovery errors for coordinate dictionaries imply a small recovery error for the KS dictionary, the other side of the statement does not necessarily hold. Hence, the latter notion is of importance when we are interested in recovering the structure of the KS dictionary.

In this section, we focus on the sample complexity of the KS-DL problem. The questions that we address in this section are as follows. (1) What are the fundamental limits of solving the statistical risk minimization problem in (5.30)? (2) Under what kind of conditions do objective functions like (5.31) recover the true coordinate dictionaries and how many samples do they need for this purpose? (3) How do these limits compare with their vectorized DL counterparts? Addressing these questions will help in understanding the benefits of KS-DL for tensor data.

5.3.3 Fundamental Limits on the Minimax Risk of KS-DL

Below, we present a result that obtains lower bounds on the minimax risk of the KS-DL problem. This result can be considered as an extension of Theorem 5.1 for the KS-DL problem for tensor data. Here, the Frobenius norm is considered as the distance metric and the result is based on the following assumption.

A7.1 (Local recovery). The true KS dictionary lies in the neighborhood of some reference dictionary, $\mathbf{D}^* \in \mathcal{D}_{\text{KS}}$, i.e., $\mathbf{D}^0 \in \widetilde{\mathcal{D}}_{\text{KS}}$, where

$$\widetilde{\mathcal{D}}_{\text{KS}} = \left\{\mathbf{D} | \mathbf{D} \in \mathcal{D}_{\text{KS}}, \left\|\mathbf{D} - \mathbf{D}^*\right\|_F \leq r\right\}. \tag{5.32}$$

THEOREM 5.7 (KS-DL minimax lower bounds [13]) *Consider a KS-DL problem with N i.i.d. observations and true KS dictionary $\mathbf{D}^0$ satisfying assumption* **A7.1** *for some $r \in (0, 2\sqrt{p}]$. Then, for any coefficient distribution with mean zero and covariance matrix $\mathbf{\Sigma}_x$, and white Gaussian noise with mean zero and variance σ^2, the minimax risk ε^* is lower-bounded as*

$$\varepsilon^* \geq \frac{t}{4} \min\left\{p, \frac{r^2}{2K}, \frac{\sigma^2}{4NK\|\mathbf{\Sigma}_x\|_2}\left(c_1\left(\sum_{k=1}^{K}(m_k - 1)p_k\right) - \frac{K}{2}\log_2(2K) - 2\right)\right\}, \tag{5.33}$$

for any $0 < t < 1$ and any $0 < c_1 < ((1 - t)/(8 \log 2))$.

Similarly to Theorem 5.1, the proof of this theorem relies on using the standard procedure for lower-bounding the minimax risk by connecting it to the maximum probability of error of a multiple-hypothesis-testing problem. Here, since the constructed hypothesis-testing class consists of KS dictionaries, the construction procedure and the minimax risk analysis are different from that in [33].

To understand this theorem, let us assume that r and p are sufficiently large that the minimizer of the left-hand side of (5.33) is the third term. In this case, Theorem 5.7 states that to achieve any error ε for the Kth-order tensor dictionary-recovery problem, we need the number of samples to be on the order of

$$N = \Omega\left(\frac{\sigma^2 \sum_k m_k p_k}{K\|\mathbf{\Sigma}_x\|_2 \varepsilon}\right).$$

Comparing this scaling with the results for the unstructured dictionary-learning problem provided in Theorem 5.1, the lower bound here is decreased from the scaling $\Omega(mp)$

to $\Omega(\sum_k m_k p_k/K)$. This reduction can be attributed to the fact that the average number of degrees of freedom in a KS-DL problem is $\sum_k m_k p_k/K$, compared with the number of degrees of the vectorized DL problem, which is mp. For the case of $K = 2$, $m_1 = m_2 = \sqrt{m}$, and $p_1 = p_2 = \sqrt{p}$, the sample complexity lower bound scales with $\Omega(mp)$ for vectorized DL and with $\Omega(\sqrt{mp})$ for KS-DL. On the other hand, when $m_1 = \alpha m, m_2 = 1/\alpha$ and $p_1 = \alpha m_1, p_2 = 1/\alpha$, where $\alpha < 1, 1/\alpha \in \mathbb{N}$, the sample complexity lower bound scales with $\Omega(mp)$ for KS-DL, which is similar to the scaling for vectorized DL.

Specializing this result to random sparse coefficient vectors and assuming that the non-zero entries of $\mathbf{x}^n$ are i.i.d. with variance σ_x^2, we get $\mathbf{\Sigma}_x = (s/p)\sigma_x^2\mathbf{I}_p$. Therefore, for sufficiently large r, the sample complexity scaling required in order to achieve any error ε for strictly sparse representations becomes

$$\Omega\left(\frac{\sigma^2 p \sum_k m_k p_k}{\sigma_x^2 s K \varepsilon}\right).$$

A very simple KS-DL algorithm is also provided in [13], which can recover a square KS dictionary that consists of the Kronecker product of two smaller dictionaries and is a perturbation of the identity matrix. It is shown that, in this case, the lower bound provided in (5.33) is order-wise achievable for the case of sparse coefficient vectors. This suggests that the obtained sample complexity lower bounds for overcomplete KS-DL are not too loose.

In the next section, we focus on achievability results for the KS dictionary recovery problem, i.e., upper bounds on the sample complexity of KS-DL.

5.3.4 Achievability Results

While the results in the previous section provide us with a lower bound on the sample complexity of the KS-DL problem, we are further interested in the sample complexity of specific KS-DL algorithms that solve (5.31). Below, we present a KS-DL achievability result that can be interpreted as an extension of Theorem 5.6 to the KS-DL problem.

Noisy Recovery

We present a result that states conditions that ensure reliable recovery of the coordinate dictionaries from noisy observations using (5.31). Shakeri *et al.* [15] solve (5.31) with $\mathcal{R}(\mathbf{x}) = \lambda\|\mathbf{x}\|_1$, where λ is a regularization parameter. Here, the coordinate dictionary error is defined as

$$\varepsilon_k = \left\|\widehat{\mathbf{D}}_k - \mathbf{D}_k^0\right\|_F, \; k \in \{1, \ldots, K\}, \tag{5.34}$$

where $\widehat{\mathbf{D}}_k$ is the recovered coordinate dictionary. The result is based on the following set of assumptions.

A8.1 (Cumulative coherence) The cumulative coherences of the true coordinate dictionaries satisfy $\mu_{s_k}(\mathbf{D}_k^0) \leq 1/4$, and the cumulative coherence of the true dictionary satisfies $\mu_s(\mathbf{D}^0) \leq 1/2$.[11]

A8.2 (Bounded random coefficients) The random support of $\mathbf{x}^n$ is generated from the separable sparsity model. Assume that non-zero entries of $\mathbf{x}^n$ are drawn i.i.d. from a distribution with absolute mean $\mathbb{E}\{|x|\}$ and variance $\mathbb{E}\{x^2\}$. Denoting $\mathbf{l}^n = \text{sign}(\mathbf{x}^n)$, and dropping the index of $\mathbf{x}^n$ and $\mathbf{l}^n$ for simplicity of notation, the following assumptions are satisfied for the coefficient vector: $\mathbb{E}\{\mathbf{x}_{\mathcal{S}}\mathbf{x}_{\mathcal{S}}^{\mathrm{T}}|\mathcal{S}\} = \mathbb{E}\{x^2\}\mathbf{I}_s$, $\mathbb{E}\{\mathbf{x}_{\mathcal{S}}\mathbf{l}_{\mathcal{S}}^{\mathrm{T}}|\mathcal{S}\} = \mathbb{E}\{|x|\}\mathbf{I}_s$, $\mathbb{E}\{\mathbf{l}_{\mathcal{S}}\mathbf{l}_{\mathcal{S}}^{\mathrm{T}}|\mathcal{S}\} = \mathbf{I}_s$, $\|\mathbf{x}\|_2 \leq M_x$, and $\min_{i\in\mathcal{S}}|\mathbf{x}_i| \geq x_{\min}$. Moreover, defining $\kappa_x \triangleq \mathbb{E}\{|x|\}/\sqrt{\mathbb{E}\{x^2\}}$ as a measure of the flatness of $\mathbf{x}$, the following inequality is satisfied:

$$\frac{\mathbb{E}\{x^2\}}{M_x\mathbb{E}\{|x|\}} > \frac{c_1}{1-2\mu_s(\mathbf{D}^0)} \max_{k\in\{1,\ldots,K\}}\left(\frac{s_k}{p_k}\left(\|\mathbf{D}_k^0\|_2 + 1\right)\|\mathbf{D}_k^{0\mathrm{T}}\mathbf{D}_k^0 - \mathbf{I}\|_{\mathrm{F}}\right), \tag{5.35}$$

where c_1 is a positive constant that is an exponential function of K.

A8.3 (Regularization parameter) The regularization parameter satisfies $\lambda \leq x_{\min}/c_2$, where c_2 is a positive constant that is an exponential function of K.

A8.4 (Bounded random noise) Assume that non-zero entries of $\mathbf{w}^n$ are drawn i.i.d. from a distribution with mean 0 and variance $\mathbb{E}\{w^2\}$. Dropping the index of vectors for simplicity of notation, $\mathbf{w}$ is a bounded random white-noise vector satisfying $\mathbb{E}\{\mathbf{w}\mathbf{w}^{\mathrm{T}}|\mathcal{S}\} = \mathbb{E}\{w^2\}\mathbf{I}_m$, $\mathbb{E}\{\mathbf{w}\mathbf{x}^{\mathrm{T}}|\mathcal{S}\} = \mathbb{E}\{\mathbf{w}\mathbf{l}^{\mathrm{T}}|\mathcal{S}\} = \mathbf{0}$, and $\|\mathbf{w}\|_2 \leq M_w$. Furthermore, denoting $\bar{\lambda} \triangleq \lambda/\mathbb{E}\{|x|\}$, we have

$$\frac{M_w}{M_x} \leq c_3\left(\bar{\lambda}Kc_{\max} - \sum_{k=1}^{K}\varepsilon_k\right), \tag{5.36}$$

where c_3 is a positive constant that is an exponential function of K and $c_{\max}$ depends on the coefficient distribution, $\mathbf{D}^0$, and K.

A8.5 (Sparsity level) The sparsity levels for each mode satisfy $s_k \leq p_k \Big/ \left(8\left(\|\mathbf{D}_k^0\|_2 + 1\right)^2\right)$ for $k \in \{1,\ldots,K\}$.

A8.6 (Radii range) The error radii $\varepsilon_k > 0$ satisfy $\varepsilon_k \in \left(\bar{\lambda}c_{k,\min}, \bar{\lambda}c_{\max}\right)$ for $k \in \{1,\ldots,K\}$, where $c_{k,\min}$ depends on s, the coefficient distribution, $\mathbf{D}^0$, and K.

THEOREM 5.8 (Noisy KS-DL, local recovery [15]) *Consider a KS-DL problem with N i.i.d. observations and suppose that assumptions **A8.1–A8.6** are satisfied. Assume*

$$N \geq \max_{k\in[K]}\Omega\left(\frac{p_k^2(\eta + m_k p_k)}{(\varepsilon_k - \varepsilon_{k,\min}(\bar{\lambda}))^2}\left(\frac{2^K(1+\bar{\lambda}^2)M_x^2}{s^2\mathbb{E}\{x^2\}^2} + \left(\frac{M_w}{s\mathbb{E}\{x^2\}}\right)^2\right)\right), \tag{5.37}$$

where $\varepsilon_{k,\min}(\bar{\lambda})$ is a function of K, $\bar{\lambda}$, and $c_{k,\min}$. Then, with probability at least $1-e^{-\eta}$, there exists a local minimum of (5.31), $\widehat{\mathbf{D}} = \bigotimes \widehat{\mathbf{D}}_k$, *such that $d(\widehat{\mathbf{D}}_k, \mathbf{D}_k^0) \leq \varepsilon_k$, for all $k \in \{1,\ldots,K\}$.*

[11] The cumulative coherence is defined in (5.18).

Table 5.2. Comparison of the scaling of vectorized DL sample complexity bounds with KS-DL, given fixed SNR

	Vectorized DL	KS-DL
Minimax lower bound	$\frac{mp^2}{\varepsilon^2}$ [33]	$\frac{p\sum_k m_k p_k}{K\varepsilon^2}$ [13]
Achievability bound	$\frac{mp^3}{\varepsilon^2}$ [32]	$\max_k \frac{m_k p_k^3}{\varepsilon_k^2}$ [15]

The proof of this theorem relies on coordinate-wise Lipschitz continuity of the objective function in (5.31) with respect to coordinate dictionaries and the use of similar sample complexity analysis arguments to those in [32]. Theorem 5.8 implies that, for fixed K and SNR, $N = \max_{k\in\{1,\ldots,K\}} \Omega\left(m_k p_k^3 \varepsilon_k^{-2}\right)$ is sufficient for the existence of a local minimum within distance ε_k of true coordinate dictionaries, with high probability. This result holds for coefficients that are generated according to the separable sparsity model. The case of coefficients generated according to the random sparsity model requires a different analysis technique that is not explored in [15].

We compare this result with the scaling in the vectorized DL problem in Theorem 5.6, which stated that $N = \Omega\left(mp^3\varepsilon^{-2}\right) = \Omega\left(\prod_k m_k p_k^3 \varepsilon^{-2}\right)$ is sufficient for the existence of $\mathbf{D}^0$ as a local minimum of (5.6) up to the predefined error ε. In contrast, $N = \max_k \Omega\left(m_k p_k^3 \varepsilon_k^{-2}\right)$ is sufficient in the case of tensor data for the existence of $\mathbf{D}_k^0$s as local minima of (5.31) up to predefined errors ε_k. This reduction in the scaling can be attributed to the reduction in the number of degrees of freedom of the KS-DL problem.

We can also compare this result with the sample complexity lower-bound scaling obtained in Theorem 5.7 for KS-DL, which stated that, given sufficiently large r and p, $N = \Omega\left(p\sum_k m_k p_k \varepsilon^{-2}/K\right)$ is necessary in order to recover the true KS dictionary $\mathbf{D}^0$ up to error ε. We can relate ε to ε_ks using the relation $\varepsilon \leq \sqrt{p}\sum_k \varepsilon_k$ [15]. Assuming all of the ε_ks are equal to each other, this implies that $\varepsilon \leq \sqrt{p}K\varepsilon_k$, and we have $N = \max_k \Omega\left(2^K K^2 p(m_k p_k^3)\varepsilon^{-2}\right)$. It can be seen from Theorem 5.7 that the sample complexity lower bound depends on the average dimension of coordinate dictionaries; in contrast, the sample complexity upper bound reported in this section depends on the maximum dimension of coordinate dictionaries. There is also a gap between the lower bound and the upper bound of order $\max_k p_k^2$. This suggests that the obtained bounds may be loose.

The sample complexity scaling results in Theorems 5.1, 5.6, 5.7, and 5.8 are shown in Table 5.2 for sparse coefficient vectors.

5.4 Extensions and Open Problems

In Sections 5.2 and 5.3, we summarized some of the key results of dictionary identification for vectorized and tensor data. In this section, we look at extensions of these works and discuss related open problems.

5.4.1 DL for Vector-Valued Data

Extensions to alternative objective functions. The works discussed in Section 5.2 all analyze variants of (5.5) and (5.6), which minimize the representation error of the dictionary. However, there do exist other works that look for a dictionary that optimizes different criteria. Schnass [56] proposed a new DL objective function called the "response maximization criterion" that extends the K-means objective function to the following:

$$\max_{\mathbf{D}\in\mathcal{D}} \sum_{n=1}^{N} \max_{|\mathcal{S}|=s} \left\| \mathbf{D}_{\mathcal{S}}^{*} \mathbf{y}^{n} \right\|_{1}. \tag{5.38}$$

Given distance metric $d(\mathbf{D}^1, \mathbf{D}^2) = \max_j \left\| \mathbf{D}_j^1 - \mathbf{D}_j^2 \right\|_2$, Schnass shows that the sample complexity needed to recover a true generating dictionary up to precision ε scales as $\mathcal{O}\left(mp^3\varepsilon^{-2}\right)$ using this objective. This sample complexity is achieved by a novel DL algorithm, called Iterative Thresholding and K-Means (ITKM), that solves (5.38) under certain conditions on the coefficient distribution, noise, and the underlying dictionary.

Efficient representations can help improve the complexity and performance of machine-learning tasks such as prediction. This means that a DL algorithm could explicitly tune the representation to optimize prediction performance. For example, some works learn dictionaries to improve classification performance [17, 25]. These works add terms to the objective function that measure the prediction performance and minimize this loss. While these DL algorithms can yield improved performance for their desired prediction task, proving sample complexity bounds for these algorithms remains an open problem.

Tightness guarantees. While dictionary identifiability has been well studied for vector-valued data, there remains a gap between the upper and lower bounds on the sample complexity. The lower bound presented in Theorem 5.1 is for the case of a particular distance metric, i.e., the Frobenius norm, whereas the presented achievability results in Theorems 5.2–5.6 are based on a variety of distance metrics. Restricting the distance metric to the Frobenius norm, we still observe a gap of order p between the sample complexity lower bound in Theorem 5.1 and the upper bound in Theorem 5.6. The partial converse result for square dictionaries that is provided in [33] shows that the lower bound is achievable for square dictionaries close to the identity matrix. For more general square matrices, however, the gap may be significant: either improved algorithms can achieve the lower bounds or the lower bounds may be further tightened. For overcomplete dictionaries the question of whether the upper bound or lower bound is tight remains open. For metrics other than the Frobenius norm, the bounds are incomparable, making it challenging to assess the tightness of many achievability results.

Finally, the works reported in Table 5.1 differ significantly in terms of the mathematical tools they use. Each approach yields a different insight into the structure of the DL problem. However, there is no unified analytic framework encompassing all of these perspectives. This gives rise to the following question. Is there a unified mathematical tool that can be used to generalize existing results on DL?

5.4.2 DL for Tensor Data

Extensions of sample complexity bounds for KS-DL. In terms of theoretical results, there are many aspects of KS-DL that have not been addressed in the literature so far. The results that are obtained in Theorems 5.7 and 5.8 are based on the Frobenius norm distance metric and provide only local recovery guarantees. Open questions include corresponding bounds for other distance metrics and global recovery guarantees. In particular, getting global recovery guarantees requires using a distance metric that can handle the inherent permutation and sign ambiguities in the dictionary. Moreover, the results of Theorem 5.8 are based on the fact that the coefficient tensors are generated according to the separable sparsity model. Extensions to coefficient tensors with arbitrary sparsity patterns, i.e., the random sparsity model, have not been explored.

Algorithmic open problems. Unlike vectorized DL problems whose sample complexity is explicitly tied to the actual algorithmic objective functions, the results in [13, 15] are not tied to an explicit KS-DL algorithm. While there exist KS-DL algorithms in the literature, none of them explicitly solve the problem in these papers. Empirically, KS-DL algorithms can outperform vectorized DL algorithms for a variety of real-world datasets [10, 11, 57–59]. However, these algorithms lack theoretical analysis in terms of sample complexity, leaving open the question of how many samples are needed in order to learn a KS dictionary using practical algorithms.

Parameter selection in KS-DL. In some cases we may not know *a priori* the parameters for which a KS dictionary yields a good model for the data. In particular, given dimension p, the problem of selecting the p_ks for coordinate dictionaries such that $p = \prod_k p_k$ has not been studied. For instance, in the case of RGB images, the selection of p_ks for the spatial modes is somewhat intuitive, as each column in the separable transform represents a pattern in each mode. However, selecting the number of columns for the depth mode, which has three dimensions (red, green, and blue), is less obvious. Given a fixed number of overall columns for the KS dictionary, how should we divide it between the number of columns for each coordinate dictionary?

Alternative structures on dictionary. In terms of DL for tensor data, the issue of extensions of identifiability results to structures other than the Kronecker product is an open problem. The main assumption in KS-DL is that the transforms for different modes of the tensor are separable from one another, which can be a limiting assumption for real-world datasets. Other structures can be enforced on the underlying dictionary to reduce sample complexity while conferring applicability to a wider range of datasets. Examples include DL using the CP decomposition [96] and the tensor t-product [61]. Characterizing the DL problem and understanding the practical benefits of these models remain interesting questions for future work.

Acknowledgments

This work is supported in part by the National Science Foundation under awards CCF-1525276 and CCF-1453073, and by the Army Research Office under award W911NF-17-1-0546.

References

[1] Y. Bengio, A. Courville, and P. Vincent, "Representation learning: A review and new perspectives," *IEEE Trans. Pattern Analysis and Machine Intelligence*, vol. 35, no. 8, pp. 1798–1828, 2013.

[2] R. N. Bracewell and R. N. Bracewell, *The Fourier transform and its applications*. McGraw-Hill, 1986.

[3] I. Daubechies, *Ten lectures on wavelets*. SIAM, 1992.

[4] E. J. Candès and D. L. Donoho, "Curvelets: A surprisingly effective nonadaptive representation for objects with edges," in *Proc. 4th International Conference on Curves and Surfaces*, 1999, vol. 2, pp. 105–120.

[5] I. T. Jolliffe, "Principal component analysis and factor analysis," in *Principal component analysis*. Springer, 1986, pp. 115–128.

[6] M. Aharon, M. Elad, and A. Bruckstein, "*K*-SVD: An algorithm for designing overcomplete dictionaries for sparse representation," *IEEE Trans. Signal Processing*, vol. 54, no. 11, pp. 4311–4322, 2006.

[7] T. G. Kolda and B. W. Bader, "Tensor decompositions and applications," *SIAM Rev.*, vol. 51, no. 3, pp. 455–500, 2009.

[8] M. F. Duarte and R. G. Baraniuk, "Kronecker compressive sensing," *IEEE Trans. Image Processing*, vol. 21, no. 2, pp. 494–504, 2012.

[9] C. F. Caiafa and A. Cichocki, "Multidimensional compressed sensing and their applications," *Wiley Interdisciplinary Rev.: Data Mining and Knowledge Discovery*, vol. 3, no. 6, pp. 355–380, 2013.

[10] S. Hawe, M. Seibert, and M. Kleinsteuber, "Separable dictionary learning," in *Proc. IEEE Conference Computer Vision and Pattern Recognition*, 2013, pp. 438–445.

[11] M. Ghassemi, Z. Shakeri, A. D. Sarwate, and W. U. Bajwa, "STARK: Structured dictionary learning through rank-one tensor recovery," in *Proc. IEEE 7th International Workshop Computational Advances in Multi-Sensor Adaptive Processing*, 2017, pp. 1–5.

[12] Z. Shakeri, W. U. Bajwa, and A. D. Sarwate, "Minimax lower bounds for Kronecker-structured dictionary learning," in *Proc. IEEE International Symposium on Information Theory*, 2016, pp. 1148–1152.

[13] Z. Shakeri, W. U. Bajwa, and A. D. Sarwate, "Minimax lower bounds on dictionary learning for tensor data," *IEEE Trans. Information Theory*, vol. 64, no. 4, pp. 2706–2726, 2018.

[14] Z. Shakeri, A. D. Sarwate, and W. U. Bajwa, "Identification of Kronecker-structured dictionaries: An asymptotic analysis," in *Proc. IEEE 7th International Workshop Computational Advances in Multi-Sensor Adaptive Processing*, 2017, pp. 1–5.

[15] Z. Shakeri, A. D. Sarwate, and W. U. Bajwa, "Identifiability of Kronecker-structured dictionaries for tensor data," *IEEE J. Selected Topics Signal Processing*, vol. 12, no. 5, pp. 1047–1062, 2018.

[16] R. Vidal, Y. Ma, and S. Sastry, "Generalized principal component analysis (GPCA)," *IEEE Trans. Pattern Analysis Machine Intelligence*, vol. 27, no. 12, pp. 1945–1959, 2005.

[17] R. Raina, A. Battle, H. Lee, B. Packer, and A. Y. Ng, "Self-taught learning: Transfer learning from unlabeled data," in *Proc. 24th International Conference on Machine Learning*, 2007, pp. 759–766.

[18] R. A. Fisher, "The use of multiple measurements in taxonomic problems," *Annals Human Genetics*, vol. 7, no. 2, pp. 179–188, 1936.

[19] A. Hyvärinen, J. Karhunen, and E. Oja, *Independent component analysis*. John Wiley & Sons, 2004.

[20] R. R. Coifman and S. Lafon, "Diffusion maps," *Appl. Comput. Harmonic Analysis*, vol. 21, no. 1, pp. 5–30, 2006.

[21] B. Schölkopf, A. Smola, and K.-R. Müller, "Kernel principal component analysis," in *Proc. International Conference on Artificial Neural Networks*, 1997, pp. 583–588.

[22] G. E. Hinton and R. R. Salakhutdinov, "Reducing the dimensionality of data with neural networks," *Science*, vol. 313, no. 5786, pp. 504–507, 2006.

[23] R. Grosse, R. Raina, H. Kwong, and A. Y. Ng, "Shift-invariance sparse coding for audio classification," in *Proc. 23rd Conference on Uncertainty in Artificial Intelligence*, 2007, pp. 149–158.

[24] J. M. Duarte-Carvajalino and G. Sapiro, "Learning to sense sparse signals: Simultaneous sensing matrix and sparsifying dictionary optimization," *IEEE Trans. Image Processing*, vol. 18, no. 7, pp. 1395–1408, 2009.

[25] J. Mairal, F. Bach, and J. Ponce, "Task-driven dictionary learning," *IEEE Trans. Pattern Analysis Machine Intelligence*, vol. 34, no. 4, pp. 791–804, 2012.

[26] M. Aharon, M. Elad, and A. M. Bruckstein, "On the uniqueness of overcomplete dictionaries, and a practical way to retrieve them," *Linear Algebra Applications*, vol. 416, no. 1, pp. 48–67, 2006.

[27] R. Remi and K. Schnass, "Dictionary identification – sparse matrix-factorization via ℓ_1-minimization," *IEEE Trans. Information Theory*, vol. 56, no. 7, pp. 3523–3539, 2010.

[28] D. A. Spielman, H. Wang, and J. Wright, "Exact recovery of sparsely-used dictionaries," in *Proc. Conference on Learning Theory*, 2012, pp. 37.11–37.18.

[29] Q. Geng and J. Wright, "On the local correctness of ℓ_1-minimization for dictionary learning," in *Proc. IEEE International Symposium on Information Theory*, 2014, pp. 3180–3184.

[30] A. Agarwal, A. Anandkumar, P. Jain, P. Netrapalli, and R. Tandon, "Learning sparsely used overcomplete dictionaries," in *Proc. 27th Annual Conference on Learning Theory*, 2014, pp. 1–15.

[31] S. Arora, R. Ge, and A. Moitra, "New algorithms for learning incoherent and overcomplete dictionaries," in *Proc. 25th Annual Conference Learning Theory*, 2014, pp. 1–28.

[32] R. Gribonval, R. Jenatton, and F. Bach, "Sparse and spurious: Dictionary learning with noise and outliers," *IEEE Trans. Information Theory*, vol. 61, no. 11, pp. 6298–6319, 2015.

[33] A. Jung, Y. C. Eldar, and N. Görtz, "On the minimax risk of dictionary learning," *IEEE Trans. Information Theory*, vol. 62, no. 3, pp. 1501–1515, 2015.

[34] O. Christensen, *An introduction to frames and Riesz bases*. Springer, 2016.

[35] K. A. Okoudjou, *Finite frame theory: A complete introduction to overcompleteness*. American Mathematical Society, 2016.

[36] W. U. Bajwa, R. Calderbank, and D. G. Mixon, "Two are better than one: Fundamental parameters of frame coherence," *Appl. Comput. Harmonic Analysis*, vol. 33, no. 1, pp. 58–78, 2012.

[37] W. U. Bajwa and A. Pezeshki, "Finite frames for sparse signal processing," in *Finite frames*, P. Casazza and G. Kutyniok, eds. Birkhäuser, 2012, ch. 10, pp. 303–335.

[38] K. Schnass, "On the identifiability of overcomplete dictionaries via the minimisation principle underlying K-SVD," *Appl. Comput. Harmonic Analysis*, vol. 37, no. 3, pp. 464–491, 2014.

[39] M. Yuan and Y. Lin, "Model selection and estimation in regression with grouped variables," *J. Roy. Statist. Soc. Ser. B*, vol. 68, no. 1, pp. 49–67, 2006.

[40] V. Vapnik, "Principles of risk minimization for learning theory," in *Proc. Advances in Neural Information Processing Systems*, 1992, pp. 831–838.

[41] K. Engan, S. O. Aase, and J. H. Husoy, "Method of optimal directions for frame design," in *Proc. IEEE International Conference on Acoustics, Speech, and Signal Processing*, vol. 5, 1999, pp. 2443–2446.

[42] J. Mairal, F. Bach, J. Ponce, and G. Sapiro, "Online learning for matrix factorization and sparse coding," *J. Machine Learning Res.*, vol. 11, no. 1, pp. 19–60, 2010.

[43] H. Raja and W. U. Bajwa, "Cloud K-SVD: A collaborative dictionary learning algorithm for big, distributed data," *IEEE Trans. Signal Processing*, vol. 64, no. 1, pp. 173–188, 2016.

[44] Z. Shakeri, H. Raja, and W. U. Bajwa, "Dictionary learning based nonlinear classifier training from distributed data," in *Proc. 2nd IEEE Global Conference Signal and Information Processing*, 2014, pp. 759–763.

[45] M. Zhou, H. Chen, J. Paisley, L. Ren, L. Li, Z. Xing, D. Dunson, G. Sapiro, and L. Carin, "Nonparametric Bayesian dictionary learning for analysis of noisy and incomplete images," *IEEE Trans. Image Processing*, vol. 21, no. 1, pp. 130–144, 2012.

[46] B. Yu, "Assouad, Fano, and Le Cam," in *Festschrift for Lucien Le Cam*. Springer, 1997, pp. 423–435.

[47] M. J. Wainwright, "Sharp thresholds for high-dimensional and noisy sparsity recovery using ℓ_1-constrained quadratic programming (lasso)," *IEEE Trans. Information Theory*, vol. 55, no. 5, pp. 2183–2202, 2009.

[48] P. Massart, *Concentration inequalities and model selection*. Springer, 2007.

[49] L. R. Tucker, "Implications of factor analysis of three-way matrices for measurement of change," in *Problems Measuring Change*. University of Wisconsin Press, 1963, pp. 122–137.

[50] C. F. Van Loan, "The ubiquitous Kronecker product," *J. Comput. Appl. Math.*, vol. 123, no. 1, pp. 85–100, 2000.

[51] R. A. Harshman, "Foundations of the PARAFAC procedure: Models and conditions for an explanatory multi-modal factor analysis," *in UCLA Working Papers in Phonetics*, vol. 16, pp. 1–84, 1970.

[52] M. E. Kilmer, C. D. Martin, and L. Perrone, "A third-order generalization of the matrix SVD as a product of third-order tensors," *Technical Report*, 2008.

[53] C. F. Caiafa and A. Cichocki, "Computing sparse representations of multidimensional signals using Kronecker bases," *Neural Computation*, vol. 25, no. 1, pp. 186–220, 2013.

[54] S. Gandy, B. Recht, and I. Yamada, "Tensor completion and low-n-rank tensor recovery via convex optimization," *Inverse Problems*, vol. 27, no. 2, p. 025010, 2011.

[55] L. De Lathauwer, B. De Moor, and J. Vandewalle, "A multilinear singular value decomposition," *SIAM J. Matrix Analysis Applications*, vol. 21, no. 4, pp. 1253–1278, 2000.

[56] K. Schnass, "Local identification of overcomplete dictionaries," *J. Machine Learning Res.*, vol. 16, pp. 1211–1242, 2015.

[57] S. Zubair and W. Wang, "Tensor dictionary learning with sparse Tucker decomposition," in *Proc. IEEE 18th International Conference on Digital Signal Processing*, 2013, pp. 1–6.

[58] F. Roemer, G. Del Galdo, and M. Haardt, "Tensor-based algorithms for learning multidimensional separable dictionaries," in *Proc. IEEE International Conference on Acoustics, Speech and Signal Processing*, 2014, pp. 3963–3967.

[59] C. F. Dantas, M. N. da Costa, and R. da Rocha Lopes, "Learning dictionaries as a sum of Kronecker products," *IEEE Signal Processing Lett.*, vol. 24, no. 5, pp. 559–563, 2017.

[60] Y. Zhang, X. Mou, G. Wang, and H. Yu, "Tensor-based dictionary learning for spectral CT reconstruction," *IEEE Trans. Medical Imaging*, vol. 36, no. 1, pp. 142–154, 2017.

[61] S. Soltani, M. E. Kilmer, and P. C. Hansen, "A tensor-based dictionary learning approach to tomographic image reconstruction," *BIT Numerical Mathe.*, vol. 56, no. 4, pp. 1–30, 2015.

6 Uncertainty Relations and Sparse Signal Recovery

Erwin Riegler and Helmut Bölcskei

Summary

This chapter provides a principled introduction to uncertainty relations underlying sparse signal recovery. We start with the seminal work by Donoho and Stark, 1989, which defines uncertainty relations as upper bounds on the operator norm of the band-limitation operator followed by the time-limitation operator, generalize this theory to arbitrary pairs of operators, and then develop out of this generalization the coherence-based uncertainty relations due to Elad and Bruckstein, 2002, as well as uncertainty relations in terms of concentration of the 1-norm or 2-norm. The theory is completed with the recently discovered set-theoretic uncertainty relations which lead to best possible recovery thresholds in terms of a general measure of parsimony, namely Minkowski dimension. We also elaborate on the remarkable connection between uncertainty relations and the "large sieve," a family of inequalities developed in analytic number theory. It is finally shown how uncertainty relations allow to establish fundamental limits of practical signal recovery problems such as inpainting, declipping, super-resolution, and denoising of signals corrupted by impulse noise or narrowband interference. Detailed proofs are provided throughout the chapter.

6.1 Introduction

The uncertainty principle in quantum mechanics says that certain pairs of physical properties of a particle, such as position and momentum, can be known to within a limited precision only [1]. Uncertainty relations in signal analysis [2–5] state that a signal and its Fourier transform cannot both be arbitrarily well concentrated; corresponding mathematical formulations exist for square-integrable or integrable functions [6, 7], for vectors in $(\mathbb{C}^m, \|\cdot\|_2)$ or $(\mathbb{C}^m, \|\cdot\|_1)$ [6–10], and for finite abelian groups [11, 12]. These results feature prominently in many areas of the mathematical data sciences. Specifically, in compressed sensing [6–9, 13, 14] uncertainty relations lead to sparse signal recovery thresholds, in Gabor and Wilson frame theory [15] they characterize limits on the time–frequency localization of frame elements, in communications [16] they play a fundamental role in the design of pulse shapes for orthogonal frequency division multiplexing (OFDM) systems [17], in the theory of partial differential equations they serve to characterize existence and smoothness properties of solutions [18], and in coding theory they help to understand questions around the existence of good cyclic codes [19].

This chapter provides a principled introduction to uncertainty relations underlying sparse signal recovery, starting with the seminal work by Donoho and Stark [6], ranging over the Elad–Bruckstein coherence-based uncertainty relation for general pairs of orthonormal bases [8], later extended to general pairs of dictionaries [10], to the recently discovered set-theoretic uncertainty relation [13] which leads to information-theoretic recovery thresholds for general notions of parsimony. We also elaborate on the remarkable connection [7] between uncertainty relations for signals and their Fourier transforms–with concentration measured in terms of support–and the "large sieve," a family of inequalities involving trigonometric polynomials, originally developed in the field of analytic number theory [20, 21].

Uncertainty relations play an important role in data science beyond sparse signal recovery, specifically in the sparse signal separation problem, which comprises numerous practically relevant applications such as (image or audio signal) inpainting, declipping, super-resolution, and the recovery of signals corrupted by impulse noise or by narrowband interference. We provide a systematic treatment of the sparse signal separation problem and develop its limits out of uncertainty relations for general pairs of dictionaries as introduced in [10]. While the flavor of these results is that beyond certain thresholds something is not possible, for example a non-zero vector cannot be concentrated with respect to two different orthonormal bases beyond a certain limit, uncertainty relations can also reveal that something unexpected is possible. Specifically, we demonstrate that signals that are sparse in certain bases can be recovered in a stable fashion from partial and noisy observations.

In practice one often encounters more general concepts of parsimony, such as, e.g., manifold structures and fractal sets. Manifolds are prevalent in the data sciences, e.g., in compressed sensing [22–27], machine learning [28], image processing [29, 30], and handwritten-digit recognition [31]. Fractal sets find application in image compression and in modeling of Ethernet traffic [32]. In the last part of this chapter, we develop an information-theoretic framework for sparse signal separation and recovery, which applies to arbitrary signals of "low description complexity." The complexity measure our results are formulated in, namely Minkowski dimension, is agnostic to signal structure and goes beyond the notion of sparsity in terms of the number of non-zero entries or concentration in 1-norm or 2-norm. The corresponding recovery thresholds are information-theoretic in the sense of applying to arbitrary signal structures and providing results of best possible nature that are, however, not constructive in terms of recovery algorithms.

To keep the exposition simple and to elucidate the main conceptual aspects, we restrict ourselves to the finite-dimensional cases $(\mathbb{C}^m, \|\cdot\|_2)$ and $(\mathbb{C}^m, \|\cdot\|_1)$ throughout. References to uncertainty relations for the infinite-dimensional case will be given wherever possible and appropriate. Some of the results in this chapter have not been reported before in the literature. Detailed proofs will be provided for most of the statements, with the goal of allowing the reader to acquire a technical working knowledge that can serve as a basis for own further research.

The chapter is organized as follows. In Sections 6.2 and 6.3, we derive uncertainty relations for vectors in $(\mathbb{C}^m, \|\cdot\|_2)$ and $(\mathbb{C}^m, \|\cdot\|_1)$, respectively, discuss the connection to the large sieve, present applications to noisy signal recovery problems, and establish a fundamental relation between uncertainty relations for sparse vectors and null-space properties of the accompanying dictionary matrices. Section 6.4 is devoted to understanding the role of uncertainty relations in sparse signal separation problems. In Section 6.5, we generalize the classical sparsity notion as used in compressed sensing to a more comprehensive concept of description complexity, namely, lower modified Minkowski dimension, which in turn leads to a set-theoretic null-space property and corresponding recovery thresholds. Section 6.6 presents a large sieve inequality in $(\mathbb{C}^m, \|\cdot\|_2)$ that one of our results in Section 6.2 is based on. Section 6.7 lists infinite-dimensional counterparts–available in the literature–to some of the results in this chapter. In Section 6.8, we provide a proof of the set-theoretic null-space property stated in Section 6.5. Finally, Section 6.9 contains results on operator norms used frequently in this chapter.

Notation. For $\mathcal{A} \subseteq \{1,\ldots,m\}$, $\mathbf{D}_{\mathcal{A}}$ denotes the $m\times m$ diagonal matrix with diagonal entries $(\mathbf{D}_{\mathcal{A}})_{i,i} = 1$ for $i \in \mathcal{A}$, and $(\mathbf{D}_{\mathcal{A}})_{i,i} = 0$ else. With $\mathbf{U} \in \mathbb{C}^{m\times m}$ unitary and $\mathcal{A} \subseteq \{1,\ldots,m\}$, we define the orthogonal projection $\mathbf{\Pi}_{\mathcal{A}}(\mathbf{U}) = \mathbf{U}\mathbf{D}_{\mathcal{A}}\mathbf{U}^*$ and set $\mathcal{W}^{\mathbf{U},\mathcal{A}} = \operatorname{range}(\mathbf{\Pi}_{\mathcal{A}}(\mathbf{U}))$. For $\mathbf{x} \in \mathbb{C}^m$ and $\mathcal{A} \subseteq \{1,\ldots,m\}$, we let $\mathbf{x}_{\mathcal{A}} = \mathbf{D}_{\mathcal{A}}\mathbf{x}$. With $\mathbf{A} \in \mathbb{C}^{m\times m}$, $|||\mathbf{A}|||_1 = \max_{\mathbf{x}:\|\mathbf{x}\|_1=1} \|\mathbf{A}\mathbf{x}\|_1$ refers to the operator 1-norm, $|||\mathbf{A}|||_2 = \max_{\mathbf{x}:\|\mathbf{x}\|_2=1} \|\mathbf{A}\mathbf{x}\|_2$ designates the operator 2-norm, $\|\mathbf{A}\|_2 = \sqrt{\operatorname{tr}(\mathbf{A}\mathbf{A}^*)}$ is the Frobenius norm, and $\|\mathbf{A}\|_1 = \sum_{i,j=1}^{m} |A_{i,j}|$. The $m\times m$ discrete Fourier transform (DFT) matrix $\mathbf{F}$ has entry $(1/\sqrt{m})e^{-2\pi jkl/m}$ in its kth row and lth column for $k,l \in \{1,\ldots,m\}$. For $x \in \mathbb{R}$, we set $[x]_+ = \max\{x,0\}$. The vector $\mathbf{x} \in \mathbb{C}^m$ is said to be s-sparse if it has at most s non-zero entries. The open ball in $(\mathbb{C}^m, \|\cdot\|_2)$ of radius ρ centered at $\mathbf{u} \in \mathbb{C}^m$ is denoted by $\mathcal{B}_m(\mathbf{u},\rho)$, and $V_m(\rho)$ refers to its volume. The indicator function on the set $\mathcal{A}$ is $\chi_{\mathcal{A}}$. We use the convention $0\cdot\infty = 0$.

6.2 Uncertainty Relations in $(\mathbb{C}^m, \|\cdot\|_2)$

Donoho and Stark [6] define uncertainty relations as upper bounds on the operator norm of the band-limitation operator followed by the time-limitation operator. We adopt this elegant concept and extend it to refer to an upper bound on the operator norm of a general orthogonal projection operator (replacing the band-limitation operator) followed by the "time-limitation operator" $\mathbf{D}_{\mathcal{P}}$ as an uncertainty relation. More specifically, let $\mathbf{U} \in \mathbb{C}^{m\times m}$ be a unitary matrix, $\mathcal{P},\mathcal{Q} \subseteq \{1,\ldots,m\}$, and consider the orthogonal projection $\mathbf{\Pi}_{\mathcal{Q}}(\mathbf{U})$ onto the subspace $\mathcal{W}^{\mathbf{U},\mathcal{Q}}$ which is spanned by $\{\mathbf{u}_i : i \in \mathcal{Q}\}$. Let[1] $\Delta_{\mathcal{P},\mathcal{Q}}(\mathbf{U}) = |||\mathbf{D}_{\mathcal{P}}\mathbf{\Pi}_{\mathcal{Q}}(\mathbf{U})|||_2$. In the setting of [6] $\mathbf{U}$ would correspond to the DFT matrix $\mathbf{F}$ and $\Delta_{\mathcal{P},\mathcal{Q}}(\mathbf{F})$ is the operator 2-norm of the band-limitation operator followed by the time-limitation operator, both in finite dimensions. By Lemma 6.12 we have

[1] We note that, for general unitary $\mathbf{A},\mathbf{B} \in \mathbb{C}^{m\times m}$, unitary invariance of $\|\cdot\|_2$ yields $|||\mathbf{\Pi}_{\mathcal{P}}(\mathbf{A})\mathbf{\Pi}_{\mathcal{Q}}(\mathbf{B})|||_2 = |||\mathbf{D}_{\mathcal{P}}\mathbf{\Pi}_{\mathcal{Q}}(\mathbf{U})|||_2$ with $\mathbf{U} = \mathbf{A}^*\mathbf{B}$. The situation where both the band-limitation operator and the time-limitation operator are replaced by general orthogonal projection operators can hence be reduced to the case considered here.

$$\Delta_{\mathcal{P},\mathcal{Q}}(\mathbf{U}) = \max_{\mathbf{x}\in\mathcal{W}^{\mathbf{U},\mathcal{Q}}\setminus\{\mathbf{0}\}} \frac{\|\mathbf{x}_{\mathcal{P}}\|_2}{\|\mathbf{x}\|_2}. \tag{6.1}$$

An uncertainty relation in $(\mathbb{C}^m, \|\cdot\|_2)$ is an upper bound of the form $\Delta_{\mathcal{P},\mathcal{Q}}(\mathbf{U}) \le c$ with $c \ge 0$, and states that $\|\mathbf{x}_{\mathcal{P}}\|_2 \le c\|\mathbf{x}\|_2$ for all $\mathbf{x} \in \mathcal{W}^{\mathbf{U},\mathcal{Q}}$. $\Delta_{\mathcal{P},\mathcal{Q}}(\mathbf{U})$ hence quantifies how well a vector supported on $\mathcal{Q}$ in the basis $\mathbf{U}$ can be concentrated on $\mathcal{P}$. Note that an uncertainty relation in $(\mathbb{C}^m, \|\cdot\|_2)$ is non-trivial only if $c < 1$. Application of Lemma 6.13 now yields

$$\frac{\|\mathbf{D}_{\mathcal{P}}\mathbf{\Pi}_{\mathcal{Q}}(\mathbf{U})\|_2}{\sqrt{\operatorname{rank}(\mathbf{D}_{\mathcal{P}}\mathbf{\Pi}_{\mathcal{Q}}(\mathbf{U}))}} \le \Delta_{\mathcal{P},\mathcal{Q}}(\mathbf{U}) \le \|\mathbf{D}_{\mathcal{P}}\mathbf{\Pi}_{\mathcal{Q}}(\mathbf{U})\|_2, \tag{6.2}$$

where the upper bound constitutes an uncertainty relation and the lower bound will allow us to assess its tightness. Next, note that

$$\|\mathbf{D}_{\mathcal{P}}\mathbf{\Pi}_{\mathcal{Q}}(\mathbf{U})\|_2 = \sqrt{\operatorname{tr}(\mathbf{D}_{\mathcal{P}}\mathbf{\Pi}_{\mathcal{Q}}(\mathbf{U}))} \tag{6.3}$$

and

$$\operatorname{rank}(\mathbf{D}_{\mathcal{P}}\mathbf{\Pi}_{\mathcal{Q}}(\mathbf{U})) = \operatorname{rank}(\mathbf{D}_{\mathcal{P}}\mathbf{U}\mathbf{D}_{\mathcal{Q}}\mathbf{U}^*) \tag{6.4}$$

$$\le \min\{|\mathcal{P}|, |\mathcal{Q}|\}, \tag{6.5}$$

where (6.5) follows from $\operatorname{rank}(\mathbf{D}_{\mathcal{P}}\mathbf{U}\mathbf{D}_{\mathcal{Q}}) \le \min\{|\mathcal{P}|, |\mathcal{Q}|\}$ and Property (c) in Section 0.4.5 of [33]. When used in (6.2) this implies

$$\sqrt{\frac{\operatorname{tr}(\mathbf{D}_{\mathcal{P}}\mathbf{\Pi}_{\mathcal{Q}}(\mathbf{U}))}{\min\{|\mathcal{P}|, |\mathcal{Q}|\}}} \le \Delta_{\mathcal{P},\mathcal{Q}}(\mathbf{U}) \le \sqrt{\operatorname{tr}(\mathbf{D}_{\mathcal{P}}\mathbf{\Pi}_{\mathcal{Q}}(\mathbf{U}))}. \tag{6.6}$$

Particularizing to $\mathbf{U} = \mathbf{F}$, we obtain

$$\sqrt{\operatorname{tr}(\mathbf{D}_{\mathcal{P}}\mathbf{\Pi}_{\mathcal{Q}}(\mathbf{F}))} = \sqrt{\operatorname{tr}(\mathbf{D}_{\mathcal{P}}\mathbf{F}\mathbf{D}_{\mathcal{Q}}\mathbf{F}^*)} \tag{6.7}$$

$$= \sqrt{\sum_{i\in\mathcal{P}}\sum_{j\in\mathcal{Q}} |\mathbf{F}_{i,j}|^2} \tag{6.8}$$

$$= \sqrt{\frac{|\mathcal{P}||\mathcal{Q}|}{m}}, \tag{6.9}$$

so that (6.6) reduces to

$$\sqrt{\frac{\max\{|\mathcal{P}|, |\mathcal{Q}|\}}{m}} \le \Delta_{\mathcal{P},\mathcal{Q}}(\mathbf{F}) \le \sqrt{\frac{|\mathcal{P}||\mathcal{Q}|}{m}}. \tag{6.10}$$

There exist sets $\mathcal{P},\mathcal{Q} \subseteq \{1,\ldots,m\}$ that saturate both bounds in (6.10), e.g., $\mathcal{P} = \{1\}$ and $\mathcal{Q} = \{1,\ldots,m\}$, which yields $\sqrt{\max\{|\mathcal{P}|,|\mathcal{Q}|\}/m} = \sqrt{|\mathcal{P}||\mathcal{Q}|/m} = 1$ and therefore $\Delta_{\mathcal{P},\mathcal{Q}}(\mathbf{F}) = 1$. An example of sets $\mathcal{P},\mathcal{Q} \subseteq \{1,\ldots,m\}$ saturating only the lower bound in (6.10) is as follows. Take n to divide m and set

$$\mathcal{P} = \left\{\frac{m}{n}, \frac{2m}{n}, \ldots, \frac{(n-1)m}{n}, m\right\} \tag{6.11}$$

and

$$\mathcal{Q} = \{l+1, \ldots, l+n\}, \tag{6.12}$$

with $l \in \{1, \ldots, m\}$ and $\mathcal{Q}$ interpreted circularly in $\{1, \ldots, m\}$. Then, the upper bound in (6.10) is

$$\sqrt{\frac{|\mathcal{P}||\mathcal{Q}|}{m}} = \frac{n}{\sqrt{m}}, \tag{6.13}$$

whereas the lower bound becomes

$$\sqrt{\frac{\max\{|\mathcal{P}|, |\mathcal{Q}|\}}{m}} = \sqrt{\frac{n}{m}}. \tag{6.14}$$

Thus, for $m \to \infty$ with fixed ratio m/n, the upper bound in (6.10) tends to infinity whereas the corresponding lower bound remains constant. The following result states that the lower bound in (6.10) is tight for $\mathcal{P}$ and $\mathcal{Q}$ as in (6.11) and (6.12), respectively. This implies a lack of tightness of the uncertainty relation $\Delta_{\mathcal{P},\mathcal{Q}}(\mathbf{F}) \leq \sqrt{|\mathcal{P}||\mathcal{Q}|/m}$ by a factor of $\sqrt{n}$. The large sieve-based uncertainty relation developed in the next section will be seen to remedy this problem.

LEMMA 6.1 *(Theorem 11 of [6]) Let n divide m and consider*

$$\mathcal{P} = \left\{\frac{m}{n}, \frac{2m}{n}, \ldots, \frac{(n-1)m}{n}, m\right\} \tag{6.15}$$

and

$$\mathcal{Q} = \{l+1, \ldots, l+n\}, \tag{6.16}$$

with $l \in \{1, \ldots, m\}$ and $\mathcal{Q}$ interpreted circularly in $\{1, \ldots, m\}$. Then, $\Delta_{\mathcal{P},\mathcal{Q}}(\mathbf{F}) = \sqrt{n/m}$.

Proof We have

$$\Delta_{\mathcal{P},\mathcal{Q}}(\mathbf{F}) = |||\mathbf{\Pi}_{\mathcal{Q}}(\mathbf{F})\mathbf{D}_{\mathcal{P}}|||_2 \tag{6.17}$$

$$= |||\mathbf{D}_{\mathcal{Q}}\mathbf{F}^*\mathbf{D}_{\mathcal{P}}|||_2 \tag{6.18}$$

$$= \max_{\mathbf{x}:\,||\mathbf{x}||_2=1} ||\mathbf{D}_{\mathcal{Q}}\mathbf{F}^*\mathbf{D}_{\mathcal{P}}\mathbf{x}||_2 \tag{6.19}$$

$$= \max_{\mathbf{x}:\,\mathbf{x}\neq\mathbf{0}} \frac{||\mathbf{D}_{\mathcal{Q}}\mathbf{F}^*\mathbf{x}_{\mathcal{P}}||_2}{||\mathbf{x}||_2} \tag{6.20}$$

$$= \max_{\substack{\mathbf{x}:\,\mathbf{x}=\mathbf{x}_{\mathcal{P}}\\ \mathbf{x}\neq\mathbf{0}}} \frac{||\mathbf{D}_{\mathcal{Q}}\mathbf{F}^*\mathbf{x}||_2}{||\mathbf{x}||_2}, \tag{6.21}$$

where in (6.17) we applied Lemma 6.12 and in (6.18) we used unitary invariance of $||\cdot||_2$. Next, consider an arbitrary but fixed $\mathbf{x} \in \mathbb{C}^m$ with $\mathbf{x} = \mathbf{x}_{\mathcal{P}}$ and define $\mathbf{y} \in \mathbb{C}^n$ according to $y_s = x_{ms/n}$ for $s = 1, \ldots, n$. It follows that

$$||\mathbf{D}_{\mathcal{Q}}\mathbf{F}^*\mathbf{x}||_2^2 = \frac{1}{m}\sum_{q\in\mathcal{Q}}\left|\sum_{p\in\mathcal{P}} x_p e^{\frac{2\pi jpq}{m}}\right|^2 \tag{6.22}$$

$$= \frac{1}{m} \sum_{q \in Q} \left| \sum_{s=1}^{n} x_{ms/n} e^{\frac{2\pi j s q}{n}} \right|^2 \tag{6.23}$$

$$= \frac{1}{m} \sum_{q \in Q} \left| \sum_{s=1}^{n} y_s e^{\frac{2\pi j s q}{n}} \right|^2 \tag{6.24}$$

$$= \frac{n}{m} \|\mathbf{F}^* \mathbf{y}\|_2^2 \tag{6.25}$$

$$= \frac{n}{m} \|\mathbf{y}\|_2^2, \tag{6.26}$$

where $\mathbf{F}$ in (6.25) is the $n \times n$ DFT matrix and in (6.26) we used unitary invariance of $\|\cdot\|_2$. With (6.22)–(6.26) and $\|\mathbf{x}\|_2 = \|\mathbf{y}\|_2$ in (6.21), we get $\Delta_{\mathcal{P},Q}(\mathbf{F}) = \sqrt{n/m}$.

6.2.1 Uncertainty Relations Based on the Large Sieve

The uncertainty relation in (6.6) is very crude as it simply upper-bounds the operator 2-norm by the Frobenius norm. For $\mathbf{U} = \mathbf{F}$ a more sophisticated upper bound on $\Delta_{\mathcal{P},Q}(\mathbf{F})$ was reported in Theorem 12 of [7]. The proof of this result establishes a remarkable connection to the so-called "large sieve," a family of inequalities involving trigonometric polynomials originally developed in the field of analytic number theory [20, 21]. We next present a slightly improved and generalized version of Theorem 12 of [7].

THEOREM 6.1 *Let $\mathcal{P} \subseteq \{1, \ldots, m\}$, $l, n \in \{1, \ldots, m\}$, and*

$$Q = \{l+1, \ldots, l+n\}, \tag{6.27}$$

with Q interpreted circularly in $\{1, \ldots, m\}$. For $\lambda \in (0, m]$, we define the circular Nyquist density $\rho(\mathcal{P}, \lambda)$ according to

$$\rho(\mathcal{P}, \lambda) = \frac{1}{\lambda} \max_{r \in [0,m)} |\widetilde{\mathcal{P}} \cap (r, r+\lambda)|, \tag{6.28}$$

where $\widetilde{\mathcal{P}} = \mathcal{P} \cup \{m + p : p \in \mathcal{P}\}$. Then,

$$\Delta_{\mathcal{P},Q}(\mathbf{F}) \leq \sqrt{\left(\frac{\lambda(n-1)}{m} + 1 \right) \rho(\mathcal{P}, \lambda)} \tag{6.29}$$

for all $\lambda \in (0, m]$.

Proof If $\mathcal{P} = \emptyset$, then $\Delta_{\mathcal{P},Q}(\mathbf{F}) = 0$ as a consequence of $\mathbf{\Pi}_\emptyset(\mathbf{F}) = \mathbf{0}$ and (6.29) holds trivially. Suppose now that $\mathcal{P} \neq \emptyset$, consider an arbitrary but fixed $\mathbf{x} \in \mathcal{W}^{\mathbf{F},Q}$ with $\|\mathbf{x}\|_2 = 1$, and set $\mathbf{a} = \mathbf{F}^* \mathbf{x}$. Then, $\mathbf{a} = \mathbf{a}_Q$ and, by unitarity of $\mathbf{F}$, $\|\mathbf{a}\|_2 = 1$. We have

$$|x_p|^2 = |(\mathbf{F}\mathbf{a})_p|^2 \tag{6.30}$$

$$= \frac{1}{m} \left| \sum_{q \in Q} a_q e^{-\frac{2\pi j p q}{m}} \right|^2 \tag{6.31}$$

$$= \frac{1}{m}\left|\sum_{k=1}^{n} a_k e^{-\frac{2\pi jpk}{m}}\right|^2 \tag{6.32}$$

$$= \frac{1}{m}\left|\psi\left(\frac{p}{m}\right)\right|^2 \quad \text{for } p \in \{1,\dots,m\}, \tag{6.33}$$

where we defined the 1-periodic trigonometric polynomial $\psi(s)$ according to

$$\psi(s) = \sum_{k=1}^{n} a_k e^{-2\pi jks}. \tag{6.34}$$

Next, let ν_t denote the unit Dirac measure centered at $t \in \mathbb{R}$ and set $\mu = \sum_{p\in\mathcal{P}} \nu_{p/m}$ with 1-periodic extension outside $[0,1)$. Then,

$$\|\mathbf{x}_{\mathcal{P}}\|_2^2 = \frac{1}{m}\sum_{p\in\mathcal{P}}\left|\psi\left(\frac{p}{m}\right)\right|^2 \tag{6.35}$$

$$= \frac{1}{m}\int_{[0,1)} |\psi(s)|^2 \,\mathrm{d}\mu(s) \tag{6.36}$$

$$\leq \left(\frac{n-1}{m} + \frac{1}{\lambda}\right) \sup_{r\in[0,1)} \mu\left(\left(r, r+\frac{\lambda}{m}\right)\right) \tag{6.37}$$

for all $\lambda \in (0,m]$, where (6.35) is by (6.30)–(6.33) and in (6.37) we applied the large sieve inequality Lemma 6.10 with $\delta = \lambda/m$ and $\|\mathbf{a}\|_2 = 1$. Now,

$$\sup_{r\in[0,1)} \mu\left(\left(r, r+\frac{\lambda}{m}\right)\right) \tag{6.38}$$

$$= \sup_{r\in[0,m)} \sum_{p\in\mathcal{P}} (\nu_p((r,r+\lambda)) + \nu_{m+p}((r,r+\lambda))) \tag{6.39}$$

$$= \max_{r\in[0,m)} |\widetilde{\mathcal{P}} \cap (r, r+\lambda)| \tag{6.40}$$

$$= \lambda\rho(\mathcal{P},\lambda) \quad \text{for all } \lambda \in (0,m], \tag{6.41}$$

where in (6.39) we used the 1-periodicity of μ. Using (6.38)–(6.41) in (6.37) yields

$$\|\mathbf{x}_{\mathcal{P}}\|_2^2 \leq \left(\frac{\lambda(n-1)}{m} + 1\right)\rho(\mathcal{P},\lambda) \quad \text{for all } \lambda \in (0,m]. \tag{6.42}$$

As $\mathbf{x} \in \mathcal{W}^{\mathbf{F},Q}$ with $\|\mathbf{x}\|_2 = 1$ was arbitrary, we conclude that

$$\Delta^2_{\mathcal{P},Q}(\mathbf{F}) = \max_{\mathbf{x}\in\mathcal{W}^{\mathbf{F},Q}\setminus\{\mathbf{0}\}} \frac{\|\mathbf{x}_{\mathcal{P}}\|_2^2}{\|\mathbf{x}\|_2^2} \tag{6.43}$$

$$\leq \left(\frac{\lambda(n-1)}{m} + 1\right)\rho(\mathcal{P},\lambda) \quad \text{for all } \lambda \in (0,m], \tag{6.44}$$

thereby finishing the proof.

Theorem 6.1 slightly improves upon Theorem 12 of [7] by virtue of applying to more general sets Q and defining the circular Nyquist density in (6.28) in terms of open intervals $(r, r+\lambda)$.

We next apply Theorem 6.1 to specific choices of $\mathcal{P}$ and $\mathcal{Q}$. First, consider $\mathcal{P} = \{1\}$ and $\mathcal{Q} = \{1,\dots,m\}$, which were shown to saturate the upper and the lower bound in (6.10), leading to $\Delta_{\mathcal{P},\mathcal{Q}}(\mathbf{F}) = 1$. Since $\mathcal{P}$ consists of a single point, $\rho(\mathcal{P},\lambda) = 1/\lambda$ for all $\lambda \in (0,m]$. Thus, Theorem 6.1 with $n = m$ yields

$$\Delta_{\mathcal{P},\mathcal{Q}}(\mathbf{F}) \le \sqrt{\frac{m-1}{m} + \frac{1}{\lambda}} \quad \text{for all } \lambda \in (0,m]. \tag{6.45}$$

Setting $\lambda = m$ in (6.45) yields $\Delta_{\mathcal{P},\mathcal{Q}}(\mathbf{F}) \le 1$.

Next, consider $\mathcal{P}$ and $\mathcal{Q}$ as in (6.11) and (6.12), respectively, which, as already mentioned, have the uncertainty relation in (6.10) lacking tightness by a factor of $\sqrt{n}$. Since $\mathcal{P}$ consists of points spaced m/n apart, we get $\rho(\mathcal{P},\lambda) = 1/\lambda$ for all $\lambda \in (0,m/n]$. The upper bound (6.29) now becomes

$$\Delta_{\mathcal{P},\mathcal{Q}}(\mathbf{F}) \le \sqrt{\frac{n-1}{m} + \frac{1}{\lambda}} \quad \text{for all } \lambda \in \left(0, \frac{m}{n}\right]. \tag{6.46}$$

Setting $\lambda = m/n$ in (6.46) yields

$$\Delta_{\mathcal{P},\mathcal{Q}}(\mathbf{F}) \le \sqrt{(2n-1)/m} \le \sqrt{2}\sqrt{n/m}, \tag{6.47}$$

which is tight up to a factor of $\sqrt{2}$ (cf. Lemma 6.1). We hasten to add, however, that the large sieve technique applies to $\mathbf{U} = \mathbf{F}$ only.

6.2.2 Coherence-Based Uncertainty Relation

We next present an uncertainty relation that is of simple form and applies to general unitary $\mathbf{U}$. To this end, we first introduce the concept of coherence of a matrix.

DEFINITION 6.1 *For* $\mathbf{A} = (\mathbf{a}_1 \dots \mathbf{a}_n) \in \mathbb{C}^{m\times n}$ *with columns* $\|\cdot\|_2$*-normalized to* 1, *the coherence is defined as* $\mu(\mathbf{A}) = \max_{i\neq j} |\mathbf{a}_i^* \mathbf{a}_j|$.

We have the following coherence-based uncertainty relation valid for general unitary $\mathbf{U}$.

LEMMA 6.2 *Let* $\mathbf{U} \in \mathbb{C}^{m\times m}$ *be unitary and* $\mathcal{P},\mathcal{Q} \subseteq \{1,\dots,m\}$. *Then,*

$$\Delta_{\mathcal{P},\mathcal{Q}}(\mathbf{U}) \le \sqrt{|\mathcal{P}||\mathcal{Q}|}\,\mu([\mathbf{I}\ \ \mathbf{U}]). \tag{6.48}$$

Proof The claim follows from

$$\begin{aligned}
\Delta^2_{\mathcal{P},\mathcal{Q}}(\mathbf{U}) &\le \operatorname{tr}(\mathbf{D}_{\mathcal{P}}\mathbf{U}\mathbf{D}_{\mathcal{Q}}\mathbf{U}^*) && (6.49)\\
&= \sum_{k\in\mathcal{P}} \sum_{l\in\mathcal{Q}} |\mathbf{U}_{k,l}|^2 && (6.50)\\
&\le |\mathcal{P}||\mathcal{Q}| \max_{k,l} |\mathbf{U}_{k,l}|^2 && (6.51)\\
&= |\mathcal{P}||\mathcal{Q}|\mu^2([\mathbf{I}\ \ \mathbf{U}]), && (6.52)
\end{aligned}$$

where (6.49) is by (6.6) and in (6.52) we used the definition of coherence.

Since $\mu([\mathbf{I}\ \ \mathbf{F}]) = 1/\sqrt{m}$, Lemma 6.2 particularized to $\mathbf{U} = \mathbf{F}$ recovers the upper bound in (6.10).

6.2.3 Concentration Inequalities

As mentioned at the beginning of this chapter, the classical uncertainty relation in signal analysis quantifies how well concentrated a signal can be in time and frequency. In the finite-dimensional setting considered here this amounts to characterizing the concentration of $\mathbf{p}$ and $\mathbf{q}$ in $\mathbf{p} = \mathbf{F}\mathbf{q}$. We will actually study the more general case obtained by replacing $\mathbf{I}$ and $\mathbf{F}$ by unitary $\mathbf{A} \in \mathbb{C}^{m \times m}$ and $\mathbf{B} \in \mathbb{C}^{m \times m}$, respectively, and will ask ourselves how well concentrated $\mathbf{p}$ and $\mathbf{q}$ in $\mathbf{A}\mathbf{p} = \mathbf{B}\mathbf{q}$ can be. Rewriting $\mathbf{A}\mathbf{p} = \mathbf{B}\mathbf{q}$ according to $\mathbf{p} = \mathbf{U}\mathbf{q}$ with $\mathbf{U} = \mathbf{A}^*\mathbf{B}$, we now show how the uncertainty relation in Lemma 6.2 can be used to answer this question. Let us start by introducing a measure for concentration in $(\mathbb{C}^m, \|\cdot\|_2)$.

DEFINITION 6.2 *Let* $\mathcal{P} \subseteq \{1, \dots, m\}$ *and* $\varepsilon_{\mathcal{P}} \in [0,1]$. *The vector* $\mathbf{x} \in \mathbb{C}^m$ *is said to be* $\varepsilon_{\mathcal{P}}$-*concentrated if* $\|\mathbf{x} - \mathbf{x}_{\mathcal{P}}\|_2 \leq \varepsilon_{\mathcal{P}} \|\mathbf{x}\|_2$.

The fraction of 2-norm an $\varepsilon_{\mathcal{P}}$-concentrated vector exhibits outside $\mathcal{P}$ is therefore no more than $\varepsilon_{\mathcal{P}}$. In particular, if $\mathbf{x}$ is $\varepsilon_{\mathcal{P}}$-concentrated with $\varepsilon_{\mathcal{P}} = 0$, then $\mathbf{x} = \mathbf{x}_{\mathcal{P}}$ and $\mathbf{x}$ is $|\mathcal{P}|$-sparse. The zero vector is trivially $\varepsilon_{\mathcal{P}}$-concentrated for all $\mathcal{P} \subseteq \{1, \dots, m\}$ and $\varepsilon_{\mathcal{P}} \in [0,1]$.

We next derive a lower bound on $\Delta_{\mathcal{P},\mathcal{Q}}(\mathbf{U})$ for unitary matrices $\mathbf{U}$ that relate $\varepsilon_{\mathcal{P}}$-concentrated vectors $\mathbf{p}$ to $\varepsilon_{\mathcal{Q}}$-concentrated vectors $\mathbf{q}$ through $\mathbf{p} = \mathbf{U}\mathbf{q}$. The formal statement is as follows.

LEMMA 6.3 *Let* $\mathbf{U} \in \mathbb{C}^{m \times m}$ *be unitary and* $\mathcal{P}, \mathcal{Q} \subseteq \{1, \dots, m\}$. *Suppose that there exist a non-zero* $\varepsilon_{\mathcal{P}}$-*concentrated* $\mathbf{p} \in \mathbb{C}^m$ *and a non-zero* $\varepsilon_{\mathcal{Q}}$-*concentrated* $\mathbf{q} \in \mathbb{C}^m$ *such that* $\mathbf{p} = \mathbf{U}\mathbf{q}$. *Then,*

$$\Delta_{\mathcal{P},\mathcal{Q}}(\mathbf{U}) \geq [1 - \varepsilon_{\mathcal{P}} - \varepsilon_{\mathcal{Q}}]_+. \tag{6.53}$$

Proof We have

$$\begin{aligned}
\|\mathbf{p} - \mathbf{\Pi}_{\mathcal{Q}}(\mathbf{U})\mathbf{p}_{\mathcal{P}}\|_2 &\leq \|\mathbf{p} - \mathbf{\Pi}_{\mathcal{Q}}(\mathbf{U})\mathbf{p}\|_2 + \|\mathbf{\Pi}_{\mathcal{Q}}(\mathbf{U})\mathbf{p}_{\mathcal{P}} - \mathbf{\Pi}_{\mathcal{Q}}(\mathbf{U})\mathbf{p}\|_2 && (6.54)\\
&\leq \|\mathbf{p} - \mathbf{\Pi}_{\mathcal{Q}}(\mathbf{U})\mathbf{p}\|_2 + \|\!|\mathbf{\Pi}_{\mathcal{Q}}(\mathbf{U})|\!\|_2 \|\mathbf{p}_{\mathcal{P}} - \mathbf{p}\|_2 && (6.55)\\
&\leq \|\mathbf{p} - \mathbf{U}\mathbf{D}_{\mathcal{Q}}\mathbf{U}^*\mathbf{p}\|_2 + \|\mathbf{p}_{\mathcal{P}} - \mathbf{p}\|_2 && (6.56)\\
&= \|\mathbf{q} - \mathbf{q}_{\mathcal{Q}}\|_2 + \|\mathbf{p}_{\mathcal{P}} - \mathbf{p}\|_2, && (6.57)\\
&\leq \varepsilon_{\mathcal{Q}}\|\mathbf{q}\|_2 + \varepsilon_{\mathcal{P}}\|\mathbf{p}\|_2 && (6.58)\\
&= (\varepsilon_{\mathcal{P}} + \varepsilon_{\mathcal{Q}})\|\mathbf{p}\|_2, && (6.59)
\end{aligned}$$

where in (6.57) we made use of the unitary invariance of $\|\cdot\|_2$. It follows that

$$\begin{aligned}
\|\mathbf{\Pi}_{\mathcal{Q}}(\mathbf{U})\mathbf{p}_{\mathcal{P}}\|_2 &\geq [\|\mathbf{p}\|_2 - \|\mathbf{p} - \mathbf{\Pi}_{\mathcal{Q}}(\mathbf{U})\mathbf{p}_{\mathcal{P}}\|_2]_+ && (6.60)\\
&\geq \|\mathbf{p}\|_2[1 - \varepsilon_{\mathcal{P}} - \varepsilon_{\mathcal{Q}}]_+, && (6.61)
\end{aligned}$$

where (6.60) is by the reverse triangle inequality and in (6.61) we used (6.54)–(6.59). Since $\mathbf{p} \neq \mathbf{0}$ by assumption, (6.60)–(6.61) implies

$$\left\| \mathbf{\Pi}_{\mathcal{Q}}(\mathbf{U})\mathbf{D}_{\mathcal{P}} \frac{\mathbf{p}}{\|\mathbf{p}\|_2} \right\|_2 \geq [1 - \varepsilon_{\mathcal{P}} - \varepsilon_{\mathcal{Q}}]_+, \tag{6.62}$$

which in turn yields $\|\!|\mathbf{\Pi}_{\mathcal{Q}}(\mathbf{U})\mathbf{D}_{\mathcal{P}}|\!\|_2 \geq [1 - \varepsilon_{\mathcal{P}} - \varepsilon_{\mathcal{Q}}]_+$. This concludes the proof as $\Delta_{\mathcal{P},\mathcal{Q}}(\mathbf{U}) = \|\!|\mathbf{\Pi}_{\mathcal{Q}}(\mathbf{U})\mathbf{D}_{\mathcal{P}}|\!\|_2$ by Lemma 6.12.

Combining Lemma 6.3 with the uncertainty relation Lemma 6.2 yields the announced result stating that a non-zero vector cannot be arbitrarily well concentrated with respect to two different orthonormal bases.

COROLLARY 6.1 *Let* $\mathbf{A},\mathbf{B} \in \mathbb{C}^{m\times m}$ *be unitary and* $\mathcal{P},\mathcal{Q} \subseteq \{1,\ldots,m\}$. *Suppose that there exist a non-zero* $\varepsilon_{\mathcal{P}}$*-concentrated* $\mathbf{p} \in \mathbb{C}^m$ *and a non-zero* $\varepsilon_{\mathcal{Q}}$*-concentrated* $\mathbf{q} \in \mathbb{C}^m$ *such that* $\mathbf{Ap} = \mathbf{Bq}$. *Then,*

$$|\mathcal{P}||\mathcal{Q}| \geq \frac{[1-\varepsilon_{\mathcal{P}}-\varepsilon_{\mathcal{Q}}]_+^2}{\mu^2([\mathbf{A}\ \ \mathbf{B}])}. \tag{6.63}$$

Proof Let $\mathbf{U} = \mathbf{A}^*\mathbf{B}$. Then, by Lemmata 6.2 and 6.3, we have

$$[1-\varepsilon_{\mathcal{P}}-\varepsilon_{\mathcal{Q}}]_+ \leq \Delta_{\mathcal{P},\mathcal{Q}}(\mathbf{U}) \leq \sqrt{|\mathcal{P}||\mathcal{Q}|}\,\mu([\mathbf{I}\ \ \mathbf{U}]). \tag{6.64}$$

The claim now follows by noting that $\mu([\mathbf{I}\ \ \mathbf{U}]) = \mu([\mathbf{A}\ \ \mathbf{B}])$.

For $\varepsilon_{\mathcal{P}} = \varepsilon_{\mathcal{Q}} = 0$, we recover the well-known Elad–Bruckstein result.

COROLLARY 6.2 *(Theorem 1 of [8]) Let* $\mathbf{A},\mathbf{B} \in \mathbb{C}^{m\times m}$ *be unitary. If* $\mathbf{Ap} = \mathbf{Bq}$ *for non-zero* $\mathbf{p},\mathbf{q} \in \mathbb{C}^m$, *then* $\|\mathbf{p}\|_0\|\mathbf{q}\|_0 \geq 1/\mu^2([\mathbf{A}\ \ \mathbf{B}])$.

6.2.4 Noisy Recovery in $(\mathbb{C}^m, \|\cdot\|_2)$

Uncertainty relations are typically employed to prove that something is not possible. For example, by Corollary 6.1 there is a limit on how well a non-zero vector can be concentrated with respect to two different orthonormal bases. Donoho and Stark [6] noticed that uncertainty relations can also be used to show that something unexpected is possible. Specifically, Section 4 of [6] considers a noisy signal recovery problem, which we now translate to the finite-dimensional setting. Let $\mathbf{p},\mathbf{n} \in \mathbb{C}^m$ and $\mathcal{P} \subseteq \{1,\ldots,m\}$, set $\mathcal{P}^c = \{1,\ldots,m\}\backslash\mathcal{P}$, and suppose that we observe $\mathbf{y} = \mathbf{p}_{\mathcal{P}^c} + \mathbf{n}$. Note that the information contained in $\mathbf{p}_{\mathcal{P}}$ is completely lost in the observation. Without structural assumptions on $\mathbf{p}$, it is therefore not possible to recover information on $\mathbf{p}_{\mathcal{P}}$ from $\mathbf{y}$. However, if $\mathbf{p}$ is sufficiently sparse with respect to an orthonormal basis and $|\mathcal{P}|$ is sufficiently small, it turns out that all entries of $\mathbf{p}$ can be recovered in a linear fashion to within a precision determined by the noise level. This is often referred to in the literature as stable recovery [6]. The corresponding formal statement is as follows.

LEMMA 6.4 *Let* $\mathbf{U} \in \mathbb{C}^{m\times m}$ *be unitary,* $\mathcal{Q} \subseteq \{1,\ldots,m\}$, $\mathbf{p} \in \mathcal{W}^{\mathbf{U},\mathcal{Q}}$, *and consider*

$$\mathbf{y} = \mathbf{p}_{\mathcal{P}^c} + \mathbf{n}, \tag{6.65}$$

where $\mathbf{n} \in \mathbb{C}^m$ *and* $\mathcal{P}^c = \{1,\ldots,m\}\backslash\mathcal{P}$ *with* $\mathcal{P} \subseteq \{1,\ldots,m\}$. *If* $\Delta_{\mathcal{P},\mathcal{Q}}(\mathbf{U}) < 1$, *then there exists a matrix* $\mathbf{L} \in \mathbb{C}^{m\times m}$ *such that*

$$\|\mathbf{Ly}-\mathbf{p}\|_2 \leq C\|\mathbf{n}_{\mathcal{P}^c}\|_2 \tag{6.66}$$

with $C = 1/(1-\Delta_{\mathcal{P},\mathcal{Q}}(\mathbf{U}))$. *In particular,*

$$|\mathcal{P}||\mathcal{Q}| < \frac{1}{\mu^2([\mathbf{I}\ \mathbf{U}])} \tag{6.67}$$

is sufficient for $\Delta_{\mathcal{P},\mathcal{Q}}(\mathbf{U}) < 1$.

Proof For $\Delta_{\mathcal{P},\mathcal{Q}}(\mathbf{U}) < 1$, it follows that (see p. 301 of [33]) $(\mathbf{I} - \mathbf{D}_{\mathcal{P}}\mathbf{\Pi}_{\mathcal{Q}}(\mathbf{U}))$ is invertible with

$$|||(\mathbf{I} - \mathbf{D}_{\mathcal{P}}\mathbf{\Pi}_{\mathcal{Q}}(\mathbf{U}))^{-1}|||_2 \le \frac{1}{1 - |||\mathbf{D}_{\mathcal{P}}\mathbf{\Pi}_{\mathcal{Q}}(\mathbf{U})|||_2} \tag{6.68}$$

$$= \frac{1}{1 - \Delta_{\mathcal{P},\mathcal{Q}}(\mathbf{U})}. \tag{6.69}$$

We now set $\mathbf{L} = (\mathbf{I} - \mathbf{D}_{\mathcal{P}}\mathbf{\Pi}_{\mathcal{Q}}(\mathbf{U}))^{-1}\mathbf{D}_{\mathcal{P}^c}$ and note that

$$\mathbf{L}\mathbf{p}_{\mathcal{P}^c} = (\mathbf{I} - \mathbf{D}_{\mathcal{P}}\mathbf{\Pi}_{\mathcal{Q}}(\mathbf{U}))^{-1}\mathbf{p}_{\mathcal{P}^c} \tag{6.70}$$

$$= (\mathbf{I} - \mathbf{D}_{\mathcal{P}}\mathbf{\Pi}_{\mathcal{Q}}(\mathbf{U}))^{-1}(\mathbf{I} - \mathbf{D}_{\mathcal{P}})\mathbf{p} \tag{6.71}$$

$$= (\mathbf{I} - \mathbf{D}_{\mathcal{P}}\mathbf{\Pi}_{\mathcal{Q}}(\mathbf{U}))^{-1}(\mathbf{I} - \mathbf{D}_{\mathcal{P}}\mathbf{\Pi}_{\mathcal{Q}}(\mathbf{U}))\mathbf{p} \tag{6.72}$$

$$= \mathbf{p}, \tag{6.73}$$

where in (6.72) we used $\mathbf{\Pi}_{\mathcal{Q}}(\mathbf{U})\mathbf{p} = \mathbf{p}$, which is by assumption. Next, we upper-bound $\|\mathbf{L}\mathbf{y} - \mathbf{p}\|_2$ according to

$$\|\mathbf{L}\mathbf{y} - \mathbf{p}\|_2 = \|\mathbf{L}\mathbf{p}_{\mathcal{P}^c} + \mathbf{L}\mathbf{n} - \mathbf{p}\|_2 \tag{6.74}$$

$$= \|\mathbf{L}\mathbf{n}\|_2 \tag{6.75}$$

$$\le |||(\mathbf{I} - \mathbf{D}_{\mathcal{P}}\mathbf{\Pi}_{\mathcal{Q}}(\mathbf{U}))^{-1}|||_2 \|\mathbf{n}_{\mathcal{P}^c}\|_2 \tag{6.76}$$

$$\le \frac{1}{1 - \Delta_{\mathcal{P},\mathcal{Q}}(\mathbf{U})}\|\mathbf{n}_{\mathcal{P}^c}\|_2, \tag{6.77}$$

where in (6.75) we used (6.70)–(6.73). Finally, Lemma 6.2 implies that (6.67) is sufficient for $\Delta_{\mathcal{P},\mathcal{Q}}(\mathbf{U}) < 1$.

We next particularize Lemma 6.4 for $\mathbf{U} = \mathbf{F}$,

$$\mathcal{P} = \left\{\frac{m}{n}, \frac{2m}{n}, \ldots, \frac{(n-1)m}{n}, m\right\} \tag{6.78}$$

and

$$\mathcal{Q} = \{l+1, \ldots, l+n\}, \tag{6.79}$$

with $l \in \{1, \ldots, m\}$ and $\mathcal{Q}$ interpreted circularly in $\{1, \ldots, m\}$. This means that $\mathbf{p}$ is n-sparse in $\mathbf{F}$ and we are missing n entries in the noisy observation $\mathbf{y}$. From Lemma 6.1, we know that $\Delta_{\mathcal{P},\mathcal{Q}}(\mathbf{F}) = \sqrt{n/m}$. Since n divides m by assumption, stable recovery of $\mathbf{p}$ is possible for $n \le m/2$. In contrast, the coherence-based uncertainty relation in Lemma 6.2 yields $\Delta_{\mathcal{P},\mathcal{Q}}(\mathbf{F}) \le n/\sqrt{m}$, and would hence suggest that $n^2 < m$ is needed for stable recovery.

6.3 Uncertainty Relations in $(\mathbb{C}^m, \|\cdot\|_1)$

We introduce uncertainty relations in $(\mathbb{C}^m, \|\cdot\|_1)$ following the same story line as in Section 6.2. Specifically, let $\mathbf{U} = (\mathbf{u}_1 \ldots \mathbf{u}_m) \in \mathbb{C}^{m \times m}$ be a unitary matrix, $\mathcal{P}, \mathcal{Q} \subseteq \{1, \ldots, m\}$, and consider the orthogonal projection $\mathbf{\Pi}_{\mathcal{Q}}(\mathbf{U})$ onto the subspace $\mathcal{W}^{\mathbf{U},\mathcal{Q}}$, which is spanned by $\{\mathbf{u}_i : i \in \mathcal{Q}\}$. Let[2] $\Sigma_{\mathcal{P},\mathcal{Q}}(\mathbf{U}) = |||\mathbf{D}_{\mathcal{P}}\mathbf{\Pi}_{\mathcal{Q}}(\mathbf{U})|||_1$. By Lemma 6.12 we have

$$\Sigma_{\mathcal{P},\mathcal{Q}}(\mathbf{U}) = \max_{\mathbf{x}\in\mathcal{W}^{\mathbf{U},\mathcal{Q}}\setminus\{\mathbf{0}\}} \frac{\|\mathbf{x}_{\mathcal{P}}\|_1}{\|\mathbf{x}\|_1}. \tag{6.80}$$

An uncertainty relation in $(\mathbb{C}^m, \|\cdot\|_1)$ is an upper bound of the form $\Sigma_{\mathcal{P},\mathcal{Q}}(\mathbf{U}) \leq c$ with $c \geq 0$ and states that $\|\mathbf{x}_{\mathcal{P}}\|_1 \leq c\|\mathbf{x}\|_1$ for all $\mathbf{x} \in \mathcal{W}^{\mathbf{U},\mathcal{Q}}$. $\Sigma_{\mathcal{P},\mathcal{Q}}(\mathbf{U})$ hence quantifies how well a vector supported on $\mathcal{Q}$ in the basis $\mathbf{U}$ can be concentrated on $\mathcal{P}$, where now concentration is measured in terms of 1-norm. Again, an uncertainty relation in $(\mathbb{C}^m, \|\cdot\|_1)$ is non-trivial only if $c < 1$. Application of Lemma 6.14 yields

$$\frac{1}{m}\|\mathbf{D}_{\mathcal{P}}\mathbf{\Pi}_{\mathcal{Q}}(\mathbf{U})\|_1 \leq \Sigma_{\mathcal{P},\mathcal{Q}}(\mathbf{U}) \leq \|\mathbf{D}_{\mathcal{P}}\mathbf{\Pi}_{\mathcal{Q}}(\mathbf{U})\|_1, \tag{6.81}$$

which constitutes the 1-norm equivalent of (6.2).

6.3.1 Coherence-Based Uncertainty Relation

We next derive a coherence-based uncertainty relation for $(\mathbb{C}^m, \|\cdot\|_1)$, which comes with the same advantages and disadvantages as its 2-norm counterpart.

LEMMA 6.5 *Let $\mathbf{U} \in \mathbb{C}^{m\times m}$ be a unitary matrix and $\mathcal{P},\mathcal{Q} \subseteq \{1,\ldots,m\}$. Then,*

$$\Sigma_{\mathcal{P},\mathcal{Q}}(\mathbf{U}) \leq |\mathcal{P}||\mathcal{Q}|\mu^2([\mathbf{I}\ \ \mathbf{U}]). \tag{6.82}$$

Proof Let $\widetilde{\mathbf{u}}_i$ denote the column vectors of $\mathbf{U}^*$. It follows from Lemma 6.14 that

$$\Sigma_{\mathcal{P},\mathcal{Q}}(\mathbf{U}) = \max_{j\in\{1,\ldots,m\}} \|\mathbf{D}_{\mathcal{P}}\mathbf{U}\mathbf{D}_{\mathcal{Q}}\widetilde{\mathbf{u}}_j\|_1. \tag{6.83}$$

With

$$\max_{j\in\{1,\ldots,m\}} \|\mathbf{D}_{\mathcal{P}}\mathbf{U}\mathbf{D}_{\mathcal{Q}}\widetilde{\mathbf{u}}_j\|_1 \leq |\mathcal{P}| \max_{i,j\in\{1,\ldots,m\}} |\widetilde{\mathbf{u}}_i^*\mathbf{D}_{\mathcal{Q}}\widetilde{\mathbf{u}}_j| \tag{6.84}$$

$$\leq |\mathcal{P}||\mathcal{Q}| \max_{i,j,k\in\{1,\ldots,m\}} |\mathbf{U}_{i,k}||\mathbf{U}_{j,k}| \tag{6.85}$$

$$\leq |\mathcal{P}||\mathcal{Q}|\mu^2([\mathbf{I}\ \ \mathbf{U}]), \tag{6.86}$$

this establishes the proof.

For $\mathcal{P} = \{1\}$, $\mathcal{Q} = \{1,\ldots,m\}$, and $\mathbf{U} = \mathbf{F}$, the upper bounds on $\Sigma_{\mathcal{P},\mathcal{Q}}(\mathbf{F})$ in (6.81) and (6.82) coincide and equal 1. We next present an example where (6.82) is sharper

[2] In contrast to the operator 2-norm, the operator 1-norm is not invariant under unitary transformations, so that we do not have $|||\mathbf{\Pi}_{\mathcal{P}}(\mathbf{A})\mathbf{\Pi}_{\mathcal{Q}}(\mathbf{B})|||_1 = |||\mathbf{D}_{\mathcal{P}}\mathbf{\Pi}_{\mathcal{Q}}(\mathbf{A}^*\mathbf{B})|||_1$ for general unitary $\mathbf{A},\mathbf{B}$. This, however, does not constitute a problem as, whenever we apply uncertainty relations in $(\mathbb{C}^m, \|\cdot\|_1)$, the case of general unitary $\mathbf{A},\mathbf{B}$ can always be reduced directly to $\mathbf{\Pi}_{\mathcal{P}}(\mathbf{I}) = \mathbf{D}_{\mathcal{P}}$ and $\mathbf{\Pi}_{\mathcal{Q}}(\mathbf{A}^*\mathbf{B})$, simply by rewriting $\mathbf{Ap} = \mathbf{Bq}$ according to $\mathbf{p} = \mathbf{A}^*\mathbf{Bq}$.

than (6.81). Let m be even, $\mathcal{P} = \{m\}$, $Q = \{1,\ldots,m/2\}$, and $\mathbf{U} = \mathbf{F}$. Then, (6.82) becomes $\Sigma_{\mathcal{P},Q}(\mathbf{F}) \leq 1/2$, whereas

$$\|\mathbf{D}_{\mathcal{P}}\mathbf{\Pi}_Q(\mathbf{F})\|_1 = \frac{1}{m}\sum_{l=1}^{m}\left|\sum_{k=1}^{m/2} e^{\frac{2\pi jlk}{m}}\right| \tag{6.87}$$

$$= \frac{1}{2} + \frac{1}{m}\sum_{l=1}^{m-1}\left|\frac{1-e^{\pi jl}}{1-e^{\frac{2\pi jl}{m}}}\right| \tag{6.88}$$

$$= \frac{1}{2} + \frac{2}{m}\sum_{l=1}^{m/2}\frac{1}{\left|1-e^{\frac{2\pi j(2l-1)}{m}}\right|} \tag{6.89}$$

$$= \frac{1}{2} + \frac{1}{m}\sum_{l=1}^{m/2}\frac{1}{\sin(\pi(2l-1)/m)}. \tag{6.90}$$

Applying Jensen's inequality, Theorem 2.6.2 of [34], to (6.90) and using $\sum_{l=1}^{m/2}(2l-1) = (m/2)^2$ then yields $\|\mathbf{D}_{\mathcal{P}}\mathbf{\Pi}_Q(\mathbf{F})\|_1 \geq 1$, which shows that (6.81) is trivial.

For $\mathcal{P}$ and Q as in (6.11) and (6.12), respectively, (6.82) becomes $\Sigma_{\mathcal{P},Q}(\mathbf{F}) \leq n^2/m$, which for fixed ratio n/m increases linearly in m and becomes trivial for $m \geq (m/n)^2$. A more sophisticated uncertainty relation based on a large sieve inequality exists for strictly bandlimited (infinite) ℓ_1-sequences, see Theorem 14 of [7]; a corresponding finite-dimensional result does not seem to be available.

6.3.2 Concentration Inequalities

Analogously to Section 6.2.3, we next ask how well concentrated a given signal vector can be in two different orthonormal bases. Here, however, we consider a different measure of concentration accounting for the fact that we are dealing with the 1-norm.

DEFINITION 6.3 *Let $\mathcal{P} \subseteq \{1,\ldots,m\}$ and $\varepsilon_{\mathcal{P}} \in [0,1]$. The vector $\mathbf{x} \in \mathbb{C}^m$ is said to be $\varepsilon_{\mathcal{P}}$-concentrated if $\|\mathbf{x}-\mathbf{x}_{\mathcal{P}}\|_1 \leq \varepsilon_{\mathcal{P}}\|\mathbf{x}\|_1$.*

The fraction of 1-norm an $\varepsilon_{\mathcal{P}}$-concentrated vector exhibits outside $\mathcal{P}$ is therefore no more than $\varepsilon_{\mathcal{P}}$. In particular, if $\mathbf{x}$ is $\varepsilon_{\mathcal{P}}$-concentrated for $\varepsilon_{\mathcal{P}} = 0$, then $\mathbf{x} = \mathbf{x}_{\mathcal{P}}$ and $\mathbf{x}$ is $|\mathcal{P}|$-sparse. The zero vector is trivially $\varepsilon_{\mathcal{P}}$-concentrated for all $\mathcal{P} \subseteq \{1,\ldots,m\}$ and $\varepsilon_{\mathcal{P}} \in [0,1]$. In the remainder of Section 6.3, concentration is with respect to the 1-norm according to Definition 6.3.

We are now ready to state the announced result on the concentration of a vector in two different orthonormal bases.

LEMMA 6.6 *Let $\mathbf{A},\mathbf{B} \in \mathbb{C}^{m\times m}$ be unitary and $\mathcal{P},Q \subseteq \{1,\ldots,m\}$. Suppose that there exist a non-zero $\varepsilon_{\mathcal{P}}$-concentrated $\mathbf{p} \in \mathbb{C}^m$ and a non-zero $\mathbf{q} \in \mathbb{C}^m$ with $\mathbf{q} = \mathbf{q}_Q$ such that $\mathbf{Ap} = \mathbf{Bq}$. Then,*

$$|\mathcal{P}||Q| \geq \frac{1-\varepsilon_{\mathcal{P}}}{\mu^2([\mathbf{A}\ \ \mathbf{B}])}. \tag{6.91}$$

Proof Rewriting $\mathbf{Ap} = \mathbf{Bq}$ according to $\mathbf{p} = \mathbf{A}^*\mathbf{Bq}$, it follows that $\mathbf{p} \in \mathcal{W}^{\mathbf{U},Q}$ with $\mathbf{U} = \mathbf{A}^*\mathbf{B}$. We have

$$1 - \varepsilon_{\mathcal{P}} \leq \frac{\|\mathbf{p}_{\mathcal{P}}\|_1}{\|\mathbf{p}\|_1} \tag{6.92}$$

$$\leq \Sigma_{\mathcal{P},Q}(\mathbf{U}) \tag{6.93}$$

$$\leq |\mathcal{P}||Q|\mu^2([\mathbf{I}\ \ \mathbf{U}]), \tag{6.94}$$

where (6.92) is by $\varepsilon_{\mathcal{P}}$-concentration of $\mathbf{p}$, (6.93) follows from (6.80) and $\mathbf{p} \in \mathcal{W}^{\mathbf{U},Q}$, and in (6.94) we applied Lemma 6.5. The proof is concluded by noting that $\mu([\mathbf{I}\ \ \mathbf{U}]) = \mu([\mathbf{A}\ \ \mathbf{B}])$.

For $\varepsilon_{\mathcal{P}} = 0$, Lemma 6.6 recovers Corollary 6.2.

6.3.3 Noisy Recovery in $(\mathbb{C}^m, \|\cdot\|_1)$

We next consider a noisy signal recovery problem akin to that in Section 6.2.4. Specifically, we investigate recovery–through 1-norm minimization–of a sparse signal corrupted by $\varepsilon_{\mathcal{P}}$-concentrated noise.

LEMMA 6.7 *Let*

$$\mathbf{y} = \mathbf{p} + \mathbf{n}, \tag{6.95}$$

where $\mathbf{n} \in \mathbb{C}^m$ *is* $\varepsilon_{\mathcal{P}}$*-concentrated to* $\mathcal{P} \subseteq \{1, \ldots, m\}$ *and* $\mathbf{p} \in \mathcal{W}^{\mathbf{U},Q}$ *for* $\mathbf{U} \in \mathbb{C}^{m\times m}$ *unitary and* $Q \subseteq \{1, \ldots, m\}$. *Denote*

$$\mathbf{z} = \underset{\mathbf{w}\in\mathcal{W}^{\mathbf{U},Q}}{\operatorname{argmin}}(\|\mathbf{y} - \mathbf{w}\|_1). \tag{6.96}$$

If $\Sigma_{\mathcal{P},Q}(\mathbf{U}) < 1/2$, *then* $\|\mathbf{z} - \mathbf{p}\|_1 \leq C\varepsilon_{\mathcal{P}}\|\mathbf{n}\|_1$ *with* $C = 2/(1 - 2\Sigma_{\mathcal{P},Q}(\mathbf{U}))$. *In particular,*

$$|\mathcal{P}||Q| < \frac{1}{2\mu^2([\mathbf{I}\ \ \mathbf{U}])} \tag{6.97}$$

is sufficient for $\Sigma_{\mathcal{P},Q}(\mathbf{U}) < 1/2$.

Proof Set $\mathcal{P}^c = \{1, \ldots, m\}\backslash\mathcal{P}$ and let $\mathbf{q} = \mathbf{U}^*\mathbf{p}$. Note that $\mathbf{q}_Q = \mathbf{q}$ as a consequence of $\mathbf{p} \in \mathcal{W}^{\mathbf{U},Q}$, which is by assumption. We have

$$\|\mathbf{n}\|_1 = \|\mathbf{y} - \mathbf{p}\|_1 \tag{6.98}$$

$$\geq \|\mathbf{y} - \mathbf{z}\|_1 \tag{6.99}$$

$$= \|\mathbf{n} - \widetilde{\mathbf{z}}\|_1 \tag{6.100}$$

$$= \|(\mathbf{n} - \widetilde{\mathbf{z}})_{\mathcal{P}}\|_1 + \|(\mathbf{n} - \widetilde{\mathbf{z}})_{\mathcal{P}^c}\|_1 \tag{6.101}$$

$$\geq \|\mathbf{n}_{\mathcal{P}}\|_1 - \|\mathbf{n}_{\mathcal{P}^c}\|_1 + \|\widetilde{\mathbf{z}}_{\mathcal{P}^c}\|_1 - \|\widetilde{\mathbf{z}}_{\mathcal{P}}\|_1 \tag{6.102}$$

$$= \|\mathbf{n}\|_1 - 2\|\mathbf{n}_{\mathcal{P}^c}\|_1 + \|\widetilde{\mathbf{z}}\|_1 - 2\|\widetilde{\mathbf{z}}_{\mathcal{P}}\|_1 \tag{6.103}$$

$$\geq \|\mathbf{n}\|_1(1 - 2\varepsilon_{\mathcal{P}}) + \|\widetilde{\mathbf{z}}\|_1(1 - 2\Sigma_{\mathcal{P},Q}(\mathbf{U})), \tag{6.104}$$

where in (6.100) we set $\widetilde{\mathbf{z}} = \mathbf{z} - \mathbf{p}$, in (6.102) we applied the reverse triangle inequality, and in (6.104) we used that $\mathbf{n}$ is $\varepsilon_{\mathcal{P}}$-concentrated and $\widetilde{\mathbf{z}} \in \mathcal{W}^{\mathbf{U},Q}$, owing to $\mathbf{z} \in \mathcal{W}^{\mathbf{U},Q}$ and $\mathbf{p} \in \mathcal{W}^{\mathbf{U},Q}$, together with (6.80). This yields

$$\|\mathbf{z} - \mathbf{p}\|_1 = \|\widetilde{\mathbf{z}}\|_1 \tag{6.105}$$

$$\leq \frac{2\varepsilon_{\mathcal{P}}}{1 - 2\Sigma_{\mathcal{P},Q}(\mathbf{U})}\|\mathbf{n}\|_1. \tag{6.106}$$

Finally, (6.97) implies $\Sigma_{\mathcal{P},Q}(\mathbf{U}) < 1/2$ thanks to (6.82).

Note that, for $\varepsilon_{\mathcal{P}} = 0$, i.e., the noise vector is supported on $\mathcal{P}$, we can recover $\mathbf{p}$ from $\mathbf{y} = \mathbf{p} + \mathbf{n}$ perfectly, provided that $\Sigma_{\mathcal{P},Q}(\mathbf{U}) < 1/2$. For the special case $\mathbf{U} = \mathbf{F}$, this is guaranteed by

$$|\mathcal{P}||Q| < \frac{m}{2}, \tag{6.107}$$

and perfect recovery of $\mathbf{p}$ from $\mathbf{y} = \mathbf{p} + \mathbf{n}$ amounts to the finite-dimensional version of what is known as Logan's phenomenon (see Section 6.2 of [6]).

6.3.4 Coherence-Based Uncertainty Relation for Pairs of General Matrices

In practice, one is often interested in sparse signal representations with respect to general (i.e., possibly redundant or incomplete) dictionaries. The purpose of this section is to provide a corresponding general uncertainty relation. Specifically, we consider representations of a given signal vector $\mathbf{s}$ according to $\mathbf{s} = \mathbf{A}\mathbf{p} = \mathbf{B}\mathbf{q}$, where $\mathbf{A} \in \mathbb{C}^{m\times p}$ and $\mathbf{B} \in \mathbb{C}^{m\times q}$ are general matrices, $\mathbf{p} \in \mathbb{C}^p$, and $\mathbf{q} \in \mathbb{C}^q$. We start by introducing the notion of mutual coherence for pairs of matrices.

DEFINITION 6.4 *For* $\mathbf{A} = (\mathbf{a}_1 \ldots \mathbf{a}_p) \in \mathbb{C}^{m\times p}$ *and* $\mathbf{B} = (\mathbf{b}_1 \ldots \mathbf{b}_q) \in \mathbb{C}^{m\times q}$*, both with columns* $\|\cdot\|_2$*-normalized to* 1*, the mutual coherence* $\bar{\mu}(\mathbf{A},\mathbf{B})$ *is defined as* $\bar{\mu}(\mathbf{A},\mathbf{B}) = \max_{i,j} |\mathbf{a}_i^*\mathbf{b}_j|$.

The general uncertainty relation we are now ready to state is in terms of a pair of upper bounds on $\|\mathbf{p}_{\mathcal{P}}\|_1$ and $\|\mathbf{q}_Q\|_1$ for $\mathcal{P} \subseteq \{1,\ldots,p\}$ and $Q \subseteq \{1,\ldots,q\}$.

THEOREM 6.2 *Let* $\mathbf{A} \in \mathbb{C}^{m\times p}$ *and* $\mathbf{B} \in \mathbb{C}^{m\times q}$*, both with column vectors* $\|\cdot\|_2$*-normalized to* 1*, and consider* $\mathbf{p} \in \mathbb{C}^p$ *and* $\mathbf{q} \in \mathbb{C}^q$*. Suppose that* $\mathbf{A}\mathbf{p} = \mathbf{B}\mathbf{q}$*. Then, we have*

$$\|\mathbf{p}_{\mathcal{P}}\|_1 \leq |\mathcal{P}|\left(\frac{\mu(\mathbf{A})\|\mathbf{p}\|_1 + \bar{\mu}(\mathbf{A},\mathbf{B})\|\mathbf{q}\|_1}{1 + \mu(\mathbf{A})}\right) \tag{6.108}$$

for all $\mathcal{P} \subseteq \{1,\ldots,p\}$ *and, by symmetry,*

$$\|\mathbf{q}_Q\|_1 \leq |Q|\left(\frac{\mu(\mathbf{B})\|\mathbf{q}\|_1 + \bar{\mu}(\mathbf{A},\mathbf{B})\|\mathbf{p}\|_1}{1 + \mu(\mathbf{B})}\right) \tag{6.109}$$

for all $Q \subseteq \{1,\ldots,q\}$.

Proof Since (6.109) follows from (6.108) simply by replacing $\mathbf{A}$ by $\mathbf{B}$, $\mathbf{p}$ by $\mathbf{q}$, and $\mathcal{P}$ by $\mathcal{Q}$, and noting that $\bar{\mu}(\mathbf{A},\mathbf{B}) = \bar{\mu}(\mathbf{B},\mathbf{A})$, it suffices to prove (6.108). Let $\mathcal{P} \subseteq \{1,\ldots,p\}$ and consider an arbitrary but fixed $i \in \{1,\ldots,p\}$. Multiplying $\mathbf{Ap} = \mathbf{Bq}$ from the left by $\mathbf{a}_i^*$ and taking absolute values results in

$$|\mathbf{a}_i^*\mathbf{Ap}| = |\mathbf{a}_i^*\mathbf{Bq}|. \tag{6.110}$$

The left-hand side of (6.110) can be lower-bounded according to

$$\begin{aligned}
|\mathbf{a}_i^*\mathbf{Ap}| &= \left| p_i + \sum_{\substack{k=1\\k\neq i}}^{p} \mathbf{a}_i^*\mathbf{a}_k p_k \right| &(6.111)\\
&\geq |p_i| - \left| \sum_{\substack{k=1\\k\neq i}}^{p} \mathbf{a}_i^*\mathbf{a}_k p_k \right| &(6.112)\\
&\geq |p_i| - \sum_{\substack{k=1\\k\neq i}}^{p} |\mathbf{a}_i^*\mathbf{a}_k||p_k| &(6.113)\\
&\geq |p_i| - \mu(\mathbf{A}) \sum_{\substack{k=1\\k\neq i}}^{p} |p_k| &(6.114)\\
&= (1+\mu(\mathbf{A}))|p_i| - \mu(\mathbf{A})\|\mathbf{p}\|_1, &(6.115)
\end{aligned}$$

where (6.112) is by the reverse triangle inequality and in (6.114) we used Definition 6.1. Next, we upper-bound the right-hand side of (6.110) according to

$$\begin{aligned}
|\mathbf{a}_i^*\mathbf{Bq}| &= \left| \sum_{k=1}^{q} \mathbf{a}_i^*\mathbf{b}_k q_k \right| &(6.116)\\
&\leq \sum_{k=1}^{q} |\mathbf{a}_i^*\mathbf{b}_k||q_k| &(6.117)\\
&\leq \bar{\mu}(\mathbf{A},\mathbf{B})\|\mathbf{q}\|_1, &(6.118)
\end{aligned}$$

where the last step is by Definition 6.4. Combining the lower bound (6.111)–(6.115) and the upper bound (6.116)–(6.118) yields

$$(1+\mu(\mathbf{A}))|p_i| - \mu(\mathbf{A})\|\mathbf{p}\|_1 \leq \bar{\mu}(\mathbf{A},\mathbf{B})\|\mathbf{q}\|_1. \tag{6.119}$$

Since (6.119) holds for arbitrary $i \in \{1,\ldots,p\}$, we can sum over all $i \in \mathcal{P}$ and get

$$\|\mathbf{p}_{\mathcal{P}}\|_1 \leq |\mathcal{P}| \left(\frac{\mu(\mathbf{A})\|\mathbf{p}\|_1 + \bar{\mu}(\mathbf{A},\mathbf{B})\|\mathbf{q}\|_1}{1+\mu(\mathbf{A})} \right). \tag{6.120}$$

For the special case $\mathbf{A} = \mathbf{I} \in \mathbb{C}^{m\times m}$ and $\mathbf{B} \in \mathbb{C}^{m\times m}$ with $\mathbf{B}$ unitary, we have $\mu(\mathbf{A}) = \mu(\mathbf{B}) = 0$ and $\bar{\mu}(\mathbf{I},\mathbf{B}) = \mu([\mathbf{I}\ \ \mathbf{B}])$, so that (6.108) and (6.109) simplify to

$$\|\mathbf{p}_{\mathcal{P}}\|_1 \leq |\mathcal{P}|\mu([\mathbf{I}\ \ \mathbf{B}])\,\|\mathbf{q}\|_1 \tag{6.121}$$

and

$$\|\mathbf{q}_{\mathcal{Q}}\|_1 \leq |\mathcal{Q}|\mu([\mathbf{I}\ \ \mathbf{B}])\|\mathbf{p}\|_1, \tag{6.122}$$

respectively. Thus, for arbitrary but fixed $\mathbf{p} \in \mathcal{W}^{\mathbf{B},\mathcal{Q}}$ and $\mathbf{q} = \mathbf{B}^*\mathbf{p}$, we have $\mathbf{q}_{\mathcal{Q}} = \mathbf{q}$ so that (6.121) and (6.122) taken together yield

$$\|\mathbf{p}_{\mathcal{P}}\|_1 \leq |\mathcal{P}||\mathcal{Q}|\mu^2([\mathbf{I}\ \ \mathbf{B}])\|\mathbf{p}\|_1. \tag{6.123}$$

As $\mathbf{p}$ was assumed to be arbitrary, by (6.80) this recovers the uncertainty relation

$$\Sigma_{\mathcal{P},\mathcal{Q}}(\mathbf{B}) \leq |\mathcal{P}||\mathcal{Q}|\mu^2([\mathbf{I}\ \ \mathbf{B}]) \tag{6.124}$$

in Lemma 6.5.

6.3.5 Concentration Inequalities for Pairs of General Matrices

We next refine the result in Theorem 6.2 to vectors that are concentrated in 1-norm according to Definition 6.3. The formal statement is as follows.

COROLLARY 6.3 *Let $\mathbf{A} \in \mathbb{C}^{m\times p}$ and $\mathbf{B} \in \mathbb{C}^{m\times q}$, both with column vectors $\|\cdot\|_2$-normalized to 1, $\mathcal{P} \subseteq \{1,\ldots,p\}$, $\mathcal{Q} \subseteq \{1,\ldots,q\}$, $\mathbf{p} \in \mathbb{C}^p$, and $\mathbf{q} \in \mathbb{C}^q$. Suppose that $\mathbf{Ap} = \mathbf{Bq}$. Then, the following statements hold.*

1. *If $\mathbf{q}$ is $\varepsilon_{\mathcal{Q}}$-concentrated, then*

$$\|\mathbf{p}_{\mathcal{P}}\|_1 \leq \frac{|\mathcal{P}|}{1+\mu(\mathbf{A})}\left(\mu(\mathbf{A}) + \frac{\bar{\mu}^2(\mathbf{A},\mathbf{B})|\mathcal{Q}|}{[(1+\mu(\mathbf{B}))(1-\varepsilon_{\mathcal{Q}})-\mu(\mathbf{B})|\mathcal{Q}|]_+}\right)\|\mathbf{p}\|_1. \tag{6.125}$$

2. *If $\mathbf{p}$ is $\varepsilon_{\mathcal{P}}$-concentrated, then*

$$\|\mathbf{q}_{\mathcal{Q}}\|_1 \leq \frac{|\mathcal{Q}|}{1+\mu(\mathbf{B})}\left(\mu(\mathbf{B}) + \frac{\bar{\mu}^2(\mathbf{A},\mathbf{B})|\mathcal{P}|}{[(1+\mu(\mathbf{A}))(1-\varepsilon_{\mathcal{P}})-\mu(\mathbf{A})|\mathcal{P}|]_+}\right)\|\mathbf{q}\|_1. \tag{6.126}$$

3. *If $\mathbf{p}$ is $\varepsilon_{\mathcal{P}}$-concentrated, $\mathbf{q}$ is $\varepsilon_{\mathcal{Q}}$-concentrated, $\bar{\mu}(\mathbf{A},\mathbf{B}) > 0$, and $(\mathbf{p}^{\mathrm{T}}\ \mathbf{q}^{\mathrm{T}})^{\mathrm{T}} \neq \mathbf{0}$, then*

$$|\mathcal{P}||\mathcal{Q}| \geq \frac{[(1+\mu(\mathbf{A}))(1-\varepsilon_{\mathcal{P}})-\mu(\mathbf{A})|\mathcal{P}|]_+[(1+\mu(\mathbf{B}))(1-\varepsilon_{\mathcal{Q}})-\mu(\mathbf{B})|\mathcal{Q}|]_+}{\bar{\mu}^2(\mathbf{A},\mathbf{B})}. \tag{6.127}$$

Proof By Theorem 6.2, we have

$$\|\mathbf{p}_{\mathcal{P}}\|_1 \leq |\mathcal{P}|\left(\frac{\mu(\mathbf{A})\|\mathbf{p}\|_1 + \bar{\mu}(\mathbf{A},\mathbf{B})\|\mathbf{q}\|_1}{1+\mu(\mathbf{A})}\right) \tag{6.128}$$

and

$$\|\mathbf{q}_{\mathcal{Q}}\|_1 \leq |\mathcal{Q}|\left(\frac{\mu(\mathbf{B})\|\mathbf{q}\|_1 + \bar{\mu}(\mathbf{A},\mathbf{B})\|\mathbf{p}\|_1}{1+\mu(\mathbf{B})}\right). \tag{6.129}$$

Suppose now that $\mathbf{q}$ is $\varepsilon_{\mathcal{Q}}$-concentrated, i.e., $\|\mathbf{q}_{\mathcal{Q}}\|_1 \geq (1-\varepsilon_{\mathcal{Q}})\|\mathbf{q}\|_1$. Then (6.129) implies that

$$\|\mathbf{q}\|_1 \leq \frac{|\mathcal{Q}|\bar{\mu}(\mathbf{A},\mathbf{B})}{[(1+\mu(\mathbf{B}))(1-\varepsilon_{\mathcal{Q}})-\mu(\mathbf{B})|\mathcal{Q}|]_+}\|\mathbf{p}\|_1. \tag{6.130}$$

Using (6.130) in (6.128) yields (6.125). The relation (6.126) follows from (6.125) by swapping the roles of $\mathbf{A}$ and $\mathbf{B}$, $\mathbf{p}$ and $\mathbf{q}$, and $\mathcal{P}$ and $\mathcal{Q}$, and upon noting that $\bar{\mu}(\mathbf{A},\mathbf{B}) = \bar{\mu}(\mathbf{B},\mathbf{A})$. It remains to establish (6.127). Using $\|\mathbf{p}_\mathcal{P}\|_1 \geq (1-\varepsilon_\mathcal{P})\|\mathbf{p}\|_1$ in (6.128) and $\|\mathbf{q}_\mathcal{Q}\|_1 \geq (1-\varepsilon_\mathcal{Q})\|\mathbf{q}\|_1$ in (6.129) yields

$$\|\mathbf{p}\|_1[(1+\mu(\mathbf{A}))(1-\varepsilon_\mathcal{P})-\mu(\mathbf{A})|\mathcal{P}|]_+ \leq \bar{\mu}(\mathbf{A},\mathbf{B})\|\mathbf{q}\|_1|\mathcal{P}| \tag{6.131}$$

and

$$\|\mathbf{q}\|_1[(1+\mu(\mathbf{B}))(1-\varepsilon_\mathcal{Q})-\mu(\mathbf{B})|\mathcal{Q}|]_+ \leq \bar{\mu}(\mathbf{A},\mathbf{B})\|\mathbf{p}\|_1|\mathcal{Q}|, \tag{6.132}$$

respectively. Suppose first that $\mathbf{p} = \mathbf{0}$. Then $\mathbf{q} \neq \mathbf{0}$ by assumption, and (6.132) becomes

$$[(1+\mu(\mathbf{B}))(1-\varepsilon_\mathcal{Q})-\mu(\mathbf{B})|\mathcal{Q}|]_+ = 0. \tag{6.133}$$

In this case (6.127) holds trivially. Similarly, if $\mathbf{q} = \mathbf{0}$, then $\mathbf{p} \neq \mathbf{0}$ again by assumption, and (6.131) becomes

$$[(1+\mu(\mathbf{A}))(1-\varepsilon_\mathcal{P})-\mu(\mathbf{A})|\mathcal{P}|]_+ = 0. \tag{6.134}$$

As before, (6.127) holds trivially. Finally, if $\mathbf{p} \neq \mathbf{0}$ and $\mathbf{q} \neq \mathbf{0}$, then we multiply (6.131) by (6.132) and divide the result by $\bar{\mu}^2(\mathbf{A},\mathbf{B})\|\mathbf{p}\|_1\|\mathbf{q}\|_1$, which yields (6.127).

Corollary 6.3 will be used in Section 6.4 to derive recovery thresholds for sparse signal separation. The lower bound on $|\mathcal{P}||\mathcal{Q}|$ in (6.127) is Theorem 1 of [9] and states that a non-zero vector cannot be arbitrarily well concentrated with respect to two different general matrices $\mathbf{A}$ and $\mathbf{B}$. For the special case $\varepsilon_\mathcal{Q} = 0$ and $\mathbf{A}$ and $\mathbf{B}$ unitary, and hence $\mu(\mathbf{A}) = \mu(\mathbf{B}) = 0$ and $\bar{\mu}(\mathbf{A},\mathbf{B}) = \mu([\mathbf{A}\ \ \mathbf{B}])$, (6.127) recovers Lemma 6.6.

Particularizing (6.127) to $\varepsilon_\mathcal{P} = \varepsilon_\mathcal{Q} = 0$ yields the following result.

COROLLARY 6.4 *(Lemma 33 of [10]) Let $\mathbf{A} \in \mathbb{C}^{m\times p}$ and $\mathbf{B} \in \mathbb{C}^{m\times q}$, both with column vectors $\|\cdot\|_2$-normalized to 1, and consider $\mathbf{p} \in \mathbb{C}^p$ and $\mathbf{q} \in \mathbb{C}^q$ with $(\mathbf{p}^\mathrm{T}\ \mathbf{q}^\mathrm{T})^\mathrm{T} \neq \mathbf{0}$. Suppose that $\mathbf{Ap} = \mathbf{Bq}$. Then, $\|\mathbf{p}\|_0\|\mathbf{q}\|_0 \geq f_{\mathbf{A},\mathbf{B}}(\|\mathbf{p}\|_0, \|\mathbf{q}\|_0)$, where*

$$f_{\mathbf{A},\mathbf{B}}(u,v) = \frac{[1+\mu(\mathbf{A})(1-u)]_+[1+\mu(\mathbf{B})(1-v)]_+}{\bar{\mu}^2(\mathbf{A},\mathbf{B})}. \tag{6.135}$$

Proof Let $\mathcal{P} = \{i \in \{1,\ldots,p\} : p_i \neq 0\}$ and $\mathcal{Q} = \{i \in \{1,\ldots,q\} : q_i \neq 0\}$, so that $\mathbf{p}_\mathcal{P} = \mathbf{p}$, $\mathbf{q}_\mathcal{Q} = \mathbf{q}$, $|\mathcal{P}| = \|\mathbf{p}\|_0$, and $|\mathcal{Q}| = \|\mathbf{q}\|_0$. The claim now follows directly from (6.127) with $\varepsilon_\mathcal{P} = \varepsilon_\mathcal{Q} = 0$.

If $\mathbf{A}$ and $\mathbf{B}$ are both unitary, then $\mu(\mathbf{A}) = \mu(\mathbf{B}) = 0$ and $\bar{\mu}(\mathbf{A},\mathbf{B}) = \mu([\mathbf{A}\ \ \mathbf{B}])$, and Corollary 6.4 recovers the Elad–Bruckstein result in Corollary 6.2.

Corollary 6.4 admits the following appealing geometric interpretation in terms of a null-space property, which will be seen in Section 6.5 to pave the way to an extension of the classical notion of sparsity to a more general concept of parsimony.

LEMMA 6.8 *Let $\mathbf{A} \in \mathbb{C}^{m\times p}$ and $\mathbf{B} \in \mathbb{C}^{m\times q}$, both with column vectors $\|\cdot\|_2$-normalized to 1. Then, the set (which actually is a finite union of subspaces)*

$$\mathcal{S} = \left\{\begin{pmatrix}\mathbf{p}\\ \mathbf{q}\end{pmatrix} : \mathbf{p} \in \mathbb{C}^p,\ \mathbf{q} \in \mathbb{C}^q,\ \|\mathbf{p}\|_0\|\mathbf{q}\|_0 < f_{\mathbf{A},\mathbf{B}}(\|\mathbf{p}\|_0, \|\mathbf{q}\|_0)\right\} \tag{6.136}$$

with $f_{\mathbf{A},\mathbf{B}}$ *defined in* (6.135) *intersects the kernel of* $[\mathbf{A}\ \ \mathbf{B}]$ *trivially, i.e.,*

$$\ker([\mathbf{A}\ \ \mathbf{B}]) \cap \mathcal{S} = \{\mathbf{0}\}. \tag{6.137}$$

Proof The statement of this lemma is equivalent to the statement of Corollary 6.4 through a chain of equivalences between the following statements:

(1) $\ker([\mathbf{A}\ \ \mathbf{B}]) \cap \mathcal{S} = \{\mathbf{0}\}$,

(2) if $(\mathbf{p}^{\mathrm{T}}\ -\mathbf{q}^{\mathrm{T}})^{\mathrm{T}} \in \ker([\mathbf{A}\ \ \mathbf{B}])\backslash\{\mathbf{0}\}$, then $\|\mathbf{p}\|_0\|\mathbf{q}\|_0 \geq f_{\mathbf{A},\mathbf{B}}(\|\mathbf{p}\|_0, \|\mathbf{q}\|_0)$,

(3) if $\mathbf{Ap} = \mathbf{Bq}$ with $(\mathbf{p}^{\mathrm{T}}\ \mathbf{q}^{\mathrm{T}})^{\mathrm{T}} \neq \mathbf{0}$, then $\|\mathbf{p}\|_0\|\mathbf{q}\|_0 \geq f_{\mathbf{A},\mathbf{B}}(\|\mathbf{p}\|_0, \|\mathbf{q}\|_0)$,

where (1) $\Leftrightarrow$ (2) is by the definition of $\mathcal{S}$, (2) $\Leftrightarrow$ (3) follows from the fact that $\mathbf{Ap} = \mathbf{Bq}$ with $(\mathbf{p}^{\mathrm{T}}\ \mathbf{q}^{\mathrm{T}})^{\mathrm{T}} \neq \mathbf{0}$ is equivalent to $(\mathbf{p}^{\mathrm{T}}\ -\mathbf{q}^{\mathrm{T}})^{\mathrm{T}} \in \ker([\mathbf{A}\ \ \mathbf{B}])\backslash\{\mathbf{0}\}$, and (3) is the statement in Corollary 6.4.

6.4 Sparse Signal Separation

Numerous practical signal recovery tasks can be cast as sparse signal separation problems of the following form. We want to recover $\mathbf{y} \in \mathbb{C}^p$ with $\|\mathbf{y}\|_0 \leq s$ and/or $\mathbf{z} \in \mathbb{C}^q$ with $\|\mathbf{z}\|_0 \leq t$ from the noiseless observation

$$\mathbf{w} = \mathbf{Ay} + \mathbf{Bz}, \tag{6.138}$$

where $\mathbf{A} \in \mathbb{C}^{m\times p}$ and $\mathbf{B} \in \mathbb{C}^{m\times q}$. Here, s and t are the sparsity levels of $\mathbf{y}$ and $\mathbf{z}$ with corresponding ambient dimensions p and q, respectively. Prominent applications include (image) inpainting, declipping, super-resolution, the recovery of signals corrupted by impulse noise, and the separation of (e.g., audio or video) signals into two distinct components (see Section I of [9]). We next briefly describe some of these problems.

1. *Clipping.* Non-linearities in power-amplifiers or in analog-to-digital converters often cause signal clipping or saturation [35]. This effect can be cast into the signal model (6.138) by setting $\mathbf{B} = \mathbf{I}$, identifying $\mathbf{s} = \mathbf{Ay}$ with the signal to be clipped, and setting $\mathbf{z} = (g_a(\mathbf{s}) - \mathbf{s})$ with $g_a(\cdot)$ realizing entry-wise clipping of the amplitude to the interval $[0, a]$. If the clipping level a is not too small, then $\mathbf{z}$ will be sparse, i.e., $t \ll q$.
2. *Missing entries.* Our framework also encompasses super-resolution [36, 37] and inpainting [38] of, e.g., images, audio, and video signals. In both these applications only a subset of the entries of the (full-resolution) signal vector $\mathbf{s} = \mathbf{Ay}$ is available and the task is to fill in the missing entries, which are accounted for by writing $\mathbf{w} = \mathbf{s} + \mathbf{z}$ with $z_i = -s_i$ if the ith entry of $\mathbf{s}$ is missing and $z_i = 0$ else. If the number of entries missing is not too large, then $\mathbf{z}$ is sparse, i.e., $t \ll q$.
3. *Signal separation.* Separation of (audio, image, or video) signals into two structurally distinct components also fits into the framework described above. A prominent example is the separation of texture from cartoon parts in images (see [39, 40] and references therein). The matrices $\mathbf{A}$ and $\mathbf{B}$ are chosen to allow sparse representations of the two distinct features. Note that here $\mathbf{Bz}$ no longer plays the role of undesired noise, and the goal is to recover both $\mathbf{y}$ and $\mathbf{z}$ from the observation $\mathbf{w} = \mathbf{Ay} + \mathbf{Bz}$.

The first two examples above demonstrate that in many practically relevant applications the locations of the possibly non-zero entries of one of the sparse vectors, say $\mathbf{z}$, may be known. This can be accounted for by removing the columns of $\mathbf{B}$ corresponding to the other entries, which results in $t = q$, i.e., the sparsity level of $\mathbf{z}$ equals the ambient dimension. We next show how Corollary 6.3 can be used to state a sufficient condition for recovery of $\mathbf{y}$ from $\mathbf{w} = \mathbf{A}\mathbf{y} + \mathbf{B}\mathbf{z}$ when $t = q$. For recovery guarantees in the case where the sparsity levels of both $\mathbf{y}$ and $\mathbf{z}$ are strictly smaller than their corresponding ambient dimensions, we refer to Theorem 8 of [9].

THEOREM 6.3 *(Theorems 4 and 7 of [9]) Let* $\mathbf{y} \in \mathbb{C}^p$ *with* $\|\mathbf{y}\|_0 \le s$, $\mathbf{z} \in \mathbb{C}^q$, $\mathbf{A} \in \mathbb{C}^{m\times p}$, *and* $\mathbf{B} \in \mathbb{C}^{m\times q}$, *both with column vectors* $\|\cdot\|_2$*-normalized to* 1 *and* $\bar{\mu}(\mathbf{A},\mathbf{B}) > 0$. *Suppose that*

$$2sq < f_{\mathbf{A},\mathbf{B}}(2s, q) \tag{6.139}$$

with

$$f_{\mathbf{A},\mathbf{B}}(u,v) = \frac{[1+\mu(\mathbf{A})(1-u)]_+[1+\mu(\mathbf{B})(1-v)]_+}{\bar{\mu}^2(\mathbf{A},\mathbf{B})}. \tag{6.140}$$

Then, $\mathbf{y}$ *can be recovered from* $\mathbf{w} = \mathbf{A}\mathbf{y} + \mathbf{B}\mathbf{z}$ *by either of the following algorithms:*

$$(P0)\ \begin{cases} \textit{minimize } \|\widetilde{\mathbf{y}}\|_0 \\ \textit{subject to } \mathbf{A}\widetilde{\mathbf{y}} \in \{\mathbf{w} + \mathbf{B}\widetilde{\mathbf{z}} : \widetilde{\mathbf{z}} \in \mathbb{C}^q\}, \end{cases} \tag{6.141}$$

$$(P1)\ \begin{cases} \textit{minimize } \|\widetilde{\mathbf{y}}\|_1 \\ \textit{subject to } \mathbf{A}\widetilde{\mathbf{y}} \in \{\mathbf{w} + \mathbf{B}\widetilde{\mathbf{z}} : \widetilde{\mathbf{z}} \in \mathbb{C}^q\}. \end{cases} \tag{6.142}$$

Proof We provide the proof for (P1) only. The proof for recovery through (P0) is very similar and can be found in Appendix B of [9].

Let $\mathbf{w} = \mathbf{A}\mathbf{y} + \mathbf{B}\mathbf{z}$ and suppose that (P1) delivers $\widetilde{\mathbf{y}} \in \mathbb{C}^p$. This implies $\|\widetilde{\mathbf{y}}\|_1 \le \|\mathbf{y}\|_1$ and the existence of a $\widetilde{\mathbf{z}} \in \mathbb{C}^q$ such that

$$\mathbf{A}\widetilde{\mathbf{y}} = \mathbf{w} + \mathbf{B}\widetilde{\mathbf{z}}. \tag{6.143}$$

On the other hand, we also have

$$\mathbf{A}\mathbf{y} = \mathbf{w} - \mathbf{B}\mathbf{z}. \tag{6.144}$$

Subtracting (6.144) from (6.143) yields

$$\mathbf{A}(\underbrace{\widetilde{\mathbf{y}} - \mathbf{y}}_{=\mathbf{p}}) = \mathbf{B}(\underbrace{\widetilde{\mathbf{z}} + \mathbf{z}}_{=\mathbf{q}}). \tag{6.145}$$

We now set

$$\mathcal{U} = \{i \in \{1,\ldots,p\} : y_i \neq 0\} \tag{6.146}$$

and

$$\mathcal{U}^c = \{1,\ldots,p\}\backslash\mathcal{U}, \tag{6.147}$$

and show that $\mathbf{p}$ is $\varepsilon_{\mathcal{U}}$-concentrated (with respect to the 1-norm) for $\varepsilon_{\mathcal{U}} = 1/2$, i.e.,

$$\|\mathbf{p}_{\mathcal{U}^c}\|_1 \leq \frac{1}{2}\|\mathbf{p}\|_1. \tag{6.148}$$

We have

$$\begin{aligned}
\|\mathbf{y}\|_1 &\geq \|\widetilde{\mathbf{y}}\|_1 && (6.149)\\
&= \|\mathbf{y}+\mathbf{p}\|_1 && (6.150)\\
&= \|\mathbf{y}_{\mathcal{U}}+\mathbf{p}_{\mathcal{U}}\|_1 + \|\mathbf{p}_{\mathcal{U}^c}\|_1 && (6.151)\\
&\geq \|\mathbf{y}_{\mathcal{U}}\|_1 - \|\mathbf{p}_{\mathcal{U}}\|_1 + \|\mathbf{p}_{\mathcal{U}^c}\|_1 && (6.152)\\
&= \|\mathbf{y}\|_1 - \|\mathbf{p}_{\mathcal{U}}\|_1 + \|\mathbf{p}_{\mathcal{U}^c}\|_1, && (6.153)
\end{aligned}$$

where (6.151) follows from the definition of $\mathcal{U}$ in (6.146), and in (6.152) we applied the reverse triangle inequality. Now, (6.149)–(6.153) implies $\|\mathbf{p}_{\mathcal{U}}\|_1 \geq \|\mathbf{p}_{\mathcal{U}^c}\|_1$. Thus, $2\|\mathbf{p}_{\mathcal{U}^c}\|_1 \leq \|\mathbf{p}_{\mathcal{U}}\|_1 + \|\mathbf{p}_{\mathcal{U}^c}\|_1 = \|\mathbf{p}\|_1$, which establishes (6.148). Next, set $\mathcal{V} = \{1,\ldots,q\}$ and note that $\mathbf{q}$ is trivially $\varepsilon_{\mathcal{V}}$-concentrated (with respect to 1-norm) for $\varepsilon_{\mathcal{V}} = 0$. Suppose, toward a contradiction, that $\mathbf{p} \neq \mathbf{0}$. Then, we have

$$\begin{aligned}
2sq &\geq 2|\mathcal{U}||\mathcal{V}| && (6.154)\\
&\geq \frac{[(1+\mu(\mathbf{A})) - 2\mu(\mathbf{A})|\mathcal{U}|]_+[1+\mu(\mathbf{B})(1-|\mathcal{V}|)]_+}{\bar{\mu}^2(\mathbf{A},\mathbf{B})} && (6.155)\\
&\geq \frac{[(1+\mu(\mathbf{A})) - 2s\mu(\mathbf{A})]_+[1+\mu(\mathbf{B})(1-q)]_+}{\bar{\mu}^2(\mathbf{A},\mathbf{B})}, && (6.156)
\end{aligned}$$

where (6.155) is obtained by applying Part 3 of Corollary 6.3 with $\mathbf{p}$ $\varepsilon_{\mathcal{U}}$-concentrated for $\varepsilon_{\mathcal{U}} = 1/2$ and $\mathbf{q}$ $\varepsilon_{\mathcal{V}}$-concentrated for $\varepsilon_{\mathcal{V}} = 0$. But (6.154)–(6.156) contradicts (6.139). Hence, we must have $\mathbf{p} = \mathbf{0}$, which yields $\widetilde{\mathbf{y}} = \mathbf{y}$.

We next provide an example showing that, as soon as (6.139) is saturated, recovery through (P0) or (P1) can fail. Take $m = n^2$ with n even, $\mathbf{A} = \mathbf{F} \in \mathbb{C}^{m\times m}$, and $\mathbf{B} \in \mathbb{C}^{m\times\sqrt{m}}$ containing every $\sqrt{m}$th column of the $m \times m$ identity matrix, i.e.,

$$B_{k,l} = \begin{cases} 1 & \text{if } k = \sqrt{m}\,l, \\ 0 & \text{else,} \end{cases} \tag{6.157}$$

for all $k \in \{1,\ldots,m\}$ and $l \in \{1,\ldots,\sqrt{m}\}$. For every $a \in \mathbb{N}$ dividing m, we define the vector $\mathbf{d}^{(a)} \in \mathbb{C}^m$ with components

$$d_l^{(a)} = \begin{cases} 1 & \text{if } l \in \{a, 2a, \ldots, (m/a-1)a, m\}, \\ 0 & \text{else.} \end{cases} \tag{6.158}$$

Straightforward calculations now yield

$$\mathbf{F}\mathbf{d}^{(a)} = \frac{\sqrt{m}}{a}\mathbf{d}^{(m/a)} \tag{6.159}$$

for all $a \in \mathbb{N}$ dividing m. Suppose that $\mathbf{w} = \mathbf{F}\mathbf{y} + \mathbf{B}\mathbf{z}$ with

$$\mathbf{y} = \mathbf{d}^{(2\sqrt{m})} - \mathbf{d}^{(\sqrt{m})} \in \mathbb{C}^m, \tag{6.160}$$

$$\mathbf{z} = (1 \dots 1)^{\mathrm{T}} \in \mathbb{C}^{\sqrt{m}}. \tag{6.161}$$

Evaluating (6.139) for $\mathbf{A} = \mathbf{F}$, $\mathbf{B}$ as defined in (6.157), and $q = \sqrt{m}$ results in $s < \sqrt{m}/2$. Now, $\mathbf{y}$ in (6.160) has $\|\mathbf{y}\|_0 = \sqrt{m}/2$ and thus just violates the threshold $s < \sqrt{m}/2$. We next show that this slender violation is enough for the existence of an alternative pair $\widetilde{\mathbf{y}} \in \mathbb{C}^m, \widetilde{\mathbf{z}} \in \mathbb{C}^{\sqrt{m}}$ satisfying $\mathbf{w} = \mathbf{F}\widetilde{\mathbf{y}} + \mathbf{B}\widetilde{\mathbf{z}}$ with $\|\widetilde{\mathbf{y}}\|_0 = \|\mathbf{y}\|_0$ and $\|\widetilde{\mathbf{y}}\|_1 = \|\mathbf{y}\|_1$. Thus, neither (P0) nor (P1) can distinguish between $\mathbf{y}$ and $\widetilde{\mathbf{y}}$. Specifically, we set

$$\widetilde{\mathbf{y}} = \mathbf{d}^{(2\sqrt{m})} \in \mathbb{C}^m, \tag{6.162}$$

$$\widetilde{\mathbf{z}} = \mathbf{0} \in \mathbb{C}^m \tag{6.163}$$

and note that $\|\widetilde{\mathbf{y}}\|_0 = \|\mathbf{y}\|_0 = \|\widetilde{\mathbf{y}}\|_1 = \|\mathbf{y}\|_1 = \sqrt{m}/2$. It remains to establish that $\mathbf{w} = \mathbf{F}\widetilde{\mathbf{y}} + \mathbf{B}\widetilde{\mathbf{z}}$. To this end, first note that

$$\mathbf{w} = \mathbf{F}\mathbf{y} + \mathbf{B}\mathbf{z} \tag{6.164}$$

$$= \frac{1}{2}\mathbf{d}^{(\sqrt{m}/2)} - \mathbf{d}^{(\sqrt{m})} + \mathbf{B}\mathbf{z} \tag{6.165}$$

$$= \frac{1}{2}\mathbf{d}^{(\sqrt{m}/2)}, \tag{6.166}$$

where (6.165) follows from (6.159) and (6.166) is by (6.157). Finally, again using (6.159), we find that

$$\mathbf{F}\widetilde{\mathbf{y}} + \mathbf{B}\widetilde{\mathbf{z}} = \frac{1}{2}\mathbf{d}^{(\sqrt{m}/2)}, \tag{6.167}$$

which completes the argument.

The threshold $s < \sqrt{m}/2$ constitutes a special instance of the so-called "square-root bottleneck" [41] all coherence-based deterministic recovery thresholds suffer from. The square-root bottleneck says that the number of measurements, m, has to scale at least quadratically in the sparsity level s. It can be circumvented by considering random models for either the signals or the measurement matrices [42–45], leading to thresholds of the form $m \propto s \log p$ and applying with high probability. Deterministic linear recovery thresholds, i.e., $m \propto s$, were, to the best of our knowledge, first reported in [46] for the DFT measurement matrix under positivity constraints on the vector to be recovered. Further instances of deterministic linear recovery thresholds were discovered in the context of spectrum-blind sampling [47, 48] and system identification [49].

6.5 The Set-Theoretic Null-Space Property

The notion of sparsity underlying the theory developed so far is that of either the number of non-zero entries or of concentration in terms of 1-norm or 2-norm. In practice, one often encounters more general concepts of parsimony, such as manifold or fractal set structures. Manifolds are prevalent in data science, e.g., in compressed sensing [22–27], machine learning [28], image processing [29, 30], and handwritten-digit recognition [31]. Fractal sets find application in image compression and in modeling of Ethernet traffic [32]. Based on the null-space property established in Lemma 6.8, we now extend the theory to account for more general notions of parsimony. To this end, we first need a suitable measure of "description complexity" that goes beyond the concepts of sparsity and concentration. Formalizing this idea requires an adequate dimension measure, which, as it turns out, is lower modified Minkowski dimension. We start by defining Minkowski dimension and modified Minkowski dimension.

DEFINITION 6.5 *(from Section 3.1 of [50])*[3] *For* $\mathcal{U} \subseteq \mathbb{C}^m$ *non-empty, the lower and upper Minkowski dimensions of* $\mathcal{U}$ *are defined as*

$$\underline{\dim}_{\mathrm{B}}(\mathcal{U}) = \liminf_{\rho \to 0} \frac{\log N_{\mathcal{U}}(\rho)}{\log(1/\rho)} \tag{6.168}$$

and

$$\overline{\dim}_{\mathrm{B}}(\mathcal{U}) = \limsup_{\rho \to 0} \frac{\log N_{\mathcal{U}}(\rho)}{\log(1/\rho)}, \tag{6.169}$$

respectively, where

$$N_{\mathcal{U}}(\rho) = \min\left\{k \in \mathbb{N} : \mathcal{U} \subseteq \bigcup_{i \in \{1,\dots,k\}} \mathcal{B}_m(\mathbf{u}_i, \rho),\ \mathbf{u}_i \in \mathcal{U}\right\} \tag{6.170}$$

is the covering number of $\mathcal{U}$ *for radius* $\rho > 0$. *If* $\underline{\dim}_{\mathrm{B}}(\mathcal{U}) = \overline{\dim}_{\mathrm{B}}(\mathcal{U})$, *this common value, denoted by* $\dim_B(\mathcal{U})$, *is the Minkowski dimension of* $\mathcal{U}$.

DEFINITION 6.6 *(from Section 3.3 of [50]) For* $\mathcal{U} \subseteq \mathbb{C}^m$ *non-empty, the lower and upper modified Minkowski dimensions of* $\mathcal{U}$ *are defined as*

$$\underline{\dim}_{\mathrm{MB}}(\mathcal{U}) = \inf\left\{\sup_{i \in \mathbb{N}} \underline{\dim}_{\mathrm{B}}(\mathcal{U}_i) : \mathcal{U} \subseteq \bigcup_{i \in \mathbb{N}} \mathcal{U}_i\right\} \tag{6.171}$$

and

$$\overline{\dim}_{\mathrm{MB}}(\mathcal{U}) = \inf\left\{\sup_{i \in \mathbb{N}} \overline{\dim}_{\mathrm{B}}(\mathcal{U}_i) : \mathcal{U} \subseteq \bigcup_{i \in \mathbb{N}} \mathcal{U}_i\right\}, \tag{6.172}$$

respectively, where in both cases the infimum is over all possible coverings $\{\mathcal{U}_i\}_{i \in \mathbb{N}}$ *of* $\mathcal{U}$ *by non-empty compact sets* $\mathcal{U}_i$. *If* $\underline{\dim}_{\mathrm{MB}}(\mathcal{U}) = \overline{\dim}_{\mathrm{MB}}(\mathcal{U})$, *this common value, denoted by* $\dim_{\mathrm{MB}}(\mathcal{U})$, *is the modified Minkowski dimension of* $\mathcal{U}$.

For further details on (modified) Minkowski dimension, we refer the interested reader to Section 3 of [50].

We are now ready to extend the null-space property in Lemma 6.8 to the following set-theoretic null-space property.

THEOREM 6.4 *Let* $\mathcal{U} \subseteq \mathbb{C}^{p+q}$ *be non-empty with* $\underline{\dim}_{\mathrm{MB}}(\mathcal{U}) < 2m$, *and let* $\mathbf{B} \in \mathbb{C}^{m \times q}$ *with* $m \geq q$ *be a full-rank matrix. Then,* $\ker[\mathbf{A}\ \ \mathbf{B}] \cap (\mathcal{U} \setminus \{\mathbf{0}\}) = \emptyset$ *for Lebesgue-a.a.* $\mathbf{A} \in \mathbb{C}^{m \times p}$.

Proof See Section 6.8.

[3] Minkowski dimension is sometimes also referred to as box-counting dimension, which is the origin of the subscript B in the notation $\dim_{\mathrm{B}}(\cdot)$ used henceforth.

The set $\mathcal{U}$ in this set-theoretic null-space property generalizes the finite union of linear subspaces $\mathcal{S}$ in Lemma 6.8. For $\mathcal{U} \subseteq \mathbb{R}^{p+q}$, the equivalent of Theorem 6.4 was reported previously in Proposition 1 of [13]. The set-theoretic null-space property can be interpreted in geometric terms as follows. If $p+q \leq m$, then $[\mathbf{A}\;\; \mathbf{B}]$ is a tall matrix so that the kernel of $[\mathbf{A}\;\; \mathbf{B}]$ is $\{\mathbf{0}\}$ for Lebesgue-a.a. matrices $\mathbf{A}$. The statement of the theorem holds trivially in this case. If $p+q > m$, then the kernel of $[\mathbf{A}\;\; \mathbf{B}]$ is a $(p+q-m)$-dimensional subspace of the ambient space $\mathbb{C}^{p+q}$ for Lebesgue-a.a. matrices $\mathbf{A}$. The theorem therefore says that, for Lebesgue-a.a. $\mathbf{A}$, the set $\mathcal{U}$ intersects the subspace $\ker([\mathbf{A}\;\; \mathbf{B}])$ at most trivially if the sum of $\dim\ker([\mathbf{A}\;\; \mathbf{B}])$ and[4] $\underline{\dim}_{\mathrm{MB}}(\mathcal{U})/2$ is strictly smaller than the dimension of the ambient space. What is remarkable here is that the notions of Euclidean dimension (for the kernel of $[\mathbf{A}\;\; \mathbf{B}]$) and of lower modified Minkowski dimension (for the set $\mathcal{U}$) are compatible. We finally note that, by virtue of the chain of equivalences in the proof of Lemma 6.8, the set-theoretic null-space property in Theorem 6.4 leads to a set-theoretic uncertainty relation, albeit not in the form of an upper bound on an operator norm; for a detailed discussion of this equivalence the interested reader is referred to [13].

We next put the set-theoretic null-space property in Theorem 6.4 in perspective with the null-space property in Lemma 6.8. Fix the sparsity levels s and t, consider the set

$$\mathcal{S}_{s,t} = \left\{ \begin{pmatrix} \mathbf{p} \\ \mathbf{q} \end{pmatrix} : \mathbf{p} \in \mathbb{C}^p, \mathbf{q} \in \mathbb{C}^q, \|\mathbf{p}\|_0 \leq s, \|\mathbf{q}\|_0 \leq t \right\}, \tag{6.173}$$

which is a finite union of $(s+t)$-dimensional linear subspaces, and, for the sake of concreteness, let $\mathbf{A} = \mathbf{I}$ and $\mathbf{B} = \mathbf{F}$ of size $q \times q$. Lemma 6.8 then states that the kernel of $[\mathbf{I}\;\; \mathbf{F}]$ intersects $\mathcal{S}_{s,t}$ trivially, provided that

$$m > st, \tag{6.174}$$

which leads to a recovery threshold in the signal separation problem that is quadratic in the sparsity levels s and t (see Theorem 8 of [9]). To see what the set-theoretic null-space property gives, we start by noting that, by Example II.2 of [26], $\dim_{\mathrm{MB}}(\mathcal{S}_{s,t}) = 2(s+t)$. Theorem 6.4 hence states that, for Lebesgue-a.a. matrices $\mathbf{A} \in \mathbb{C}^{m \times p}$, the kernel of $[\mathbf{A}\;\; \mathbf{B}]$ intersects $\mathcal{S}_{s,t}$ trivially, provided that

$$m > s+t. \tag{6.175}$$

This is striking as it says that, while the threshold in (6.174) is quadratic in the sparsity levels s and t and, therefore, suffers from the square-root bottleneck, the threshold in (6.175) is linear in s and t.

To understand the operational implications of the observation just made, we demonstrate how the set-theoretic null-space property in Theorem 6.4 leads to a sufficient condition for the recovery of vectors in sets of small lower modified Minkowski dimension.

[4] The factor $1/2$ stems from the fact that (modified) Minkowski dimension "counts real dimensions." For example, the modified Minkowski dimension of an n-dimensional linear subspace of $\mathbb{C}^m$ is $2n$ (see Example II.2 of [26]).

LEMMA 6.9 *Let* $\mathcal{S} \subseteq \mathbb{C}^{p+q}$ *be non-empty with* $\underline{\dim}_{MB}(\mathcal{S} \ominus \mathcal{S}) < 2m$, *where* $\mathcal{S} \ominus \mathcal{S} = \{\mathbf{u} - \mathbf{v} : \mathbf{u}, \mathbf{v} \in \mathcal{S}\}$, *and let* $\mathbf{B} \in \mathbb{C}^{m \times q}$, *with* $m \geq q$, *be a full-rank matrix. Then,* $[\mathbf{A}\ \mathbf{B}]$ *is one-to-one on* $\mathcal{S}$ *for Lebesgue-a.a.* $\mathbf{A} \in \mathbb{C}^{m \times p}$.

Proof The proof follows immediately from the set-theoretic null-space property in Theorem 6.4 and the linearity of matrix-vector multiplication.

To elucidate the implications of Lemma 6.9, consider $\mathcal{S}_{s,t}$ defined in (6.173). Since $\mathcal{S}_{s,t} \ominus \mathcal{S}_{s,t}$ is again a finite union of linear subspaces of dimensions no larger than $\min\{p, 2s\} + \min\{q, 2t\}$, where the $\min\{\cdot, \cdot\}$ operation accounts for the fact that the dimension of a linear subspace cannot exceed the dimension of its ambient space, we have (see Example II.2 of [26])

$$\dim_{\mathrm{MB}}(\mathcal{S}_{s,t} \ominus \mathcal{S}_{s,t}) = 2(\min\{p, 2s\} + \min\{q, 2t\}). \tag{6.176}$$

Application of Lemma 6.9 now yields that, for Lebesgue-a.a. matrices $\mathbf{A} \in \mathbb{C}^{m \times p}$, we can recover $\mathbf{y} \in \mathbb{C}^p$ with $\|\mathbf{y}\|_0 \leq s$ and $\mathbf{z} \in \mathbb{C}^q$ with $\|\mathbf{z}\|_0 \leq t$ from $\mathbf{w} = \mathbf{A}\mathbf{y} + \mathbf{B}\mathbf{z}$, provided that $m > \min\{p, 2s\} + \min\{q, 2t\}$. This qualitative behavior (namely, linear in s and t) is best possible as it cannot be improved even if the support sets of $\mathbf{y}$ and $\mathbf{z}$ were known prior to recovery. We emphasize, however, that the statement in Lemma 6.9 guarantees injectivity of $[\mathbf{A}\ \mathbf{B}]$ only absent computational considerations for recovery.

6.6 A Large Sieve Inequality in $(\mathbb{C}^m, \|\cdot\|_2)$

We present a slightly improved and generalized version of the large sieve inequality stated in Equation (32) of [7].

LEMMA 6.10 *Let* μ *be a* 1*-periodic,* σ*-finite measure on* $\mathbb{R}$, $n \in \mathbb{N}$, $\varphi \in [0,1)$, $\mathbf{a} \in \mathbb{C}^n$, *and consider the* 1*-periodic trigonometric polynomial*

$$\psi(s) = e^{2\pi j\varphi} \sum_{k=1}^{n} a_k e^{-2\pi jks}. \tag{6.177}$$

Then,

$$\int_{[0,1)} |\psi(s)|^2 \, d\mu(s) \leq \left(n - 1 + \frac{1}{\delta}\right) \sup_{r \in [0,1)} \mu((r, r+\delta)) \|\mathbf{a}\|_2^2 \tag{6.178}$$

for all $\delta \in (0,1]$.

Proof Since

$$|\psi(s)| = \left| \sum_{k=1}^{n} a_k e^{-2\pi jks} \right|, \tag{6.179}$$

we can assume, without loss of generality, that $\varphi = 0$. The proof now follows closely the line of argumentation on pp. 185–186 of [51] and in the proof of Lemma 5 of [7]. Specifically, we make use of the result on p. 185 of [51] saying that, for every $\delta > 0$, there exists a function $g \in L^2(\mathbb{R})$ with Fourier transform

$$G(s) = \int_{-\infty}^{\infty} g(t) e^{-2\pi j s t}\, dt \tag{6.180}$$

such that $\|G\|_2^2 = n - 1 + 1/\delta$, $|g(t)|^2 \geq 1$ for all $t \in [1, n]$, and $G(s) = 0$ for all $s \notin [-\delta/2, \delta/2]$. With this g, consider the 1-periodic trigonometric polynomial

$$\theta(s) = \sum_{k=1}^{n} \frac{a_k}{g(k)} e^{-2\pi j k s} \tag{6.181}$$

and note that

$$\int_{-\delta/2}^{\delta/2} G(r)\theta(s-r)\, dr = \sum_{k=1}^{n} \frac{a_k}{g(k)} e^{-2\pi j k s} \int_{-\infty}^{\infty} G(r) e^{2\pi j k r} dr \tag{6.182}$$

$$= \sum_{k=1}^{n} a_k e^{-2\pi j k s} \tag{6.183}$$

$$= \psi(s) \quad \text{for all } s \in \mathbb{R}. \tag{6.184}$$

We now have

$$\int_{[0,1)} |\psi(s)|^2\, d\mu(s) = \int_{[0,1)} \left| \int_{-\delta/2}^{\delta/2} G(r)\theta(s-r) dr \right|^2 d\mu(s) \tag{6.185}$$

$$\leq \|G\|_2^2 \int_{[0,1)} \left(\int_{-\delta/2}^{\delta/2} |\theta(s-r)|^2\, dr \right) d\mu(s) \tag{6.186}$$

$$= \|G\|_2^2 \int_{[0,1)} \left(\int_{s-\delta/2}^{s+\delta/2} |\theta(r)|^2\, dr \right) d\mu(s) \tag{6.187}$$

$$= \|G\|_2^2 \int_{-1}^{2} \mu((r-\delta/2, r+\delta/2) \cap [0,1)) |\theta(r)|^2\, dr \tag{6.188}$$

$$= \|G\|_2^2 \sum_{i=-1}^{1} \int_{0+i}^{1+i} \mu((r-\delta/2, r+\delta/2) \cap [0,1)) |\theta(r)|^2\, dr \tag{6.189}$$

$$= \|G\|_2^2 \sum_{i=-1}^{1} \int_{0}^{1} \mu((r-\delta/2, r+\delta/2) \cap [i, 1+i)) |\theta(r)|^2\, dr \tag{6.190}$$

$$= \|G\|_2^2 \int_{0}^{1} \mu((r-\delta/2, r+\delta/2) \cap [-1, 2)) |\theta(r)|^2\, dr \tag{6.191}$$

$$= \|G\|_2^2 \int_{0}^{1} \mu((r-\delta/2, r+\delta/2)) |\theta(r)|^2\, dr \tag{6.192}$$

for all $\delta \in (0, 1]$, where (6.185) follows from (6.182)–(6.184), in (6.186) we applied the Cauchy–Schwartz inequality (Theorem 1.37 of [52]), (6.188) is by Fubini's theorem (Theorem 1.14 of [53]) (recall that μ is σ-finite by assumption) upon noting that

$$\{(r,s) : s \in [0,1), r \in (s-\delta/2, s+\delta/2)\} = \{(r,s) : r \in [-1,2), s \in (r-\delta/2, r+\delta/2) \cap [0,1)\} \tag{6.193}$$

for all $\delta \in (0,1]$, in (6.190) we used the 1-periodicity of μ and θ, and (6.191) is by σ-additivity of μ. Now,

$$\int_0^1 \mu((r-\delta/2, r+\delta/2))|\theta(r)|^2\, dr \leq \sup_{r\in[0,1)} \mu((r,r+\delta)) \int_0^1 |\theta(r)|^2\, dr \tag{6.194}$$

$$= \sup_{r\in[0,1)} \mu((r,r+\delta)) \sum_{k=1}^{n} \frac{|a_k|^2}{|g(k)|^2} \tag{6.195}$$

$$\leq \sup_{r\in[0,1)} \mu((r,r+\delta))\|\mathbf{a}\|_2^2 \tag{6.196}$$

for all $\delta > 0$, where (6.196) follows from $|g(t)|^2 \geq 1$ for all $t \in [1,n]$. Using (6.194)–(6.196) and $\|G\|_2^2 = n - 1 + 1/\delta$ in (6.192) establishes (6.178).

Lemma 6.10 is a slightly strengthened version of the large sieve inequality (Equation (32) of [7]). Specifically, in (6.178) it is sufficient to consider open intervals $(r, r+\delta)$, whereas Equation (32) of [7] requires closed intervals $[r, r+\delta]$. Thus, the upper bound in Equation (32) of [7] can be strictly larger than that in (6.178) whenever μ has mass points.

6.7 Uncertainty Relations in L_1 and L_2

Table 6.1 contains a list of infinite-dimensional counterparts–available in the literature–to results in this chapter. Specifically, these results apply to bandlimited L_1- and L_2-functions and hence correspond to $\mathbf{A} = \mathbf{I}$ and $\mathbf{B} = \mathbf{F}$ in our setting.

Table 6.1. Infinite-dimensional counterparts to results in this chapter

	L_2 analog	L_1 analog
Upper bound in (6.10)	Lemma 2 of [6]	
Corollary 6.1	Theorem 2 of [6]	
Lemma 6.4	Theorem 4 of [6]	
Lemma 6.5		Lemma 3 of [6]
Lemma 6.7		Lemma 2 of [7]
Lemma 6.10		Theorem 4 of [7]

6.8 Proof of Theorem 6.4

By definition of lower modified Minkowski dimension, there exists a covering $\{\mathcal{U}_i\}_{i\in\mathbb{N}}$ of $\mathcal{U}$ by non-empty compact sets $\mathcal{U}_i$ satisfying $\underline{\dim}_{\mathrm{B}}(\mathcal{U}_i) < 2m$ for all $i \in \mathbb{N}$. The countable sub-additivity of Lebesgue measure λ now implies

$$\lambda(\{\mathbf{A} \in \mathbb{C}^{m\times p} : \ker[\mathbf{A}\ \ \mathbf{B}] \cap (\mathcal{U}\backslash\{\mathbf{0}\}) \neq \emptyset\}) \leq \sum_{i=1}^{\infty} \lambda(\{\mathbf{A} \in \mathbb{C}^{m\times p} : \ker[\mathbf{A}\ \ \mathbf{B}] \cap (\mathcal{U}_i\backslash\{\mathbf{0}\}) \neq \emptyset\}). \tag{6.197}$$

We next establish that every term in the sum on the right-hand side of (6.197) equals zero. Take an arbitrary but fixed $i \in \mathbb{N}$. Repeating the steps in Equations (10)–(14) of [13] shows that it suffices to prove that

$$\mathbb{P}[\ker([\mathbf{A}\ \ \mathbf{B}]) \cap \mathcal{V} \neq \emptyset] = 0, \tag{6.198}$$

with

$$\mathcal{V} = \left\{ \begin{pmatrix} \mathbf{u} \\ \mathbf{v} \end{pmatrix} : \mathbf{u} \in \mathbb{C}^p, \mathbf{v} \in \mathbb{C}^q, \|\mathbf{u}\|_2 > 0 \right\} \cap \mathcal{U}_i \tag{6.199}$$

and $\mathbf{A} = (\mathbf{A}_1 \dots \mathbf{A}_m)^*$, where the random vectors $\mathbf{A}_i$ are independent and uniformly distributed on $\mathcal{B}_p(\mathbf{0}, r)$ for arbitrary but fixed $r > 0$. Suppose, toward a contradiction, that (6.198) is false. This implies

$$0 = \liminf_{\rho \to 0} \frac{\log \mathbb{P}[\ker([\mathbf{A}\ \ \mathbf{B}]) \cap \mathcal{V} \neq \emptyset]}{\log(1/\rho)} \tag{6.200}$$

$$\leq \liminf_{\rho \to 0} \frac{\log \sum_{i=1}^{N_{\mathcal{V}}(\rho)} \mathbb{P}[\ker([\mathbf{A}\ \ \mathbf{B}]) \cap \mathcal{B}_{p+q}(\mathbf{c}_i, \rho) \neq \emptyset]}{\log(1/\rho)}, \tag{6.201}$$

where we have chosen $\{\mathbf{c}_i : i = 1, \dots, N_{\mathcal{V}}(\rho)\} \subseteq \mathcal{V}$ such that

$$\mathcal{V} \subseteq \bigcup_{i=1}^{N_{\mathcal{V}}(\rho)} \mathcal{B}_{p+q}(\mathbf{c}_i, \rho), \tag{6.202}$$

with $N_{\mathcal{V}}(\rho)$ denoting the covering number of $\mathcal{V}$ for radius $\rho > 0$ (cf. (6.170)). Now let $i \in \{1, \dots, N_{\mathcal{V}}(\rho)\}$ be arbitrary but fixed and write $\mathbf{c}_i = (\mathbf{u}_i^{\mathrm{T}}\ \mathbf{v}_i^{\mathrm{T}})^{\mathrm{T}}$. It follows that

$$\|\mathbf{A}\mathbf{u}_i + \mathbf{B}\mathbf{v}_i\|_2 = \|[\mathbf{A}\ \ \mathbf{B}]\mathbf{c}_i\|_2 \tag{6.203}$$

$$\leq \|[\mathbf{A}\ \ \mathbf{B}](\mathbf{x} - \mathbf{c}_i)\|_2 + \|[\mathbf{A}\ \ \mathbf{B}]\mathbf{x}\|_2 \tag{6.204}$$

$$\leq \|[\mathbf{A}\ \ \mathbf{B}]\|_2 \|(\mathbf{x} - \mathbf{c}_i)\|_2 + \|[\mathbf{A}\ \ \mathbf{B}]\mathbf{x}\|_2 \tag{6.205}$$

$$\leq (\|[\mathbf{A}\|_2 + \|\mathbf{B}\|_2)\rho + \|[\mathbf{A}\ \ \mathbf{B}]\mathbf{x}\|_2 \tag{6.206}$$

$$\leq (r\sqrt{m} + \|\mathbf{B}\|_2)\rho + \|[\mathbf{A}\ \ \mathbf{B}]\mathbf{x}\|_2 \quad \text{for all } \mathbf{x} \in \mathcal{B}_{p+q}(\mathbf{c}_i, \rho), \tag{6.207}$$

where in the last step we made use of $\|\mathbf{A}_i\|_2 \leq r$ for $i = 1, \dots, m$. We now have

$$\mathbb{P}[\ker([\mathbf{A}\ \ \mathbf{B}]) \cap \mathcal{B}_{p+q}(\mathbf{c}_i, \rho) \neq \emptyset] \tag{6.208}$$

$$\leq \mathbb{P}[\exists \mathbf{x} \in \mathcal{B}_{p+q}(\mathbf{c}_i, \rho) : \|[\mathbf{A}\ \ \mathbf{B}]\mathbf{x}\|_2 < \rho] \tag{6.209}$$

$$\leq \mathbb{P}[\|\mathbf{A}\mathbf{u}_i + \mathbf{B}\mathbf{v}_i\|_2 < \rho(1 + r\sqrt{m} + \|\mathbf{B}\|_2)] \tag{6.210}$$

$$\leq \rho^{2m} \frac{C(p, m, r)}{\|\mathbf{u}_i\|_2^{2m}} (1 + r\sqrt{m} + \|\mathbf{B}\|_2)^{2m}, \tag{6.211}$$

where (6.210) is by (6.203)–(6.207), and in (6.211) we applied the concentration of measure result Lemma 6.11 below (recall that $\mathbf{c}_i = (\mathbf{u}_i^{\mathrm{T}}\ \mathbf{v}_i^{\mathrm{T}})^{\mathrm{T}} \in \mathcal{V}$ implies $\mathbf{u}_i \neq \mathbf{0}$) with $C(p,m,r)$ as in (6.216) below. Inserting (6.208)–(6.211) into (6.201) yields

$$0 \leq \liminf_{\rho\to\infty} \frac{\log(N_{\mathcal{V}}(\rho)\rho^{2m})}{\log(1/\rho)} \tag{6.212}$$

$$= \underline{\dim}_{\mathrm{B}}(\mathcal{V}) - 2m \tag{6.213}$$

$$< 0, \tag{6.214}$$

where (6.214) follows from $\underline{\dim}_{\mathrm{B}}(\mathcal{V}) \leq \underline{\dim}_{\mathrm{B}}(\mathcal{U}_i) < 2m$, which constitutes a contradiction. Therefore, (6.198) must hold.

LEMMA 6.11 *Let $\mathbf{A} = (\mathbf{A}_1 \ldots \mathbf{A}_m)^*$ with independent random vectors $\mathbf{A}_i$ uniformly distributed on $\mathcal{B}_p(\mathbf{0}, r)$ for $r > 0$. Then*

$$\mathbb{P}[\|\mathbf{A}\mathbf{u} + \mathbf{v}\|_2 < \delta] \leq \frac{C(p,m,r)}{\|\mathbf{u}\|_2^{2m}} \delta^{2m}, \tag{6.215}$$

with

$$C(p,m,r) = \left(\frac{\pi V_{p-1}(r)}{V_p(r)}\right)^m \tag{6.216}$$

for all $\mathbf{u} \in \mathbb{C}^p \backslash \{\mathbf{0}\}$, $\mathbf{v} \in \mathbb{C}^m$, and $\delta > 0$.

Proof Since

$$\mathbb{P}[\|\mathbf{A}\mathbf{u} + \mathbf{v}\|_2 < \delta] \leq \prod_{i=1}^{m} \mathbb{P}[|\mathbf{A}_i^*\mathbf{u} + v_i| < \delta] \tag{6.217}$$

owing to the independence of the $\mathbf{A}_i$ and as $\|\mathbf{A}\mathbf{u} + \mathbf{v}\|_2 < \delta$ implies $|\mathbf{A}_i^*\mathbf{u} + v_i| < \delta$ for $i = 1, \ldots, m$, it is sufficient to show that

$$\mathbb{P}[|\mathbf{B}^*\mathbf{u} + v| < \delta] \leq \frac{D(p,r)}{\|\mathbf{u}\|_2^2} \delta^2 \tag{6.218}$$

for all $\mathbf{u} \in \mathbb{C}^p \backslash \{\mathbf{0}\}$, $v \in \mathbb{C}$, and $\delta > 0$, where the random vector $\mathbf{B}$ is uniformly distributed on $\mathcal{B}_p(\mathbf{0}, r)$ and

$$D(p,r) = \frac{\pi V_{p-1}(r)}{V_p(r)}. \tag{6.219}$$

We have

$$\mathbb{P}[|\mathbf{B}^*\mathbf{u} + v| < \delta] = \mathbb{P}\left[\frac{|\mathbf{B}^*\mathbf{u} + v|}{\|\mathbf{u}\|_2} < \frac{\delta}{\|\mathbf{u}\|_2}\right] \tag{6.220}$$

$$= \mathbb{P}[|\mathbf{B}^*\mathbf{U}^*\mathbf{e}_1 + \widetilde{v}| < \widetilde{\delta}] \tag{6.221}$$

$$= \mathbb{P}[|\mathbf{B}^*\mathbf{e}_1 + \widetilde{v}| < \widetilde{\delta}] \tag{6.222}$$

$$= \frac{1}{V_p(r)} \int_{\mathcal{B}_p(\mathbf{0},r)} \chi_{\{b_1 : |b_1 + \widetilde{v}| < \widetilde{\delta}\}}(b_1)\, \mathrm{d}\mathbf{b} \tag{6.223}$$

$$\leq \frac{1}{V_p(r)} \int_{|b_1+\widetilde{\nu}|\leq\widetilde{\delta}} db_1 \int_{\mathcal{B}_{p-1}(\mathbf{0},r)} d(b_2 \dots b_p)^{\mathrm{T}} \tag{6.224}$$

$$= \frac{V_{p-1}(r)}{V_p(r)} \int_{|b_1+\widetilde{\nu}|<\widetilde{\delta}} db_1 \tag{6.225}$$

$$= \frac{V_{p-1}(r)}{V_p(r)} \pi\widetilde{\delta}^2 \tag{6.226}$$

$$= \frac{\pi V_{p-1}(r)}{V_p(r)\|\mathbf{u}\|^2} \delta^2, \tag{6.227}$$

where the unitary matrix $\mathbf{U}$ in (6.221) has been chosen such that $\mathbf{U}(\mathbf{u}/\|\mathbf{u}\|_2) = \mathbf{e}_1 = (1\ 0\ \dots\ 0)^{\mathrm{T}} \in \mathbb{C}^p$ and we set $\widetilde{\delta} := \delta/\|\mathbf{u}\|_2$ and $\widetilde{\nu} := \nu/\|\mathbf{u}\|_2$. Further, (6.222) follows from the unitary invariance of the uniform distribution on $\mathcal{B}_p(\mathbf{0}, r)$, and in (6.223) the factor $1/V_p(r)$ is owing to the assumption of a uniform probability density function on $\mathcal{B}_p(\mathbf{0}, r)$.

6.9 Results for $|||\cdot|||_1$ and $|||\cdot|||_2$

LEMMA 6.12 *Let $\mathbf{U} \in \mathbb{C}^{m\times m}$ be unitary, $\mathcal{P}, \mathcal{Q} \subseteq \{1, \dots, m\}$, and consider the orthogonal projection $\mathbf{\Pi}_{\mathcal{Q}}(\mathbf{U}) = \mathbf{U}\mathbf{D}_{\mathcal{Q}}\mathbf{U}^*$ onto the subspace $\mathcal{W}^{\mathbf{U},\mathcal{Q}}$. Then*

$$|||\mathbf{\Pi}_{\mathcal{Q}}(\mathbf{U})\mathbf{D}_{\mathcal{P}}|||_2 = |||\mathbf{D}_{\mathcal{P}}\mathbf{\Pi}_{\mathcal{Q}}(\mathbf{U})|||_2. \tag{6.228}$$

Moreover, we have

$$|||\mathbf{D}_{\mathcal{P}}\mathbf{\Pi}_{\mathcal{Q}}(\mathbf{U})|||_2 = \max_{\mathbf{x}\in\mathcal{W}^{\mathbf{U},\mathcal{Q}}\setminus\{\mathbf{0}\}} \frac{\|\mathbf{x}_{\mathcal{P}}\|_2}{\|\mathbf{x}\|_2} \tag{6.229}$$

and

$$|||\mathbf{D}_{\mathcal{P}}\mathbf{\Pi}_{\mathcal{Q}}(\mathbf{U})|||_1 = \max_{\mathbf{x}\in\mathcal{W}^{\mathbf{U},\mathcal{Q}}\setminus\{\mathbf{0}\}} \frac{\|\mathbf{x}_{\mathcal{P}}\|_1}{\|\mathbf{x}\|_1}. \tag{6.230}$$

Proof The identity (6.228) follows from

$$|||\mathbf{D}_{\mathcal{P}}\mathbf{\Pi}_{\mathcal{Q}}(\mathbf{U})|||_2 = |||(\mathbf{D}_{\mathcal{P}}\mathbf{\Pi}_{\mathcal{Q}}(\mathbf{U}))^*|||_2 \tag{6.231}$$

$$= |||\mathbf{\Pi}_{\mathcal{Q}}^*(\mathbf{U})\mathbf{D}_{\mathcal{P}}^*|||_2 \tag{6.232}$$

$$= |||\mathbf{\Pi}_{\mathcal{Q}}(\mathbf{U})\mathbf{D}_{\mathcal{P}}|||_2, \tag{6.233}$$

where in (6.231) we used that $|||\cdot|||_2$ is self-adjoint (see p. 309 of [33]), $\mathbf{\Pi}_{\mathcal{Q}}(\mathbf{U})^* = \mathbf{\Pi}_{\mathcal{Q}}(\mathbf{U})$, and $\mathbf{D}_{\mathcal{P}}^* = \mathbf{D}_{\mathcal{P}}$. To establish (6.229), we note that

$$|||\mathbf{D}_{\mathcal{P}}\mathbf{\Pi}_{\mathcal{Q}}(\mathbf{U})|||_2 = \max_{\mathbf{x}:\|\mathbf{x}\|_2=1} \|\mathbf{D}_{\mathcal{P}}\mathbf{\Pi}_{\mathcal{Q}}(\mathbf{U})\mathbf{x}\|_2 \tag{6.234}$$

$$= \max_{\substack{\mathbf{x}:\mathbf{\Pi}_{\mathcal{Q}}(\mathbf{U})\mathbf{x}\neq\mathbf{0}\\ \|\mathbf{x}\|_2=1}} \|\mathbf{D}_{\mathcal{P}}\mathbf{\Pi}_{\mathcal{Q}}(\mathbf{U})\mathbf{x}\|_2 \tag{6.235}$$

$$\leq \max_{\substack{\mathbf{x}:\mathbf{\Pi}_{\mathcal{Q}}(\mathbf{U})\mathbf{x}\neq\mathbf{0}\\ \|\mathbf{x}\|_2=1}} \left\|\mathbf{D}_{\mathcal{P}} \frac{\mathbf{\Pi}_{\mathcal{Q}}(\mathbf{U})\mathbf{x}}{\|\mathbf{\Pi}_{\mathcal{Q}}(\mathbf{U})\mathbf{x}\|_2}\right\|_2 \tag{6.236}$$

$$\leq \max_{\mathbf{x}:\,\mathbf{\Pi}_\mathcal{Q}(\mathbf{U})\mathbf{x}\neq\mathbf{0}} \left\| \mathbf{D}_\mathcal{P} \frac{\mathbf{\Pi}_\mathcal{Q}(\mathbf{U})\mathbf{x}}{\|\mathbf{\Pi}_\mathcal{Q}(\mathbf{U})\mathbf{x}\|_2} \right\|_2 \tag{6.237}$$

$$= \max_{\mathbf{x}\in\mathcal{W}^{\mathbf{U},\mathcal{Q}}\setminus\{\mathbf{0}\}} \frac{\|\mathbf{x}_\mathcal{P}\|_2}{\|\mathbf{x}\|_2} \tag{6.238}$$

$$= \max_{\mathbf{x}:\,\mathbf{\Pi}_\mathcal{Q}(\mathbf{U})\mathbf{x}\neq\mathbf{0}} \left\| \mathbf{D}_\mathcal{P}\mathbf{\Pi}_\mathcal{Q}(\mathbf{U}) \frac{\mathbf{\Pi}_\mathcal{Q}(\mathbf{U})\mathbf{x}}{\|\mathbf{\Pi}_\mathcal{Q}(\mathbf{U})\mathbf{x}\|_2} \right\|_2 \tag{6.239}$$

$$\leq \max_{\mathbf{x}:\,\|\mathbf{x}\|_2=1} \|\mathbf{D}_\mathcal{P}\mathbf{\Pi}_\mathcal{Q}(\mathbf{U})\mathbf{x}\|_2 \tag{6.240}$$

$$= |||\mathbf{D}_\mathcal{P}\mathbf{\Pi}_\mathcal{Q}(\mathbf{U})|||_2, \tag{6.241}$$

where in (6.236) we used $\|\mathbf{\Pi}_\mathcal{Q}(\mathbf{U})\mathbf{x}\|_2 \leq \|\mathbf{x}\|_2$, which implies $\|\mathbf{\Pi}_\mathcal{Q}(\mathbf{U})\mathbf{x}\|_2 \leq 1$ for all $\mathbf{x}$ with $\|\mathbf{x}\|_2 = 1$. Finally, (6.230) follows by repeating the steps in (6.234)–(6.241) with $\|\cdot\|_2$ replaced by $\|\cdot\|_1$ at all occurrences.

LEMMA 6.13 *Let* $\mathbf{A} \in \mathbb{C}^{m\times n}$. *Then*

$$\frac{\|\mathbf{A}\|_2}{\sqrt{\operatorname{rank}(\mathbf{A})}} \leq |||\mathbf{A}|||_2 \leq \|\mathbf{A}\|_2. \tag{6.242}$$

Proof The proof is trivial for $\mathbf{A} = \mathbf{0}$. If $\mathbf{A} \neq \mathbf{0}$, set $r = \operatorname{rank}(\mathbf{A})$ and let $\sigma_1, \ldots, \sigma_r$ denote the non-zero singular values of $\mathbf{A}$ organized in decreasing order. Unitary invariance of $|||\cdot|||_2$ and $\|\cdot\|_2$ (see Problem 5, on p. 311 of [33]) yields $|||\mathbf{A}|||_2 = \sigma_1$ and $\|\mathbf{A}\|_2 = \sqrt{\sum_{i=1}^r \sigma_i^2}$. The claim now follows from

$$\sigma_1 \leq \sqrt{\sum_{i=1}^r \sigma_i^2} \leq \sqrt{r}\sigma_1. \tag{6.243}$$

LEMMA 6.14 *For* $\mathbf{A} = (\mathbf{a}_1 \ldots \mathbf{a}_n) \in \mathbb{C}^{m\times n}$, *we have*

$$|||\mathbf{A}|||_1 = \max_{j\in\{1,\ldots,n\}} \|\mathbf{a}_j\|_1 \tag{6.244}$$

and

$$\frac{1}{n}\|\mathbf{A}\|_1 \leq |||\mathbf{A}|||_1 \leq \|\mathbf{A}\|_1. \tag{6.245}$$

Proof The identity (6.244) is established on p. 294 of [33], and (6.245) follows directly from (6.244).

References

[1] W. Heisenberg, *The physical principles of the quantum theory*. University of Chicago Press, 1930.

[2] W. G. Faris, "Inequalities and uncertainty principles," *J. Math. Phys.*, vol. 19, no. 2, pp. 461–466, 1978.

[3] M. G. Cowling and J. F. Price, "Bandwidth versus time concentration: The Heisenberg–Pauli–Weyl inequality," *SIAM J. Math. Anal.*, vol. 15, no. 1, pp. 151–165, 1984.

[4] J. J. Benedetto, *Wavelets: Mathematics and applications.* CRC Press, 1994, ch. Frame decompositions, sampling, and uncertainty principle inequalities.

[5] G. B. Folland and A. Sitaram, "The uncertainty principle: A mathematical survey," *J. Fourier Analysis and Applications*, vol. 3, no. 3, pp. 207–238, 1997.

[6] D. L. Donoho and P. B. Stark, "Uncertainty principles and signal recovery," *SIAM J. Appl. Math.*, vol. 49, no. 3, pp. 906–931, 1989.

[7] D. L. Donoho and B. F. Logan, "Signal recovery and the large sieve," *SIAM J. Appl. Math.*, vol. 52, no. 2, pp. 577–591, 1992.

[8] M. Elad and A. M. Bruckstein, "A generalized uncertainty principle and sparse representation in pairs of bases," *IEEE Trans. Information Theory*, vol. 48, no. 9, pp. 2558–2567, 2002.

[9] C. Studer, P. Kuppinger, G. Pope, and H. Bölcskei, "Recovery of sparsely corrupted signals," *IEEE Trans. Information Theory*, vol. 58, no. 5, pp. 3115–3130, 2012.

[10] P. Kuppinger, G. Durisi, and H. Bölcskei, "Uncertainty relations and sparse signal recovery for pairs of general signal sets," *IEEE Trans. Information Theory*, vol. 58, no. 1, pp. 263–277, 2012.

[11] A. Terras, *Fourier analysis on finite groups and applications.* Cambridge University Press, 1999.

[12] T. Tao, "An uncertainty principle for cyclic groups of prime order," *Math. Res. Lett.*, vol. 12, no. 1, pp. 121–127, 2005.

[13] D. Stotz, E. Riegler, E. Agustsson, and H. Bölcskei, "Almost lossless analog signal separation and probabilistic uncertainty relations," *IEEE Trans. Information Theory*, vol. 63, no. 9, pp. 5445–5460, 2017.

[14] S. Foucart and H. Rauhut, *A mathematical introduction to compressive sensing.* Birkhäuser, 2013.

[15] K. Gröchenig, *Foundations of time–frequency analysis.* Birkhäuser, 2001.

[16] D. Gabor, "Theory of communications," *J. Inst. Elec. Eng.*, vol. 96, pp. 429–457, 1946.

[17] H. Bölcskei, *Advances in Gabor analysis.* Birkhäuser, 2003, ch. Orthogonal frequency division multiplexing based on offset QAM, pp. 321–352.

[18] C. Fefferman, "The uncertainty principle," *Bull. Amer. Math. Soc.*, vol. 9, no. 2, pp. 129–206, 1983.

[19] S. Evra, E. Kowalski, and A. Lubotzky, "Good cyclic codes and the uncertainty principle," L'Enseignement Mathématique, vol. 63, no. 2, pp. 305–332, 2017.

[20] E. Bombieri, *Le grand crible dans la théorie analytique des nombres.* Société Mathématique de France, 1974.

[21] H. L. Montgomery, *Twentieth century harmonic analysis – A celebration.* Springer, 2001, ch. Harmonic analysis as found in analytic number theory.

[22] R. G. Baraniuk and M. B. Wakin, "Random projections of smooth manifolds," *Found. Comput. Math.*, vol. 9, no. 1, pp. 51–77, 2009.

[23] E. J. Candès and B. Recht, "Exact matrix completion via convex optimization," *Found. Comput. Math.*, vol. 9, no. 6, pp. 717–772, 2009.

[24] Y. C. Eldar, P. Kuppinger, and H. Bölcskei, "Block-sparse signals: Uncertainty relations and efficient recovery," *IEEE Trans. Signal Processing*, vol. 58, no. 6, pp. 3042–3054, 2010.

[25] E. J. Candès and Y. Plan, "Tight oracle inequalities for low-rank matrix recovery from a minimal number of noisy random measurements," *IEEE Trans. Information Theory*, vol. 4, no. 57, pp. 2342–2359, 2011.

[26] G. Alberti, H. Bölcskei, C. De Lellis, G. Koliander, and E. Riegler, "Lossless analog compression," *IEEE Trans. Information Theory*, vol. 65, no. 11, pp. 7480–7513, 2019.

[27] E. Riegler, D. Stotz, and H. Bölcskei, "Information-theoretic limits of matrix completion," in *Proc. IEEE International Symposium on Information Theory*, 2015, pp. 1836–1840.

[28] A. J. Izenman, "Introduction to manifold learning," *WIREs Comput. Statist.*, vol. 4, pp. 439–446, 2012.

[29] H. Lu, Y. Fainman, and R. Hecht-Nielsen, "Image manifolds," in *Proc. SPIE*, 1998, vol. 3307, pp. 52–63.

[30] N. Sochen and Y. Y. Zeevi, "Representation of colored images by manifolds embedded in higher dimensional non-Euclidean space," in *Proc. IEEE International Conference on Image Processing*, 1998, pp. 166–170.

[31] G. E. Hinton, P. Dayan, and M. Revow, "Modeling the manifolds of images of handwritten digits," *IEEE Trans. Neural Networks*, vol. 8, no. 1, pp. 65–74, 1997.

[32] W. E. Leland, M. S. Taqqu, W. Willinger, and D. V. Wilson, "On the self-similar nature of Ethernet traffic (extended version)," *IEEE/ACM Trans. Networks*, vol. 2, no. 1, pp. 1–15, 1994.

[33] R. A. Horn and C. R. Johnson, *Matrix analysis*. Cambridge University Press, 1990.

[34] T. M. Cover and J. A. Thomas, *Elements of information theory*, 2nd edn. Wiley, 2006.

[35] J. S. Abel and J. O. Smith III, "Restoring a clipped signal," in *Proc. IEEE International Conference on Acoustics, Speech and Signal Processing*, 1991, pp. 1745–1748.

[36] S. G. Mallat and G. Yu, "Super-resolution with sparse mixing estimators," *IEEE Trans. Image Processing*, vol. 19, no. 11, pp. 2889–2900, 2010.

[37] M. Elad and Y. Hel-Or, "Fast super-resolution reconstruction algorithm for pure translational motion and common space-invariant blur," *IEEE Trans. Image Processing*, vol. 10, no. 8, pp. 1187–1193, 2001.

[38] M. Bertalmio, G. Sapiro, V. Caselles, and C. Ballester, "Image inpainting," in *Proc. 27th Annual Conference on Computer Graphics and Interactive Techniques*, 2000, pp. 417–424.

[39] M. Elad, J.-L. Starck, P. Querre, and D. L. Donoho, "Simultaneous cartoon and texture image inpainting using morphological component analysis (MCA)," *Appl. Comput. Harmonic Analysis*, vol. 19, pp. 340–358, 2005.

[40] D. L. Donoho and G. Kutyniok, "Microlocal analysis of the geometric separation problem," *Commun. Pure Appl. Math.*, vol. 66, no. 1, pp. 1–47, 2013.

[41] J. A. Tropp, "On the conditioning of random subdictionaries," *Appl. Comput. Harmonic Analysis*, vol. 25, pp. 1–24, 2008.

[42] E. J. Candès, J. Romberg, and T. Tao, "Robust uncertainty principles: Exact signal reconstruction from highly incomplete frequency information," *IEEE Trans. Information Theory*, vol. 52, no. 2, pp. 489–509, 2006.

[43] G. Pope, A. Bracher, and C. Studer, "Probabilistic recovery guarantees for sparsely corrupted signals," *IEEE Trans. Information Theory*, vol. 59, no. 5, pp. 3104–3116, 2013.

[44] D. L. Donoho, "Compressed sensing," *IEEE Trans. Information Theory*, vol. 52, no. 4, pp. 1289–1306, 2006.

[45] J. A. Tropp, "On the linear independence of spikes and sines," *J. Fourier Analysis and Applications*, vol. 14, no. 5, pp. 838–858, 2008.

[46] D. L. Donoho, I. M. Johnstone, J. C. Hoch, and A. S. Stern, "Maximum entropy and the nearly black object," *J. Roy. Statist. Soc. Ser. B.*, vol. 54, no. 1, pp. 41–81, 1992.

[47] P. Feng and Y. Bresler, "Spectrum-blind minimum-rate sampling and reconstruction of multiband signals," in *Proc. IEEE International Conference on Acoustics, Speech and Signal Processing*, 1996, pp. 1688–1691.

[48] M. Mishali and Y. C. Eldar, "From theory to practice: Sub-Nyquist sampling of sparse wideband analog signals," *IEEE J. Selected Areas Commun.*, vol. 4, no. 2, pp. 375–391, 2010.

[49] R. Heckel and H. Bölcskei, "Identification of sparse linear operators," *IEEE Trans. Information Theory*, vol. 59, no. 12, pp. 7985–8000, 2013.

[50] K. Falconer, *Fractal geometry*, 1st edn. Wiley, 1990.

[51] J. D. Vaaler, "Some extremal functions in Fourier analysis," *Bull. Amer. Math. Soc.*, vol. 2, no. 12, pp. 183–216, 1985.

[52] C. Heil, *A basis theory primer*. Springer, 2011.

[53] P. Mattila, *Geometry of sets and measures in Euclidean space: Fractals and rectifiability*. Cambridge University Press, 1999.

7 Understanding Phase Transitions via Mutual Information and MMSE

Galen Reeves and Henry D. Pfister

Summary

The ability to understand and solve high-dimensional inference problems is essential for modern data science. This chapter examines high-dimensional inference problems through the lens of information theory and focuses on the standard linear model as a canonical example that is both rich enough to be practically useful and simple enough to be studied rigorously. In particular, this model can exhibit phase transitions where an arbitrarily small change in the model parameters can induce large changes in the quality of estimates. For this model, the performance of optimal inference can be studied using the replica method from statistical physics but, until recently, it was not known whether the resulting formulas were actually correct. In this chapter, we present a tutorial description of the standard linear model and its connection to information theory. We also describe the replica prediction for this model and outline the authors' recent proof that it is exact.

7.1 Introduction

7.1.1 What Can We Learn from Data?

Given a probabilistic model, this question can be answered succinctly in terms of the difference between the prior distribution (what we know before looking at the data) and the posterior distribution (what we know after looking at the data). Throughout this chapter we will focus on high-dimensional inference problems where the posterior distribution may be complicated and difficult to work with directly. We will show how techniques rooted in information theory and statistical physics can provide explicit characterizations of the statistical relationship between the data and the unknown quantities of interest.

In his seminal paper, Shannon showed that *mutual information* provides an important measure of the difference between the prior and posterior distributions [1]. Since its introduction, mutual information has played a central role for applications in engineering, such as communication and data compression, by describing the fundamental constraints imposed solely by the *statistical* properties of the problems.

Traditional problems in information theory assume that all statistics are known and that certain system parameters can be chosen to optimize performance. In contrast, data-science problems typically assume that the important distributions are either given by the problem or must be estimated from the data. When the distributions are unknown, the implied inference problems are more challenging and their analysis can become intractable. Nevertheless, similar behavior has also been observed in more general high-dimensional inference problems such as Gaussian mixture clustering [2].

In this chapter, we use the standard linear model as a simple example to illustrate phase transitions in high-dimensional inference. In Section 7.2, basic properties of the standard linear model are described and examples are given to describe its behavior. In Section 7.3, a number of connections to information theory are introduced. In Section 7.4, we present an overview of the authors' proof that the replica formula for mutual information is exact. In Section 7.5, connections between posterior correlation and phase transitions are discussed. Finally, in Section 7.6, we offer concluding remarks.

Notation The probability $\mathbb{P}[\mathbf{Y} = \mathbf{y} | \mathbf{X} = \mathbf{x}]$ is denoted succinctly by $p_{\mathbf{Y}|\mathbf{X}}(\mathbf{y}|\mathbf{x})$ and shortened to $p(\mathbf{y}|\mathbf{x})$ when the meaning is clear. Similarly, the distinction between discrete and continuous distributions is neglected when it is inconsequential.

7.1.2 High-Dimensional Inference

Suppose that the relationship between the unobserved random vector $\mathbf{X} = (X_1, \ldots, X_N)$ and the observed random vector $\mathbf{Y} = (Y_1, \ldots, Y_M)$ is modeled by the joint probability distribution $p(\mathbf{x}, \mathbf{y})$. The central problem of Bayesian inference is to answer questions about the unobserved variables in terms of the posterior distribution:

$$p(\mathbf{x} \,|\, \mathbf{y}) = \frac{p(\mathbf{y} \,|\, \mathbf{x}) p(\mathbf{x})}{p(\mathbf{y})}. \tag{7.1}$$

In the high-dimensional setting where both M and N are large, direct evaluation of the posterior distribution can become intractable, and one often resorts to summary statistics such as the posterior mean/covariance or the marginal posterior distribution of a small subset of the variables.

The analysis of high-dimensional inference problems focuses on two questions.

- What is the fundamental limit of inference without computational constraints?
- What can be inferred from data using computationally efficient methods?

It is becoming increasingly common for the answers to these questions to be framed in terms of *phase diagrams*, which provide important information about fundamental trade-offs involving the amount and quality of data. For example, the phase diagram in Fig. 7.1 shows that increasing the amount of data not only provides more information, but also moves the problem into a regime where efficient methods are optimal. By contrast, increasing the SNR may lead to improvements that can be attained only with significant computational complexity.

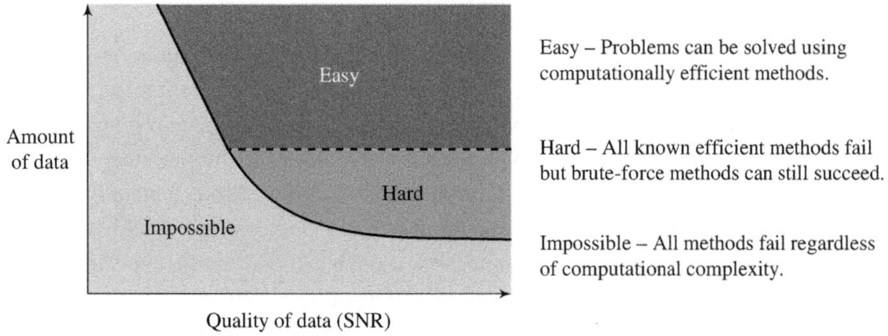
Easy
Hard
Impossible
Amount of data
Quality of data (SNR)
Easy – Problems can be solved using computationally efficient methods.
Hard – All known efficient methods fail but brute-force methods can still succeed.
Impossible – All methods fail regardless of computational complexity.

Figure 7.1 Example phase diagram for a high-dimensional inference problem such as signal estimation for the standard linear model. The parameter regions indicate the difficulty of inference of some fixed quality.

7.1.3 Three Approaches to Analysis

For the standard linear model, the qualitative behavior shown in Fig. 7.1 is correct and can be made quantitatively precise in the large-system limit. In general, there are many different approaches that can be used to analyze high-dimensional inference problems. Three popular approaches are described below.

Information-Theoretic Analysis

The standard approach taken in information theory is to first obtain precise characterizations of the fundamental limits, without any computational constraints, and then use these limits to inform the design and analysis of practical methods. In many cases, the fundamental limits can be understood by studying macroscopic system properties in the large-system limit, such as the mutual information and the minimum mean-squared error (MMSE). There are a wide variety of mathematical tools to analyze these quantities in the context of compression and communication [3, 4]. Unfortunately, these tools alone are often unable to provide simple descriptions for the behavior of high-dimensional statistical inference problems.

The Replica Method from Statistical Physics

An alternative approach for analyzing the fundamental limits is provided by the powerful but heuristic *replica method* from statistical physics [5, 6]. This method, which was developed originally in the context of disordered magnetic materials known as spin glasses, has been applied successfully to a wide variety of problems in science and engineering. At a high level, the replica method consists of a sequence of derivations that provide explicit formulas for the mutual information in the large-system limit. The main limitation, however, is that the validity of these formulas relies on certain assumptions that are unproven in general. A common progression in the statistical physics literature is that results are first conjectured using the replica method and then proven using very different techniques. For example, formulas for the Sherrington–Kirkpatrick model were conjectured using the replica method in 1980 by Parisi [7], but were not rigorously proven until 2006 by Talagrand [8].

Analysis of Approximate Inference

Significant work has focused on tractable methods for computing summary statistics of the posterior distribution. Variational inference [9–11] refers to a large class of methods where the inference problem is recast as an optimization problem. The well-known mean-field variational approach typically refers to minimizing the Kullback–Leibler divergence with respect to product distributions. More generally, however, the variational formulation also encompasses the Bethe and Kikuchi methods for sparsely connected or highly decomposable models as well as the expectation-consistent (EC) approximate inference framework of Opper and Winther [12].

A variety of methods can be used to solve or approximately solve the variational optimization problem, including message-passing algorithms such as belief propagation [13], expectation propagation [14], and approximate message passing [15]. In some cases, the behavior of these algorithms can be characterized precisely via density evolution (coding theory) or state evolution (compressed sensing), which leads to single-letter characterizations of the behavior in the large-system limit.

7.2 Problem Setup and Characterization

7.2.1 Standard Linear Model

Consider the inference problem implied by an unobserved random vector $\mathbf{X} \in \mathbb{R}^N$ and an observed random vector $\mathbf{Y} \in \mathbb{R}^M$. An important special case is the *Gaussian channel*, where $M = N$ and the unknown variables are related to the observations via

$$Y_n = \sqrt{s} X_n + W_n, \qquad n = 1, \dots, N, \tag{7.2}$$

where $\{W_n\}$ is i.i.d. standard Gaussian noise and $s \in [0, \infty)$ parameterizes the signal-to-noise ratio. The Gaussian channel, which is also known as the Gaussian sequence model in the statistics literature [16], provides a useful first-order approximation for a wide variety of applications in science and engineering.

The *standard linear model* is an important generalization of the Gaussian channel in which the observations consist of noisy linear measurements:

$$Y_m = \langle \mathbf{A}_m, \mathbf{X} \rangle + W_m, \qquad m = 1, \dots, M, \tag{7.3}$$

where $\langle \cdot, \cdot \rangle$ denotes the standard Euclidean inner product, $\{\mathbf{A}_m\}$ is a known sequence of N-length measurement vectors, and $\{W_m\}$ is i.i.d. Gaussian noise. Unlike for the Gaussian channel, the number of observations M may be different from the number of unknown variables N. For this reason the measurement indices are denoted by m instead of n. In matrix form, the standard linear model can be expressed as

$$\mathbf{Y} = \mathbf{A}\mathbf{X} + \mathbf{W}, \tag{7.4}$$

where $\mathbf{A} \in \mathbb{R}^{M \times N}$ is a known matrix and $\mathbf{W} \sim \mathcal{N}(0, \mathbf{I}_M)$. Inference problems involving the standard linear model include linear regression in statistics, both channel and symbol estimation in wireless communications, and sparse signal recovery in compressed sensing [17, 18]. In the standard linear model, the matrix $\mathbf{A}$ induces dependences between the unknown variables which make the inference problem significantly more difficult.

Typical inference questions for the standard linear model include the following.

- **Estimation** of unknown variables. The performance of an estimator $\mathbf{Y} \mapsto \widehat{\mathbf{X}}$ is often measured using its *mean-squared error* (MSE),
$$\mathbb{E}\left[\|\mathbf{X} - \widehat{\mathbf{X}}\|^2\right].$$
The optimal MSE, computed by minimizing over all possible estimators, is called the *minimum mean-squared error* (MMSE),
$$\mathbb{E}\left[\|\mathbf{X} - \mathbb{E}[\mathbf{X} \mid \mathbf{Y}]\|^2\right].$$
The MMSE is also equivalent to the Bayes risk under squared-error loss.
- **Prediction** of a new observation $Y_{\text{new}} = \langle \mathbf{A}_{\text{new}}, \mathbf{X}\rangle + W_{\text{new}}$. The performance of an estimator $(\mathbf{Y}, \mathbf{A}_{\text{new}}) \mapsto \widehat{Y}_{\text{new}}$ is often measured using the *prediction mean-squared error*,
$$\mathbb{E}\left[(Y_{\text{new}} - \widehat{Y}_{\text{new}})^2\right].$$
- **Detection** of whether the ith entry belongs to a subset K of the real line. For example, $K = \mathbb{R} \setminus \{0\}$ tests whether entries are non-zero. In practice, one typically defines a test statistic
$$T(\mathbf{y}) = \ln\left(\frac{p(\mathbf{Y} = \mathbf{y} | X_i \in K)}{p(\mathbf{Y} = \mathbf{y} | X_i \notin K)}\right)$$
and then uses a threshold rule that chooses $X_i \in K$ if $T(\mathbf{y}) \geq \lambda$ and $X_i \notin K$ otherwise. The performance of this detection rule can be measured using the true-positive rate (TPR) and the false-positive rate (FPR) given by
$$\mathsf{TPR} = p(T(\mathbf{Y}) \geq \lambda | X_i \in K), \qquad \mathsf{FPR} = p(T(\mathbf{Y}) \geq \lambda | X_i \notin K).$$
The receiver operating characteristic (ROC) curve for this binary decision problem is obtained by plotting the TPR versus the FPR as a parametric function of the threshold λ. An example is given in Fig. 7.3 below.
- **Posterior marginal approximation** of a subset S of unknown variables. The goal is to compute an approximation $\widehat{p}(\mathbf{x}_S \mid \mathbf{Y})$ of the marginal distribution of entries in S, which can be used to provide summary statistics and measures of uncertainty. In some cases, accurate approximation of the posterior for small subsets is possible even though the full posterior distribution is intractable.

Analysis of Fundamental Limits

To understand the fundamental and practical limits of inference with the standard linear model, a great deal of work has focused on the setting where (1) the entries of $\mathbf{X}$ are drawn i.i.d. from a known prior distribution, and (2) the matrix $\mathbf{A}$ is an $M \times N$ random matrix whose entries $\mathbf{A}_{ij}$ are drawn i.i.d. from $\mathcal{N}(0, 1/N)$. Sparsity in $\mathbf{X}$ can be modeled by using a spike–slab signal distribution (i.e., a mixture of a very narrow distribution with a very wide distribution). Consider a sequence of problems where the number of measurements per signal dimension converges to δ. In this case, the normalized mutual information and MMSE corresponding to the large-system limit[1] are given by

$$\mathcal{I}(\delta) \triangleq \lim_{\substack{M,N\to\infty \\ M/N\to\delta}} \frac{1}{N} I(\mathbf{X};\mathbf{Y}\,|\,\mathbf{A}), \qquad \mathcal{M}(\delta) \triangleq \lim_{\substack{M,N\to\infty \\ M/N\to\delta}} \frac{1}{N}\,\mathsf{mmse}(\mathbf{X}\,|\,\mathbf{Y},\mathbf{A}).$$

Part of what makes this problem interesting is that the MMSE can have discontinuities, which are referred to as *phase transitions*. The values of δ at which these discontinuities occur are of significant interest because they correspond to problem settings in which a small change in the number of measurements makes a large difference in the ability to estimate the unknown variables. In the above limit, the value of $\mathcal{M}(\delta)$ is undefined at these points.

Replica-Symmetric Formula

Using the heuristic replica method from statistical physics, Guo and Verdú [19] derived single-letter formulas for the mutual information and MMSE in the standard linear model with i.i.d. variables and an i.i.d. Gaussian matrix:

$$\mathcal{I}(\delta) = \min_{s\geq 0} \underbrace{\left\{ I\left(X;\, \sqrt{s}X + W\right) + \frac{\delta}{2}\left(\log\left(\frac{\delta}{s}\right) + \frac{s}{\delta} - 1\right)\right\}}_{\mathcal{F}(s)} \tag{7.5}$$

and $$\mathcal{M}(\delta) = \mathsf{mmse}\left(X \,|\, \sqrt{s^*(\delta)}X + W\right). \tag{7.6}$$

In these expressions, X is a univariate random variable drawn according to the prior p_X, $W \sim \mathcal{N}(0,1)$ is independent Gaussian noise, and $s^*(\delta)$ is a minimizer of the objective function $\mathcal{F}(s)$. Precise definitions of the mutual information and the MMSE are provided in Section 7.3 below. By construction, the replica mutual information (7.5) is a continuous function of the measurement rate δ. However, the replica MMSE prediction (7.6) may have discontinuities when the global minimizer $s^*(\delta)$ jumps from one minimum to another, and $\mathcal{M}(\delta)$ is well defined only if $s^*(\delta)$ is the unique minimizer. In [20–22], the authors prove these expressions are exact for the standard linear model with an i.i.d. Gaussian measurement matrix. An overview of this proof is presented in Section 7.4.

Approximate Message Passing

An algorithmic breakthrough for the standard linear model with i.i.d. Gaussian matrices was provided by the approximate message-passing (AMP) algorithm [15, 23] and its generalizations [24–29]. For CDMA waveforms, the same idea was applied earlier in [30].

An important property of this class of algorithms is that the performance for large i.i.d. Gaussian matrices is characterized precisely via a state-evolution formalism [23, 31]. Remarkably, the fixed points of the state evolution correspond to the stationary points

[1] Under reasonable conditions, one can show that these limits are well defined for almost all non-negative values of δ and that $\mathcal{I}(\delta)$ is continuous.

of the objective function $\mathcal{F}(s)$ in (7.5). For cases where the replica formulas are exact, this means that AMP-type algorithms can be optimal with respect to marginal inference problems whenever the largest local minimizer of $\mathcal{F}(s)$ is also the global minimizer [32].

The Generalized Linear Model

In the generalized linear model, the observations $\mathbf{Y} \in \mathbb{R}^M$ are related to the unknown variables $\mathbf{X} \in \mathbb{R}^N$ by way of the $M \times N$ matrix $\mathbf{A}$, the random variable $\mathbf{Z} = \mathbf{AX}$, and the conditional distribution

$$p_{\mathbf{Y}|\mathbf{X}}(\mathbf{y} \mid \mathbf{x}) \triangleq p_{\mathbf{Y}|\mathbf{Z}}(\mathbf{y} \mid \mathbf{AX}), \tag{7.7}$$

where $p_{\mathbf{Y}|\mathbf{Z}}(\mathbf{y}|\mathbf{z})$ defines a memoryless (i.e., separable) channel. The generalized linear model is fundamental to generalized linear regression in statistics. It is also used to model different sensing architectures (e.g., Poisson channels, phase retrieval) and the effects of scalar quantization. The AMP algorithm was introduced by Donoho, Maleki, and Montanari in [15] and extended to the GLM by Rangan in [24]. More recent work has focused on AMP-style algorithms for rotationally invariant random matrices [33–38].

7.2.2 Illustrative Examples

We now consider some examples that illustrate the similarities and differences between the Gaussian channel and the standard linear model. In these examples, the unknown variables are drawn i.i.d. from the Bernoulli-Gaussian distribution, which corresponds to the product of independent Bernoulli and Gaussian random variables and is given by

$$\mathsf{BG}(x \mid \mu, \sigma^2, \gamma) \triangleq (1-\gamma)\delta_0(x) + \gamma \mathcal{N}(x \mid \mu, \sigma^2). \tag{7.8}$$

Here, δ_0 denotes a Dirac distribution with all probability mass at zero and $\mathcal{N}(x|\mu,\sigma^2)$ denotes the Gaussian p.d.f. $(2\pi\sigma^2)^{-1/2}e^{-(x-\mu)^2/(2\sigma^2)}$ with mean μ and variance σ^2. The parameter $\gamma \in (0,1)$ determines the expected fraction of non-zero entries. The mean and variance of a random variable $X \sim \mathsf{BG}(x \mid \mu, \sigma^2, \gamma)$ are given by

$$\mathbb{E}[X] = \gamma\mu, \qquad \mathsf{Var}(X) = \gamma(1-\gamma)\mu^2 + \gamma\sigma^2. \tag{7.9}$$

Gaussian Channel with Bernoulli–Gaussian Prior

If the unknown variables are i.i.d. $\mathsf{BG}(x \mid \mu, \sigma^2, \gamma)$ and the observations are generated according to the Gaussian channel (7.2), then the posterior distribution *decouples* into the product of its marginals:

$$p(\mathbf{x} \mid \mathbf{y}) = \prod_{n=1}^{N} p(x_n \mid y_n). \tag{7.10}$$

Furthermore, the posterior marginal $p(x_n \mid y_n)$ is also a Bernoulli Gaussian distribution but with new parameters $(\mu_n, \sigma_n^2, \gamma_n)$ that depend on y_n:

$$\mu_n = \mu + \frac{\sqrt{s}\sigma^2}{1+s\sigma^2}(y_n - \sqrt{s}\mu), \tag{7.11}$$

$$\sigma_n^2 = \frac{\sigma^2}{1+s\sigma^2}, \tag{7.12}$$

$$\gamma_n = \left[1+(1/\gamma-1)\sqrt{1+s\sigma^2}\exp\left(\frac{s\mu^2-2\sqrt{s}\mu y_n - s\sigma^2 y_n^2}{2(1+s\sigma^2)}\right)\right]^{-1}. \tag{7.13}$$

Given these parameters, the posterior mean $\mathbb{E}[X_n \mid Y_n]$ and the posterior variance $\mathsf{Var}(X_n \mid Y_n)$ can be computed using (7.9). The parameter γ_n is the conditional probability that X_n is non-zero given Y_n. This parameter is often called the *posterior inclusion probability* in the statistics literature.

The decoupling of the posterior distribution makes it easy to characterize the fundamental limits of performance measures. For example the MMSE is the expectation of the posterior variance $\mathsf{Var}(X_n \mid Y_n)$ and the optimal trade-off between the true-positive rate and false-positive rate for detecting the event $\{X_n \neq 0\}$ is characterized by the distribution of γ_n.

To investigate the statistical properties of the posterior distribution we perform a numerical experiment. First, we draw $N = 10000$ variables according to the Bernoulli–Gaussian variables with $\mu = 0$, $\sigma^2 = 10^6$, and prior inclusion probability $\gamma = 0.2$. Then, for various values of the signal-to-noise-ratio parameter s, we evaluate the posterior distribution corresponding to the output of the Gaussian channel.

In Fig. 7.2 (left panel), we plot three quantities associated with the estimation error:

$$\text{average squared error:} \quad \frac{1}{N}\sum_{n=1}^{N}(X_n - \mathbb{E}[X \mid Y_n])^2,$$

$$\text{average posterior variance:} \quad \frac{1}{N}\sum_{n=1}^{N}\mathsf{Var}(X_n \mid Y_n),$$

$$\text{average MMSE:} \quad \frac{1}{N}\sum_{n=1}^{N}\mathbb{E}[\mathsf{Var}(X_n \mid Y_n)].$$

Note that the squared error and posterior variance are both random quantities because they are functions of the data. This means that the corresponding plots would look

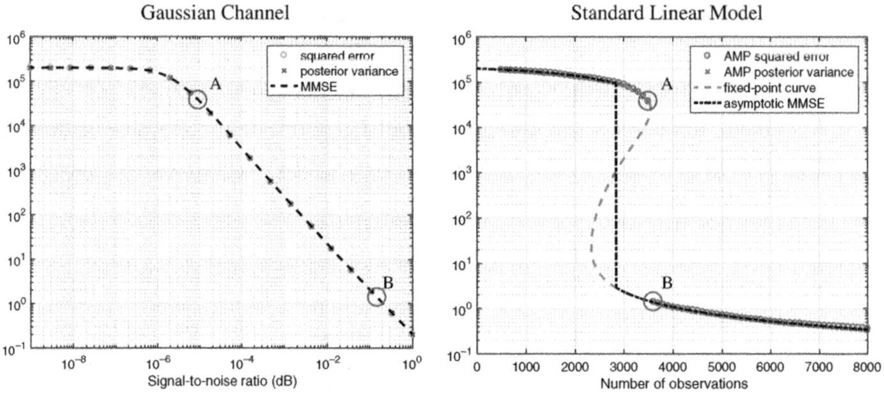

Figure 7.2 Comparison of average squared error for the Gaussian channel as a function of the signal-to-noise ratio (left panel) and the standard linear model as a function of the number of observations (right panel). In both cases, the unknown variables are i.i.d. Bernoulli–Gaussian with zero mean and a fraction $\gamma = 0.20$ of non-zero entries.

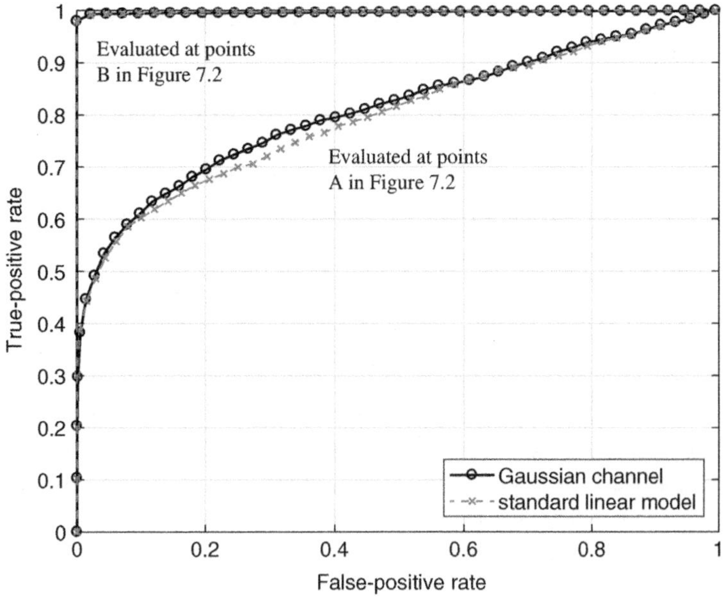

Figure 7.3 ROC curves for detecting the non-zero variables in the parameter regimes labeled A and B in Fig. 7.2. The curves for the Gaussian channel are obtained by thresholding the true posterior inclusion probabilities $\{\gamma_n\}$. The curves for the standard linear model are obtain by thresholding the AMP approximations of the posterior inclusion probabilities.

slightly different if the experiment were repeated multiple times. The MMSE, however, is a function of the joint distribution of $(\mathbf{X}, \mathbf{Y})$ and is thus non-random. In this setting, the fact that there is little difference between the *averages* of these quantities can be seen as a consequence of the decoupling of the posterior distribution and the law of large numbers.

In Fig. 7.3, we plot the ROC curve for the problem of detecting the non-zero variables. The curves are obtained by thresholding the posterior inclusion probabilities $\{\gamma_n\}$ associated with the values of the signal-to-noise ratio at the points A and B in Fig. 7.2.

Standard Linear Model with Bernoulli–Gaussian Prior

Next, we consider the setting where the observations are generated by the standard linear model (7.4). In this case, the measurement matrix introduces dependence in the posterior distribution, and the decoupling seen in (7.10) does not hold in general.

To characterize the posterior distribution, one can use the fact that $\mathbf{X}$ is conditionally Gaussian given the support vector $\mathbf{U} \in \{0,1\}^N$, where $X_n = 0$ if $U_n = 0$ and $X_n \neq 0$ with probability one if $U_n = 1$. Consequently, the posterior distribution can be expressed as a Gaussian mixture model of the form

$$p(\mathbf{x} \mid \mathbf{y}, \mathbf{A}) = \sum_{\mathbf{u} \in \{0,1\}^N} p(\mathbf{u} \mid \mathbf{y}, \mathbf{A}) \mathcal{N}(\mathbf{x} \mid \mathbb{E}[\mathbf{X} \mid \mathbf{y}, \mathbf{A}, \mathbf{u}], \mathrm{Cov}(\mathbf{X} \mid \mathbf{y}, \mathbf{A}, \mathbf{u})),$$

where the summation is over all possible support sets. The posterior probability of the support set is given by

$$p(\mathbf{u} \mid \mathbf{y}, \mathbf{A}) \propto p(\mathbf{u}) \mathcal{N}\big(\mathbf{y} \mid \mu \mathbf{A}_{\mathbf{u}} \mathbf{1}, \mathbf{I} + \sigma^2 \mathbf{A}_{\mathbf{u}}^{\mathrm{T}} \mathbf{A}_{\mathbf{u}}\big),$$

where $\mathbf{A}_{\mathbf{u}}$ is the submatrix of $\mathbf{A}$ formed by removing columns where $u_n = 0$ and $\mathbf{1}$ denotes a vector of ones. The posterior marginal is obtained by integrating out the other variables to get

$$p(x_n \mid \mathbf{y}, \mathbf{A}) = \int p(\mathbf{x} \mid \mathbf{y}, \mathbf{A}) \, d\mathbf{x}_{\sim n},$$

where $\mathbf{x}_{\sim n}$ denotes all the entries except for x_n.

Here, the challenge is that the number of terms in the summation over $\mathbf{u}$ grows exponentially with the signal dimension N. Since it is difficult to compute the posterior distribution in general, we use AMP to compute approximations to the posterior marginal distributions. The marginals of the approximation, which belong to the Bernoulli–Gaussian family of distributions, are given by

$$\widehat{p}(x_n \mid \mathbf{y}, \mathbf{A}) = \mathsf{BG}(x_n \mid \mu_n, \sigma_n^2, \gamma_n), \tag{7.14}$$

where the parameters $(\mu_n, \sigma_n^2, \gamma_n)$ are the outputs of the AMP algorithm.

Similarly to the previous example, we perform a numerical experiment to investigate the statistical properties of the marginal approximations. First, we draw $N = 10\,000$ variables according to the Bernoulli–Gaussian variables with $\mu = 0$, $\sigma^2 = 10^6$, and prior inclusion probability $\gamma = 0.2$. Then, for various values of M, we obtain measurements from the standard linear model with i.i.d. Gaussian measurement vectors $\mathbf{A}_m \sim \mathcal{N}(0, N^{-1}\mathbf{I})$ and use AMP to compute the parameters $(\mu_n, \sigma_n^2, \gamma_n)$ used in the marginal posterior approximations.

In Fig. 7.2 (right panel), we plot the squared error and the approximation of the posterior variance associated with the AMP marginal approximations:

$$\text{average AMP squared error:} \quad \frac{1}{N} \sum_{n=1}^{N} \big(X_n - \mathbb{E}_{\widehat{p}}[X_n \mid \mathbf{Y}, \mathbf{A}]\big)^2,$$

$$\text{average AMP posterior variance:} \quad \frac{1}{N} \sum_{n=1}^{N} \mathsf{Var}_{\widehat{p}}(X_n \mid \mathbf{Y}, \mathbf{A}).$$

In these expressions, the expectation and the variance are computed with respect to the marginal approximation in (7.14). Because these quantities are functions of the random data, one expects that they would look slightly different if the experiment were repeated multiple times.

At this point, there are already some interesting observations that can be made. First, we note that the AMP approximation of the mean can be viewed as a point-estimate of the unknown variables. Similarly, the AMP approximation of the posterior variance (which depends on the observations but not the ground truth) can be viewed as a point-estimate of the squared error. From this perspective, the close correspondence between the squared error and the AMP approximation of the variance seen in Fig. 7.2 suggests that AMP is self-consistent in the sense that it provides an accurate estimate of its square error.

Another observation is that the squared error undergoes an abrupt change at around 3500 observations, between the points labeled A and B. Before this, the squared error is within an order of magnitude of the prior variance. After this, the squared error drops discontinuously. This illustrates that the estimator provided by AMP is quite accurate in this setting.

However, there are still some important questions that remain. For example, how accurate are the AMP posterior marginal approximations? Is it possible that a different algorithm (e.g., one that computes the true posterior marginals) would lead to estimates with significantly smaller squared error? Further questions concern how much information is lost in focusing only on the marginals of the posterior distribution as opposed to the full posterior distribution.

Unlike the Gaussian channel, it is not possible to evaluate the MMSE directly because the summation over all 2^{10000} support vectors is prohibitively large. For comparison, we plot the large-system MMSE predicted by (7.6), which corresponds to the large-N limit where the fraction of observations is parameterized by $\delta = M/N$.

The behavior of the large-system MMSE is qualitatively similar to the AMP squared error because it has a single jump discontinuity (or phase transition). However, the jump occurs after only 2850 observations for the MMSE as opposed to after 3600 observations for AMP. By comparing the AMP squared error with the asymptotic MMSE, we see that the AMP marginal approximations are accurate in some cases (e.g., when the number of observations is fewer than 2840 or greater than 3600) but highly inaccurate in others (e.g., when the number of observations is between 2840 and 3600).

In Fig. 7.3, we plot the ROC curve for the problem of detecting the non-zero variables in the standard linear model. In this case, the curves are obtained by thresholding AMP approximations of the posterior inclusion probabilities $\{\gamma_n\}$. It is interesting to note that the ROC curves corresponding to the two different observation models (the Gaussian channel and the standard linear model) have similar shapes when they are evaluated in problem settings with matched squared error.

7.3 The Role of Mutual Information and MMSE

The amount one learns about an unknown vector $\mathbf{X}$ from an observation $\mathbf{Y}$ can be quantified in terms of the difference between the prior and posterior distributions. For a particular realization of the observations $\mathbf{y}$, a fundamental measure of this difference is provided by the relative entropy

$$D(p_{\mathbf{X}|\mathbf{Y}}(\cdot \mid \mathbf{y}) \,\|\, p_{\mathbf{X}}(\cdot)).$$

This quantity is non-negative and equal to zero if and only if the posterior is the same as the prior almost everywhere.

For real vectors, another way to assess the difference between the prior and posterior distributions is to compare the first and second moments of their distributions. These moments are summarized by the mean $\mathbb{E}[\mathbf{X}]$, the conditional mean $\mathbb{E}[\mathbf{X} \mid \mathbf{Y} = \mathbf{y}]$, the covariance matrix

$$\mathsf{Cov}(\mathbf{X}) \triangleq \mathbb{E}\Big[(\mathbf{X} - \mathbb{E}[\mathbf{X}])(\mathbf{X} - \mathbb{E}[\mathbf{X}])^{\mathrm{T}}\Big],$$

and the conditional covariance matrix

$$\mathsf{Cov}(\mathbf{X} \mid \mathbf{Y} = \mathbf{y}) \triangleq \mathbb{E}\Big[(\mathbf{X} - \mathbb{E}[\mathbf{X} \mid \mathbf{Y}])(\mathbf{X} - \mathbb{E}[\mathbf{X} \mid \mathbf{Y}])^{\mathrm{T}} \mid \mathbf{Y} = \mathbf{y}\Big].$$

Together, these provide some measure of how much "information" there is in the data.

One of the difficulties of working with the posterior distribution directly is that it can depend non-trivially on the particular realization of the data. It can be much easier to focus on the behavior for *typical* realizations of the data by studying the distribution of the relative entropy when $\mathbf{Y}$ is drawn according to the marginal distribution $p(\mathbf{y})$. For example, the expectation of the relative entropy is the mutual information

$$I(\mathbf{X}; \mathbf{Y}) \triangleq \mathbb{E}[D(p_{\mathbf{X}|\mathbf{Y}}(\cdot \mid \mathbf{Y}) \,\|\, p_{\mathbf{X}}(\cdot))].$$

Similarly, the expected value of the conditional covariance matrix[2] is

$$\mathbb{E}[\mathsf{Cov}(\mathbf{X} \mid \mathbf{Y})] = \mathbb{E}\Big[(\mathbf{X} - \mathbb{E}[\mathbf{X} \mid \mathbf{Y}])(\mathbf{X} - \mathbb{E}[\mathbf{X} \mid \mathbf{Y}])^{\mathrm{T}}\Big].$$

The trace of this matrix equals the Bayes risk for squared-error loss, which is more commonly called the MMSE and defined by

$$\mathsf{mmse}(\mathbf{X} \mid \mathbf{Y}) \triangleq \mathrm{tr}(\mathbb{E}[\mathsf{Cov}(\mathbf{X} \mid \mathbf{Y})]) = \mathbb{E}\Big[\|\mathbf{X} - \mathbb{E}[\mathbf{X} \mid \mathbf{Y}]\|_2^2\Big],$$

where $\|\cdot\|_2$ denotes the Euclidean norm. Part of the appeal of working with the mutual information and the MMSE is that they satisfy a number of useful functional properties, including chain rules and data-processing inequalities.

The prudence of focusing on the expectation with respect to the data depends on the extent to which the random quantities of interest deviate from their expectations. In the statistical physics literature, the concentration of the relative entropy and squared error around their expectations is called the *self-averaging property* and is often assumed for large systems [5].

7.3.1 I-MMSE Relationships for the Gaussian Channel

Given an N-dimensional random vector $\mathbf{X} = (X_1, \ldots, X_N)$, the output of the Gaussian channel with signal-to-noise-ratio parameter $s \in [0, \infty)$ is denoted by

$$\mathbf{Y}(s) = \sqrt{s}\mathbf{X} + \mathbf{W},$$

where $\mathbf{W} \sim \mathcal{N}(0, \mathbf{I}_N)$ is independent Gaussian noise. Two important functionals of the joint distribution of $(\mathbf{X}, \mathbf{Y}_s)$ are the *mutual information function*

$$I_{\mathbf{X}}(s) = \frac{1}{N} I(\mathbf{X}; \mathbf{Y}(s))$$

[2] Observe that $\mathsf{Cov}(\mathbf{X} \mid \mathbf{Y})$ is a random variable in the same sense as $\mathbb{E}[\mathbf{X} \mid \mathbf{Y}]$.

and the *MMSE function*

$$M_{\mathbf{X}}(s) = \frac{1}{N}\mathbb{E}\left[\|\mathbf{X} - \mathbb{E}[\mathbf{X} \mid \mathbf{Y}(s)]\|_2^2\right].$$

In some cases, these functions can be computed efficiently using numerical integration or Monte Carlo approximation. For example, if the entries of $\mathbf{X}$ are i.i.d. copies of a scalar random variable X then the mutual information and MMSE depend only on the marginal distribution:

$$I_{\mathbf{X}}(s) = I(X; \sqrt{s}X + W),$$
$$M_{\mathbf{X}}(s) = \mathbb{E}\left[\left(X - \mathbb{E}\left[X \mid \sqrt{s}X + W\right]\right)^2\right].$$

Another example is if $\mathbf{X}$ is drawn according to a Gaussian mixture model with a small number of mixture components. For general high-dimensional distributions, however, direct computation of these functions can be intractable due to the curse of dimensionality. Instead, one often resorts to asymptotic approximations.

The I-MMSE relationship [39] asserts that the derivative of the mutual information is one-half of the MMSE,

$$\frac{d}{ds} I_{\mathbf{X}}(s) = \frac{1}{2} M_{\mathbf{X}}(s). \tag{7.15}$$

This result is equivalent to the classical De Bruijn identity [40], which relates the derivative of differential entropy to the Fisher information. Part of the significance of the I-MMSE relationship is that it provides a link between an information-theoretic quantity and an estimation-theoretic quantity.

Another important property of the MMSE function is that its derivative is

$$M'_{\mathbf{X}}(s) = -\frac{1}{N}\mathbb{E}\left[\|\operatorname{Cov}(\mathbf{X} \mid \sqrt{s}\mathbf{X} + \mathbf{W})\|_{\mathrm{F}}^2\right], \tag{7.16}$$

where $\|\cdot\|_{\mathrm{F}}$ denotes the Frobenious norm. Since the derivative is non-positive, it follows that the MMSE function is non-increasing and the mutual information function is concave.

The relationship between the MMSE function and its derivative imposes some useful constraints on the MMSE. One example is the so-called *single-crossing property* [41], which asserts that, for any random vector $\mathbf{X}$ and isotropic Gaussian random vector $\mathbf{Z}$, the MMSE functions $M_{\mathbf{X}}(s)$ and $M_{\mathbf{Z}}(s)$ cross at most once.

The following result states a monotonicity property concerning a transformation of the MMSE function. A matrix generalization of this result is given in [42].

THEOREM 7.1 (Monotonicity of MMSE) *For any random vector* $\mathbf{X}$ *that is not almost-surely constant, the function*

$$k_{\mathbf{X}}(s) \triangleq \frac{1}{M_{\mathbf{X}}(s)} - s \tag{7.17}$$

is well defined and non-decreasing on $(0, \infty)$.

Proof The MMSE function is real analytic, and hence infinitely differentiable, on $(0,\infty)$ [41]. By differentiation, one finds that

$$\frac{d}{ds} k_{\mathbf{X}}(s) = -M'_{\mathbf{X}}(s)/M^2_{\mathbf{X}}(s) - 1. \tag{7.18}$$

Let $\lambda_1, \ldots, \lambda_N \in [0,\infty)$ be the eigenvalues of the $N \times N$ matrix $\mathbb{E}[\mathsf{Cov}(\mathbf{X} \mid \mathbf{Y}))]$, where $\mathbf{Y} = \sqrt{s}\mathbf{X} + \mathbf{W}$. Starting with (7.16), we find that

$$\begin{aligned} -M'_{\mathbf{X}}(s) &= \frac{1}{N}\mathbb{E}\left[\|\mathsf{Cov}(\mathbf{X} \mid \mathbf{Y})\|_{\mathrm{F}}^2\right] \geq \frac{1}{N}\|\mathbb{E}[\mathsf{Cov}(\mathbf{X} \mid \mathbf{Y})]\|_{\mathrm{F}}^2 \\ &= \frac{1}{N}\sum_{n=1}^{N} \lambda_n^2 \geq \left(\frac{1}{N}\sum_{n=1}^{N} \lambda_n\right)^2 = M^2_{\mathbf{X}}(s), \end{aligned}$$

where both inequalities are due to Jensen's inequality. Combining this inequality with (7.18) establishes that the derivative of $k_{\mathbf{X}}(s)$ is non-negative and hence $k_{\mathbf{X}}(s)$ is non-decreasing. □

We remark that Theorem 7.1 implies the single-crossing property. To see this, note that if $\mathbf{Z} \sim \mathcal{N}(0, \sigma^2 I)$ then $k_{\mathbf{Z}}(s) = \sigma^{-2}$ is a constant and thus $k_{\mathbf{X}}(s)$ and $k_{\mathbf{Z}}(s)$ cross at most once. Furthermore, Theorem 7.1 shows that, for many problems, the Gaussian distribution plays an extremal role for distributions with finite second moments. For example, if we let $\mathbf{Z}$ be a Gaussian random vector with the same mean and covariance as $\mathbf{X}$, then we have

$$\begin{aligned} I_{\mathbf{X}}(s) &\leq I_{\mathbf{Z}}(s), \\ M_{\mathbf{X}}(s) &\leq M_{\mathbf{Z}}(s), \\ M'_{\mathbf{X}} &\leq M'_{\mathbf{Z}}(s), \end{aligned}$$

where equality holds if and only if $\mathbf{X}$ is Gaussian. The importance of these inequalities follows from the fact that the Gaussian distribution is easy to analyze and often well behaved.

7.3.2 Analysis of Good Codes for the Gaussian Channel

This section provides an example of how the properties described in Section 7.3.1 can be applied in the context of a high-dimensional inference problem. The focus is on the channel coding problem for the Gaussian channel. A code for the Gaussian channel is a collection $\mathcal{X} = \{\mathbf{x}(1), \ldots, \mathbf{x}(L)\}$ of L codewords in $\mathbb{R}^N$ such that $\mathbf{X} = \mathbf{x}(J)$, where J is drawn uniformly from $\{1, 2, \ldots, L\}$. The output of the channel is given by

$$\mathbf{Y} = \sqrt{\mathsf{snr}}\,\mathbf{X} + \mathbf{W}, \tag{7.19}$$

where $\mathbf{W} \sim \mathcal{N}(0, \mathbf{I})$ is independent Gaussian noise.

The code is called η-good if it satisfies three conditions.

- *Power Constraint.* The codewords satisfy the average power constraint

$$\frac{1}{N}\mathbb{E}\left[\|\mathbf{X}\|_2^2\right] \leq 1. \tag{7.20}$$

- *Low Error Probability.* For the MAP decoding decision $\widehat{\mathbf{X}}$, we have
$$\Pr\left(\widehat{\mathbf{X}} = \mathbf{X}\right) \geq \mathbb{P}\left[p_{\mathbf{X}|\mathbf{Y}}(\mathbf{x}(J)|\mathbf{Y}) > \max_{\ell \neq J} p_{\mathbf{X}|\mathbf{Y}}(\mathbf{x}(\ell)|\mathbf{Y})\right] \geq 1 - \eta.$$
- *Sufficient Rate.* The number of codewords satisfies $L \geq (1 + \mathsf{snr})^{(1-\eta)N/2}$.

LEMMA 7.1 (Corollary of the channel coding theorem) *For every* $\mathsf{snr} > 0$ *and* $\epsilon > 0$ *there exist an integer* N *and random vector* $\mathbf{X} = (X_1, \ldots, X_N)$ *satisfying the average power constraint* (7.20) *as well as the following inequalities:*

$$I_{\mathbf{X}}(\mathsf{snr}) \geq \frac{1}{2}\log(1 + \mathsf{snr}) - \epsilon, \tag{7.21}$$
$$M_{\mathbf{X}}(\mathsf{snr}) \leq \epsilon. \tag{7.22}$$

The distribution on $\mathbf{X}$ induced by a good code is fundamentally different from the i.i.d. Gaussian distribution that maximizes the mutual information. For example, a good code defines a discrete distribution that has finite entropy, whereas the Gaussian distribution is continuous and hence has infinite entropy. Furthermore, while the MMSE of a good code can be made arbitrarily small (in the large-N limit), the MMSE of the Gaussian channel is lower-bounded by $1/(1 + \mathsf{snr})$ for all N.

Nevertheless, the distribution induced by a good code and the Gaussian distribution are similar in the sense that their mutual information functions must become arbitrarily close for large N. It is natural to ask whether this closeness implies other similarities between the good code and the Gaussian distribution. The next result shows that closeness in mutual information also implies closeness in MMSE.

THEOREM 7.2 *For any N-dimensional random vector* $\mathbf{X}$ *satisfying the average power constraint* (7.20) *and the mutual information lower bound* (7.21) *the MMSE function satisfies*

$$\frac{e^{-2\epsilon}}{1+s} - \frac{1 - e^{-2\epsilon}}{\mathsf{snr} - s} \leq M_{\mathbf{X}}(s) \leq \frac{1}{1+s} \tag{7.23}$$

for all $0 \leq s < \mathsf{snr}$.

Proof For the upper bound, we have

$$M_{\mathbf{X}}(s) = \frac{1}{k_{\mathbf{X}}(s) + s} \leq \frac{1}{k_{\mathbf{X}}(0) + s} \leq \frac{1}{1+s}, \tag{7.24}$$

where the first inequality follows from Theorem 7.1 and the second inequality holds because the assumption $(1/N)\mathbb{E}\left[\|\mathbf{X}\|_2^2\right] \leq 1$ implies that $M_{\mathbf{X}}(0) \leq 1$, and hence $k_{\mathbf{X}}(0) \geq 1$. For the lower bound, we use the following chain of inequalities:

$$\log\left(\frac{(1+\mathsf{snr})(k_{\mathbf{X}}(s) + s)}{(1+s)(k_{\mathbf{X}}(s) + \mathsf{snr})}\right) = \int_s^{\mathsf{snr}} \left(\frac{1}{1+t} - \frac{1}{k_{\mathbf{X}}(s) + t}\right) dt \tag{7.25}$$
$$\leq \int_s^{\mathsf{snr}} \left(\frac{1}{1+t} - \frac{1}{k_{\mathbf{X}}(t) + t}\right) dt \tag{7.26}$$
$$= \int_s^{\mathsf{snr}} \left(\frac{1}{1+t} - M_{\mathbf{X}}(t)\right) dt \tag{7.27}$$

$$\leq \int_0^{\mathsf{snr}} \left(\frac{1}{1+t} - M_{\mathbf{X}}(t) \right) dt \tag{7.28}$$

$$= \log(1 + \mathsf{snr}) - 2I_{\mathbf{X}}(\mathsf{snr}) \tag{7.29}$$

$$\leq 2\epsilon, \tag{7.30}$$

where (7.26) follows from Theorem 7.1, (7.28) holds because the upper bound in (7.23) ensures that the integrand is non-negative, and (7.30) follows from the assumed lower bound on the mutual information. Exponentiating both sides, rearranging terms, and recalling the definition of $k_{\mathbf{X}}(s)$ leads to the stated lower bound.

An immediate consequence of Theorem 7.2 is that the MMSE function associated with a sequence of good codes undergoes a phase transition in the large-N limit. In particular,

$$\lim_{N\to\infty} M_{\mathbf{X}}(s) = \begin{cases} \dfrac{1}{1+s}, & 0 \leq s < \mathsf{snr}, \\ 0, & \mathsf{snr} < s. \end{cases} \tag{7.31}$$

The case $s \in [\mathsf{snr}, \infty)$ follows from the definition of a good code and the monotonicity of the MMSE function. The case $s \in [0, \mathsf{snr})$ follows from Theorem 7.2 and the fact that ϵ can be arbitrarily small.

An analogous result for binary linear codes on the Gaussian channel can be found in [43]. The characterization of the asymptotic MMSE for good Gaussian codes, described by (7.31), was also obtained previously using ideas from statistical physics [44]. The derivation presented in this chapter, which relies only on the monotonicity of $k_{\mathbf{X}}(s)$, bypasses some technical difficulties encountered in the previous approach.

7.3.3 Incremental-Information Sequence

We now consider a different approach for decomposing the mutual information between random vectors $\mathbf{X} = (X_1, \ldots, X_N)$ and $\mathbf{Y} = (Y_1, \ldots, Y_M)$. The main idea is to study the increase in information associated with new observations. In order to make general statements, we average over all possible presentation orders for the elements of $\mathbf{Y}$. To this end, we define the *information sequence* $\{I_m\}$ according to

$$I_m \triangleq \frac{1}{M!} \sum_{\pi} I(\mathbf{X}; Y_{\pi(1)}, \ldots Y_{\pi(m)}), \qquad m = 1, \ldots, M,$$

where the sum is over all permutations $\pi : [M] \to [M]$. Note that each summand on the right-hand side is the mutual information between $\mathbf{X}$ and an m-tuple of the observations. The average over all possible permutations can be viewed as the expected mutual information when the order of observations is chosen uniformly at random.

Owing to the random ordering, we will find that I_m is an increasing sequence with $0 = I_0 \leq I_1 \leq \cdots \leq I_{M-1} \leq I_M = I(\mathbf{X}; \mathbf{Y})$. To study the increase in information with additional observations, we focus on the first- and second-order difference sequences, which are defined as follows:

$$I'_m \triangleq I_{m+1} - I_m,$$
$$I''_m \triangleq I'_{m+1} - I'_m.$$

Using the chain rule for mutual information, it is straightforward to show that the first- and second-order difference sequences can also be expressed as

$$\begin{aligned} I'_m &= \frac{1}{M!} \sum_{\pi} I(\mathbf{X}; Y_{\pi(m+1)} \mid Y_{\pi(1)}, \ldots Y_{\pi(m)}), \\ I''_m &= \frac{1}{M!} \sum_{\pi} I(Y_{\pi(m+2)}; Y_{\pi(m+1)} \mid \mathbf{X}, Y_{\pi(1)}, \ldots Y_{\pi(m)}) \\ &\quad - \frac{1}{M!} \sum_{\pi} I(Y_{\pi(m+2)}; Y_{\pi(m+1)} \mid Y_{\pi(1)}, \ldots Y_{\pi(m)}). \end{aligned} \tag{7.32}$$

The incremental-information approach is well suited to observation models in which the entries of $\mathbf{Y}$ are conditionally independent given $\mathbf{X}$, that is

$$p_{\mathbf{Y}|\mathbf{X}}(\mathbf{y} \mid \mathbf{x}) = \prod_{m=1}^{M} p_{Y_k|\mathbf{X}}(y_k \mid \mathbf{x}). \tag{7.33}$$

The class of models satisfying this condition is quite broad and includes memoryless channels and generalized linear models as special cases. The significance of the conditional independence assumption is summarized in the following result.

THEOREM 7.3 (Monotonicity of incremental information) *The first-order difference sequence $\{I'_m\}$ is monotonically decreasing for any observation model satisfying the conditional independence condition in* (7.33).

Proof Under assumption (7.33), two new observations $Y_{\pi(m+1)}$ and $Y_{\pi(m+2)}$ are conditionally independent given $\mathbf{X}$, and thus the first term on the right-hand side of (7.33) is zero. This means that the second-order difference sequence is non-positive, which implies monotonicity of the first-order difference.

The monotonicity in Theorem 7.3 can also be seen as a consequence of the subset inequalities studied by Han; see Chapter 17 of [3]. Our focus on the incremental information is also related to prior work in coding theory that uses an integral–derivative relationship for the mutual information called the area theorem [45].

Similarly to the monotonicity properties studied in Section 7.3.1, the monotonicity of the first-order difference imposes a number of constraints on the mutual information sequence. Some examples illustrating the usefulness of these constraints will be provided in the following sections.

7.3.4 Standard Linear Model with i.i.d. Measurement Vectors

We now provide an example of how the incremental-information sequences can be used in the context of the standard linear model (7.3). We focus on the setting where the measurement vectors $\{\mathbf{A}_m\}$ are drawn i.i.d. from a distribution on $\mathbb{R}^N$. In this setting, the entire observation consists of the pair $(\mathbf{Y}, \mathbf{A})$ and the mutual information sequences defined in Section 7.3.3 can be expressed compactly as

$$I_m = I(\mathbf{X}; Y^m \mid \mathbf{A}^m),$$

$$I'_m = I(\mathbf{X}; Y_{m+1} \mid Y^m, \mathbf{A}^{m+1}),$$

$$I''_m = -I(Y_{m+1}; Y_{m+2} \mid Y^m, \mathbf{A}^{m+2}),$$

where $Y^m = (Y_1, \ldots, Y_m)$ and $\mathbf{A}^m = (\mathbf{A}_1, \ldots, \mathbf{A}_m)$. In these expressions, we do not average over permutations of measurement indices because the distribution of the observations is permutation-invariant. Furthermore, the measurement vectors appear only as conditional variables in the mutual information because they are independent of all other random variables.

The sequence perspective can also be applied to other quantities of interest. For example, the MMSE sequence $\{M_m\}$ is defined by

$$M_m \triangleq \frac{1}{N}\mathbb{E}\left[\left\|\mathbf{X} - \mathbb{E}[\mathbf{X} \mid Y^m, A^m]\right\|_2^2\right], \tag{7.34}$$

where $M_0 = (1/N)\operatorname{tr}(\mathsf{Cov}(\mathbf{X}))$ and $M_M = (1/N)\,\mathsf{mmse}(\mathbf{X} \mid \mathbf{Y}, \mathbf{A})$. By the data-processing inequality for MMSE, it follows that M_m is a decreasing sequence.

Motivated by the I-MMSE relations in Section 7.3.1, one might wonder whether there also exists a relationship between the mutual information and MMSE sequences. For simplicity, consider the setting where the measurement vectors are i.i.d. with mean zero and covariance proportional to the identity matrix:

$$\mathbb{E}[\mathbf{A}_m] = 0, \qquad \mathbb{E}\left[\mathbf{A}_m \mathbf{A}_m^{\mathrm{T}}\right] = N^{-1}\mathbf{I}_N. \tag{7.35}$$

One example of a distribution satisfying these constraints is when the entries of $\mathbf{A}_m$ are i.i.d. with mean zero and variance $1/N$. Another example is when $\mathbf{A}_m$ is drawn uniformly from a collection of N mutually orthogonal unit vectors.

THEOREM 7.4 *Consider the standard linear model* (7.3) *with i.i.d. measurement vectors* $\{\mathbf{A}_m\}$ *satisfying* (7.35). *If* $\mathbf{X}$ *has finite covariance, then the sequences* $\{I'_m\}$ *and* $\{M_m\}$ *satisfy*

$$I'_m \le \frac{1}{2}\log(1 + M_m) \tag{7.36}$$

for all integers m.

Proof Conditioned on the observations $(Y^m, \mathbf{A}^{m+1})$, the variance of a new measurement can be expressed as

$$\mathsf{Var}(\langle \mathbf{A}_{m+1}, \mathbf{X}\rangle \mid Y^m, \mathbf{A}^{m+1}) = \mathbf{A}_{m+1}^{\mathrm{T}}\,\mathsf{Cov}(\mathbf{X} \mid Y^m, \mathbf{A}^m)\mathbf{A}_{m+1}.$$

Taking the expectation of both sides and leveraging the assumptions in (7.35), we see that

$$\mathbb{E}\left[\mathsf{Var}(\langle \mathbf{A}_{m+1}, \mathbf{X}\rangle \mid Y^m, \mathbf{A}^{m+1})\right] = M_m. \tag{7.37}$$

Next, starting with the fact that the mutual information in a Gaussian channel is maximized when the input (i.e., $\langle \mathbf{A}_{m+1}, \mathbf{X}\rangle$) is Gaussian, we have

$$
\begin{aligned}
I'_m &= I(\mathbf{X}; Y_{m+1} \mid Y^m, \mathbf{A}^{m+1}) \\
&\le \mathbb{E}\left[\frac{1}{2}\log\left(1 + \mathsf{Var}(\langle \mathbf{A}_{m+1}, \mathbf{X}\rangle \mid Y^m, \mathbf{A}^{m+1})\right)\right] \\
&\le \mathbb{E}\left[\frac{1}{2}\log\left(1 + \mathbb{E}\left[\mathsf{Var}(\langle \mathbf{A}_{m+1}, \mathbf{X}\rangle \mid Y^n, \mathbf{A}^{m+1})\right]\right)\right],
\end{aligned}
\tag{7.38}
$$

where the second step follows from Jensen's inequality and the concavity of the logarithm. Combining (7.37) and (7.38) gives the stated inequality.

Theorem 7.4 is reminiscent of the I-MMSE relation for Gaussian channels in the sense that it relates a change in mutual information to an MMSE estimate. One key difference, however, is that (7.36) is an inequality instead of an equality. The difference between the right- and left-hand sides of (7.36) can be viewed as a measure of the difference between the posterior distribution of a new observation Y_{m+1} given observations $(Y^m, \mathbf{A}^{m+1})$ and the Gaussian distribution with matched first and second moments [20–22, 46, 47].

Combining Theorem 7.4 with the monotonicity of the first-order difference sequence (Theorem 7.3) leads to a lower bound on the MMSE in terms of the total mutual information.

THEOREM 7.5 *Under the assumptions of Theorem 7.4, we have*

$$
M_k \ge \left(\exp\left(\frac{2I_m - k\log(1 + M_0)}{m-k}\right) - 1\right) \tag{7.39}
$$

for all integers $0 \le k < m$.

Proof For any $0 \le k < m$, the monotonicity of I'_m (Theorem 7.3) allows us to write $I_m = \sum_{\ell=0}^{m-1} I'_\ell = \sum_{\ell=0}^{k-1} I'_k + \sum_{\ell=k}^{m-1} I'_\ell \le kI'_0 + (m-k)I'_k$. Using Theorem 7.4 to upper-bound the terms I'_0 and I'_k and then rearranging terms leads to the stated result.

Theorem 7.5 is particularly meaningful when the mutual information is large. For example, if the mutual information satisfies the lower bound

$$
I_m \ge (1-\epsilon)\frac{n}{2}\log(1 + M_0)
$$

for some $\epsilon \in [0, 1)$, then Theorem 7.5 implies that

$$
M_k \ge \left((1 + M_0)^{1 - \frac{\epsilon}{1-k/m}} - 1\right)
$$

for all integers $0 \le k < m$. As ϵ converges to zero, the right-hand side of this inequality increases to M_0. In other words, a large value of I_m after m observations implies that the MMSE sequence is nearly constant for all k that are sufficiently small relative to m.

7.4 Proving the Replica-Symmetric Formula

The authors' prior work [20–22] provided the first rigorous proof of the replica formulas (7.5) and (7.6) for the standard linear model with an i.i.d. signal and a Gaussian sensing matrix.

In this section, we give an overview of the proof. It begins by focusing on the increase in mutual information associated with adding a new observation as described in Sections 7.3.3 and 7.3.4. Although this approach is developed formally using finite-length sequences, we describe the large-system limit first for simplicity.

7.4.1 Large-System Limit and Replica Formulas

For the large-system limit, the increase in mutual information $\mathcal{I}(\delta)$ with additional measurement is characterized by its derivative $\mathcal{I}'(\delta)$. The main technical challenge is to establish the following relationships:

$$\text{fixed-point formula} \qquad \mathcal{M}(\delta) = M_X\left(\frac{\delta}{1+\mathcal{M}(\delta)}\right), \tag{7.40}$$

$$\text{I-MMSE formula} \qquad \mathcal{I}'(\delta) = \frac{1}{2}\log(1+\mathcal{M}(\delta)), \tag{7.41}$$

where these equalities hold almost everywhere but not at phase transitions.

The next step is to use these two relationships to prove that the replica formulas, (7.5) and (7.6), are exact. First, by solving the minimization over s in (7.5), one finds that any local minimizer must satisfy the fixed-point formula (7.40). In addition, by differentiating $\mathcal{I}(\delta)$ in (7.5), one can show that $\mathcal{I}'(\delta)$ must satisfy the I-MMSE formula (7.41). Thus, if the fixed-point formula (7.40) defines $\mathcal{M}(\delta)$ uniquely, then the mutual information $\mathcal{I}(\delta)$ can be computed by integrating (7.41) and the proof is complete. However, this happens only if there are no phase transitions. Later, we will discuss how to handle the case of multiple solutions and phase transitions.

7.4.2 Information and MMSE Sequences

To establish the large-limit formulas, (7.40) and (7.41), the authors' proof focuses on functional properties of the incremental mutual information as well as the MMSE sequence $\{M_m\}$ defined by (7.34). In particular, the results in [20, 21] first establish the following approximate relationships between the mutual information and MMSE sequences:

$$\text{fixed-point formula} \qquad M_m \approx M_X\left(\frac{m/N}{1+M_m}\right), \tag{7.42}$$

$$\text{I-MMSE formula} \qquad I'_m \approx \frac{1}{2}\log(1+M_m). \tag{7.43}$$

The fixed-point formula (7.42) shows that the MMSE M_m corresponds to a scalar estimation problem whose signal-to-noise ratio is a function of the number of observations m as well as M_m. In Section A7.1, it is shown how the standard linear model can be related to a scalar estimation problem of one signal entry. Finally, the I-MMSE formula (7.43) implies that the increase in information with a new measurement corresponds to a single use of a Gaussian channel with a Gaussian input whose variance is matched to the

MMSE. The following theorem, from [20, 21], quantifies the precise sense in which these approximations hold.

THEOREM 7.6 *Consider the standard linear model* (7.3) *with i.i.d. Gaussian measurement vectors* $\mathbf{A}_m \sim \mathcal{N}(0, N^{-1}\mathbf{I}_N)$. *If the entries of* $\mathbf{X}$ *are i.i.d. with bounded fourth moment* $\mathbb{E}\left[X_n^4\right] \le B$, *then the sequences* $\{I'_m\}$ *and* $\{M_m\}$ *satisfy*

$$\sum_{m=1}^{\lceil \delta N \rceil} \left| I'_m - \frac{1}{2}\log(1 + M_m) \right| \le C_{B,\delta} N^{\alpha}, \tag{7.44}$$

$$\sum_{m=1}^{\lceil \delta N \rceil} \left| M_m - M_X\left(\frac{m/N}{1 + M_m}\right) \right| \le C_{B,\delta} N^{\alpha} \tag{7.45}$$

for every integer N *and* $\delta > 0$, *where* $\alpha \in (0,1)$ *is a universal constant and* $C_{B,\delta}$ *is a constant that depends only on the pair* (B, δ).

Theorem 7.6 shows that the normalized sum of the cumulative absolute error in approximations (7.42) and (7.43) grows sub-linearly with the vector length N. Thus, if one normalizes these sums by $M = \delta N$, then the resulting expressions converge to zero as $N \to \infty$. This is sufficient to establish (7.40) and (7.41).

7.4.3 Multiple Fixed-Point Solutions

At this point, the remaining difficulty is that the MMSE fixed-point formula (7.40) can have *multiple solutions*, as is illustrated by the information fixed-point curve in Fig. 7.4. In this case, the formulas (7.40) and (7.41) alone are not sufficient to uniquely define the actual mutual information $\mathcal{I}(\delta)$.

For many signal distributions with a single phase transition, the curve (see Fig. 7.4) defined by the fixed-point formula (7.40) has the following property. For each δ, there are at most two solutions where the slope of the curve is non-increasing. Since $\mathcal{M}(\delta)$ is

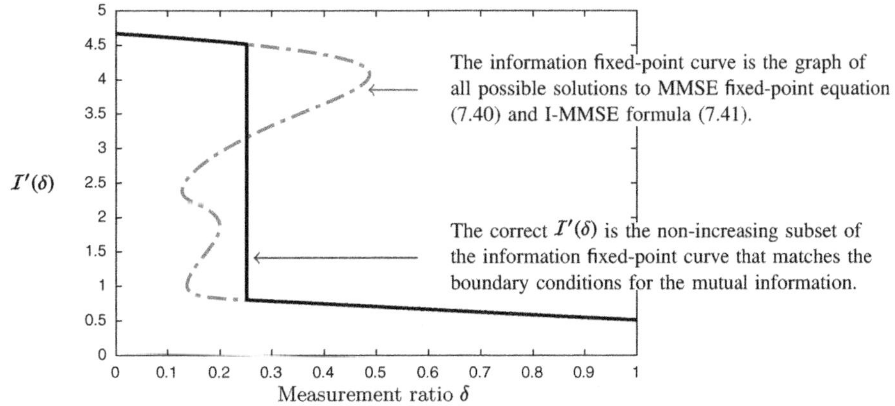

$\mathcal{I}'(\delta)$
Measurement ratio δ
The information fixed-point curve is the graph of all possible solutions to MMSE fixed-point equation (7.40) and I-MMSE formula (7.41).
The correct $\mathcal{I}'(\delta)$ is the non-increasing subset of the information fixed-point curve that matches the boundary conditions for the mutual information.

Figure 7.4 The derivative of the mutual information $\mathcal{I}'(\delta)$ as a function of the measurement rate δ for linear estimation with i.i.d. Gaussian matrices. A phase transition occurs when the derivative jumps from one branch of the information fixed-point curve to another.

non-increasing, in this case, it must jump from the upper solution branch to the lower solution branch (see Fig. 7.4) at the phase transition.

The final step in the authors' proof technique is to resolve the location of the phase transition using boundary conditions on the mutual information for $\delta = 0$ and $\delta \to \infty$, which can be obtained directly using different arguments. Under the signal property stated below in Definition 7.1, it is shown that the only solution consistent with the boundary conditions is the one predicted by the replica method. A graphical illustration of this argument is provided in Fig. 7.4.

7.4.4 Formal Statement

In this section, we formally state the main theorem in [21]. To do this, we need the following definition.

DEFINITION 7.1 *A signal distribution p_X has the* single-crossing property[3] *if its replica MMSE* (7.6) *crosses its fixed-point curve* (7.40) *at most once.*

For any $\delta > 0$, consider a sequence of standard linear models indexed by N where the number of measurements is $M = \lceil \delta N \rceil$, the signal $\mathbf{X} \in \mathbb{R}^N$ is an i.i.d. vector with entries drawn from p_X, the $M \times N$ measurement matrix $\mathbf{A}$ has i.i.d. entries drawn from $\mathcal{N}(0, 1/N)$, and the observed vector is $\mathbf{Y} = \mathbf{AX} + \mathbf{W}$, where $\mathbf{W} \in \mathbb{R}^M$ is a standard Gaussian vector. For this sequence, we can also define a sequence of mutual information and MMSE functions:

$$\mathcal{I}_N(\delta) \triangleq = I(X^N; Y^{\lceil \delta M \rceil}), \tag{7.46}$$

$$\mathcal{M}_N(\delta) \triangleq = \frac{1}{N} \mathsf{mmse}(X^N \mid Y^{\lceil \delta M \rceil}). \tag{7.47}$$

THEOREM 7.7 *Consider the sequence of problems defined above and assume that p_X has a bounded fourth moment (i.e., $\mathbb{E}\left[X^4\right] \le B < \infty$) and satisfies the single-crossing property. Then the following two statements hold.*

1. *The sequence of mutual information functions $\mathcal{I}_N(\delta)$ converges to the replica prediction* (7.5). *In other words, for all $\delta > 0$,*
$$\lim_{N \to \infty} \mathcal{I}_N(\delta) = \mathcal{I}(\delta).$$
2. *The sequence of MMSE functions $\mathcal{M}_N(\delta)$ converges almost everywhere to the replica prediction* (7.6). *In other words, at all continuity points of $\mathcal{M}(\delta)$,*
$$\lim_{N \to \infty} \mathcal{M}_N(\delta) = \mathcal{M}(\delta).$$

Relationship with Other Methods

The use of an integral relationship defining the mutual information is reminiscent of the generalized area theorems introduced by Méasson *et al.* in coding theory [45]. However,

[3] Regrettably, this is unrelated to the "single-crossing property" described earlier that says the MMSE function $M_{\mathbf{X}}(s)$ may cross the matched Gaussian MMSE $M_{\mathbf{Z}}(s)$ at most once.

one of the key differences in the compressed sensing problem is that the conditional entropy of the signal does not drop to zero after the phase transition.

The authors' proof technique also differs from previous approaches that use system-wide interpolation methods to obtain one-sided bounds [5, 48] or that focus on special cases, such as sparse matrices [49], Gaussian mixture models [50], or the detection problem of support recovery [51, 52]. After [20, 22], Barbier *et al.* obtained similar results using a substantially different method [53, 54]. More recent work has provided rigorous results for the generalized linear model [55].

7.5 Phase Transitions and Posterior Correlation

A *phase transition* refers to an abrupt change in the macroscopic properties of a system. In the context of thermodynamic systems, a phase transition may correspond to the transition from one state of matter to another (e.g., from solid to liquid or from liquid to gas). In the context of inference problems, a phase transition can be used to describe a sharp change in the quality of inference. For example, the channel coding problem undergoes a phase transition as the signal-to-noise ratio crosses a threshold because the decoder error probability transitions from ≈ 1 to ≈ 0 over a very small range of signal-to-noise ratios. In the standard linear model, the asymptotic MMSE may also contain a jump discontinuity with respect to the fraction of observations.

In many cases, the existence of phase transitions in inference problems can be related to the emergence of significant correlation in the posterior distribution [5]. In these cases, a small change in the uncertainty for one variable (e.g., reduction in the posterior variance with a new observation) corresponds to a change in the uncertainty for a large number of other variables as well. The net effect is a large change in the overall system properties, such as the MMSE.

In this section, we show how the tools introduced in Section 7.3 can be used to provide a link between a measure of the average correlation in the posterior distribution and second-order differences in the mutual information for both the Gaussian channel and the standard linear model.

7.5.1 Mean-Squared Covariance

Let us return to the general inference problem of estimating a random vector $\mathbf{X} = (X_1, \ldots, X_N)$ from observations $\mathbf{Y} = (Y_1, \ldots, Y_M)$. As discussed in Section 7.3, the posterior covariance matrix $\mathsf{Cov}(\mathbf{X} \mid \mathbf{Y})$ provides a geometric measure of the amount of uncertainty in the posterior distribution. One important function of this matrix is the MMSE, which corresponds to the expected posterior variance:

$$\mathsf{mmse}(\mathbf{X} \mid \mathbf{Y}) = \sum_{n=1}^{N} \mathbb{E}[\mathsf{Var}(X_n \mid \mathbf{Y})]. \tag{7.48}$$

Going beyond the MMSE, there is also important information contained in the off-diagonal entries, which describe the pair-wise correlations. A useful measure of this correlation is provided by the *mean-squared covariance*:

$$\mathbb{E}\left[\|\mathsf{Cov}(\mathbf{X} \mid \mathbf{Y})\|_F^2\right] = \sum_{k=1}^{N}\sum_{n=1}^{N} \mathbb{E}\left[\mathsf{Cov}^2(X_k, X_n \mid \mathbf{Y})\right]. \tag{7.49}$$

Note that, while the MMSE corresponds to N terms, the mean-squared covariance corresponds to N^2 terms. If the entries in $\mathbf{X}$ have bounded fourth moments (i.e., $\mathbb{E}\left[X_i^4\right] \le B$), then it follows from the Cauchy–Schwarz inequality that each summand on the right-hand side of (7.49) is upper-bounded by B, and it can be verified that

$$\frac{1}{N}(\mathsf{mmse}(\mathbf{X} \mid \mathbf{Y}))^2 \le \mathbb{E}\left[\|\mathsf{Cov}(\mathbf{X} \mid \mathbf{Y})\|_F^2\right] \le N^2 B. \tag{7.50}$$

The left inequality is tight when the posterior distribution is uncorrelated and hence the off-diagonal terms of the conditional covariance are zero. The right inequality is tight when the off-diagonal terms are of the same order as the variance.

Another way to view the relationship between the MMSE and mean-squared covariance its to consider the spectral decomposition of the covariance matrix. Let $\Lambda_1 \ge \Lambda_2 \ge \cdots \ge \Lambda_N$ denote the (random) eigenvalues of $\mathsf{Cov}(\mathbf{X} \mid \mathbf{Y})$. Then we can write

$$\mathsf{tr}(\mathsf{Cov}(\mathbf{X} \mid \mathbf{Y})) = \sum_{n=1}^{N} \Lambda_n, \qquad \|\mathsf{Cov}(\mathbf{X} \mid \mathbf{Y})\|_F^2 = \sum_{n=1}^{N} \Lambda_n^2.$$

Taking the expectations of these random quantaties and rearranging terms, one finds that the mean-squared covariance can be decomposed into three non-negative terms:

$$\mathbb{E}\left[\|\mathsf{Cov}(\mathbf{X} \mid \mathbf{Y})\|_F^2\right] = N\left(\mathbb{E}\left[\bar{\Lambda}\right]\right)^2 + N\mathsf{Var}(\bar{\Lambda}) + \sum_{n=1}^{N} \mathbb{E}\left[\left(\Lambda_n - \bar{\Lambda}\right)^2\right], \tag{7.51}$$

where $\bar{\Lambda} = (1/N)\sum_{n=1}^{N} \Lambda_n$ denotes the arithmetic mean of the eigenvalues. The first term on the right-hand side corresponds to the square of the MMSE and is equal to the lower bound in (7.50). The second term on the right-hand side corresponds to the variance of $(1/N)\,\mathsf{tr}(\mathsf{Cov}(\mathbf{X} \mid \mathbf{Y}))$ with respect to the randomness in $\mathbf{Y}$. This term is equal to zero if $\mathbf{X}$ and $\mathbf{Y}$ are jointly Gaussian. The last term corresponds to the expected variation in the eigenvalues. If a small number of eigenvalues are significantly larger than the others then it is possible for this term to be N times larger than the first term. When this occurs, most of the uncertainty in the posterior distribution is concentrated on a low-dimensional subspace.

7.5.2 Conditional MMSE Function and Its Derivative

The relationship between phase transitions and correlation in the posterior distribution can be made precise using the properties of the Gaussian channel discussed in Section 7.3.1. Given a random vector $\mathbf{X} \in \mathbb{R}^N$ and a random observation $\mathbf{Y} \in \mathbb{R}^M$, the *conditional MMSE function* is defined by

$$M_{\mathbf{X}|\mathbf{Y}}(s) \triangleq \frac{1}{N}\mathbb{E}\left[\|\mathbf{X} - \mathbb{E}[\mathbf{X} \mid \mathbf{Y}, \mathbf{Z}(s)]\|^2\right], \tag{7.52}$$

where $\mathbf{Z}(s) = \sqrt{s}\mathbf{X} + \mathbf{W}$ is a new observation of $\mathbf{X}$ from an independent Gaussian noise channel [32]. From the expression for the derivative of the MMSE in (7.16), it can be verified that the derivative of the conditional MMSE function is

$$M'_{\mathbf{X}|\mathbf{Y}}(s) = -\frac{1}{N}\mathbb{E}\left[\|\mathsf{Cov}(\mathbf{X} \mid \mathbf{Y}, \mathbf{Z}(s))\|_{\mathrm{F}}^2\right]. \tag{7.53}$$

Here, we recognize that the right-hand side is proportional to the mean-squared covariance associated with the pair of observations $(\mathbf{Y}, \mathbf{Z}(s))$. Meanwhile, the left-hand side describes the change in the MMSE associated with a small increase in the signal-to-noise ratio of the Gaussian channel.

In Section 7.3.2, we saw that a phase transition in the channel coding problem for the Gaussian channel corresponds to a jump discontinuity in the MMSE. More generally, one can say that the inference problem defined by the pair $(\mathbf{Y}, \mathbf{Z}(s))$ undergoes a phase transition whenever $M_{\mathbf{X}|\mathbf{Y}}(s)$ has a jump discontinuity in the large-N limit. If such a phase transition occurs, then it implies that the magnitude of $M'_{\mathbf{X}|\mathbf{Y}}(s)$ is increasing without bound. From (7.53), we see that this also implies significant correlation in the posterior distribution.

Evaluating the conditional MMSE function and its derivative at $s = 0$ provides expressions for the MMSE and mean-squared covariance associated with the orignal observation model:

$$M_{\mathbf{X}|\mathbf{Y}}(0) = \frac{1}{N}\,\mathsf{mmse}(\mathbf{X} \mid \mathbf{Y}), \tag{7.54}$$

$$M'_{\mathbf{X}|\mathbf{Y}}(0) = -\frac{1}{N}\mathbb{E}\left[\|\mathsf{Cov}(\mathbf{X} \mid \mathbf{Y})\|_{\mathrm{F}}^2\right]. \tag{7.55}$$

In light of the discussion above, the mean-squared covariance can be interpreted as the rate of MMSE change with s that occurs when one is presented with an independent observation of $\mathbf{X}$ from a Gaussian channel with infinitesimally small signal-to-noise ratio. Furthermore, we see that significant correlation in the posterior distribution corresponds to a jump discontinuity in the large-N limit of $M_{\mathbf{X}|\mathbf{Y}}(s)$ at the point $s = 0$.

7.5.3 Second-Order Differences of the Information Sequence

Next, we consider some further properties of the incremental-mutual information sequence introduced in Section 7.3.3. For any observation model satisfying the conditional independence condition in (7.33) the second-order difference sequence given in (7.33) can be expressed as

$$I''_m = -\frac{1}{M!}\sum_{\pi} I(Y_{\pi(m+2)}; Y_{\pi(m+1)} \mid Y_{\pi(1)}, \dots Y_{\pi(m)}). \tag{7.56}$$

Note that each summand is a measure of the *pair-wise dependence* in the posterior distribution of new measurements. If the pair-wise dependence is large (on average), then this means that there is a significant decrease in the first-order difference sequence I'_m.

The monotonicity and non-negativity of the first-order difference sequence imposes some important constraints on the second-order difference sequence. For example, the number of terms for which $|I''_m|$ is "large" can be upper-bounded in terms of the information provided by a single observation. The following result provides a quantitative description of this constraint.

THEOREM 7.8 *For any observation model satisfying the conditional independence condition in* (7.33) *and positive number T, the second-order difference sequence $\{I''_m\}$ satisfies*

$$|\{m : |I''_m| \geq T\}| \leq I_1/T, \tag{7.57}$$

where $I_1 = (1/M)\sum_{m=1}^{M} I(\mathbf{X}; Y_m)$ is the first term in the information sequence.

Proof The monotonicity of the first-order difference (Theorem 7.3) means that I''_m is non-positive, and hence the indicator function of the event $\{|I''_m| \geq T\}$ is upper-bounded by $-I''_m/T$. Summing this inequality over m, we obtain

$$|\{m : |I''_m| \geq T\}| = \sum_{m=1}^{M-2} \mathbf{1}_{[T,\infty)}(|I''_m|) \leq \sum_{m=1}^{M-2} -I''_m/T = \left(I'_0 - I'_{M-1}\right)/T.$$

Noting that $I'_0 = I_1$ and $I'_{M-1} \geq 0$ completes the proof.

An important property of Theorem 7.8 is that, for many problems of interest, the term I_1 does not depend on the total number of observations M. For example, in the standard linear model with i.i.d. measurement vectors, the upper bound in Theorem 7.4 gives $I_1 \leq \frac{1}{2}\log(1 + M_{\mathbf{X}}(0))$. Some implications of these results are discussed in the next section.

Implications for the Standard Linear Model

One of the key steps in the authors' prior work on the standard linear model [20–22] is the following inequality, which relates the second-order difference sequence and the mean-squared covariance.

THEOREM 7.9 *Consider the standard linear model* (7.3) *with i.i.d. Gaussian measurement vectors $\mathbf{A}_m \sim \mathcal{N}(0, N^{-1}\mathbf{I}_N)$. If the entries of $\mathbf{X}$ are independent with bounded fourth-moment $\mathbb{E}\left[X_n^4\right] \leq B$, then the mean-squared covariance satisfies*

$$\frac{1}{N^2}\mathbb{E}\left[\left\|\mathsf{Cov}(\mathbf{X} \mid Y^m, \mathbf{A}^m)\right\|_{\mathrm{F}}^2\right] \leq C_B \left|I''_m\right|^{1/4}, \tag{7.58}$$

for all integers N and $m = 1, \ldots, M$, where C_B is a constant that depends only on the fourth-moment upper bound B.

Theorem 7.9 shows that significant correlation in the posterior distribution implies pair-wise dependence in the joint distribution of new measurements and, hence, a significant decrease in the first-order difference sequence I'_m. In particular, if the mean-squared covariance is order N^2 (corresponding to the upper bound in (7.50)), then $|I''_m|$ is lower-bounded by a constant. If we consider the large-N limit in which the number of observations is parameterized by the fraction $\delta = m/N$, then an order-one difference in I'_m corresponds to a jump discontinuity with respect to δ. In other words, significant pair-wise correlation implies a phase transition with respect to the fraction of observations.

Viewed in the other direction, Theorem 7.9 also shows that small changes in the first-order difference sequence imply that the average pair-wise correlation is small. From

Theorem 7.8, we see that this is, in fact, the typical situation. Under the assumptions of Theorem 7.9, it can be verified that

$$\left|\left\{m : \mathbb{E}\left[\|\mathrm{Cov}(\mathbf{X} \mid Y^m, \mathbf{A}^m)\|_F^2\right] \geq N^{2-\epsilon/4}\right\}\right| \leq \widetilde{C}_B N^{\epsilon} \tag{7.59}$$

for all $0 \leq \epsilon \leq 1$, where $\widetilde{C}_B$ is a constant that depends only on the fourth-moment bound B. In other words, the number of m-values for which the mean-squared covariance has the same order as the upper bound in (7.50) must be sub-linear in N. This fact plays a key role in the proof of Theorem 7.6; see [20, 21] for details.

7.6 Conclusion

This chapter provides a tutorial introduction to high-dimensional inference and its connection to information theory. The standard linear model is analyzed in detail and used as a running example. The primary goal is to present intuitive links between phase transitions, mutual information, and estimation error. To that end, we show how general functional properties (e.g., the chain rule, data-processing inequality, and I-MMSE relationship) of mutual information and MMSE can imply meaningful constraints on the solutions of challenging problems. In particular, the replica prediction of the mutual information and MMSE is described and an outline is given for the authors' proof that it is exact in some cases. We hope that the approach described here will make this material accessible to a wider audience.

7.6.1 Further Directions

Beyond the standard linear model, there are other interesting high-dimensional inference problems that can be addressed using the ideas in this chapter. For example, one recent line of work has focused on multilayer networks, which consist of multiple stages of a linear transform followed by a nonlinear (possibly random) function [32, 56, 57]. There has also been significant work on bilinear estimation problems, matrix factorization, and community detection [58–61]. Finally, there has been some initial progress on optimal quantization for the standard linear model [62, 63].

A7.1 Subset Response for the Standard Linear Model

This appendix describes the mapping between the standard linear model and a signal-plus-noise response model for a subset of the observations. Recall the problem formulation

$$\mathbf{Y} = \mathbf{A}\mathbf{X} + \mathbf{W} \tag{7.60}$$

where $\mathbf{A}$ is an $M \times N$ matrix. Suppose that we are interested in the posterior distribution of a subset $S \subset \{1, \ldots, N\}$ of the signal entries where the size of the subset $K = |S|$ is small relative to the signal length N and the number of measurements M. Letting $S^{\mathrm{c}} = \{1, \ldots, N\} \backslash S$ denote the complement of S, the measurements can be decomposed as

$$\mathbf{Y} = \mathbf{A}_S \mathbf{X}_S + \mathbf{A}_{S^c} \mathbf{X}_{S^c} + \mathbf{W}, \tag{7.61}$$

where $\mathbf{A}_S$ is an $M \times K$ matrix corresponding to the columns of $\mathbf{A}$ indexed by S and $\mathbf{A}_{S^c}$ is an $M \times (N-K)$ matrix corresponding to the columns indexed by the complement of S.

This decomposition suggests an alternative interpretation of the linear model in which $\mathbf{X}_S$ is a low-dimensional signal of interest and $\mathbf{X}_{S^c}$ is a high-dimensional interference term. Note that $\mathbf{A}_S$ is a tall skinny matrix, and thus the noiseless measurements of $\mathbf{X}_S$ lie in a K-dimensional subspace of the M-dimensional measurement space.

Next, we introduce a linear transformation of the problem that attempts to separate the signal of interest from the interference term. The idea is to consider the QR decomposition of the tall skinny matrix $\mathbf{A}_S$ of the form

$$\mathbf{A}_S = \underbrace{[\mathbf{Q}_1, \mathbf{Q}_2]}_{\mathbf{Q}} \begin{bmatrix} \mathbf{R} \\ \mathbf{0} \end{bmatrix},$$

where $\mathbf{Q}$ is an $M \times M$ orthogonal matrix ($\mathbf{Q}\mathbf{Q}^{\mathrm{T}} = I$), $\mathbf{Q}_1$ is $M \times K$, $\mathbf{Q}_2$ is $M \times (M-K)$, and $\mathbf{R}$ is an $K \times K$ upper triangular matrix whose diagonal entries are non-negative. If $\mathbf{A}_S$ has full column rank, then the pair $(\mathbf{Q}_1, \mathbf{R})$ is uniquely defined. The matrix $\mathbf{Q}_2$ can be chosen arbitrarily subject to the constraint $\mathbf{Q}_2^{\mathrm{T}} \mathbf{Q}_2 = I - \mathbf{Q}_1^{\mathrm{T}} \mathbf{Q}_1$. To facilitate the analysis, we will assume that $\mathbf{Q}_2$ is chosen uniformly at random over the set of matrices satisfying this constraint.

Multiplication by $\mathbf{Q}^{\mathrm{T}}$ is a one-to-one linear transformation. The transformed problem parameters are defined as

$$\widetilde{\mathbf{Y}} \triangleq \mathbf{Q}^{\mathrm{T}} \mathbf{Y}, \qquad \mathbf{B} \triangleq \mathbf{Q}^{\mathrm{T}} \mathbf{A}_{S^c}, \qquad \widetilde{\mathbf{W}} \triangleq \mathbf{Q}^{\mathrm{T}} \mathbf{W}.$$

At this point, it is important to note that the isotropic Gaussian distribution is invariant to orthogonal transformations. Consequently, the transformed noise $\widetilde{\mathbf{W}}$ has the same distribution as $\mathbf{W}$ and is independent of everything else. Using the transformed parameters, the linear model can be expressed as

$$\begin{bmatrix} \widetilde{\mathbf{Y}}_1 \\ \widetilde{\mathbf{Y}}_2 \end{bmatrix} = \begin{bmatrix} \mathbf{R} & \mathbf{B}_1 \\ \mathbf{0} & \mathbf{B}_2 \end{bmatrix} \begin{bmatrix} \mathbf{X}_S \\ \mathbf{X}_{S^c} \end{bmatrix} + \begin{bmatrix} \widetilde{\mathbf{W}}_1 \\ \widetilde{\mathbf{W}}_2 \end{bmatrix}, \tag{7.62}$$

where $\widetilde{\mathbf{Y}}_1$ corresponds to the first K measurements and $\widetilde{\mathbf{Y}}_2$ corresponds to the remaining $(M-K)$ measurements.

A useful property of the transformed model is that $\widetilde{\mathbf{Y}}_2$ is independent of the signal of interest $\mathbf{X}_S$. This decomposition motivates a two-stage approach in which one first estimates $\mathbf{X}_{S^c}$ from the data $(\widetilde{\mathbf{Y}}_2, \mathbf{B}_2)$ and then uses this estimate to "subtract out" the interference term in $\widetilde{\mathbf{Y}}_1$. To be more precise, we define

$$\mathbf{Z} \triangleq \widetilde{\mathbf{Y}}_1 - \mathbf{B}_1 \mathbb{E}\left[\mathbf{X}_{S^c} \mid \widetilde{\mathbf{Y}}_2, \mathbf{B}_2\right]$$

to be the measurements $\widetilde{\mathbf{Y}}_1$ after subtracting the conditional expectation of the interference term. Rearranging terms, one finds that the relationship between $\mathbf{Z}$ and $\mathbf{X}_S$ can be expressed succinctly as

$$\mathbf{Z} = \mathbf{R}\mathbf{X}_S + \mathbf{V}, \quad \mathbf{V} \sim p(\mathbf{v} \mid \widetilde{\mathbf{Y}}_2, \mathbf{B}), \tag{7.63}$$

where

$$\mathbf{V} \triangleq \mathbf{B}_1\left(\mathbf{X}_{S^c} - \mathbb{E}\left[\mathbf{X}_{S^c} \mid \widetilde{\mathbf{Y}}_2, \mathbf{B}_2\right]\right) + \widetilde{\mathbf{W}}_1 \tag{7.64}$$

is the error due to both the interference and the measurement noise.

Thus far, this decomposition is quite general in the sense that it can be applied for any matrix $\mathbf{A}$ and subset S of size less than M. The key question at this point is whether the error term $\mathbf{V}$ is approximately Gaussian.

References

[1] C. E. Shannon, "A mathematical theory of communication," *Bell System Technical J.*, vol. 27, pp. 379–423, 623–656, 1948.

[2] T. Lesieur, C. De Bacco, J. Banks, F. Krzakala, C. Moore, and L. Zdeborová, "Phase transitions and optimal algorithms in high-dimensional Gaussian mixture clustering," in *Proc. Allerton Conference on Communication, Control, and Computing*, 2016.

[3] T. M. Cover and J. A. Thomas, *Elements of information theory*, 2nd edn. Wiley-Interscience, 2006.

[4] T. S. Han, *Information-spectrum methods in information theory*. Springer, 2004.

[5] M. Mézard and A. Montanari, *Information, physics, and computation*. Oxford University Press, 2009.

[6] L. Zdeborová and F. Krzakala, "Statistical physics of inference: Thresholds and algorithms," *Adv. Phys.*, vol. 65, no. 5, pp. 453–552, 2016.

[7] G. Parisi, "A sequence of approximated solutions to the S–K model for spin glasses," *J. Phys. A: Math. and General*, vol. 13, no. 4, pp. L115–L121, 1980.

[8] M. Talagrand, "The Parisi formula," *Annals Math.*, vol. 163, no. 1, pp. 221–263, 2006.

[9] D. J. MacKay, *Information theory, inference, and learning algorithms*. Cambridge University Press, 2003.

[10] M. J. Wainwright and M. I. Jordan, *Graphical models, exponential families, and variational inference*. Now Publisher Inc., 2008.

[11] M. Pereyra, P. Schniter, E. Chouzenoux, J.-C. Pesquet, J.-Y. Tourneret, A. O. Hero, and S. McLaughlin, "A survey of stochastic simulation and optimization methods in signal processing," *IEEE J. Selected Topics Signal Processing*, vol. 10, no. 2, pp. 224–241, 2016.

[12] M. Opper and O. Winther, "Expectation consistent approximate inference," *J. Machine Learning Res.*, vol. 6, pp. 2177–2204, 2005.

[13] J. Pearl, *Probabilistic reasoning in intelligent systems: Networks of plausible inference*. Morgan Kaufmann Publishers Inc., 1998.

[14] T. P. Minka, "Expectation propagation for approximate Bayesian inference," in *Proc. 17th Conference in Uncertainty in Artificial Intelligence*, 2001, pp. 362–369.

[15] D. L. Donoho, A. Maleki, and A. Montanari, "Message-passing algorithms for compressed sensing," *Proc. Natl. Acad. Sci. USA*, vol. 106, no. 45, pp. 18 914–18 919, 2009.

[16] I. M. Johnstone, "Gaussian estimation: Sequence and wavelet models," 2015, http://statweb.stanford.edu/~imj/.

[17] S. Foucart and H. Rauhut, *A mathematical introduction to compressive sensing*. Birkhäuser, 2013.

[18] Y. C. Eldar and G. Kutyniok, *Compressed sensing theory and applications*. Cambridge University Press, 2012.

[19] D. Guo and S. Verdú, "Randomly spread CDMA: Asymptotics via statistical physics," *IEEE Trans. Information Theory*, vol. 51, no. 6, pp. 1983–2010, 2005.

[20] G. Reeves and H. D. Pfister, "The replica-symmetric prediction for compressed sensing with Gaussian matrices is exact," in *Proc. IEEE International Symposium on Information Theory (ISIT)*, 2016, pp. 665–669.

[21] G. Reeves and H. D. Pfister, "The replica-symmetric prediction for compressed sensing with Gaussian matrices is exact," 2016, https://arxiv.org/abs/1607.02524.

[22] G. Reeves, "Understanding the MMSE of compressed sensing one measurement at a time," presented at the Institut Henri Poincaré Spring 2016 Thematic Program on the Nexus of Information and Computation Theories, Paris, 2016, https://youtu.be/vmd8-CMv04I.

[23] M. Bayati and A. Montanari, "The dynamics of message passing on dense graphs, with applications to compressed sensing," *IEEE Trans. Information Theory*, vol. 57, no. 2, pp. 764–785, 2011.

[24] S. Rangan, "Generalized approximate message passing for estimation with random linear mixing," in *Proc. IEEE International Symposium on Information Theory (ISIT)*, 2011, pp. 2174–2178.

[25] J. P. Vila and P. Schniter, "Expectation-maximization Gaussian-mixture approximate message passing," *IEEE Trans. Signal Processing*, vol. 61, no. 19, pp. 4658–4672, 2013.

[26] Y. Ma, J. Zhu, and D. Baron, "Compressed sensing via universal denoising and approximate message passing," *IEEE Trans. Signal Processing*, vol. 64, no. 21, pp. 5611–5622, 2016.

[27] C. A. Metzler, A. Maleki, and R. G. Baraniuk, "From denoising to compressed sensing," *IEEE Trans. Information Theory*, vol. 62, no. 9, pp. 5117–5144, 2016.

[28] P. Schniter, S. Rangan, and A. K. Fletcher, "Vector approximate message passing for the generalized linear model," in *Asilomar Conference on Signals, Systems and Computers*, 2016.

[29] S. Rangan, P. Schniter, and A. K. Fletcher, "Vector approximate message passing," in *Proc. IEEE International Symposium on Information Theory (ISIT)*, 2017, pp. 1588–1592.

[30] Y. Kabashima, "A CDMA multiuser detection algorithm on the basis of belief propagation," *J. Phys. A: Math. General*, vol. 36, no. 43, pp. 11 111–11 121, 2003.

[31] M. Bayati, M. Lelarge, and A. Montanari, "Universality in polytope phase transitions and iterative algorithms," in *IEEE International Symposium on Information Theory*, 2012.

[32] G. Reeves, "Additivity of information in multilayer networks via additive Gaussian noise transforms," in *Proc. Allerton Conference on Communication, Control, and Computing*, 2017, https://arxiv.org/abs/1710.04580.

[33] B. Çakmak, O. Winther, and B. H. Fleury, "S-AMP: Approximate message passing for general matrix ensembles," 2014, http://arxiv.org/abs/1405.2767.

[34] A. Fletcher, M. Sahree-Ardakan, S. Rangan, and P. Schniter, "Expectation consistent approximate inference: Generalizations and convergence," in *Proc. IEEE International Symposium on Information Theory (ISIT)*, 2016.

[35] S. Rangan, P. Schniter, and A. K. Fletcher, "Vector approximate message passing," 2016, https://arxiv.org/abs/1610.03082.

[36] P. Schniter, S. Rangan, and A. K. Fletcher, "Vector approximate message passing for the generalized linear model," 2016, https://arxiv.org/abs/1612.01186.

[37] B. Çakmak, M. Opper, O. Winther, and B. H. Fleury, "Dynamical functional theory for compressed sensing," 2017, https://arxiv.org/abs/1705.04284.

[38] H. He, C.-K. Wen, and S. Jin, "Generalized expectation consistent signal recovery for nonlinear measurements," in *Proc. IEEE International Symposium on Information Theory (ISIT)*, 2017.

[39] D. Guo, S. Shamai, and S. Verdú, "Mutual information and minimum mean-square error in Gaussian channels," *IEEE Trans. Information Theory*, vol. 51, no. 4, pp. 1261–1282, 2005.

[40] A. J. Stam, "Some inequalities satisfied by the quantities of information of Fisher and Shannon," *Information and Control*, vol. 2, no. 2, pp. 101–112, 1959.

[41] D. Guo, Y. Wu, S. Shamai, and S. Verdú, "Estimation in Gaussian noise: Properties of the minimum mean-square error," *IEEE Trans. Information Theory*, vol. 57, no. 4, pp. 2371–2385, 2011.

[42] G. Reeves, H. D. Pfister, and A. Dytso, "Mutual information as a function of matrix SNR for linear Gaussian channels," in *Proc. IEEE International Symposium on Information Theory (ISIT)*, 2018.

[43] K. Bhattad and K. R. Narayanan, "An MSE-based transfer chart for analyzing iterative decoding schemes using a Gaussian approximation," *IEEE Trans. Information Theory*, vol. 58, no. 1, pp. 22–38, 2007.

[44] N. Merhav, D. Guo, and S. Shamai, "Statistical physics of signal estimation in Gaussian noise: Theory and examples of phase transitions," *IEEE Trans. Information Theory*, vol. 56, no. 3, pp. 1400–1416, 2010.

[45] C. Méasson, A. Montanari, T. J. Richardson, and R. Urbanke, "The generalized area theorem and some of its consequences," *IEEE Trans. Information Theory*, vol. 55, no. 11, pp. 4793–4821, 2009.

[46] G. Reeves, "Conditional central limit theorems for Gaussian projections," in *Proc. IEEE International Symposium on Information Theory (ISIT)*, 2017, pp. 3055–3059.

[47] G. Reeves, "Two-moment inequailties for Rényi entropy and mutual information," in *Proc. IEEE International Symposium on Information Theory (ISIT)*, 2017, pp. 664–668.

[48] S. B. Korada and N. Macris, "Tight bounds on the capacity of binary input random CDMA systems," *IEEE Trans. Information Theory*, vol. 56, no. 11, pp. 5590–5613, 2010.

[49] A. Montanari and D. Tse, "Analysis of belief propagation for non-linear problems: The example of CDMA (or: How to prove Tanaka's formula)," in *Proc. IEEE Information Theory Workshop (ITW)*, 2006, pp. 160–164.

[50] W. Huleihel and N. Merhav, "Asymptotic MMSE analysis under sparse representation modeling," *Signal Processing*, vol. 131, pp. 320–332, 2017.

[51] G. Reeves and M. Gastpar, "The sampling rate–distortion trade-off for sparsity pattern recovery in compressed sensing," *IEEE Trans. Information Theory*, vol. 58, no. 5, pp. 3065–3092, 2012.

[52] G. Reeves and M. Gastpar, "Approximate sparsity pattern recovery: Information-theoretic lower bounds," *IEEE Trans. Information Theory*, vol. 59, no. 6, pp. 3451–3465, 2013.

[53] J. Barbier, M. Dia, N. Macris, and F. Krzakala, "The mutual information in random linear estimation," in *Proc. Allerton Conference on Communication, Control, and Computing*, 2016.

[54] J. Barbier, F. Krzakala, N. Macris, L. Miolane, and L. Zdeborová, "Phase transitions, optimal errors and optimality of message-passing in generalized linear models," 2017, https://arxiv.org/abs/1708.03395.

[55] J. Barbier, F. Krzakala, N. Macris, L. Miolane, and L. Zdeborová, "Optimal errors and phase transitions in high-dimensional generalized linear models," in *Conference on Learning Theory*, 2018, pp. 728–731.

[56] A. Manoel, F. Krzakala, M. Mézard, and L. Zdeborová, "Multi-layer generalized linear estimation," in *Proc. IEEE International Symposium on Information Theory (ISIT)*, 2017, pp. 2098–2102.

[57] A. K. Fletcher, S. Rangan, and P. Schniter, "Inference in deep networks in high dimensions," in *Proc. IEEE International Symposium on Information Theory (ISIT)*, 2018.

[58] J. Barbier, M. Dia, N. Macris, F. Krzakala, T. Lesieur, and L. Zdeborová, "Mutual information for symmetric rank-one matrix estimation: A proof of the replica formula," in *Advances in Neural Information Processing Systems (NIPS)*, 2016, pp. 424–432.

[59] M. Lelarge and L. Miolane, "Fundamental limits of symmetric low-rank matrix estimation," 2016, https://arxiv.org/abs/1611.03888.

[60] T. Lesieur, F. Krzakala, and L. Zdeborová, "Constrained low-rank matrix estimation: Phase transitions, approximate message passing and applications," 2017, https://arxiv.org/abs/1701.00858.

[61] E. Abbe, "Community detection and stochastic block models: Recent developments," 2017, https://arxiv.org/abs/1703.10146.

[62] A. Kipnis, G. Reeves, Y. C. Eldar, and A. Goldsmith, "Compressed sensing under optimal quantization," in *Proc. IEEE International Symposium on Information Theory (ISIT)*, 2017, pp. 2153–2157.

[63] A. Kipnis, G. Reeves, and Y. C. Eldar, "Single letter formulas for quantized compressed sensing with Gaussian codebooks," in *Proc. IEEE International Symposium on Information Theory (ISIT)*, 2018.

8 Computing Choice: Learning Distributions over Permutations

Devavrat Shah

Summary

We discuss the question of learning distributions over permutations of a given set of choices, options, or items on the basis of partial observations. This is central to capturing the so-called "choice" in a variety of contexts: understanding preferences of consumers over a collection of products from purchasing and browsing data in the setting of retail and e-commerce, learning public opinion amongst a collection of socio-economic issues from sparse polling data, deciding a ranking of teams or players from outcomes of games, electing leaders according to votes, and more generally collaborative decision-making employing collective judgment such as accepting papers in a competitive academic conference. The question of learning distributions over permutations arises beyond capturing "choice" as well, for example, tracking a collection of objects using noisy cameras, or aggregating ranking of web-pages using outcomes of multiple search engines. It is only natural that this topic has been studied extensively in economics, political science, and psychology for more than a century, and more recently in computer science, electrical engineering, statistics, and operations research.

Here we shall focus on the task of learning distributions over permutations from marginal distributions of two types: first-order marginals and pair-wise comparisons. A lot of progress has been made on this topic in the last decade. The ideal goal is to provide a comprehensive overview of the state-of-the-art on this topic. We shall provide a detailed overview of selective aspects, biased by the author's perspective of the topic, and provide sufficient pointers to aspects not covered here. We shall emphasize the ability to identify the entire distribution over permutations as well as the "best ranking."

8.1 Background

8.1.1 Learning from Comparisons

Consider a grocery store around the corner from your home. The owner of the store would like to have the ability to identify exactly what every customer would purchase (or not) given the options available in the store. If such an ability exists, then, for example, optimal stocking decisions can be made by the store operator or the net worth of the store can be evaluated. This ability is what one would call the "choice model" of consumers of the store. More precisely, such a "choice model" can be viewed as a black-box that spits out the probability of purchase of a particular option when the customer is presented with a collection of options.

A canonical fine-grained representation for such a "choice model" is the distribution over permutations of all the possible options (including the *no-purchase* option). Then, probability of purchasing a particular option when presented with a collection of options is simply the probability that this particular option has the highest (relative) order or rank amongst all the presented options (including the *no-purchase* option).

Therefore, one way to operationalize such a "choice model" is to learn the distribution over permutations of all options that a store owner can stock in the store. Clearly, such a distribution needs to be learned from the observations or data. The data available to the store owner are the historical transactions as well as what was stocked in the store when each transaction happened. Such data effectively provide a bag of pair-wise comparisons between options: consumer exercises or purchases option A over option B corresponds to a pair-wise comparison A > B or "A is preferred to B."

In summary, to model consumer choice, we wish to learn the distribution over permutations of all possible options using observations in terms of a collection of pair-wise comparisons that are consistent with the learned distribution.

In the context of sports, we wish to go a step further, to obtain a ranking of sports teams or players that is based on outcomes of games, which are simply pair-wise comparisons (between teams or players). Similarly, for the purpose of data-driven policy-making, we wish to aggregate people's opinions about socio-economic issues such as modes of transportation according to survey data; for designing online recommendation systems that are based on historical online activity of individuals, we wish to recommend the top few options; or to sort objects on the basis of noisy outcomes of pair-wise comparisons.

8.1.2 Learning from First-Order Marginals

The task of learning distributions over permutations of options, using different types of partial information comes up in other scenarios. To that end, now suppose that the store owner wants to track each consumer's journey within the store with the help of cameras. The consumers constantly move within the store as they search through the aisles. Naturally, when multiple consumers are in the store, their paths are likely to cross. When the paths of two (or more) consumers cross, and they subsequently follow different trajectories, confusion can arise regarding which of the multiple trajectories maps to which of the consumers. That is, at each instant of time we need to continue mapping physical locations of consumers observed by cameras with the trajectories of consumers who are being tracked. Equivalently, it's about keeping track of a "matching" between locations and individuals in a bipartite graph or keeping track of permutations!

The authors of [1] proposed distribution over permutations as the canonical model where a permutation corresponds to "matching" of consumers or trajectories to locations. In such a scenario, due to various constraints and tractability reasons, the information that is available is the likelihoods of each consumer or trajectory being in a specific location. In the context of distribution over permutations, this corresponds to knowing the "first-order" marginal distribution information that states the probability of a given option being in a certain position in the permutation. Therefore, to track consumers in the store, we wish to learn the distribution over consumer trajectories that is consistent with this first-order marginal information over time.

In summary, the model to track trajectories of individuals boils down to continually learning the distribution over permutations that is consistent with the first-order marginal information and subsequently finding the *most likely* ranking or permutation from the learned distribution. It is the very same question that arises in the context of aggregating web-page rankings obtained through results of search from multiple search engines in a computationally efficient manner.

8.1.3 Historical Remarks

This fine-grained representation for choice, distribution over permutations, is ancient. Here, we provide a brief historical overview of the use of distribution over permutations as a model for choice and other applications. We also refer to Chapter 9 of the monograph by Diaconis [2] for a nice historical overview from a statistician's perspective.

One of the earliest references regarding how to model and learn choice using (potentially inconsistent) comparisons is the seminal work by Thurstone [3]. It presents "a law of comparative judgement" or more precisely a simple parametric model to capture the outcomes of a collection of pair-wise comparisons between given options (or *stimuli* in the language of [3]). This model can be rephrased as an instance of the *random utility model* (RUM) as follows (also see [4, 5]): given N options, let each option, say i, have inherent utility u_i associated with it; when two options i and j are compared, random variables Y_i, Y_j are sampled and i is preferred over j iff $Y_i > Y_j$; here $Y_i = u_i + \varepsilon_i$, $Y_j = u_j + \varepsilon_j$, with $\varepsilon_i, \varepsilon_j$ independent random variables with identical mean.

A specialization of the above model when the ε_is are assumed to be Gaussian with mean 0 and variance 1 for all i is known as the Thurstone–Mosteller model. It is also known as the *probit* model. Another specialization of the Thurstone model is realized when the ε_is are assumed to have the Gumbel distribution (one of the extreme value distributions). This model has been credited differently across communities. Holman and Marley established that this model is equivalent (see [6] for details) to a generative model described in detail in Section 8.3.2. It is known as the Luce model [7] and the Plackett model [8]. In the context when the partial observations are choice observations (i.e., the observation that an item is chosen from an offered subset of items), this model is called the multinomial logit model (MNL) after McFadden [9] called it *conditional logit*; also see [10]. It is worth remarking that, when restricted to pair-wise comparisons only, this model matches the Bradley–Terry [11] model, but the Bradley–Terry model did not consider the requirement that the pair-wise comparison marginals need to be consistent with an underlying distribution over permutations.

The MNL model is of central importance for various reasons. It was introduced by Luce to be consistent with the axiom of *independence from irrelevant alternatives* (IIA). The model was shown to be consistent with the induced preferences assuming a form of random utility maximization framework whose inquiry was started by [4, 5]. Very early on, simple statistical tests as well as simple estimation procedures were developed to fit such a model to observed data [9]. Now the IIA property possessed by the MNL model is not necessarily desirable as evidenced in many empirical scenarios. Despite such structural limitations, the MNL model has been widely utilized across application areas primarily due to the ability to learn the model parameters easily from observed data. For example, see [12–14] for applications in transportation and [15, 16] for applications in operations management and marketing.

With a view to addressing the structural limitations of the MNL model, a number of generalizations to this model have been proposed over the years. Notable among these are the so-called "nested" MNL model, as well as mixtures of MNL models (or MMNL models). These generalizations avoid the IIA property and continue to be consistent with the random utility maximization framework at the expense of increased model complexity; see [13, 17–20] for example. The interested reader is also referred to an overview article on this line of research [14]. While generalized models of this sort are in principle attractive, their complexity makes them difficult to learn while avoiding the risk of overfitting. More generally, specifying an appropriate parametric model is a difficult task, and the risks associated with mis-specification are costly in practice. For an applied view of these issues, see [10, 21, 22].

As an alternative to the MNL model (and its extensions), one might also consider the parametric family of choice models induced by the exponential family of distributions over permutations. These may be viewed as the models that have maximum entropy among those models that satisfy the constraints imposed by the observed data. The number of parameters in such a model is equal to the number of constraints in the maximum-entropy optimization formulation, or equivalently the effective dimension of the underlying data, see the Koopman–Pitman–Darmois theorem [23]. This scaling of the number of parameters with the effective data dimension makes the exponential family obtained via the maximum-entropy principle very attractive. Philosophically, this approach imposes on the model only those constraints implied by the observed data. On the flip side, learning the parameters of an exponential family model is a computationally challenging task (see [24–26]) as it requires computing a "partition function," possibly over a complex state space.

Very recently, Jagabathula and Shah [27, 28] introduced a *non-parametric* sparse model. Here the distribution over permutations is assumed to have sparse (or small) support. While this may not be exactly true, it can be an excellent approximation to the reality and can provide computationally efficient ways to both infer the model [27, 28] in a manner consistent with observations and utilize it for effective decision-making [29, 30].

8.2 Setup

Given N objects or items denoted as $[N] = \{1, \ldots, N\}$, we are interested in the distribution over permutations of these N items. A permutation $\sigma : [N] \to [N]$ is one-to-one and onto mapping, with $\sigma(i)$ denoting the position or ordering of element $i \in [N]$.

Let S_N denote the space of $N!$ permutations of these N items. The set of distributions over S_N is denoted as $\mathcal{M}(S_N) = \{\nu : S_N \to [0,1] : \sum_{\sigma \in S_N} \nu(\sigma) = 1\}$. Given $\nu \in \mathcal{M}(S_N)$, the first-order marginal information, $\mathbf{M}(\nu) = [M_{ij}(\nu)]$, is an $N \times N$ doubly stochastic matrix with non-negative entries defined as

$$M_{ij}(\nu) = \sum_{\sigma \in S_N} \nu(\sigma) \mathbf{1}_{\{\sigma(i)=j\}}, \tag{8.1}$$

where, for $\sigma \in S_N$, $\sigma(i)$ denotes the rank of item i under permutation σ, and $\mathbf{1}_{\{x\}}$ is the standard indicator with $\mathbf{1}_{\{\text{true}\}} = 1$ and $\mathbf{1}_{\{\text{false}\}} = 0$. The comparison marginal information, $\mathbf{C}(\nu) = [C_{ij}(\nu)]$, is an $N \times N$ matrix with non-negative entries defined as

$$C_{ij}(\nu) = \sum_{\sigma \in S_N} \nu(\sigma) \mathbf{1}_{\{\sigma(i)>\sigma(j)\}}. \tag{8.2}$$

By definition, the diagonal entries of $\mathbf{C}(\nu)$ are all 0s, and $C_{ij}(\nu) + C_{ji}(\nu) = 1$ for all $1 \leq i \neq j \leq N$. We shall abuse the notation by using $\mathbf{M}(\sigma)$ and $\mathbf{C}(\sigma)$ to denote the matrices obtained by applying them to the distribution where the support of the distribution is simply $\{\sigma\}$.

Throughout, we assume that there is a *ground-truth* model ν. We observe marginal information $\mathbf{M}(\nu)$ or $\mathbf{C}(\nu)$, or their noisy versions.

8.2.1 Questions of Interest

We are primarily interested in two questions: recovering the distribution and producing the ranking, on the basis of the distribution.

QUESTION 1 (Recover the distribution.) *The primary goal is to recover ν from observations. Precisely, we observe $\mathbf{D} = \mathbf{P}(\nu) + \boldsymbol{\eta}$, where $\mathbf{P}(\nu) \in \{\mathbf{M}(\nu), \mathbf{C}(\nu)\}$ and there is a potentially noisy perturbation $\boldsymbol{\eta}$. The precise form of noisy perturbation that allows recovery of the model is explained later in detail. Intuitively, the noisy perturbation may represent the finite sample error introduced due to forming an empirical estimation of $\mathbf{P}(\nu)$ or an inability to observe data associated with certain components.*

A generic ν has $N! - 1$ unknowns, while the dimension of $\mathbf{D}$ is at most N^2. Learning ν from $\mathbf{D}$ boils down to finding the solution to a set of linear equations where there are at most N^2 linear equations involving $N! - 1$ unknowns. This is a highly under-determined system of equations and hence, without imposing structural conditions on ν, it is unlikely that we will be able to recover ν faithfully. Therefore, the basic "information" question would be to ask under what structural assumption on ν it is feasible to recover it from $\mathbf{P}(\nu)$ (i.e., when $\boldsymbol{\eta} = \mathbf{0}$). The next question concerns the "robustness" of such a recovery condition when we have non-trivial noise, $\boldsymbol{\eta}$. And finally, we would like to answer the "computational" question associated with it, which asks whether such recovery is possible using computation that scales polynomially in N.

QUESTION 2 (*Produce the ranking.*) *An important associated decision question is that of finding the "ranking" or most "relevant" permutation for the underlying ν. To begin with, what is the most "relevant" permutation assuming we know the ν perfectly? This,* *in a sense, is ill-defined due to the impossibility result of Arrow [31]:* there is no ranking algorithm that works for all ν and satisfies certain basic hypotheses expected from any ranking algorithm even when $N = 3$.

For this reason, like in the context of recovering ν, we will have to impose structure on ν. In particular, the structure that we shall impose (e.g., the sparse model or the multinomial logit model) seems to suggest a natural answer for ranking or the most "relevant" permutation: find the σ that has maximal probability, i.e., find $\sigma^*(\nu)$, where

$$\sigma^*(\nu) \in \arg\max_{\sigma \in S_N} \nu(\sigma). \tag{8.3}$$

Again, the goals would include the ability to recover $\sigma^*(\nu)$ (exactly or approximately) using observations (a) when $\boldsymbol{\eta} = \mathbf{0}$ and (b) when there is a non-trivial $\boldsymbol{\eta}$, and (c) the ability to do this in a computationally efficient manner.

8.3 Models

We shall consider two types of model here: the non-parametric sparse model and the parametric random utility model. As mentioned earlier, a large number of models have been studied in the literature and are not discussed in detail here.

8.3.1 Sparse Model

The support of distribution ν, denoted as $\mathsf{supp}(\nu)$ is defined as

$$\mathsf{supp}(\nu) \triangleq \{\sigma \in S_N : \nu(\sigma) > 0\}. \tag{8.4}$$

The ℓ_0-norm of ν, denoted as $||\nu||_0$, is defined as

$$||\nu||_0 \triangleq \left|\mathsf{supp}(\nu)\right|. \tag{8.5}$$

We say that ν has sparsity K if $K = ||\nu||_0$. Naturally, by varying K, all possible $\nu \in \mathcal{M}(S_N)$ can be captured. In that sense, this is a *non-parametric* model. This model was introduced in [27, 28].

The goal would be to learn the sparsest possible ν that is consistent with observations. Formally, this corresponds to solving

$$\begin{aligned} &\text{minimize} \quad ||\mu||_0 \qquad \text{over} \quad \mu \in \mathcal{M}(S_N) \\ &\text{such that} \quad \mathbf{P}(\mu) \approx \mathbf{D}, \end{aligned} \tag{8.6}$$

where $\mathbf{P}(\mu) \in \{\mathbf{M}(\mu), \mathbf{C}(\mu)\}$ depending upon the type of information considered.

8.3.2 Random Utility Model (RUM)

We consider the random utility model (RUM) that in effect was considered in the "law of comparative judgement" by Thurstone [3]. Formally, each option $i \in [N]$ has a deterministic utility u_i associated with it. The random utility Y_i associated with option $i \in [N]$ obeys the form

$$Y_i = u_i + \varepsilon_i, \tag{8.7}$$

where ε_i are independent random variables across all $i \in [N]$ – they represent "random perturbation" of the "inherent utility" u_i. We assume that all of the ε_i have identical mean across all $i \in [N]$, but can have varying distribution. The specific form of the distribution gives rise to different types of models. We shall describe a few popular examples of this in what follows. Before we do that, we explain how this setup gives rise to the distribution over permutations by describing the generative form of the distribution. Specifically, to generate a random permutation over the N options, we first sample random variable $Y_i,\ i \in [N]$, independently. Then we sort $Y_1, \dots, Y_N$ in decreasing order,[1] and this sorted order of indices $[N]$ provides the permutation. Now we describe two popular examples of this model.

Probit model. Let ε_i have a Gaussian distribution with mean 0 and variance σ_i^2 for $i \in [N]$. Then, the resulting model is known as the Probit model. In the homogeneous setting, we shall assume that $\sigma_i^2 = \sigma^2$ for all $i \in [N]$.

Multinomial logit (MNL) model. Let ε_i have a Gumbel distribution with mode μ_i and scaling parameter $\beta_i > 0$, i.e., the PDF of ε_i is given by

$$f(x) = \frac{1}{\beta_i} \exp(-(z + \exp(-z))), \text{ where } z = \frac{x - \mu_i}{\beta_i}, \text{ for } x \in \mathbb{R}. \tag{8.8}$$

In the homogeneous setting, $\mu_i = \mu$ and $\beta_i = \beta$ for all $i \in [N]$. In this scenario, the resulting distribution over permutations turns out to be equivalent to the following generative model.

Let $w_i > 0$ be a parameter associated with $i \in [N]$. Then the probability of permutation $\sigma \in S_N$ is given by (for example, see [32])

$$\mathbb{P}(\sigma) = \prod_{j=1}^{N} \frac{w_{\sigma^{-1}(j)}}{w_{\sigma^{-1}(j)} + w_{\sigma^{-1}(j+1)} + \cdots + w_{\sigma^{-1}(N)}}. \tag{8.9}$$

Above, $\sigma^{-1}(j) = i$ iff $\sigma(i) = j$. Specifically, for $i \neq j \in [N]$,

$$\mathbb{P}(\sigma(i) > \sigma(j)) = \frac{w_i}{w_i + w_j}. \tag{8.10}$$

We provide a simple explanation of the above, seemingly mysterious, relationship between two very different descriptions of the MNL model.

[1] We shall assume that the distribution of $\varepsilon_i,\ i \in [N]$, has a *density* and hence ties never happen between Y_i, Y_j for any $i \neq j \in [N]$.

LEMMA 8.1 *Let $\varepsilon_i, \varepsilon_j$ be independent random variables with Gumbel distributions with mode μ_i, μ_j, respectively, with scaling parameters $\beta_i = \beta_j = \beta > 0$. Then, $\Delta_{ij} = \varepsilon_i - \varepsilon_j$ has a logistic distribution with parameters $\mu_i - \mu_j$ (location) and β (scale).*

The proof of Lemma 8.1 follows by, for example, using the characteristic function associated with the Gumbel distribution along with the property of the gamma function ($\Gamma(1+z)\Gamma(1-z) = z\pi/\sin(\pi z)$) and then identifying the characteristic function of the logistic distribution.

Returning to our model, when we compare the random utilities associated with options i and j, Y_i and Y_j, respectively, we assume the corresponding random perturbation to be homogeneous, i.e., $\mu_i = \mu_j = \mu$ and $\beta_i = \beta_j = \beta > 0$. Therefore, Lemma 8.1 suggests that

$$
\begin{aligned}
\mathbb{P}(Y_i > Y_j) &= \mathbb{P}(\varepsilon_i - \varepsilon_j > u_j - u_i) \\
&= \mathbb{P}(\mathsf{Logistic}(0,\beta) > u_j - u_i) \\
&= 1 - \mathbb{P}(\mathsf{Logistic}(0,\beta) < u_j - u_i) \\
&= 1 - \frac{1}{1 + \exp(-u_j - u_i)/\beta} \\
&= \frac{\exp(u_i/\beta)}{\exp(u_i/\beta) + \exp(u_j/\beta)} \\
&= \frac{w_i}{w_i + w_j},
\end{aligned}
\tag{8.11}
$$

where $w_i = \exp(u_i/\beta)$, $w_j = \exp(u_j/\beta)$. It is worth remarking that the property of Lemma 8.9 relates the MNL model to the Bradley–Terry model.

Learning the model and ranking. For the random utility model, the question of learning the model from data effectively boils down to learning the model parameters from observations. In the context of a homogeneous model, i.e., ε_i in (8.7), for which we have an identical distribution across all $i \in [N]$, the primary interest is in learning the inherent utility parameters u_i, for $i \in [N]$. The question of recovering the ranking, on the other hand, is about recovering $\sigma \in S_N$, which is the sorted (decreasing) order of the inherent utilities $u_i, i \in [N]$: for example, if $u_1 \geq u_2 \geq \cdots \geq u_N$, then the ranking is the identity permutation.

8.4 Sparse Model

In this section, we describe the conditions under which we can learn the underlying sparse distribution using first-order marginal and comparison marginal information. We divide the presentation into two parts: first, we consider access to exact or noise-less marginals for exact recovery; and, then, we discuss its robustness.

We can recover a ranking in terms of the most likely permutation once we have recovered the sparse model by simply sorting the likelihoods of the permutations in the support of the distribution, which requires time $O(K \log K)$, where K is the sparsity of the model. Therefore, the key question in the context of the sparse model is the recovery of the distribution, on which we shall focus in the remainder of this section.

8.4.1 Exact Marginals: Infinite Samples

We are interested in understanding when it is feasible to recover the underlying distribution ν given access to its marginal information $\mathbf{M}(\nu)$ or $\mathbf{C}(\nu)$. As mentioned earlier, one way to recover siuch a distribution using exactly marginal information is to solve (8.6) with the equality constraint of $\mathbf{P}(\nu) = D$, where $\mathbf{P}(\nu) \in \{\mathbf{M}(\nu), \mathbf{C}(\nu)\}$ depending upon the type of marginal information.

We can view the unknown ν as a high-dimensional vector in $\mathbb{R}^{N!}$ which is sparse. That is, $\|\nu\|_0 \ll N!$. The observations are marginals of ν, either first-order marginals $\mathbf{M}(\nu)$ or comparison marginals $\mathbf{C}(\nu)$. They can be viewed as linear projections of the ν vector of dimension N^2 or $N(N-1)$. Therefore, recovering the sparse model from marginal information boils down to recovering a sparse vector in high-dimensional space (here $N!$ dimensional) from a small number of linear measurements of the sparse vector. That is, we wish to recover $\mathbf{x} \in \mathbb{R}^n$ from observation $\mathbf{y} = \mathbf{A}\mathbf{x}$, where $\mathbf{y} \in \mathbb{R}^m$, $\mathbf{A} \in \mathbb{R}^{m \times n}$ with $m \ll n$. In the best case, one can hope to recover $\mathbf{x}$ uniquely as long as $m \sim \|\mathbf{x}\|_0$.

This question has been well studied in the context of sparse model learning from linear measurements of the signal in signal processing and it has been popularized under the umbrella term of *compressed sensing*, see for example [33–37]. It has been argued that such recovery is possible as long as A satisfies certain conditions, for example the *restricted isoperimetric property* (RIP) (see [34, 38]) with $m \sim K \log n/K$, in which case the ℓ_0 optimization problem,

$$\min_{\mathbf{z}} \|\mathbf{z}\|_0 \quad \text{over} \quad \mathbf{y} = \mathbf{A}\mathbf{z},$$

recovers the true signal $\mathbf{x}$ as long as the sparsity of $\mathbf{x}$, $\|\mathbf{x}\|_0$, is at most K. The remarkable fact about an RIP-like condition is that it not only recovers the sparse signal using the ℓ_0 optimization, but also can be done using a computationally efficient procedure, a linear program.

Impossibility of recovering the distribution even for $N = 4$. Returning to our setting, $n = N!$, $m = N^2$ and we wish to understand up to what level of sparsity of ν we can recover it. The key difference here is in the fact that $\mathbf{A}$ is *not designed* but *given*. Therefore, the question is whether A has a *nice* property such as RIP that can allow sparse recovery. To that end, consider the following simple counterexample that shows that it is impossible to recover a sparse model uniquely even with support size 3 using the ℓ_0 optimization [27, 28].

Example 8.1 (Impossibility) For $N = 4$, consider the four permutations $\sigma_1 = [1 \to 2, 2 \to 1, 3 \to 3, 4 \to 4]$, $\sigma_2 = [1 \to 1, 2 \to 2, 3 \to 4, 4 \to 3]$, $\sigma_3 = [1 \to 2, 2 \to 1, 3 \to 4, 4 \to 3]$, and $\sigma_4 = \mathsf{id} = [1 \to 1, 2 \to 2, 3 \to 3, 4 \to 4]$, i.e., the identity permutation. It is easy to check that

$$\mathbf{M}(\sigma_1) + \mathbf{M}(\sigma_2) = \mathbf{M}(\sigma_3) + \mathbf{M}(\sigma_4).$$

Now suppose that $\nu(\sigma_i) = p_i$, where $p_i \in [0,1]$ for $1 \leq i \leq 3$, and $\nu(\sigma) = 0$ for all other $\sigma \in S_N$. Without loss of generality, let $p_1 \leq p_2$. Then

$$p_1\mathbf{M}(\sigma_1) + p_2\mathbf{M}(\sigma_2) + p_3\mathbf{M}(\sigma_3) = (p_2 - p_1)\mathbf{M}(\sigma_2) + (p_3 + p_1)\mathbf{M}(\sigma_3) + p_1\mathbf{M}(\sigma_4).$$

Here, note that $\{\mathbf{M}(\sigma_1), \mathbf{M}(\sigma_2), \mathbf{M}(\sigma_3)\}$ are linearly independent, yet the sparsest solution is not unique. Therefore, it is not feasible to recover the sparse model uniquely.

Note that the above example can be extended for any $N \geq 4$ by simply having identity permutation for all elements larger than 4 in the above example. Therefore, for any N with support size 3, we cannot always recover them uniquely.

Signature condition for recovery. Example 8.1 suggests that it is not feasible to expect an RIP-like condition for the "projection matrix" corresponding to the first-order marginals or comparison marginals so that *any* sparse probability distribution can be recovered. The next best thing we can hope for is the ability to recover *almost all* of the sparse probability distributions. This leads us to the *signature condition* of the matrix for a given sparse vector, which, as we shall see, allows recovery of the particular sparse vector [27, 28].

CONDITION 8.1 (Signature conditions) A given matrix $\mathbf{A} \in \mathbb{R}^{m\times n}$ is said to satisfy a *signature condition* with respect to index set $S \subset \{1,\ldots,n\}$ if, for each $i \in S$, there exists $j(i) \in [m]$ such that $\mathbf{A}_{j(i)i} \neq 0$ and $\mathbf{A}_{j(i)i'} = 0$ for all $i' \neq i, i' \in S$.

The signature condition allows recovery of sparse vector using a simple "peeling" algorithm. We summarize the recovery result followed by an algorithm that will imply the result.

THEOREM 8.1 *Let $\mathbf{A} \in \{0,1\}^{m\times n}$ with all of its columns being distinct. Let $\mathbf{x} \in \mathbb{R}^n_{\geq 0}$ be such that $\mathbf{A}$ satisfies a signature condition with respect to the set $\mathsf{supp}(\mathbf{x})$. Let the non-zero components of $\mathbf{x}$, i.e., $\{\mathbf{x}_i : i \in \mathsf{supp}(\mathbf{x})\}$, be such that, for any two distinct $S_1, S_2 \subset \mathsf{supp}(\mathbf{x})$, $\sum_{i\in S_1} \mathbf{x}_i \neq \sum_{i'\in S_2} \mathbf{x}_{i'}$. Then $\mathbf{x}$ can be recovered from $\mathbf{y}$, where $\mathbf{y} = \mathbf{A}\mathbf{x}$.*

Proof To establish Theorem 8.1, we shall describe the algorithm that recovers $\mathbf{x}$ under the conditions of the theorem and simultaneously argue for its correctness. To that end, the algorithm starts by sorting components of $\mathbf{y}$. Since $\mathbf{A} \in \{0,1\}^{m\times n}$, for each $j \in [m]$, $\mathbf{y}_j = \sum_{i\in S_j} \mathbf{x}_i$, with $S_j \subseteq \mathsf{supp}(\mathbf{x})$. Owing to the signature condition, for each $i \in \mathsf{supp}(\mathbf{x})$, there exists $j(i) \in [m]$ such that $\mathbf{y}_{j(i)} = \mathbf{x}_i$. If we can identify $j(i)$ for each i, we recover the values $\mathbf{x}_i$, but not necessarily the position i. To identify the position, we identify the ith column of matrix $\mathbf{A}$, and, since the columns of matrix $\mathbf{A}$ are all distinct, this will help us identify the position. This will require use of the property that, for any $S_1 \neq S_2 \subset \mathsf{supp}(\mathbf{x})$, $\sum_{i\in S_1} \mathbf{x}_i \neq \sum_{i'\in S_2} \mathbf{x}_{i'}$. This implies, to begin with, that all of the non-zero elements of $\mathbf{x}$ are distinct. Without loss of generality, let the non-zero elements of $\mathbf{x}$ be $\mathbf{x}_1, \ldots, \mathbf{x}_K$, with $K = |\mathsf{supp}(\mathbf{x})|$ such that $0 < \mathbf{x}_1 < \cdots < \mathbf{x}_K$.

Now consider the smallest non-zero element of $\mathbf{y}$. Let it be $\mathbf{y}_{j_1}$. From the property of $\mathbf{x}$, it follows that $\mathbf{y}_{j_1}$ must be the smallest non-zero element of $\mathbf{x}$, $\mathbf{x}_1$. The second smallest component of $\mathbf{y}$ that is distinct from $\mathbf{x}_1$, let it be $\mathbf{y}_{j_2}$, must be $\mathbf{x}_2$. The third distinct smallest component, $\mathbf{y}_{j_3}$, however, could be $\mathbf{x}_1 + \mathbf{x}_2$ or $\mathbf{x}_3$. Since we know $\mathbf{x}_1, \mathbf{x}_2$, and the property of $\mathbf{x}$, we can identify whether $\mathbf{y}_{j_3}$ is $\mathbf{x}_3$ or not. Iteratively, we consider the kth distinct smallest value of $\mathbf{y}$, say $\mathbf{y}_{j_k}$. Then, it equals either the sum of the subset of already identified components of $\mathbf{x}$ or the next smallest unidentified component of $\mathbf{x}$ due to the signature property and non-negativity of $\mathbf{x}$. In summary, by the time we have gone through all of the non-zero components of $\mathbf{y}$ in the increasing order as described above, we will recover all the non-zero elements of $\mathbf{x}$ in increasing order as well as the corresponding columns of $\mathbf{A}$. This is because iteratively we identify for each $\mathbf{y}_j$ the set $S_j \subset \mathsf{supp}(\mathbf{x})$ such that $\mathbf{y}_j = \sum_{i \in S_j} \mathbf{x}_i$. That is, $\mathbf{A}_{ji} = 1$ for all $i \in S_j$ and 0 otherwise. This completes the proof.

Now we remark on the computation complexity of the "peeling" algorithm described above. It runs for at most m iterations. In each iteration, it tries to effectively solve a subset sum problem whose computation cost is at most $O(2^K)$, where $K = \|\mathbf{x}\|_0$ is the sparsity of $\mathbf{x}$. Then the additional step for sorting the components of $\mathbf{y}$ costs $O(m \log m)$. In summary, the computation cost is $O(2^K + m \log m)$. Notice that, somewhat surprisingly, this does not depend on n at all. In contrast, for the linear programming-based approach used for sparse signal recovery in the context of compressed sensing literature, the computational complexity scales at least linearly in n, the ambient dimension of the signal. For example, if this were applicable to our setting, it would scale as $N!$, which is simply prohibitive.

Recovering the distribution using first-order marginals via the signature condition. We shall utilize the signature condition, Condition 8.1, in the context of recovering the distribution over permutations from its first-order marginals. Again, given the counterexample, Example 8.1, it is not feasible to recover *all* sparse models even with sparsity 3 from first-order marginals uniquely. However, with the aid of the signature condition, we will argue that it is feasible to recover *most* sparse models with reasonably large sparsity.

To that end, let $\mathbf{A}^{\mathrm{f}} \in \{0,1\}^{N^2 \times N!}$ denote the first-order marginal matrix that maps the $N!$-dimensional vector corresponding to the distribution over permutations to an N^2-dimensional vector corresponding to the first-order marginals of the distribution. We state the signature property of $\mathbf{A}^{\mathrm{f}}$ next.

LEMMA 8.2 *Let S be a randomly chosen subset of $\{1,\ldots,N!\}$ of size K. Then the first-order marginal matrix $\mathbf{A}^{\mathrm{f}}$ satisfies the signature condition with respect to S with probability $1 - o(1)$ as long as $K \le (1-\epsilon) N \log N$ for any $\epsilon > 0$.*

The proof of the above lemma can be found in [27, 28]. Lemma 8.2 and Theorem 8.1 immediately imply the following result.

THEOREM 8.2 *Let $S \subset S_N$ be a randomly chosen subset of S_N of size K, denoted as $S = \{\sigma_1,\ldots,\sigma_K\}$. Let $p_1,\ldots,p_K$ be chosen from a joint distribution with a continuous density over subspace of $[0,1]^K$ corresponding to $p_1 + \cdots + p_K = 1$. Let ν be the distribution over S_N such that*

$$\nu(\sigma) = \begin{cases} p_k & \text{if } \sigma = \sigma_k,\ k \in [K], \\ 0 & \text{otherwise}. \end{cases} \tag{8.12}$$

Then ν can be recovered from its first-order marginal distribution with probability $1 - o(1)$ as long as $K \leq (1 - \epsilon)N \log N$ for a fixed $\epsilon > 0$.

The proof of Theorem 8.2 can be found in [27, 28]. In a nutshell, it states that the most sparse distribution over permutations with sparsity up to $N \log N$ can be recovered from its first-order marginals. This is in sharp contrast with the counterexample, Example 8.1, which states that for any N, a distribution with sparsity 3 cannot be recovered uniquely.

Recovering the distribution using comparison marginals via the signature condition. Next, we utilize the signature condition, Condition 8.1, in the context of recovering the distribution over permutations from its comparison marginals. Let $\mathbf{A}^c \in \{0,1\}^{N(N-1)\times N!}$ denote the comparison marginal matrix that maps the $N!$-dimensional vector corresponding to the distribution over permutations to an N^2-dimensional vector corresponding to the comparison marginals of the distribution. We state the signature property of $\mathbf{A}^c$ next.

LEMMA 8.3 *Let S be a randomly chosen subset of $\{1,\dots,N!\}$ of size K. Then the comparison marginal matrix $\mathbf{A}^c$ satisfies the signature condition with respect to S with probability $1 - o(1)$ as long as $K = o(\log N)$.*

The proof of the above lemma can be found in [29, 30]. Lemma 8.3 and Theorem 8.1 immediately imply the following result.

THEOREM 8.3 *Let $S \subset S_N$ be a randomly chosen subset of S_N of size K, denoted as $S = \{\sigma_1,\dots,\sigma_K\}$. Let $p_1,\dots,p_K$ be chosen from a joint distribution with a continuous density over the subspace of $[0,1]^K$ corresponding to $p_1 + \cdots + p_K = 1$. Let ν be a distribution over S_N such that*

$$\nu(\sigma) = \begin{cases} p_k & \text{if } \sigma = \sigma_k,\ k \in [K], \\ 0 & \text{otherwise.} \end{cases} \tag{8.13}$$

Then ν can be recovered from its comparison marginal distribution with probability $1 - o(1)$ as long as $K = o(\log N)$.

The proof of Theorem 8.3 can be found in [29, 30]. This proof suggests that it is feasible to recover a sparse model with growing support size with N as long as it is $o(\log N)$. However, it is exponentially smaller than the recoverable support size compared with the first-order marginal. This seems to be related to the fact that the first-order marginal is relatively information-rich compared with the comparison marginal.

8.4.2 Noisy Marginals: Finite Samples

Thus far, we have considered a setup where we had access to exact marginal distribution information. Instead, suppose we have access to marginal distributions formed on the basis of an empirical distribution of finite samples from the underlying distribution. This can be viewed as access to a "noisy" marginal distribution. Specifically, given a distribution ν, we observe $\mathbf{D} = \mathbf{P}(\nu) + \eta$, where $\mathbf{P}(\nu) \in \{\mathbf{M}(\nu), \mathbf{C}(\nu)\}$ depending upon the type of marginal information, with η being noise such that some norm of η, e.g., $||\eta||_2$ or $||\eta||_\infty$, is bounded above by δ, with $\delta > 0$ being small if we have access to enough samples. Here δ represents the error observed due to access to finitely many samples and is assumed to be known.

For example, if we have access to n independent samples for each marginal entry (e.g., i ranked in position j for a first-order marginal or i found to be better than j for a comparison marginal) according to ν, and we create an empirical estimation of each entry in $\mathbf{M}(\nu)$ or $\mathbf{C}(\nu)$, then, using the Chernoff bound for the binomial distribution and the union bound over a collection of events, it can be argued that $||\eta||_\infty \leq \delta$ with probability $1-\delta$ as long as $n \sim (1/\delta^2)\log(4N/\delta)$ for each entry. Using more sophisticated methods from the matrix-estimation literature, it is feasible to obtain better estimations of $\mathbf{M}(\nu)$ or $\mathbf{C}(\nu)$ from fewer samples of entries and even when some of the entries are entirely unobserved as long as $\mathbf{M}(\nu)$ or $\mathbf{C}(\nu)$ has *structure*. This is beyond the scope of this exposition; however, we refer the interested reader to [39–42] as well as the discussion in Section 8.6.4.

Given this, the goal is to recover a *sparse* distribution whose marginals are close to the observations. More precisely, we wish to find a distribution $\widehat{\nu}$ such that $||\mathbf{P}(\widehat{\nu})-\mathbf{D}||_2 \leq f(\delta)$ and $||\widehat{\nu}||_0$ is small. Here, ideally we would like $f(\delta) = \delta$ but we may settle for any f such that $f(\delta) \to 0$ as $\delta \to 0$.

Following the line of reasoning in Section 8.4.1, we shall assume that there is a sparse model ν^s with respect to which the marginal matrix satisfies the *signature condition*, Condition 8.1, and $||\mathbf{P}(\nu^s)-\mathbf{D}||_2 \leq \delta$. The goal would be to produce an estimate $\widehat{\nu}$ so that $||\widehat{\nu}||_0 = ||\nu||_0$ and $||\mathbf{P}(\widehat{\nu})-\mathbf{D}||_2 \leq f(\delta)$.

This is the exact analog of the robust recovery of a sparse signal in the context of compressed sensing where the RIP-like condition allowed recovery of a sparse approximation to the original signal from linear projections through linear optimization. The computational complexity of such an algorithm scales at least linearly in n, the ambient dimension of the signal. As discussed earlier, in our context this would lead to the computation cost scaling as $N!$, which is prohibitive. The exact recovery algorithm discussed in Section 8.4.1 has computation cost $O(2^K + N^2 \log N)$ in the context of recovering a sparse model satisfying the signature condition. The brute-force search for the sparse model will lead to cost at least $\binom{N!}{K} \approx (N!)^K$ or $\exp(\Theta(KN \log N))$ for $K \ll N!$. The question is whether it is possible to get rid of the dependence on $N!$, and ideally to achieve a scaling of $O(2^K + N^2 \log N)$, as in the case of exact model recovery.

In what follows, we describe the conditions on noise under which the algorithm described in Section 8.4.1 is robust. This requires an assumption that the underlying ground-truth distribution is sparse and satisfies the signature condition. This recovery result requires noise to be *small*. How to achieve such a recovery in a higher-noise regime remains broadly unknown; initial progress toward it has been made in [43].

Robust recovery under the signature condition: low-noise regime. Recall that the "peeling" algorithm recovers the sparse model when the *signature condition* is satisfied using exact marginals. Here, we discuss the robustness of the "peeling" algorithm under noise. Specifically, we argue that the "peeling" algorithm as described is robust as long as the noise is "low." We formalize this in the statement below.

THEOREM 8.4 *Let $\mathbf{A} \in \{0,1\}^{m\times n}$, with all of its columns being distinct. Let $\mathbf{x} \in \mathbb{R}^n_{\geq 0}$ be such that $\mathbf{A}$ satisfies the signature condition with respect to the set $\mathsf{supp}(\mathbf{x})$. Let the non-zero components of $\mathbf{x}$, i.e., $\{\mathbf{x}_i : i \in \mathsf{supp}(\mathbf{x})\}$, be such that, for any $S_1 \neq S_2 \subset \mathsf{supp}(\mathbf{x})$,*

$$\left|\sum_{i\in S_1} \mathbf{x}_i - \sum_{i'\in S_2} \mathbf{x}_{i'}\right| > 2\delta K, \tag{8.14}$$

for some $\delta > 0$. Then, given $\mathbf{y} = \mathbf{A}\mathbf{x} + \boldsymbol{\eta}$ with $\|\boldsymbol{\eta}\|_\infty < \delta$, it is feasible to find $\widehat{\mathbf{x}}$ so that $\|\widehat{\mathbf{x}} - \mathbf{x}\|_\infty \leq \delta$.

Proof To establish Theorem 8.4, we shall utilize effectively the same algorithm as that utilized for establishing Theorem 8.1 in the proof of Theorem 8.1. However, we will have to deal with the "error" in measurement $\mathbf{y}$ delicately.

To begin with, according to arguments in the proof of Theorem 8.1, it follows that all non-zero elements of $\mathbf{x}$ are distinct. Without loss of generality, let the non-zero elements of $\mathbf{x}$ be $\mathbf{x}_1, \ldots, \mathbf{x}_K$, with $K = |\mathsf{supp}(\mathbf{x})| \geq 2$ such that $0 < \mathbf{x}_1 < \cdots < \mathbf{x}_K$; $\mathbf{x}_i = 0$ for $K+1 \leq i \leq n$. From (8.14), it follows that $\mathbf{x}_{i+1} \geq \mathbf{x}_i + 4\delta$ for $1 \leq i < K$ and $\mathbf{x}_1 \geq 2\delta$. Therefore,

$$\mathbf{x}_k \geq (k-1)4\delta + 2\delta, \tag{8.15}$$

for $1 \leq k \leq K$. Next, we shall argue that, inductively, it is feasible to find $\widehat{\mathbf{x}}_i$, $1 \leq i \leq n$, so that $|\widehat{\mathbf{x}}_i - \mathbf{x}_i| \leq \delta$ for $1 \leq i \leq K$ and $\widehat{\mathbf{x}}_i = 0$ for $K+1 \leq i \leq n$.

Now, since $\mathbf{A} \in \{0,1\}^{m\times n}$, for each $j \in [m]$, $\mathbf{y}_j = \sum_{i\in S_j} \mathbf{x}_i + \boldsymbol{\eta}_j$, with $S_j \subseteq \mathsf{supp}(\mathbf{x})$ and $|\boldsymbol{\eta}_j| \leq \delta$. From (8.14), it follows for $\mathbf{x}_1 > 2\delta$. Therefore, if $S_j \neq \emptyset$ then $\mathbf{y}_j > \delta$. That is, we will start by restricting the discussion to indices $J^1 = \{j \in [m] : \mathbf{y}_j > \delta\}$.

Let j_1 be an index in J such that $\mathbf{y}_{j_1} \in \arg\min_{j\in J^1} \mathbf{y}_j$. We set $\widehat{\mathbf{x}}_1 = \mathbf{y}_{j_1}$ and $\mathbf{A}_{j_1 1} = 1$. To justify this, we next argue that $\mathbf{y}_{j_1} = \mathbf{x}_1 + \boldsymbol{\eta}_{j_1}$ and hence $|\widehat{\mathbf{x}}_1 - \mathbf{x}_1| < \delta$. By virtue of the signature condition, for each $i \in \mathsf{supp}(\mathbf{x})$, there exists $j(i) \in [m]$ such that $|\mathbf{y}_{j(i)} - \mathbf{x}_i| \leq \delta$ and hence $j(i) \in J^1$, since $\mathbf{y}_{j(i)} \geq \mathbf{x}_i - \delta > \delta$. Let $J(1) = \{j \in J^1 : \mathbf{y}_j = \mathbf{x}_1 + \boldsymbol{\eta}_j\}$. Clearly, $J(1) \neq \emptyset$. Effectively, we want to argue that $j_1 \in J(1)$. To that end, suppose that this does not hold. Then there exists $S \subset \mathsf{supp}(\mathbf{x})$, such that $S \neq \emptyset$, $S \neq \{1\}$, and $\mathbf{y}_{j_1} = \sum_{i\in S} \mathbf{x}_i + \boldsymbol{\eta}_{j_1}$. Then

$$\begin{aligned}
\mathbf{y}_{j_1} &> \sum_{i\in S} \mathbf{x}_i - \delta, \quad \text{since } |\boldsymbol{\eta}_{j_1}| < \delta, \\
&> \mathbf{x}_1 + 2K\delta - \delta, \quad \text{by (8.14)}, \\
&\geq \mathbf{x}_1 + \delta, \quad \text{since } K \geq 1, && (8.16) \\
&\geq \mathbf{y}_j, && (8.17)
\end{aligned}$$

for any $j \in J(1) \subset J^1$. But this is a contradiction, since $\mathbf{y}_{j_1} \leq \mathbf{y}_j$ for all $j \in J$. That is, $S = \{1\}$ or $\mathbf{y}_{j_1} = \mathbf{x}_1 + \boldsymbol{\eta}_{j_1}$. Thus, we have found $\widehat{\mathbf{x}}_1 = \mathbf{y}_{j_1}$ such that $|\widehat{\mathbf{x}}_1 - \mathbf{x}_1| < \delta$.

Now, for any $j \in J^1$ with $\mathbf{y}_j = \sum_{i\in S} \mathbf{x}_i + \boldsymbol{\eta}_j$ with $S \cap \{2, \ldots, K\} \neq \emptyset$, we have, with the notation $\mathbf{x}(S) = \sum_{i\in S} \mathbf{x}_i$,

$$\begin{aligned}
|\widehat{\mathbf{x}}_1 - \mathbf{y}_j| &= |\mathbf{x}_1 - \mathbf{y}_j + \widehat{\mathbf{x}}_1 - \mathbf{x}_1| \\
&= |\mathbf{x}_1 - \mathbf{x}(S) - \boldsymbol{\eta}_j + \widehat{\mathbf{x}}_1 - \mathbf{x}_1|
\end{aligned}$$

$$\geq |\mathbf{x}_1 - \mathbf{x}(S)| - |\boldsymbol{\eta}_j| - |\widehat{\mathbf{x}}_1 - \mathbf{x}_1|$$
$$> 2K\delta - \delta - \delta$$
$$\geq 2\delta. \tag{8.18}$$

Furthermore, if $S = \{1\}$, then $|\widehat{\mathbf{x}}_1 - \mathbf{y}_j| < 2\delta$. Therefore, we set

$$\mathbf{A}_{j1} = 1 \quad \text{if} \quad |\widehat{\mathbf{x}}_1 - \mathbf{y}_j| < 2\delta, \text{ for } j \in J^1,$$

and

$$J^2 \leftarrow J^1 \backslash \{j \in J^1 : |\mathbf{y}_j - \widehat{\mathbf{x}}_1| < 2\delta\}.$$

Clearly,

$$j \in J^2 \;\Leftrightarrow\; j \in [m],\ \mathbf{y}_j = x(S) + \boldsymbol{\eta}_j, \text{ such that } S \cap \{2, \ldots, K\} \neq \emptyset.$$

Now suppose, inductively, that we have found $\widehat{\mathbf{x}}_1, \ldots, \widehat{\mathbf{x}}_k$, $1 \leq k < K$, so that $|\widehat{\mathbf{x}}_i - \mathbf{x}_i| < \delta$ for $1 \leq i \leq k$ and

$$j \in J^{k+1} \;\Leftrightarrow j \in [m],\ \mathbf{y}_j = \mathbf{x}(S) + \boldsymbol{\eta}_j, \text{ such that } S \cap \{k+1, \ldots, K\} \neq \emptyset.$$

To establish the inductive step, we suggest that one should set $j_{k+1} \in \arg\min_{j \in J^{k+1}} \{\mathbf{y}_j\}$, $\widehat{\mathbf{x}}_{k+1} = \mathbf{y}_{j_{k+1}}$ and

$$J^{k+2} \leftarrow J^{k+1} \backslash \{j \in J^{k+1} : |\mathbf{y}_j - \widehat{\mathbf{x}}_1| < (k+1)\delta\}.$$

We shall argue that $|\widehat{\mathbf{x}}_{k+1} - \mathbf{x}_{k+1}| < \delta$ by showing that $\mathbf{y}_{j_{k+1}} = \mathbf{x}_{k+1} + \boldsymbol{\eta}_{j_{k+1}}$ and establishing

$$j \in J^{k+2} \;\Leftrightarrow\; j \in [m],\ \mathbf{y}_j = \mathbf{x}(S) + \boldsymbol{\eta}_j, \text{ such that } S \cap \{k+2, \ldots, K\} \neq \emptyset.$$

To that end, let $\mathbf{y}_{j_{k+1}} = \mathbf{x}(S) + \boldsymbol{\eta}_{j_{k+1}}$. By the inductive hypothesis, $S \subset \mathsf{supp}(\mathbf{x})$ and $S \cap \{k+1, \ldots, K\} \neq \emptyset$. Suppose $S \neq \{k+1\}$. Then

$$\begin{aligned}
\mathbf{y}_{j_{k+1}} &> \mathbf{x}(S) - \delta \\
&= (\mathbf{x}(S) - \mathbf{x}_{k+1}) + \mathbf{x}_{k+1} - \delta \\
&\geq 2\delta + \mathbf{x}_{k+1} - \delta \\
&= \delta + \mathbf{x}_{k+1}, \\
&> \mathbf{y}_j,
\end{aligned}$$

for any $j \in J(k+1) \equiv \{j \in J^{k+1} : \mathbf{y}_j = \mathbf{x}_{k+1} + \boldsymbol{\eta}_j\}$. In the above, we have used the fact that, since $S \cap \{k+1, \ldots, K\} \neq \emptyset$ and $S \neq \{k+1\}$, it must hold that $\mathbf{x}(S) \geq \min\{\mathbf{x}_1 + \mathbf{x}_{k+1}, \mathbf{x}_{k+2}\}$. In either case, using (8.15), it follows that $\mathbf{x}(S) - \mathbf{x}_{k+1} \geq 2\delta$. We note that, due to the signature condition and the inductive hypothesis about J^{k+1}, it follows that $J(k+1) \neq \emptyset$. But $\mathbf{y}_{j_{k+1}}$ is the minimal value of $\mathbf{y}_j$ for $j \in J^{k+1}$. This is a contradiction. Therefore, $S = \{k+1\}$. That is, $\widehat{\mathbf{x}}_{k+1} = \mathbf{y}_{j_{k+1}}$ satisfies $|\widehat{\mathbf{x}}_{k+1} - \mathbf{x}_{k+1}| < \delta$.

Now, consider any set $S \subset \{1, \ldots, k+1\}$ and any $j \in J^{k+2}$ such that $\mathbf{y}_j = \sum_{i \in S'} \mathbf{x}_i + \boldsymbol{\eta}_j$ with $S' \cap \{k+2, \ldots, K\} \neq \emptyset$. Using the notation $\widehat{\mathbf{x}}(S) = \sum_{i \in S} \widehat{\mathbf{x}}_i$, we have

$$\begin{aligned}
|\widehat{\mathbf{x}}(S) - \mathbf{y}_j| &= |\mathbf{x}(S) - \mathbf{y}_j + \widehat{\mathbf{x}}(S) - \mathbf{x}(S)| \\
&= |\mathbf{x}(S) - \mathbf{x}(S') - \boldsymbol{\eta}_j + \widehat{\mathbf{x}}(S) - \mathbf{x}(S)|
\end{aligned}$$

$$
\begin{aligned}
&\geq |\mathbf{x}(S) - \mathbf{x}(S')| - |\boldsymbol{\eta}_j| - |\widehat{\mathbf{x}}(S) - \mathbf{x}(S)| \\
&> 2K\delta - (1 + |S|)\delta \\
&\geq (|S| + 1)\delta, \qquad (8.19)
\end{aligned}
$$

where we used the fact that $|S| + 1 \leq K$. Therefore, if we set

$$J^{k+2} \leftarrow J^{k+1} \backslash \{j \in J^{k+1} : |\nu y_j - \widehat{\mathbf{x}}(S)| \leq (|S| + 1)\delta, \text{ for some } S \subset \{1, \ldots, k+1\}\},$$

it follows that

$$j \in J^{k+2} \Leftrightarrow j \in [m], \ \mathbf{y}_j = \mathbf{x}(S) + \boldsymbol{\eta}_j, \text{ such that } S \cap \{k+2, \ldots, K\} \neq \emptyset.$$

This completes the induction step. It also establishes the desired result that we can recover $\widehat{\mathbf{x}}$ such that $\|\widehat{\mathbf{x}} - \mathbf{x}\|_\infty \leq \delta$.

Naturally, as before, Theorem 8.4 implies *robust* versions of Theorems 8.2 and 8.3. In particular, if we are forming an empirical estimation of $\mathbf{M}(\nu)$ or $\mathbf{C}(\nu)$ that is based on independently drawn samples, then simple application of the Chernoff bound along with a union bound will imply that it may be sufficient to have samples that scale as $\delta^{-2} \log N$ in order to have $\widehat{\mathbf{M}}(\nu)$ or $\widehat{\mathbf{C}}(\nu)$ so that $\|\widehat{\mathbf{M}}(\nu) - \mathbf{M}(\nu)\|_\infty < \delta$ or $\|\widehat{\mathbf{C}}(\nu) - \mathbf{C}(\nu)\|_\infty < \delta$ with high probability (i.e., $1 - o_N(1)$). Then, as long as ν satisfies condition (8.14) in addition to the signature condition, Theorem 8.4 guarantees approximate recovery as discussed above.

Robust recovery under the signature condition: high-noise regime. Theorem 8.4 provides conditions under which the "peeling" algorithm manages to recover the distribution as long as the elements in the support are far enough apart. Putting it another way, for a given $\mathbf{x}$, the error tolerance needs to be *small enough* compared with the *gap* that is implicitly defined by (8.14) for recovery to be feasible.

Here, we make an attempt to go beyond such restrictions. In particular, supposing we can view observations as a noisy version of a model that satisfies the *signature condition*, then we will be content if we can recover *any* model that satisfies the *signature condition* and is consistent with observations (up to noise). For this, we shall assume knowledge of the sparsity K.

Now, we need to learn $\mathsf{supp}(\mathbf{x})$, i.e., positions of $\mathbf{x}$ that are non-zero and the non-zero values at those positions. The determination of $\mathsf{supp}(\mathbf{x})$ corresponds to selecting the columns of $\mathbf{A}$. Now, if $\mathbf{A}$ satisfies the signature condition with respect to $\mathsf{supp}(\mathbf{x})$, then we can simply choose the entries in the positions of $\mathbf{y}$ corresponding to the signature component. If the choice of $\mathsf{supp}(\mathbf{x})$ is correct, then this will provide an estimate $\widehat{\mathbf{x}}$ so that $\|\widehat{\mathbf{x}} - \mathbf{x}\|_2 \leq \delta$. In general, if we assume that there exists x such that $\mathbf{A}$ satisfies the signature condition with respect to $\mathsf{supp}(\mathbf{x})$ with $K = \|\mathbf{x}\|_0$ and $\|\mathbf{y} - \mathbf{A}\mathbf{x}\|_2 \leq \delta$, then a suitable approach would be to find $\widehat{\mathbf{x}}$ such that $\|\widehat{\mathbf{x}}\|_0 = K$, $\mathbf{A}$ satisfies the signature condition with respect to $\mathsf{supp}(\widehat{x})$, and it minimizes $\|\mathbf{y} - \mathbf{A}\widehat{\mathbf{x}}\|_2$.

In summary, we are solving a combinatorial optimization problem over the space of columns of $\mathbf{A}$ that collectively satisfy the signature condition. Formally, the space of subsets of columns of $\mathbf{A}$ of size K can be encoded through a binary-valued matrix $\mathbf{Z} \in \{0,1\}^{m \times m}$ as follows: all but K columns of $\mathbf{Z}$ are zero, and the non-zero columns of $\mathbf{Z}$ are distinct columns of A collectively satisfying the signature condition. More precisely, for any $1 \le i_1 < i_2 < \cdots < i_K \le m$, representing the signature columns, the variable $\mathbf{Z}$ should satisfy

$$\mathbf{Z}_{i_j i_j} = 1, \text{ for } 1 \le j \le K, \tag{8.20}$$

$$\mathbf{Z}_{i_j i_k} = 0, \text{ for } 1 \le j \ne k \le K, \tag{8.21}$$

$$[\mathbf{Z}_{a i_j}]_{a \in [m]} \in \mathsf{col}(\mathbf{A}), \text{ for } 1 \le j \le K, \tag{8.22}$$

$$\mathbf{Z}_{ab} = 0, \text{ for } a \in [m],\ b \notin \{i_1, \ldots, i_K\}. \tag{8.23}$$

In the above, $\mathsf{col}(\mathbf{A}) = \{[\mathbf{A}_{ij}]_{i \in [m]} : 1 \le j \le n\}$ represents the set of columns of matrix $\mathbf{A}$. Then the optimization problem of interest is

$$\begin{aligned} &\mathsf{minimize}\ \|\mathbf{y} - \mathbf{Z}\mathbf{y}\|_2 \quad \mathsf{over} \quad \mathbf{Z} \in \{0,1\}^{m \times m} \\ &\mathsf{such\ that}\ \mathbf{Z}\ \mathsf{satisfies\ constraints}\ (8.20) - (8.23). \end{aligned} \tag{8.24}$$

The constraint set (8.20)–(8.23) can be viewed as a disjoint union of $\binom{m}{K}$ sets, each one corresponding to a choice of $1 \le i_1 < \cdots < i_K \le m$. For each such choice, we can solve the optimization (8.24) and choose the best solution across all of them. That is, the computation cost is $O(m^K)$ times the cost of solving the optimization problem (8.24). The complexity of solving (8.24) fundamentally depends on the constraint (8.22) – it captures the structural complexity of describing the column set of matrix $\mathbf{A}$.

A natural convex relaxation of the optimization problem (8.24) involves replacing (8.22) and $\mathbf{Z} \in \{0,1\}^{m \times m}$ by

$$[\mathbf{Z}_{a i_j}]_{a \in [m]} \in \mathsf{convex\text{-}hull}(\mathsf{col}(\mathbf{A})), \text{ for } 1 \le j \le K; \quad \mathbf{Z} \in [0,1]^{m \times m}. \tag{8.25}$$

In the above, for any set S,

$$\mathsf{convex\text{-}hull}(S) \equiv \left\{ \sum_{\ell=1}^{Q} a_\ell \mathbf{x}_\ell : a_\ell \ge 0, \mathbf{x}_\ell \in S \text{ for } \ell \in [Q],\ \sum_\ell a_\ell = 1, \text{ for } Q \ge 2 \right\}.$$

In the best case, it may be feasible to solve the optimization with the convex relaxation efficiently. However, the relaxation may not yield a solution that is achieved at the extreme points of convex-hull(col($\mathbf{A}$)), which is what we desire. This is due to the fact that the objective ℓ_2-norm of the error we are considering is strictly convex. To overcome this challenge, we can replace ℓ_2 by ℓ_∞. The constraints of interest are, for a given $\varepsilon > 0$,

$$\mathbf{Z}_{i_j i_j} = 1, \text{ for } 1 \le j \le K, \tag{8.26}$$

$$\mathbf{Z}_{i_j i_k} = 0, \text{ for } 1 \le j \ne k \le K, \tag{8.27}$$

$$\mathbf{y}_i - (\mathbf{Z}\mathbf{y})_i \le \varepsilon, \text{ for } 1 \le i \le m, \tag{8.28}$$

$$\mathbf{y}_i - (\mathbf{Z}\mathbf{y})_i \ge -\varepsilon, \text{ for } 1 \le i \le m, \tag{8.29}$$

$$[\mathbf{Z}_{a i_j}]_{a \in [m]} \in \mathsf{convex\text{-}hull}(\mathsf{col}(\mathbf{A})), \text{ for } 1 \le j \le K, \tag{8.30}$$

$$\mathbf{Z}_{ab} = 0, \text{ for } a \in [m],\ b \notin \{i_1, \ldots, i_K\}. \tag{8.31}$$

This results in the linear program

$$\mathsf{minimize} \sum_{i,j=1}^{m} \zeta_{ij} \mathbf{Z}_{ij} \quad \mathsf{over} \quad \mathbf{Z} \in [0,1]^{m \times m}$$

$$\text{such that} \quad \mathbf{Z} \text{ satisfies constraints (8.26)} - (8.31). \tag{8.32}$$

In the above, $\zeta = [\zeta_{ij}] \in [0,1]^{m\times m}$ is a random vector with each of its components chosen by drawing a number from $[0,1]$ uniformly at random. The purpose of choosing ζ is to obtain a unique solution, if it is feasible. Note that, when this is feasible, the solution is achieved at the extreme point, which happens to be the valid solution of interest. We can solve (8.32) iteratively for choices of $\varepsilon = 2^{-q}$ for $q \geq 0$ until we fail to find a feasible solution. The value of ε before that will be the smallest (within a factor of 2) error tolerance that is feasible within the signature family. Therefore, the cost of finding such a solution is within $O(\log 1/\varepsilon)$ times the cost of solving the linear program (8.32). The cost of the linear program (8.32) depends on the complexity of the convex relaxation of the set col($\mathbf{A}$). If it is indeed simple enough, then we can solve (8.32) efficiently.

As it turns out, for the case of *first-order* marginals, the convex hull of col($\mathbf{A}$) is succinct due to the classical result by Birkhoff and von Neumann which characterizes the convex relaxation of permutation matrices through a number of equalities that is linear in the size of the permutation, here N. Each of the elements in col($\mathbf{A}$) corresponds to a (flattened) permutation matrix. Therefore, its convex hull is simply that of (flattened) permutation matrices, leading to a succinct description. This results in a polynomial-time algorithm for solving (8.32). In summary, we conclude the following (see [43] for details).

THEOREM 8.5 *For a given observation vector* $\mathbf{y} \in [0,1]^m$*, if there exists a distribution* μ *in a* signature familty *of support size* K *such that the corresponding projection is within* $\varepsilon \in (0,1]$ *of* $\mathbf{y}$ *in terms of the* ℓ_∞*-norm, then it can be found through an algorithm with computation cost* $O(N^{\Theta(K)} \log(1/\varepsilon))$.

Open question. It is not known how to perform an efficient computation equivalent to Theorem 8.5 (or whether one can do so) for finding the approximate distribution in the signature family for the pair-wise comparison marginals.

On the universality of the signature family. Thus far, we have focused on developing algorithms for learning a sparse model with a signature condition. The sparse model is a natural approximation for a generic distribution over permutations. In Theorems 8.2 and 8.3, we effectively argued that a model with randomly chosen sparse support satisfies the signature condition as long as the sparsity is not too large. However, it is not clear whether a sparse model with a signature condition is a good approximation beyond such a setting. For example, is there a sparse model with a signature condition that approximates the marginal information of a simple parametric model such as the multinomial logit (MNL) model well?

To that end, recently Farias *et al.* [43] have established the following representation result which we state without proof here.

THEOREM 8.6 *Let* ν *be an MNL model with parameters* $w_1, \ldots, w_N$ *(and, without loss of generality, let* $0 < w_1 < \cdots < w_N$*) such that*

$$\frac{w_N}{\sum_{k=1}^{N-L} w_k} \leq \frac{\sqrt{\log N}}{N}, \tag{8.33}$$

for some $L = N^{\delta}$ *for some* $\delta \in (0,1)$. *Then there exists* $\widehat{\nu}$ *such that* $|\mathrm{supp}(\widehat{\nu})| = O(N/\varepsilon^2)$, $\widehat{\nu}$ *satisfies the signature condition with respect to the first-order marginals, and* $\|\mathbf{M}(\nu) - \mathbf{M}(\widehat{\nu})\|_2 \leq \varepsilon$.

8.5 Random Utility Model (RUM)

We discuss recovery of an exact model for the MNL model and recovery of the ranking for a generic random utility model with homogeneous random perturbation.

8.5.1 Exact Marginals: Infinite Samples

Given the exact marginal information $\mathbf{M}(\nu)$ or $\mathbf{C}(\nu)$ for ν, we wish to recover the parameters of the model when ν is MNL, and we wish to recover the ranking when ν is a generic random utility model. We first discuss recovery of the MNL model for both types of marginal information and then discuss recovery of ranking for a generic model.

Recovering MNL: first-order marginals. Without loss of generality, let us assume that the parameters $w_1, \ldots, w_N$ are normalized so that $\sum_i w_i = 1$. Then, under the MNL model according to (8.9),

$$\mathbb{P}(\sigma(i) = 1) = w_i. \tag{8.34}$$

That is, the first column of the first-order marginal matrix $\mathbf{M}(\nu) = [\mathbf{M}_{ij}(\nu)]$ precisely provides the parameters of the MNL model!

Recovering the MNL model: comparison marginals. Under the MNL model, according to (8.11), for any $i \neq j \in [N]$,

$$\mathbb{P}(\sigma(i) > \sigma(j)) = \frac{w_i}{w_i + w_j}. \tag{8.35}$$

The comparison marginals, $\mathbf{C}(\nu)$ provide access to $\mathbb{P}(\sigma(i) > \sigma(j))$ for all $i \neq j \in [N]$. Using these, we wish to recover the parameters $w_1, \ldots, w_N$.

Next, we describe a reversible Markov chain over N states whose stationary distribution is precisely the parameters of our interest, and its transition kernel utilizes the $\mathbf{C}(\nu)$. This alternative representation provides an intuitive algorithm for recovering the MNL parameters, and more generally what is known as the *rank centrality* [44, 45].

To that end, the Markov chain of interest has N states. The transition kernel or transition probability matrix $\mathbf{Q} = [\mathbf{Q}_{ij}] \in [0,1]^{N \times N}$ of the Markov chain is defined using comparison marginals $\mathbf{C} = \mathbf{C}(\nu)$ as follows:

$$\mathbf{Q}_{ij} = \begin{cases} \mathbf{C}_{ji}/2N, & \text{if } i \neq j, \\ 1 - \sum_{j \neq i} \mathbf{C}_{ji}/2N, & \text{if } i = j. \end{cases} \tag{8.36}$$

The Markov chain has a unique stationary distribution because (a) $\mathbf{Q}$ is irreducible, since $\mathbf{C}_{ij}, \mathbf{C}_{ji} > 0$ for all $i \neq j$ as long as $w_i > 0$ for all $i \in [N]$, and (b) $\mathbf{Q}_{ii} > 0$ by definition for all $i \in [N]$ and hence it is aperiodic. Further, $\mathbf{w} = [w_i]_{i\in[N]} \in [0,1]^N$ is a stationary distribution since it satisfies the detailed-balance condition, i.e., for any $i \neq j \in [N]$,

$$\begin{aligned} w_i \mathbf{Q}_{ij} = w_i \frac{\mathbf{C}_{ji}}{2N} &= w_i \frac{w_j}{2N(w_i + w_j)} \\ = w_j \frac{w_i}{2N(w_i + w_j)} &= w_j \frac{\mathbf{C}_{ij}}{2N} \\ = w_j \mathbf{Q}_{ji}. \end{aligned} \tag{8.37}$$

Thus, by finding the stationary distribution of a Markov chain as defined above, we can find the parameters of the MNL model. This boils down to finding the largest eigenvector of $\mathbf{Q}$, which can be done using various efficient algorithms, including the standard power-iteration method.

We note that the algorithm employed to find the parameters of the MNL model does not need to have access to *all* entries of $\mathbf{C}$. Suppose $E \subset \{(i,j) : i \neq j \in [N]\}$ to be a subset of all possible $\binom{N}{2}$ pairs for which we have access to $\mathbf{C}$. Let us define a Markov chain with $\mathbf{Q}$ such that, for $i \neq j \in [N]$, $\mathbf{Q}_{ij}$ is defined according to (8.36) if $(i,j) \in E$ (we assume $(i,j) \in E$ and then $(j,i) \in E$ because $\mathbf{C}_{ji} = 1 - \mathbf{C}_{ij}$ by definition), else $\mathbf{Q}_{ij} = 0$; and $\mathbf{Q}_{ii} = 1 - \sum_{j \neq i} \mathbf{Q}_{ij}$. The resulting Markov chain is aperiodic, since by definition $\mathbf{Q}_{ii} > 0$. Therefore, as long as the resulting Markov chain is irreducible, it has a unique stationary distribution. Now the Markov chain is irreducible if effectively all N states are reachable from each other via transitions $\{(i,j),(j,i) : (i,j) \in E\}$. That is, there are data that compare any two $i \neq j \in [N]$ through, potentially, chains of comparisons, which, in a sense, is a minimal requirement in order to have consistent ranking across all $i \in [N]$. Once we have this, again it follows that the stationary distribution is given by $\mathbf{w} = [w_i]_{i\in[N]} \in [0,1]^N$, since the detailed-balance equation (8.37) holds for all $i \neq j \in [N]$ with $(i,j) \in E$.

Recovering ranking for a homogeneous RUM. As mentioned in Section 8.3.2, we wish to recover the ranking or ordering of inherent utilities for a homogeneous RUM. That is, if $u_1 \geq \cdots \geq u_N$, then the ranking of interest is identity, i.e., $\sigma \in S_N$ such that $\sigma(i) = i$ for all $i \in [N]$. Recall that the homogeneous RUM random perturbations ε_i in (8.7) have an identical distribution for all $i \in [N]$. We shall assume that the distribution of the random perturbation is absolutely continuous with respect to the Lebesgue measure on $\mathbb{R}$. Operationally, for any $t_1 < t_2 \in \mathbb{R}$,

$$\mathbb{P}(\varepsilon_1 \in (t_1, t_2)) > 0. \tag{8.38}$$

The following is the key characterization of a homogeneous RUM with (8.38) that will enable recovery of ranking from marginal data (both comparison and first-order); also see [46, 47].

LEMMA 8.4 *Consider a homogeneous RUM with property* (8.38). *Then, for* $i \neq j \in [N]$,

$$u_i > u_j \Leftrightarrow \mathbb{P}(Y_i > Y_j) > \frac{1}{2}. \tag{8.39}$$

Further, for any $k \neq i, j \in [N]$,

$$u_i > u_j \Leftrightarrow \mathbb{P}(Y_i > Y_k) > \mathbb{P}(Y_j > Y_k). \tag{8.40}$$

Proof By definition,

$$\mathbb{P}(Y_i > Y_j) = \mathbb{P}(\varepsilon_i - \varepsilon_j > u_j - u_i). \tag{8.41}$$

Since $\varepsilon_i, \varepsilon_j$ are independent and identically distributed (i.i.d.) with propert (8.38), their difference random variable $\varepsilon_i - \varepsilon_j$ has 0 mean, symmetric and with property (8.38). That is, 0 is its unique median as well, and, for any $t > 0$,

$$\mathbb{P}(\varepsilon_i - \varepsilon_j > t) = \mathbb{P}(\varepsilon_i - \varepsilon_j < -t) < \frac{1}{2}. \tag{8.42}$$

This leads to the conclusion that

$$u_i > u_j \Leftrightarrow \mathbb{P}(Y_i > Y_j) > \frac{1}{2}.$$

Similarly,

$$\mathbb{P}(Y_i > Y_k) = \mathbb{P}(\varepsilon_i - \varepsilon_k > u_k - u_i), \tag{8.43}$$

$$\mathbb{P}(Y_j > Y_k) = \mathbb{P}(\varepsilon_j - \varepsilon_k > u_k - u_j). \tag{8.44}$$

Now $\varepsilon_i - \varepsilon_k$ and $\varepsilon_j - \varepsilon_k$ are identically distributed with property (8.38). That is, one has a strictly monotonically increasing cumulative distribution function (CDF). Therefore, (8.40) follows immediately.

Recovering the ranking: comparison marginals. From (8.39) of Lemma 8.4, using comparison marginals $C(v)$, we can recover a ranking of $[N]$ that corresponds to the ranking of their inherent utility for a generic homogeneous RUM as follows. For each $i \in [N]$, assign rank as

$$\mathsf{rank}(i) = N - \left|\left\{j \in [N] : j \neq i, \, \mathbf{C}_{ij} > \frac{1}{2}\right\}\right|. \tag{8.45}$$

From Lemma 8.4, it immediately follows that the rank provides the ranking of $[N]$ as desired.

We also note that (8.40) of Lemma 8.4 suggests an alternative way (which will turn out to be robust and more useful) to find the same ranking. To that end, for each $i \in [N]$, define the score as

$$\mathsf{score}(i) = \frac{1}{N-1} \sum_{k \neq i} \mathbf{C}_{ik}. \tag{8.46}$$

From (8.40) of Lemma 8.4, it follows that, for any $i \neq j \in [N]$,

$$\mathsf{score}(i) > \mathsf{score}(j) \Leftrightarrow u_i > u_j. \tag{8.47}$$

That is, by ordering $[N]$ in decreasing order of score values, we obtain the desired ranking.

Recovering the ranking: first-order marginals. We are given a first-order marginal data matrix, $M = M(v) \in [0,1]^{N \times N}$, where the M_{ij} represent $\mathbb{P}(\sigma(i) = j)$ under v for $i, j \in [N]$. To recover the ranking under a generic homogeneous RUM using M, we shall introduce the notion of the *Borda count*, see [48]. More precisely, for any $i \in [N]$,

$$\mathsf{borda}(i) = \mathbb{E}[\sigma(i)] = \sum_{j \in [N]} \mathbb{P}(\sigma(i) = j) j = \sum_{j \in [N]} j M_{ij}. \tag{8.48}$$

That is, borda(i) can be computed using $\mathbf{M}$ for any $i \in [N]$. Recall that we argued earlier that the score($\cdot$) (in decreasing order) provides the desired ordering or ranking of $[N]$. However, computing score required access to comparison marginals $\mathbf{C}$. And it's not feasible to recover $\mathbf{C}$ from $\mathbf{M}$.

On the other hand, intuitively it seems that borda (in increasing order) provides an ordering over $[N]$ that might be what we want. Next, we state a simple invariant that ties score(i) and borda(i), which will lead to the conclusion that we can recover the desired ranking by sorting $[N]$ in increasing order of borda count [46, 47].

LEMMA 8.5 *For any $i \in [N]$ and any distribution over permutations,*

$$\text{borda}(i) + (N-1)\text{score}(i) = N. \tag{8.49}$$

Proof Consider any permutation $\sigma \in S_N$. For any $i \in [N]$, $\sigma(i)$ denotes the position in $[N]$ that i is ranked to. That is, $N - \sigma(i)$ is precisely the number of elements in $[N]$ (and not equal to i) that are ranked below i. Formally,

$$N - \sigma(i) = \sum_{j \neq i} \mathbf{1}(\sigma(i) > \sigma(j)). \tag{8.50}$$

Taking the expectation on both sides with respect to the underlying distribution over permutations and rearranging terms, we obtain

$$N = \mathbb{E}[\sigma(i)] + \sum_{j \neq i} \mathbb{P}(\sigma(i) > \sigma(j)). \tag{8.51}$$

Using definitions from (8.46) and (8.48), we have

$$N = \text{borda}(i) + (N-1)\text{score}(i). \tag{8.52}$$

This completes the proof.

8.5.2 Noisy Marginals: Finite Samples

Now we consider a setup where we have access to marginal distributions formed on the basis of an empirical distribution of finite samples from the underlying distribution. This can be viewed as access to a "noisy" marginal distribution. Specifically, given a distribution ν, we observe $\mathbf{D} = \mathbf{P}(\nu) + \boldsymbol{\eta}$, where $\mathbf{P}(\nu) \in \{\mathbf{M}(\nu), \mathbf{C}(\nu)\}$ depending upon the type of marginal information, with $\boldsymbol{\eta}$ being noise such that $\|\boldsymbol{\eta}\|_2 \leq \delta$, with $\delta > 0$ being small if we have access to enough samples. Here δ represents the error observed due to access to finitely many samples and is assumed to be known.

As before, we wish to recover the parameters of the model when ν is MNL, and we wish to recover the ranking when ν is a generic RUM. We first discuss recovery of the MNL model for both types of marginal information and then discuss recovery of the ranking for the generic model.

Recovering the MNL model: first-order marginals. As discussed in Section 8.5.1, we can recover the parameters of the MNL model, $\mathbf{w} = [w_i]_{i \in [N]}$, using the first row of the first-order marginal matrix, $\mathbf{M} = \mathbf{M}(\nu)$, by simply setting $w_i = \mathbf{M}_{i1}$. Since we have access to $\mathbf{M}_{i1} + \boldsymbol{\eta}_{i1}$, a simple estimator is to set $\widehat{w}_i = \mathbf{M}_{i1} + \boldsymbol{\eta}_{i1}$. Then it follows that

$$\|\widehat{\mathbf{w}} - \mathbf{w}\|_2 = \|\boldsymbol{\eta}_{\cdot 1}\|_2 \leq \|\boldsymbol{\eta}\|_2 \leq \delta. \tag{8.53}$$

That is, using the same algorithm for estimating a parameter as in the case of access to the exact marginals, we obtain an estimator which seems to have reasonably good properties.

Recovering the MNL model: comparison marginals. We shall utilize the noisy comparison data to create a Markov chain as in Section 8.5.1. The stationary distribution of this noisy or perturbed Markov chain will be a good approximation of the original Markov chain, i.e., the true MNL parameters. This will lead to a good estimator of the MNL model using noisy comparison data.

To that end, we have access to noisy comparison marginals $\widehat{\mathbf{C}} = \mathbf{C} + \boldsymbol{\eta}$. To keep things generic, we shall assume that we have access to comparisons for a subset of pairs. Let $E \subset \{(i,j) : i \neq j \in [N]\}$ denote the subset of all possible $\binom{N}{2}$ pairs for which we have access to noisy comparison marginals, and we shall assume that $(i,j) \in E$ iff $(j,i) \in E$. Define $d_i = \{j \in [N] : j \neq i, (i,j) \in E\}$ and $d_{\max} = \max_i d_i$. Define a noisy Markov chain with transition matrix $\widehat{\mathbf{Q}}$ as

$$\widehat{\mathbf{Q}}_{ij} = \begin{cases} \widehat{\mathbf{C}}_{ji}/2d_{\max} & \text{if } i \neq j,\ (i,j) \in E, \\ 1 - (1/2d_{\max}) \sum_{j:(i,j)\in E} \widehat{\mathbf{C}}_{ji} & \text{if } i = j, \\ 0, & \text{if } (i,j) \notin E. \end{cases} \tag{8.54}$$

We shall assume that E is such that the resulting Markov chain with transition matrix $\widehat{\mathbf{Q}}$ is irreducible; it is aperiodic since $\widehat{\mathbf{Q}}_{ii} > 0$ for all $i \in [N]$ by definition (8.54). As before, it can be verified that this noisy Markov chain is reversible and has a unique stationary distribution that satisfies the detailed-balance condition. Let $\widehat{\pi}$ denote this stationary distribution. The corresponding ideal Markov chain has transition matrix $\mathbf{Q}$ defined as

$$\mathbf{Q}_{ij} = \begin{cases} \mathbf{C}_{ji}/2d_{\max} & \text{if } i \neq j,\ (i,j) \in E, \\ 1 - (1/2d_{\max}) \sum_{j:(i,j)\in E} \mathbf{C}_{ji} & \text{if } i = j, \\ 0, & \text{if } (i,j) \notin E. \end{cases} \tag{8.55}$$

It is reversible and has a unique stationary distribution $\pi = \mathbf{w}$. We want to bound $\|\widehat{\pi} - \pi\|$.

By definition of $\widehat{\pi}$ being the stationary distribution of $\widehat{\mathbf{Q}}$, we have that

$$\widehat{\pi}^{\mathrm{T}} \widehat{\mathbf{Q}} = \widehat{\pi}^{\mathrm{T}}. \tag{8.56}$$

We can find $\widehat{\pi}$ using a power-iteration algorithm. More precisely, let $\nu_0 \in [0,1]^N$ be a probability distribution as our initial guess. Iteratively, for iteration $t \geq 0$,

$$\nu_{t+1}^{\mathrm{T}} = \nu_t^{\mathrm{T}} \widehat{\mathbf{Q}}. \tag{8.57}$$

We make the following claim, see [44, 45].

LEMMA 8.6 *For any $t \geq 1$,*

$$\frac{\|\nu_t - \pi\|}{\|\pi\|} \leq \left(\rho^t \frac{\|\nu_0 - \pi\|}{\|\pi\|} + \frac{1}{1-\rho} \|\Delta\|_2 \right) \sqrt{\frac{\pi_{\max}}{\pi_{\min}}}. \tag{8.58}$$

Here $\Delta = \widehat{\mathbf{Q}} - \mathbf{Q}$, $\pi_{\max} = \max_i \pi_i$, $\pi_{\min} = \min_i \pi_i$; *let* $\lambda_{\max}(\mathbf{Q})$ *be the second largest (in norm) eigenvalue of* $\mathbf{Q}$; $\rho = \lambda_{\max}(\mathbf{Q}) + \|\Delta\|_2 \sqrt{\pi_{\max}/\pi_{\min}}$; *and it is assumed that* $\rho < 1$.

Before we provide the proof of this lemma, let us consider its implications. It is quantifying the robustness of our approach for identifying the parameters of the MNL model using comparison data. Specifically, since $\lim_{t\to\infty} \nu_t \to \widehat{\pi}$, (8.58) implies

$$\frac{\|\widehat{\pi} - \pi\|}{\|\pi\|} \leq \frac{1}{1-\rho}\|\Delta\|_2 \sqrt{\frac{\pi_{\max}}{\pi_{\min}}}. \tag{8.59}$$

Thus, the operator or spectral norm of the perturbation matrix $\Delta = \widehat{\mathbf{Q}} - \mathbf{Q}$ determines the error in our ability to learn parameters using the above-mentioned *rank centrality* algorithm. By definition

$$|\widehat{\mathbf{Q}}_{ij} - \mathbf{Q}_{ij}| \leq \begin{cases} \left|\widehat{\mathbf{C}}_{ij} - \mathbf{C}_{ji}\right|/2d_{\max} & \text{if } i \neq j,\ (i,j) \in E, \\ (1/2d_{\max})\left|\sum_{j:(i,j)\in E}(\widehat{\mathbf{C}}_{ij} - \mathbf{C}_{ji})\right| & \text{if } i = j, \\ 0 & \text{if } (i,j) \notin E. \end{cases} \tag{8.60}$$

Therefore, it follows that

$$\begin{aligned} \|\Delta\|_{\mathrm{F}}^2 = \sum_{i,j} \Delta_{ij}^2 &= \sum_{i,j} (\widehat{\mathbf{Q}}_{ij} - \mathbf{Q}_{ij})^2 \\ &= \frac{1}{4d_{\max}^2} \sum_{i\neq j} (\widehat{\mathbf{C}}_{ij} - \mathbf{C}_{ji})^2 + \frac{1}{4d_{\max}^2} \sum_i \Big| \sum_{j:(i,j)\in E} (\widehat{\mathbf{C}}_{ij} - \mathbf{C}_{ji}) \Big|^2 \\ &\leq \frac{1}{4d_{\max}^2} \sum_{i,j} (\widehat{\mathbf{C}}_{ij} - \mathbf{C}_{ji})^2 + \frac{1}{4d_{\max}^2} \sum_i d_{\max} \sum_{j:(i,j)\in E} (\widehat{\mathbf{C}}_{ij} - \mathbf{C}_{ji})^2 \\ &\leq \frac{d_{\max}+1}{4d_{\max}^2} \sum_{i,j} (\widehat{\mathbf{C}}_{ij} - \mathbf{C}_{ji})^2 \\ &= \frac{d_{\max}+1}{4d_{\max}^2} \|\boldsymbol{\eta}\|_{\mathrm{F}}^2 \leq \frac{1}{2d_{\max}} \|\boldsymbol{\eta}\|_{\mathrm{F}}^2, \end{aligned} \tag{8.61}$$

where $\boldsymbol{\eta}$ is the error in comparison marginals, i.e., $\boldsymbol{\eta} = \widehat{\mathbf{C}} - \mathbf{C}$. Thus, if $\|\boldsymbol{\eta}\|_{\mathrm{F}}^2 \leq \delta^2$, then, since $\|\Delta\|_2 \leq \|\Delta\|_{\mathrm{F}}$, we have that

$$\frac{\|\widehat{\pi} - \pi\|}{\|\pi\|} \leq \frac{1}{1-\rho} \frac{\delta}{\sqrt{2d_{\max}}} \sqrt{\frac{\pi_{\max}}{\pi_{\min}}}, \tag{8.62}$$

with

$$|\rho - \lambda_{\max}(\mathbf{Q})| \leq \frac{\delta}{\sqrt{2d_{\max}}} \sqrt{\pi_{\max}/\pi_{\min}}. \tag{8.63}$$

Therefore, if $\mathbf{Q}$ has a good spectral gap, i.e., $1 - \lambda_{\max}(\mathbf{Q})$ is large enough, and δ is small enough, then the estimate $\widehat{\pi}$ is a good proxy of π, the true parameters. The precise role of the "graph structure" induced by the observed entries, E, comes into play in determining $\lambda_{\max}(\mathbf{Q})$. This, along with implications regarding the sample complexity for the random sampling model, is discussed in detail in [44, 45].

Proof of Lemma 8.6. Define the inner-product space induced by π. For any $u, v \in \mathbb{R}^N$, define the inner product $\langle \cdot, \cdot \rangle_\pi$ as

$$\langle u, v \rangle_\pi = \sum_i u_i v_i \pi_i. \tag{8.64}$$

This defines norm $||u||_\pi = \sqrt{\langle u, u \rangle_\pi}$ for all $u \in \mathbb{R}$. Let $L^2(\pi)$ denote the space of all vectors with finite $||\cdot||_\pi$ norm endowed with inner product $\langle \cdot, \cdot \rangle_\pi$. Then, for any $u, v \in L^2(\pi)$,

$$\begin{aligned}\langle u, \mathbf{Q}v \rangle_\pi &= \sum_i u_i \left(\sum_j \mathbf{Q}_{ij} v_j \right) \pi_i \\ &= \sum_{i,j} u_i v_j \pi_i \mathbf{Q}_{ij} = \sum_{i,j} u_i v_j \pi_j \mathbf{Q}_{ji} \\ &= \sum_j \pi_j v_j \left(\sum_i \mathbf{Q}_{ji} u_i \right) = \langle \mathbf{Q}u, v \rangle_\pi. \end{aligned} \tag{8.65}$$

That is, $\mathbf{Q}$ is *self-adjoint* over $L^2(\pi)$. For a self-adjoint matrix $\mathbf{Q}$ over $L^2(\pi)$, define the norm

$$||\mathbf{Q}||_{2,\pi} = \max_u \frac{||\mathbf{Q}u||_\pi}{||u||_\pi}. \tag{8.66}$$

It can be verified that, for any $u \in \mathbb{R}^N$ and $\mathbf{Q}$,

$$\sqrt{\pi_{\min}} ||u||_2 \le ||u||_\pi \le \sqrt{\pi_{\max}} ||u||_2, \tag{8.67}$$

$$\sqrt{\frac{\pi_{\min}}{\pi_{\max}}} ||\mathbf{Q}||_2 \le ||\mathbf{Q}||_{2,\pi} \le \sqrt{\frac{\pi_{\max}}{\pi_{\min}}} ||\mathbf{Q}||_2. \tag{8.68}$$

Consider a symmetrized version of $\mathbf{Q}$ as $\mathbf{S} = \Pi^{\frac{1}{2}} \mathbf{Q} \Pi^{-\frac{1}{2}}$, where $\Pi^{\pm\frac{1}{2}}$ is the $N \times N$ diagonal matrix with the ith entry on the diagonal being $\pi_i^{\pm\frac{1}{2}}$. The symmetry of $\mathbf{S}$ follows due to the detailed-balance property of $\mathbf{Q}$, i.e., $\pi_i \mathbf{Q}_{ij} = \pi_j \mathbf{Q}_{ji}$ for all i, j. Since $\mathbf{Q}$ is a probability transition matrix, by the Perron–Frobenius theorem we have that its eigenvalues are in $[-1, 1]$, with the top eigenvalue being 1 and unique. Let the eigenvalues be $1 = \lambda_1 > \lambda_2 \ge \cdots \ge \lambda_N > -1$. Let the corresponding (left) eigenvectors of $\mathbf{Q}$ be $v_1, \ldots, v_N$. By definition $v_1 = \pi$. Therefore, $u_i = \Pi^{-\frac{1}{2}} v_i$ are (left) eigenvectors of S with eigenvalue λ_i for $1 \le i \le N$, since

$$\begin{aligned} u_i^{\mathrm{T}} \mathbf{S} &= (\Pi^{-\frac{1}{2}} v_i)^{\mathrm{T}} \Pi^{\frac{1}{2}} \mathbf{Q} \Pi^{-\frac{1}{2}} = v_i^{\mathrm{T}} \mathbf{Q} \Pi^{-\frac{1}{2}} \\ &= \lambda_i v_i^{\mathrm{T}} \Pi^{-\frac{1}{2}} = \lambda_i (\Pi^{-\frac{1}{2}} v_i)^{\mathrm{T}} = \lambda_i u_i^{\mathrm{T}}. \end{aligned} \tag{8.69}$$

That is, $u_1 = \pi^{\frac{1}{2}}$ or $\Pi^{-\frac{1}{2}} u_1 = \mathbf{1}$. By singular value decomposition, we can write $\mathbf{S} = \mathbf{S}_1 + \mathbf{S}_{\backslash 1}$, where $\mathbf{S}_1 = \lambda_1 u_1 u_1^{\mathrm{T}}$ and $\mathbf{S}_{\backslash 1} = \sum_{i=2}^N \lambda_i u_i u_i^{\mathrm{T}}$. That is,

$$\mathbf{Q} = \Pi^{-\frac{1}{2}} \mathbf{S} \Pi^{\frac{1}{2}} = \Pi^{-\frac{1}{2}} \mathbf{S}_1 \Pi^{\frac{1}{2}} + \Pi^{-\frac{1}{2}} \mathbf{S}_{\backslash 1} \Pi^{\frac{1}{2}} = \mathbf{1}\pi^{\mathrm{T}} + \Pi^{-\frac{1}{2}} \mathbf{S}_{\backslash 1} \Pi^{\frac{1}{2}}. \tag{8.70}$$

Recalling the notation $\Delta = \widehat{\mathbf{Q}} - \mathbf{Q}$, we can write (8.57) as

$$v_{t+1}^{\mathrm{T}} - \pi^{\mathrm{T}} = v_t^{\mathrm{T}} \widehat{\mathbf{Q}} - \pi^{\mathrm{T}} \mathbf{Q} = (v_t - \pi)^{\mathrm{T}} (\mathbf{Q} + \Delta) + \pi^{\mathrm{T}} \Delta$$

$$= (v_t - \pi)^{\mathrm{T}}(\mathbf{1}\pi^{\mathrm{T}} + \Pi^{-\frac{1}{2}}\mathbf{S}_{\backslash 1}\Pi^{\frac{1}{2}}) + (v_t - \pi)^{\mathrm{T}}\Delta + \pi^{\mathrm{T}}\Delta$$

$$= (v_t - \pi)^{\mathrm{T}}\Pi^{-\frac{1}{2}}\mathbf{S}_{\backslash 1}\Pi^{\frac{1}{2}} + (v_t - \pi)^{\mathrm{T}}\Delta + \pi^{\mathrm{T}}\Delta, \tag{8.71}$$

where we used the fact that $(v_t - \pi)^{\mathrm{T}}\mathbf{1} = 0$ since both v_t and π are probability vectors. Now, for any matrix $\mathbf{W}$, $||\Pi^{-\frac{1}{2}}\mathbf{W}\Pi^{\frac{1}{2}}||_{2,\pi} = ||\mathbf{W}||_2$. Therefore,

$$||v_{t+1}^{\mathrm{T}} - \pi^{\mathrm{T}}||_\pi \leq ||v_t - \pi||_\pi\Big(||\mathbf{S}_{\backslash 1}||_2 + ||\Delta||_{\pi,2}\Big) + ||\pi^{\mathrm{T}}\Delta||_\pi. \tag{8.72}$$

By definition $||\mathbf{S}_{\backslash 1}||_2 = \max\big(\lambda_2, |\lambda_N|\big) = \lambda_{\max}(\mathbf{Q})$. Let $\gamma = \Big(\lambda_{\max}(\mathbf{Q}) + ||\Delta||_{\pi,2}\Big)$. Then

$$||v_t^{\mathrm{T}} - \pi^{\mathrm{T}}||_\pi \leq \gamma^t ||v_0 - \pi||_\pi + \left(\sum_{s=0}^{t-1} \gamma^s\right)||\pi^{\mathrm{T}}\Delta||_\pi. \tag{8.73}$$

Using the bounds in (8.67) and (8.68), we have

$$\gamma \leq \lambda_{\max}(\mathbf{Q}) + ||\Delta||_2 \sqrt{\frac{\pi_{\max}}{\pi_{\min}}} \equiv \rho, \tag{8.74}$$

$$||v_t^{\mathrm{T}} - \pi^{\mathrm{T}}||_2 \leq \frac{1}{\sqrt{\pi_{\min}}}||v_t^{\mathrm{T}} - \pi^{\mathrm{T}}||_{\pi,2}, \tag{8.75}$$

$$||v_0 - \pi||_\pi \leq \sqrt{\pi_{\max}}||v_0 - \pi||_2, \tag{8.76}$$

$$||\pi^{\mathrm{T}}\Delta||_\pi \leq ||\pi||_2||\Delta||_2 \sqrt{\pi_{\max}}. \tag{8.77}$$

Therefore, we conclude that

$$\frac{||v_t^{\mathrm{T}} - \pi^{\mathrm{T}}||}{||\pi||} \leq \left[\rho^t \frac{||v_0 - \pi||}{||\pi||} + \left(\sum_{s=0}^{t-1} \rho^s\right)||\Delta||_2\right]\sqrt{\frac{\pi_{\max}}{\pi_{\min}}}. \tag{8.78}$$

This completes the proof by bounding $\sum_{s=0}^{t-1}\rho^s = (1-\rho^t)/(1-\rho) \leq 1/(1-\rho)$ for $\rho < 1$.

Recovering the ranking: comparison marginals. We will consider recovering the ranking from noisy comparison marginals $\widehat{\mathbf{C}} = \mathbf{C} + \eta$, using the scores as in (8.46) defined using noisy marginals. That is, for $i \in [N]$ define

$$\widehat{\mathsf{score}}(i) = \frac{1}{N-1}\sum_{k \neq i} \widehat{\mathbf{C}}_{ik}. \tag{8.79}$$

Then the error in the score for i is

$$\mathsf{error}(i) = |\widehat{\mathsf{score}}(i) - \mathsf{score}(i)| = \frac{1}{N-1}\left|\sum_{k \neq i} \widehat{\mathbf{C}}_{ik} - \mathbf{C}_{ik}\right|$$

$$\leq \frac{1}{N-1}\sum_{k \neq i}|\widehat{\mathbf{C}}_{ik} - \mathbf{C}_{ik}| \leq \frac{1}{N-1}||\boldsymbol{\eta}_{i\cdot}||_1, \tag{8.80}$$

where $\boldsymbol{\eta}_{i\cdot} = [\boldsymbol{\eta}_{ik}]_{k \in [N]}$. Therefore, the relative order of any pair of $i, j \in [N]$ is preserved under a noisy score as long as

$$\mathsf{error}(i) + \mathsf{error}(j) < |\mathsf{score}(i) - \mathsf{score}(j)|. \tag{8.81}$$

That is,

$$\|\boldsymbol{\eta}_{i\cdot}\|_1 + \|\boldsymbol{\eta}_{j\cdot}\|_1 < (N-1)|\mathsf{score}(i) - \mathsf{score}(j)|. \tag{8.82}$$

In summary, (8.82) provides the robustness property of a ranking algorithm based on noisy comparison being able to recover the true relative order for each pair of i, j; and subsequently for the entire ranking.

Recovering the ranking: first-order marginals. We consider using a Borda count for finding the ranking using noisy first-order marginals. More precisely, given noisy first-order marginals $\widehat{\mathbf{M}} = \mathbf{M} + \eta$, we define the noisy Borda count for $i \in [N]$ as

$$\widehat{\mathsf{borda}}(i) = \sum_{k \in [N]} k\widehat{\mathbf{M}}_{ik}. \tag{8.83}$$

Then, the error in the Borda count for i is

$$\begin{aligned} \mathsf{error}(i) = |\widehat{\mathsf{borda}}(i) - \mathsf{borda}(i)| &\le \sum_{k \in [N]} k|\widehat{\mathbf{M}}_{ik} - \mathbf{M}_{ik}| \\ &= \sum_{k \in [N]} k|\boldsymbol{\eta}_{ik}| = \mathsf{borda}^{\eta^+}(i). \end{aligned} \tag{8.84}$$

That is, $\mathsf{error}(i)$ is like computing the Borda count for i using $\eta^+ \equiv [|\boldsymbol{\eta}_{ik}|]$, which we define as $\mathsf{borda}^{\eta^+}(i)$. Then the relative order of any pair $i, j \in [N]$ per noisy Borda count is preserved if

$$\mathsf{borda}^{\eta^+}(i) + \mathsf{borda}^{\eta^+}(j) < |\mathsf{borda}(i) - \mathsf{borda}(j)|. \tag{8.85}$$

In summary, (8.85) provides the robustness property of a ranking algorithm based on noisy first-order marginal being able to recover the relative order for a given pair i, j; and subsequently for the entire ranking.

8.6 Discussion

We discussed learning the distribution over permutations from marginal information. In particular, we focused on marginal information of two types: first-order marginals and pair-wise marginals. We discussed the conditions for recovering the distribution as well as the ranking associated with the distribution under two model classes: *sparse* models and *random utility models* (RUM). For all of these settings, we discussed settings where we had access to *exact* marginal information as well as *noisy* marginals. A lot of progress has been made, especially in the past decade, on this topic in various directions. Here, we point out some of the prominent directions and provide associated references.

8.6.1 Beyond First-Order and Pair-Wise Marginals

To start with, learning the distribution for different types of marginal information has been discussed in [28], where Jagabathula and Shah discuss the relationship between the level of sparsity and the type of marginal information available. Specifically, through connecting marginal information with the spectral representation of the permutation group, Jagabathula and Shah find that, as the higher-order marginal information is made available, the distribution with larger support size can be recovered with a tantalizingly similar relationship between the dimensionality of the higher-order information and the recoverability of the support size just like in the first-order marginal information. The reader is referred to [28] for more details and some of the open questions. We note that this collection of results utilizes the *signature* condition discussed in Section 8.4.1.

8.6.2 Learning MNL beyond Rank Centrality

The work in [44, 45] on recovering the parameters of the MNL model from noisy observations has led to exciting subsequent work in recent years. Maximum-likelihood estimation (MLE) for the MNL model is particularly nice – it boils down to solving a simple, convex minimization problem. For this reason, historically it has been well utilized in practice. However, only recently has its sample complexity been well studied. In particular, in [49] Hajek *et al.* argue that the sample complexity requirement of MLE is similar to that of *rank centrality*, discussed in Section 8.5.2.

There has been work to find refined estimation of parameters, for example, restricting the analysis to the top few parameters, see [50–52]. We also note an interesting algorithmic generalization of rank centrality that has been discussed in [53] through a connection to a continuous-time representation of the reversible Markov chain considered in rank centrality.

8.6.3 Mixture of MNL Models

The RUM discussed in detail here has the weakness that all options are parameterized by one parameter. This does not allow for heterogeneity in options in terms of multiple "modes" of preferences or rankings. Putting it another way, the RUM captures a sliver of the space of all possible distributions over permutations. A natural way to generalize such a model is to consider a mixture of RUM models. A specific instance is a mixture of MNL models, which is known as the mixture MNL or mixed MNL model. It can be argued that such a mixed MNL model can approximate any distribution over permutations with enough mixture components. This is because we can approximate a distribution with unit mass on a permutation by an MNL model by choosing parameters appropriately. Therefore, it makes sense to understand when we can learn the mixed MNL model from pair-wise rankings or more generally partial rankings. In [54], Ammar *et al.* considered this question and effectively identified an impossibility result that suggests that pair-wise information is not sufficient to learn mixtures in general. They provided lower bounds that related the number of mixture components, the number of choices (here N), and the length of partial rankings observed. For a *separable* mixture MNL model, they provide a natural clustering-based solution for recovery. Such a recovery approach has been further refined in the context of collaborative ranking [55, 56] through use of convex optimization-based methods and imposing a "low-rank" structure on the model parameter matrix to enable recovery. In another line of work, using higher-moment information for a *separable* mixture model, in [57] Oh and Shah provided a tensor-decomposition-based approach for recovering the mixture MNL model.

8.6.4 Matrix Estimation for Denoising Noisy Marginals

In a very different view, the first-order marginal and pair-wise marginal information considered here can be viewed as a matrix of observations with an underlying structure. Or, more generally, we have access to a noisy observation of an underlying ground matrix with structure. The structure is implied by the underlying distribution over permutations generating it. Therefore, recovering *exact* marginal information from *noisy* marginal information can be viewed as recovering a matrix, with structure, employing a noisy and potentially partial view of it. In [39], this view was considered for denoising pair-wise comparison marginal data. In [39], it was argued that, when the ground-truth comparison marginal matrix satisfies a certain *stochastic transitivity* condition, for example that implied by the MNL model, the true pair-wise marginal matrix can be recovered from noisy, partial observations. This approach has been further studied in a sequence of recent works including [42, 58].

8.6.5 Active Learning and Noisy Sorting

Active learning with a view to ranking using pair-wise comparisons with noisy observations has been well studied for a long time. For example, in [59] Adler *et al.* considered the design of an adaptive "tournament," i.e., the draw of games is not decided upfront, but is decided as more and more games are played. The eventual goal is to find the "true" winner with a high probability. It is assumed that there is an underlying ranking which corresponds to the ordering implied by the MNL or Bradley–Terry–Luce model. The work of Adler *et al.* provides an adaptive algorithm that leads to finding the true winner with high probability efficiently.

When there is "geometry" imposed on the space of preferences, very efficient adaptive algorithms can be designed for searching using comparisons, for example [60, 61]. Another associated line of works is that of noisy sorting. For example, see [62].

It is worth noting that the variation of online learning in the context of a "bandit" setting is known as "dueling bandits" wherein comparisons between pairs of arms are provided and the goal is to find the top arm. This, again can be viewed as finding the top element from pair-wise comparisons in an online setting, with the goal being to minimize "regret." For example, see [63–65].

In our view, the dueling-bandits model using distributions over permutations, i.e., the outcome of pair-wise comparisons of arms are consistent with the underlying distribution over permutations, provides an exciting direction for making progress toward online matrix estimation.

8.6.6 And It Continues ...

There is a lot more that is tangentially related to this topic. For example, the question of ranking or selecting the winner in an election is fundamental to so many disciplines and each (sub)discipline brings to the question a different perspective that makes this topic rich and exciting. The statistical challenges related to learning the distribution over permutations are very recent, as is clear from the exposition provided here. The scale and complexity of the distribution over permutations ($N!$ for N options) makes it challenging from a computational point of view and thus provides a fertile ground for emerging interactions between statistics and computation. The rich group structure embedded within the permutation group makes it an exciting arena for the development of algebraic statistics, see [66].

The statistical philosophy of max-entropy model learning leads to learning a parametric distribution from an exponential family. This brings in rich connections to the recent advances in learning and inference on graphical models. For example, fitting such a model using first-order marginals boils down to computing a partition function over the space of matchings or permutations; which can be computationally efficiently solved due to fairly recent progress in computing the permanent [67]. In contrast, learning such a model efficiently in the context of pair-wise marginals is not easy due to a connection to the feedback arc set problem.

We note an interesting connection: a mode computation heuristic based on maximum weight matching in a bipartite graph using a first-order marginal turns out to be a "first-order" approximation of the mode of such a distribution, see [47]. On the other hand, using pair-wise comparison marginals, there is a large number of heuristics to compute ranking, including the rank centrality algorithm discussed in detail here, or more generally a variety of spectral methods considered for ranking, including in [68, 69], and more recently [70, 71].

This exciting list of work continues to grow even as the author completes these final keystrokes and as you get inspired to immerse yourself in this fascinating topic of *computing choice*.

References

[1] J. Huang, C. Guestrin, and L. Guibas, "Fourier theoretic probabilistic inference over permutations," *J. Machine Learning Res.*, vol. 10, no. 5, pp. 997–1070, 2009.

[2] P. Diaconis, *Group representations in probability and statistics*. Institute of Mathematical Statistics, 1988.

[3] L. Thurstone, "A law of comparative judgement," *Psychol. Rev.*, vol. 34, pp. 237–286, 1927.

[4] J. Marschak, "Binary choice constraints on random utility indicators," Cowles Foundation Discussion Paper, 1959.

[5] J. Marschak and R. Radner, *Economic theory of teams*. Yale University Press, 1972.

[6] J. I. Yellott, "The relationship between Luce's choice axiom, Thurstone's theory of comparative judgment, and the double exponential distribution," *J. Math. Psychol.*, vol. 15, no. 2, pp. 109–144, 1977.

[7] R. Luce, *Individual choice behavior: A theoretical analysis*. Wiley, 1959.

[8] R. Plackett, "The analysis of permutations," *Appl. Statist.*, vol. 24, no. 2, pp. 193–202, 1975.

[9] D. McFadden, "Conditional logit analysis of qualitative choice behavior," in *Frontiers in Econometrics*, P. Zarembka, ed. Academic Press, 1973, pp. 105–142.

[10] G. Debreu, "Review of R. D. Luce, 'individual choice behavior: A theoretical analysis,'" *Amer. Economic Rev.*, vol. 50, pp. 186–188, 1960.

[11] R. A. Bradley, "Some statistical methods in taste testing and quality evaluation," *Biometrics*, vol. 9, pp. 22–38, 1953.

[12] D. McFadden, "Econometric models of probabilistic choice," in *Structural analysis of discrete data with econometric applications*, C. F. Manski and D. McFadden, eds. MIT Press, 1981.

[13] M. E. Ben-Akiva and S. R. Lerman, *Discrete choice analysis: Theory and application to travel demand.* CMIT Press, 1985.

[14] D. McFadden, "Disaggregate behavioral travel demand's RUM side: A 30-year retrospective," *Travel Behaviour Res.*, pp. 17–63, 2001.

[15] P. M. Guadagni and J. D. C. Little, "A logit model of brand choice calibrated on scanner data," *Marketing Sci.*, vol. 2, no. 3, pp. 203–238, 1983.

[16] S. Mahajan and G. J. van Ryzin, "On the relationship between inventory costs and variety benefits in retail assortments," *Management Sci.*, vol. 45, no. 11, pp. 1496–1509, 1999.

[17] M. E. Ben-Akiva, "Structure of passenger travel demand models," Ph.D. dissertation, Department of Civil Engineering, MIT, 1973.

[18] J. H. Boyd and R. E. Mellman, "The effect of fuel economy standards on the U.S. automotive market: An hedonic demand analysis," *Transportation Res. Part A: General*, vol. 14, nos. 5–6, pp. 367–378, 1980.

[19] N. S. Cardell and F. C. Dunbar, "Measuring the societal impacts of automobile downsizing," *Transportation Res. Part A: General*, vol. 14, nos. 5–6, pp. 423–434, 1980.

[20] D. McFadden and K. Train, "Mixed MNL models for discrete response," *J. Appl. Econometrics*, vol. 15, no. 5, pp. 447–470, 2000.

[21] K. Bartels, Y. Boztug, and M. M. Muller, "Testing the multinomial logit model," 1999, unpublished working paper.

[22] J. L. Horowitz, "Semiparametric estimation of a work-trip mode choice model," *J. Econometrics*, vol. 58, pp. 49–70, 1993.

[23] B. Koopman, "On distributions admitting a sufficient statistic," *Trans. Amer. Math. Soc.*, vol. 39, no. 3, pp. 399–409, 1936.

[24] B. Crain, "Exponential models, maximum likelihood estimation, and the Haar condition," *J. Amer. Statist. Assoc.*, vol. 71, pp. 737–745, 1976.

[25] R. Beran, "Exponential models for directional data," *Annals Statist.*, vol. 7, no. 6, pp. 1162–1178, 1979.

[26] M. Wainwright and M. Jordan, "Graphical models, exponential families, and variational inference," *Foundations and Trends Machine Learning*, vol. 1, nos. 1–2, pp. 1–305, 2008.

[27] S. Jagabathula and D. Shah, "Inferring rankings under constrained sensing," in *Advances in Neural Information Processing Systems*, 2009, pp. 753–760.

[28] S. Jagabathula and D. Shah, "Inferring rankings under constrained sensing," *IEEE Trans. Information Theory*, vol. 57, no. 11, pp. 7288–7306, 2011.

[29] V. Farias, S. Jagabathula, and D. Shah, "A data-driven approach to modeling choice," in *Advances in Neural Information Processing Systems*, 2009.

[30] V. Farias, S. Jagabathula, and D. Shah, "A nonparametric approach to modeling choice with limited data," *Management Sci.*, vol. 59, no. 2, pp. 305–322, 2013.

[31] K. J. Arrow, "A difficulty in the concept of social welfare," *J. Political Economy*, vol. 58, no. 4, pp. 328–346, 1950.

[32] J. Marden, *Analyzing and modeling rank data*. Chapman & Hall/CRC, 1995.

[33] E. J. Candès and T. Tao, "Decoding by linear programming," *IEEE Trans. Information Theory*, vol. 51, no. 12, pp. 4203–4215, 2005.

[34] E. J. Candès, J. K. Romberg, and T. Tao, "Stable signal recovery from incomplete and inaccurate measurements," *Communications Pure Appl. Math.*, vol. 59, no. 8, pp. 1207–1223, 2006.

[35] E. J. Candès and J. Romberg, "Quantitative robust uncertainty principles and optimally sparse decompositions," *Foundations Computational Math.*, vol. 6, no. 2, pp. 227–254, 2006.

[36] E. J. Candès, J. Romberg, and T. Tao, "Robust uncertainty principles: Exact signal reconstruction from highly incomplete frequency information," *IEEE Trans. Information Theory*, vol. 52, no. 2, pp. 489–509, 2006.

[37] D. L. Donoho, "Compressed sensing," *IEEE Trans. Information Theory*, vol. 52, no. 4, pp. 1289–1306, 2006.

[38] R. Berinde, A. C. Gilbert, P. Indyk, H. Karloff, and M. J. Strauss, "Combining geometry and combinatorics: A unified approach to sparse signal recovery," in *Proc. 46th Annual Allerton Conference on Communications, Control, and Computation*, 2008, pp. 798–805.

[39] S. Chatterjee, "Matrix estimation by universal singular value thresholding," *Annals Statist.*, vol. 43, no. 1, pp. 177–214, 2015.

[40] D. Song, C. E. Lee, Y. Li, and D. Shah, "Blind regression: Nonparametric regression for latent variable models via collaborative filtering," in *Advances in Neural Information Processing Systems*, 2016, pp. 2155–2163.

[41] C. Borgs, J. Chayes, C. E. Lee, and D. Shah, "Thy friend is my friend: Iterative collaborative filtering for sparse matrix estimation," in *Advances in Neural Information Processing Systems*, 2017, pp. 4715–4726.

[42] N. Shah, S. Balakrishnan, A. Guntuboyina, and M. Wainwright, "Stochastically transitive models for pairwise comparisons: Statistical and computational issues," in *International Conference on Machine Learning*, 2016, pp. 11–20.

[43] V. F. Farias, S. Jagabathula, and D. Shah, "Sparse choice models," in *46th Annual Conference on Information Sciences and Systems (CISS)*, 2012, pp. 1–28.

[44] S. Negahban, S. Oh, and D. Shah, "Iterative ranking from pair-wise comparisons," in *Advances in Neural Information Processing Systems*, 2012, pp. 2474–2482.

[45] S. Negahban, S. Oh, and D. Shah, "Rank centrality: Ranking from pairwise comparisons," *Operations Res.*, vol. 65, no. 1, pp. 266–287, 2016.

[46] A. Ammar and D. Shah, "Ranking: Compare, don't score," in *Proc. 49th Annual Allerton Conference on Communication, Control, and Computing*, 2011, pp. 776–783.

[47] A. Ammar and D. Shah, "Efficient rank aggregation using partial data," in *ACM SIGMETRICS Performance Evaluation Rev.*, vol. 40, no. 1, 2012, pp. 355–366.

[48] P. Emerson, "The original Borda count and partial voting," *Social Choice and Welfare*, vol. 40, no. 2, pp. 353–358, 2013.
[49] B. Hajek, S. Oh, and J. Xu, "Minimax-optimal inference from partial rankings," in *Advances in Neural Information Processing Systems*, 2014, pp. 1475–1483.
[50] Y. Chen and C. Suh, "Spectral mle: Top-k rank aggregation from pairwise comparisons," in *International Conference on Machine Learning*, 2015, pp. 371–380.
[51] Y. Chen, J. Fan, C. Ma, and K. Wang, "Spectral method and regularized mle are both optimal for top-k ranking," *arXiv:1707.09971*, 2017.
[52] M. Jang, S. Kim, C. Suh, and S. Oh, "Top-k ranking from pairwise comparisons: When spectral ranking is optimal," *arXiv:1603.04153*, 2016.
[53] L. Maystre and M. Grossglauser, "Fast and accurate inference of Plackett–Luce models," in *Advances in Neural Information Processing Systems*, 2015, pp. 172–180.
[54] A. Ammar, S. Oh, D. Shah, and L. F. Voloch, "What's your choice?: Learning the mixed multi-nomial," in *ACM SIGMETRICS Performance Evaluation Rev.*, vol. 42, no. 1, 2014, pp. 565–566.
[55] S. Oh, K. K. Thekumparampil, and J. Xu, "Collaboratively learning preferences from ordinal data," in *Advances in Neural Information Processing Systems*, 2015, pp. 1909–1917.
[56] Y. Lu and S. N. Negahban, "Individualized rank aggregation using nuclear norm regularization," in *Proc. 53rd Annual Allerton Conference on Communication, Control, and Computing*, 2015, pp. 1473–1479.
[57] S. Oh and D. Shah, "Learning mixed multinomial logit model from ordinal data," in *Advances in Neural Information Processing Systems*, 2014, pp. 595–603.
[58] N. B. Shah, S. Balakrishnan, and M. J. Wainwright, "Feeling the bern: Adaptive estimators for Bernoulli probabilities of pairwise comparisons," in *IEEE International Symposium on Information Theory (ISIT)*, 2016, pp. 1153–1157.
[59] M. Adler, P. Gemmell, M. Harchol-Balter, R. M. Karp, and C. Kenyon, "Selection in the presence of noise: The design of playoff systems," in *SODA*, 1994, pp. 564–572.
[60] K. G. Jamieson and R. Nowak, "Active ranking using pairwise comparisons," in *Advances in Neural Information Processing Systems*, 2011, pp. 2240–2248.
[61] A. Karbasi, S. Ioannidis, and L. Massoulié, "From small-world networks to comparison-based search," *IEEE Trans. Information Theory*, vol. 61, no. 6, pp. 3056–3074, 2015.
[62] M. Braverman and E. Mossel, "Noisy sorting without resampling," in *Proc. 19th Annual ACM–SIAM Symposium on Discrete Algorithms*, 2008, pp. 268–276.
[63] Y. Yue and T. Joachims, "Interactively optimizing information retrieval systems as a dueling bandits problem," in *Proc. 26th Annual International Conference on Machine Learning*, 2009, pp. 1201–1208.
[64] K. G. Jamieson, S. Katariya, A. Deshpande, and R. D. Nowak, "Sparse dueling bandits," in *AISTATS*, 2015.
[65] M. Dudík, K. Hofmann, R. E. Schapire, A. Slivkins, and M. Zoghi, "Contextual dueling bandits," in *Proc. 28th Conference on Learning Theory*, 2015, pp. 563–587.
[66] R. Kondor, A. Howard, and T. Jebara, "Multi-object tracking with representations of the symmetric group," in *Artificial Intelligence and Statistics*, 2007, pp. 211–218.
[67] M. Jerrum, A. Sinclair, and E. Vigoda, "A polynomial-time approximation algorithm for the permanent of a matrix with nonnegative entries," *J. ACM*, vol. 51, no. 4, pp. 671–697, 2004.
[68] T. L. Saaty and G. Hu, "Ranking by eigenvector versus other methods in the analytic hierarchy process," *Appl. Math. Lett.*, vol. 11, no. 4, pp. 121–125, 1998.

[69] C. Dwork, R. Kumar, M. Naor, and D. Sivakumar, "Rank aggregation methods for the web," in *Proc. 10th International Conference on World Wide Web*, 2001, pp. 613–622.

[70] A. Rajkumar and S. Agarwal, "A statistical convergence perspective of algorithms for rank aggregation from pairwise data," in *International Conference on Machine Learning*, 2014, pp. 118–126.

[71] H. Azari, D. Parks, and L. Xia, "Random utility theory for social choice," in *Advances in Neural Information Processing Systems*, 2012, pp. 126–134.

9 Universal Clustering

Ravi Kiran Raman and Lav R. Varshney

Summary

Clustering is a general term for the set of techniques that, given a set of objects, aim to select those that are closer to one another than to the rest of the objects, according to a chosen notion of closeness. It is an unsupervised learning problem since objects are not externally labeled by category. Much research effort has been expended on finding natural mathematical definitions of *closeness* and then developing/evaluating algorithms in these terms [1]. Many have argued that there is no domain-independent mathematical notion of similarity, but that it is context-dependent [2]; categories are perhaps natural in that people can evaluate them when they see them [3, 4]. Some have dismissed the problem of unsupervised learning in favor of supervised learning, saying it is not a powerful natural phenomenon (see p. 159 of [5]).

Yet, most of the learning carried out by people and animals is unsupervised. We largely learn how to think through categories by observing the world in its unlabeled state. This is a central problem in data science. Whether grouping behavioral traces into categories to understand their neural correlates [6], grouping astrophysical data to understand galactic potentials [7], or grouping light spectra into color names [8], clustering is crucial to data-driven science.

Drawing on insights from universal information theory, in this chapter we ask whether there are universal approaches to unsupervised clustering. In particular, we consider instances wherein the ground-truth clusters are defined by the unknown statistics governing the data to be clustered. By universality, we mean that the system does not have prior access to such statistical properties of the data to be clustered (as is standard in machine learning), nor does it have a strong sense of the appropriate notion of similarity to measure which objects are close to one another.

9.1 Unsupervised Learning In Machine Learning

In an age with explosive data-acquisition rates, we have seen a drastic increase in the amount of raw, unlabeled data in the form of pictures, videos, text, and voice. By 2023, it is projected that the per capita amount of data stored in the world will exceed the entire Library of Congress (10^{14} bits) at the time Shannon first estimated it in 1949 [9, 10].

Making sense of such large volumes of data, e.g., for statistical inference, requires extensive, efficient pre-processing to transform the data into meaningful forms. In addition to the increasing volumes of known data forms, we are also collecting new varieties of data with which we have little experience [10].

In astronomy, GalaxyZoo uses the general intelligence of people in crowdsourcing systems to classify celestial imagery (and especially point out unknown and anomalous objects in the universe); supervised learning is not effective since there is a limited amount of training data available for this task, especially given the level of noise in the images [11]. Analyses of such new data forms often need to be performed with minimal or no contextual information.

Crowdsourcing is also popular for collecting (noisy) labeled training data for machine-learning algorithms [12, 13]. As an example, impressive classification performance of images has been attained using deep convolutional networks on the ImageNet database [14], but this is after training over a set of 1000 classes of images with 1000 labeled samples for each class, i.e., one million labeled samples, which were obtained from costly crowdsourcing [15]. Similarly, Google Translate has achieved near-human-level translation using deep recurrent neural networks (RNNs) [16] at the cost of large labeled training corpora on the order of millions of examples [17]. For instance, the training dataset for English to French consisted of 36 million pairs – far more than what the average human uses. On the other hand, reinforcement learning methods in their current form rely on exploring state–action pairs to learn the underlying cost function by Q-learning. While this might be feasible in learning games such as chess and Go, in practice such exploration through reinforcement might be expensive. For instance, the recently successful Alpha Go algorithm uses a combination of supervised and reinforcement learning of the optimal policy, by learning from a dataset consisting of 29.4 million positions learned from 160000 games [18]. Beside this dataset, the algorithm learned from the experience of playing against other Go programs and players. Although collecting such a database and the games to learn from might be feasible for a game of Go, in general inferring from experience is far more expensive.

Although deep learning has given promising solutions for supervised learning problems, results for unsupervised learning pale in comparison [19]. Is it possible to make sense of (training) data without people? Indeed, one of the main goals of artificial general intelligence has always been to develop solutions that work with minimal supervision [20]. To move beyond data-intensive supervised learning methods, there is growing interest in unsupervised learning problems such as density/support estimation, clustering, and independent-component analysis. Careful choice and application of unsupervised methods often reveal structure in the data that is otherwise not evident. Good clustering solutions help sort unlabeled data into simpler sub-structures, not only allowing for subsequent inferential learning, but also adding insights into the characteristics of the data being studied. This chapter studies information-theoretic formulations of unsupervised clustering, emphasizing their strengths under limited contextual information.

9.2 Universality in Information Theory

Claude Shannon identified the fundamental mathematical principles governing data compression and transmission [21], introducing the field of information theory. Much work in understanding the fundamental limits of and in designing compression and communication systems followed, typically under the assumption that statistical knowledge of the source and channel is given, respectively. Information theory has also subsequently found fundamental limits and optimal algorithms for statistical inference.

However, such contextual knowledge of the statistics governing the system is not always available. Thus, it is important to be able to design solutions that operate reliably under an appropriate performance criterion, for any model in a large family. This branch of exploration, more commonly known as *universal information theory*, has also helped understand worst-case fundamental performance limits of systems.

Here we give a very brief overview of prior work in the field, starting with compression and communication, and then highlighting results from universal statistical inference. As part of our development, we use block-diagrammatic reasoning [22] to provide graphical summaries.

9.2.1 Universality in Compression and Communication

Compression and communication are the two most extensively studied problems in information theory. However, besides being of historical significance the two topics also shed light on problems in statistical inference, such as the clustering problem that we are interested in.

In his Shannon lecture, Robert Gray described the inherent connection between data compression and simulation [23]. The problem of simulation aims to design a coder that takes a random bitstream as input and generates a sequence that has the statistical properties of a desired distribution [24, 25]. In particular, it is notable that a data-compression decoder that operates close to the Shannon limit also serves as the nearly optimal coder for the corresponding simulation problem, highlighting the underlying relation between lossy data compression and this statistical inference problem. Similarly, the connection between data compression and the problem of gambling is well studied, in that a sequence of outcomes over which a gambler can make significant profits is also a sequence that can be compressed well [26].

Contrarily in his Shannon lecture, Jorma Rissanen discussed the philosophical undertones of similarity between data compression and estimation theory with a disclaimer [27]:

> in data compression the shortest code length cannot be achieved without taking advantage of the regular features in data, while in estimation it is these regular features, the underlying mechanism, that we want to learn ... like a jigsaw puzzle where the pieces almost fit but not quite, and, moreover, vital pieces were missing.

This observation is significant not only in reaffirming the connection between the two fields, but also in highlighting the fact that the unification/translation of ideas often requires closer inspection and some additional effort.

Compression, by definition, is closely related to the task of data clustering. Consider the k-means clustering algorithm where each data point is represented by one of k possible centroids, chosen according to Euclidean proximity (and thereby similar to lossy data compression under the squared loss function where the compressed representation can be viewed as a cluster center). Explicit clustering formulations for communication under the universal setting, wherein the transmitted messages represent cluster labels, have also been studied [28]; this work and its connection to clustering are elaborated in Section 9.5.2.

It is evident from history and philosophy that universal compression and communication may yield much insight into both the design and the analysis of inference algorithms, especially of clustering. Thus, we give some brief insight into prominent prior work in these areas.

Compression

Compression, as originally introduced by Shannon [21], considers the task of efficiently representing discrete data samples, $X_1, \ldots, X_n$, generated by a source independently and identically according to a distribution $P_X(\cdot)$, as shown in Fig. 9.1. Shannon showed that the best rate of lossless compression for such a case is given by the entropy, $H(P_X)$, of the source.

A variety of subsequently developed methods, such as arithmetic coding [29], Huffman coding [30], and Shannon–Fano–Elias coding [26], approach this limit to perform optimal compression of memoryless sources. However, they often require knowledge of the source distribution for encoding and decoding.

On the other hand, Lempel and Ziv devised a universal encoding scheme for compressing unknown sources [31], which has proven to be asymptotically optimal for strings generated by stationary, ergodic sources [32, 33]. This result is impressive as it highlights the feasibility of optimal compression (in the asymptotic sense), without contextual knowledge, by using a simple encoding scheme. Unsurprisingly, this work also inspired the creation of practical compression algorithms such as gzip. The method has subsequently been generalized to random fields (such as images) with lesser storage requirements as well [34]. Another approach to universal compression is to first approximate the source distribution from the symbol stream, and then perform optimal compression for this estimated distribution [35–37]. Davisson identifies the necessary and sufficient condition for noiseless universal coding of ergodic sources as being that the per-letter average mutual information between the parameter that defines the source distribution and the message is zero [38].

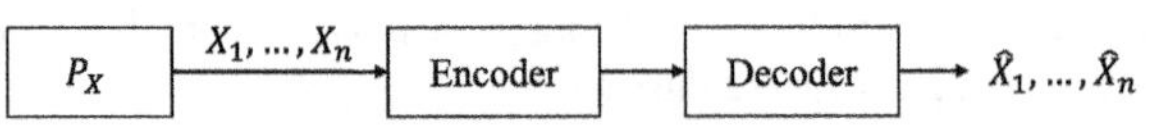

Figure 9.1 Model of data compression. The minimum rate of compression for a memoryless source is given by the entropy of the source.

Beyond the single-terminal case, the problem of separately compressing a correlated pair of sources was addressed in the seminal work of Slepian and Wolf [39] and later generalized to jointly stationary ergodic sources [40]. Universal analogs for this problem have also been derived by Csiszár [41].

Through rate-distortion theory, Shannon also established theoretical foundations of lossy compression, where the aim is to compress a source subject to a constraint on the distortion experienced by the decoded bitstream, instead of exact recovery [42, 43]. Universal analogs for lossy compression, albeit limited in comparison with the lossless setting, have also been explored [44–46]. These are particularly important from the universal clustering standpoint due to the underlying similarity between the two problems noted earlier. A comprehensive survey of universal compression can be found in [47] and references therein.

Communication

In the context of communicating messages across a noisy, discrete memoryless channel (DMC), W, Shannon established that the rate of communication is limited by the *capacity* of the channel [21, 48], which is given by

$$C(W) = \max_{P_X} I(X;Y). \tag{9.1}$$

Here, the maximum is over all information sources generating the codeword symbol X that is transmitted through the channel W to receive the noisy symbol Y, as shown in Fig. 9.2.

Shannon showed that using a random codebook for encoding the messages and joint typicality decoding achieves the channel capacity. Thus, the decoder requires knowledge of the channel. Goppa later defined the maximum mutual information (MMI) decoder to perform universal decoding [49]. The decoder estimates the message as the one with the codeword that maximizes the empirical mutual information with the received vector. This decoder is universally optimal in the error exponent [50] and has also been generalized to account for erasures using constant-composition random codes [51].

Universal communication over channels subject to conditions on the decoding error probability has also been studied [52]. Beside DMCs, the problem of communication over channels with memory has also been extensively studied both with channel knowledge [53–55] and in the universal sense [56, 57].

Note that the problem of universal communication translates to one of classification of the transmitted codewords, on the basis of the message, using noisy versions without knowledge of the channel. In addition, if the decoder is unaware of the transmission codebook, then the best one could hope for is to cluster the messages. This problem has been studied in [28, 58], which demonstrate that the MMI decoder is sub-optimal

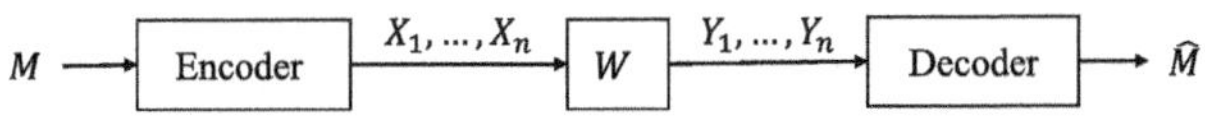

Figure 9.2 Model of data communication. The maximum rate of communication for discrete memoryless channels is given by the capacity of the channel.

for this context, and therefore introduces minimum partition information (MPI) decoding to cluster the received noisy codewords. More recently, universal communication in the presence of noisy codebooks has been considered [59]. A comprehensive survey of universal communication under a variety of channel models can be found in [60].

9.2.2 Universality in Statistical Inference

Beside compression and communication, a variety of statistical inference problems have also been studied from the perspective of universality. Let us review a few such problems.

The likelihood ratio test is well known to be Bayes-optimal for hypothesis testing when the conditional distributions defining the hypotheses are known. Composite hypothesis testing, shown in Fig. 9.3, considers null and alternate hypotheses defined by classes of distributions parameterized by families Θ_0 and Θ_1, respectively. In such contexts, the generalized likelihood ratio, defined by

$$L_{\mathrm{G}}(y) = \frac{\sup_{\theta\in\Theta_1} p_\theta(y)}{\sup_{\theta\in\Theta_0} p_\theta(y)},$$

is used in place of the likelihood ratio. The generalized likelihood ratio test (GLRT) is asymptotically optimal under specific hypothesis classes [61].

A universal version of composite hypothesis testing is given by the competitive minimax formulation [62]:

$$\min_{\delta} \max_{(\theta_0,\theta_1)\in\Theta_0\times\Theta_1} \frac{P_{\mathrm{e}}(\delta|\theta_0,\theta_1)}{\inf_{\widetilde{\delta}} P_{\mathrm{e}}(\widetilde{\delta}|\theta_0,\theta_1)}.$$

That is, the formulation compares the worst-case error probabilities with the corresponding Bayes error probability of the binary hypothesis test. The framework highlights the loss from the composite nature of the hypotheses and from the lack of knowledge of these families, proving to be a benchmark. This problem has also been studied in the Neyman–Pearson formulation [63].

Universal information theory has also been particularly interested in the problem of prediction. In particular, given the sequence $X_1,\ldots,X_n$, the aim is to predict X_{n+1} such

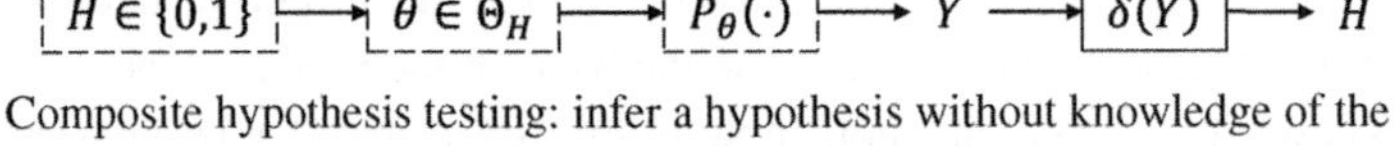

Figure 9.3 Composite hypothesis testing: infer a hypothesis without knowledge of the conditional distribution of the observations. In this and all other subsequent figures in this chapter we adopt the convention that the decoders and algorithms are aware of the system components enclosed in solid lines and unaware of those enclosed by dashed lines.

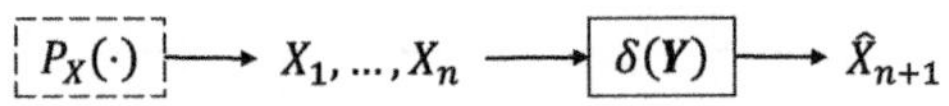

Figure 9.4 Prediction: estimate the next symbol given all past symbols without knowledge of the stationary, ergodic source distribution.

$$\boxed{P_X(\cdot)} \longmapsto X_1, \ldots, X_n \longrightarrow \boxed{W(Y|X)} \longmapsto Y_1, \ldots, Y_n \longrightarrow \boxed{\delta(Y)} \longmapsto \hat{X}_1, \ldots, \hat{X}_n$$

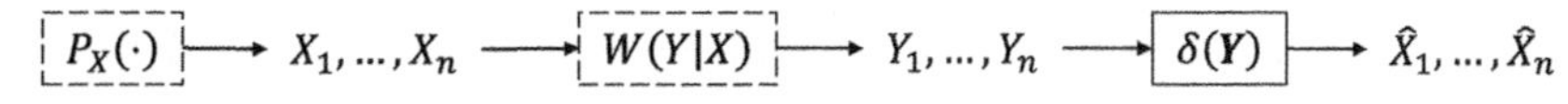

Figure 9.5 Denoising: estimate a transmitted sequence of symbols without knowledge of one or both of the channel and source distributions.

$$X_1, \ldots, X_n \longrightarrow \boxed{W(Y|X)} \longmapsto \boxed{\pi \in \Pi} \longmapsto Y_{\pi_1}, \ldots, Y_{\pi_n} \longrightarrow \boxed{\delta(Y)} \longmapsto \hat{X}_{c+1}, \ldots, \hat{X}_{c+n}$$

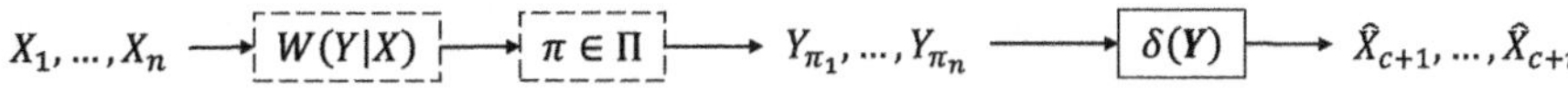

Figure 9.6 Multireference alignment: estimate the transmitted sequence up to a cyclic rotation, without knowledge of the channel distribution and permutation.

that a distortion function is minimized, as shown in Fig. 9.4. A universal prediction algorithm is designed using the parsing idea of Lempel–Ziv compression in [64] and has been proven to achieve finite-state predictability in the asymptotic sense. A similar mechanism is also used for determining wagers in a sequential gambling framework using finite-state machines [65].

Another problem of interest in the universal information theory framework is that of denoising. This considers the context of estimating a sequence of symbols $X_1, \ldots, X_n$ transmitted through a channel from the noisy received sequence $Y_1, \ldots, Y_n$, without channel coding to add redundancy (as against communication). This is shown in Fig. 9.5. Denoising in the absence of knowledge of the source distribution, for a known, discrete memoryless channel, has been studied in [66]. It has been shown that universal denoising algorithms asymptotically perform as well as the optimal algorithm that has knowledge of the source distribution.

For a memoryless source, estimating the transmitted sequence translates to a symbol-wise m-ary hypothesis test for each symbol without knowledge of a prior distribution. By restricting their consideration to a DMC with a full-rank transition matrix in this work, the authors use the statistics of the noisy symbols to estimate the source distribution. Once the source distribution has been estimated, the sequence is denoised.

The twice universal version of the problem wherein both source and channel distributions are unknown has also been explored [67]. Given a parameter k, the loss between the universal denoiser and the optimal kth-order sliding window denoiser is shown to vanish asymptotically. These problems try to exploit the inherent redundancy (if any) in the source to predict the input sequences.

A problem closely related to the denoising problem is that of multireference alignment [68]. Here an input sequence is corrupted by a channel and permuted from a given class of feasible permutations as shown in Fig. 9.6. The aim here is to recover the input sequence, up to a cyclic shift, given the noisy signal. It finds application in signal alignment and image registration. The sample complexity of the model-aware, Bayes-optimal maximum *a posteriori* (MAP) decoder for the Boolean version of the problem was recently identified in [69]. Multireference alignment with a linear observation model and sub-Gaussian additive noise was considered in [70]. Here, denoising methods agnostic to the noise model parameters were defined with order-optimal minimax performance rates.

The problem of image registration involves identifying the transformation that aligns a noisy copy of an image to its reference, as shown in Fig. 9.7. Note that the aim here is

Figure 9.7 Image registration: align a corrupted copy **Y** to reference **X** without knowledge of the channel distribution.

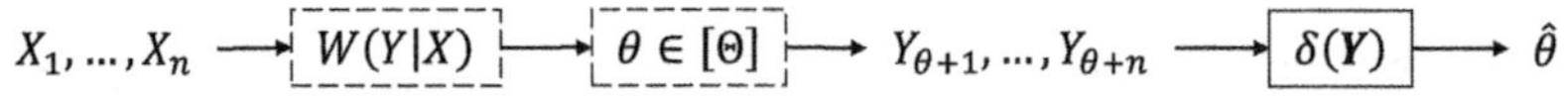

Figure 9.8 Delay estimation: estimate the finite cyclic delay from a noisy, cyclically rotated sequence without channel knowledge.

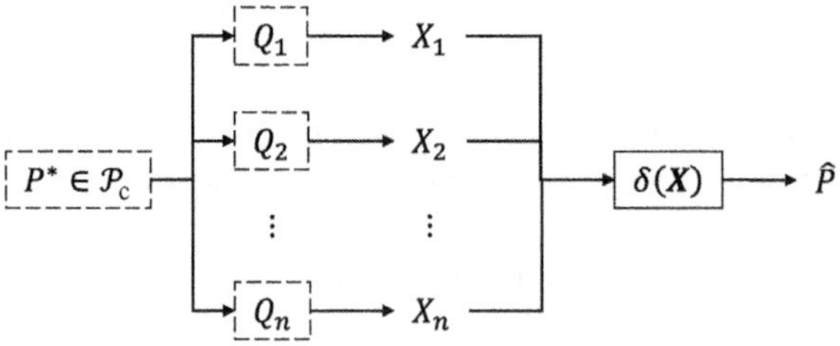

Figure 9.9 Clustering: determine the correct partition without knowledge of the conditional distributions of observations.

only to align the copy to the reference and not necessarily to denoise the signal. The max mutual information method has been considered for universal image registration [71]. In the universal version of the registration problem with unknown discrete memoryless channels, the decoder has been shown to be universally asymptotically optimal in the error exponent for registering two images (but not for more than two images) [72]. The registration approach is derived from that used for universal delay estimation for discrete channels and memoryless sources, given cyclically shifted, noisy versions of the transmitted signal, in [73]. The model for the delay estimation problem is shown in Fig. 9.8.

Classification and clustering of data sources have also been considered from the universal setting and can broadly be summarized by Fig. 9.9. As described earlier, even the communication problem can be formulated as one of universal clustering of messages, encoded according to an unknown random codebook, and transmitted across an unknown discrete memoryless channel, by using the MPI decoder [58].

Note that the task of clustering such objects is strictly stronger than the universal binary hypothesis test of identifying whether two objects are drawn from the same source or not. Naturally, several ideas from the hypothesis literature translate to the design of clustering algorithms. Binary hypothesis testing, formulated as a classification problem using empirically observed statistics, and the relation of the universal discriminant functions to universal data compression are elaborated in [74]. An asymptotically optimal decision rule for the problem under hidden Markov models with unknown statistics in the Neyman–Pearson formulation is designed and compared with the GLRT in [75].

Unsupervised clustering of discrete objects under universal crowdsourcing was studied in [76]. In the absence of knowledge of the crowd channels, we define budget-optimal universal clustering algorithms that use distributional identicality and response

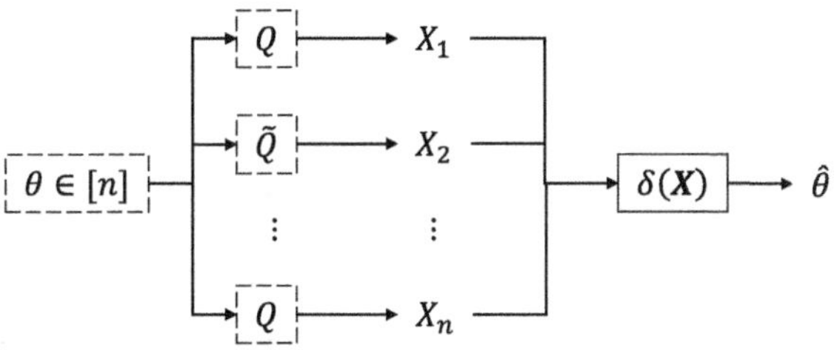

Figure 9.10 Outlier hypothesis testing: determine outlier without knowledge of typical and outlier hypothesis distributions.

dependence across similar objects. Exponentially consistent universal image clustering algorithms using multivariate information functionals were also defined in [72].

A problem that is similar to clustering data sources is that of outlier hypothesis testing, where the aim is to identify data sources that are drawn according to an outlier distribution. The system model for the outlier hypothesis testing problem is shown in Fig. 9.10. Note that this forms a strict subset of the clustering problems. The GLRT is studied for outlier testing in the universal setting where the distributions of the outlier distribution and typical distribution are unknown [77]. The error exponents of the test and converse results on the error exponent are also computed.

Note the strong similarities between the various problems of compression, communication, and statistical inference and the intuition they add to the problem of universal clustering. We next define information-theoretic formulations of clustering for both distance- and dependence-based methods, and present some results therein. Before proceeding, however, we formally define the problem of clustering and discuss various notions of similarity that are central to the definition of any clustering algorithm.

9.3 Clustering and Similarity

We now formally define the problem of clustering. Consider a set of n objects $\{X_1,\ldots,X_n\}$. The term object here is defined broadly to encompass a variety of data types. For instance, the objects could be random variables drawn according to an underlying joint distribution, or a set of hypotheses drawn from a family of distributions, or even points in a vector space. The definition of objects is application-specific and is often evident from usage. In general, they may be viewed as random variables, taking values dictated by the space they belong to. For instance, the objects could be images of dogs and cats, with each image being a collection of pixels as determined by the kind of animal in it.

In the following discussion, let $\mathcal{P}$ be the set of all partitions of the index set $[n] = \{1,\ldots,n\}$. Given a set of constraints to be followed by the clusters, as defined by the problem, let $\mathcal{P}_c \subseteq \mathcal{P}$ be the subset of viable partitions satisfying the system constraints.

DEFINITION 9.1 (Clustering) *A* clustering *of a set of n objects is a partition $P \in \mathcal{P}_c$ of the set $[n]$ that satisfies the set of all constraints. A* cluster *is a set in the partition and objects X_i and X_j are in the same cluster when $i, j \in C$ for some cluster $C \in P$.*

In this chapter we restrict the discussion to hard, non-overlapping clustering models and methods. Some applications consider soft versions of clustering wherein each object is assigned a probability of belonging to a set rather than a deterministic assignment. Other versions also consider clusters such that an object can belong to multiple overlapping clusters owing to varying similarity criteria. In our discussion, each object is assigned to a unique cluster.

Data sources have an underlying label associated with them depending on either the class they are sampled from or the statistics governing their generation. However, in the absence of training data to identify the optimal label assignment, clustering the data according to the labels is desirable.

DEFINITION 9.2 (Correct clustering) *A clustering, $P \in \mathcal{P}_{\mathrm{c}}$, is said to be* correct *if all objects with the same label are clustered together, and no two objects with differing labels are clustered together.*

The space of partitions includes a natural ordering to compare the partitions, and is defined as follows.

DEFINITION 9.3 (Partition ordering) *For $P, P' \in \mathcal{P}$, P is* finer *than P', if the following ordering holds:*

$$P \preceq P' \Leftrightarrow \textit{for all } C \in P, \textit{ there exists } C' \in P' : C \subseteq C'.$$

Similarly, P is denser *than P', $P \succeq P' \Leftrightarrow P' \preceq P$. If $P \npreceq P'$ and $P \nsucceq P'$, then the partitions are* comparable, *$P \sim P'$.*

As is evident from Definition 9.2, learning algorithms aim to identify the finest clustering maximizing the similarity criterion. In the context of designing hierarchical clustering structures (also known as taxonomies), algorithms identify sequentially finer partitions, defined by the similarity criterion, to construct the phylogenetic tree.

The performance of a clustering algorithm is characterized by distance metrics such as the symbol or block-wise error probabilities. Depending on context, a unique correct clustering may not exist and, in such cases, we might be interested in a hierarchical clustering of the objects using relative similarity structures.

9.3.1 Universal Clustering Statistical Model

In this chapter we focus on universal clustering of objects drawn according to a statistical model defined according to the correct clustering of the data sources. Let us first formalize the data model under consideration. Let the maximum number of labels be ℓ, and the set of labels, without loss of generality, be $\mathcal{L} = [\ell]$. Let the true label of object i be $L_i \in \mathcal{L}$. Each label has an associated source distribution defined by the label, which for object i is $Q_{L_i} \in \{Q_1, \ldots, Q_\ell\}$. According to this source distribution, an m-dimensional latent vector representation, $\mathbf{X}_i = (X_{i,1}, \ldots, X_{i,m}) \in \mathcal{X}^m$, of the source is drawn. Depending on the source distribution, $\mathcal{X}$ could be a finite or infinite alphabet.

Latent vector representations define the class that the source belongs to. However, we often get to observe only noisy versions of these data sources. In this respect,

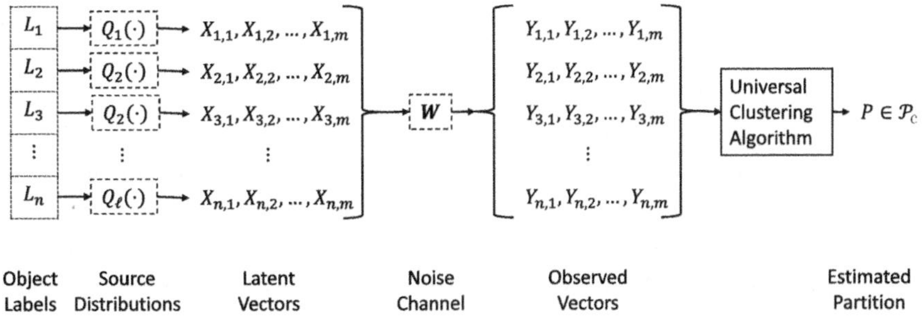

Figure 9.11 Data-generation model for universal clustering. Here we cluster n data sources with labels $(L_1,\ldots,L_n) = (1,2,2,\ldots,\ell)$. True labels define unknown source distributions. Latent vectors are drawn from the source distributions and corrupted by the unknown noise channel, resulting in m-dimensional observation vectors for each source. Universal clustering algorithms cluster the sources using the observation vectors generated according to this model.

let $W(\mathbf{Y}_1,\ldots,\mathbf{Y}_n|\mathbf{X}_1,\ldots,\mathbf{X}_n)$ be the channel that corrupts the latent vectors to generate observation vectors $\mathbf{Y}_1,\ldots,\mathbf{Y}_n \in \mathcal{Y}^m$. The specific source or channel models that define the data samples are application-specific.

A clustering algorithm infers the correct clustering of the data sources using the observation vectors, such that any two objects are in the same cluster if and only if they share the same label. The statistical model is depicted in Fig. 9.11.

DEFINITION 9.4 (Universal clustering) *A clustering algorithm Φ is* universal *if it performs clustering in the absence of knowledge of the source and/or channel distributions, $\{P_1,\ldots,P_\ell\}, W$, and other parameters defining the data samples.*

Example 9.1 Let us now consider a simple example to illustrate the statistical model under consideration. Let $\mu_1,\ldots,\mu_\ell \in \mathbb{R}^n$ be the latent vector distributions corresponding to the labels, i.e., the source distribution is given by $Q_j(\mu_j) = 1$. That is, if source i has label j, i.e., $L_i = j$, then $\mathbf{X}_i = \mu_j$ with probability one. Let W be a memoryless additive white Gaussian noise (AWGN) channel, i.e., $\mathbf{Y}_i = \mathbf{X}_i + \mathbf{Z}_i$, where $\mathbf{Z}_i \sim \mathcal{N}(0,\mathbf{I})$.

Since the latent vector representation is dictated by the mean of the observation vectors corresponding to the label, and since the distribution of each observation vector is Gaussian, the maximum-likelihood estimation of the correct cluster translates to the partition that minimizes the intra-cluster distance of points from the centroids of the cluster. That is, it is the same problem as that which the k-means algorithm attempts to solve.

9.3.2 Similarity Scores and Comparing Clusters

As is evident from the example, the optimal clustering algorithm is dependent on the statistical model (or family) defining the data, the similarity criterion, and the performance metric. The three defining characteristics often depend on each other. For instance, in the example, the Gaussian statistical model in conjunction with the Hamming loss (probability of clustering error) will result in the Euclidean distance between the vectors being the distance criterion, and the ML estimate as obtained from the k-means algorithm (with appropriate initializations) being the optimal clustering algorithm. One can extrapolate the example to see that a Laplacian model for data generation with Hamming loss would translate to an L_1-distance criterion among the data vectors.

In the absence of knowledge of the statistics governing data generation, we require appropriate notions of similarity to define insightful clustering algorithms. Let us define the similarity score of the set of objects $\{\mathbf{Y}_1, \ldots, \mathbf{Y}_n\}$ clustered according to a partition P as $S(\mathbf{Y}_1, \ldots, \mathbf{Y}_n; P)$, where $S : (\mathcal{Y}^m)^n \times \mathcal{P} \to \mathbb{R}$. Clustering can equivalently be defined by a notion of distance between objects, where the similarity can be treated as just the negative/inverse of the distance.

Identifying the best clustering for a task is thus equivalent to maximizing the intra-cluster (or minimizing the inter-cluster) similarity of objects over the space of all viable partitions, i.e.,

$$P^* \in \arg\max_{P \in \mathcal{P}_c} S(\mathbf{Y}_1, \ldots, \mathbf{Y}_n; P). \tag{9.2}$$

Such optimization may not be computationally efficient except for certain similarity functions, as the partition space is discrete and exponentially large.

In the absence of a well-defined similarity criterion, S, defining the "true" clusters that satisfy (9.2), the task may be more art than science [2]. Developing contextually appropriate similarity functions is thus of standalone interest, and we outline some popular similarity functions used in the literature.

Efforts have long been made at identifying universal discriminant functions that define the notion of similarity in classification [74]. Empirical approximation of the Kullback–Leibler (KL) divergence has been considered as a viable candidate for a universal discriminant function to perform binary hypothesis testing in the absence of knowledge of the alternate hypothesis, such that it incurs an error exponent arbitrarily close to the optimal exponent.

The Euclidean distance is the most common measure of pair-wise distance between observations used to cluster objects, as evidenced in the k-means clustering algorithm. A variety of other distances such as the Hamming distance, edit distance, Lempel–Ziv distance, and more complicated formulations such as those in [78, 79] have also been considered to quantify similarity in a variety of applications. The more general class of Bregman divergences have been considered as inter-object distance functions for clustering [80]. On the other hand, a fairly broad class of f-divergence functionals were used as the notion of distance between two objects in [76]. Both of these studies estimate the distributions from the data samples $\mathbf{Y}_1, \ldots, \mathbf{Y}_n$ and use a notion of pair-wise distance between the sources on the probability simplex to cluster them.

Clusters can also be viewed as alternate efficient representations of the data, comprising sufficient information about the objects. The Kolmogorov complexity is widely accepted as a measure of the information content in an object [81]. The Kolomogorov complexity, $K(x)$, of x is the length of the shortest binary program to compute x using a universal Turing machine. Similarly, $K(x|y)$, the Kolmogorov complexity of x given y, is the length of the shortest binary program to compute x, given y, using a universal Turing machine. Starting from the Kolmogorov complexity, in [82] Bennett *et al.* introduce the notion of the *information distance* between two objects X, Y as

$$E_1(X,Y) = \max\{K(X|Y), K(Y|X)\},$$

and argue that this is a universal cognitive similarity distance. The smaller the information distance between two sources, the easier it is to represent one by the other, and hence the more similar they are. A corresponding normalized version of this distance, called the *normalized information distance*, is obtained by normalizing the information distance by the maximum Kolmogorov complexity of the individual objects [83] and is a useful metric in clustering sources according to the pair-wise normalized cognitive similarity.

Salient properties of this notion of universal similarity have been studied [84], and heuristically implemented using word-similarity scores, computed using Google page counts, as empirical quantifications of the score [85]. While such definitions are theoretically universal and generalize a large class of similarity measures, it is typically not feasible to compute the Kolmogorov complexity. They do, however, inspire other practical notions of similarity. For instance, the normalized mutual information between objects has been used as the notion of similarity to maximize the inter-cluster correlation [86]. The normalized mutual information has also been used for feature selection [87], a problem that translates to clustering in the feature space.

Whereas a similarity score guides us toward the design of a clustering algorithm, it is also important to be able to quantitatively compare and evaluate the results of such methods. One standard notion of the quality of a clustering scheme is comparing it against the benchmark set by random clustering ensembles. A variety of metrics based on this notion have been defined.

Arguably the most popular such index was introduced by Rand [1]. Let $P_1, P_2 \in \mathcal{P}_c$ be two viable partitions of a set of objects $\{\mathbf{X}_1, \ldots, \mathbf{X}_n\}$. Let a be the number of pairs of objects that are in the same cluster in both partitions and b be the number of pairs of objects that are in different clusters in both partitions. Then, the Rand index for deterministic clustering algorithms is given by

$$R = \frac{a+b}{\binom{n}{2}},$$

quantifying the similarity of the two partitions as observed through the fraction of pairwise comparisons they agree on. For randomized algorithms, this index is adjusted using expectations of these quantities.

A quantitative comparison of clustering methods can also be made through the normalized mutual information [88]. Specifically, for two partitions P_1, P_2, if $\widehat{p}(C_1, C_2)$ is the fraction of objects that are in cluster C_1 in P_1 and in cluster C_2 in P_2, and if $\widehat{p}_1, \widehat{p}_2$ are the corresponding marginal distributions, then the normalized mutual information is defined as

$$\text{NMI}(P_1, P_2) = \frac{2I(\widehat{p})}{H(\widehat{p}_1) + H(\widehat{p}_2)}.$$

Naturally the NMI value is higher for similar clustering algorithms.

Whereas comparing against the baselines established by random clustering ensembles helps identify structure in the result, the permutation model is often used to generate these ensembles. Drawbacks of using the same method across applications and the variation in the evaluations across different models have been identified [89].

Apart from intuitive quantifications such as these, it has long been argued that the task of clustering is performed only as well as it is evaluated to be done by the humans who evaluate it [2]. The inherently subjective nature of the task has evoked calls for human-based evaluation of clustering methods [90]; some of the standard clustering quality measures are at times in sharp contrast with human evaluations.

Clustering is thus best understood when the appropriate notions of similarity, or equivalently the statistical class generating the data, are available, and when the methods can be benchmarked and evaluated using rigorous mathematical notions or human test subjects. We now describe clustering algorithms defined under distance- and dependence-based similarity notions.

9.4 Distance-Based Clustering

The most prominent criterion for clustering objects is based on a notion of distance between the samples. Conventional clustering tools in machine learning such as k-means, support vector machines, linear discriminant analysis, and k-nearest-neighbor classifiers [91–93] have used Euclidean distances as the notion of distance between objects represented as points in a feature space. As in Example 9.1, these for instance translate to optimal universal clustering methods under Gaussian observation models and Hamming loss. Other similar distances in the feature-space representation of the observations can be used to perform clustering universally and optimally under varying statistical and loss models.

In this section, however, we restrict the focus to universal clustering of data sources in terms of the notions of distance in the conditional distributions generating them. That is, according to the statistical model of Fig. 9.11, we cluster according to the distance between the estimated conditional distributions of $\widehat{p}(Y_i|L_i)$. We describe clustering under sample independence in detail first, and then highlight some distance-based methods used for clustering sources with memory.

9.4.1 Clustering under Independence and Identicality

Let us first consider universal clustering under the following restrictions of independence in data sources. Let us assume that the latent vectors corresponding to the same label are generated as independent and identical samples of a source distribution and that the noise channel is memoryless across sources and samples of the vector. That is, each sample of an observation vector is independent and identically distributed (i.i.d.), and observations of the same label are drawn according to identical distributions. The effective model is depicted in Fig. 9.12.

The iterative clustering mechanism of the k-means algorithm is, for instance, used by the Linde–Buzo–Gray algorithm [94] under the Itakura–Saito distance. Note that this

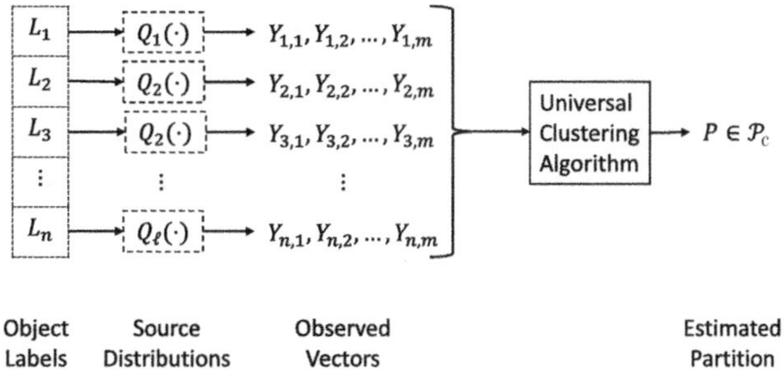

Figure 9.12 Data-generation model for universal distance-based clustering. Here we cluster n sources with labels $(L_1, \dots, L_n) = (1, 2, 2, \dots, \ell)$. True labels define the source distributions and, for each source, m i.i.d. samples are drawn to generate the observation vectors. Source identicality is used to cluster the sources.

is equivalent to data clustering using pair-wise distances between the spectral densities of the sources. A similar algorithm using KL divergences between probability distributions was considered in [95]. More generally, the k-means clustering algorithm can be extended directly to solve the clustering problem under a variety of distortion functions between the distributions corresponding to the sources, in particular, the class of Bregman divergences [96].

DEFINITION 9.5 (Bregman divergence) *Let $f : \mathbb{R}^+ \to \mathbb{R}$ be a strictly convex function, and let p_0, p_1 be two probability mass functions (p.m.f.s), represented by vectors. The Bregman divergence is defined as*

$$B_f(p_0 \| p_1) = f(p_1) - f(p_2) - \langle \nabla f(p_2), p_1 - p_2 \rangle. \tag{9.3}$$

The class of Bregman divergences includes a variety of distortion functions such as the squared loss, KL divergence, Mahalanobis distance, I-divergence, and Itakuro–Saito distance. Clustering according to various such distortion functions translates to varying notions of similarity between the sources. Each of these similarity notions proves to be an optimal universal clustering method for corresponding families of statistical models and loss functions.

Hard and soft clustering algorithms based on minimizing an intra-cluster Bregman divergence functional were defined in [80]. In particular, the clustering task is posed as a quantization problem aimed at minimizing the loss in Bregman information, which is a generalized notion of mutual information defined with respect to Bregman divergences. Iterative solutions are developed for the clustering problem, as in the k-means method. Co-clustering of rows and columns of a doubly stochastic matrix, with each row and column treated as a p.m.f., has also been considered under the same criterion [97].

In [76], we consider the task of clustering n discrete objects according to their labels using the responses of temporary crowd workers. Here, for each object, we collect a set of m independent responses from unknown independent workers drawn at random from a worker pool. Thus, the m responses to a given object are i.i.d., conditioned on a label. Clustering is performed using the empirical response distribution for each object. Consistent clustering involves grouping objects that have the same conditional response distributions.

In particular, we cluster objects that have empirical distributions that are close to each other in the sense of an f-divergence functional.

DEFINITION 9.6 (f-divergence) *Let p,q be discrete probability distributions defined on a space of m alphabets. Given a convex function $f : [0,\infty) \to \mathbb{R}$, the* f-divergence *is defined as*

$$D_f(p\|q) = \sum_{i=1}^{m} q_i f\left(\frac{p_i}{q_i}\right). \tag{9.4}$$

The function f is said to be normalized *if $f(1) = 0$.*

We consider clustering according to a large family of pair-wise distance functionals defined by f-divergence functionals with convex functions f such that

(1) f is twice differentiable on $[r,R]$, and
(2) there exist real constants $c, C < \infty$ such that

$$c \le x f''(x) \le C, \text{ for all } x \in (r,R).$$

Using empirical estimates of conditional distributions of observations, we define a weighted graph G such that the weights are defined by $w(i,j) = d_{ij}\mathbf{1}\{d_{ij} > \gamma_m\}$, where $d_{ij} = \widehat{D}_f(Y_i\|Y_j)$ is the plug-in estimate of the f-divergence estimates. Here, a sufficiently small threshold is chosen according to the number of samples m per object, decaying polynomially with m. Then the clustering is performed by identifying the minimum-weight partition such that the clusters are maximal cliques in the graph, as shown in Fig. 9.13.

This universal clustering algorithm is not only statistically consistent, but also incurs order-optimal sample complexity [98]. In particular, the number of responses per object, m, for reliable clustering scales as $m = O(\log n)$ with the number of objects n to cluster, which is order-optimal. That is, in the absence of the statistics defining worker responses to objects and their reliabilities, the universal clustering algorithm performs budget-optimal reliable clustering of the discrete set of objects.

Outlier hypothesis testing is a problem closely related to clustering and has been considered extensively in the universal context. In this problem, the set of labels is $\mathcal{L} = \{0\} \cup \mathcal{S}$, where $\mathcal{S}$ is a set of parameters. We have a collection of n sources of i.i.d. data, of which a majority follow a typical distribution π (label 0), and the rest are outliers

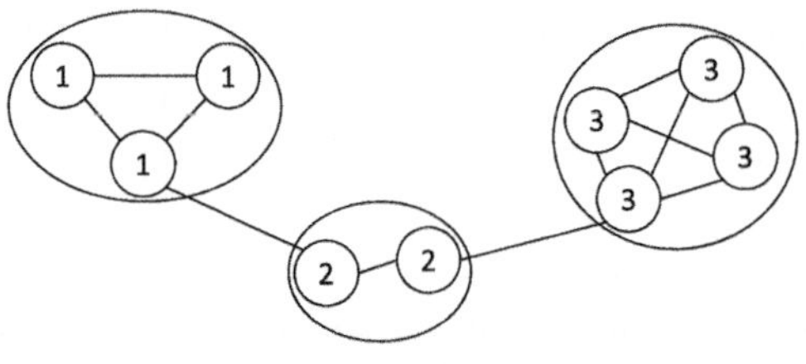

Figure 9.13 Distance-based clustering of nine objects of three types according to [76]. The graph is obtained by thresholding the f-divergences of empirical response distributions. The clustering is then done by identifying the maximal clusters in the thresholded graph.

whose data are generated according to a distribution p_s for some $s \in \mathcal{S}$. In general the set of parameters can be uncountably infinite, and in this sense the set of labels need not necessarily be discrete. However, this does not affect our study of clustering of a finite collection of sources. We are interested in separating the outliers from the typical sources, and thus we wish to cluster the sources into two clusters. In the presence of knowledge of the typical and outlier distributions, the optimal detection rule is characterized by the generalized likelihood ratio test (GLRT) wherein each sequence can be tested for being an outlier as

$$\delta(y^m): \frac{\max_{s \in \mathcal{S}} \prod_{j \in [m]} p_s(y_j)}{\prod_{j \in [m]} \pi(y_j)} \underset{0}{\overset{1}{\gtrless}} \eta,$$

where the threshold η is chosen depending on the priors of typical and outlier distributions, and decision 1 implies that the sequence is drawn from an outlier.

In universal outlier hypothesis testing [77], we are unaware of the typical and outlier hypothesis distributions. The aim is then to design universal tests such that we can detect the outlier sequences universally. We know that, for any distribution p and a sequence of i.i.d. observations y^m drawn according to this distribution, if $\widehat{p}$ is the empirical distribution of the observed sequence, then

$$p(y^m) = \exp(-m(H(\widehat{p}) + D(p\|\widehat{p}))). \tag{9.5}$$

Then any likelihood ratio for two distributions essentially depends on the difference in KL divergences of the corresponding distributions from the empirical estimate. Hence, the outlier testing problem is formulated as one of clustering typical and outlier sequences according to the KL divergences of the empirical distributions from the cluster centroids by using (9.5) in a variety of settings of interest [99–101]. Efficient clustering-based outlier detection methods, with a linear computational complexity in the number of sources, with universal exponential consistency have also been devised [102]. Here the problem is addressed by translating it into one of clustering according to the empirical distributions, with the KL divergence as the similarity measure.

A sequential testing variation of the problem has also been studied [103], where the sequential probability ratio test (SPRT) is adapted to incorporate (9.5) in the likelihood ratios. In the presence of a unique outlier distribution μ, the universal test is exponentially consistent with the optimal error exponent given by $2B(\pi,\mu)$, where $B(p,q)$ is the Bhattacharya distance between the distributions p and q. Further, this approach is also exponentially consistent when there exists at least one outlier, and it is consistent in the case of the null hypothesis (no outlier distributions) [77].

9.4.2 Clustering Sources with Memory

Let us now consider universal clustering models using sources with memory. That is, for each label, the latent vector is generated by a source with memory and thus the components of the vector are dependent. Let us assume that the channel W is the same – memoryless. For such sources, we now describe some novel universal clustering methods.

As noted earlier, compressed streams represent cluster centers either as themselves or as clusters in the space of the codewords. In this sense, compression techniques have been used in defining clustering algorithms for images [104]. In particular, a variety of image segmentation algorithms have been designed starting from compression techniques that fundamentally employ distance-based clustering mechanisms over the sources generating the image features [105–107].

Compression also inspires more distance-based clustering algorithms, using the compression length of each sequence when compressed by a selected compressor C. The normalized compression distance between sequences X, Y is defined as

$$\mathrm{NCD}(X\|Y) = \frac{C(X,Y) - \min\{C(X), C(Y)\}}{\max\{C(X), C(Y)\}}, \tag{9.6}$$

where $C(X,Y)$ is the length of compressing the pair. Thus, the normalized compression distance represents the closeness of the compressed representations of the sequence, therein translating to closeness of the sources as dictated through the compressor C. Thus, the choice of compression scheme here characterizes the notion of similarity and the corresponding universal clustering model it applies to. Clustering schemes based on the NCD minimizing (maximizing) the intra-cluster (inter-cluster) distance are considered in [108].

Clustering of stationary, ergodic random processes according to the source distribution has also been studied, wherein two objects are in the same cluster if and only if they are drawn from the same distribution [109]. The algorithm uses empirical estimates of a weighted L_1 distance between the source distributions as obtained from the data streams corresponding to each source. Statistical consistency of the algorithm is established for generic ergodic sources. This method has also been used for clustering time-series data [110].

The works we have summarized here represent a small fraction of the distance-based clustering methods in the literature. It is important to notice the underlying thread of universality in each of these methods as they do not assume explicit knowledge of the statistics that defines the objects to be clustered. Thus we observe that, in the universal framework, one is able to perform clustering reliably, and often with strong guarantees such as order-optimality and exponential consistency using information-theoretic methods.

9.5 Dependence-Based Clustering

Another large class of clustering algorithms consider the dependence among random variables to cluster them. Several applications such as epidemiology and meteorology, for instance, generate temporally or spatially correlated sources of information with an element of dependence across sources belonging to the same label. We study the task of clustering such sources in this section, highlighting some of these formulations and universal solutions under two main classes – independence clustering using graphical models and clustering using multivariate information functionals.

9.5.1 Independence Clustering by Learning Graphical Models

In this section, we consider the problem of independence clustering of a set of random variables, defined according to an underlying graphical model, in a universal sense.

DEFINITION 9.7 (Independence clustering) *Consider a collection of random variables* $(X_1,\ldots,X_n) \sim Q_G$*, where* Q_G *is the joint distribution defined according to a graphical model defined on the graph* $G = ([n], E)$. Independence clustering *of this set of random variables identifies the finest partition* $P^* \in \mathcal{P}$ *such that the clusters of random variables are mutually independent [111].*

Before we study the problem in detail, it is important to note that this problem is fundamentally different from independent component analysis (ICA) [112], where the idea is to learn a linear transformation such that the components are mutually independent. Several studies have also focused on learning the dependence structures [113] or their optimal tree-structured approximations [114]. In its most general form, it is known that recovering the Bayesian network structure, from data even for just Ising models, for a collection of random variables is NP-hard [115, 116]. Independence clustering essentially requires us to identify the connected components of the Bayesian network and hence, though simpler, cannot always be efficiently learned from data. Additionally, as we are interested in mutual independence among clusters, pair-wise measurements are insufficient, translating to prohibitive sample complexities.

It has been observed that under constraints like bounded in-degrees, or under restricted statistical families, the recovery of a graphical model can in fact be performed efficiently [117–119].

Example 9.2 Consider a collection of n jointly Gaussian random variables $(X_1,\ldots,X_n) \sim \mathcal{N}(0,\Sigma)$, represented by the Gaussian graphical model defined on the graph $G = ([n], E)$. For the Gaussian graphical model, the adjacency matrix of the graph is equivalent to the support of the precision matrix Σ^{-1}. Hence, learning the graphical model structure is equivalent to identifying the support of the precision matrix.

Since independence clustering seeks only to identify mutual independence, for the Gaussian graphical model, we are interested in identifying the block diagonal decomposition of the precision matrix, i.e., elements of each block diagonal form the clusters. Identifying the non-zero blocks of the precision matrix (independence clusters) is exactly equivalent to identifying the non-zero blocks of the covariance matrix, therein making the task of independence clustering for Gaussian graphical models trivial in terms of both computational $\left(O\left(nk - \binom{k+1}{2}\right) \text{ for } k \text{ clusters}\right)$ and sample complexity (depends on the exact loss function, but the worst-case complexity is $O(n)$).

In this section we study the problem of independence clustering under similar constraints on the statistical models to obtain insightful results and algorithms. A class of graphical models that has been well understood is that of the Ising model [120–122]. Most generally, a fundamental threshold for conditional mutual information has been used in devising an iterative algorithm for structure recovery in Ising models [122]. For the independence clustering problem, we can use the same threshold to design a simple extension of this algorithm to define an iterative independence clustering algorithm for Ising models.

Beside such restricted distribution classes, a variety of other techniques involving greedy algorithms [123], maximum-likelihood estimation [124], variational methods [125], locally tree-like grouping strategies [126], and temporal sampling [127] have been studied for dependence structure recovery using samples, for all possible distributions under specific families of graphs and alphabets. The universality of these methods in recovering the graphical model helps define direct extensions of these algorithms to perform universal independence clustering.

Clustering using Bayesian network recovery has been studied in a universal crowdsourcing context [76], where the conditional dependence structure is used to identify object similarity. We consider object clustering using responses of a crowdsourcing system with long-term employees. Here, the system employs m workers such that each worker labels each of the n objects. The responses of a given worker are assumed to be dependent in a Markov fashion on the responses to the previous object and the most recent object of the same class as shown in Fig. 9.14.

More generally, relating to the universal clustering model of Fig. 9.11, this version of independence clustering reduces to Fig. 9.15. That is, the crowd workers observe the object according to their true label, and provide a label appropriately, depending on the observational noise introduced by the channel W. Thus, the latent vector representation is a simple repetition code of the true label. The responses of the crowd workers are defined by the Markov channel W, such that the response to an object is dependent on that offered for the most recent object of the same label, and the most recent object.

Here, we define a clustering algorithm that identifies the clusters by recovering the Bayesian network from the responses of workers. In particular, the algorithm computes the maximum-likelihood (ML) estimates of the mutual information between the responses to pairs of objects and, using the data-processing inequality, reconstructs the Bayesian network defining the worker responses. As elaborated in [98] the number of responses per object required for reliable clustering of n objects is $m = O(\log n)$, which is order-optimal. The method and algorithm directly extend to graphical models with any finite-order Markov dependences.

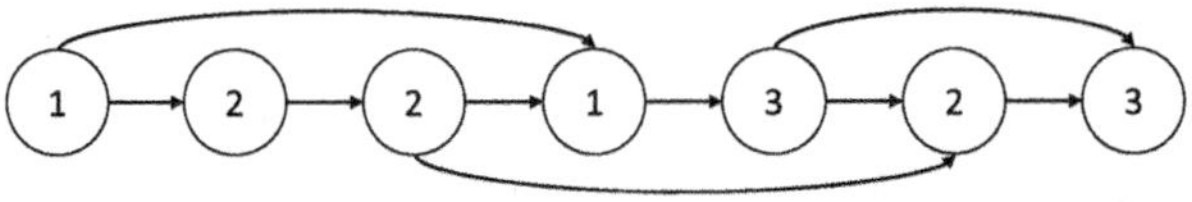

Figure 9.14 Bayesian network model of responses to a set of seven objects chosen from a set of three types in [76]. The most recent response and the response to the most recent object of the same type influence a response.

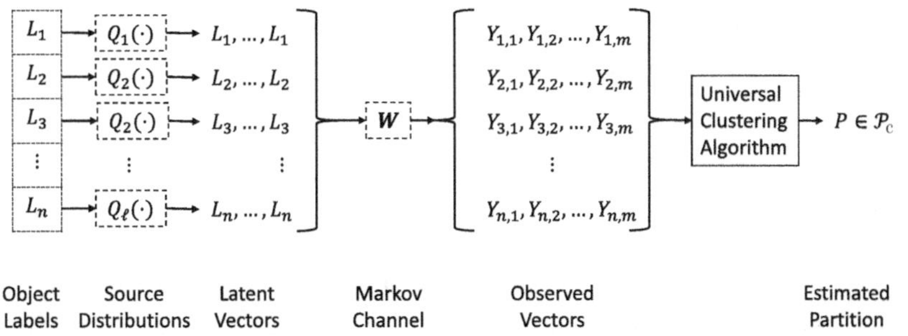

Figure 9.15 Block diagram of independence clustering under a temporal dependence structure. Here we cluster n objects with labels $(L_1, \ldots, L_n) = (1, 2, 2, \ldots, \ell)$. The latent vectors are m-dimensional repetitions of the true label. The observations are drawn according to the Markov channel that introduces dependences in observations across objects depending on the true labels. Clustering is performed according to the inferred conditional dependences in the observations.

A hierarchical approach toward independence clustering is adopted by Ryabko in [111]. Here Ryabko first designs an algorithm to identify the clusters when the joint distribution is given, by recursively dividing each cluster into two, as long as it is feasible. Then the idea is extended to clustering from data samples, wherein the criterion for splitting into clusters can be obtained through a statistical test for conditional independence [128–131]. The universal method is proven to be consistent for stationary, ergodic, unknown sources when the number of clusters is given, as the conditional independence test is reliable, given a sufficient number of samples.

The relationship between compression and clustering within the independence clustering framework emerges through the information-bottleneck principle [132, 133]. This problem considers quantization of an observation in the presence of latent variables, such that the random variable is maximally compressed up to a requirement of a minimum amount of information on the latent variable. Specifically, consider the Markov chain

$$Y \to X \to \widetilde{X}.$$

Let Y be the latent variable representing the object label, and let X be the observation. The quantized representation of X, $\widetilde{X}$, represents the cluster it is associated with. If the (randomized) clustering is performed according to the conditional distribution $p(\widetilde{x}|x)$, then the information-bottleneck principle aims to minimize the Lagrangian

$$\mathcal{L}(p(\widetilde{x}|x)) = I(X; \widetilde{X}) - \beta I(\widetilde{X}; Y), \tag{9.7}$$

where $\beta \geq 0$ is the Lagrange multiplier, as against the unconstrained maximization of $I(\widetilde{X}; Y)$. It has been observed that the optimal assignment minimizing (9.7) is given by the exponential family

$$p(\widetilde{x}|x) = \frac{p(x)}{Z(x;\beta)} \exp(-\beta D(p(Y|x) \| p(Y|\widetilde{x}))), \tag{9.8}$$

and can be computed iteratively using a method similar to the Blahut–Arimoto algorithm [134]. This framework extends naturally to the clustering of multiple random variables according to the latent variable, and corresponding universal clustering methods can be defined according to the information-bottleneck constraints.

The rate of information bottleneck (hypercontractivity coefficient) for random variables X, Y is defined as

$$s(X;Y) = \sup_{U-X-Y} \frac{I(U;Y)}{I(U;X)},$$

where $U-X-Y$ is a Markov chain and U is viewed as a summary of X. This functional has been used as a measure of correlation between random variables and has been used in performing correlation-based clustering [135]. Other variants of the information-bottleneck method including multivariate [136], Gaussian [137], and agglomerative formulations [138, 139] have also been studied extensively, forming information-constrained universal clustering methods.

Thus we saw in this section that universal clustering algorithms can be defined for random variables drawn according to a graphical model, using reliable, and often efficient, structure recovery methods.

9.5.2 Multivariate Information-Based Clustering

Whereas independence clustering aims to recover mutually independent components, such components rarely exist in practice. By analogy to the distance clustering framework, we can consider a relaxed, and more practical notion of clustering that clusters components such that the intra-cluster information is maximized.

Mutual information has been considered as a valid measure of pair-wise similarity to perform clustering of random variables according to independence [140]. In particular, it has been prominent in the area of genomic clustering and gene-expression evaluation [141, 142]. Hierarchical clustering algorithms using mutual information as the pair-wise similarity functional have also been defined [143, 144]. Mutual information is also a prominent tool used in clustering features to perform feature extraction in big data analysis [87].

However, mutual information quantifies only pair-wise dependence and can miss non-linear dependences in the data. For instance, consider objects $X_1, X_2 \sim \text{Bern}(1/2)$ drawn independently of each other, and let $X_3 = X_1 \oplus X_2$. While any two of the three are pair-wise independent, collectively the three objects in the set are dependent on each other. Hence a meaningful cluster might require all three to be grouped in the same set, which would not be the case if mutual information were used as the similarity score.

In applications that require one to account for such dependences, multivariate information functionals prove to be an effective alternative. One such information functional that is popular for clustering is the *minimum partition information* [145].

DEFINITION 9.8 (Partition information) *Let $X_1, \ldots, X_n$ be a set of random variables, and let $P \in \mathcal{P}$ be a partition of $[n]$ with $|P| > 1$. Then, the* partition information *is defined as*

$$I_P(X_1; \ldots; X_n) = \frac{1}{|P| - 1} \left(\sum_{C \in P} H(X_C) - H(X_1, \ldots, X_n) \right), \tag{9.9}$$

where $H(X_C)$ is the joint entropy of all random variables in the cluster C.

The partition information quantifies the inter-cluster information, normalized by the size of the partition.

LEMMA 9.1 *For any partition $P \in \mathcal{P}$, $I_P(X_1;\ldots;X_n) \geq 0$, with equality if and only if the clusters of P are mutually independent.*

Proof Let $(X_1,\ldots,X_n) \sim Q$. Then, the results follow from the observation that

$$I_P(X_1;\ldots;X_n) = \frac{1}{|P|-1} D\left(Q \| \prod_{C \in P} Q_C\right) \geq 0,$$

where Q_C is the marginal distribution of X_C.

Example 9.3 Consider a set of jointly Gaussian random variables $(X_1,\ldots,X_n) \sim \mathcal{N}(0,\Sigma)$. Then, for any partition $P \in \mathcal{P}$ and any cluster $C \in C$,

$$H(X_C) = \frac{1}{2}(|C|\log(2\pi e) + \log(|\Sigma_C|)),$$

where Σ_C is the covariance submatrix of Σ corresponding to the variables in the cluster, and $|\Sigma_C|$ is the determinant of the matrix. Here of course we use the differential entropy.

Thus, the partition information is given by

$$I_P(X_1;\ldots;X_n) = \frac{1}{|P|-1}\left(\sum_{C \in P} \log(|\Sigma_C|) - \log(|\Sigma|)\right) \tag{9.10}$$

$$= \frac{1}{|P|-1} \log\left(\frac{|\Sigma_P|}{|\Sigma|}\right), \tag{9.11}$$

where $[\Sigma_P]_{i,j} = [\Sigma]_{i,j}\mathbf{1}\{i,j \in C \text{ for some } C \in P\}$, for all $i,j \in [n]^2$. This is essentially the block diagonal matrix obtained according to the partition P. The clusters of P are mutually independent if and only if $\Sigma_P = \Sigma$.

From the definition of the partition information, it is evident that the average inter-cluster (intra-cluster) dependence is minimized (maximized) by a partition that minimizes the partition information.

DEFINITION 9.9 (Minimum partition information) *The minimum partition information (MPI) of a set of random variables $X_1,\ldots,X_n$ is defined as*

$$\underline{I}(X_1;\ldots;X_n) = \min_{P \in \mathcal{P}, |P|>1} I_P(X_1;\ldots;X_n). \tag{9.12}$$

The finest partition minimizing the partition information is referred to as the fundamental partition, i.e., if $\mathcal{P}^* = \arg\min_{P \in \mathcal{P}, |P|>1} I_P(X_1;\ldots;X_n)$, then $P^* \in \mathcal{P}^*$ such that $P^* \preceq P$, for any $P \in \mathcal{P}^*$.

The MPI finds operational significance as the capacity of the multiterminal secret-key agreement problem [146]. More recently, the change in MPI with the addition or removal of sources of common randomness was identified [147], giving us a better understanding of the multivariate information functional.

The MPI finds functional use in universal clustering under communication, in the absence of knowledge of the codebook (also thought of as communicating with aliens) [28, 58]. It is of course infeasible to recover the messages when the codebook is not available, and so the focus here is on clustering similar messages according to the transmitted code stream. In particular, the empirical MPI was used as the universal decoder for clustering messages transmitted through an unknown discrete memoryless channel, when the decoder is unaware of the codebook used for encoding the messages. By identifying the received codewords that are most dependent on each other, the decoder optimally clusters the messages up to the capacity of the channel.

Clustering random variables under the Chow–Liu tree approximation by minimizing the partition information is studied in [148]. A more general version of the clustering problem that is obtained by minimizing the partition information is considered in [149], where the authors describe the clustering algorithm, given the joint distribution.

Identifying the optimal partition is often computationally expensive as the number of partitions is exponential in the number of objects. An efficient method to cluster the random variables is identified in [149]. We know that entropy is a submodular function as for any set of indices A, B,

$$H(X_{A\cup B}) + H(X_{A\cap B}) \leq H(X_A) + H(X_B),$$

as conditioning reduces entropy. The equality holds if and only if $X_A \perp X_B$. For any $A \subseteq [n]$, let

$$h_\gamma(A) = H(X_A) - \gamma, \quad \widehat{h}_\gamma(A) = \min_{P\in\mathcal{P}(A)} \sum_{C\in P} h_\gamma(C), \tag{9.13}$$

where $\mathcal{P}(A)$ is the set of all partitions of the index set A. Then, we have

$$\begin{aligned} \widehat{h}_\gamma(A_1 \cup A_2) &= \min_{P\in\mathcal{P}(A_1\cup A_2)} \sum_{C\in P} h_\gamma(C) \\ &\leq \min_{P\in\mathcal{P}(A_1)} \sum_{C\in P} h_\gamma(C) + \min_{P\in\mathcal{P}(A_2)} \sum_{C\in P} h_\gamma(C) \\ &= \widehat{h}_\gamma(A_1) + \widehat{h}_\gamma(A_2). \end{aligned}$$

Thus, $\widehat{h}_\gamma(\cdot)$ is intersection submodular.

Further, the MPI is given by the value of γ satisfying (9.13) for $A = [n]$. Thus, identifying the partition that minimizes the intersection submodular function $\widehat{h}_\gamma$ yields the clustering solution (fundamental partition), and the value of γ minimizing the function corresponds to the MPI.

Since submodular function minimization can be performed efficiently, the clustering can be performed in strongly polynomial time by using the principal sequence of partitions (PSP) based on the Dilworth truncation lattice formed by $\widehat{h}_\gamma$ [145, 149]. From the partition information and the PSP, a hierarchical clustering solution can be obtained that creates the phylogenetic tree with increasing partition information [150]. However, the submodular minimization requires knowledge of the joint distribution of the random variables. Defining universal equivalents of these clustering algorithms would require not only efficient methods for estimating the partition information, but also a minimization algorithm that is robust to the errors in the estimate.

Example 9.4 Continuing with the example of the multivariate Gaussian random variable in Example 9.3, the minimum partition function is determined by submodular minimization over the functions $f : 2^{[n]} \to \mathbb{R}$, defined by $f(C) = \log(|\Sigma_C|)$, since the optimal partition is obtained as

$$P^* = \arg\min_{P \in \mathcal{P}} \frac{1}{|P| - 1} \sum_{C \in P} \log(|\Sigma_C|).$$

Alternatively, the minimization problem can also be viewed as

$$P^* = \arg\min_{P \in \mathcal{P}} \frac{1}{|P| - 1} \left(\sum_{C \in P} \sum_{i \in [|C|]} \log \lambda_i^C - \sum_{i \in [n]} \log \lambda_i \right),$$

where λ_i^C are the eigenvalues of the covariance submatrix Σ_C and λ_i are the eigenvalues of Σ. In this sense, clustering according to the multivariate information translates to a clustering similar to spectral clustering.

For another view on the optimal clustering solution to this problem, relating to the independence clustering equivalent, let us compare this case with Example 9.2. The independence clustering solution for Gaussian graphical models recovered the block-diagonal decomposition of the covariance matrix. Here, we perform a relaxed version of the same retrieval wherein we return the block-diagonal decomposition with minimum normalized entropy difference from the given distribution.

Thus, the partition information functional proves to be a useful multivariate information functional to cluster random variables according to the dependence structure. This approach is particularly conducive for the design of efficient optimization frameworks to perform the clustering, given a sufficiently large number of samples of the sources.

Multiinformation is another multivariate information functional that has been used extensively to study multivariate dependence [151].

DEFINITION 9.10 (Multiinformation) *Given a set of random variables $X_1, \ldots, X_n$, the multiinformation is defined as*

$$I_M(X_1; \ldots; X_n) = \sum_{i-1}^{n} H(X_i) - H(X_1, \ldots, X_n). \tag{9.14}$$

The multiinformation has, for instance, been used in identifying function transformations that maximize the multivariate correlation within the set of random variables, using a greedy search algorithm [86]. It has also been used in devising unsupervised methods for identifying abstract structure in data from areas like genomics and natural language [152]. In both of these studies the multiinformation proves to be a more robust measure of correlation between data sources and hence a stronger notion to characterize the relationship between them. The same principle is exploited in devising meaningful universal clustering algorithms.

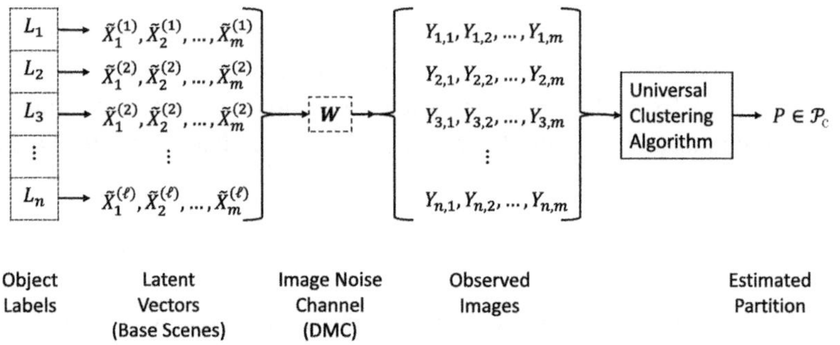

Figure 9.16 Source model for multi-image clustering according to mutual independence of images. Here we cluster n images with labels $(L_1,\ldots,L_n) = (1,2,2,\ldots,\ell)$. The labels define the base scene being imaged (latent vectors). The scenes are subject to an image-capture noise which is independent across scenes and pixels. Clustering is performed according to these captured images universally.

The role of multiinformation in independence-based clustering is best elucidated through the example of joint clustering and registration of images under rigid-body transformations [153]. The problem of clustering of images is defined by the source model depicted in Fig. 9.16. Consider a collection of labels drawn i.i.d. and assume an n-dimensional latent vector for each label. Let the noise model be a discrete memoryless channel resulting in corrupted versions of the latent vectors. In the case of multi-image clustering, the latent vectors represent the true scene being captured and the observations are the resulting images to be clustered. The channel represents the image-capture noise from the imaging process.

Under this model, observations corresponding to the same label (scene) are dependent on each other, whereas those of different labels (scenes) are mutually independent. Thus the sources can be clustered using a multiinformation-based clustering algorithm [72]. In particular, the algorithm considers the intra-cluster multiinformation, called cluster information, defined as

$$I_C^{(P)}(X_1;\ldots;X_n) = \sum_{C\subset P} I_M(X_C). \tag{9.15}$$

Here $I_M(X_C)$ is the multiinformation of the random variables in cluster C.

LEMMA 9.2 ([72]) *If $\underline{I}(X_1;\ldots;X_n) = 0$, then the optimal clustering that minimizes the partition information is given by the finest partition that maximizes the cluster information.*

Proof For any partition P,

$$I_P(X_1;\ldots;X_n) = \frac{1}{|P|-1}\Big(I_M(X_1;\ldots;X_n) - I_C^{(P)}(X_1;\ldots;X_n)\Big).$$

Thus,

$$\begin{aligned}\max_{P\in\mathcal{P}} I_C^{(P)}(X_1;\ldots;X_n) &= I_M(X_1;\ldots;X_n) - \min_{P\in\mathcal{P}}(|P|-1)I_P(X_1;\ldots;X_n)\\ &= I_M(X_1;\ldots;X_n).\end{aligned}$$

From the non-negativity of partition information, and since $\underline{I}(X_1;\ldots;X_n) = 0$,

$$\arg\max_{P\in\mathcal{P}} I_C^{(P)}(X_1;\ldots;X_n) = \arg\min_{P\in\mathcal{P}} I_P(X_1;\ldots;X_n).$$

This attribute characterizes the above-defined optimal universal clustering algorithm for the model in Fig. 9.16. This algorithm is exponentially consistent given the number of clusters [153]. We can also use the cluster information for hierarchical clustering, constructing finer partitions of increasing inter-cluster dependence.

The cluster information is also intersection supermodular, and thus its maximization to identify the optimal partition can also be done efficiently given the joint distribution, much like for the partition information. Robust estimates of the multiinformation and cluster information [154] can thus be used to perform efficient universal clustering.

Example 9.5 For a jointly Gaussian random vector $(X_1,\ldots,X_n) \sim \mathcal{N}(0,\Sigma)$, the multi-information is given by

$$I_M(X_1;\ldots;X_n) = \frac{1}{2}\log\left(\frac{\prod_{i\in[n]}\sigma_i^2}{|\Sigma|}\right), \tag{9.16}$$

where σ_i^2 is the variance of X_i.

Then, the cluster information is given by

$$I_C^{(P)}(X_1;\ldots;X_n) = \frac{1}{2}\log\left(\frac{\prod_{i\in[n]}\sigma_i^2}{|\Sigma_P|}\right), \tag{9.17}$$

where again $[\Sigma_P]_{i,j} = [\Sigma]_{i,j}\mathbf{1}\{i,j\in C \text{ for some } C\in P\}$, for all $i,j\in[n]^2$. If $\widetilde{P} = \{\{1\},\ldots,\{n\}\}$, i.e., the partition of singletons, then the cluster information is essentially given by

$$I_C^{(P)}(X_1;\ldots;X_n) = \frac{1}{2}\log\left(\frac{|\Sigma_{\widetilde{P}}|}{|\Sigma_P|}\right).$$

Thus the cluster information represents the information in the clustered form as compared against the singletons.

Another useful multivariate information functional, which is inspired by the multi-information, is the illum information [155]. This is the Csiszár conjugate of the multi-information and the multivariate extension of the lautum information [156]:

$$L(X_1;\ldots;X_n) = D\left(\prod_{i\in[n]} Q_i \| Q\right), \tag{9.18}$$

where $(X_1,\ldots,X_n)\sim Q$, and Q_i is the marginal distribution of X_i for all $i\in[n]$. The cluster version of the illum information, correspondingly defined as

$$L_C^{(P)} = D\left(\prod_{C\in P} Q_C \| Q\right),$$

is minimized by the partition that corresponds to the independence clustering solution, yielding another similarity function to perform clustering. Whereas the multi-information is upper-bounded by the sum of marginal entropies, the illum information has no such upper bound. This proves particularly useful at recovering dependence structures and in clustering low-entropy data sources.

Example 9.6 For the jointly Gaussian random vector $(X_1,\dots,X_n) \sim \mathcal{N}(0,\Sigma)$, let $\Sigma = \sum_{i=1}^{n} \lambda_i u_i u_i^{\mathrm{T}}$ be the orthonormal eigendecomposition. Then, $|\Sigma| = \prod_{i=1}^{n} \lambda_i$, and the illum information is given by

$$L(X_1;\dots;X_n) = \frac{1}{2}\sum_{i=1}^{n}\left[\frac{u_i^{\mathrm{T}}\widehat{\Sigma}u_i}{\lambda_i} - \log\left(\frac{\sigma_i^2}{\lambda_i}\right) - 1\right], \tag{9.19}$$

where $\widehat{\Sigma}$ is the diagonal matrix of variance values. In comparison, the multiinformation is given by

$$I_M(X_1;\dots;X_n) = \frac{1}{2}\sum_{i=1}^{n}\log\left(\frac{\sigma_i^2}{\lambda_i}\right).$$

Example 9.7 For a pair-wise Markov random field (MRF) defined on a graph $G = (V,E)$ as

$$Q_G(X_1;\dots;X_n) = \exp\left(\sum_{i\in V}\psi_i(X_i) + \sum_{(i,j)\in E}\psi_{ij}(X_i,X_j) - A(\psi)\right), \tag{9.20}$$

where $A(\psi)$ is the log partition function, the *sum information* [155], which is defined as

$$S(X_1;\dots;X_n) = I(X_1;\dots;X_n) + L(X_1;\dots;X_n), \tag{9.21}$$

reduces to

$$S(X_1;\dots;X_n) = \sum_{(i,j)\in E}\mathbb{E}_{Q_G}\left[\psi_{ij}(X_i,X_j)\right] - \mathbb{E}_{Q_iQ_j}\left[\psi_{ij}(X_i,X_j)\right].$$

That is, the sum information is dependent only on the expectation of the pair-wise potential functions defining the MRF. In particular, it is independent of the partition function and thus statistically and computationally easier to estimate when the potential functions are known. This also gives us an alternative handle on information-based dependence structure recovery.

From Example 9.7, we see the advantages of the illum information and other correspondingly defined multivariate information functionals in universal clustering of data sources. Thus, depending on the statistical families defining the data sources, we see that a variety of multivariate information functionals, serving as natural notions of similarity, can be used to recover clusters. The bottlenecks for most of these methods are the sample and/or the computational complexity (moving away from universality, costs may be reduced for restricted families).

9.6 Applications

Data often include all the information we require for processing them, and the difficulty in extracting useful information typically lies in understanding their fundamental structure and in designing effective methods to process such data. Universal methods aim to learn this information and perform unsupervised learning tasks such as clustering and provide a lot of insight into problems we have little prior understanding on. In this section we give a very brief overview of some of the applications of information-theoretic clustering algorithms.

For instance, in [28] Misra hypothesizes the ability to communicate with aliens by clustering messages under a universal communication setting without knowledge of the codebook. That is, if one presumes that aliens communicate using a random codebook, then the codewords can be clustered reliably, albeit without the ability to recover the mapping between message and codeword.

The information-bottleneck method has been used for document clustering [157] and for image clustering [158] applications. Similarly, clustering of pieces of music through universal compression techniques has been studied [159, 160].

Clustering using the minimum partition information has been considered in biology [149]. The information content in gene-expression patterns is exploited through info-clustering for genomic studies of humans. Similarly, firing patterns of neurons over time are used as data to better understand neural stimulation of the human brain by identifying modular structures through info-clustering.

Clustering of images in the absence of knowledge of image and noise models along with joint registration has been considered [153]. It is worth noting that statistically consistent clustering and registration of images can be performed with essentially no information about the source and channel models.

Multivariate information functionals in the form of multiinformation have also been utilized in clustering data sources according to their latent factors in a hierarchical fashion [152, 161]. This method, which is related to ICA in the search for latent factors using information, has been found to be efficient in clustering statements according to personality types, in topical characterization of news, and in learning hierarchical clustering from DNA strands. The method has in fact been leveraged in social-media content categorization with the motivation of identifying useful content for disaster response [162].

Beside direct clustering applications, the study of universal clustering is significant for understanding fundamental performance limits of systems. As noted earlier, understanding the task of clustering using crowdsourced responses, without knowledge of the crowd channels, helps establish fundamental benchmarks for practical systems [76].

Establishing tight bounds on the cost of clustering or the performance of universal algorithms helps in choosing appropriate parameters for practical adaptations of the algorithms. Moreover, universal methods highlight the fundamental structural properties inherent in the data, with minimal contextual knowledge.

9.7 Conclusion

Clustering is ubiquitous in data pre-processing and structure learning in unsupervised settings. Hence it is imperative that we have a stronger understanding of the problem. Although it is often context-sensitive, the task of clustering can also be viewed in the broader sense of universality. Here we have reviewed the strength of these context-independent techniques for clustering.

In our review, we have seen universal information-theoretic methods yielding novel techniques that not only translate to practice, but also establish fundamental limits in the space. Practical heuristics are often built either using these techniques or by using the fundamental limits as performance benchmarks.

We also noticed the integral role of classical information theory in inference problems in unsupervised clustering. Design principles observed for the problems of communication and compression also prove to be useful guidelines for the data clustering framework. Further, information theory also helps develop rigorous theoretical insight into the problem and the methods therein, helping us devise systematic development pipelines for tackling the challenge.

9.7.1 Future Directions

While many tools and techniques have been explored, devised, and analyzed, much is still left to be understood on universal clustering in particular, and unsupervised learning in general. A careful exploration of the information-theory literature could very well be a strong guiding principle for solving learning problems.

Efficient, robust, universal learning could very well reduce the sample complexities of supervised learning methods. It has already been observed that efficient unsupervised learning of abstract data representations significantly reduces the training complexity in visual object recognition [163]. Universal clustering in crowdsourcing has been proposed to reduce the labeling complexity and the domain expert requirement [76]. Similarly, statistically grounded, information-theoretic methods addressing universal clustering could very well generate efficient representations of unlabeled information, thereby facilitating better learning algorithms.

The focus of most information-theoretic methods addressing universal clustering has largely been on statistical consistency, in the process often neglecting the associated computational complexity. An important avenue of study for the future would be to design computationally efficient clustering algorithms. More generally, we require a concrete understanding of the fundamental trade-offs between computational and statistical optimalities [164]. This would also help in translating such methods to practical datasets and in developing new technologies and applications.

Most analyses performed within this framework have considered first-order optimalities in the performance of the algorithms such as error exponents. These results are useful when we have sufficiently large (asymptotic) amounts of data. However, when we have finite (moderately large) data, stronger theoretical results focusing on second- and third-order characteristics of the performance through CLT-based finite-blocklength analyses are necessary [165].

Information functionals, which are often used in universal clustering algorithms, have notoriously high computational and/or sample complexities of estimation and thus an important future direction of research concerns the design of robust and efficient estimators [166].

These are just some of the open problems in the rich and burgeoning area of universal clustering.

References

[1] W. M. Rand, "Objective criteria for the evaluation of clustering methods," *J. Am. Statist. Assoc.*, vol. 66, no. 336, pp. 846–850, 1971.

[2] U. von Luxburg, R. C. Williamson, and I. Guyon, "Clustering: Science or art?" in *Proc. 29th International Conference on Machine Learning (ICML 2012)*, 2012, pp. 65–79.

[3] G. C. Bowker and S. L. Star, *Sorting things out: Classification and its consequences*. MIT Press, 1999.

[4] D. Niu, J. G. Dy, and M. I. Jordan, "Iterative discovery of multiple alternative clustering views," *IEEE Trans. Pattern Analysis Machine Intelligence*, vol. 36, no. 7, pp. 1340–1353, 2014.

[5] L. Valiant, *Probably approximately correct: Nature's algorithms for learning and prospering in a complex world*. Basic Books, 2013.

[6] J. T. Vogelstein, Y. Park, T. Ohyama, R. A. Kerr, J. W. Truman, C. E. Priebe, and M. Zlatic, "Discovery of brainwide neural-behavioral maps via multiscale unsupervised structure learning," *Science*, vol. 344, no. 6182, pp. 386–392, 2014.

[7] R. E. Sanderson, A. Helmi, and D. W. Hogg, "Action-space clustering of tidal streams to infer the Galactic potential," *Astrophys. J.*, vol. 801, no. 2, 18 pages, 2015.

[8] E. Gibson, R. Futrell, J. Jara-Ettinger, K. Mahowald, L. Bergen, S. Ratnasingam, M. Gibson, S. T. Piantadosi, and B. R. Conway, "Color naming across languages reflects color use," *Proc. Natl. Acad. Sci. USA*, vol. 114, no. 40, pp. 10 785–10 790, 2017.

[9] C. E. Shannon, "Bits storage capacity," Manuscript Division, Library of Congress, handwritten note, 1949.

[10] M. Weldon, *The Future X Network: A Bell Labs perspective*. CRC Press, 2015.

[11] C. Lintott, K. Schawinski, S. Bamford, A. Slosar, K. Land, D. Thomas, E. Edmondson, K. Masters, R. C. Nichol, M. J. Raddick, A. Szalay, D. Andreescu, P. Murray, and J. Vandenberg, "Galaxy Zoo 1: Data release of morphological classifications for nearly 900 000 galaxies," *Monthly Notices Roy. Astron. Soc.*, vol. 410, no. 1, pp. 166–178, 2010.

[12] A. Kittur, E. H. Chi, and B. Suh, "Crowdsourcing user studies with Mechanical Turk," in *Proc. SIGCHI Conference on Human Factors in Computational Systems (CHI 2008)*, 2008, pp. 453–456.

[13] P. G. Ipeirotis, F. Provost, and J. Wang, "Quality management on Amazon Mechanical Turk," in *Proc. ACM SIGKDD Workshop Human Computation (HCOMP '10)*, 2010, pp. 64–67.

[14] A. Krizhevsky, I. Sutskever, and G. E. Hinton, "ImageNet classification with deep convolutional neural networks," in *Proc. Advances in Neural Information Processing Systems 25*, 2012, pp. 1097–1105.

[15] O. Russakovsky, J. Deng, H. Su, J. Krause, S. Satheesh, S. Ma, Z. Huang, A. Karpathy, A. Khosla, M. Bernstein, A. C. Berg, and L. Fei-Fei, "ImageNet large scale visual recognition challenge," *arXiv:1409.0575* [cs.CV], 2014.

[16] T. Simonite, "Google's new service translates languages almost as well as humans can," *MIT Technol. Rev.*, Sep. 27, 2016.

[17] Y. Wu, M. Schuster, Z. Chen, Q. V. Le, M. Norouzi, W. Macherey, M. Krikun, Y. Cao, Q. Gao, K. Macherey, J. Klingner, A. Shah, M. Johnson, X. Liu, L. Kaiser, S. Gouws, Y. Kato, T. Kudo, H. Kazawa, K. Stevens, G. Kurian, N. Patil, W. Wang, C. Young, J. Smith, J. Riesa, A. Rudnick, O. Vinyals, G. Corrado, M. Hughes, and J. Dean, "Google's neural machine translation system: Bridging the gap between human and machine translation," *arXiv:1609.08144* [cs.CL], 2017.

[18] D. Silver, A. Huang, C. J. Maddison, A. Guez, L. Sifre, G. Van Den Driessche, J. Schrittwieser, I. Antonoglou, V. Panneershelvam, M. Lanctot, S. Dieleman, D. Grewe, J. Nham, N. Kalchbrenner, I. Sutskever, T. Lillicrap, M. Leach, K. Kavukcuoglu, T. Graepel, and D. Hassabis, "Mastering the game of Go with deep neural networks and tree search," *Nature*, vol. 529, no. 7587, pp. 484–489, 2016.

[19] Y. LeCun, Y. Bengio, and G. Hinton, "Deep learning," *Nature*, vol. 521, no. 7553, pp. 436–444, 2015.

[20] T. Simonite, "The missing link of artificial intelligence," *MIT Technol. Rev.*, Feb. 18, 2016.

[21] C. E. Shannon, "A mathematical theory of communication," *Bell Systems Technical J.*, vol. 27, nos. 3–4, pp. 379–423, 623–656, 1948.

[22] L. R. Varshney, "Block diagrams in information theory: Drawing things closed," in *SHOT Special Interest Group on Computers, Information, and Society Workshop 2014*, 2014.

[23] R. M. Gray, "Source coding and simulation," *IEEE Information Theory Soc. Newsletter*, vol. 58, no. 4, pp. 1/5–11, 2008 (2008 Shannon Lecture).

[24] R. M. Gray, "Time-invariant trellis encoding of ergodic discrete-time sources with a fidelity criterion," *IEEE Trans. Information Theory*, vol. 23, no. 1, pp. 71–83, 1977.

[25] Y. Steinberg and S. Verdú, "Simulation of random processes and rate-distortion theory," *IEEE Trans. Information Theory*, vol. 42, no. 1, pp. 63–86, 1996.

[26] T. M. Cover and J. A. Thomas, *Elements of information theory*. John Wiley & Sons, 1991.

[27] J. Rissanen, "Optimal estimation," *IEEE Information Theory Soc. Newsletter*, vol. 59, no. 3, pp. 1/6–7, 2009 (2009 Shannon Lecture).

[28] V. Misra, "Universal communication and clustering," Ph.D. dissertation, Stanford University, 2014.

[29] J. J. Rissanen, "Generalized Kraft inequality and arithmetic coding," *IBM J. Res. Development*, vol. 20, no. 3, pp. 198–203, 1976.

[30] D. A. Huffman, "A method for the construction of minimum-redundancy codes," *Proc. IRE*, vol. 40, no. 9, pp. 1098–1101, 1952.

[31] J. Ziv and A. Lempel, "Compression of individual sequences via variable-rate coding," *IEEE Trans. Information Theory*, vol. 24, no. 5, pp. 530–536, 1978.

[32] J. Ziv, "Coding theorems for individual sequences," *IEEE Trans. Information Theory*, vol. 24, no. 4, pp. 405–412, 1978.

[33] A. D. Wyner and J. Ziv, "The rate-distortion function for source coding with side information at the decoder," *IEEE Trans. Information Theory*, vol. 22, no. 1, pp. 1–10, 1976.

[34] J. J. Rissanen, "A universal data compression system," *IEEE Trans. Information Theory*, vol. 29, no. 5, pp. 656–664, 1983.

[35] R. Gallager, "Variations on a theme by Huffman," *IEEE Trans. Information Theory*, vol. 24, no. 6, pp. 668–674, 1978.

[36] J. C. Lawrence, "A new universal coding scheme for the binary memoryless source," *IEEE Trans. Information Theory*, vol. 23, no. 4, pp. 466–472, 1977.

[37] J. Ziv, "Coding of sources with unknown statistics – Part I: Probability of encoding error," *IEEE Trans. Information Theory*, vol. 18, no. 3, pp. 384–389, 1972.

[38] L. D. Davisson, "Universal noiseless coding," *IEEE Trans. Information Theory*, vol. 19, no. 6, pp. 783–795, 1973.

[39] D. Slepian and J. K. Wolf, "Noiseless coding of correlated information sources," *IEEE Trans. Information Theory*, vol. 19, no. 4, pp. 471–480, 1973.

[40] T. M. Cover, "A proof of the data compression theorem of Slepian and Wolf for ergodic sources," *IEEE Trans. Information Theory*, vol. 21, no. 2, pp. 226–228, 1975.

[41] I. Csiszár, "Linear codes for sources and source networks: Error exponents, universal coding," *IEEE Trans. Information Theory*, vol. 28, no. 4, pp. 585–592, 1982.

[42] C. E. Shannon, "Coding theorems for a discrete source with a fidelity criterion," *IRE National Convention Record*, vol. 4, no. 1, pp. 142–163, 1959.

[43] T. Berger, *Rate distortion theory: A mathematical basis for data compression.* Prentice-Hall, 1971.

[44] J. Ziv, "Coding of sources with unknown statistics – Part II: Distortion relative to a fidelity criterion," *IEEE Trans. Information Theory*, vol. 18, no. 3, pp. 389–394, May 1972.

[45] J. Ziv, "On universal quantization," *IEEE Trans. Information Theory*, vol. 31, no. 3, pp. 344–347, 1985.

[46] E. Hui Yang and J. C. Kieffer, "Simple universal lossy data compression schemes derived from the Lempel–Ziv algorithm," *IEEE Trans. Information Theory*, vol. 42, no. 1, pp. 239–245, 1996.

[47] A. D. Wyner, J. Ziv, and A. J. Wyner, "On the role of pattern matching in information theory," *IEEE Trans. Information Theory*, vol. 44, no. 6, pp. 2045–2056, 1998.

[48] C. E. Shannon, "Communication in the presence of noise," *Proc. IRE*, vol. 37, no. 1, pp. 10–21, 1949.

[49] V. D. Goppa, "Nonprobabilistic mutual information without memory," *Problems Control Information Theory*, vol. 4, no. 2, pp. 97–102, 1975.

[50] I. Csiszár, "The method of types," *IEEE Trans. Information Theory*, vol. 44, no. 6, pp. 2505–2523, 1998.

[51] P. Moulin, "A Neyman–Pearson approach to universal erasure and list decoding," *IEEE Trans. Information Theory*, vol. 55, no. 10, pp. 4462–4478, 2009.

[52] N. Merhav, "Universal decoding for arbitrary channels relative to a given class of decoding metrics," *IEEE Trans. Information Theory*, vol. 59, no. 9, pp. 5566–5576, 2013.

[53] C. E. Shannon, "Certain results in coding theory for noisy channels," *Information Control*, vol. 1, no. 1, pp. 6–25, 1957.

[54] A. Feinstein, "On the coding theorem and its converse for finite-memory channels," *Information Control*, vol. 2, no. 1, pp. 25–44, 1959.

[55] I. Csiszár and P. Narayan, "Capacity of the Gaussian arbitrarily varying channel," *IEEE Trans. Information Theory*, vol. 37, no. 1, pp. 18–26, 1991.

[56] J. Ziv, "Universal decoding for finite-state channels," *IEEE Trans. Information Theory*, vol. 31, no. 4, pp. 453–460, 1985.

[57] M. Feder and A. Lapidoth, "Universal decoding for channels with memory," *IEEE Trans. Information Theory*, vol. 44, no. 5, pp. 1726–1745, 1998.

[58] V. Misra and T. Weissman, "Unsupervised learning and universal communication," in *Proc. 2013 IEEE International Symposium on Information Theory*, 2013, pp. 261–265.

[59] N. Merhav, "Universal decoding using a noisy codebook," *arXiv:1609:00549* [cs.IT], in *IEEE Trans. Information Theory* , vol. 64. no. 4, pp. 2231–2239, 2018.

[60] A. Lapidoth and P. Narayan, "Reliable communication under channel uncertainty" *IEEE Trans. Information Theory*, vol. 44, no. 6, pp. 2148–2177, 1998.

[61] O. Zeitouni, J. Ziv, and N. Merhav, "When is the generalized likelihood ratio test optimal?" *IEEE Trans. Information Theory*, vol. 38, no. 5, pp. 1597–1602, 1992.

[62] M. Feder and N. Merhav, "Universal composite hypothesis testing: A competitive minimax approach," *IEEE Trans. Information Theory*, vol. 48, no. 6, pp. 1504–1517, 2002.

[63] E. Levitan and N. Merhav, "A competitive Neyman–Pearson approach to universal hypothesis testing with applications," *IEEE Trans. Information Theory*, vol. 48, no. 8, pp. 2215–2229, 2002.

[64] M. Feder, N. Merhav, and M. Gutman, "Universal prediction of individual sequences," *IEEE Trans. Information Theory*, vol. 38, no. 4, pp. 1258–1270, 1992.

[65] M. Feder, "Gambling using a finite state machine," *IEEE Trans. Information Theory*, vol. 37, no. 5, pp. 1459–1465, 1991.

[66] T. Weissman, E. Ordentlich, G. Seroussi, S. Verdú, and M. J. Weinberger, "Universal discrete denoising: Known channel," *IEEE Trans. Information Theory*, vol. 51, no. 1, pp. 5–28, 2005.

[67] E. Ordentlich, K. Viswanathan, and M. J. Weinberger, "Twice-universal denoising," *IEEE Trans. Information Theory*, vol. 59, no. 1, pp. 526–545, 2013.

[68] T. Bendory, N. Boumal, C. Ma, Z. Zhao, and A. Singer, "Bispectrum inversion with application to multireference alignment," vol. 66, no. 4, pp. 1037–1050, 2018.

[69] E. Abbe, J. M. Pereira, and A. Singer, "Sample complexity of the Boolean multireference alignment problem," in *Proc. 2017 IEEE International Symposium on Information Theory*, 2017, pp. 1316–1320.

[70] A. Pananjady, M. J. Wainwright, and T. A. Courtade, "Denoising linear models with permuted data," in *Proc. 2017 IEEE International Symposium on Information Theory*, 2017, pp. 446–450.

[71] P. Viola and W. M. Wells III, "Alignment by maximization of mutual information," *Int. J. Computer Vision*, vol. 24, no. 2, pp. 137–154, 1997.

[72] R. K. Raman and L. R. Varshney, "Universal joint image clustering and registration using partition information," in *Proc. 2017 IEEE International Symposium on Information Theory*, 2017, pp. 2168–2172.

[73] J. Stein, J. Ziv, and N. Merhav, "Universal delay estimation for discrete channels," *IEEE Trans. Information Theory*, vol. 42, no. 6, pp. 2085–2093, 1996.

[74] J. Ziv, "On classification with empirically observed statistics and universal data compression," *IEEE Trans. Information Theory*, vol. 34, no. 2, pp. 278–286, 1988.

[75] N. Merhav, "Universal classification for hidden Markov models," *IEEE Trans. Information Theory*, vol. 37, no. 6, pp. 1586–1594, Nov. 1991.

[76] R. K. Raman and L. R. Varshney, "Budget-optimal clustering via crowdsourcing," in *Proc. 2017 IEEE International Symposium on Information Theory*, 2017, pp. 2163–2167.

[77] Y. Li, S. Nitinawarat, and V. V. Veeravalli, "Universal outlier hypothesis testing," *IEEE Trans. Information Theory*, vol. 60, no. 7, pp. 4066–4082, 2014.

[78] G. Cormode, M. Paterson, S. C. Sahinalp, and U. Vishkin, "Communication complexity of document exchange," in *Proc. 11th Annual ACM-SIAM Symposium on Discrete Algorithms (SODA '00)*, 2000, pp. 197–206.

[79] S. Muthukrishnan and S. C. Sahinalp, "Approximate nearest neighbors and sequence comparison with block operations," in *Proc. 32nd Annual ACM Symposium on Theory Computation (STOC '00)*, 2000, pp. 416–424.

[80] A. Banerjee, S. Merugu, I. S. Dhillon, and J. Ghosh, "Clustering with Bregman divergences," *J. Machine Learning Res.*, vol. 6, pp. 1705–1749, 2005.

[81] M. Li and P. Vitányi, *An introduction to Kolmogorov complexity and its applications*, 3rd edn. Springer, 2008.

[82] C. H. Bennett, P. Gács, M. Li, P. M. B. Vitányi, and W. H. Zurek, "Information distance," *IEEE Trans. Information Theory*, vol. 44, no. 4, pp. 1407–1423, 1998.

[83] M. Li, X. Chen, X. Li, B. Ma, and P. M. B. Vitányi, "The similarity metric," *IEEE Trans. Information Theory*, vol. 50, no. 12, pp. 3250–3264, 2004.

[84] P. Vitanyi, "Universal similarity," in *Proc. IEEE Information Theory Workshop (ITW '05)*, 2005, pp. 238–243.

[85] R. L. Cilibrasi and P. M. B. Vitányi, "The Google similarity distance," *IEEE Trans. Knowledge Data Engineering*, vol. 19, no. 3, pp. 370–383, 2007.

[86] H. V. Nguyen, E. Müller, J. Vreeken, P. Efros, and K. Böhm, "Multivariate maximal correlation analysis," in *Proc. 31st Internatinal Conference on Machine Learning (ICML 2014)*, 2014, pp. 775–783.

[87] P. A. Estévez, M. Tesmer, C. A. Perez, and J. M. Zurada, "Normalized mutual information feature selection," *IEEE Trans. Neural Networks*, vol. 20, no. 2, pp. 189–201, 2009.

[88] L. Danon, A. Diaz-Guilera, J. Duch, and A. Arenas, "Comparing community structure identification," *J. Statist. Mech.*, vol. 2005, p. P09008, 2005.

[89] A. J. Gates and Y.-Y. Ahn, "The impact of random models on clustering similarity," *J. Machine Learning Res.*, vol. 18, no. 87, pp. 1–28, 2017.

[90] J. Lewis, M. Ackerman, and V. de Sa, "Human cluster evaluation and formal quality measures: A comparative study," in *Proc. 34th Annual Conference on Cognitive Science in Society*, 2012.

[91] J. MacQueen, "Some methods for classification and analysis of multivariate observations," in *Proc. 5th Berkeley Symposium on Mathematics Statistics and Probability*, 1967, pp. 281–297.

[92] R. O. Duda, P. E. Hart, and D. G. Stork, *Pattern classification*, 2nd edn. Wiley, 2001.

[93] C. M. Bishop, *Pattern recognition and machine learning*. Springer, 2006.

[94] Y. Linde, A. Buzo, and R. M. Gray, "An algorithm for vector quantizer design," *IEEE Trans. Communication*, vol. 28, no. 1, pp. 84–95, 1980.

[95] I. S. Dhillon, S. Mallela, and R. Kumar, "A divisive information-theoretic feature clustering algorithm for text classification," *J. Machine Learning Res.*, vol. 3, pp. 1265–1287, 2003.

[96] A. Banerjee, X. Guo, and H. Wang, "On the optimality of conditional expectation as a Bregman predictor," *IEEE Trans. Information Theory*, vol. 51, no. 7, pp. 2664–2669, 2005.

[97] I. S. Dhillon, S. Mallela, and D. S. Modha, "Information-theoretic co-clustering," in *Proc. 9th ACM SIGKDD International Conference on Knowledge Discovery and Data Mining (KDD '03)*, 2003, pp. 89–98.

[98] R. K. Raman and L. R. Varshney, "Universal clustering via crowdsourcing," *arXiv:1610.02276* [cs.IT], 2016.

[99] Y. Li, S. Nitinawarat, and V. V. Veeravalli, "Universal outlier detection," in *Proc. 2013 Information Theory Applications Workshop*, 2013.

[100] Y. Li, S. Nitinawarat, and V. V. Veeravalli, "Universal outlier hypothesis testing," in *Proc. 2014 IEEE Internatinal Symposium on Information Theory*, 2014, pp. 4066–4082.

[101] Y. Li, S. Nitinawarat, Y. Su, and V. V. Veeravalli, "Universal outlier hypothesis testing: Application to anomaly detection," in *Proc. IEEE International Conference on Acoustics, Speech, Signal Process. (ICASSP 2015)*, 2015, pp. 5595–5599.

[102] Y. Bu, S. Zou, and V. V. Veeravalli, "Linear-complexity exponentially-consistent tests for universal outlying sequence detection," in *Proc. 2017 IEEE International Symposium on Information Theory*, 2017, pp. 988–992.

[103] Y. Li, S. Nitinawarat, and V. V. Veeravalli, "Universal sequential outlier hypothesis testing," *Sequence Analysis*, vol. 36, no. 3, pp. 309–344, 2017.

[104] J. Wright, Y. Tao, Z. Lin, Y. Ma, and H.-Y. Shum, "Classification via minimum incremental coding length (MICL)," in *Proc. Advances in Neural Information Processing Systems 20*. MIT Press, 2008, pp. 1633–1640.

[105] Y. Ma, H. Derksen, W. Hong, and J. Wright, "Segmentation of multivariate mixed data via lossy data coding and compression," *IEEE Trans. Pattern Analysis Machine Intelligence*, vol. 29, no. 9, pp. 1546–1562, 2007.

[106] A. Y. Yang, J. Wright, Y. Ma, and S. S. Sastry, "Unsupervised segmentation of natural images via lossy data compression," *Comput. Vision Image Understanding*, vol. 110, no. 2, pp. 212–225, 2008.

[107] S. R. Rao, H. Mobahi, A. Y. Yang, S. S. Sastry, and Y. Ma, "Natural image segmentation with adaptive texture and boundary encoding," in *Computer Vision – ACCV 2009*. Springer.

[108] R. Cilibrasi and P. M. B. Vitányi, "Clustering by compression," *IEEE Trans. Information Theory*, vol. 51, no. 4, pp. 1523–1545, 2005.

[109] D. Ryabko, "Clustering processes," in *27th International Conference on Machine Learning*, 2010, pp. 919–926.

[110] A. Khaleghi, D. Ryabko, J. Mary, and P. Preux, "Consistent algorithms for clustering time series," *J. Machine Learning Res.*, vol. 17, no. 3, pp. 1–32, 2016.

[111] D. Ryabko, "Independence clustering (without a matrix)," in *Proc. Advances in Neural Information Processing Systems 30*, 2017, pp. 4016–4026.

[112] A. J. Bell and T. J. Sejnowski, "An information-maximization approach to blind separation and blind deconvolution," *Neural Computation*, vol. 7, no. 6, pp. 1129–1159, 1995.

[113] F. R. Bach and M. I. Jordan, "Beyond independent components: Trees and clusters," *J. Machine Learning Res.*, vol. 4, no. 12, pp. 1205–1233, 2003.

[114] C. K. Chow and C. N. Liu, "Approximating discrete probability distributions with dependence trees," *IEEE Trans. Information Theory*, vol. 14, no. 3, pp. 462–467, 1968.

[115] D. M. Chickering, "Learning Bayesian networks is NP-complete," in *Learning from data*, D. Fisher and H.-J. Lenz, eds. Springer, 1996, pp. 121–130.

[116] A. Montanari and J. A. Pereira, "Which graphical models are difficult to learn?" in *Proc. Advances in Neural Information Processing Systems 22*, 2009, pp. 1303–1311.

[117] P. Abbeel, D. Koller, and A. Y. Ng, "Learning factor graphs in polynomial time and sample complexity," *J. Machine Learning Res.*, vol. 7, pp. 1743–1788, 2006.

[118] Z. Ren, T. Sun, C.-H. Zhang, and H. H. Zhou, "Asymptotic normality and optimalities in estimation of large Gaussian graphical models," *Annals Statist.*, vol. 43, no. 3, pp. 991–1026, 2015.

[119] P.-L. Loh and M. J. Wainwright, "Structure estimation for discrete graphical models: Generalized covariance matrices and their inverses," in *Proc. Advances in Neural Information Processing Systems 25*, 2012, pp. 2087–2095.

[120] N. P. Santhanam and M. J. Wainwright, "Information-theoretic limits of selecting binary graphical models in high dimensions," *IEEE Trans. Information Theory*, vol. 58, no. 7, pp. 4117–4134, 2012.

[121] L. Bachschmid-Romano and M. Opper, "Inferring hidden states in a random kinetic Ising model: Replica analysis," *J. Statist. Mech.*, vol. 2014, no. 6, p. P06013, 2014.

[122] G. Bresler, "Efficiently learning Ising models on arbitrary graphs," in *Proc. 47th Annual ACM Symposium Theory of Computation (STOC '15)*, 2015, pp. 771–782.

[123] P. Netrapalli, S. Banerjee, S. Sanghavi, and S. Shakkottai, "Greedy learning of Markov network structure," in *Proc. 48th Annual Allerton Conference on Communication Control Computation*, 2010, pp. 1295–1302.

[124] V. Y. F. Tan, A. Anandkumar, L. Tong, and A. S. Willsky, "A large-deviation analysis of the maximum-likelihood learning of Markov tree structures," *IEEE Trans. Information Theory*, vol. 57, no. 3, pp. 1714–1735, 2011.

[125] M. J. Beal and Z. Ghahramani, "Variational Bayesian learning of directed graphical models with hidden variables," *Bayesian Analysis*, vol. 1, no. 4, pp. 793–831, 2006.

[126] A. Anandkumar and R. Valluvan, "Learning loopy graphical models with latent variables: Efficient methods and guarantees," *Annals Statist.*, vol. 41, no. 2, pp. 401–435, 2013.

[127] G. Bresler, D. Gamarnik, and D. Shah, "Learning graphical models from the Glauber dynamics," *arXiv:1410.7659* [cs.LG], 2014, to be published in *IEEE Trans. Information Theory*.

[128] A. P. Dawid, "Conditional independence in statistical theory," *J. Roy. Statist. Soc. Ser. B. Methodol.*, vol. 41, no. 1, pp. 1–31, 1979.

[129] T. Batu, E. Fischer, L. Fortnow, R. Kumar, R. Rubinfeld, and P. White, "Testing random variables for independence and identity," in *Proc. 42nd Annual Symposium on the Foundations Computer Science*, 2001, pp. 442–451.

[130] A. Gretton and L. Györfi, "Consistent non-parametric tests of independence," *J. Machine Learning Res.*, vol. 11, no. 4, pp. 1391–1423, 2010.

[131] R. Sen, A. T. Suresh, K. Shanmugam, A. G. Dimakis, and S. Shakkottai, "Model-powered conditional independence test," in *Proc. Advances in Neural Information Processing Systems 30*, 2017, pp. 2955–2965.

[132] N. Tishby, F. C. Pereira, and W. Bialek, "The information bottleneck method," in *Proc. 37th Annual Allerton Conference on Communation Control Computication*, 1999, pp. 368–377.

[133] R. Gilad-Bachrach, A. Navot, and N. Tishby, "An information theoretic trade-off between complexity and accuracy," in *Learning Theory and Kernel Machines*, B. Schölkopf and M. K. Warmuth, eds. Springer, 2003, pp. 595–609.

[134] N. Slonim, "The information bottleneck: Theory and applications," Ph.D. dissertation, The Hebrew University of Jerusalem, 2002.

[135] H. Kim, W. Gao, S. Kannan, S. Oh, and P. Viswanath, "Discovering potential correlations via hypercontractivity," in *Proc. 30th Annual Conference on Neural Information Processing Systems (NIPS)*, 2017, pp. 4577–4587.

[136] N. Slonim, N. Friedman, and N. Tishby, "Multivariate information bottleneck," *Neural Comput.*, vol. 18, no. 8, pp. 1739–1789, 2006.

[137] G. Chechik, A. Globerson, N. Tishby, and Y. Weiss, "Information bottleneck for Gaussian variables," *J. Machine Learning Res.*, vol. 6, no. 1, pp. 165–188, 2005.

[138] N. Slonim and N. Tishby, "Agglomerative information bottleneck," in *Proc. Advances in Neural Information Processing Systems 12*, 1999, pp. 617–625.

[139] N. Slonim, N. Friedman, and N. Tishby, "Agglomerative multivariate information bottleneck," in *Proc. Advances in Neural Information Processing Systems 14*, 2002, pp. 929–936.

[140] J. S. Bridle, A. J. R. Heading, and D. J. C. MacKay, "Unsupervised classifiers, mutual information and 'phantom targets,'" in *Proc. Advances in Neural Information Processing Systems 4*, 1992, pp. 1096–1101.

[141] A. J. Butte and I. S. Kohane, "Mutual information relevance networks: Functional genomic clustering using pairwise entropy measurements," in *Biocomputing 2000*, 2000, pp. 418–429.

[142] I. Priness, O. Maimon, and I. Ben-Gal, "Evaluation of gene-expression clustering via mutual information distance measure," *BMC Bioinformatics*, vol. 8, no. 1, p. 111, 2007.

[143] A. Kraskov, H. Stögbauer, R. G. Andrzejak, and P. Grassberger, "Hierarchical clustering using mutual information," *Europhys. Lett.*, vol. 70, no. 2, p. 278, 2005.

[144] M. Aghagolzadeh, H. Soltanian-Zadeh, B. Araabi, and A. Aghagolzadeh, "A hierarchical clustering based on mutual information maximization," in *Proc. IEEE International Conference on Image Processing (ICIP 2007)*, vol. 1, 2007, pp. I-277–I-280.

[145] C. Chan, A. Al-Bashabsheh, J. B. Ebrahimi, T. Kaced, and T. Liu, "Multivariate mutual information inspired by secret-key agreement," *Proc. IEEE*, vol. 103, no. 10, pp. 1883–1913, 2015.

[146] I. Csiszár and P. Narayan, "Secrecy capacities for multiple terminals," *IEEE Trans. Information Theory*, vol. 50, no. 12, pp. 3047–3061, 2004.

[147] C. Chan, A. Al-Bashabsheh, and Q. Zhou, "Change of multivariate mutual information: From local to global," *IEEE Trans. Information Theory*, vol. 64, no. 1, pp. 57–76, 2018.

[148] C. Chan and T. Liu, "Clustering by multivariate mutual information under Chow–Liu tree approximation," in *Proc. 53rd Annual Allerton Conference on Communication Control Computation*, 2015, pp. 993–999.

[149] C. Chan, A. Al-Bashabsheh, Q. Zhou, T. Kaced, and T. Liu, "Info-clustering: A mathematical theory for data clustering," *IEEE Trans. Mol. Biol. Multi-Scale Commun.*, vol. 2, no. 1, pp. 64–91, 2016.

[150] C. Chan, A. Al-Bashabsheh, and Q. Zhou, "Agglomerative info-clustering," *arXiv:1701.04926* [cs.IT], 2017.

[151] M. Studený and J. Vejnarová, "The multiinformation function as a tool for measuring stochastic dependence," in *Learning in Graphical Models*, M. I. Jordan, ed. Kluwer Academic Publishers, 1998, pp. 261–297.

[152] G. V. Steeg and A. Galstyan, "Discovering structure in high-dimensional data through correlation explanation," in *Proc. 28th Annual Conference on Neural Information Processing Systems (NIPS)*, 2014, pp. 577–585.

[153] R. K. Raman and L. R. Varshney, "Universal joint image clustering and registration using multivariate information measures," *IEEE J. Selected Topics Signal Processing*, vol. 12, no. 5, pp. 928–943, 2018.

[154] M. Studený, "Asymptotic behaviour of empirical multiinformation," *Kybernetika*, vol. 23, no. 2, pp. 124–135, 1987.

[155] R. K. Raman, H. Yu, and L. R. Varshney, "Illum information," in *Proc. 2017 Information Theory Applications Workshop*, 2017.

[156] D. P. Palomar and S. Verdú, "Lautum information," *IEEE Trans. Information Theory*, vol. 54, no. 3, pp. 964–975, 2008.

[157] N. Slonim and N. Tishby, "Document clustering using word clusters via the information bottleneck method," in *Proc. 23rd Annual International ACM SIGIR Conference on Research and Development in Information Retrieval (SIGIR '00)*, 2000, pp. 208–215.

[158] J. Goldberger, H. Greenspan, and S. Gordon, "Unsupervised image clustering using the information bottleneck method," in *Pattern Recognition*, L. Van Gool, ed. Springer, 2002, pp. 158–165.

[159] S. Dubnov, G. Assayag, O. Lartillot, and G. Bejerano, "Using machine-learning methods for musical style modeling," *IEEE Computer*, vol. 36, no. 10, pp. 73–80, 2003.

[160] R. Cilibrasi, P. Vitányi, and R. de Wolf, "Algorithmic clustering of music based on string compression," *Czech. Math. J.*, vol. 28, no. 4, pp. 49–67, 2004.

[161] G. V. Steeg and A. Galstyan, "The information sieve," in *Proc. 33rd International Conference on Machine Learning (ICML 2016)*, 2016, pp. 164–172.

[162] N. O. Hodas, G. V. Steeg, J. Harrison, S. Chikkagoudar, E. Bell, and C. D. Corley, "Disentangling the lexicons of disaster response in twitter," in *Proc. 24th International Conference on the World Wide Web (WWW '15)*, 2015, pp. 1201–1204.

[163] F. Anselmi, J. Z. Leibo, L. Rosasco, J. Mutch, A. Tacchetti, and T. Poggio, "Unsupervised learning of invariant representations," *Theoretical Computer Sci.*, vol. 633, pp. 112–121, 2016.

[164] M. I. Jordan, "On statistics, computation and scalability," *Bernoulli*, vol. 19, no. 4, pp. 1378–1390, 2013.

[165] V. Y. Tan, "Asymptotic estimates in information theory with non-vanishing error probabilities," *Foundations Trends Communication Information Theory*, vol. 11, nos. 1–2, pp. 1–184, 2014.

[166] W. Gao, S. Oh, and P. Viswanath, "Demystifying fixed *k*-nearest neighbor information estimators," *arXiv:1604.03006* [cs.LG], to be published in *IEEE Trans. Information Theory*.

10 Information-Theoretic Stability and Generalization

Maxim Raginsky, Alexander Rakhlin, and Aolin Xu

Summary

Machine-learning algorithms can be viewed as stochastic transformations that map training data to hypotheses. Following Bousquet and Elisseeff, we say that such an algorithm is stable if its output does not depend too much on any individual training example. Since stability is closely connected to the generalization capabilities of learning algorithms, it is of theoretical and practical interest to obtain sharp quantitative estimates on the generalization bias of machine-learning algorithms in terms of their stability properties. This chapter describes several information-theoretic measures of algorithmic stability and illustrates their use for upper-bounding the generalization bias of learning algorithms.

Specifically, we relate the expected generalization error of a learning algorithm (i.e., the expected difference between the empirical risk and the population risk of the hypothesis generated by the algorithm) to several information-theoretic quantities that capture the amount of statistical dependence between the training data and the hypothesis. These include the mutual information and the erasure mutual information, as well as their counterparts induced by the total variation distance. We illustrate the general theory through a number of examples, including the Gibbs algorithm and differentially private algorithms, and discuss various strategies for controlling the generalization error, such as pre- and post-processing, as well as adaptive composition.

10.1 Introduction

In the standard framework of statistical learning theory (see, e.g., [1]), we are faced with the stochastic optimization problem

$$\text{minimize} \qquad L_\mu(w) := \int_{\mathsf{Z}} \ell(w,z)\mu(\mathrm{d}z),$$

where w takes values in some *hypothesis space* W, μ is an unknown probability law on an *instance space* Z, and $\ell : \mathsf{W} \times \mathsf{Z} \to \mathbb{R}^+$ is a given *loss function*. The quantity $L_\mu(w)$ defined above is referred to as the *expected* (or *population*) *risk* of a hypothesis $w \in \mathsf{W}$.

We are given a *training sample* of size n, i.e., an n-tuple $\mathbf{Z} = (Z_1, \dots, Z_n)$ of i.i.d. random elements of Z drawn from μ. A (possibly randomized) *learning algorithm*[1] is a Markov kernel $P_{W|\mathbf{Z}}$ that maps the training data $\mathbf{Z}$ to a random element W of W, and the objective is to generate W with a suitably small population risk

$$L_\mu(W) = \int_{\mathsf{Z}} \ell(W, z)\mu(\mathrm{d}z).$$

Since $L_\mu(W)$ is a random variable, we require it to be small either in expectation or with high enough probability.

This framework is sufficiently rich to cover a wide variety of learning problems (see [2] for additional examples).

- *Binary classification* – let $\mathsf{Z} = \mathsf{X} \times \{0,1\}$, where X is an arbitrary space, let W be a class of functions $w : \mathsf{X} \to \{0,1\}$, and let $\ell(w,z) = \ell(w,(x,y)) = \mathbf{1}_{\{y \neq w(x)\}}$. The elements of W are typically referred to as (binary) *classifiers*.
- *Regression with quadratic loss* – let $\mathsf{Z} = \mathsf{X} \times \mathsf{Y}$, where X is arbitrary and $\mathsf{Y} \subseteq \mathbb{R}$, let W be a class of functions $w : \mathsf{X} \to \mathbb{R}$, and let $\ell(w,z) = \ell(w,(x,y)) = (y - w(x))^2$. The elements of W are typically referred to as *predictors*.
- *Clustering* – let Z be a Hilbert space, let W be the class of all k-element subsets of Z, and let $\ell(w,z) = \min_{a \in w} \|a - z\|_{\mathsf{Z}}^2$, where $\|\cdot\|_{\mathsf{Z}}$ denotes the norm induced by the inner product on Z.
- *Density estimation* – let Z be a Borel set in $\mathbb{R}^d$, let W be a class of bounded probability densities on Z, and let $\ell(w,z) = -\log w(z)$, the negative log-likelihood of z according to the hypothesis density w. It is typically assumed that the densities $w \in \mathsf{W}$ are all lower-bounded by some constant $c > 0$.

The first two examples describe *supervised* learning problems, where each instance $z \in \mathsf{Z}$ splits naturally into a "feature" x and a "label" y, and the objective is to learn to predict the label corresponding to a given feature. The last two examples fall into the class of *unsupervised* learning problems.

Since the data-generating distribution μ is unknown, we do not have enough information to compute $L_\mu(W)$. However, the *empirical loss*

$$L_{\mathbf{Z}}(W) := \frac{1}{n}\sum_{i=1}^{n} \ell(W, Z_i)$$

is a natural proxy that can be computed from the data $\mathbf{Z}$ and from the output of the algorithm W. The *generalization error* of $P_{W|\mathbf{Z}}$ is the difference $L_\mu(W) - L_{\mathbf{Z}}(W)$, and we are interested in its expected value:

$$\mathrm{gen}(\mu, P_{W|\mathbf{Z}}) := \mathbb{E}\big[L_\mu(W) - L_{\mathbf{Z}}(W)\big],$$

where the expectation is w.r.t. the joint probability law $\mathbf{P} := \mu^{\otimes n} \otimes P_{W|\mathbf{Z}}$ of the training data $\mathbf{Z}$ and the algorithm's output W.

[1] We are using the term "algorithm" as a synonym for "rule" or "procedure," without necessarily assuming computational efficiency.

One motivation to study this quantity is as follows. Let us assume, for simplicity, that the infimum $\inf_{w \in \mathsf{W}} L_\mu(w)$ exists, and is achieved by some $w_\circ \in \mathsf{W}$. We can decompose the expected value of the *excess risk* $L_\mu(W) - L_\mu(w_\circ)$ as

$$\begin{aligned}\mathbb{E}[L_\mu(W) - L_\mu(w_\circ)] &= \mathbb{E}\big[L_\mu(W) - L_Z(W)\big] + \mathbb{E}[L_Z(W)] - L_\mu(w_\circ) \\ &= \mathbb{E}\big[L_\mu(W) - L_Z(W)\big] + \mathbb{E}[L_Z(W) - L_Z(w_\circ)] \\ &= \mathrm{gen}(\mu, P_{W|Z}) + \mathbb{E}[L_Z(W) - L_Z(w_\circ)], \end{aligned} \tag{10.1}$$

where in the second line we have used the fact that, for any fixed $w \in \mathsf{W}$, the empirical risk $L_Z(w)$ is an unbiased estimate of the population risk $L_\mu(w)$: $\mathbb{E}L_Z(w) = L_\mu(w)$. This decomposition shows that the expected excess risk of a learning algorithm will be small if its expected generalization error is small and if the expected difference of the empirical risks of W and $w_\circ$ is bounded from above by a small non-negative quantity.

For example, we can consider the empirical risk minimization (ERM) algorithm [1] that returns any minimizer of the empirical risk:

$$W \in \arg\min_{w \in \mathsf{W}} L_Z(w).$$

Evidently, the second term in (10.1) is non-positive, so the expected excess risk of ERM is upper-bounded by its expected generalization error. A crude upper bound on the latter is given by

$$\mathrm{gen}(\mu, P_{W|Z}) \le \mathbb{E}\left[\sup_{w \in \mathsf{W}} |L_\mu(w) - L_Z(w)|\right], \tag{10.2}$$

and it can be shown that, under some restrictions on the complexity of W, the expected supremum on the right-hand side decays to zero as the sample $n \to \infty$, i.e., empirical risks converge to population risks uniformly over the hypothesis class. However, in many cases it is possible to attain asymptotically vanishing excess risk without uniform convergence (see the article by Shalev-Shwartz *et al.* [2] for several practically relevant examples); moreover, the bound in (10.2) is oblivious to the interaction between the data Z and the algorithm's output W, and may be rather loose in some settings (for example, if the algorithm has a fixed computational budget and therefore cannot be expected to explore the entire hypothesis space W).

10.1.1 Algorithmic Stability

As discussed above, machine-learning algorithms are stochastic transformations (or channels, in information-theoretic terms) that map training data to hypotheses. To gauge the quality of the resulting hypothesis, one should ideally evaluate its performance on a "fresh" independent dataset. In practice, this additional verification step is not always performed. In the era of "small" datasets, such a procedure would have been deemed wasteful, while these days it is routine to test many hypotheses and tune many parameters on the same dataset [3]. One may ask whether the quality of the hypothesis can be estimated simply by the *resubstitution estimate* $L_Z(W)$ or by some other statistic, such as the leave-one-out estimate. Algorithmic stability, which quantifies the sensitivity of learning algorithms to local modifications of the training dataset, was introduced in the 1970s by Devroye and Wagner [4] and Rogers and Wagner [5] as a tool for controlling the error of these estimates. More recently, it was studied by Ron and Kearns [6], Bousquet and Elisseeff [7], Poggio *et al.* [8], and Shalev-Shwartz *et al.* [2]. Kutin and Niyogi [9] provided a taxonomy of at least *twelve* different notions of stability, and Rakhlin *et al.* [10] studied another handful.

Recently, the interest in algorithmic stability was renewed through the work on differential privacy [11], which quantifies the sensitivity of the *distribution* of the algorithm's output to local modifications in the data, and can therefore be viewed as a form of algorithmic stability. Once the connection to generalization bounds had been established, it was used to study adaptive methods that choose hypotheses by interacting with the data over multiple rounds [12]. Quantifying stability via differential privacy is an inherently information-theoretic idea, as it imposes limits on the amount of information the algorithm can glean from the observed data. Moreover, differential privacy behaves nicely under composition of algorithms [13, 14], which makes it particularly amenable to information-theoretic analysis.

In this chapter, which is based in part on preliminary results from [15, 16], we focus on information-theoretic definitions of stability, which capture the idea that the output of a stable learning algorithm cannot depend "too much" on any particular example from the training dataset. These notions of stability are weaker (i.e., less restrictive) than differential privacy.

10.2 Preliminaries

10.2.1 Information-Theoretic Background

The *relative entropy* between two probability measures μ and ν on a common measurable space $(\Omega, \mathcal{F})$ is given by

$$D(\mu \| \nu) := \begin{cases} \mathbb{E}_\mu\left[\log\left(\frac{d\mu}{d\nu}\right)\right], & \text{if } \mu \ll \nu, \\ +\infty, & \text{otherwise,} \end{cases}$$

where $\ll$ denotes absolute continuity of measures, and $d\mu/d\nu$ is the Radon–Nikodym derivative of μ with respect to ν. Here and in what follows, we work with natural logarithms. The *total variation distance* between μ and ν is defined as

$$d_{\mathrm{TV}}(\mu, \nu) := \sup_{E \in \mathcal{F}} |\mu(E) - \nu(E)|,$$

where the supremum is over all measurable subsets of Ω. An equivalent representation is

$$d_{\mathrm{TV}}(\mu, \nu) = \sup_{f:\Omega \to [0,1]} \left| \int_\Omega f \, d\mu - \int_\Omega f \, d\nu \right|. \tag{10.3}$$

Moreover, if $\mu \ll \nu$, then

$$d_{\mathrm{TV}}(\mu,\nu) = \frac{1}{2}\int_\Omega \mathrm{d}\nu \left|\frac{\mathrm{d}\mu}{\mathrm{d}\nu} - 1\right|. \tag{10.4}$$

For these and related results, see Section 1.2 of [17].

If (U,V) is a random couple with joint probability law P_{UV}, the mutual information between U and V is defined as $I(U;V) := D(P_{UV} \| P_U \otimes P_V)$. The conditional mutual information between U and V given a third random element Y jointly distributed with (U,V) is defined as

$$I(U;V|Y) := \int P_Y(\mathrm{d}y) D(P_{UV|Y=y} \| P_{U|Y=y} \otimes P_{V|Y=y}).$$

If we use the total variation distance instead of the relative entropy, we obtain the *T-information* $T(U;V) := d_{\mathrm{TV}}(P_{UV}, P_U \otimes P_V)$ (see, e.g., [18]) and the *conditional T-information*

$$T(U;V|Y) := \int P_Y(\mathrm{d}y) d_{\mathrm{TV}}(P_{UV|Y=y}, P_{U|Y=y} \otimes P_{V|Y=y}).$$

The *erasure mutual information* [19] between jointly distributed random objects U and $V = (V_1,\dots,V_m)$ is

$$I^-(U;\mathbf{V}) = I^-(U;V_1,\dots,V_m) := \sum_{k=1}^m I(U;V_k|\mathbf{V}^{-k}),$$

where $\mathbf{V}^{-k} := (V_1,\dots,V_{k-1},V_{k+1},\dots,V_m)$. By analogy, we define the *erasure T-information* as

$$T^-(U;\mathbf{V}) = T^-(U;V_1,\dots,V_m) := \sum_{k=1}^m T(U;V_k|\mathbf{V}^{-k}).$$

The erasure mutual information $I^-(U;\mathbf{V})$ is related to the usual mutual information $I(U;\mathbf{V}) = I(U;V_1,\dots,V_m)$ via the identity

$$I^-(U;\mathbf{V}) = mI(U;\mathbf{V}) - \sum_{k=1}^m I(U;\mathbf{V}^{-k}) \tag{10.5}$$

(Theorem 7 of [19]). Moreover, $I^-(U;\mathbf{V})$ may be larger or smaller than $I(U;\mathbf{V})$, as can be seen from the following examples [19]:

- if U takes at most countably many values and $V_1 = \ldots = V_m = U$, then $I(U;\mathbf{V}) = H(U)$, the Shannon entropy of U, while $I^-(U;\mathbf{V}) = 0$;
- if $V_1,\dots,V_m \overset{\text{i.i.d.}}{\sim} \mathrm{Bern}(\frac{1}{2})$ and $U = V_1 \oplus V_2 \oplus \cdots \oplus V_m$, then $I(U;\mathbf{V}) = \log 2$, while $I^-(U;\mathbf{V}) = n \log 2$;
- if $U \sim \mathcal{N}(0,\sigma^2)$ and $V_m = U + N_m$, where $N_1,\dots,N_m$ are i.i.d. $\mathcal{N}(0,1)$ independent of U, then

$$I(U;\mathbf{V}) = \frac{1}{2}\log(1+n\sigma^2),$$

$$I^{-}(U;V) = \frac{n}{2}\log\left(1 + \frac{\sigma^2}{1+(n-1)\sigma^2}\right).$$

We also have the following.

PROPOSITION 10.1 If $V_1,\ldots,V_m$ are independent, then, for an arbitrary U jointly distributed with $\boldsymbol{V}$,

$$I(U;\boldsymbol{V}) \le I^{-}(U;\boldsymbol{V}). \tag{10.6}$$

Proof For any $k \in [m]$, we can write

$$\begin{aligned} I(U;V_k|V_1,\ldots,V_{k-1}) &= I(U,V_1,\ldots,V_{k-1};V_k) \\ &\le I(U,\boldsymbol{V}^{-k};V_k) \\ &= I(U;V_k|\boldsymbol{V}^{-k}), \end{aligned}$$

where the first and third steps use the chain rule and the independence of $V_1,\ldots,V_m$, while the second step is provided by the data-processing inequality. Summing over all k, we get (10.6).

10.2.2 Sub-Gaussian Random Variables and a Decoupling Estimate

A real-valued random variable U with $\mathbb{E}U < \infty$ is *σ^2-sub-Gaussian* if, for every $\lambda \in \mathbb{R}$,

$$\mathbb{E}[e^{\lambda(U-\mathbb{E}U)}] \le e^{\lambda^2\sigma^2/2}. \tag{10.7}$$

A classic result due to Hoeffding states that any almost surely bounded random variable is sub-Gaussian.

LEMMA 10.1 (Hoeffding [20]) *If $a \le U \le b$ almost surely, for some $-\infty < a \le b < \infty$, then*

$$\mathbb{E}[e^{\lambda(U-\mathbb{E}U)}] \le e^{\lambda^2(b-a)^2/8}, \tag{10.8}$$

i.e., U is $\left((b-a)^2/4\right)$-sub-Gaussian.

Consider a pair of random elements U and V of some spaces U and V, respectively, with joint distribution P_{UV}. Let $\bar{U}$ and $\bar{V}$ be independent copies of U and V, such that $P_{\bar{U}\bar{V}} = P_U \otimes P_V$. For an arbitrary real-valued function $f : \mathsf{U} \times \mathsf{V} \to \mathbb{R}$, we have the following upper bound on the absolute difference between $\mathbb{E}[f(U,V)]$ and $\mathbb{E}[f(\bar{U},\bar{V})]$.

LEMMA 10.2 *If $f(u,V)$ is σ^2-sub-Gaussian under P_V for every u, then*

$$\left|\mathbb{E}[f(U,V)] - \mathbb{E}[f(\bar{U},\bar{V})]\right| \le \sqrt{2\sigma^2 I(U;V)}. \tag{10.9}$$

Proof We exploit the Donsker–Varadhan variational representation of the relative entropy, Corollary 4.15 of [21]: for any two probability measures π, ν on a common measurable space $(\Omega,\mathcal{F})$,

$$D(\pi\|\nu) = \sup_F \left\{ \int_\Omega F\, d\pi - \log \int_\Omega e^F\, d\nu \right\}, \tag{10.10}$$

where the supremum is over all measurable functions $F : \Omega \to \mathbb{R}$, such that $\int e^F \, d\nu < \infty$. From (10.10), we know that, for any $\lambda \in \mathbb{R}$,

$$\begin{aligned} D(P_{V|U=u} \| P_V) &\geq \mathbb{E}[\lambda f(u,V)|U=u] - \log \mathbb{E}[e^{\lambda f(u,V)}] \\ &\geq \lambda(\mathbb{E}[f(u,V)|U=u] - \mathbb{E}[f(u,V)]) - \frac{\lambda^2\sigma^2}{2}, \end{aligned} \tag{10.11}$$

where the second step follows from the sub-Gaussian assumption on $f(u,V)$:

$$\log \mathbb{E}[e^{\lambda(f(u,V) - \mathbb{E}[f(u,V)])}] \leq \frac{\lambda^2\sigma^2}{2} \qquad \forall \lambda \in \mathbb{R}.$$

By maximizing the right-hand side of (10.11) over all $\lambda \in \mathbb{R}$ and rearranging we obtain

$$\left|\mathbb{E}[f(u,V)|U=u] - \mathbb{E}[f(u,V)]\right| \leq \sqrt{2\sigma^2 D(P_{V|U=u} \| P_V)}.$$

Then, using the law of iterated expectation and Jensen's inequality,

$$\begin{aligned} \left|\mathbb{E}[f(U,V)] - \mathbb{E}[f(\bar{U},\bar{V})]\right| &= \left|\mathbb{E}[\mathbb{E}[f(U,V) - f(U,\bar{V})|U]]\right| \\ &\leq \int_{\mathsf{U}} P_U(du) |\mathbb{E}[f(u,V)|U=u] - \mathbb{E}[f(u,V)]| \\ &\leq \int_{\mathsf{U}} P_U(du) \sqrt{2\sigma^2 D(P_{V|U=u} \| P_V)} \\ &\leq \sqrt{2\sigma^2 D(P_{UV} \| P_U \otimes P_V)}. \end{aligned}$$

The result follows by noting that $I(U;V) = D(P_{UV} \| P_U \otimes P_V)$.

10.3 Learning Algorithms and Stability

The magnitude of $\text{gen}(\mu, P_{W|\mathbf{Z}})$ is determined by the stability properties of $P_{W|\mathbf{Z}}$, i.e., by the sensitivity of $P_{W|\mathbf{Z}}$ to local modifications of the training data $\mathbf{Z}$. We wish to quantify this variability in information-theoretic terms. Let $P_{W|\mathbf{Z}=z}$ denote the probability distribution of the output of the algorithm in response to a deterministic $z = (z_1, \ldots, z_n) \in \mathsf{Z}^n$. This coincides with the conditional distribution of W given $\mathbf{Z} = z$, i.e., $\mathbf{P}_{W|\mathbf{Z}=z}(\cdot) = P_{W|\mathbf{Z}=z}(\cdot)$. Recalling the notation $z^{-i} := (z_1, \ldots, z_{i-1}, z_{i+1}, \ldots, z_n)$, we can write the conditional distribution of W given $\mathbf{Z}^{-i} = z^{-i}$ in the following form:

$$\mathbf{P}_{W|\mathbf{Z}^{-i}=z^{-i}}(\cdot) = \int_{\mathsf{Z}} \mu(dz_i) P_{W|\mathbf{Z}=(z_1,\ldots,z_i,\ldots,z_n)}(\cdot); \tag{10.12}$$

unlike $P_{W|\mathbf{Z}=z}$, this conditional distribution is determined by both μ and $P_{W|\mathbf{Z}}$. We put forward the following definition.

DEFINITION 10.1 *Given the data-generating distribution μ, we say that a learning algorithm $P_{W|\mathbf{Z}}$ is (ε,μ)-stable in erasure T-information if*

$$T^-(W;\mathbf{Z}) = T^-(W;Z_1,\ldots,Z_n) \leq n\varepsilon, \tag{10.13}$$

and (ε,μ)-stable in erasure mutual information *if*

$$I^-(W;\mathbf{Z}) = I^-(W;Z_1,\ldots,Z_n) \leq n\varepsilon, \tag{10.14}$$

where all expectations are w.r.t. $\mathbf{P}$. *We say that* $P_{W|Z}$ *is* ε*-stable (in erasure* T*-information or in erasure mutual information) if it is* (ε,μ)*-stable in the appropriate sense for every* μ.

These two notions of stability are related.

LEMMA 10.3 *Consider a learning algorithm* $P_{W|Z}$ *and a data-generating distribution* μ.

1. *If* $P_{W|Z}$ *is* (ε,μ)*-stable in erasure mutual information, then it is* $(\sqrt{\varepsilon/2},\mu)$*-stable in erasure* T*-information.*
2. *If* $P_{W|Z}$ *is* (ε,μ)*-stable in erasure* T*-information with* $\varepsilon \le 1/4$ *and the hypothesis class* W *is finite, i.e.,* $|\mathsf{W}| < \infty$*, then* $P_{W|Z}$ *is* $(\varepsilon\log(|\mathsf{W}|/\varepsilon),\mu)$*-stable in erasure mutual information.*

Proof For Part 1, using Pinsker's inequality and the concavity of the square root, we have

$$\begin{aligned}
\frac{1}{n}T^{-}(W;\mathbf{Z}) &= \frac{1}{n}\sum_{i=1}^{n} T(W;Z_i|\mathbf{Z}^{-i}) \\
&\le \frac{1}{n}\sum_{i=1}^{n}\sqrt{\frac{1}{2}I(W;Z_i|\mathbf{Z}^{-i})} \\
&\le \sqrt{\frac{1}{2n}\sum_{i=1}^{n} I(W;Z_i|\mathbf{Z}^{-i})} \\
&\le \sqrt{\frac{\varepsilon}{2}}.
\end{aligned}$$

For Part 2, since the output W is finite-valued, we can express the conditional mutual information $I(W;Z_i|\mathbf{Z}^{-i})$ as the difference of two conditional entropies:

$$\begin{aligned}
I(W;Z_i|\mathbf{Z}^{-i}) &= H(W|\mathbf{Z}^{-i}) - H(W|Z_i,\mathbf{Z}^{-i}) \\
&= H(W|\mathbf{Z}^{-i}) - H(W|\mathbf{Z}) \\
&= \int \mu^{\otimes n}(\mathrm{d}z)\Big(H(\mathbf{P}_{W|\mathbf{Z}^{-i}=z^{-i}}) - H(\mathbf{P}_{W|\mathbf{Z}=z})\Big).
\end{aligned} \tag{10.15}$$

We now recall the following inequality (Theorem 17.3.3 of [22]): for any two probability distributions μ and ν on a common finite set Ω with $d_{\mathsf{TV}}(\mu,\nu) \le 1/4$,

$$|H(\mu) - H(\nu)| \le d_{\mathsf{TV}}(\mu,\nu)\log\left(\frac{|\Omega|}{d_{\mathsf{TV}}(\mu,\nu)}\right). \tag{10.16}$$

Applying (10.16) to (10.15), we get

$$\begin{aligned}
I(W;Z_i|\mathbf{Z}^{-i}) \le \int \mu^{\otimes n}(\mathrm{d}z)\, & d_{\mathsf{TV}}(\mathbf{P}_{W|\mathbf{Z}^{-i}=z^{-i}},\mathbf{P}_{W|\mathbf{Z}=z}) \\
&\cdot \log\left(\frac{|\mathsf{W}|}{d_{\mathsf{TV}}(\mathbf{P}_{W|\mathbf{Z}^{-i}=z^{-i}},\mathbf{P}_{W|\mathbf{Z}=z})}\right).
\end{aligned} \tag{10.17}$$

Since the function $u \mapsto u\log(1/u)$ is concave, an application of Jensen's inequality to (10.17) gives

$$I(W;Z_i|\mathbf{Z}^{-i}) \le T(W;Z_i|\mathbf{Z}^{-i})\log\left(\frac{|\mathsf{W}|}{T(W;Z_i|\mathbf{Z}^{-i})}\right).$$

Summing from $i = 1$ to $i = n$, using the fact that the function $u \mapsto u\log(1/u)$ is concave and monotone increasing on $[0,1/4]$, and the stability assumption on $P_{W|\mathbf{Z}}$, we obtain the claimed stability result.

The next lemma gives a sufficient condition for stability by comparing the distributions $P_{W|\mathbf{Z}=z}$ and $P_{W|\mathbf{Z}=z'}$ for any two training sets that differ in one instance.

LEMMA 10.4 *A learning algorithm $P_{W|\mathbf{Z}}$ is ε-stable in erasure mutual information if $D(P_{W|\mathbf{Z}=z}\|P_{W|\mathbf{Z}=z'}) \le \varepsilon$ and in erasure T-information if $d_{\mathrm{TV}}(P_{W|\mathbf{Z}=z},P_{W|\mathbf{Z}=z'}) \le \varepsilon$ for all $z,z' \in \mathsf{Z}^n$ with*

$$d_{\mathrm{H}}(z,z') := \sum_{i=1}^{n} \mathbf{1}\{z_i \ne z'_i\} \le 1,$$

where $d_{\mathrm{H}}(\cdot,\cdot)$ is the Hamming distance on Z^n.

REMARK 10.1 Note that the sufficient conditions of Lemma 10.4 are distribution-free; that is, they do not involve μ. These notions of stability were introduced recently by Bassily *et al.* [23] under the names of KL- and TV-stability.

Proof We give the proof for the erasure mutual information; the case of the erasure T-information is analogous. Fix $\mu \in \mathcal{P}(\mathsf{Z})$, $s \in \mathsf{Z}^n$, and $i \in [n]$. For $z \in \mathsf{Z}$, let $z^{i,z}$ denote the n-tuple obtained by replacing z_i in z with z. Then

$$D(\mathbf{P}_{W|\mathbf{Z}=z}\|\mathbf{P}_{W|\mathbf{Z}^{-i}=z^{-i}}) \le \int_{\mathsf{Z}} \mu(\mathrm{d}z')D(\mathbf{P}_{W|\mathbf{Z}=z}\|\mathbf{P}_{W|\mathbf{Z}=z^{i,z'}}) \le \varepsilon,$$

where the first inequality follows from the convexity of the relative entropy. The claimed result follows by substituting this estimate into the expression

$$I(W;Z_i|\mathbf{Z}^{-i}) = \int_{\mathsf{Z}^n} \mu^{\otimes n}(\mathrm{d}z)D(\mathbf{P}_{W|\mathbf{Z}=z}\|\mathbf{P}_{W|\mathbf{Z}^{-i}=z^{-i}})$$

for the conditional mutual information.

Another notion of stability results if we consider plain T-information and mutual information.

DEFINITION 10.2 *Given a pair $(\mu, P_{W|\mathbf{Z}})$, we say that $P_{W|\mathbf{Z}}$ is (ε,μ)*-stable in T-information *if*

$$T(W;\mathbf{Z}) = T(W;Z_1,\ldots,Z_n) \le n\varepsilon, \tag{10.18}$$

and (ε,μ)-stable in mutual information *if*

$$I(W;\mathbf{Z}) = I(W;Z_1,\ldots,Z_n) \le n\varepsilon, \tag{10.19}$$

where all expectations are w.r.t. $\mathbf{P}$. We say that $P_{W|\mathbf{Z}}$ is ε-stable (in T*-information or in mutual information) if it is (ε,μ)-stable in the appropriate sense for every μ.*

LEMMA 10.5 *Consider a learning algorithm $P_{W|Z}$ and a data-generating distribution μ.*

1. *If $P_{W|Z}$ is (ε,μ)-stable in mutual information, then it is $(\sqrt{\varepsilon/2},\mu)$-stable in T-information.*
2. *If $P_{W|Z}$ is (ε,μ)-stable in T-information with $\varepsilon \le 1/4$ and the hypothesis class W is finite, i.e., $|\mathsf{W}| < \infty$, then $P_{W|Z}$ is $(\varepsilon \log(|\mathsf{W}|/\varepsilon),\mu)$-stable in mutual information.*
3. *If $P_{W|Z}$ is (ε,μ)-stable in erasure mutual information, then it is (ε,μ)-stable in mutual information.*
4. *If $P_{W|Z}$ is (ε,μ)-stable in mutual information, then it is $(n\varepsilon,\mu)$-stable in erasure mutual information.*

Proof Parts 1 and 2 are proved essentially in the same way as in Lemma 10.3. For Part 3, since the Z_is are independent, $I(W;\mathbf{Z}) \le I^-(W;\mathbf{Z})$ by Proposition 10.1. Hence, $I^-(W;\mathbf{Z}) \le n\varepsilon$ implies $I(W;\mathbf{Z}) \le n\varepsilon$. For Part 4, $I^-(W;\mathbf{Z}) \le nI(W;\mathbf{Z})$ by (10.5), so $I(W;\mathbf{Z}) \le n\varepsilon$ implies $I^-(W;\mathbf{Z}) \le n^2\varepsilon$.

10.4 Information Stability and Generalization

In this section, we will relate the generalization error of an arbitrary learning algorithm to its information-theoretic stability properties. We start with the following simple, but important, result.

THEOREM 10.1 *If the loss function ℓ takes values in $[0,1]$, then, for any pair $(\mu, P_{W|Z})$,*

$$|\mathrm{gen}(\mu, P_{W|Z})| \le \frac{1}{n} T^-(W;\mathbf{Z}). \tag{10.20}$$

In particular, if $P_{W|Z}$ is (ε,μ)-stable in erasure T-information, then $|\mathrm{gen}(\mu, P_{W|Z})| \le 2\varepsilon$.

Proof The proof technique is standard in the literature on algorithmic stability (see, e.g., the proof of Lemma 7 in [7], the discussion at the beginning of Section 3.1 in [10], or the proof of Lemma 11 in [2]); note, however, that we do not require $P_{W|Z}$ to be symmetric. Introduce an auxiliary sample $\mathbf{Z}' = (Z'_1,\dots,Z'_n) \sim \mu^{\otimes n}$ that is independent of $(\mathbf{Z}, W) \sim \mathbf{P}$. Since $\mathbb{E}[L_\mu(W)] = \mathbb{E}[\ell(W,Z'_i)]$ for each $i \in [n]$, we write

$$\mathrm{gen}(\mu, P_{W|Z}) = \frac{1}{n}\sum_{i=1}^{n} \mathbb{E}[\ell(W,Z'_i) - \ell(W,Z_i)].$$

Now, for each $i \in [n]$ let us denote by $W^{(i)}$ the output of the algorithm when the input is equal to $\mathbf{Z}^{i,Z'_i}$. Then, since the joint probability law of $(W,\mathbf{Z},Z'_i)$ evidently coincides with the joint probability law of $(W^{(i)},\mathbf{Z}^{i,Z'_i},Z_i)$, we have

$$\mathbb{E}[\ell(W,Z'_i) - \ell(W,Z_i)] = \mathbb{E}[\ell(W^{(i)},Z_i) - \ell(W^{(i)},Z'_i)]. \tag{10.21}$$

Moreover,

$$\begin{aligned}\mathbb{E}\ell(W^{(i)},Z_i) &= \int \mu^{\otimes(n-1)}(\mathrm{d}z^{-i})\mu(\mathrm{d}z_i)\mu(\mathrm{d}z'_i)P_{W|\mathbf{Z}=z^{i,z'_i}}(\mathrm{d}w)\ell(w,z_i)\\ &= \int \mu^{\otimes n}(\mathrm{d}z)\mathbf{P}_{W|Z^{-i}=z^{-i}}(\mathrm{d}w)\ell(w,z_i),\end{aligned} \tag{10.22}$$

where in the second line we have used (10.12), and

$$\begin{aligned}\mathbb{E}\ell(W^{(i)},Z_i') &= \int \mu^{\otimes(n-1)}(\mathrm{d}z^{-i})\mu(\mathrm{d}z_i)\mu(\mathrm{d}z_i')P_{W|\mathbf{Z}=z^{i,z_i'}}(\mathrm{d}w)\ell(w,z_i') \\ &= \int \mu^{\otimes(n-1)}(\mathrm{d}z^{-i})\mu(\mathrm{d}z_i)\mu(\mathrm{d}z_i')P_{W|\mathbf{Z}=z}(\mathrm{d}w)\ell(w,z_i) \\ &= \int \mu^{\otimes n}(\mathrm{d}z)\mathbf{P}_{W|\mathbf{Z}=z}\ell(w,z_i), \qquad (10.23)\end{aligned}$$

where in the second line we used the fact that $Z_1,\ldots,Z_n$ and Z_i' are i.i.d. draws from μ. Using (10.22) and (10.23), we obtain

$$\begin{aligned}&\left|\mathbb{E}\ell(W^{(i)},Z_i) - \mathbb{E}\ell(W^{(i)},Z_i')\right| \\ &\leq \int_{\mathsf{Z}^n} \mu^{\otimes n}(\mathrm{d}z)\left|\int_{\mathsf{W}} \mathbf{P}_{W|\mathbf{Z}^{-i}=z^{-i}}(\mathrm{d}w)\ell(w,z_i) - \int_{\mathsf{W}} \mathbf{P}_{W|\mathbf{Z}=z}(\mathrm{d}w)\ell(w,z_i)\right| \\ &\leq \int_{\mathsf{Z}^n} \mu^{\otimes n}(\mathrm{d}z) d_{\mathrm{TV}}(\mathbf{P}_{W|\mathbf{Z}=z}, \mathbf{P}_{W|\mathbf{Z}^{-i}=z^{-i}}) \\ &= T(W;Z_i|\mathbf{Z}^{-i}),\end{aligned}$$

where we have used (10.3) and the fact that $T(U;V|Y) = \mathbb{E}d_{\mathrm{TV}}(P_{U|VY}, P_{U|Y})$ [18]. Summing over $i \in [n]$ and using the definition of $T^-(W;\mathbf{Z})$, we get the claimed bound.

The following theorem replaces the assumption of bounded loss with a less restrictive sub-Gaussianity condition.

THEOREM 10.2 *Consider a pair $(\mu, P_{W|\mathbf{Z}})$, where $\ell(w,Z)$ is σ^2-sub-Gaussian under μ for every $w \in \mathsf{W}$. Then*

$$|\mathrm{gen}(\mu, P_{W|\mathbf{Z}})| \leq \sqrt{\frac{2\sigma^2}{n} I(W;\mathbf{Z})}. \qquad (10.24)$$

In particular, if $P_{W|\mathbf{Z}}$ is (ε,μ)-stable in mutual information, then $|\mathrm{gen}(\mu, P_{W|\mathbf{Z}})| \leq \sqrt{2\sigma^2\varepsilon}$.

REMARK 10.2 Upper bounds on the expected generalization error in terms of the mutual information $I(\mathbf{Z};W)$ go back to earlier results of McAllester on PAC-Bayes methods [24] (see also the tutorial paper [25]).

REMARK 10.3 For a bounded loss function $\ell(\cdot,\cdot) \in [a,b]$, $\ell(w,Z)$ is $(b-a)^2/4$-sub-Gaussian for all μ and all $w \in \mathsf{W}$, by Hoeffding's lemma.

REMARK 10.4 Since $Z_1,\ldots,Z_n$ are i.i.d., $I(W;\mathbf{Z}) \leq I^-(W;\mathbf{Z})$, by Proposition 10.1. Thus, if $P_{W|\mathbf{Z}}$ is ε-stable in erasure mutual information, then it is automatically ε-stable in mutual information, and therefore the right-hand side of (10.24) is bounded by $\sqrt{2\sigma^2\varepsilon}$. On the other hand, proving stability in erasure mutual information is often easier than proving stability in mutual infornation.

Proof For each w, $L_{\mathbf{Z}}(w) = (1/n)\sum_{i=1}^n \ell(w,Z_i)$ is (σ^2/n)-sub-Gaussian. Thus, we can apply Lemma 10.2 to $U = W$, $V = \mathbf{Z}$, $f(U,V) = L_{\mathbf{Z}}(W)$.

Theorem 10.2 allows us to control the generalization error of a learning algorithm by the erasure mutual information or by the plain mutual information between the input and the output of the algorithm. Russo and Zou [26] considered the same setup, but with the restriction to finite hypothesis spaces, and showed that $|\mathrm{gen}(\mu, P_{W|\mathbf{Z}})|$ can be upper-bounded in terms of $I(\Lambda_{\mathsf{W}}(\mathbf{Z}); W)$, where

$$\Lambda_{\mathsf{W}}(\mathbf{Z}) := (L_{\mathbf{Z}}(w))_{w \in \mathsf{W}} \tag{10.25}$$

is the collection of empirical risks of the hypotheses in W. Using Lemma 10.2 by setting $U = \Lambda_{\mathsf{W}}(\mathbf{Z})$, $V = W$, and $f(\Lambda_{\mathsf{W}}(\mathbf{z}), w) = L_{\mathbf{z}}(w)$, we immediately recover the result obtained by Russo and Zou even when W is uncountably infinite.

THEOREM 10.3 (*Russo and Zou [26]*) *Suppose $\ell(w, Z)$ is σ^2-sub-Gaussian under μ for all $w \in \mathsf{W}$, then*

$$\left|\mathrm{gen}(\mu, P_{W|\mathbf{Z}})\right| \le \sqrt{\frac{2\sigma^2}{n} I(W; \Lambda_{\mathsf{W}}(\mathbf{Z}))}. \tag{10.26}$$

It should be noted that Theorem 10.2 can also be obtained as a consequence of Theorem 10.3 because

$$I(W; \Lambda_{\mathsf{W}}(\mathbf{Z})) \le I(W; \mathbf{Z}), \tag{10.27}$$

by the data-processing inequality for the Markov chain $\Lambda_{\mathsf{W}}(\mathbf{Z}) - \mathbf{Z} - W$. The latter holds since, for each $w \in \mathsf{W}$, $L_{\mathbf{Z}}(w)$ is a function of $\mathbf{Z}$. However, if the output W depends on $\mathbf{Z}$ only through the empirical risks $\Lambda_{\mathsf{W}}(\mathbf{Z})$ (i.e., the Markov chain $\mathbf{Z} - \Lambda_{\mathsf{W}}(\mathbf{Z}) - W$ holds), then Theorems 10.2 and 10.3 are equivalent. The advantage of Theorem 10.2 is that $I(W; \mathbf{Z})$ is often much easier to evaluate than $I(W; \Lambda_{\mathsf{W}}(\mathbf{Z}))$. We will elaborate on this when we discuss the Gibbs algorithm and adaptive composition of learning algorithms. Recent work by Jiao *et al.* [27] extends the results of Russo and Zou by introducing a generalization of mutual information that can handle the cases when $\ell(w, Z)$ is not sub-Gaussian.

10.4.1 A Concentration Inequality for $|L_\mu(W) - L_{\mathbf{Z}}(W)|$

Theorems 10.2 and 10.3 merely give upper bounds on the expected generalization error. We are often interested in analyzing the expected value or the tail behavior of the absolute generalization error $|L_\mu(W) - L_{\mathbf{Z}}(W)|$. This is the subject of the present section.

For any fixed $w \in \mathsf{W}$, if $\ell(w, Z)$ is σ^2-sub-Gaussian, the Chernoff–Hoeffding bound gives $\mathbf{P}[|L_\mu(w) - L_{\mathbf{Z}}(w)| > \alpha] \le 2e^{-\alpha^2 n/2\sigma^2}$. Thus, if $\mathbf{Z}$ and W are independent, a sample size of

$$n \ge \frac{2\sigma^2}{\alpha^2} \log\left(\frac{2}{\beta}\right) \tag{10.28}$$

suffices to guarantee

$$\mathbf{P}[|L_\mu(W) - L_{\mathbf{Z}}(W)| > \alpha] \le \beta. \tag{10.29}$$

The following results pertain to the case when W and $\mathbf{Z}$ are dependent, but the mutual information $I(W;\mathbf{Z})$ is sufficiently small. The tail probability now is taken with respect to the joint distribution $\mathbf{P} = \mu^{\otimes n} \otimes P_{W|\mathbf{Z}}$.

THEOREM 10.4 *Suppose $\ell(w,Z)$ is σ^2-sub-Gaussian under μ for all $w \in \mathsf{W}$. If a learning algorithm satisfies $I(W;\Lambda_{\mathsf{W}}(\mathbf{Z})) \leq C$, then for any $\alpha > 0$ and $0 < \beta \leq 1$,* (10.29) *can be guaranteed by a sample complexity of*

$$n \geq \frac{8\sigma^2}{\alpha^2}\left(\frac{C}{\beta} + \log\left(\frac{2}{\beta}\right)\right). \tag{10.30}$$

In view of (10.27), any learning algorithm that is (ε,μ)-stable in mutual information satisfies the condition $I(W;\Lambda_{\mathsf{W}}(\mathbf{Z})) \leq n\varepsilon$. We also have the following corollary.

COROLLARY 10.1 *Under the conditions in Theorem 10.4, if $C \leq (g(n)-1)\beta\log(2/\beta)$ for some function $g(n) \geq 1$, then a sample complexity that satisfies $n/g(n) \geq (8\sigma^2/\alpha^2)\log(2/\beta)$ guarantees* (10.29).

For example, taking $g(n) = 2$, Corollary 10.1 implies that, if $C \leq \beta\log(2/\beta)$, then (10.29) can be guaranteed by a sample complexity of $n = (16\sigma^2/\alpha^2)\log(2/\beta)$, which is on the same order as the sample complexity when $\mathbf{Z}$ and W are independent as in (10.28). As another example, taking $g(n) = \sqrt{n}$, Corollary 10.1 implies that, if $C \leq (\sqrt{n}-1)\beta\log(2/\beta)$, then a sample complexity of $n = (64\sigma^4/\alpha^4)(\log(2/\beta))^2$ guarantees (10.29).

Recent papers of Dwork *et al.* [28, 29] give "high-probability" bounds on the absolute generalization error of differentially private algorithms with bounded loss functions, i.e., the tail bound $\mathbf{P}[|L_\mu(W) - L_{\mathbf{Z}}(W)| > \alpha] \leq \beta$ is guaranteed to hold whenever $n \gtrsim (1/\alpha^2)\log(1/\beta)$. By contrast, Theorem 10.4 does not require differential privacy and assumes that $\ell(w,Z)$ is sub-Gaussian. Bassily *et al.* [30] obtain a concentration inequality on the absolute generalization error on the same order as the bound of Theorem 10.4 and show that this bound is sharp – they give an example of a learning problem (μ,W,ℓ) and a learning algorithm $\mathbf{P}_{W|\mathbf{Z}}$ that satisfies $I(W;\mathbf{Z}) \leq O(1)$ and

$$\mathbf{P}\left(|L_{\mathbf{Z}}(W) - L_\mu(W)| \geq \frac{1}{2}\right) \geq \frac{1}{n}.$$

They also give an example where a sufficient amount of mutual information is necessary in order for the ERM algorithm to generalize.

The proof of Theorem 10.4 is based on Lemma 10.2 and an adaptation of the "monitor technique" of Bassily *et al.* [23]. We first need the following two lemmas. The first lemma is a simple consequence of the tensorization property of mutual information.

LEMMA 10.6 *Consider the parallel execution of m independent copies of $P_{W|\mathbf{Z}}$ on independent datasets $\mathbf{Z}_1,\ldots,\mathbf{Z}_m$: for $t = 1,\ldots,m$, an independent copy of $P_{W|\mathbf{Z}}$ takes $\mathbf{Z}_t \sim \mu^{\otimes n}$ as input and outputs W_t. Let $\mathbf{Z}^m := (\mathbf{Z}_1,\ldots,\mathbf{Z}_m)$ be the overall dataset. If under μ, $P_{W|\mathbf{Z}}$ satisfies $I(W;\Lambda_{\mathsf{W}}(\mathbf{Z})) \leq C$, then the overall algorithm $P_{W^m|\mathbf{Z}^m}$ satisfies $I(W^m;\Lambda_{\mathsf{W}}(\mathbf{Z}_1),\ldots,\Lambda_{\mathsf{W}}(\mathbf{Z}_m)) \leq mC$.*

Proof The proof is by the independence among $(\mathbf{Z}_t, W_t)$, $t = 1,\ldots,m$, and the chain rule of mutual information.

The next lemma is the key piece. It will be used to construct a procedure that executes m copies of a learning algorithm in parallel and then selects the one with the largest absolute generalization error.

LEMMA 10.7 *Let $\mathbf{Z}^m = (\mathbf{Z}_1,\ldots,\mathbf{Z}_m)$, where each $\mathbf{Z}_t \sim \mu^{\otimes n}$ is independent of all of the others. If an algorithm $P_{W,T,R|\mathbf{Z}^m} : \mathsf{Z}^{m\times n} \to \mathsf{W} \times [m] \times \{\pm 1\}$ satisfies $I(W,T,R;(\Lambda_{\mathsf{W}}(\mathbf{Z}_1),\ldots,\Lambda_{\mathsf{W}}(\mathbf{Z}_m))) \le C$, and if $\ell(w,Z)$ is σ^2-sub-Gaussian for all $w \in \mathsf{W}$, then*

$$\mathbb{E}[R(L_{\mathbf{Z}_T}(W) - L_\mu(W))] \le \sqrt{\frac{2\sigma^2 C}{n}}.$$

Proof The proof is based on Lemma 10.2. Let $U = (\Lambda_{\mathsf{W}}(\mathbf{Z}_1),\ldots,\Lambda_{\mathsf{W}}(\mathbf{Z}_m))$, $V = (W,T,R)$, and

$$f((\Lambda_{\mathsf{W}}(z_1),\ldots,\Lambda_{\mathsf{W}}(z_m)),(w,t,r)) = rL_{z_t}(w).$$

If $\ell(w,Z)$ is σ^2-sub-Gaussian under $Z \sim \mu$ for all $w \in \mathsf{W}$, then $(r/n)\sum_{i=1}^n \ell(w,Z_{t,i})$ is (σ^2/n)-sub-Gaussian for all $w \in \mathsf{W}$, $t \in [m]$ and $r \in \{\pm 1\}$, and hence $f(u,V)$ is (σ^2/n)-sub-Gaussian for every u. Lemma 10.2 implies that

$$\mathbb{E}[RL_{\mathbf{Z}_T}(W)] - \mathbb{E}[RL_\mu(W)] \le \sqrt{\frac{2\sigma^2 I(W,T,R;\Lambda_{\mathsf{W}}(\mathbf{Z}_1),\ldots,\Lambda_{\mathsf{W}}(\mathbf{Z}_m))}{n}},$$

which proves the claim.

Note that the upper bound in Lemma 10.7 does not depend on m. With these lemmas, we can prove Theorem 10.4.

Proof of Theorem 10.4 First, let $P_{W^m|\mathbf{Z}^m}$ be the parallel execution of m independent copies of $P_{W|\mathbf{Z}}$, as in Lemma 10.6. Given $\mathbf{Z}^m$ and W^m, let the output of the "monitor" be a sample (W^*,T^*,R^*) drawn from $\mathsf{W} \times [m] \times \{\pm 1\}$ according to

$$(T^*,R^*) = \underset{t\in[m],\, r\in\{\pm 1\}}{\arg\max}\ r(L_\mu(W_t) - L_{\mathbf{Z}_t}(W_t)) \quad \text{and} \quad W^* = W_{T^*}. \tag{10.31}$$

This gives

$$R^*(L_\mu(W^*) - L_{\mathbf{Z}_{T^*}}(W^*)) = \max_{t\in[m]} \left|L_\mu(W_t) - L_{\mathbf{Z}_t}(W_t)\right|.$$

Taking the expectation of both sides, we have

$$\mathbb{E}[R^*(L_\mu(W^*) - L_{\mathbf{Z}_{T^*}}(W^*))] = \mathbb{E}\Big[\max_{t\in[m]} \left|L_\mu(W_t) - L_{\mathbf{Z}_t}(W_t)\right|\Big]. \tag{10.32}$$

Note that, conditionally on W^m, the tuple (W^*,T^*,R^*) can take only $2m$ values, which means that

$$I(W^*,T^*,R^*;\Lambda_{\mathsf{W}}(\mathbf{Z}_1),\ldots,\Lambda_{\mathsf{W}}(\mathbf{Z}_m)|W^m) \le \log(2m). \tag{10.33}$$

In addition, since $P_{W|\mathbf{Z}}$ is assumed to satisfy $I(W;\Lambda_{\mathsf{W}}(\mathbf{Z})) \le C$, Lemma 10.6 implies that

$$I(W^m;\Lambda_{\mathsf{W}}(\mathbf{Z}_1),\ldots,\Lambda_{\mathsf{W}}(\mathbf{Z}_m)) \le mC.$$

Therefore, by the chain rule of mutual information and the data-processing inequality, we have

$$\begin{aligned} I(W^*, T^*, R^*; \Lambda_{\mathsf{W}}(\mathbf{Z}_1), \ldots, \Lambda_{\mathsf{W}}(\mathbf{Z}_m)) \\ \leq I(W^m, W^*, T^*, R^*; \Lambda_{\mathsf{W}}(\mathbf{Z}_1), \ldots, \Lambda_{\mathsf{W}}(\mathbf{Z}_m)) \\ \leq mC + \log(2m). \end{aligned}$$

By Lemma 10.7 and the assumption that $\ell(w, Z)$ is σ^2-sub-Gaussian,

$$\mathbb{E}[R^*(L_{\mathbf{Z}_{T^*}}(W^*) - L_\mu(W^*))] \leq \sqrt{\frac{2\sigma^2}{n}(mC + \log(2m))}. \tag{10.34}$$

Combining (10.34) and (10.32) gives

$$\mathbb{E}\Big[\max_{t\in[m]} \big|L_{\mathbf{Z}_t}(W_t) - L_\mu(W_t)\big|\Big] \leq \sqrt{\frac{2\sigma^2}{n}(mC + \log(2m))}. \tag{10.35}$$

The rest of the proof is by contradiction. Choose $m = \lfloor 1/\beta \rfloor$. Suppose the algorithm $P_{W|\mathbf{Z}}$ does not satisfy the claimed generalization property, namely

$$\mathbf{P}\big[\big|L_{\mathbf{Z}}(W) - L_\mu(W)\big| > \alpha\big] > \beta. \tag{10.36}$$

Then, by the independence among the pairs $(\mathbf{Z}_t, W_t)$, $t = 1, \ldots, m$,

$$\mathbf{P}\Big[\max_{t\in[m]} \big|L_{\mathbf{Z}_t}(W_t) - L_\mu(W_t)\big| > \alpha\Big] > 1 - (1-\beta)^{\lfloor 1/\beta \rfloor} > \frac{1}{2}.$$

Thus

$$\mathbb{E}\Big[\max_{t\in[m]} \big|L_{\mathbf{Z}_t}(W_t) - L_\mu(W_t)\big|\Big] > \frac{\alpha}{2}. \tag{10.37}$$

Combining (10.35) and (10.37) gives

$$\frac{\alpha}{2} < \sqrt{\frac{2\sigma^2}{n}\left(\frac{C}{\beta} + \log\left(\frac{2}{\beta}\right)\right)}. \tag{10.38}$$

The above inequality implies that

$$n < \frac{8\sigma^2}{\alpha^2}\left(\frac{C}{\beta} + \log\left(\frac{2}{\beta}\right)\right), \tag{10.39}$$

which contradicts the condition in (10.30). Therefore, under the condition in (10.30), the assumption in (10.36) cannot hold. This completes the proof.

A byproduct of the proof of Theorem 10.4 (setting $m = 1$ in (10.35)) is an upper bound on the expected absolute generalization error.

THEOREM 10.5 *Suppose $\ell(w, Z)$ is σ^2-sub-Gaussian under μ for all $w \in \mathsf{W}$. If a learning algorithm satisfies $I(W; \Lambda_{\mathsf{W}}(\mathbf{Z})) \leq C$, then*

$$\mathbb{E}\big|L_\mu(W) - L_{\mathbf{Z}}(W)\big| \leq \sqrt{\frac{2\sigma^2}{n}(C + \log 2)}. \tag{10.40}$$

This result improves on Proposition 3.2 of [26], which states that $\mathbb{E}\left|L_Z(W) - L_\mu(W)\right| \leq \sigma/\sqrt{n} + 36\sqrt{2\sigma^2 C/n}$. Theorem 10.5 together with Markov's inequality implies that (10.29) can be guaranteed by $n \geq \left(2\sigma^2/\alpha^2\beta^2\right)(C + \log 2)$, but it has a worse dependence on β than does the sample complexity given by Theorem 10.4.

10.5 Information-Theoretically Stable Learning Algorithms

In this section, we illustrate the framework of information-theoretic stability in the context of several learning problems and algorithms. We first consider two cases where upper bounds on the input–output mutual information of a learning algorithm can be obtained in terms of the geometric or combinatorial properties of the hypothesis space. Then we show that one can obtain learning algorithms with controlled input–output mutual information by regularizing the ERM algorithm. We also discuss alternative methods to induce input–output mutual information stability, as well as stability of learning algorithms built by adaptive composition.

10.5.1 Countable Hypothesis Spaces

When the hypothesis space is countable, the input–output mutual information can be upper-bounded by $H(W)$, the entropy of W. In that case, if $\ell(w, Z)$ is σ^2-sub-Gaussian for all $w \in \mathsf{W}$, any learning algorithm $P_{W|Z}$ satisfies

$$\left|\text{gen}(\mu, P_{W|Z})\right| \leq \sqrt{\frac{2\sigma^2 H(W)}{n}} \tag{10.41}$$

by Theorem 10.2. If $|\mathsf{W}| = k < \infty$, we have $H(W) \leq \log k$, and the bound (10.41) can be weakened to

$$\left|\text{gen}(\mu, P_{W|Z})\right| \leq \sqrt{\frac{2\sigma^2 \log k}{n}}.$$

However, a simple argument based on the union bound also shows that, when $|S| = k < \infty$, we have

$$\mathbb{E}\left[\max_{w \in \mathsf{W}} |L_Z(w) - L_\mu(w)|\right] \leq \sqrt{\frac{2\sigma^2 \log k}{n}}.$$

Thus, the information-theoretic bound (10.41) is useful only when the learning algorithm $P_{W|Z}$ is such that $H(W) \ll \log k$.

An uncountable hypothesis space can often be replaced with a countable one by quantizing (or discretizing) the output hypothesis. For example, suppose W is a bounded subset of $\mathbb{R}^d$ equipped with some norm $\|\cdot\|$, i.e., $\mathsf{W} \subseteq \{w \in \mathbb{R}^d : \|w\| \leq B\}$ for some $B < \infty$. Let $P_{W|Z}$ be an arbitrary learning algorithm. Let $N(r, \|\cdot\|, \mathsf{W})$ denote the *covering number* of W, defined as the cardinality of the smallest set $\mathsf{W}' \subset \mathbb{R}^m$, such that for every $w \in \mathsf{W}$ there exists some $w' \in \mathsf{W}'$ with $\|w - w'\| \leq r$. Consider the Markov chain $\mathbf{Z}$–W–W', where $W' = \arg\min_{v \in \mathsf{W}'} \|v - W\|$, with ties broken arbitrarily. If $r < B$, then, using standard estimates for covering numbers in finite-dimensional Banach spaces [31], we can write

$$I(\mathbf{Z}; W') \le \log N(\mathsf{W}, \cdot\|\cdot\|, r) \le d \log\left(\frac{3B}{r}\right),$$

and therefore, under the sub-Gaussian assumption on ℓ, the composite learning algoritm $P_{W'|\mathbf{Z}}$ satisfies

$$\left|\text{gen}(\mu, P_{W'|\mathbf{Z}})\right| \le \sqrt{\frac{2\sigma^2 d}{n} \log\left(\frac{3B}{r}\right)}. \tag{10.42}$$

For example, if we set $r = 3B/\sqrt{n}$, the above bound on the generalization error will scale as $\sqrt{(\sigma^2 d \, \log \, n)/n}$. If $\ell(\cdot, z)$ is Lipschitz, i.e., $|\ell(w,z) - \ell(w',z)| \le \|w - w'\|_2$, then we can use (10.42) to obtain the following generalization bound for the original algorithm $P_{W|\mathbf{Z}}$:

$$\left|\text{gen}(\mu, P_{W|\mathbf{Z}})\right| \le \inf_{r \ge 0}\left(2r + \sqrt{\frac{2\sigma^2 d}{n} \log\left(\frac{3B}{r}\right)}\right).$$

Again, taking $r = 3B/\sqrt{n}$, we get

$$\left|\text{gen}(\mu, P_{W|\mathbf{Z}})\right| \le \sqrt{\frac{2\sigma^2 d \log n}{n}} + \frac{6B}{\sqrt{n}}.$$

10.5.2 Binary Classification

Recall the problem of binary classification, briefly described in Section 16.1: $\mathsf{Z} = \mathsf{X} \times \mathsf{Y}$, where $\mathsf{Y} = \{0, 1\}$, W is a collection of classifiers $w : \mathsf{X} \to \mathsf{Y}$ (which could be uncountably infinite), and $\ell(w, z) = \mathbf{1}\{w(x) \ne y\}$. Before proceeding, we need some basic facts from Vapnik–Chervonenkis theory [1, 32]. We say that W *shatters* a set $\mathsf{S} \subset \mathsf{X}$ if for every $\mathsf{S}' \subseteq \mathsf{S}$ there exists some $w \in \mathsf{W}$ such that, for any $x \in \mathsf{S}$,

$$w(x) = \mathbf{1}_{\{x \in \mathsf{S}'\}}.$$

The Vapnik–Chervonenkis dimension (or VC dimension) of W is defined as

$$\text{vc}(\mathsf{W}) := \sup\{|\mathsf{S}| : \mathsf{S} \text{ is shattered by } \mathsf{W}\}.$$

If $\text{vc}(\mathsf{W})$ is finite, we say that W is a *VC class*. For example, if $\mathsf{X} \subseteq \mathbb{R}^d$ and W consists of indicators of all halfspaces of $\mathbb{R}^d$, then $\text{vc}(\mathsf{W}) = d + 1$. A fundamental combinatorial estimate, known as the *Sauer–Shelah lemma*, states that, for any finite set $\mathsf{S} \subseteq \mathsf{X}$ and any VC class W with $\text{vc}(\mathsf{W}) = V$,

$$|\{(w(x) : x \in \mathsf{S}) : w \in \mathsf{W}\}| \le \left(\frac{e|\mathsf{S}|}{V}\right)^V \le (|\mathsf{S}| + 1)^V.$$

With these preliminaries out of the way, we can use Theorem 10.2 to perform a simple analysis of the following two-stage algorithm [32, 33] that can achieve the same performance as ERM. Given the dataset $\mathbf{Z}$ of size n, split it into $\mathbf{Z}_1$ and $\mathbf{Z}_2$ with sizes n_1 and n_2. First, pick a subset of hypotheses $\mathsf{W}_1 \subset \mathsf{W}$ based on $\mathbf{Z}_1$, such that the binary strings $(w(X_1),\ldots,w(X_{n_1}))$ for $w \in \mathsf{W}_1$ are all distinct and $\{(w(X_1),\ldots,w(X_{n_1})) : w \in \mathsf{W}_1\} = \{(w(X_1),\ldots,w(X_{n_1})) : w \in \mathsf{W}\}$. In other words, W_1 forms an *empirical cover* of W with respect to $\mathbf{Z}_1$. Then pick a hypothesis from W_1 with the minimal empirical risk on $\mathbf{Z}_2$, i.e.,

$$W = \arg\min_{w \in \mathsf{W}_1} L_{\mathbf{Z}_2}(w). \tag{10.43}$$

Let V denote the VC dimension of W. We can upper-bound the expected generalization error of W with respect to $\mathbf{Z}_2$ as

$$\mathbb{E}[L_\mu(W)] - \mathbb{E}[L_{\mathbf{Z}_2}(W)] = \mathbb{E}[\mathbb{E}[L_\mu(W) - L_{\mathbf{Z}_2}(W)|\mathbf{Z}_1]] \leq \sqrt{\frac{V\log(n_1+1)}{2n_2}}, \tag{10.44}$$

where we have used Theorem 10.2 and the fact that $I(W;\mathbf{Z}_2|\mathbf{Z}_1 = z_1) \leq H(W|\mathbf{Z}_1 = z_1) \leq V\log(n_1+1)$, by the Sauer–Shelah lemma. It can also be shown that [32, 33]

$$\mathbb{E}[L_{\mathbf{Z}_2}(W)] \leq \mathbb{E}\Big[\inf_{w \in \mathsf{W}_1} L_\mu(w)\Big] \leq \inf_{w \in \mathsf{W}} L_\mu(w) + c\sqrt{\frac{V}{n_1}}, \tag{10.45}$$

where the second expectation is taken with respect to W_1, which depends on $\mathbf{Z}_1$, and c is an absolute constant. Combining (10.44) and (10.45) and setting $n_1 = n_2 = n/2$, we have, for some constant c,

$$\mathbb{E}[L_\mu(W)] \leq \inf_{w \in \mathsf{W}} L_\mu(w) + c\sqrt{\frac{V\log n}{n}}. \tag{10.46}$$

From an information-theoretic point of view, the above two-stage algorithm effectively controls the conditional mutual information $I(W;\mathbf{Z}_2|\mathbf{Z}_1)$ by extracting an empirical cover of W using $\mathbf{Z}_1$, while maintaining a small empirical risk using $\mathbf{Z}_2$.

10.5.3 Gibbs Algorithm

Theorem 10.2 upper-bounds the generalization error of a learning algorithm by the mutual information $I(W;\mathbf{Z})$. Hence, it is natural to consider an idealized algorithm that minimizes the empirical risk regularized by $I(W;\mathbf{Z})$:

$$P^\circ_{W|\mathbf{Z}} = \arg\min_{P_{W|\mathbf{Z}}} \left(\mathbb{E}[L_{\mathbf{Z}}(W)] + \frac{1}{\beta} I(\mathbf{Z};W)\right), \tag{10.47}$$

where $\beta > 0$ is a parameter that balances data fit and generalization. Since μ is unknown to the learning algorithm, we relax the above optimization problem by appealing to the so-called *golden formula* for the mutual information:

$$I(U;V) = D(P_{UV}\|P_U \otimes Q_V) - D(P_V\|Q_V), \tag{10.48}$$

which is valid for any Q_V satisfying $D(P_V\|Q_V) < \infty$. In particular, applying (10.48) with $U = \mathbf{Z}$ and $V = W$, we see that we can choose an arbitrary probability distribution Q on W and upper-bound $I(W;\mathbf{Z})$ as follows:

$$\begin{aligned} I(W;\mathbf{Z}) &\leq D(P_{W\mathbf{Z}}\|P_{\mathbf{Z}} \otimes Q) \\ &= \int_{\mathsf{Z}^n} \mu^{\otimes n}(\mathrm{d}z) D(P_{W|\mathbf{Z}=z}\|Q). \end{aligned}$$

Using this, we relax the minimization problem (10.47) to

$$P^*_{W|Z} = \arg\min_{P_{W|Z}} \left(\mathbb{E}[L_Z(W)] + \frac{1}{\beta} D(P_{WZ} \| Q \otimes P_Z) \right) \tag{10.49a}$$

$$= \arg\min_{P_{W|Z}} \int_{Z^n} \mu^{\otimes n}(dz) \left[\int_W P_{W|Z=z}(dw) L_z(w) + \frac{1}{\beta} D(P_{W|Z=z} \| Q) \right]. \tag{10.49b}$$

It is evident that the solution of the relaxed optimization problem does not depend on μ, and it is straightforward to obtain the following.

THEOREM 10.6 *The solution to the optimization problem* (10.49) *is given by the Gibbs algorithm*

$$P^*_{W|Z=z}(dw) = \frac{e^{-\beta L_z(w)} Q(dw)}{\mathbb{E}_Q[e^{-\beta L_z(W)}]} \quad \text{for each } z \in Z^n. \tag{10.50}$$

REMARK 10.5 We would not have been able to arrive at the Gibbs algorithm had we used $I(W; \Lambda_W(\mathbf{Z}))$ as the regularization term instead of $I(W; \mathbf{Z})$ in (10.47), even if we had chosen to upper-bound $I(W; \Lambda_W(\mathbf{Z}))$ by $D(P_{W|\Lambda_W(Z)} \| Q | P_{\Lambda_W(Z)})$.

Using Theorems 10.1 and 10.2, we can obtain the following distribution-free bounds on the generalization error of the Gibbs algorithm.

THEOREM 10.7 *If the function ℓ takes values in* $[0,1]$, *then, for any data-generating distribution μ, the generalization error of the Gibbs algorithm defined in* (10.50) *is bounded as*

$$|\mathrm{gen}(\mu, P^*_{W|Z})| \le \frac{1}{2}\left(1 - e^{-2\beta/n}\right) \wedge \frac{\beta}{2n} \wedge \sqrt{\frac{\beta}{n}}. \tag{10.51}$$

Proof Fix any two $z, z' \in Z^n$ with $d_H(z, z') = 1$ and let $i \in [n]$ be the coordinate where $z_i \neq z'_i$. Then, from the identity

$$L_{z'}(w) = \frac{1}{n}(\ell(w, z'_i) - \ell(w, z_i)) + L_z(w),$$

it follows that

$$\frac{\mathbb{E}_Q[e^{-\beta L_{z'}(W)}]}{\mathbb{E}_Q[e^{-\beta L_z(W)}]} = \frac{\mathbb{E}_Q\left[\exp\left(-(\beta/n)(\ell(W, z'_i) - \ell(W, z_i)) - \beta L_z(W)\right)\right]}{\mathbb{E}_Q[e^{-\beta L_z(W)}]}$$
$$\le e^{\beta/n}.$$

Consequently,

$$\frac{dP^*_{W|Z=z}}{dP^*_{W|Z=z'}}(w) = \exp\left(\frac{\beta}{n}(\ell(w, z'_i) - \ell(w, z_i))\right) \frac{\mathbb{E}_Q[e^{-\beta L_{z'}(W)}]}{\mathbb{E}_Q[e^{-\beta L_z(W)}]} \le e^{2\beta/n},$$

and, interchanging the roles of z and z', we see that $e^{-2\beta/n} \le \left((dP^*_{W|Z=z})/(dP^*_{W|Z=z'})\right) \le e^{2\beta/n}$. It follows that

$$D(P^*_{W|Z=z} \| P^*_{W|Z=z'}) = \int_W dP^*_{W|Z=z} \log\left(\frac{dP^*_{W|Z=z}}{dP^*_{W|Z=z'}}\right) \le \frac{2\beta}{n}$$

and

$$d_{\mathsf{TV}}(P^*_{W|\mathbf{Z}=z}, P^*_{W|\mathbf{Z}=z'}) = \frac{1}{2}\int_{\mathsf{W}} \mathrm{d}P^*_{W|\mathbf{Z}=z'}\left|\frac{\mathrm{d}P^*_{W|\mathbf{Z}=z}}{\mathrm{d}P^*_{W|\mathbf{Z}=z'}} - 1\right| \leq \frac{1}{2}\left(1 - e^{-2\beta/n}\right),$$

where we have used (10.4). Another bound on the relative entropy between $P^*_{W|\mathbf{Z}=z}$ and $P^*_{W|\mathbf{Z}=z'}$ can be obtained as follows. We start with

$$D(P^*_{W|\mathbf{Z}=z}\|P^*_{W|\mathbf{Z}=z'}) = \frac{\beta}{n}\int_{\mathsf{W}} P^*_{W|\mathbf{Z}=z}(\mathrm{d}w)(\ell(w,z_i')-\ell(w,z_i))+\log\left(\frac{\mathbb{E}_Q[e^{-\beta L_{z'}(W)}]}{\mathbb{E}_Q[e^{-\beta L_z(W)}]}\right). \quad (10.52)$$

Then

$$\begin{aligned}\log\left(\frac{\mathbb{E}_Q[e^{-\beta L_{z'}(W)}]}{\mathbb{E}_Q[e^{-\beta L_z(W)}]}\right) &= \log \int_{\mathsf{W}} \exp\left(\frac{\beta}{n}(\ell(w,z_i) - \ell(w,z_i'))\right) P^*_{W|\mathbf{Z}=z}(\mathrm{d}w)\\ &\leq \frac{\beta}{n}\int_{\mathsf{W}} P^*_{W|\mathbf{Z}=z}(\mathrm{d}w)(\ell(w,z_i) - \ell(w,z_i')) + \frac{\beta^2}{2n^2}, \end{aligned} \quad (10.53)$$

where the inequality follows by applying Hoeffding's lemma to the random variable $\ell(W,z_i) - \ell(W,z_i')$ with $W \sim P^*_{W|\mathbf{Z}=z}$, which takes values in $[-1,1]$ almost surely. Combining (10.52) and (10.53), we get the bound

$$D(P^*_{W|\mathbf{Z}=z}\|P^*_{W|\mathbf{Z}=z'}) \leq \frac{\beta^2}{2n^2}$$

for any $z,z' \in \mathsf{Z}^n$ with $d_{\mathsf{H}}(z,z') = 1$. Invoking Lemma 10.4, we see that $P^*_{W|\mathbf{Z}}$ is $(1 - e^{-2\beta/n})$-stable in erasure T-information and $((2\beta/n) \wedge (\beta^2/2n^2))$-stable in erasure mutual information.

Therefore, Theorem 10.1 gives $|\mathrm{gen}(\mu, P^*_{W|\mathbf{Z}})| \leq 1 - e^{-2\beta/n}$. Moreover, since ℓ takes values in $[0,1]$, the sub-Gaussian assumption of Theorem 10.2 is satisfied with $\sigma^2 = 1/4$ by Hoeffding's lemma, and thus $|\mathrm{gen}(\mu, P^*_{W|\mathbf{Z}})| \leq \sqrt{\beta/n} \wedge (\beta/2n)$. Taking the minimum of the two bounds, we obtain (10.51).

With the above guarantees on the generalization error, we can analyze the population risk of the Gibbs algorithm. We first present a result for countable hypothesis spaces.

COROLLARY 10.2 *Suppose* W *is countable. Let* W *denote the output of the Gibbs algorithm applied to* $\mathbf{Z}$*, and let* w_o *denote the hypothesis that achieves the minimum population risk among* W*. If* ℓ *takes values in* $[0,1]$*, the population risk of* W *satisfies*

$$\mathbb{E}[L_\mu(W)] \leq \inf_{w \in \mathsf{W}} L_\mu(w) + \frac{1}{\beta}\log\left(\frac{1}{Q(w_o)}\right) + \frac{\beta}{2n}. \quad (10.54)$$

Proof We can bound the expected empirical risk of the Gibbs algorithm $P^*_{W|\mathbf{Z}}$ as

$$\begin{aligned}\mathbb{E}[L_{\mathbf{Z}}(W)] &\leq \mathbb{E}[L_{\mathbf{Z}}(W)] + \frac{1}{\beta}D(P^*_{W|\mathbf{Z}}\|Q|P_{\mathbf{Z}}) & (10.55)\\ &\leq \mathbb{E}[L_{\mathbf{Z}}(w)] + \frac{1}{\beta}D(\delta_w\|Q) \qquad \text{for all } w \in \mathsf{W}, & (10.56)\end{aligned}$$

where δ_w is the point mass at w. The second inequality is obtained via the optimality of the Gibbs $P^*_{W|\mathbf{Z}}$ in (10.49), since one can view δ_w as a sub-optimal learning algorithm that simply ignores the dataset and always outputs w. Taking $w = w_o$, noting that $\mathbb{E}[L_{\mathbf{Z}}(w_o)] = L_\mu(w_o)$, and combining this with the upper bound of Theorem 10.7 we obtain

$$\mathbb{E}[L_\mu(W)] \leq \inf_{w \in \mathsf{W}} L_\mu(w) + \frac{1}{\beta} D(\delta_{w_o} \| Q) + \frac{\beta}{2n}. \tag{10.57}$$

This leads to (10.54), as $D(\delta_{w_o} \| Q) = -\log Q(w_o)$ when W is countable.

The auxiliary distribution Q in the Gibbs algorithm can encode any prior knowledge of the population risks of the hypotheses in W. For example, we can order the hypotheses according to our (possibly imperfect) prior knowledge of their population risks, and set $Q(w_m) = 6/(\pi^2 m^2)$ for the mth hypothesis in the order.[2] Then, setting $\beta = \sqrt{n}$, (10.54) becomes

$$\mathbb{E}[L_\mu(W)] \leq \inf_{w \in \mathsf{W}} L_\mu(w) + \frac{2\log m_o + 1}{\sqrt{n}}, \tag{10.58}$$

where m_o is the index of w_o in the order. Thus, better prior knowledge of the population risks leads to a smaller sample complexity to achieve a certain expected excess risk. As another example, if $|\mathsf{W}| = k < \infty$ and we do not have any *a priori* preferences among the hypotheses, then we can take Q to be the uniform distribution on W. Upon setting $\beta = \sqrt{2n \log k}$, (10.54) becomes

$$\mathbb{E}[L_\mu(W)] \leq \inf_{w \in \mathsf{W}} L_\mu(w) + \sqrt{\frac{2\log k}{n}}.$$

For uncountable hypothesis spaces, we can proceed in an analogous fashion to analyze the population risk under the assumption that the loss function is Lipschitz.

COROLLARY 10.3 *Suppose* $\mathsf{W} = \mathbb{R}^d$ *with the Euclidean norm* $\|\cdot\|_2$. *Let* w_o *be the hypothesis that achieves the minimum population risk among* W. *Suppose that* ℓ *takes values in* $[0,1]$, *and* $\ell(\cdot, z)$ *is* ρ*-Lipschitz for all* $z \in \mathsf{Z}$. *Let* W *denote the output of the Gibbs algorithm applied to* $\mathbf{Z}$. *The population risk of* W *satisfies*

$$\mathbb{E}[L_\mu(W)] \leq \inf_{w \in \mathsf{W}} L_\mu(w) + \frac{\beta}{2n} + \inf_{a>0} \left(a\rho\sqrt{d} + \frac{1}{\beta} D(\mathcal{N}(w_o, a^2 \mathbf{I}_d) \| Q) \right), \tag{10.59}$$

where $\mathcal{N}(v, \Sigma)$ *denotes the d-dimensional normal distribution with mean v and covariance matrix* Σ.

Proof Just as we did in the proof of Corollary 10.2, we first bound the expected empirical risk of the Gibbs algorithm $P^*_{W|\mathbf{Z}}$. For any $a > 0$, we can view $\mathcal{N}(w_o, a^2 \mathbf{I}_d)$ as a learning algorithm that ignores the dataset and always draws a hypothesis from this Gaussian distribution. Denote by γ_d the standard normal density on $\mathbb{R}^d$. The non-negativity of relative entropy and (10.49) imply that

$$\mathbb{E}[L_{\mathbf{Z}}(W)] \leq \mathbb{E}[L_{\mathbf{Z}}(W)] + \frac{1}{\beta} D(P^*_{W|\mathbf{Z}} \| Q | P_{\mathbf{Z}}) \tag{10.60}$$

[2] Recall that $\sum_{m=1}^{\infty}(1/m^2) = (\pi^2/6) < 2$.

$$\leq \int_{\mathsf{W}} \mathbb{E}[L_{\mathbf{Z}}(w)]\gamma_d\left(\frac{w-w_{\mathrm{o}}}{a}\right)\mathrm{d}w + \frac{1}{\beta}D(\mathcal{N}(w_{\mathrm{o}},a^2\mathbf{I}_d)\|Q) \tag{10.61}$$

$$= \int_{\mathsf{W}} L_\mu(w)\gamma_d\left(\frac{w-w_{\mathrm{o}}}{a}\right)\mathrm{d}w + \frac{1}{\beta}D(\mathcal{N}(w_{\mathrm{o}},a^2\mathbf{I}_d)\|Q). \tag{10.62}$$

Combining this with the upper bound on the expected generalization error of Theorem 10.7, we obtain

$$\mathbb{E}[L_\mu(W)] \leq \inf_{a>0}\left(\int_{\mathsf{W}} L_\mu(w)\gamma_d\left(\frac{w-w_{\mathrm{o}}}{a}\right)\mathrm{d}w + \frac{1}{\beta}D(\mathcal{N}(w_{\mathrm{o}},a^2\mathbf{I}_d)\|Q)\right) + \frac{\beta}{2n}. \tag{10.63}$$

Since $\ell(\cdot,z)$ is ρ-Lipschitz for all $z \in \mathsf{Z}$, we have that, for any $w \in \mathsf{W}$,

$$|L_\mu(w) - L_\mu(w_{\mathrm{o}})| \leq \mathbb{E}[|\ell(w,Z) - \ell(w_{\mathrm{o}},Z)|] \leq \rho\|w - w_{\mathrm{o}}\|_2. \tag{10.64}$$

Then

$$\int_{\mathsf{W}} L_\mu(w)\gamma_d\left(\frac{w-w_{\mathrm{o}}}{a}\right)\mathrm{d}w \leq \int_{\mathsf{W}} (L_\mu(w_{\mathrm{o}}) + \rho\|w-w_{\mathrm{o}}\|_2)\gamma_d\left(\frac{w-w_{\mathrm{o}}}{a}\right)\mathrm{d}w \tag{10.65}$$

$$\leq L_\mu(w_{\mathrm{o}}) + \rho a\sqrt{d}. \tag{10.66}$$

Substituting this into (10.63), we obtain (10.59).

Again, we can use the distribution Q to express our preference of the hypotheses in W. For example, we can choose $Q = \mathcal{N}(w_Q, b^2\mathbf{I}_d)$ with $b = n^{-1/4}d^{-1/4}\rho^{-1/2}$ and choose $\beta = n^{3/4}d^{1/4}\rho^{1/2}$. Then, setting $a = b$ in (10.59), we have

$$\mathbb{E}[L_\mu(W)] \leq \inf_{w\in\mathsf{W}} L_\mu(w) + \frac{d^{1/4}\rho^{1/2}}{2n^{1/4}}\left(\|w_Q - w_{\mathrm{o}}\|_2^2 + 3\right). \tag{10.67}$$

This result places essentially no restrictions on W, which could be unbounded, and only requires the Lipschitz condition on $\ell(\cdot,z)$, which could be non-convex. The sample complexity decreases with better prior knowledge of the optimal hypothesis.

10.5.4 Differentially Private Learning Algorithms

As mentioned in Section 10.1.1, differential privacy can be interpreted as a stability property. In this section, we investigate this in detail. A learning algorithm $P_{W|\mathbf{Z}}$ is (ε,δ)-*differentially private* [13] if, for any measurable set $F \subseteq \mathsf{W}$,

$$d_{\mathsf{H}}(z,z') \leq 1 \implies P_{W|\mathbf{Z}=z}(F) \leq e^{\varepsilon}P_{W|\mathbf{Z}=z'}(F) + \delta. \tag{10.68}$$

THEOREM 10.8 *Suppose that the loss function ℓ takes values in* $[0,1]$. *If $P_{W|\mathbf{Z}}$ is (ε,δ)-differentially private for some $\varepsilon \geq 0$ and $\delta \in [0,1]$, then*

$$|\,\mathrm{gen}(\mu, P_{W|\mathbf{Z}})| \leq 1 - e^{-\varepsilon}(1-\delta) \tag{10.69}$$

for any data-generating distribution μ. If ℓ is arbitrary, but the hypothesis space W is finite, the pair $(\mu, P_{W|\mathbf{Z}})$ satisfies the conditions of Theorem 10.2, and $e^{\varepsilon} \leq 4(1-\delta)/3$, then

$$|\,\mathrm{gen}(\mu, P_{W|\mathbf{Z}})| \leq \sqrt{2\sigma^2(1-e^{-\varepsilon}(1-\delta))\log\left(\frac{|\mathsf{W}|}{1-e^{-\varepsilon}(1-\delta)}\right)}. \tag{10.70}$$

Proof We will prove that any (ε,δ)-differentially private learning algorithm is $(1 - e^{-\varepsilon}(1-\delta))$-stable in total variation. To that end, let us rewrite the differential privacy condition (10.68) as follows. For any z,z' with $d_{\mathsf{H}}(z,z') \le 1$,

$$E_{e^{\varepsilon}}(P_{W|Z=z} \| P_{W|Z=z'}) \le \delta, \tag{10.71}$$

where the E_γ-divergence (with $\gamma \ge 1$) between two probability measures μ and ν on a measurable space $(\Omega,\mathcal{F})$ is defined as

$$E_\gamma(\mu\|\nu) := \max_{E\in\mathcal{F}}(\mu(E) - \gamma\nu(E))$$

(see, e.g., [34] or p. 28 of [35]). It satisfies the following inequality (see Section VII.A of [36]):

$$E_\gamma(\mu\|\nu) \ge 1 - \gamma(1 - d_{\mathsf{TV}}(\mu,\nu)). \tag{10.72}$$

Using (10.72) in (10.71) with $\gamma = e^{\varepsilon}$, we get

$$d_{\mathsf{H}}(z,z') \le 1 \quad \Longrightarrow \quad d_{\mathsf{TV}}(P_{W|Z=z}, P_{W|Z=z'}) \le 1 - e^{-\varepsilon}(1-\delta),$$

and therefore A is $(1 - e^{-\varepsilon}(1-\delta))$-stable in erasure T-information by Lemma 10.4. Using Theorem 10.1, we obtain Eq. (10.69); the inequality (10.70) follows from (10.72), Lemma 10.3, and Theorem 10.2.

For example, the Gibbs algorithm is $(2\beta/n, 0)$-differentially private [37], so Theorem 10.7 is a special case of Theorem 10.8.

10.5.5 Noisy Empirical Risk Minimization

Another algorithm that provides control on the input–output mutual information is noisy ERM, where we add independent scalar noise N_w to the empirical risk of each hypothesis $w \in \mathsf{W}$ and then output any hypothesis that minimizes the noise-perturbed empirical risk:

$$W = \arg\min_{w\in\mathsf{W}} (L_Z(w) + N_w). \tag{10.73}$$

Just as in the case of the Gibbs algorithm, we can encode our preferences for (or prior knowledge about) various hypotheses by controlling the amount of noise added to each hypothesis. The following result formalizes this idea.

COROLLARY 10.4 *Suppose $\mathsf{W} = \{w_m\}_{m=1}^{\infty}$ is countably infinite, and the ordering is such that a hypothesis with a lower index is preferred over one with a higher index. Also suppose $\ell \in [0,1]$. For the noisy ERM algorithm in* (10.73)*, choosing N_m to be an exponential random variable with mean b_m, we have*

$$\mathbb{E}[L_\mu(W)] \le \min_m L_\mu(w_m) + b_{m_o} + \sqrt{\frac{1}{2n}\sum_{m=1}^{\infty}\frac{L_\mu(w_m)}{b_m} - \left(\sum_{m=1}^{\infty}\frac{1}{b_m}\right)^{-1}}, \tag{10.74}$$

where $m_o = \arg\min_m L_\mu(w_m)$. In particular, choosing $b_m = m^{1.1}/n^{1/3}$, we have

$$\mathbb{E}[L_\mu(W)] \le \min_m L_\mu(w_m) + \frac{m_o^{1.1} + 3}{n^{1/3}}. \tag{10.75}$$

Without adding noise, the ERM algorithm applied to the above case when card (W) = $k < \infty$ can achieve $\mathbb{E}[L_\mu(W_{\text{ERM}})] \le \min_{m\in[k]} L_\mu(w_m) + \sqrt{(1/2n)\log k}$. Compared with (10.75), we see that performing noisy ERM may be beneficial when we have high-quality prior knowledge of w_{o} and when k is large.

Proof We prove the result assuming $\text{card}(\mathsf{W}) = k < \infty$. When W is countably infinite, the proof carries over by taking the limit as $k \to \infty$.

First, we upper-bound the expected generalization error via $I(W;\mathbf{Z})$. We have the following chain of inequalities:

$$\begin{aligned}
I(W;\mathbf{Z}) &\le I((L_{\mathbf{Z}}(w_m)+N_m)_{m\in[k]};(L_{\mathbf{Z}}(w_m))_{m\in[k]}) \\
&\le \sum_{m=1}^{k} I(L_{\mathbf{Z}}(w_m);L_{\mathbf{Z}}(w_m)+N_m) \\
&\le \sum_{m=1}^{k} \log\left(1+\frac{\mathbb{E}[L_{\mathbf{Z}}(w_m)]}{b_m}\right) \\
&= \sum_{m=1}^{k} \log\left(1+\frac{L_\mu(w_m)}{b_m}\right),
\end{aligned} \tag{10.76}$$

where we have used the data-processing inequality for mutual information; the fact that, for product channels, the overall input–output mutual information is upper-bounded by the sum of the input–output mutual information of individual channels [38]; the formula for the capacity of the additive exponential noise channel under an input mean constraint [39]; and the fact that $\mathbb{E}[L_{\mathbf{Z}}(w_m)] = L_\mu(w_m)$. The assumption that ℓ takes values in $[0,1]$ implies that $\ell(w,Z)$ is $1/4$-sub-Gaussian for all $w \in \mathsf{W}$, and, as a consequence of (10.76),

$$\text{gen}(\mu, P_{W|\mathbf{Z}}) \le \sqrt{\frac{1}{2n}\sum_{m=1}^{k}\log\left(1+\frac{L_\mu(w_m)}{b_m}\right)}. \tag{10.77}$$

We now upper-bound the expected empirical risk. By the construction of the algorithm, almost surely

$$\begin{aligned}
L_{\mathbf{Z}}(W) &= L_{\mathbf{Z}}(W) + N_W - N_W \\
&\le L_{\mathbf{Z}}(w_{m_{\text{o}}}) + N_{m_{\text{o}}} - N_W \\
&\le L_{\mathbf{Z}}(w_{m_{\text{o}}}) + N_{m_{\text{o}}} - \min\{N_m, m \in [k]\}.
\end{aligned}$$

Taking the expectation on both sides, we get

$$\mathbb{E}[L_{\mathbf{Z}}(W)] \le L_\mu(w_{m_{\text{o}}}) + b_{m_{\text{o}}} - \left(\sum_{m=1}^{k}\frac{1}{b_m}\right)^{-1}. \tag{10.78}$$

Combining (10.77) and (10.78), we have

$$\mathbb{E}[L_\mu(W)] \le \min_{m\in[k]} L_\mu(w_m) + \sqrt{\frac{1}{2n}\sum_{m=1}^{k}\log\left(1+\frac{L_\mu(w_m)}{b_m}\right)} + b_{m_{\text{o}}} - \left(\sum_{m=1}^{k}\frac{1}{b_m}\right)^{-1},$$

which leads to (10.74) with the fact that $\log(1+x) \le x$.

When $b_m = m^{1.1}/n^{1/3}$, using the fact that

$$\sum_{m=1}^{k} \frac{1}{m^{1.1}} \le 11 - 10k^{-1/10}$$

and upper-bounding $L_\mu(w_m)$s by 1, we get

$$\begin{aligned}\mathbb{E}[L_\mu(W)] &\le \min_{m\in[k]} L_\mu(w_m) + \frac{1}{n^{1/3}}\left(\sqrt{\frac{1}{2}(11-10k^{-1/10}) + m_o^{1.1}} - \frac{1}{11-10k^{-1/10}}\right)\\ &\le \min_{m\in[k]} L_\mu(w_m) + \frac{3+m_o^{1.1}}{n^{1/3}},\end{aligned}$$

which proves (10.75).

10.5.6 Other Methods to Induce Input–Output Mutual Information Stability

There is a wide variety of methods for controlling the input–output mutual information of a learning algorithm. One method is to pre-process the dataset $\mathbf{Z}$ and then run a learning algorithm on the transformed dataset $\widetilde{\mathbf{Z}}$. The pre-processing step can consist of adding noise to the data or removing some of the instances from the dataset. For any choice of the pre-processing mechanism and the learning algorithm, we have the Markov chain $\mathbf{Z}-\widetilde{\mathbf{Z}}-W$, which implies $I(W;\mathbf{Z}) \le \min\{I(\mathbf{Z};\widetilde{\mathbf{Z}}), I(W;\widetilde{\mathbf{Z}})\}$. Another method is to postprocess the output of a learning algorithm. For instance, the entries of the weight matrix $\widetilde{W}$ generated by a neural-network training algorithm can be quantized or perturbed by noise. Any choice of the learning algorithm and the post-processing mechanism induces the Markov chain $\mathbf{Z}-\widetilde{W}-W$, which implies $I(W;\mathbf{Z}) \le \min\{I(\widetilde{W};W), I(\widetilde{W};\mathbf{Z})\}$. One can use strong data-processing inequalities [40] to sharpen these upper bounds on $I(W;\mathbf{Z})$. Pre-processing of the dataset and post-processing of the output hypothesis are among the numerous regularization methods used in deep learning (see Chapter 7.5 of [41]). Other regularization methods may also be interpreted as strategies to induce the input–output mutual information stability of a learning algorithm; this would be an interesting direction of future research.

10.5.7 Adaptive Composition of Learning Algorithms

Apart from controlling the generalization error of individual learning algorithms, the input–output mutual information can also be used for analyzing the generalization capability of complex learning algorithms obtained by adaptively composing simple constituent algorithms. Under a k-fold adaptive composition, a common $\mathbf{Z}$ is shared by k learning algorithms that are executed sequentially. Specifically, for each $j \in [k]$, the output W_j of the jth algorithm is drawn from a hypothesis space W_j that is based on the data $\mathbf{Z}$ and on the outputs W^{j-1} of the previously executed algorithms, according to $P_{W_j|\mathbf{Z},W^{j-1}}$. An example when $k=2$ is the simplest case of selective inference [3]: data-driven model selection followed by a learning algorithm executed the same dataset.

Various boosting techniques in machine learning can also be viewed as instances of adaptive composition. From the data-processing inequality and the chain rule of mutual information,

$$I(W_k;\mathbf{Z}) \le I(W^k;\mathbf{Z}) = \sum_{j=1}^{k} I(W_j;\mathbf{Z}|W^{j-1}). \tag{10.79}$$

If the Markov chain $\mathbf{Z}-\Lambda_{\mathsf{W}_j}(\mathbf{Z})-W_j$ holds conditional on W^{j-1} for $j = 1,\dots,k$, then the upper bound in (10.79) can be sharpened to $\sum_{j=1}^{k} I(W_j;\Lambda_{\mathsf{W}_j}(\mathbf{Z})|W^{j-1})$. Thus, we can control the generalization error of the final output by controlling the conditional mutual information at each step of the composition. This also gives us a way to analyze the generalization error of the composite learning algorithm using local information-theoretic properties of the constituent algorithms. Recent work by Feldman and Steinke [42] proposes a stronger notion of stability in erasure mutual information (uniformly with respect to all, not necessarily product, distributions of $\mathbf{Z}$) and applies it to adaptive data analysis. Armed with this notion, they design and analyze a noise-adding algorithm that calibrates the noise variance to the empirical variance of the data. A related information-theoretic notion of stability was also proposed by Cuff and Yu [43] in the context of differential privacy.

Acknowledgments

The authors would like to thank the referees and Nir Weinberger for reading the manuscript and for suggesting a number of corrections, and Vitaly Feldman for valuable general comments and for pointing out the connection to the work of McAllester on PAC-Bayes bounds. The work of A. Rakhlin is supported in part by the NSF under grant no. CDS&E-MSS 1521529 and by the DARPA Lagrange program. The work of M. Raginsky and A. Xu is supported in part by NSF CAREER award CCF–1254041 and in part by the Center for Science of Information (CSoI), an NSF Science and Technology Center, under grant agreement CCF-0939370.

References

[1] V. N. Vapnik, *Statistical learning theory*. Wiley, 1998.

[2] S. Shalev-Shwartz, O. Shamir, N. Srebro, and K. Sridharan, "Learnability, stability, and uniform convergence," *J. Machine Learning Res.*, vol. 11, pp. 2635–2670, 2010.

[3] J. Taylor and R. J. Tibshirani, "Statistical learning and selective inference," *Proc. Natl. Acad. Sci. USA*, vol. 112, no. 25, pp. 7629–7634, 2015.

[4] L. Devroye and T. Wagner, "Distribution-free performance bounds for potential function rules," *IEEE Trans. Information Theory*, vol. 25, no. 5, pp. 601–604, 1979.

[5] W. H. Rogers and T. J. Wagner, "A finite sample distribution-free performance bound for local discrimination rules," *Annals Statist.*, vol. 6, no. 3, pp. 506–514, 1978.

[6] D. Ron and M. Kearns, "Algorithmic stability and sanity-check bounds for leave-one-out crossvalidation," *Neural Computation*, vol. 11, no. 6, pp. 1427–1453, 1999.

[7] O. Bousquet and A. Elisseeff, "Stability and generalization," *J. Machine Learning Res.*, vol. 2, pp. 499–526, 2002.

[8] T. Poggio, R. Rifkin, S. Mukherjee, and P. Niyogi, "General conditions for predictivity in learning theory," *Nature*, vol. 428, no. 6981, pp. 419–422, 2004.

[9] S. Kutin and P. Niyogi, "Almost-everywhere algorithmic stability and generalization error," in *Proc.* 18*th Conference on Uncertainty in Artificial Intelligence* (*UAI 2002*), 2002, pp. 275–282.

[10] A. Rakhlin, S. Mukherjee, and T. Poggio, "Stability results in learning theory," *Analysis Applications*, vol. 3, no. 4, pp. 397–417, 2005.

[11] C. Dwork, F. McSherry, K. Nissim, and A. Smith, "Calibrating noise to sensitivity in private data analysis," in *Proc. Theory of Cryptography Conference*, 2006, pp. 265–284.

[12] C. Dwork, V. Feldman, M. Hardt, T. Pitassi, O. Reingold, and A. Roth, "Generalization in adaptive data analysis and holdout reuse," *arXiv:1506.02629*, 2015.

[13] C. Dwork and A. Roth, "The algorithmic foundatons of differential privacy," *Foundations and Trends in Theoretical Computer Sci.*, vol. 9, nos. 3–4, pp. 211–407, 2014.

[14] P. Kairouz, S. Oh, and P. Viswanath, "The composition theorem for differential privacy," in *Proc. 32nd International Conference on Machine Learning (ICML)*, 2015, pp. 1376–1385.

[15] M. Raginsky, A. Rakhlin, M. Tsao, Y. Wu, and A. Xu, "Information-theoretic analysis of stability and bias of learning algorithms," in *Proc. IEEE Information Theory Workshop*, 2016.

[16] A. Xu and M. Raginsky, "Information-theoretic analysis of generalization capability of learning algorithms," in *Proc. Conference on Neural Information Processing Systems*, 2017.

[17] H. Strasser, *Mathematical theory of statistics: Statistical experiments and asymptotic decision Theory*. Walter de Gruyter, 1985.

[18] Y. Polyanskiy and Y. Wu, "Dissipation of information in channels with input constraints," *IEEE Trans. Information Theory*, vol. 62, no. 1, pp. 35–55, 2016.

[19] S. Verdú and T. Weissman, "The information lost in erasures," *IEEE Trans. Information Theory*, vol. 54, no. 11, pp. 5030–5058, 2008.

[20] W. Hoeffding, "Probability inequalities for sums of bounded random variables," *J. Amer. Statist. Soc.*, vol. 58, no. 301, pp. 13–30, 1963.

[21] S. Boucheron, G. Lugosi, and P. Massart, *Concentration inequalities: A nonasymptotic theory of independence*. Oxford University Press, 2013.

[22] T. M. Cover and J. A. Thomas, *Elements of information theory*, 2nd edn. Wiley, 2006.

[23] R. Bassily, K. Nissim, A. Smith, T. Steinke, U. Stemmer, and J. Ullman, "Algorithmic stability for adaptive data analysis," in *Proc. 48th ACM Symposium on Theory of Computing (STOC)*, 2016, pp. 1046–1059.

[24] D. McAllester, "PAC-Bayesian model averaging," in *Proc. 1999 Conference on Learning Theory*, 1999.

[25] D. McAllester, "A PAC-Bayesian tutorial with a dropout bound," *arXiv:1307.2118*, 2013, http://arxiv.org/abs/1307.2118.

[26] D. Russo and J. Zou, "Controlling bias in adaptive data analysis using information theory," in *Proc. 19th International Conference on Artificial Intelligence and Statistics (AISTATS)*, 2016, pp. 1232–1240.

[27] J. Jiao, Y. Han, and T. Weissman, "Dependence measures bounding the exploration bias for general measurements," in *Proc. IEEE International Symposium on Information Theory*, 2017, pp. 1475–1479.

[28] C. Dwork, V. Feldman, M. Hardt, T. Pitassi, O. Reingold, and A. Roth, "Preserving statistical validity in adaptive data analysis," in *Proc. 47th ACM Symposium on Theory of Computing (STOC)*, 2015.

[29] C. Dwork, V. Feldman, M. Hardt, T. Pitassi, O. Reingold, and A. Roth, "Generalization in adaptive data analysis and holdout reuse," in *28th Annual Conference on Neural Information Processing Systems (NIPS)*, 2015.

[30] R. Bassily, S. Moran, I. Nachum, J. Shafer, and A. Yehudayoff, "Learners that use little information," in *Proc. Conference on Algorithmic Learning Theory (ALT)*, 2018.

[31] R. Vershynin, *High-dimensional probability: An introduction with applications in data science*. Cambridge University Press, 2018.

[32] L. Devroye, L. Györfi, and G. Lugosi, *A probabilistic theory of pattern recognition*. Springer, 1996.

[33] K. L. Buescher and P. R. Kumar, "Learning by canonical smooth estimation. I. Simultaneous estimation," *IEEE Tran. Automatic Control*, vol. 41, no. 4, pp. 545–556, 1996.

[34] J. E. Cohen, J. H. B. Kemperman, and G. Zbăganu, *Comparisons of stochastic matrices, with applications in information theory, statistics, economics, and population sciences*. Birkhäuser, 1998.

[35] Y. Polyanskiy, "Channel coding: Non-asymptotic fundamental limits," Ph.D. dissertation, Princeton University, 2010.

[36] I. Sason and S. Verdú, "f-divergence inequalities," *IEEE Trans. Information Theory*, vol. 62, no. 11, pp. 5973–6006, 2016.

[37] F. McSherry and K. Talwar, "Mechanism design via differential privacy," in *Proc. 48th Annual IEEE Symposium on Foundations of Computer Science (FOCS)*, 2007.

[38] Y. Polyanskiy and Y. Wu, "Lecture notes on information theory," Lecture Notes for ECE563 (UIUC) and 6.441 (MIT), 2012–2016, http://people.lids.mit.edu/yp/homepage/data/itlectures_v4.pdf.

[39] S. Verdú, "The exponential distribution in information theory," *Problems of Information Transmission*, vol. 32, no. 1, pp. 86–95, 1996.

[40] M. Raginsky, "Strong data processing inequalities and Φ-Sobolev inequalities for discrete channels," *IEEE Trans. Information Theory*, vol. 62, no. 6, pp. 3355–3389, 2016.

[41] I. Goodfellow, Y. Bengio, and A. Courville, *Deep learning*. MIT Press, 2016.

[42] V. Feldman and T. Steinke, "Calibrating noise to variance in adaptive data analysis," in *Proc. 2018 Conference on Learning Theory*, 2018.

[43] P. Cuff and L. Yu, "Differential privacy as a mutual information constraint," in *Proc. 2016 ACM SIGSAC Conference on Computer and Communication Security (CCS)*, 2016, pp. 43–54.

11 Information Bottleneck and Representation Learning

Pablo Piantanida and Leonardo Rey Vega

A grand challenge in representation learning is the development of computational algorithms that learn the different explanatory factors of variation behind high-dimensional data. Representation models (usually referred to as encoders) are often determined for optimizing performance on training data when the real objective is to generalize well to other (unseen) data. The first part of this chapter is devoted to providing an overview of and introduction to fundamental concepts in statistical learning theory and the information-bottleneck principle. It serves as a mathematical basis for the technical results given in the second part, in which an upper bound to the *generalization gap* corresponding to the *cross-entropy risk* is given. When this penalty term times a suitable multiplier and the *cross-entropy empirical risk* are minimized jointly, the problem is equivalent to optimizing the information-bottleneck objective with respect to the empirical data distribution. This result provides an interesting connection between mutual information and generalization, and helps to explain why noise injection during the training phase can improve the generalization ability of encoder models and enforce invariances in the resulting representations.

11.1 Introduction and Overview

Information theory aims to characterize the fundamental limits for data compression, communication, and storage. Although the coding techniques used to prove these fundamental limits are impractical, they provide valuable insight, highlighting key properties of good codes and leading to designs approaching the theoretical optimum (e.g., turbo codes, ZIP and JPEG compression algorithms). On the other hand, statistical models and machine learning are used to acquire knowledge from data. Models identify relationships between variables that allow one to make predictions and assess their accuracy. A good choice of data representation is paramount for performing large-scale data processing in a computationally efficient and statistically meaningful manner [1], allowing one to decrease the need for storage, or to reduce inter-node communication if the data are distributed.

Shannon's abstraction of information merits careful study [2]. While a layman might think that the problem of communication is to convey meaning, Shannon clarified that "*the fundamental problem of communication is that of reproducing at one point a message selected at another point.*" Shannon further argued that the meaning of a message is subjective, i.e., dependent on the observer, and irrelevant to the engineering problem of communication. However, what does matter for the theory of communication is finding suitable representations for given data. In source coding, for example, one generally aims at distilling the relevant information from the data by removing unnecessary redundancies. This can be cast in information-theoretic terms, as higher redundancy makes data more predictable and lowers their information content.

In the context of learning [3, 4], we propose to distinguish between these two rather different aspects of data: *information* and *knowledge*. *Information* contained in data is unpredictable and random, while additional structure and redundancy in the data stream constitute *knowledge* about the data-generation process, which a learner must acquire. Indeed, according to connectionist models [5], the redundancy contained within messages enables the brain to build up its cognitive maps and the statistical regularities in these messages are being used for this purpose. Hence, this *knowledge*, provided by redundancy [6, 7] in the data, must be what drives unsupervised learning. While information theory is a unique success story, from its birth, it discarded *knowledge* as being irrelevant to the engineering problem of communication. However, knowledge is recognized as being a critical – almost central – component of representation learning. The present text provides an information-theoretic treatment of this problem.

Knowledge representation. The data deluge of recent decades has led to new expectations for scientific discoveries from massive data. While mankind is drowning in data, a significant part of the data is unstructured, and it is difficult to discover relevant information. A common denominator in these novel scenarios is the challenge of representation learning: how to extract salient features or statistical relationships from data in order to build meaningful representations of the relevant content. In many ways, *deep neural networks* have turned out to be very good at discovering structures in high-dimensional data and have dramatically improved the state-of-the-art in several pattern-recognition tasks [8]. The global learning task is decomposed into a hierarchy of layers with non-linear processing, a method achieving great success due to its ability not only to fit different types of datasets but also to generalize incredibly well. The representational capabilities of neural networks [9] have attracted significant interest from the machine-learning community. These networks seem to be able to learn multi-level abstractions, with a capability to harness unlabeled data, multi-task learning, and multiple inputs, while learning from distributed and hierarchical data, to represent context at multiple levels.

The actual goal of representation learning is neither accurate estimation of model parameters [10] nor compact representation of the data themselves [11, 12]; rather, we are mostly interested in the generalization capabilities, meaning the ability to successfully apply rules extracted from previously seen data to characterize unseen data. According to the statistical learning theory [13], models with many parameters tend to *overfit* by representing the learned data too accurately, therefore diminishing their ability to generalize to unseen data. In order to reduce this "generalization gap," i.e., the difference between the "training error" and the "test error" (a measure of how well the learner has learned), several regularization methods were proposed in the literature [13].

A recent breakthrough in this area has been the development of *dropout* [14] for training deep neural networks. This consists in randomly dropping units during training to prevent their co-adaptation, including some information-based regularization [15] that yields a slightly more general form of the variational auto-encoder [16].

Why is that we succeed in learning high-dimensional representations? Recently there has been much interest in understanding the importance of implicit regularization. Numerical experiments in [17] demonstrate that network size may not be the main form of capacity control for deep neural networks and, hence, some other, unknown, form of capacity control plays a central role in learning multilayer feed-forward networks. From a theoretical perspective, regularization seems to be an indispensable component in order to improve the final misclassification probability, while convincing experiments support the idea that the absence of all regularization does not necessarily induce a poor generalization gap. Possible explanations were approached via rate-distortion theory [18, 19] by exploring heuristic connections with the celebrated information-bottleneck principle [20]. Within the same line of work, in [21, 22] Russo and Zou and Xu and Raginsky have proven bounds showing that the square root of the mutual information between the training inputs and the parameters inferred from the training algorithm provides a concise bound on the generalization gap. These bounds crucially depend on the Markov operator that maps the training set into the network parameters and whose characterization could not be an easy task. Similarly, in [23] Achille and Soatto explored how the use of an information-bottleneck objective on the network parameters (and not on the representations) may help to avoid overfitting while enforcing invariant representations.

The interplay between information and complexity. The goal of data representation may be cast as trying to find regularity in the data. Regularity may be identified with the "ability to compress" by viewing representation learning as lossy data compression: this tells us that, for a given set of encoder models and dataset, we should try to find the encoder or combination of encoders that compresses the data most. In this sense, we may speak of the information complexity of a structure, meaning the minimum amount of information (number of bits) we need to store enough information about the structure that allows us to achieve its reconstruction. The central result in this chapter states that good representation models should squeeze out as much regularity as possible from the given data. In other words, representations are expected to distill the *meaningful information* present in the data, i.e., to separate structure as seeing the regularity from noise, interpreted as the *accidental information*.

The structure of this chapter. This chapter can be read without any prior knowledge of information theory and statistical learning theory. In the first part, the basic learning framework for analysis is developed and an accessible overview of basic concepts in statistical learning theory and the information-bottleneck principle are presented. The second part introduces an upper bound to the generalization gap corresponding to the cross-entropy loss and shows that, when this penalty term times a suitable multiplier and the cross-entropy empirical risk are minimized jointly, the problem is equivalent to optimizing the information-bottleneck objective with respect to the empirical data distribution. The notion of information complexity is introduced and intuitions behind it are developed.

11.2 Representation and Statistical Learning

We introduce the framework within which learning from examples is to be studied. We develop precise notions of risk and the generalization gap, and discuss the mathematical factors upon which these depend.

11.2.1 Basic Definitions

In this text we are concerned with the problem of pattern classification, which is about predicting the unknown class of an example or observation. An example can be modeled as an information source $X \in \mathcal{X}$ presented to the learner about a target concept $Y \in \mathcal{Y}$ (the concept class). In our model we simply assume $(\mathcal{X}, \mathcal{Y})$ are abstract discrete spaces equipped with a σ-algebra. In the problem of pattern classification, one searches for a function $c : \mathcal{X} \longrightarrow \mathcal{Y}$ which represents one's guess of Y given X. Although there is much to say about statistical learning, this section does not cover the field extensively (an overview can be found in [13]). Besides this, we limit ourselves to describing the key ideas in a simple way, often sacrificing generality.

DEFINITION 11.1 (Misclassification probability) *An $|\mathcal{Y}|$-ary classifier is defined by a (possibly stochastic) decision rule $Q_{\widehat{Y}|X} : \mathcal{X} \to \mathcal{P}(\mathcal{Y})$, where $\widehat{Y} \in \mathcal{Y}$ denotes the random variable associated with the classifier output and X is the information source. The probability of misclassification of a rule $Q_{\widehat{Y}|X}$ with respect to a data distribution P_{XY} is given by*

$$P_{\mathcal{E}}(Q_{\widehat{Y}|X}) := 1 - \mathbb{E}_{P_{XY}}\left[Q_{\widehat{Y}|X}(Y|X)\right]. \tag{11.1}$$

Minimizing over all possible classifiers $Q_{\widehat{Y}|X}$ gives the smallest average probability of misclassification. An optimum classifier $c^{\star}(\cdot)$ chooses the hypothesis $\widehat{y} \in \mathcal{Y}$ with largest posterior probability $P_{Y|X}$ given the observation x, that is the *maximum a posteriori (MAP)* decision. The MAP test that breaks ties randomly with equal probability is given by[1]

$$Q^{\text{MAP}}_{\widehat{Y}|X}(y|x) := \begin{cases} \dfrac{1}{|\mathcal{B}(x)|}, & \text{if } y \in \mathcal{B}(x), \\ 0, & \text{otherwise}, \end{cases} \tag{11.2}$$

where the set $\mathcal{B}(x)$ is defined as

$$\mathcal{B}(x) := \left\{y \in \mathcal{Y} : P_{Y|X}(y|x) = \max_{y' \in \mathcal{Y}} P_{Y|X}(y'|x)\right\}.$$

This classification rule is called the *Bayes decision rule*. The Bayes decision rule is optimal in the sense that no other decision rule has a smaller probability of misclassification. It is straightforward to obtain the following lemma.

[1] In general, the optimum classifier given in (11.2) is not unique. Any conditional p.m.f. with support in $\mathcal{B}(x)$ for each $x \in \mathcal{X}$ will be equally good.

LEMMA 11.1 (Bayes error) *The misclassification error rate of the Bayes decision rule is given by*

$$P_{\mathcal{E}}(Q^{\mathrm{MAP}}_{\widehat{Y}|X}) \;= 1 - \mathbb{E}_{P_X}\left[\max_{y' \in \mathcal{Y}} P_{Y|X}(y'|X)\right]. \tag{11.3}$$

Finding the Bayes decision rule requires knowledge of the underlying distribution P_{XY}, but typically in applications these distributions are not known. In fact, even a parametric form or an approximation to the true distribution is unknown. In this case, the *learner* tries to overcome the lack of knowledge by resorting to labeled examples. In addition, the probability of misclassification using the labeled examples has the particularity that it is mathematically hard to solve for the optimal decision rule. As a consequence, it is common to work with a surrogate (information measure) given by the *average logarithmic loss or cross-entropy loss*. This loss is used when a probabilistic interpretation of the scores is desired by measuring the dissimilarity between the true label distribution $P_{Y|X}$ and the predicted label distribution $Q_{\widehat{Y}|X}$, and is defined below.

LEMMA 11.2 (Surrogate based on the average logarithmic loss) *A natural surrogate for the probability of misclassification $P_{\mathcal{E}}(Q_{\widehat{Y}|X})$ corresponding to a classifier $Q_{\widehat{Y}|X}$ is given by the average logarithmic loss $\mathbb{E}_{P_{XY}}\left[-\log Q_{\widehat{Y}|X}(Y|X)\right]$, which satisfies*

$$P_{\mathcal{E}}(Q_{\widehat{Y}|X}) \leq 1 - \exp\left(-\mathbb{E}_{P_{XY}}\left[-\log Q_{\widehat{Y}|X}(Y|X)\right]\right). \tag{11.4}$$

A lower bound for the average logarithmic loss can be computed as

$$\mathbb{E}_{P_{XY}}\left[-\log Q_{\widehat{Y}|X}(Y|X)\right] \geq H(P_{Y|X}|P_X). \tag{11.5}$$

The average logarithmic loss can provide an effective and better-behaved surrogate for the particular problem of minimizing the probability of misclassification [9]. Evidently, the optimal decision rule for the average logarithmic loss is $Q_{\widehat{Y}|X} \equiv P_{Y|X}$. This does not match in general with the optimal decision rule for the probability of misclassification $Q^{\mathrm{MAP}}_{\widehat{Y}|X}$ in expression (11.2). Although the average logarithmic loss may induce an irreducible gap with respect to the probability of misclassification, it is clear that when the true $P_{Y|X}$ concentrates around a particular value $y(x)$ for each $x \in \mathcal{X}$ (which is necessary for a statistical model $P_{Y|X}$ to induce low probability of misclassification) this gap could be significantly reduced.

11.2.2 Learning Data Representations

We will concern ourselves with learning representation models (randomized encoders) and self-classifiers (randomized decoders) from labeled examples, in other words, learning target probability distributions which are assumed to belong to some class of distributions. The motivation behind this paradigm relies on a view of the brain as an information processor that in solving certain problems (e.g., object recognition) builds a series of internal representations starting with the sensory (external) input from which it computes a function (e.g., detecting the orientations of edges in an image or learning to recognize individual faces).

The problem of finding a good classifier can be divided into that of simultaneously finding a (possibly randomized) encoder $Q_{U|X} : \mathcal{X} \to \mathcal{P}(\mathcal{U})$ that maps raw data to a representation, possibly living in a higher-dimensional (feature) space $\mathcal{U}$, and a soft-decoder $Q_{\widehat{Y}|U} : \mathcal{U} \to \mathcal{P}(\mathcal{Y})$, which maps the representation to a probability distribution on the label space $\mathcal{Y}$. Although these mappings induce an equivalent classifier,

$$Q_{\widehat{Y}|X}(y|x) = \sum_{u\in\mathcal{U}} Q_{U|X}(u|x)Q_{\widehat{Y}|U}(y|u), \,. \tag{11.6}$$

the computation of the latter expression requires marginalizing out $u \in \mathcal{U}$, which is in general computationally hard due to the exponential number of atoms involved in the representations. A variational upper bound is commonly used to rewrite this intractable problem into

$$\mathbb{E}_{P_{XY}}\Big[-\log Q_{\widehat{Y}|X}(Y|X)\Big] \le \mathbb{E}_{P_{XY}}\mathbb{E}_{Q_{U|X}}\Big[-\log Q_{\widehat{Y}|U}(Y|U)\Big], \tag{11.7}$$

which simply follows by applying the Jensen inequality [24]. This bound induces the well-known *cross-entropy risk* defined below.

DEFINITION 11.2 (Cross-entropy loss and risk) *Given two randomized mappings $Q_{U|X}$: $\mathcal{X} \to \mathcal{P}(\mathcal{U})$ and $Q_{\widehat{Y}|U} : \mathcal{U} \to \mathcal{P}(\mathcal{Y})$, we define the average (over representations)* cross-entropy loss *as*

$$\ell(Q_{U|X}(\cdot|x), Q_{\widehat{Y}|U}(y|\cdot)) := \langle Q_{U|X}(\cdot|x), -\log Q_{\widehat{Y}|U}(y|\cdot)\rangle \tag{11.8}$$

$$= -\sum_{u\in\mathcal{U}} Q_{U|X}(u|x)\log Q_{\widehat{Y}|U}(y|u). \tag{11.9}$$

We measure the expected performance of $(Q_{U|X}, Q_{\widehat{Y}|U})$ via the risk *function:*

$$(Q_{U|X}, Q_{\widehat{Y}|U}) \mapsto \mathcal{L}(Q_{U|X}, Q_{\widehat{Y}|U}) := \mathbb{E}_{P_{XY}}[\ell(Q_{U|X}(\cdot|X), Q_{\widehat{Y}|U}(Y|\cdot))]. \tag{11.10}$$

In addition to the points noted earlier, another crucial component of knowledge representation is the use of deep representations. Formally speaking, we consider Kth-layer randomized encoders $\{Q_{U_k|U_{k-1}}\}_{k=1}^{K}$ with $U_0 \equiv X$ instead of one randomized encoder $Q_{U|X}$. Although this appears at first to be more general, it can be casted using the one-layer randomized encoder formulation induced by the marginal distribution that relates the input layer and the output layer of the network. Therefore any result for the one-layer formulation immediately implies a result for the Kth-layer formulation, and for this reason we shall focus on the one-layer case without loss of generality.

LEMMA 11.3 (Optimal decoders) *The minimum cross-entropy loss risk satisfies*

$$\inf_{Q_{\widehat{Y}|U}:\mathcal{U}\to\mathcal{P}(\mathcal{Y})} \mathcal{L}(Q_{\widehat{Y}|U}, Q_{U|X}) = H(Q_{Y|U}|Q_U), \tag{11.11}$$

where

$$Q_{Y|U}(y|u) = \frac{\sum_{x\in\mathcal{X}} Q_{U|X}(u|x)P_{XY}(x,y)}{\sum_{x\in\mathcal{X}} Q_{U|X}(u|x)P_X(x)}. \tag{11.12}$$

Proof The proof follows from the positivity of the relative entropy by noticing that $\mathcal{L}(Q_{U|X}, Q_{\widehat{Y}|U}) = D\big(Q_{Y|U}\|Q_{\widehat{Y}|U}|Q_U\big) + H(Q_{Y|U}|Q_U)$.

The associated risk to the optimal decoder is

$$\mathcal{L}(Q_{U|X}, Q_{Y|U}) := \mathbb{E}_{P_{XY}}\left[-\sum_{u\in\mathcal{U}} Q_{U|X}(u|x)\log Q_{Y|U}(Y|U)\right], \tag{11.13}$$

which is only a function of the encoder model $Q_{U|X}$. However, the optimal decoder cannot be determined since P_{XY} is unknown.

The learner's goal is to select $Q_{U|X}$ and $Q_{\widehat{Y}|U}$ by minimizing the risk (11.10). However, since P_{XY} is unknown the learner cannot directly measure the risk, and it is common to measure the agreement of a pair of candidates with a finite training dataset in terms of the *empirical risk*.

DEFINITION 11.3 (Empirical risk) *Let $\widehat{P}_{XY}$ denote the empirical distribution through the training dataset $\mathcal{S}_n := \{(x_1,y_1),\ldots,(x_n,y_n)\}$. The empirical risk is*

$$\mathcal{L}_{\text{emp}}(Q_{U|X}, Q_{\widehat{Y}|U}) := \mathbb{E}_{\widehat{P}_{XY}}\left[\ell(Q_{U|X}(\cdot|X), Q_{\widehat{Y}|U}(Y|\cdot))\right] \tag{11.14}$$

$$= \frac{1}{n}\sum_{i=1}^{n} \ell(Q_{U|X}(\cdot|x_i), Q_{\widehat{Y}|U}(y_i|\cdot)). \tag{11.15}$$

LEMMA 11.4 (Optimality of empirical decoders) *Given a randomized encoder $Q_{U|X} : \mathcal{X} \to \mathcal{P}(\mathcal{U})$, define the empirical decoder with respect to the empirical distribution $\widehat{P}_{XY}$ as*

$$\widehat{Q}_{Y|U}(y|u) := \frac{\sum_{x\in\mathcal{X}} Q_{U|X}(u|x)\widehat{P}_{XY}(x,y)}{\sum_{x\in\mathcal{X}} Q_{U|X}(u|x)\widehat{P}_{X}(x)}. \tag{11.16}$$

Then, the risk can be lower-bounded uniformly over $Q_{\widehat{Y}|U} : \mathcal{U} \to \mathcal{P}(\mathcal{Y})$ as

$$\mathcal{L}_{\text{emp}}(Q_{U|X}, Q_{\widehat{Y}|U}) \geq \mathcal{L}_{\text{emp}}(Q_{U|X}, \widehat{Q}_{Y|U}), \tag{11.17}$$

where equality holds provided that $Q_{\widehat{Y}|U} \equiv \widehat{Q}_{Y|U}$, i.e., the optimal decoder is computed from the encoder and the empirical distribution as done in (11.16).

Proof The inequality follows along the lines of Lemma (11.3) by noticing that $\mathcal{L}_{\text{emp}}(Q_{U|X}, Q_{\widehat{Y}|U}) = D\big(\widehat{Q}_{Y|U}\|Q_{\widehat{Y}|U}|\widehat{Q}_{U}\big) + \mathcal{L}_{\text{emp}}(\widehat{Q}_{Y|U}, Q_{U|X})$. Finally, the non-negativity of the relative conditional entropy completes the proof.

Since the empirical risk is evaluated on finite samples, its evaluation may be sensitive to sampling (noise) error, thus giving rise to the issue of generalization. It can be argued that a key component of learning is not just the development of a representation model on the basis of a finite training dataset, but its use in order to generalize to unseen data. Clearly, successful generalization necessitates the closeness (in some sense) of the selected representation and decoder models. Therefore, successful representation learning would involve successful generalization. This chapter deals with the information complexity of successful generalization. The generalization gap defined below is a measure of how an algorithm could perform on new data, i.e., data that are not available during the training phase. In the light of Lemmas 11.3 and 11.4, we will restrict our analysis to encoders only and assume that the optimal empirical decoder has been selected, i.e., $Q_{\widehat{Y}|U} \equiv \widehat{Q}_{Y|U}$ both in the empirical risk (11.14) and in the true risk (11.10). This is reasonable given the fact that the true P_{XY} is not known, and the only decoder that can be implemented in practice is the empirical one.

DEFINITION 11.4 (Generalization gap) *Given a stochastic mapping $Q_{U|X} : \mathcal{X} \to \mathcal{P}(\mathcal{U})$, the* generalization gap *is defined as*

$$(Q_{U|X}, \mathcal{S}_n) \mapsto \mathcal{E}_{\text{gap}}(Q_{U|X}, \mathcal{S}_n) := \left| \mathcal{L}_{\text{emp}}(Q_{U|X}, \widehat{Q}_{Y|U}) - \mathcal{L}(Q_{U|X}, \widehat{Q}_{Y|U}) \right|, \tag{11.18}$$

which represents the error incurred by the selected $Q_{U|X}$ when the rule $\mathcal{L}_{\text{emp}}(Q_{U|X}, \widehat{Q}_{Y|U})$ is used instead of the true risk $\mathcal{L}(Q_{U|X}, \widehat{Q}_{Y|U})$.

11.2.3 Optimizing on Restricted Classes of Randomized Encoders

We have already introduced the notions of representation and inference models and risk functions from which these candidates are chosen. Another related question of interest is as follows: how do we define the encoder class? A simple approach is to model classes in a parametric fashion. We first introduce the Bayes risk and then the restricted classes of randomized encoders and decoders.

DEFINITION 11.5 (Bayes risk) *The minimum cross-entropy risk over all possible candidates is called the* Bayes risk *and will be denoted by $\mathcal{L}^\star$. In this case,*

$$\mathcal{L}^\star := \inf_{Q_{U|X}: \mathcal{X} \to \mathcal{P}(\mathcal{U})} \mathcal{L}(Q_{U|X}, \widehat{Q}_{Y|U}) = H(P_{Y|X} | P_X). \tag{11.19}$$

DEFINITION 11.6 (Learning model) *The encoder functions are defined by $f_\theta : \mathcal{X}^d \times \mathcal{Z} \longrightarrow \mathcal{U}_\theta^m$, where $\mathcal{X}$ is the finite input alphabet with cardinality $|\mathcal{X}|$ and d is a positive integer, $\theta \in \Theta \subset \mathbb{R}^{d_\Theta}$ denotes the unknown parameters to be optimized, Z is a random variable taking values on a finite alphabet $\mathcal{Z}$ with probability P_Z, whose role is to randomize encoders, and $\mathcal{U}_\theta \subset [0,1]$ is the alphabet corresponding to the hidden representation which satisfies $|\mathcal{U}_\theta| \le |\mathcal{X}|^d \cdot |\mathcal{Z}|$. For notational convenience, we let $\mathcal{X} \equiv \mathcal{X}^d$ and $\mathcal{U} \equiv \mathcal{U}_\theta^m$, and denote this class as*

$$\mathcal{F} := \{Q_{U|X}(u|x) = \mathbb{E}_{P_Z}[\mathbb{1}[u = f_\theta(x, Z)]] \ : \ \theta \in \Theta\}.$$

It is clear that, for every θ, $\theta \mapsto Q_{U|X} \in \mathcal{F}$ induces a randomized encoder.

In order to simplify subsequent analysis we will assume the following conditions over the possible data p.m.f. and over the family $\mathcal{F}$ of encoders.

DEFINITION 11.7 (Restricted model class) *We assume that alphabets $\mathcal{X}$, $\mathcal{Y}$ are of arbitrarily large size but finite. Furthermore, there exists $\eta > 0$ such that the unknown data-generating distribution P_{XY} satisfies $P_X(x_{\min}) := \min_{x \in \mathcal{X}} P_X(x) \ge \eta$ and $P_Y(y_{\min}) := \min_{y \in \mathcal{Y}} P_Y(y) \ge \eta$.*

DEFINITION 11.8 (Empirical risk minimization) *The methodology of empirical risk minimization is one of the most straightforward approaches, yet it is usually efficient,* *provided that the chosen model class $\mathcal{F}$ is restricted [25]. The learner chooses a pair $\widehat{Q}^{\star}_{U|X} \in \mathcal{F}$ that minimizes the empirical risk:*

$$\mathcal{L}_{\text{emp}}(\widehat{Q}^{\star}_{U|X}, \widehat{Q}^{\star}_{Y|U}) \leq \mathcal{L}_{\text{emp}}(Q_{U|X}, \widehat{Q}_{Y|U}), \text{ for all } Q_{U|X} \in \mathcal{F}. \tag{11.20}$$

Moreover, it is possible to minimize a surrogate of the true risk:

$$\mathcal{L}(Q_{U|X}, \widehat{Q}_{Y|U}) \leq \mathcal{L}_{\text{emp}}(Q_{U|X}, \widehat{Q}_{Y|U}) + \mathcal{E}_{\text{gap}}(Q_{U|X}, \mathcal{S}_n), \tag{11.21}$$

which depends on the empirical risk and the so-called generalization gap, respectively. Expression (11.21) states that an adequate selection of the encoder should be performed in order to minimize the empirical risk and the generalization gap simultaneously. It is reasonable to expect that the optimal encoder achieving the minimal risk in (11.19) does not belong to our restricted class of models $\mathcal{F}$, so the learner may want to enlarge the model classes $\mathcal{F}$ as much as possible. However, this could induce a larger value of the generalization gap, which could lead to a trade-off between these two fundamental quantities.

DEFINITION 11.9 (Approximation and estimation error) *The sub-optimality of the model class $\mathcal{F}$ is measured in terms of the approximation error:*

$$\mathcal{E}_{\text{app}}(\mathcal{F}) := \inf_{Q_{U|X} \in \mathcal{F}} \mathcal{L}(Q_{U|X}, \widehat{Q}_{Y|U}) - \mathcal{L}^{\star}. \tag{11.22}$$

The induced risk of the selected pair of encoders is given by

$$\mathcal{E}_{\text{est}}(\mathcal{F}, \widehat{Q}^{\star}_{U|X}, \widehat{Q}^{\star}_{Y|U}) := \mathcal{L}(\widehat{Q}^{\star}_{U|X}, \widehat{Q}^{\star}_{Y|U}) - \inf_{Q_{U|X} \in \mathcal{F}} \mathcal{L}(Q_{U|X}, \widehat{Q}_{Y|U}), \tag{11.23}$$

where $\widehat{Q}^{\star}_{U|X}$ denotes the minimizer of expression (11.20).

DEFINITION 11.10 (Excess risk) *The* excess risk *of the algorithm* (11.20) *selecting an optimal pair $(\widehat{Q}^{\star}_{Y|U}, \widehat{Q}^{\star}_{U|X})$ can be decomposed as*

$$\begin{aligned}\mathcal{E}_{\text{exc}}(\mathcal{F}, \widehat{Q}^{\star}_{U|X}, \widehat{Q}^{\star}_{Y|U}) &:= \mathbb{E}[\mathcal{L}(\widehat{Q}^{\star}_{U|X}, \widehat{Q}^{\star}_{Y|U})] - \mathcal{L}^{\star} \\ &= \mathcal{E}_{\text{app}}(\mathcal{F}) + \mathbb{E}[\mathcal{E}_{\text{est}}(\mathcal{F}, \widehat{Q}^{\star}_{U|X}, \widehat{Q}^{\star}_{Y|U})],\end{aligned}$$

where the expectation is taken with respect to the random choice $\mathcal{S}_n$ of dataset which induces the optimal pair $(\widehat{Q}^{\star}_{Y|U}, \widehat{Q}^{\star}_{U|X})$.

The *approximation error* $\mathcal{E}_{\text{app}}(\mathcal{F})$ measures how closely encoders in the model class $\mathcal{F}$ can approximate the optimal solution $\mathcal{L}^{\star}$. On the other hand, the *estimation error* $\mathcal{E}_{\text{est}}(\mathcal{F}, \widehat{Q}^{\star}_{U|X}, \widehat{Q}^{\star}_{Y|U})$ measures the effect of minimizing the empirical risk instead of the true risk, which is caused by the finite size of the training data. The estimation error is determined by the number of training samples and by the complexity of the model class, i.e., large models have smaller approximation error but lead to higher estimation errors, and it is also related to the generalization error [25]. However, for the sake of simplicity, in this chapter we restrict our attention only to the generalization gap.

11.3 Information-Theoretic Principles and Information Bottleneck

11.3.1 Lossy Source Coding

The problem of source coding is, jointly with that of channel coding, one of the two most important and relevant problems in information theory [24, 26]. In the source coding problem, we face the fundamental question of how to represent in the most compact way possible a given stochastic source such that we can be able to reconstruct, with a given level of fidelity, the original source. Shannon was the first to formalize and completely solve this problem [2, 27] in the asymptotic regime,[2] establishing the optimal trade-off between the compactness of the representations and the level of fidelity in the reconstruction. In the *lossless source coding* problem, the level of fidelity required is maximal: it is desired to have a short-length representation which can be used to reconstruct almost exactly the original source. According to the more general *lossy source coding* setup, we look for more compact representations of the original source by dropping the requirement of almost-exact reconstruction. The level of fidelity required is measured by using a predefined *distortion measure*, which is an essential part of the problem. Interestingly enough, the problem of lossy source coding can be solved completely for any well-defined distortion measure. Let us formulate this problem mathematically.

DEFINITION 11.11 (Lossy source coding problem) *Consider a discrete and finite alphabet $\mathcal{X}$ and a stochastic source X, which generates identically and independently distributed samples according to $P_X \in \mathcal{P}(\mathcal{X})$. Consider an alternative alphabet $\widehat{\mathcal{X}}$ and a distortion function $d : \mathcal{X} \times \widehat{\mathcal{X}} \to \mathbb{R}_{\geq 0}$. Consider also a realization X^n of the source, an* encoder *function $f_n : \mathcal{X}^n \to \{1, \dots M_n\}$, where $M_n \in \mathbb{N}$, and a* decoder *function $g_n : \{1, \dots, M_n\} \to \widehat{\mathcal{X}}^n$. We say that $C_n = (f_n, g_n)$ is an n-code for X and $d(\cdot;\cdot)$.*

DEFINITION 11.12 (Achievable rate and fidelity) *A pair (R, D) is said to be achievable if, for every $\epsilon > 0$, there exist $n \geq 1$ and an n-code C_n such that*

$$\frac{\log M_n}{n} \leq R + \epsilon, \tag{11.24}$$

$$\mathbb{E}_{P_X^n}\left[\bar{d}(X^n; g_n(f_n(X^n)))\right] \leq D + \epsilon, \tag{11.25}$$

where $\bar{d}(X^n; \widehat{X}^n) \equiv (1/n) \sum_{i=1}^n d(x_i; \widehat{x}_i)$.

The set of all achievable pairs (R, D) contains the complete characterization of all the possible trade-offs between the *rate* R (which quantifies the level of compression of the source X measuring the necessary number of bits per symbol) and the *distortion* D (which quantifies the average fidelity level per symbol in the reconstruction using the

[2] In the asymptotic regime one considers that the number of realizations of the stochastic source to be compressed tends to infinity. Although this could be questionable in practice, the asymptotic problem reflects accurately the important trade-offs of the problem. In this presentation, our focus will be on the asymptotic problem originally solved by Shannon.

distortion function $d(\cdot;\cdot)$ symbol by symbol). An equivalent characterization of the set of achievable pairs (R,D) is given by the *rate-distortion* function defined by

$$R(D) = \inf\{R : (R,D) \text{ is achievable}\}. \tag{11.26}$$

It is the great achievement of Shannon [27] to have obtained the following result.

THEOREM 11.1 (Rate-distortion function) *The rate-distortion function for source X with reconstruction alphabet $\widehat{\mathcal{X}}$ and with distortion function $d(\cdot;\cdot)$ is given by*

$$\mathcal{R}_{X,d}(D) = \inf_{\substack{P_{\widehat{X}|X}:\mathcal{X}\to\mathcal{P}(\widehat{\mathcal{X}})\\ \mathbb{E}_{P_{\widehat{X}X}}[d(X;\widehat{X})]\le D}} I(P_X; P_{\widehat{X}|X}). \tag{11.27}$$

This function depends solely on the distribution P_X and the distortion function $d(\cdot;\cdot)$ and contains the exact trade-off between compression and fidelity that can be expected for the particular source and distortion function. It is easy to establish that this function is positive, non-increasing in D, and convex. Moreover, there exists $D > 0$ such that $\mathcal{R}_{X,d}(D)$ is finite and we denote the minimum of such values of D by $D_{\min}$ with $R_{\max} := \lim_{D\to D_{\min}^+} \mathcal{R}_{X,d}(D)$. Although $\mathcal{R}_{X,d}(D)$ could be hard to compute in closed form for a particular P_X and $d(\cdot;\cdot)$, the problem in (11.27) is a convex optimization one, for which there exist efficient numerical techniques. However, several important cases admit closed-form expressions, such as the Gaussian case with quadratic distortion[3] [24].

Another important function related to the rate-distortion function is the *distortion-rate* function. This function can be defined independently from the rate-distortion function and directly from information-theoretic principles. Intuitively, this function is the infimum value of the distortion D as a function of the rate R for all (R,D) achievable pairs. We will define it directly from the rate-distortion function:

$$\mathcal{R}_{X,d}^{-1}(I) := \inf\{D \in \mathbb{R}_{\ge 0} : \mathcal{R}_{X,d}(D) \le I\}. \tag{11.28}$$

It is not hard to show the following.[4]

LEMMA 11.5 *Consider the distortion-rate function defined according to (11.28). This function is positive, non-increasing in I and convex.*

Proof The proof follows easily from (11.28).

Besides their obvious importance in the problem of source coding, the definitions of the rate-distortion and distortion-rate functions will be useful for the problem of learning as presented in the previous section. They will permit one to establish connections between the misclassification probability, the cross-entropy, and the mutual information

[3] Although the Gaussian case does not correspond to a finite cardinality set $\mathcal{X}$, the result in (11.27) can easily be extended to that case using quantization arguments.

[4] It is worth mentioning that by using $\mathcal{R}_{X,d}^{-1}(I)$ we are abusing notation. This is because in general it is not true that $\mathcal{R}_{X,d}(D)$ is injective for every $D \ge 0$. However, when $I \in [R_{\min}, R_{\max})$ with $R_{\min} := \mathcal{R}_{X,d}(D_{\max})$ and $D_{\max} := \min_{\widehat{x}\in\widehat{\mathcal{X}}} \mathbb{E}_{P_X}[d(X;\widehat{x})]$, under some very mild conditions on P_X and $d(\cdot;\cdot)$, $\mathcal{R}_{X,d}^{-1}(I)$ is the true inverse of $\mathcal{R}_{X,d}(D)$, which is guaranteed to be injective in the interval $D \in (D_{\min}, D_{\max}]$.

between the input X and the output of the encoder $Q_{U|X}$. These connections will be conceptually important for the rest of the chapter, at least from a qualitative point of view.

11.3.2 Misclassification Probability and Cross-Entropy Loss

It is easy to show that the proposed learning framework can be set up as a lossy source coding problem. This formulation, however, is not an operational one as was the case for the information-theoretic one presented in Definitions 11.11 and 11.12. The reason for this comes from the fact that for our learning framework we do not have the same type of scaling with n as in the source coding problem in information theory. While, in the typical source coding problem, encoders and decoders act upon the entire sequence of observed samples $x^n = (x_1, \ldots, x_n)$, in the learning framework, the encoder $Q_{U|X}$ acts on a sample-by-sample basis. Nevertheless, the definition of the rate-distortion (w.r.t. distortion-rate) function is relevant for the learning framework as well, provided that we avoid any operational interpretation and concentrate on its strictly mathematical meaning.

Consider alphabets $\mathcal{U}$, $\mathcal{X}$, and $\mathcal{Y}$, corresponding to the descriptions generated by the encoder $Q_{U|X}$ and to the examples and their corresponding labels. From (11.1) and (11.6), we can write the misclassification probability as

$$
\begin{aligned}
P_{\mathcal{E}}(Q_{U|X}, Q_{\widehat{Y}|U}) &= 1 - \mathbb{E}_{P_{XY}}\left[\sum_{u\in\mathcal{U}} Q_{U|X}(u|X)Q_{\widehat{Y}|U}(Y|u)\right] \\
&= 1 - \mathbb{E}_{P_Y}\left[\sum_{u\in\mathcal{U}} Q_{\widehat{Y}|U}(Y|u)\mathbb{E}_{P_{X|Y}}\left[Q_{U|X}(u|X)\right]\right] \\
&= 1 - \mathbb{E}_{P_Y}\left[\sum_{u\in\mathcal{U}} Q_{\widehat{Y}|U}(Y|u)Q_{U|Y}(u|Y)\right] \\
&= \mathbb{E}_{P_{UY}}\left[1 - Q_{\widehat{Y}|U}(Y|U)\right]. \qquad (11.29)
\end{aligned}
$$

From the above derivation, we can set a distortion measure: $d(u;y) := 1 - Q_{\widehat{Y}|U}(y|u)$. In this way, the probability of misclassification can be written as an average over the outcomes of Y (taken as the source) and U (taken as the reconstruction) of the distortion measure: $1 - Q_{\widehat{Y}|U}(y|u)$. In this manner, we can consider the following rate-distortion function:

$$
\mathcal{R}_{Y,Q_{\widehat{Y}|U}}(D) := \inf_{\substack{P_{U|Y}:\mathcal{Y}\to\mathcal{P}(\mathcal{U}) \\ \mathbb{E}_{P_{UY}}[1-Q_{\widehat{Y}|U}(Y|U)]\le D}} I(P_Y; P_{U|Y}), \qquad (11.30)
$$

which provides a connection between the misclassification probability and the mutual information $I(P_Y; P_{U|Y})$.

From this formulation, we are able to obtain the following lemma, which provides an upper and a lower bound on the probability of misclassification via the distortion-rate function and the cross-entropy loss.

LEMMA 11.6 (Probability of misclassification and cross-entropy loss) *The probability of misclassification* $P_{\mathcal{E}}(Q_{\widehat{Y}|U}, Q_{U|X})$ *induced by a randomized encoder* $Q_{U|X} : \mathcal{X} \to \mathcal{P}(\mathcal{U})$ *and decoder* $Q_{\widehat{Y}|U} : \mathcal{U} \to \mathcal{P}(\mathcal{Y})$ *is bounded by*

$$\mathcal{R}^{-1}_{Y,Q_{\widehat{Y}|U}}(I(P_X; Q_{U|X})) \leq \mathcal{R}^{-1}_{Y,Q_{\widehat{Y}|U}}(I(P_Y; Q_{U|Y})) \tag{11.31}$$

$$\leq P_{\mathcal{E}}(Q_{\widehat{Y}|U}, Q_{U|X}) \tag{11.32}$$

$$\leq 1 - \exp\Big(-\mathcal{L}(Q_{\widehat{Y}|U}, Q_{U|X})\Big), \tag{11.33}$$

where $Q_{U|Y}(u|y) = \sum_{x \in \mathcal{X}} Q_{U|X}(u|x) P_{X|Y}(x|y)$ *for* $(u, y) \in \mathcal{U} \times \mathcal{Y}$.

Proof The upper bound simply follows by using the Jensen inequality from [24], while the lower bound is a consequence of the definition of the rate-distortion and distortion-rate functions. The probability of misclassification corresponding to the classifier can be expressed by the expected distortion $\mathbb{E}_{P_{XY}Q_{U|X}}[d(Y, U)] = P_{\mathcal{E}}(Q_{\widehat{Y}|U}, Q_{U|X})$, which is based on the fidelity function $d(y, u) := 1 - Q_{\widehat{Y}|U}(y|u)$ as shown in (11.29). Because of the Markov chain $Y -\!\!\circ\!\!- X -\!\!\circ\!\!- U$, we can use the data-processing inequality [24] and the definition of the rate-distortion function, obtaining the following bound for the classification error:

$$I(P_X; Q_{U|X}) \geq I(P_Y; Q_{U|Y}) \tag{11.34}$$

$$\geq \inf_{\substack{P_{\widehat{U}|Y} : \mathcal{Y} \to \mathcal{P}(\mathcal{U}) \\ \mathbb{E}_{P_{\widehat{U}Y}}[d(Y, \widehat{U})] \leq \mathbb{E}_{P_{XY}Q_{U|X}}[d(Y,U)]}} I(P_Y; P_{\widehat{U}|Y}) \tag{11.35}$$

$$= \mathcal{R}_{Y,Q_{\widehat{Y}|U}}(P_{\mathcal{E}}(Q_{\widehat{Y}|U}, Q_{U|X})). \tag{11.36}$$

For $\mathbb{E}_{P_{XY}Q_{U|X}}[d(Y, U)]$, we can use the definition of $\mathcal{R}^{-1}_{Y,Q_{\widehat{Y}|U}}(\cdot)$, and thus obtain from (11.34) the fundamental bound

$$\mathcal{R}^{-1}_{Y,Q_{\widehat{Y}|U}}(I(P_X; Q_{U|X})) \leq \mathcal{R}^{-1}_{Y,Q_{\widehat{Y}|U}}(I(P_Y; Q_{U|Y})) \leq P_{\mathcal{E}}(Q_{\widehat{Y}|U}, Q_{U|X}).$$

The lower bound in the above expression states that any limitation in terms of the mutual information between the raw data and their representation will bound from below the probability of misclassification while the upper bound shows that the cross-entropy loss introduced in (11.10) can be used as a surrogate to optimize the probability of misclassification, as was also pointed out in Lemma 11.2. As a matter of fact, it appears that the probability of misclassification is controlled by two fundamental information quantities: the mutual information $I(P_X; Q_{U|X})$ and the cross-entropy loss $\mathcal{L}(Q_{\widehat{Y}|U}, Q_{U|X})$.

11.3.3 Noisy Lossy Source Coding and the Information Bottleneck

A more subtle variant of the lossy source coding problem is the *noisy lossy source coding problem*, which was first introduced in [28]. The main difference with respect to Shannon's original problem is that the source Y is not observed directly at the encoder. Instead, a noisy version of Y denoted by X is observed and appropriately compressed.

More precisely, we have a memoryless source with single-letter distribution P_Y observed through a noisy channel with single-input transition probability $P_{X|Y}$. From the compressed version of X, it is desired to reconstruct, with a predetermined level of fidelity, the realization of the unobserved source Y. The fidelity is measured, similarly to the usual lossy source coding problem, with distortion function $d : \mathcal{Y} \times \mathcal{U} \to \mathbb{R}_{\geq 0}$, where $\mathcal{U}$ is the alphabet in which we generate the reconstructions. The operational information-theoretic definitions for this problem are analogous to Definitions 11.11 and 11.12, and for this reason are omitted. The rate distortion in this case is given by

$$\mathcal{R}_{XY,d}(D) = \inf_{\substack{P_{U|X}:\mathcal{X}\to\mathcal{P}(\mathcal{U}) \\ \mathbb{E}_{P_{UY}}[d(Y;U)]\leq D}} I(P_X; P_{U|X}). \tag{11.37}$$

Consider the case of logarithmic distortion $d(y;u) = -\log P_{Y|U}(y|u)$, where

$$P_{Y|U}(y|u) = \frac{\sum_{x\in\mathcal{X}} P_{U|X}(u|x)P_{XY}(x,y)}{\sum_{x\in\mathcal{X}} P_{U|X}(u|x)P_X(x)}. \tag{11.38}$$

The noisy lossy source coding with this choice of distortion function gives rise to the celebrated *information bottleneck* [20]. In precise terms,

$$\mathcal{R}_{XY,d}(D) = \inf_{\substack{P_{U|X}:\mathcal{X}\to\mathcal{P}(\mathcal{U}) \\ H(P_{Y|U}|P_U)\leq D}} I(P_X; P_{U|X}). \tag{11.39}$$

Noticing that $H(P_{Y|U}|P_U) = -I(P_Y; P_{U|Y}) + H(P_Y)$ and defining $\mu := H(P_Y) - D$, we can write (11.39) as

$$\bar{\mathcal{R}}_{XY}(\mu) = \inf_{\substack{P_{U|X}:\mathcal{X}\to\mathcal{P}(\mathcal{U}) \\ I(P_Y;P_{U|Y})\geq\mu}} I(P_X; P_{U|X}). \tag{11.40}$$

Equation (11.40) summarizes the trade-off that exists between the level of compression of the observable source X, using representation U, and the level of information about the *hidden* source Y preserved by this representation. This function is called the *rate-relevance* function, where μ is the minimum level of relevance we expect from representation U when the rate used for the compression of X is $\bar{\mathcal{R}}_{XY}(\mu)$. Notice that in the information-bottleneck case the distortion $d(y;u)$ depends on the optimal conditional distribution $P^*_{U|X}$ through (11.38). This makes the problem of characterizing $\bar{\mathcal{R}}_{XY}(\mu)$ more difficult than (11.37), in which the distortion function is fixed. In fact, although $\bar{\mathcal{R}}_{XY}(\mu)$ is positive, non-decreasing, and convex, the problem in (11.40) is not convex, which leads to the need for more sophisticated tools for its solution. Moreover, from the corresponding operational definition for the lossy source coding problem (analogous to Definitions 11.11 and 11.12), it is clear that the distortion function for sequences Y^n and U^n is applied symbol-by-symbol $\bar{d}(Y^n; U^n) = -(1/n)\sum_{i=1} \log P_{Y|U}(Y_i|U_i)$, implying a memoryless condition between hidden source realization Y^n and description $U^n = f_n(X^n)$. It is possible to show [29, 30] that, if we apply a full logarithmic distortion $\bar{d}(Y^n; U^n) = -(1/n)\log P_{Y^n|U^n}(Y^n|U^n)$, not necessarily additive as in the previous case, the rate-relevance function in (11.40) remains unchanged, where relevance is measured by the non-additive multi-letter mutual information:

$$\bar{d}(Y^n;U^n) \equiv \frac{1}{n} I\Big(P_{Y^n};P_{f_n(X^n)|Y^n}\Big). \tag{11.41}$$

As a simple example in which the rate-relevance function in (11.40) can be calculated in closed form, we can consider the case in which X and Y are jointly Gaussian with zero mean, variances σ_X^2 and σ_Y^2, and Pearson correlation coefficient given by ρ_{XY}. Using standard information-theoretic arguments [30], it can be shown that the optimal distribution $P_{U|X}$ is also Gaussian, with mean X and variance given by

$$\sigma_{U|X}^2 = \sigma_X^2 \frac{2^{-2\mu} - (1-\rho_{XY}^2)}{1-2^{-2\mu}}. \tag{11.42}$$

With this choice for $P_{U|X}$ we easily obtain that $I(P_Y;P_{U|Y}) = \mu$ and that

$$\bar{\mathcal{R}}_{XY}(\mu) = \frac{1}{2}\log\left(\frac{\rho_{XY}^2}{2^{-2\mu}-(1-\rho_{XY}^2)}\right), \quad 0 \le \mu \le \frac{1}{2}\log\left(\frac{1}{1-\rho_{XY}^2}\right). \tag{11.43}$$

It is interesting to observe that $\bar{\mathcal{R}}_{XY}(\mu)$ depends only on the structure of the sources X and Y through the correlation coefficient ρ_{XY} and not on their variances. It should also be noted that the level of relevance μ is constrained to lie in a bounded interval. This is not surprising because of the Markov chain $U -\!\!\circ\!\!- X -\!\!\circ\!\!- Y$, the maximum value for the relevance level is $I(P_X;P_{Y|X})$, which is easily shown to be equal to $\frac{1}{2}\log(1/(1-\rho_{XY}^2))$. The maximum level of relevance is achievable only as long as the rate $R \to \infty$, that is, when the source X is minimally compressed. The trade-off between rate and relevance for this simple example can be appreciated in Fig. 11.1 for $\rho_{XY} = 0.9$.

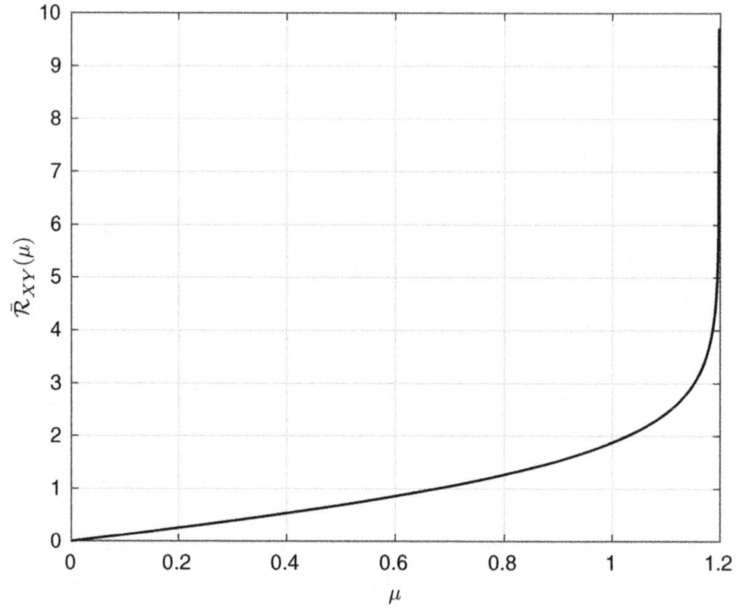

Figure 11.1 $\bar{\mathcal{R}}_{XY}(\mu)$ for $\rho_{XY} = 0.9$.

11.3.4 The Information-Bottleneck Method

Noisy lossy source coding with logarithmic loss can be used as a general principle for learning problems leading to the *information-bottleneck method*. This method was successfully used in several learning problems with considerable success (see [31, 32] and references therein). Consider the classification problem introduced in Section 11.2.1 and encoder–decoder pairs $(Q_{U|X}, Q_{\widehat{Y}|U})$, as was explained in Section 11.2.2. The information-bottleneck method can be introduced through the following optimization problem:

$$\inf_{Q_{U|X}, Q_{\widehat{Y}|U}} \left\{ \mathbb{E}_{P_{XY}Q_{U|X}}[-\log Q_{\widehat{Y}|U}(Y|U)] + \beta \cdot I(P_X; Q_{U|X}) \right\}. \tag{11.44}$$

Expression (11.44) can be interpreted as a cross-entropy loss with a regularization term given by $\beta \cdot I(P_X; Q_{U|X})$, where β is a positive number. The regularization term can be interpreted as penalization on the complexity of the descriptions generated from the examples X using the encoder $Q_{U|X}$. The smaller the term $I(P_X; Q_{U|X})$, the simpler the descriptions U will be. Moreover, the simpler the descriptions U, the less information they share with labels X and Y (because of the Markov chain $U -\!\!\circ\!\!- X -\!\!\circ\!\!- Y$). As the information content in U with respect to Y is naturally decreased, the value of the cross-entropy $\mathbb{E}_{P_{XY}Q_{U|X}}[-\log Q_{\widehat{Y}|U}(Y|U)]$ increases. In this way, a trade-off between the cross-entropy loss and the complexity of the descriptions extracted from X is established. It can happen, though, that the regularization term given $I(P_X; Q_{U|X})$ penalizes very complex descriptions that could provide a low cross-entropy value at the cost of poor generalization and overfitting.

From the result in Lemma 11.3 and the fact that the regularization term $I(P_X; Q_{U|X})$ does not depend on the decoder $Q_{\widehat{Y}|U}$, problem (11.44) can be written as

$$\inf_{Q_{U|X}: \mathcal{X} \to \mathcal{P}(\mathcal{U})} \left\{ \mathbb{E}_{P_{XY}Q_{U|X}}[-\log Q_{Y|U}(Y|U)] + \beta \cdot I(P_X; Q_{U|X}) \right\}, \tag{11.45}$$

where the decoder can be written as a function of the encoder as follows:

$$Q_{Y|U}(y|u) = \frac{\sum_{x \in \mathcal{X}} Q_{U|X}(u|x) P_{XY}(x, y)}{\sum_{x \in \mathcal{X}} Q_{U|X}(u|x) P_X(x)}. \tag{11.46}$$

On recognizing that $\mathbb{E}_{P_{XY}Q_{U|X}}[-\log Q_{Y|U}(Y|U)] = H(Q_{Y|U}|Q_U)$, where $Q_U(u) = \sum_{x \in \mathcal{X}} Q_{U|X}(u|x) P_X(x)$, we see that (11.45) is closely related to the information bottleneck and to the rate-relevance function defined in (11.40). In fact, the problem in (11.45) can be equivalently written as

$$\sup_{Q_{U|X}: \mathcal{X} \to \mathcal{P}(\mathcal{U})} [I(P_Y; Q_{U|Y}) - \beta \cdot I(P_X; Q_{U|X})], \tag{11.47}$$

with $Q_{U|Y}(u|y) = \sum_{x \in \mathcal{X}} Q_{U|X}(u|x) P_{X|Y}(x|y)$. We can easily see that in (11.47) we are considering the dual problem to (11.40), looking for the supremum of relevance μ subject to a given rate R. The value of β (which can be thought of as a typical Lagrange multiplier [33]) can be thought of as a hyperparameter which controls the trade-off between $I(P_Y; Q_{U|Y})$ (relevance) and $I(P_X; Q_{U|X})$ (rate). In more precise terms, consider the following set:

$$\begin{aligned}\mathcal{R} := \{(\mu, R) \in \mathbb{R}^2_{\geq 0} : \exists\, Q_{U|X} : \mathcal{X} \to \mathcal{P}(\mathcal{U}) \text{ s.t.}\\ R \geq I(P_X; Q_{U|X}),\\ \mu \leq I(P_Y; Q_{U|Y}), \quad U -\!\!\circ\!\!- X -\!\!\circ\!\!- Y\}.\end{aligned} \tag{11.48}$$

It is easy to show that this region corresponds to the set of achievable values of relevance and rate (μ, R) for the corresponding noisy lossy source coding problem with logarithmic distortion as was defined in Section 11.3.3. This set is closed and convex and it is not difficult to show that [34]

$$\sup_{\substack{Q_{U|X}:\mathcal{X}\to\mathcal{P}(\mathcal{U})\\ I(P_X;Q_{U|X})\leq R}} I(P_Y; Q_{U|Y}) = \sup\{\mu : (\mu, R) \in \mathcal{R}\}. \tag{11.49}$$

Using convex optimization theory [33], we can easily conclude that (11.47) corresponds to obtaining the *supporting hyperplane* of region $\mathcal{R}$ with slope β. As any convex and closed set is characterized by all of its supporting hyperplanes, by varying β and solving (11.47) we are reconstructing the upper boundary of $\mathcal{R}$ which coincides with (11.49). In other words, the hyperparameter β is directly related to the value of R at which we are considering the maximum possible value of the redundancy μ, or, which amounts to the same thing, the value of β controls the complexity of representations of X, as was pointed out above.

It remains only to discuss the implementation of a procedure for solving (11.47). Unfortunately, although the set $\mathcal{R}$ characterizing the solutions of (11.47) is convex, it is not true that (11.47) is itself a convex optimization problem. However, the structure of the problem allows the use of efficient numerical optimization procedures that guarantee convergence to local optimum solutions. These numerical procedures are basically Blahut–Arimoto (BA)-type algorithms. These are often used to refer to a class of algorithms for numerically computing the capacity of a noisy channel and the rate-distortion function for given channel and source distributions, respectively [35, 36]. For these reasons, these algorithms can be applied with minor changes to the problem (11.47), as was done in [20].

Clearly, for the solution of (11.47) we need as input the distribution P_{XY}. When only training samples and labels $\mathcal{S}_n := \{(x_1, y_1), \ldots, (x_n, y_n)\}$ are available, we use the empirical distribution $\widehat{P}_{XY}$ instead of the true distribution P_{XY}.

In Fig. 11.2, we plot what we call the excess risk (as presented in Definition 11.10), rewritten as

$$\text{Excess risk} := H(Q^{*\beta}_{Y|U} | Q^{*\beta}_{U}) - H(P_{Y|X} | P_X), \tag{11.50}$$

where $Q^{*\beta}_{Y|U}$, $Q^{*\beta}_{U}$ are computed by using the optimal solution $Q^{*\beta}_{U|X}$ in (11.47) and the empirical distribution $\widehat{P}_{XY}$. As β defines unequivocally the value of $I(P_X; Q^{*\beta}_{U|X})$, which is basically the rate or complexity associated with the chosen encoder, we choose the horizontal axis to be labeled by rate R. Experiments were performed by using synthetic data with alphabets $|\mathcal{X}| = 128$ and $|\mathcal{Y}| = 4$. The excess-risk curve as a function of the

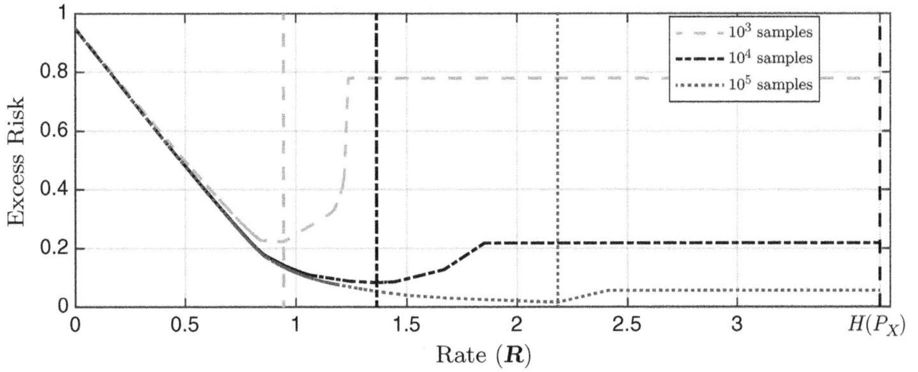

Figure 11.2 Excess risk (11.50) as a function of rate R being the mutual information between the representation U and the corresponding input X.

rate constraint for different sizes of training samples is plotted. With dashed vertical lines, we denote the rate for which the excess risk achieves its minimum. When the number of training samples increases, the optimal rate R approaches its maximum possible value: $H(P_X)$ (black vertical dashed line on the far right). We emphasize that for every curve there exists a different limiting rate R_{lim}, such that, for each $R \geq R_{\text{lim}}$, the excess risk remains constant for that value. It is not difficult to check that $R_{\text{lim}} = H(\widehat{P}_X)$. Furthermore, for every size of the training samples, there is an optimal value of R_{opt} which provides the lowest excess risk in (11.50). In a sense, this is indicating that the rate R can be interpreted as an effective regularization term and, thus, it can provide robustness for learning in practical scenarios in which the true input distribution is not known and the empirical data distribution is used. It is worth mentioning that when more data are available the optimal value of the regularizing rate R becomes less critical. This fact was expected since, when the amount of training data increases, the empirical distribution approaches the data-generating distribution.

In the next section, we provide a formal mathematical proof of the explicit relation between the generalization gap and the rate constraint, which explains the heuristic observations presented in Fig. 11.2.

11.4 The Interplay between Information and Generalization

In the following, we will denote $\mathcal{L}(Q_{U|X}) = \mathcal{L}(Q_{U|X}, \widehat{Q}_{Y|U})$ and $\mathcal{L}_{\text{emp}}(Q_{U|X}) = \mathcal{L}_{\text{emp}}(Q_{U|X}, \widehat{Q}_{Y|U})$. We will study informational bounds on the generalization gap (11.18). More precisely, the goal is to find the learning rate $\epsilon_n(Q, \mathcal{S}_n, \gamma_n)$ such that

$$\mathbb{P}\Big(\mathcal{E}_{\text{gap}}(Q_{U|X}, \mathcal{S}_n) > \epsilon_n(Q_{U|X}, \mathcal{S}_n, \gamma_n)\Big) \leq \gamma_n, \tag{11.51}$$

for a given $Q_{U|X} \in \mathcal{F}$ and some $\gamma_n \to 0$ as $n \to \infty$. We will further comment on the implications for practical algorithms minimizing the surrogate of the risk function,

$$\mathcal{L}(Q_{U|X}) \leq \mathcal{L}_{\text{emp}}(Q_{U|X}) + \mathcal{E}_{\text{gap}}(Q_{U|X}), \tag{11.52}$$

which depends on the empirical risk and the so-called generalization gap. Expression (11.52) states that a suitable selection of the encoder can be obtained by minimizing the empirical risk and the generalization gap simultaneously, that is

$$\mathcal{L}_{\text{emp}}(Q_{U|X}) + \lambda \cdot \epsilon_n(Q_{U|X}, S_n, \gamma_n), \tag{11.53}$$

for some suitable multiplier $\lambda \geq 0$. It is reasonable to expect that the optimal encoder achieving the minimal risk in (11.10) does not belong to $\mathcal{F}$, so we may want to enlarge the model classes as much as possible. However, as usual, we expect a sensitive trade-off between these two fundamental quantities.

11.4.1 Bounds on the Generalization Gap

We first present the main technical result in Theorem 11.2, that is a sample-dependent bound on the generalization gap (11.18) with probability of at least $1-\delta$, as a function of a selected randomized encoder $Q_{U|X}$ and the data probability distribution P_{XY}. In particular, we will show that the mutual information between the raw data and their representation controls the learning rate with an order $O\big(\log(n)/\sqrt{n}\big)$, which leads to an informational PAC-style generalization error bound. From this perspective, we discuss the implications for model selection, variational auto-encoders, and the information-bottleneck method.

THEOREM 11.2 (Informational bound) *Let $\mathcal{F}$ be a class of encoders. Then, for every P_{XY} and every $\delta \in (0,1)$, with probability at least $1-\delta$ over the choice of $S_n \sim P^n_{XY}$ the following inequality holds $\forall\ Q_{U|X} \in \mathcal{F}$:*

$$\mathcal{E}_{\text{gap}}(Q_{U|X}, S_n) \leq A_\delta \sqrt{I(\widehat{P}_X; Q_{U|X})} \cdot \frac{\log(n)}{\sqrt{n}} + \frac{C_\delta}{\sqrt{n}} + O\left(\frac{\log(n)}{n}\right), \tag{11.54}$$

where $A_\delta, B_\delta, C_\delta$ are universal constants:

$$A_\delta := \frac{\sqrt{2}B_\delta}{P_X(x_{\min})}\left(1 + 1/\sqrt{|\mathcal{X}|}\right), \quad B_\delta := 2 + \sqrt{\log\left(\frac{|\mathcal{Y}|+3}{\delta}\right)}, \tag{11.55}$$

$$C_\delta := 2|\mathcal{U}|e^{-1} + B_\delta \sqrt{|\mathcal{Y}|} \log\left(\frac{|\mathcal{U}|}{P_Y(y_{\min})}\right). \tag{11.56}$$

The importance of this result is that the main quantity involves the empirical mutual information between the data X and their randomized representation $U(X)$. This can be understood as a "measure of information complexity" scaling with rate $n^{-1/2}\log(n)$. The remaining issue is merely how to interpret this information-theoretic bound and its implication in the learning problem.

By combining Theorem 11.2 with inequality (11.52) we obtain the following corollary.

COROLLARY 11.1 (PAC-style generalization error bound) *Let $\mathcal{F}$ be the class of randomized encoders. Then, for every P_{XY} and every $\delta \in (0,1)$, with probability at least $1-\delta$ over the choice of $S_n \sim P^n_{XY}$ the following inequality holds:*

$$\mathcal{L}(\widehat{Q}_{Y|U}, Q_{U|X}) \leq H(\widehat{Q}_{Y|U}|\widehat{Q}_U) + A_\delta \sqrt{I(\widehat{P}_X; Q_{U|X})}\frac{\log(n)}{\sqrt{n}} + \frac{C_\delta}{\sqrt{n}} + O\left(\frac{\log(n)}{n}\right). \quad (11.57)$$

An interesting connection between the empirical risk minimization of the cross-entropy loss and the information-bottleneck method presented in the previous section arises which motivates formally the following algorithm [15, 20, 37].

DEFINITION 11.13 (Information-bottleneck algorithm) *A representation learning algorithm inspired by the information-bottleneck principle [20] consists in finding an encoder $Q_{U|X} \in \mathcal{F}$ that minimizes over the random choice $\mathcal{S}_n \sim P^n_{XY}$ the functional*

$$\mathcal{L}^{(\lambda)}_{IB}(Q_{U|X}) := H(\widehat{Q}_{Y|U}|\widehat{Q}_U) + \lambda \cdot I(\widehat{P}_X; Q_{U|X}), \quad (11.58)$$

for a suitable multiplier $\lambda > 0$, where $\widehat{Q}_{Y|U}$ is given by (11.16) *and $\widehat{Q}_U$ is its denominator.*

This algorithm optimizes a trade-off between $H(\widehat{Q}_{Y|U}|\widehat{Q}_U)$ and the information-based regularization term $I(\widehat{P}_X; Q_{U|X})$. Interestingly, the resulting regularized empirical risk suggested by (11.57) can be seen as an optimization of the information-bottleneck method from the empirical distribution (11.58) but based on the square root of the mutual information in expression (11.58). Additionally, we observe that, on selecting an arbitrary $\widetilde{Q}_U \in \mathcal{P}(\mathcal{U})$ in (11.58) and using the information-radius identity [24], the following inequality holds:

$$\mathcal{L}^{(\lambda)}_{\text{IB}}(Q_{U|X}) \leq H(\widehat{Q}_{Y|U}|\widehat{Q}_U) + \lambda \cdot D(Q_{U|X}(\cdot|X)\|\widetilde{Q}_U|\widehat{P}_X) \quad (11.59)$$

$$\equiv \mathcal{L}^{(\lambda)}_{\text{VA}}(Q_{U|X}, \widetilde{Q}_U). \quad (11.60)$$

The new surrogate function (11.60), denoted by $\mathcal{L}^{(\lambda)}_{\text{VA}}(Q_{U|X})$, shares a lot in common with a slightly more general form of the variational auto-encoders (VAEs) [16] and the recently introduced information dropout (ID) [15, 37], where the latent space is regularized using a prior $\widetilde{Q}_U$. Therefore, the information-theoretic bound in Theorem 11.2 shows that the algorithm in Definition 11.13 as well as VAEs and ID are slightly different but related information-theoretic ways to control the generalization gap. In all of them the mutual information $I(\widehat{P}_X; Q_{U|X})$ (or its upper bound given by $D(Q_{U|X}(\cdot|X)\|\widetilde{Q}_U|\widehat{P}_X)$) plays the fundamental role, although the specific way in which this term controls the generalization gap could be different for each case.

11.4.2 Information Complexity of Representations

We could think of the most significative term in the upper bound (11.54) as an information-complexity cost of data representations, which depends only on the data samples and on the selected randomized encoder from the restricted model. Suppose we are given a set of different model classes for the randomized encoders $k = [1 : K]$:

$$\mathcal{F}^{(k)}_E := \left\{Q_{U|X} \equiv \mathbb{E}_{P_Z}[\mathbb{1}[u = f_\theta(x, Z)]] : \boldsymbol{\theta} = (\theta_1, \ldots, \theta_k) \in \Theta_k, P_Z \in \mathcal{P}_k(\mathcal{Z})\right\},$$

where there are two kinds of parameters: a structure parameter k and real-value parameters θ, whose parameters depend on the structure, e.g., Θ_k may account for different number of layers or nonlinearities, while $\mathcal{P}_k(\mathcal{Z})$ indicates different kinds of noise distribution. Theorem 11.2 motivates the following model-selection principle for learning compact representations.

Find a parameter k and real-value parameters $\boldsymbol{\theta}$ for the observed data $\mathcal{S}_n$ with which the corresponding data representation can be encoded with the shortest code length:

$$\inf_{\boldsymbol{\theta}\in\Theta_k, k=[1:K]} \left[\mathcal{L}_{\mathrm{emp}}\left(Q_{U|X}^{(\boldsymbol{\theta},k)}, \mathcal{S}_n\right) + \lambda \cdot \sqrt{I\left(\widehat{P}_X; Q_{U|X}^{(\boldsymbol{\theta},k)}\right)} \right], \tag{11.61}$$

where the mutual information penalty term indicates the minimum of the expected redundancy between the minimum code length[5] *(measured in bits)* $-\log Q_{U|X}^{(\boldsymbol{\theta},k)}(\cdot|x)$ *to encode representations under a known data source and the best code length* $-\log Q_U(\cdot)$ *chosen to encode the data representations without knowing the input samples:*

$$I\left(\widehat{P}_X; Q_{U|X}^{(\boldsymbol{\theta},k)}\right) = \min_{Q_U \in \mathcal{P}(\mathcal{U})} \mathbb{E}_{\widehat{P}_X} \mathbb{E}_{Q_{U|X}^{(\boldsymbol{\theta},k)}} \left[-\log Q_U(U) + \log Q_{U|X}^{(\boldsymbol{\theta},k)}(U|X) \right]. \tag{11.62}$$

This information principle combines the empirical cross-entropy risk (11.14) with the "information complexity" of the selected encoder (11.62) as being a regularization that acts as a sample-dependent penalty against *overfitting*. One may view (11.62) as a possible means of comparing the appropriateness of distinct representation models (e.g., number of layers or amount of noise), after a parametric choice has been selected.

The coding interpretation of the penalty term in (11.61) is that the length of the description of the representations themselves can be quantified in the same units as the code length in data compression, namely, bits. In other words, for each data sample x, a randomized encoder can induce different types of representations $U(x)$ with expected information length given by $H(Q_{U|X}(\cdot|x))$. When this representation has to be encoded without knowing $Q_{U|X}$ since x is not given to us (e.g., in a communication problem where the sender wishes to communicate the representations only), the required average length of an encoding distribution Q_U results in $\mathbb{E}_{Q_{U|X}}[-\log Q_U(U)]$. In this sense, expression (11.61) suggests that we should select encoders that allow us to then encode representations efficiently. Interestingly, this is closely related to the celebrated minimum-description-length (MDL) method for density estimation [38, 39]. However, the fundamental difference between these principles is that the information complexity (11.62) follows from the generalization gap and measures the amount of information conveyed by the representations relative to an encoder model, as opposed to the model parameters of the encoder itself.

The information-theoretic significance of (11.62) goes beyond simply a regularization term, since it leads us to introduce the fundamental notion of *encoder capacity*. This key idea of encoder capacity is made possible thanks to Theorem 11.2 that connects mathematically the generalization gap to the information complexity, which is intimately related to the number of distinguishable samples from the representations. Notice that the information complexity can be upper-bounded as

[5] As is well known in information theory, the shortest expected code length is achievable by a uniquely decodable code under a known data source [24].

$$I\left(\widehat{P}_X; Q_{U|X}\right) = \frac{1}{n}\sum_{i=1}^{n} D\left(Q_{U|X}(\cdot|x_i) \middle\| \frac{1}{n}\sum_{j=1}^{n} Q_{U|X}(\cdot|x_j)\right) \tag{11.63}$$

$$\leq \frac{1}{n^2}\sum_{i=1}^{n}\sum_{j=1}^{n} D\left(Q_{U|X}(\cdot|x_i) \| Q_{U|X}(\cdot|x_j)\right), \tag{11.64}$$

where $\{x_i\}_{i=1}^n$ are the training examples from the dataset $\mathcal{S}_n$ and the last inequality follows from the convexity of the relative entropy. This bound is measuring the average degree of closeness between the corresponding representations for the different sample inputs. When two distributions, $Q_{U|X}(\cdot|x_i)$ and $Q_{U|X}(\cdot|x_j)$, are very close to each other, i.e., $Q_{U|X}$ assigns high likelihood to similar representations corresponding to different inputs $x_i \neq x_j$, they do not contribute so much to the complexity of the overall representations. In other words, the more sample inputs an encoder can differentiate, the more patterns it can fit well, and hence the larger the mutual information and thus the risk of overfitting. This observation suggests that the complexity of a representation model with respect to a sample dataset can be related to the number of data samples that essentially yield different (distinguishable) representations. Inspired by the concept of *stochastic complexity* [39], we introduce below the notion of *encoder capacity* to measure the complexity of a representation model.

DEFINITION 11.14 (Capacity of randomized encoders) *The encoder capacity* C_e *of a randomized encoder* $Q_{U|X}$ *with respect to a sample set* $\mathcal{A} \subseteq \mathcal{X}$ *is defined as*

$$C_e(\mathcal{A}, Q_{U|X}) := \max_{\psi:\mathcal{U}\to\mathcal{A}} \log\left(\sum_{u\in\mathcal{U}} Q_{U|X}(u|\psi(u))\right) = \log|\mathcal{A}| - \log\left(\frac{1}{1-\varepsilon}\right), \tag{11.65}$$

$$\varepsilon := \min_{\psi:\mathcal{U}\to\mathcal{A}} \frac{1}{|\mathcal{A}|}\sum_{x\in\mathcal{A}}\sum_{u\in\mathcal{U}} Q_{U|X}(u|x)\mathbb{1}[\psi(u) \neq x] \leq 1 - \frac{1}{|\mathcal{A}|}. \tag{11.66}$$

The argument of the logarithm in the second term of (11.65) represents the probability of being able to distinguish samples from their representations $1-\varepsilon$, i.e., the average probability that estimated samples via the maximum-likelihood estimator $\psi(\cdot)$ from $Q_{U|X}$ are equal to the true samples. Therefore, the encoder capacity is the logarithm of the total number of samples minus a term that depends on the probability of misclassification of the input samples from their representations. When ε is small, then $C_e(\mathcal{A}, Q_{U|X}) \approx \log|\mathcal{A}| - \varepsilon$ and thus all samples are perfectly distinguishable. The following proposition gives simple bounds[6] on the encoder capacity from the information complexity (11.62), which, as we already know, has a close relation to the generalization gap.

PROPOSITION 11.1 Let $Q_{U|X}$ be an encoder distribution and let $\widehat{P}_X$ be an empirical distribution with support $\mathcal{A}_n \equiv \operatorname{supp}(\widehat{P}_X)$. Then, the information complexity and the encoder capacity satisfy

[6] Notice that it is possible to provide better bounds on ε by relying on the results in [40]. However, we preferred simplicity to "tightness" since the purpose of Proposition 11.1 is to link the encoder capacity and the information complexity.

$$C_{\mathrm{e}}(\mathcal{A}_n, Q_{U|X}) = \log|\mathcal{A}_n| - \log\left(\frac{1}{1-\varepsilon}\right) \tag{11.67}$$

and

$$g^{-1}\Big(\log|\mathcal{A}_n| - I(\widehat{P}_X; Q_{U|X})\Big) \leq \varepsilon \leq \frac{1}{2}\Big(\log|\mathcal{A}_n| - I(\widehat{P}_X; Q_{U|X})\Big), \tag{11.68}$$

where ε is defined by (11.66) with respect to $\mathcal{A}_n$ and, for $0 \leq t \leq 1$,

$$g(t) := t \cdot \log(|\mathcal{A}_n| - 1) + h(t), \tag{11.69}$$

with $h(t) := -t\log(t) - (1-t)\log(1-t)$ and $0\log 0 := 0$. Furthermore,

$$I(\widehat{P}_X; Q_{U|X}) \leq C_{\mathrm{e}}. \tag{11.70}$$

Proof We begin with the lower bound (11.70). Consider the inequalities

$$I(\widehat{P}_X; Q_{U|X}) = \min_{Q_U \in \mathcal{P}(\mathcal{U})} D(Q_{U|X} \| Q_U | \widehat{P}_X) \tag{11.71}$$

$$\leq \min_{Q_U \in \mathcal{P}(\mathcal{U})} \mathbb{E}_{\widehat{P}_X} \mathbb{E}_{Q_{U|X}}\left[\max_{x \in \mathcal{A}_n} \log\left(\frac{Q_{U|X}(U|x)}{Q_U(U)}\right)\right] \tag{11.72}$$

$$\leq \min_{Q_U \in \mathcal{P}(\mathcal{U})} \max_{u \in \mathcal{U}} \log\left(\frac{Q_{U|X}(u|\psi^\star(u))}{Q_U(u)}\right) \tag{11.73}$$

$$= \log\left(\sum_{u \in \mathcal{U}} Q_{U|X}(u|\psi^\star(u))\right) = C_{\mathrm{e}}(Q_{U|X}, \mathcal{A}_n), \tag{11.74}$$

where (11.73) follows by letting $\psi^\star$ be the mapping maximizing $C_{\mathrm{e}}(Q_{U|X}, \mathcal{A}_n)$, and (11.74) follows by noticing that (11.73) is the smallest worst-case regret, known as the *minimax regret*, and thus by choosing Q_U to be the normalized maximum-likelihood distribution on the restricted set $\mathcal{A}_n$ the claim is a consequence of the remarkable result of Shtarkov [41].

It remains to show the bounds in (11.68). In order to show the lower bound, we can simply apply Fano's lemma (Lemma 2.10 of [42]), from which we can bound from below the error probability (11.66) that is based on $\mathcal{A}_n$. As for the upper bound,

$$\log|\mathcal{A}_n| - I(\widehat{P}_X; Q_{U|X}) \geq H(\widehat{P}_X) - I(\widehat{P}_X; Q_{U|X}) \tag{11.75}$$

$$= \sum_{u \in \mathcal{U}} \widehat{Q}_U(u) H(\widehat{Q}_{X|U}(\cdot|u)) \tag{11.76}$$

$$\geq 2 \sum_{u \in \mathcal{U}} \widehat{Q}_U(u)\left(1 - \max_{x' \in \mathcal{X}} \widehat{Q}_{X|U}(x'|u)\right) \tag{11.77}$$

$$= 2\varepsilon, \tag{11.78}$$

where (11.75) follows from the assumption $\mathcal{A}_n = \mathrm{supp}(\widehat{P}_X)$ and the fact that the entropy is maximal over the uniform distribution; (11.77) follows by using Equation (7) of [43] and (11.78) by the definition of ε in (11.66). This concludes the proof.

REMARK 11.1 In Proposition 11.1, the function $g^{-1}(t) := 0$ for $t < 0$ and, for $0 < t < \log|\mathcal{A}_n|$, $g^{-1}(t)$ is a solution of the equation $g(\varepsilon) = t$ with respect to

$\varepsilon \in [0, 1 - 1/|\mathcal{A}_n|]$; this solution exists since the function g is continuous and increasing on $[0, 1 - 1/|\mathcal{A}_n|]$ and $g(0) = 0$, $g(1 - 1/|\mathcal{A}_n|) = \log|\mathcal{A}_n|$.

REMARK 11.2 (Generalization requires learning invariant representations) An important consequence of the lower bound in (11.68) in Proposition 11.1 is that by limiting the information complexity, i.e., by controlling the generalization gap according to the criterion (11.61), we bound from below the error probability of distinguishing input samples from their representations. In other words, from expression (11.67) and Theorem 11.2 we can conclude that encoders inducing a large misclassification probability on input samples from their representations, i.e., different inputs must share similar representations, are expected to achieve better generalization. Specifically, this also implies formally that we need only enforce invariant representations to control the encoder capacity (e.g., injecting noise during training), from which the generalization is upper-bounded naturally thanks to Theorem 11.2 and the connection with the information complexity. However, there is a sensitive trade-off between the amount of noise (enforcing both invariance and generalization) and the minimization of the cross-entropy loss. Additionally, it is not difficult to show from the data-processing inequality that stacking noisy encoder layers reinforces increasingly invariant representations since distinguishing inputs from their representations becomes harder – or equivalently the encoder capacity decreases – the deeper the network.

11.4.3 Sketch of the Proofs

We begin by observing that the generalization gap can easily be bounded as

$$\begin{aligned} \mathcal{E}_{\text{gap}}(Q_{U|X}, \mathcal{S}_n) \le{}& \widetilde{\mathcal{E}}_{\text{gap}}(Q_{U|X}, \mathcal{S}_n) \\ &+ \left| \sum_{(u,y)\in\mathcal{U}\times\mathcal{Y}} \left[Q_{YU}(y,u) - \widehat{Q}_{YU}(y,u) \right] \log\left(\frac{Q_{Y|U}(y|u)}{\widehat{Q}_{Y|U}(y|u)} \right) \right|, \end{aligned} \tag{11.79}$$

where we define

$$\begin{aligned} \widetilde{\mathcal{E}}_{\text{gap}}(Q_{U|X}, \mathcal{S}_n) = \Bigg| & \mathbb{E}_{P_{XY}}\left[-\sum_{u\in\mathcal{U}} Q_{U|X}(u|x) \log Q_{Y|U}(Y|U) \right] \\ & - \mathbb{E}_{\widehat{P}_{XY}}\left[-\sum_{u\in\mathcal{U}} Q_{U|X}(u|x) \log Q_{Y|U}(Y|U) \right] \Bigg|. \end{aligned} \tag{11.80}$$

That is, $\widetilde{\mathcal{E}}_{\text{gap}}(Q_{U|X}, \mathcal{S}_n)$ is the gap corresponding to the optimal decoder selecting, which depends on the true P_{XY}, according to Lemma 11.3. It is not difficult to show that

$$\widetilde{\mathcal{E}}_{\text{gap}}(Q_{U|X}, \mathcal{S}_n) \le \left| H(Q_{Y|U}|Q_U) - H(\widehat{Q}_{Y|U}|\widehat{Q}_U) \right| + \mathbb{E}_{\widehat{Q}_U}\left[D(\widehat{Q}_{Y|U} \| Q_{Y|U}) \right], \tag{11.81}$$

where the second term can be bounded as $\mathbb{E}_{\widehat{Q}_U}\left[D(\widehat{Q}_{Y|U} \| Q_{Y|U}) \right] \le D(\widehat{P}_{XY} \| P_{XY})$. The first term of (11.81) is bounded as

$$\begin{aligned} \left| H(Q_{Y|U}|Q_U) - H(\widehat{Q}_{Y|U}|\widehat{Q}_U) \right| \le{}& \left| H(Q_U) - H(\widehat{Q}_U) \right| + \left| H(P_Y) - H(\widehat{P}_Y) \right| \\ &+ \left| H(Q_{U|Y}|P_Y) - H(\widehat{Q}_{U|Y}|\widehat{P}_Y) \right|. \end{aligned} \tag{11.82}$$

To obtain an upper bound, we use the following bounds [18]:

$$\left|H(Q_U) - H(\widehat{Q}_U)\right| \leq \sum_{u\in\mathcal{U}} \phi\Big(\|\mathbf{p}_X - \widehat{\mathbf{p}}_X\|_2 \cdot \sqrt{\mathbb{V}(\{Q_{U|X}(u|x)\}_{x\in\mathcal{X}})}\Big), \tag{11.83}$$

$$\begin{aligned}\left|H(Q_{U|Y}|P_Y) - H(\widehat{Q}_{U|Y}|\widehat{P}_Y)\right| &\leq \|\mathbf{p}_Y - \widehat{\mathbf{p}}_Y\|_2 \sqrt{|\mathcal{Y}|}\log|\mathcal{U}| \\ &+ \mathbb{E}_{P_Y}\Bigg[\sum_{u\in\mathcal{U}} \phi\Big(\big\|\mathbf{p}_{X|Y}(\cdot|Y) - \widehat{\mathbf{p}}_{X|Y}(\cdot|Y)\big\|_2 \\ &\cdot \sqrt{\mathbb{V}(\{\mathbf{q}_{U|X}(u|x)\}_{x\in\mathcal{X}})}\Big)\Bigg],\end{aligned} \tag{11.84}$$

where

$$\phi(x) = \begin{cases} 0 & x \leq 0, \\ -x\log(x) & 0 < x < e^{-1}, \\ e^{-1} & x \geq e^{-1} \end{cases} \tag{11.85}$$

and $\mathbb{V}(\mathbf{a}) = \|\mathbf{a} - \bar{a}\mathbb{1}_d\|_2^2$, with $\mathbf{a} \in \mathbb{R}^d$, $d \in \mathbb{N}_+$, $\bar{a} = (1/d)\sum_{i=1}^d a_i$, and $\mathbb{1}_d$ is the vector of ones of length d.

It is clear that $P_Y \mapsto H(P_Y)$ is a differentiable function and, thus, we can apply a first-order Taylor expansion to obtain

$$H(P_Y) - H(\widehat{P}_Y) = \left\langle \frac{\partial H(P_Y)}{\partial \mathbf{p}_Y}, \mathbf{p}_Y - \widehat{\mathbf{p}}_Y \right\rangle + o(\|\mathbf{p}_Y - \widehat{\mathbf{p}}_Y\|_2), \tag{11.86}$$

where $\partial H(P_Y)/\partial P_Y(y) = -\log P_Y(y) - \log(e)$ for each $y \in \mathcal{Y}$. Then, using the Cauchy–Schwartz inequality, we have

$$\left|H(P_Y) - H(\widehat{P}_Y)\right| \leq \sqrt{\mathbb{V}(\{\log \mathbf{p}_Y(y)\}_{y\in\mathcal{Y}})}\big\|\mathbf{p}_Y - \widehat{\mathbf{p}}_Y\big\|_2 + o(\|\mathbf{p}_Y - \widehat{\mathbf{p}}_Y\|_2). \tag{11.87}$$

McDiarmid's concentration inequality and Theorem 12.2.1 of [24] allow us to bound with an arbitrary probability close to one the terms $D(\widehat{P}_{XY}\|P_{XY})$, $\|\mathbf{p}_X - \widehat{\mathbf{p}}_X\|_2$, $\|\mathbf{p}_Y - \widehat{\mathbf{p}}_Y\|_2$, and $\|\mathbf{p}_{X|Y}(\cdot|y) - \widehat{\mathbf{p}}_{X|Y}(\cdot|y)\|_2$, $\forall\, y \in \mathcal{Y}$ simultaneously. To make sure the bounds hold simultaneously over these $\mathcal{Y}+3$ quantities, we replace δ with $\delta/(|\mathcal{Y}|+3)$ in each concentration inequality. Then, with probability at least $1-\delta$, the following bounds hold:

$$\max\Big\{\big\|\mathbf{p}_X - \widehat{\mathbf{p}}_X\big\|_2, \big\|\mathbf{p}_{X|Y}(\cdot|y) - \widehat{\mathbf{p}}_{X|Y}(\cdot|y)\big\|_2, \big\|\mathbf{p}_Y - \widehat{\mathbf{p}}_Y\big\|_2\Big\} \leq \frac{B_\delta}{\sqrt{n}} \tag{11.88}$$

and

$$D(\widehat{P}_{XY}\|P_{XY}) \leq |\mathcal{X}||\mathcal{Y}|\frac{\log(n+1)}{n} + \frac{1}{n}\log\left(\frac{|\mathcal{Y}|+3}{\delta}\right). \tag{11.89}$$

Then, with probability at least $1-\delta$, we have

$$\begin{aligned}\widetilde{\mathcal{E}}_{\text{gap}}(Q_{U|X}, \mathcal{S}_n) &\leq 2\sum_{u\in\mathcal{U}} \phi\left(\frac{B_\delta}{\sqrt{n}}\sqrt{\mathbb{V}(\{\mathbf{q}_{U|X}(u|x)\}_{x\in\mathcal{X}})}\right) + \frac{B_\delta}{\sqrt{n}}\sqrt{|\mathcal{Y}|}\log|\mathcal{U}| \\ &+ \sqrt{\mathbb{V}(\{\log \mathbf{p}_Y(y)\}_{y\in\mathcal{Y}})}\frac{B_\delta}{\sqrt{n}} + O\left(\frac{\log(n)}{n}\right)\end{aligned} \tag{11.90}$$

$$\leq \frac{\log(n)}{\sqrt{n}} B_\delta \sum_{u\in\mathcal{U}} \sqrt{\mathbb{V}(\{\mathbf{q}_{U|X}(u|x)\}_{x\in\mathcal{X}})} + \frac{B_\delta \sqrt{|\mathcal{Y}|}\log|\mathcal{U}|}{\sqrt{n}}$$
$$+ \frac{2|\mathcal{U}|e^{-1}}{\sqrt{n}} + \sqrt{\mathbb{V}(\{\log \mathbf{p}_Y(y)\}_{y\in\mathcal{Y}})}\frac{B_\delta}{\sqrt{n}} + O\left(\frac{\log(n)}{n}\right), \tag{11.91}$$

where we use $n \geq a^2e^2$ and $\phi\left(a/\sqrt{n}\right) \leq ((a/2)\log(n)/\sqrt{n}) + (e^{-1}/\sqrt{n})$. By combining this result with the next inequality [18]:

$$\sum_{u\in\mathcal{U}} \sqrt{\mathbb{V}(\{\mathbf{q}_{U|X}(u|x)\}_{x\in\mathcal{X}})} \leq \frac{\sqrt{2}}{p_X(x_{\min})}\left(1+\sqrt{\frac{1}{|\mathcal{X}|}}\right)\sqrt{I(P_X;Q_{U|X})}, \tag{11.92}$$

we relate to the mutual information. Finally, using Taylor arguments as above, we can easily write

$$\left|\sqrt{I(P_X;Q_{U|X})} - \sqrt{I\left(\widehat{P}_X;Q_{U|X}\right)}\right| \equiv O(\|\mathbf{p}_X - \widehat{\mathbf{p}}_X\|_2) \leq O(n^{-1/2}) \tag{11.93}$$

with probability $1-\delta$. It only remains to analyze the second term on the right-hand side of (11.79). Using standard manipulations, we can easily show that this term can be equivalently written as

$$\left|\sum_{(x,y)\in\mathcal{X}\times\mathcal{Y}} \left[P_{XY}(x,y) - \widehat{P}_{XY}(x,y)\right]\sum_{u\in\mathcal{U}} Q_{U|X}(u|x)\log\left(\frac{Q_{Y|U}(y|u)}{\widehat{Q}_{Y|U}(y|u)}\right)\right|. \tag{11.94}$$

It is not difficult to see that given $Q_{U|X}$, $P_{XY} \mapsto \log(Q_{Y|U}(y|u))$ is a differentiable function and, thus, we can apply a first-order Taylor expansion to obtain

$$\sum_{u\in\mathcal{U}} Q_{U|X}(u|x)\log\left(\frac{Q_{Y|U}(y|u)}{\widehat{Q}_{Y|U}(y|u)}\right) = -\sum_{u\in\mathcal{U}} Q_{U|X}(u|x)\left\langle\frac{\partial \log Q_{Y|U}(y|u)}{\partial \mathbf{p}_{XY}}, \mathbf{p}_{XY} - \widehat{\mathbf{p}}_{XY}\right\rangle$$
$$+ o(\|\mathbf{p}_{XY} - \widehat{\mathbf{p}}_{XY}\|_2) \tag{11.95}$$

and

$$\frac{\partial \log Q_{Y|U}(y|u)}{\partial P_{XY}(x',y')} = \frac{Q_{U|X}(u|x')[\mathbb{1}\{y'=y\} - Q_{Y|U}(y|u)]}{Q_{UY}(u,y)}. \tag{11.96}$$

With the assumption that every encoder $Q_{U|X}(u|x)$ in the family $\mathcal{F}$ satisfies that $Q_{U|X}(u|x) > \alpha$ for every $(u,x) \in \mathcal{U}\times\mathcal{X}$ with $\alpha > 0$, we obtain that

$$\left|\frac{\partial \log Q_{Y|U}(y|u)}{\partial P_{XY}(x',y')}\right| < \frac{2}{\alpha}, \ \forall (x,x',y',u) \in \mathcal{X}\times\mathcal{X}\times\mathcal{Y}\times\mathcal{U}. \tag{11.97}$$

From simple algebraic manipulations, we can bound the term in (11.94) as

$$\left|\sum_{(x,y)\in\mathcal{X}\times\mathcal{Y}} \left[P_{XY}(x,y) - \widehat{P}_{XY}(x,y)\right]\sum_{u\in\mathcal{U}} Q_{U|X}(u|x)\log\left(\frac{Q_{Y|U}(y|u)}{\widehat{Q}_{Y|U}(y|u)}\right)\right|$$
$$\leq \frac{2}{\alpha}\left(\sum_{(x,y)\in\mathcal{X}\times\mathcal{Y}} |P_{XY}(x,y) - \widehat{P}_{XY}(x,y)|\right)^2. \tag{11.98}$$

Again, using McDiarmid's concentration inequality, it can be shown that with probability close to one this term is $O(1/n)$, which can be neglected compared with the other terms calculated previously. This concludes the proof of the theorem.

11.5 Summary and Outlook

We discussed how generalization in representation learning that is based on the cross-entropy loss is related to the notion of information complexity, and how this connection is employed to view learning in terms of the information-bottleneck principle. The resulting information-complexity penalty is a sample-dependent bound on the generalization gap that crucially depends on the mutual information between the inputs and the randomized (representation) outputs of the selected encoder, revealing an interesting connection between the generalization capabilities of representation models and the information carried by the representations. Furthermore, we have shown that the information complexity is closely related to the so-called encoder capacity, revealing the well-known fact that enforcing invariance in the representations is a critical aspect to control the generalization gap. Among other things, the results of this chapter present a new viewpoint on the foundations of representation learning, showing the usefulness of information-theoretic concepts and tools in the comprehension of fundamental learning problems. This survey provided a summary of some useful links between information theory and representation learning from which we expect to see advances in years to come.

In the present analysis, the number of samples is the most useful resource for the reduction of the generalization gap. Nevertheless, we have not considered other important ingredients of the problem related to the computational-complexity aspect of learning representation models. One of them is the particular optimization problem that has to be solved in order to find an appropriate encoder. It is well known that the specific "landscape" of the cost function (as a function of the parameters of the family of encoders) to be optimized and the particular optimization algorithm used (e.g., stochastic gradient-descent algortihms) could have some major effects such that performance may not be improved by increasing the number of samples. Additional constraints imposed by real-world applications such as that computations must be performed with a limited time budget could also be relevant from a more practical perspective. Evidently, it is pretty clear that many challenges still remain in this exciting research area.

References

[1] National Research Council, *Frontiers in massive data analysis*. National Academies Press, 2013.

[2] C. Shannon, "A mathematical theory of communication," *Bell System Technical J.*, vols. 3, 4, 27, pp. 379–423, 623–656, 1948.

[3] V. Vapnik, *The nature of statistical learning theory*, 2nd edn. Springer, 2000.

[4] G. I. Hinton, "Connectionist learning procedures," in *Machine learning*, Y. Kodratoff and R. S. Michalski, eds. Elsevier, 1990, pp. 555–610.

[5] H. B. Barlow, "Unsupervised learning," *Neural Computation*, vol. 1, no. 3, pp. 295–311, 1989.

[6] A. Pouget, J. M. Beck, W. J. Ma, and P. E. Latham, "Probabilistic brains: Knowns and unknowns," *Nature Neurosci.*, vol. 16, no. 9, pp. 1170–1178, 2013.

[7] H. Barlow, "The exploitation of regularities in the environment by the brain," *Behav. Brain Sci.*, vol. 24, no. 8, pp. 602–607, 2001.

[8] Y. LeCun, Y. Bengio, and G. Hinton, "Deep learning," *Nature*, vol. 521, no. 7553, pp. 436–444, May 2015.

[9] Y. Bengio, A. Courville, and P. Vincent, "Representation learning: A review and new perspectives," *IEEE Trans. Pattern Analysis Machine Intelligence*, vol. 35, no. 8, pp. 1798–1828, 2013.

[10] A. R. Barron, "Approximation and estimation bounds for artificial neural networks," *Machine Learning*, vol. 14, no. 1, pp. 115–133, 1994.

[11] J. Rissanen, "Modeling by shortest data description," *Automatica*, vol. 14, no. 5, pp. 465–471, 1978.

[12] A. R. Barron and T. M. Cover, "Minimum complexity density estimation," *IEEE Trans. Information Theory*, vol. 37, no. 4, pp. 1034–1054, 1991.

[13] S. Boucheron, O. Bousquet, and G. Lugosi, "Theory of classification: A survey of some recent advances," *ESAIM: Probability Statist.*, vol. 9, no. 11, pp. 323–375, 2005.

[14] N. Srivastava, G. E. Hinton, A. Krizhevsky, I. Sutskever, and R. Salakhutdinov, "Dropout: A simple way to prevent neural networks from overfitting," *J. Machine Learning Res.*, vol. 15, no. 1, pp. 1929–1958, 2014.

[15] A. Achille and S. Soatto, "Information dropout: Learning optimal representations through noisy computation," *arXiv:1611.01353* [stat.ML], 2016.

[16] D. P. Kingma and M. Welling, "Auto-encoding variational Bayes," in *Proc. 2nd International Conference on Learning Representations (ICLR)*, 2013.

[17] C. Zhang, S. Bengio, M. Hardt, B. Recht, and O. Vinyals, "Understanding deep learning requires rethinking generalization," *CoRR*, vol. abs/1611.03530, 2016.

[18] O. Shamir, S. Sabato, and N. Tishby, "Learning and generalization with the information bottleneck," *Theor. Comput. Sci.*, vol. 411, nos. 29–30, pp. 2696–2711, 2010.

[19] R. Shwartz-Ziv and N. Tishby, "Opening the black box of deep neural networks via information," *CoRR*, vol. abs/1703.00810, 2017.

[20] N. Tishby, F. C. Pereira, and W. Bialek, "The information bottleneck method," in *Proc. 37th Annual Allerton Conference on Communication, Control and Computing*, 1999, pp. 368–377.

[21] D. Russo and J. Zou, "How much does your data exploration overfit? Controlling bias via information usage," *arXiv:1511.05219* [CS, stat], 2015.

[22] A. Xu and M. Raginsky, "Information-theoretic analysis of generalization capability of learning algorithms," in *Proc. Advances in Neural Information Processing Systems 30*, 2017, pp. 2524–2533.

[23] A. Achille and S. Soatto, "Emergence of invariance and disentangling in deep representations," *arXiv:1706.01350* [CS, stat], 2017.

[24] T. M. Cover and J. A. Thomas, *Elements of information theory*. Wiley-Interscience, 2006.

[25] V. N. Vapnik, *Statistical learning theory*. Wiley, 1998.

[26] A. E. Gamal and Y.-H. Kim, *Network information theory*. Cambridge University Press, 2012.

[27] C. E. Shannon, "Coding theorems for a discrete source with a fidelity criterion," *IRE National Convention Record*, vol. 4, no. 1, pp. 142–163, 1959.

[28] R. Dobrushin and B. Tsybakov, "Information transmission with additional noise," *IEEE Trans. Information Theory*, vol. 8, no. 5, pp. 293–304, 1962.

[29] T. Courtade and T. Weissman, "Multiterminal source coding under logarithmic loss," *IEEE Trans. Information Theory*, vol. 60, no. 1, pp. 740–761, 2014.

[30] M. Vera, L. R. Vega, and P. Piantanida, "Collaborative representation learning," *arXiv:1604.01433* [cs.IT], 2016.

[31] N. Slonim and N. Tishby, "Document clustering using word clusters via the information bottleneck method," in *Proc. 23rd Annual International ACM SIGIR Conference on Research and Development in Information Retrieval*, 2000, pp. 208–215.

[32] L. Wang, M. Chen, M. Rodrigues, D. Wilcox, R. Calderbank, and L. Carin, "Information-theoretic compressive measurement design," *IEEE Trans. Pattern Analysis Machine Intelligence*, vol. 39, no. 6, pp. 1150–1164, 2017.

[33] S. Boyd and L. Vandenberghe, *Convex optimization*. Cambridge University Press, 2004.

[34] M. Vera, L. R. Vega, and P. Piantanida, "Compression-based regularization with an application to multi-task learning," *IEEE J. Selected Topics Signal Processing*, vol. 5, no. 12, pp. 1063–1076, 2018.

[35] S. Arimoto, "An algorithm for computing the capacity of arbitrary discrete memoryless channels," *IEEE Trans. Information Theory*, vol. 18, no. 1, pp. 14–20, 1972.

[36] R. Blahut, "Computation of channel capacity and rate-distortion functions," *IEEE Trans. Information Theory*, vol. 18, no. 4, pp. 460–473, 1972.

[37] A. A. Alemi, I. Fischer, J. V. Dillon, and K. Murphy, "Deep variational information bottleneck," *CoRR*, vol. abs/1612.00410, 2016.

[38] J. Rissanen, "Paper: Modeling by shortest data description," *Automatica*, vol. 14, no. 5, pp. 465–471, 1978.

[39] P. D. Grünwald, I. J. Myung, and M. A. Pitt, *Advances in minimum description length: Theory and applications*. MIT Press, 2005.

[40] S. Arimoto, "On the converse to the coding theorem for discrete memoryless channels (corresp.)," *IEEE Trans. Information Theory*, vol. 19, no. 3, pp. 357–359, 1973.

[41] Y. M. Shtarkov, "Universal sequential coding of single messages," *Problems Information Transmission*, vol. 23, no. 3, pp. 175–186, 1987.

[42] A. B. Tsybakov, *Introduction to nonparametric estimation*, 1st edn. Springer, 2008.

[43] D. Tebbe and S. Dwyer, "Uncertainty and the probability of error (corresp.)," *IEEE Trans. Information Theory*, vol. 14, no. 3, pp. 516–518, 1968.

12 Fundamental Limits in Model Selection for Modern Data Analysis

Jie Ding, Yuhong Yang, and Vahid Tarokh

Summary

With rapid development in hardware storage, precision instrument manufacturing, and economic globalization etc., data in various forms have become ubiquitous in human life. This enormous amount of data can be a double-edged sword. While it provides the possibility of modeling the world with a higher fidelity and greater flexibility, improper modeling choices can lead to false discoveries, misleading conclusions, and poor predictions. Typical data-mining, machine-learning, and statistical-inference procedures learn from and make predictions on data by fitting parametric or non-parametric models (in a broad sense). However, there exists no model that is universally suitable for all datasets and goals. Therefore, a crucial step in data analysis is to consider a set of postulated candidate models and learning methods (referred to as the model class) and then select the most appropriate one. In this chapter, we provide integrated discussions on the fundamental limits of inference and prediction that are based on model-selection principles from modern data analysis. In particular, we introduce two recent advances of model-selection approaches, one concerning a new information criterion and the other concerning selection of the modeling procedure.

12.1 Introduction

Model selection is the task of selecting a statistical model or learning method from a model class, given a set of data. Some common examples are selecting the variables for low- or high-dimensional linear regression, basis terms such as polynomials, splines, or wavelets in function estimation, the order of an autoregressive process, the best machine-learning techniques for solving real-data challenges on an online competition platform, etc. There has been a long history of model-selection techniques that arise from fields such as statistics, information theory, and signal processing. A considerable number of methods have been proposed, following different philosophies and with sometimes drastically different performances. Reviews of the literature can be found in [1–6] and references therein. In this chapter, we aim to provide an integrated understanding of the properties of various approaches, and introduce two recent advances leading to the improvement of classical model-selection methods.

We first introduce some notation. We use $\mathcal{M}_m = \{p_\theta, \theta \in \mathcal{H}_m\}$ to denote a model which is a set of probability density functions, where $\mathcal{H}_m$ is the parameter space and p_θ is short for $p_m(Z_1,\ldots,Z_n \mid \theta)$, the probability density function of data $Z_1,\ldots,Z_n \in \mathbb{Z}$. A model class, $\{\mathcal{M}_m\}_{m\in\mathbb{M}}$, is a collection of models indexed by $m \in \mathbb{M}$. We denote by n the sample size, and by d_m the size/dimension of the parameter in model $\mathcal{M}_m$. We use p_* to denote the true-data generating distribution, and $\mathbb{E}_*$ for the expectation associated with it. In the *parametric* framework, there exists some $m \in \mathbb{M}$ and some $\theta_* \in \mathcal{H}_m$ such that p_* is equivalent to p_{θ_*} almost surely. Otherwise it is in the *non-parametric* framework. We use $\rightarrow_p$ to denote convergence in probability (under p_*), and $\mathcal{N}(\mu,\sigma^2)$ to denote a Gaussian distribution of mean μ and variance σ^2. We use capital and lower-case letters to denote random variables and the realized values.

The chapter's content can be outlined as follows. In Section 12.2, we review the two statistical/machine-learning goals (i.e., prediction and inference) and the fundamental limits associated with each of them. In Section 12.3, we explain how model selection is the key to reliable data analysis through a toy example. In Section 12.4, we introduce the background and theoretical properties of the Akaike information criterion (AIC) and the Bayesian information criterion (BIC), as two fundamentally important model-selection criteria (and other principles sharing similar asymptotic properties). We shall discuss their conflicts in terms of large-sample performances in relation to two statistical goals. In Section 12.5 we introduce a new information criterion, referred to as the bridge criterion (BC), that bridges the conflicts of the AIC and the BIC. We provide its background, theoretical properties, and a related quantity referred to as the parametricness index that is practically very useful to describe how likely it is that the selected model can be practically trusted as the "true model." In Section 12.6, we review recent developments in modeling-procedure selection, which, differently from model selection in the narrow sense of choosing among parametric models, aims to select the better statistical or machine-learning procedure.

12.2 Fundamental Limits in Model Selection

Data analysis usually consists of two steps. In the first step, candidate models are postulated, and for each candidate model $\mathcal{M}_m = \{p_\theta, \theta \in \mathcal{H}_m\}$ we estimate its parameter $\theta \in \mathcal{H}_m$. In the second step, from the set of estimated candidate models $p_{\widehat{\theta}_m}$ $(m \in \mathbb{M})$ we select the most appropriate one (for either interpretation or prediction purposes). We note that not every data analysis and its associated model-selection procedure formally rely on probability distributions. An example is nearest-neighbor learning with the neighbor size chosen by cross-validation, which requires only that the data splitting is meaningful and that the predictive performance of each candidate model/method can be assessed in terms of some measure (e.g., quadratic loss, hinge loss, or perceptron loss). Also, motivated by computational feasibility, there are methods (e.g., LASSO[1]) that combine the two steps into a single step.

Before we proceed, we introduce the concepts of "fitting" and "optimal model" relevant to the above two steps. The fitting procedure given a certain candidate model $\mathcal{M}_m$ is usually achieved by minimizing the negative log-likelihood function

$$\boldsymbol{\theta} \mapsto \ell_{n,m}(\boldsymbol{\theta}),$$

with $\widehat{\boldsymbol{\theta}}_m$ being the maximum-likelihood estimator (MLE) (under model $\mathcal{M}_m$). The maximized log-likelihood value is defined by $\ell_{n,m}(\widehat{\boldsymbol{\theta}}_m)$. The above notion applies to non-i.i.d. data as well. Another function often used in time-series analysis is the quadratic loss function $\{z_t - \mathbb{E}_p(Z_t \mid z_1,\ldots,z_{t-1})\}^2$ instead of the negative log-likelihood. (Here the expectation is taken over the distribution p.) Since there can be a number of other variations, the notion of the negative log-likelihood function could be extended as a general loss function $\ell(p,z)$ involving a density function $p(\cdot)$ and data z. Likewise, the notion of the MLE could be thought of as a specific form of M-estimator.

To define what the "optimal model" means, let $\widehat{p}_m = p_{\widehat{\boldsymbol{\theta}}_m}$ denote the estimated distribution under model $\mathcal{M}_m$. The predictive performance may be assessed via the out-sample prediction loss: $\mathbb{E}_*(\ell(\widehat{p}_m, Z')|Z) = \int_{\mathbb{Z}} \ell(\widehat{p}_m, z')p_*(z')dz'$, where Z' is independent from and identically distributed to the data used to obtain $\widehat{p}_m$, and $\mathbb{Z}$ is the data domain. A sample analog of $\mathbb{E}_*(\ell(\widehat{p}_m, Z')|Z)$ may be the in-sample prediction loss (also referred to as the empirical loss), defined as $\mathbb{E}_n(\ell(\widehat{p}_m, z)) = n^{-1}\sum_{t=1}^{n} \ell(\widehat{p}_m, z_t)$, to measure the fitness of the observed data to the model $\mathcal{M}_m$. In view of this definition, the optimal model can be naturally defined as the candidate model with the smallest out-sample prediction loss, i.e., $m_0 = \arg\min_{m\in\mathbb{M}} \mathbb{E}_*(\ell(\widehat{p}_m, Z')|Z)$. In other words, $\mathcal{M}_{m_0}$ is the model whose predictive power is the best offered by the candidate models given the observed data and the specified model class. It can be regarded as the theoretical limit of learning given the current data and the model list.

In a parametric framework, the true data-generating model is usually the optimal model for sufficiently large sample size [8]. In this vein, if the true density function p_* belongs to some model $\mathcal{M}_m$, or equivalently $p_* = p_{\boldsymbol{\theta}_*}$ for some $\boldsymbol{\theta}_* \in \mathcal{H}_m$ and $m \in \mathbb{M}$, then we aim to select such $\mathcal{M}_m$ (from $\{\mathcal{M}_m\}_{m\in\mathbb{M}}$) with probability going to one as the sample size increases. This is called *consistency in model selection*. In addition, the MLE of $p_{\boldsymbol{\theta}}$ for $\boldsymbol{\theta} \in \mathcal{H}_m$ is known to be an asymptotically efficient estimator of the true parameter $\boldsymbol{\theta}_*$ [9]. In a non-parametric framework, the optimal model depends on the sample size: for a larger sample size, the optimal model tends to be larger since more observations can help reveal weak effects that are out of reach at a small sample size. In that situation, it can be statistically unrealistic to pursue selection consistency [10]. We note that the aforementioned equivalence between the optimal model and the true model may not

[1] Least absolute shrinkage and selection operator (LASSO) [7] is a penalized regression method, whose penalty term is in the form of $\lambda\|\boldsymbol{\beta}\|_1$, where $\boldsymbol{\beta}$ is the regression coefficient vector and λ is a tuning parameter that controls how many (and which) variables are selected. In practice, data analysts often select an appropriate λ that is based on e.g., five-fold cross-validation.

hold for high-dimensional regression settings where the number of independent variables is large relative to the sample size [8]. Here, even if the true model is included as a candidate, its dimension may be too high to be appropriately identified on the basis of a relatively small amount of data. Then the (literally) parametric setting becomes virtually non-parametric.

There are two main objectives in learning from data. One is to understand the data-generation process for scientific discoveries. Under this objective, a notion of *fundamental limits* is concerned with the consistency of selecting the optimal model. For example, a scientist may use the data to support his/her physics model or identify genes that clearly promote early onset of a disease. Another objective of learning from data is for prediction, where the data scientist does not necessarily care about obtaining an accurate probabilistic description of the data (e.g., which covariates are independent of the response variable given a set of other covariates). Under this objective, a notion of *fundamental limits* is taken in the sense of achieving optimal predictive performance. Of course, one may also be interested in both directions. For example, scientists may be interested in a physical model that explains well the causes of precipitation (inference), and at the same time they may also want to have a good model for predicting the amount of precipitation on the next day or during the next year (prediction).

In line with the two objectives above, model selection can also have two directions: *model selection for inference* and *model selection for prediction*. The first is intended to identify the optimal model for the data, in order to provide a reliable characterization of the sources of uncertainty for scientific interpretation. The second is to choose a model as a vehicle to arrive at a model/method that offers a satisfying top performance. For the former goal, it is crucial that the selected model is stable for the data, meaning that a small perturbation of the data does not affect the selection result. For the latter goal, however, the selected model may be simply the lucky winner among a few close competitors whose predictive performance can be nearly optimal. If so, the model selection is perfectly fine for prediction, but the use of the selected model for insight and interpretation may be misleading. For instance, in linear regression, because the covariates are often correlated, it is quite possible that two very different sets of covariates may offer nearly identical top predictive performances yet neither can justify its own explanation of the regression relationship against that by the other.

Associated with the first goal of model selection for *inference* or identifying the best candidate is the concept of selection consistency. *The selection consistency means that the optimal model/method is selected with probability going to one as the sample size goes to infinity*. This idealization that the optimal model among the candidates can be practically deemed the "true model" is behind the derivations of several model-selection methods. In the context of variable selection, in practical terms, model-selection consistency is intended to mean that the useful variables are identified and their statistical significance can be ascertained in a follow-up study while the rest of the variables cannot. However, in reality, with limited data and large noise the goal of model-selection consistency may not be reachable. Thus, to certify the selected model as the "true" model for reliable statistical inference, the data scientist must conduct a proper selection of model diagnostic assessment (see [11] and references therein). Otherwise, the use of the selected model for drawing conclusions on the data-generation process may give irreproducible results, which is a major concern in the scientific community [12].

In various applications where prediction accuracy is the dominating consideration, the optimal model as defined earlier is the target. When it can be selected with high probability, the selected model can not only be trusted for optimal prediction but also comfortably be declared the best. However, even when the optimal model is out of reach in terms of selection with high confidence, other models may provide asymptotically equivalent predictive performance. In this regard, asymptotic efficiency is a natural consideration for the second goal of model selection. When *prediction* is the goal, obviously prediction accuracy is the criterion to assess models. For theoretical examination, the convergence behavior of the loss of the prediction based on the selected model characterizes the performance of the model-selection criterion. Two properties are often used to describe good model-selection criteria. The asymptotic efficiency property demands that the loss of the selected model/method is asymptotically equivalent to the smallest among all the candidates. *The asymptotic efficiency is technically defined by*

$$\frac{\min_{m \in \mathbb{M}} \mathcal{L}_m}{\mathcal{L}_{\widehat{m}}} \to_p 1 \tag{12.1}$$

as $n \to \infty$, where $\widehat{m}$ denotes the selected model. Here, $\mathcal{L}_m = \mathbb{E}_*(\ell(\widehat{p}_m, Z)) - \mathbb{E}_*(\ell(p_*, Z))$ is the adjusted prediction loss, where $\widehat{p}_m$ denotes the estimated distribution under model m. The subtraction of $\mathbb{E}_*(\ell(p_*, Z))$ makes the definition more refined, which allows one to make a better comparison of competing model-selection methods. Overall, the goal of *prediction* is to select a model that is comparable to the optimal model regardless of whether it is stable or not as the sample size varies. This formulation works both for parametric and for non-parametric settings.

12.3 An Illustration on Fitting and the Optimal Model

We provide a synthetic experiment to demonstrate that better fitting does not imply better predictive performance due to *inflated variances in parameter estimation.*

Example 12.1 Linear regression Suppose that we generated synthetic data from a regression model $Y = f(X) + \varepsilon$, where each item of data is in the form of $z_i = (y_i, x_i)$. Each response y_i ($i = 0, \ldots, n-1$) is observed at $x_i = i/n$ (fixed design points), namely $y_i = f(i/n) + \varepsilon_i$. Suppose that the ε_is are independent standard Gaussian noises. Suppose that we use polynomial regression, and the specified models are in the form of $f(x) = \sum_{j=0}^{m} \beta_j x^j$ ($0 \leq x < 1$, m being a positive integer). The candidate models are specified to be $\{\mathcal{M}_m, m = 1, \ldots, d_n\}$, with $\mathcal{M}_m$ corresponding to $f(x) = \sum_{j=0}^{m} \beta_j x^j$. Clearly, the dimension of $\mathcal{M}_m$ is $d_m = m + 1$.

The prediction loss for regression is calculated as $\mathcal{L}_m = n^{-1} \sum_{i=1}^{n} (f(x_i) - \widehat{f}_m(x_i))^2$, where $\widehat{f}$ is the least-squares estimate of f using model $\mathcal{M}_m$ (see, for example, [8]). The efficiency, as before, is defined as $\min_{m \in \mathbb{M}} \mathcal{L}_m / \mathcal{L}_{\widehat{m}}$.

When the data-generating model is unknown, one critical problem is the identification of the degree of the polynomial model fitted to the data. We need to first estimate polynomial coefficients with different degrees $1,\dots,d_n$, and then select one of them according to a certain principle.

In an experiment, we first generate independent data using each of the following true data-generating models, with sample sizes $n = 100, 500, 2000, 3000$. We then fit the data using the model class as given above, with the maximal order $d_n = 15$.

(1) Parametric framework. The data are generated by $f(x) = 10(1 + x + x^2)$.

Suppose that we adopt the quadratic loss in this example. Then we obtain the in-sample prediction loss $\widehat{e}_m = n^{-1}\sum_{i=1}^{n}(Y_i - f_m(x_i))^2$. Suppose that we plot $\widehat{e}_m$ against d_m, then the curve must be monotonically decreasing, because a larger model fits the same data better. We then compute the out-sample prediction loss $\mathbb{E}_*(\ell(\widehat{p}_m, Z))$, which is equivalent to

$$\mathbb{E}_*(\ell(\widehat{p}_m, Z)) = n^{-1}\sum_{i=1}^{n}(f(x_i) - \widehat{f}_m(x_i))^2 + \sigma^2 \tag{12.2}$$

in this example. The above expectation is taken over the true distribution of an independent future data item Z_t. Instead of showing the out-sample prediction loss of each candidate model, we plot its rescaled version (on $[0,1]$). Recall the *asymptotic efficiency* as defined in (12.1). Under quadratic loss, we have $\mathbb{E}_*(\ell(p_*, Z)) = \sigma^2$, and the asymptotic efficiency requires

$$\frac{\min_{m\in\mathbb{M}} n^{-1}\sum_{i=1}^{n}(f(x_i) - \widehat{f}_m(x_i))^2}{n^{-1}\sum_{i=1}^{n}(f(x_i) - \widehat{f}_{\widehat{m}}(x_i))^2} \to_p 1, \tag{12.3}$$

where $\widehat{m}$ denotes the selected model. In order to describe how the predictive performance of each model deviates from the best possible, we define the *efficiency* of each model as the term on the left-hand side in (12.1), or (12.3) in this example.

We now plot the efficiency of each candidate model on the left-hand side of Fig. 12.1. The curves show that the predictive performance is optimal only for the true model. We note that the minus-σ^2 adjustment of the out-sample prediction loss in the numerator and denominator of (12.3), compared with (12.2), makes it highly non-trivial to achieve the property (see, for example, [8, 13–15]). Consider, for example, the comparison between two nested polynomial models with degrees $d = 2$ and $d = 3$, the former being the true data-generating model. It can be proved that, without subtracting σ^2, the ratio (of the mean-square prediction errors) for each candidate model approaches 1; on subtracting σ^2, the ratio for the former model still approaches 1, while the ratio for the latter model approaches $2/3$.

(2) Non-parametric framework. The data are generated by $f(x) = 20x^{1/3}$.

As for framework (1), we plot the efficiency on the right-hand side of Fig. 12.1. Differently from the case (1), the predictive performance is optimal at increasing model dimensions (as the sample size n increases). As mentioned before, in such a non-parametric framework (i.e., the true f is not in any of the candidate models), the optimal model is highly unstable as the sample size varies, so that pursuing an inference of a fixed good model becomes improper. Intuitively, this is because, in a non-parametric

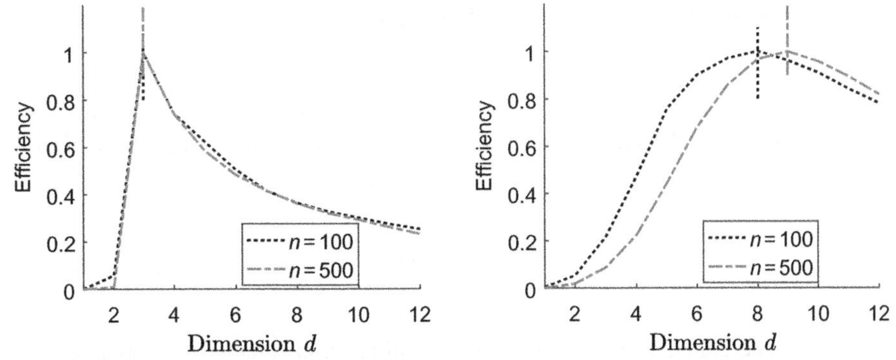

Figure 12.1 The efficiency of each candidate model under two different data-generating processes. The best-performing model is the true model of dimension 3 in the parametric framework (left figure), whereas the best-performing model varies with sample size in the non-parametric framework (right figure).

framework, more complex models are needed to accommodate more observed data in order to strike an appropriate trade-off between the estimation bias (i.e., the smallest approximation error between the data-generating model and a model in the model space) and the variance (i.e., the variance due to parameter estimation) so that the prediction loss can be reduced. Thus, in the non-parametric framework, the optimal model changes, and the model-selection task aims to select a model that is optimal for prediction (e.g., asymptotically efficient), while recognizing it is not tangible to identify the true/optimal model for the inference purpose. Note that Fig. 12.1 is drawn using information of the underlying true model, but that information is unavailable in practice, hence the need for a model-selection method so that the asymptotically best efficiency can still be achieved.

This toy experiment illustrates the general rules that (1) a larger model tends to fit data better, and (2) the predictive performance is optimal with a candidate model that typically depends both on the sample size and on the true data-generating process (which is unknown in practice). In a virtually (or practically) parametric scenario, the optimal model is stable around the present sample size and it may be practically treated as the true model. In contrast, in a virtually (or practically) non-parametric scenario, the optimal model changes sensitively to the sample size (around the present sample size) and the task of identifying the elusive optimal model for reliable inference is unrealistic. With this understanding, an appropriate model-selection technique is called for so as to single out the optimal model for inference and prediction in a strong practically parametric scenario, or to strike a good balance between the *goodness of fit* and *model complexity* (i.e., the number of free unknown parameters) on the observed data to facilitate optimal prediction in a practically non-parametric scenario.

12.4 The AIC, the BIC, and Other Related Criteria

Various model-selection criteria have been proposed in the literature. Though each approach was motivated by a different consideration, many of them originally aimed to select either the order in an autoregressive model or a subset of variables in a regression model. We shall revisit an important class of them referred to as information criteria.

12.4.1 Information Criteria and Related Methods

Several model-selection methods that are based on likelihood functions can be put within the framework of information criteria. They are generally applicable to parametric model-based problems, and their asymptotic performances have been well studied in various settings. A general form of information criterion is to select $\mathcal{M}_m$ such that

$$\widehat{m} = \arg\min_{m\in\mathbb{M}} n^{-1} \sum_{t=1}^{n} \ell(p_{\widehat{\theta}_m}, z_t) + f_{n,d_m},$$

where the objective function is the estimated in-sample loss plus a penalty $f_{n,d}$ (indexed by the sample size n and model dimension d).

The *Akaike information criterion* (AIC) [16, 17] is a model-selection principle that was originally derived by minimizing the Kullback–Leibler (KL) divergence from a candidate model to the true data-generating model p_*. Equivalently, the idea is to approximate the out-sample prediction loss by the sum of the in-sample prediction loss and a correction term. In the typical setting where the loss is logarithmic, the AIC procedure is to select the model $\mathcal{M}_m$ that minimizes

$$\mathrm{AIC}_m = -2\ell_{n,m} + 2d_m, \tag{12.4}$$

where $\ell_{n,m}$ is the maximized log-likelihood of model $\mathcal{M}_m$ given n observations and d_m is the dimension of model $\mathcal{M}_m$. In the task of selecting the appropriate order in autoregressive models or subsets of variables in regression models, it is also common to use the alternative version $\mathrm{AIC}_k = n \log \widehat{e}_k + 2k$ for the model of order k, where $\widehat{e}_k$ is the average in-sample prediction error based on the quadratic loss.

The *Bayesian information criterion* (BIC) [18] is a principle that selects the model m that minimizes

$$\mathrm{BIC}_m = -2\ell_{n,m} + d_m \log n. \tag{12.5}$$

The only difference from the AIC is that the constant 2 in the penalty term is replaced with the logarithm of the sample size. Its original derivation by Schwarz was only for an exponential family from a frequentist perspective. But it turned out to have a nice Bayesian interpretation, as its current name suggests. Recall that, in Bayesian data analysis, marginal likelihood is commonly used for model selection [19]. In a Bayesian setting, we would introduce a prior with density $\boldsymbol{\theta} \mapsto p_m(\boldsymbol{\theta})$ ($\boldsymbol{\theta} \in \mathcal{H}_m$), and a likelihood of data $p_m(Z \mid \boldsymbol{\theta})$, where $Z = [Z_1, \ldots, Z_n]$, for each $m \in \mathbb{M}$. We first define the marginal likelihood of model $\mathcal{M}_m$ by

$$p(Z \mid \mathcal{M}_m) = \int_{\mathcal{H}_m} p_m(Z \mid \boldsymbol{\theta}) p_m(\boldsymbol{\theta}) d\boldsymbol{\theta}. \tag{12.6}$$

The candidate model with the largest marginal likelihood should be selected. Interestingly, this Bayesian principle is asymptotically equivalent to the BIC in selecting models. To see the equivalence, we assume that $Z_1, \ldots, Z_n$ are i.i.d., and $\pi(\cdot)$ is any prior distribution on θ which has dimension d. We let $\ell_n(\theta) = \sum_{i=1}^{n} \log p_\theta(z_i)$ be the log-likelihood function, and $\widehat{\boldsymbol{\theta}}_n$ the MLF of $\boldsymbol{\theta}$. Note that ℓ_n implicitly depends on the model. A proof of the Bernstein–von Mises theorem (see Chapter 10.2 of [20]) implies (under regularity conditions)

$$p(Z_1,\ldots,Z_n)\exp\left(-\ell_n(\widehat{\theta}_n)+\frac{d}{2}\log n\right)\to_p c_* \tag{12.7}$$

as $n \to \infty$, for some constant c_* that does not depend on n. Therefore, selecting a model with the largest marginal likelihood $p(Z_1,\ldots,Z_n)$ (as advocated by Bayesian model comparison) is asymptotically equivalent to selecting a model with the smallest BIC in (12.5). It is interesting to see that the marginal likelihood of a model does not depend on the imposed prior at all, given a sufficiently large sample size. We note that, in many cases of practical data analysis, especially when likelihoods cannot be written analytically, the BIC is implemented less than the Bayesian marginal likelihood, because the latter can easily be implemented by utilizing Monte Carlo-based computation methods [21].

Cross-validation (CV) [22–26] is a class of model-selection methods that are widely used in machine-learning practice. CV does not require the candidate models to be parametric, and it works as long as data splittings make sense and one can assess the predictive performance in terms of some measure. A specific type of CV is the delete-1 CV method [27] (or leave-one-out, LOO). The idea is explained as follows. For brevity, let us consider a parametric model class as before. Recall that we wish to select a model $\mathcal{M}_m$ with as small an out-sample loss $\mathbb{E}_*(\ell(p_{\widehat{\theta}_m}, Z))$ as possible. Its computation involves an unknown true-data-generating process, but we may approximate it by $n^{-1}\sum_{i=1}^{n} \ell(p_{\widehat{\theta}_{m,-i}}, z_i)$, where $\widehat{\theta}_{m,-i}$ is the MLE under model $\mathcal{M}_m$ using all the observations except z_i. In other words, given n observations, we leave each observation out and attempt to predict that data point by using the $n-1$ remaining observations, and record the average prediction loss over n rounds. It is worth mentioning that LOO is asymptotically equivalent to the AIC under some regularity conditions [27].

The general practice of CV works in the following manner. It randomly splits the original data into a training set of n_t data and a validation set of $n_v = n - n_t$ data; each candidate model is trained from the n_t data and validated on the remaining data (i.e., to record the average validation loss); the above procedure is replicated a few times, each with a different validation set, in order to alleviate the variance caused by splitting; in the end, the model with the least average validation loss is selected, and the model is re-trained using the complete data for future use. The v-fold CV (with v being a positive integer) is a specific version of CV. It randomly partitions data into v subsets of (approximately) equal size; each model is trained on $v-1$ folds and validated on the remaining 1 fold; the procedure is repeated v times, and the model with the smallest average validation loss is selected. The v-fold CV is perhaps more commonly used than LOO, partly due to the large computational complexity involved in LOO. The holdout method, which is often used in data competitions (e.g., Kaggle competition), may be viewed as a special case of CV: it does data splitting only once, producing one part as the training set and the remaining part as the validation set.

We have mentioned that LOO was asymptotically equivalent to the AIC. How about a general CV with n_t training data and n_v validation data? For regression problems, it has been proved that CV is asymptotically similar to the AIC when $n_v/n_t \to 0$ (including LOO as a special case), and to the BIC when $n_v/n_t \to \infty$ (see, e.g., [8]). Additional comments on CV and corrections of some misleading folklore will be elaborated in Section 12.6.

Note that many other model-selection methods are closely related to the AIC and the BIC. For example, methods that are asymptotically equivalent to the AIC include finite-sample-corrected AIC [28], which was proposed as a corrected version of the AIC, generalized cross-validation [29], which was proposed for selecting the degree of smoothing in spline regression, the final-prediction error criterion which was proposed as a predecessor of the AIC [30, 31], and Mallows' C_p method [32] for regression-variable selection. Methods that share the selection-consistency property of the BIC include the Hannan and Quinn criterion [33], which was proposed as the smallest penalty that achieves strong consistency (meaning that the best-performing model is selected almost surely for sufficiently large sample size), the predictive least-squares (PLS) method based on the minimum-description-length principle [34, 35], and the use of Bayes factors, which is another form of Bayesian marginal likelihood. For some methods such as CV and the generalized information criterion (GIC, or written as GIC_{λ_n}) [8, 36, 37], their asymptotic behavior usually depends on the tuning parameters.

In general, the AIC and the BIC have served as the golden rules for model selection in statistical theory since their coming into existence. Their asymptotic properties have been rigorously established for autoregressive models and regression models, among many other models. Though cross-validations or Bayesian procedures have also been widely used, their asymptotic justifications are still rooted in frequentist approaches in the form of the AIC, the BIC, etc. Therefore, understanding the asymptotic behavior of the AIC and the BIC is of vital value both in theory and in practice. We therefore focus on the properties of the AIC and the BIC in the rest of this section and Section 12.5.

12.4.2 Theoretical Properties of the Model-Selection Criteria

Theoretical examinations of model-selection criteria have centered on several properties: selection consistency, asymptotic efficiency (both defined in Section 12.2), and minimax-rate optimality (defined below). The selection consistency targets the goal of identifying the optimal model or method on its own for scientific understanding, statistical inference, insight, or interpretation. Asymptotic efficiency and minimax-rate optimality are in tune with the goal of prediction. Before we introduce these properties, it is worth mentioning that many model-selection methods can also be categorized into two classes according to their large-sample performances, respectively represented by the AIC and the BIC.

First of all, the AIC has been proved to be *minimax-rate optimal* (defined below) for a range of variable-selection tasks, *including the usual subset-selection and order-selection problems in linear regression, and non-parametric regression based on series expansion* (with bases such as polynomials, splines, or wavelets) (see, e.g., [38] and the references therein). For example, consider the minimax risk of estimating the regression function $f \in \mathcal{F}$ under the squared error

$$\inf_{\widehat{f}} \sup_{f \in \mathcal{F}} n^{-1} \sum_{i=1}^{n} \mathbb{E}_*(\widehat{f}(x_i) - f(x_i))^2, \tag{12.8}$$

where $\widehat{f}$ is over all estimators based on the observations, and $f(x_i)$ equals the expectation of the ith response variable (or the ith value of the regression function) conditional on the ith vector of variables x_i. Each x_i can refer to a vector of explanatory variables, or polynomial basis terms, etc. For a model-selection method δ, its worst-case risk is $\sup_{f \in \mathcal{F}} R(f, \delta, n)$, where $R(f, \delta, n) = n^{-1} \sum_{i=1}^{n} \mathbb{E}_*\{\widehat{f_\delta}(x_i) - f(x_i)\}^2$, and $\widehat{f_\delta}$ is the least-squares estimate of f under the variables selected by δ. *The method δ is said to be minimax-rate optimal over $\mathcal{F}$ if $\sup_{f \in \mathcal{F}} R(f, \delta, n)$ converges (as $n \to \infty$) at the same rate as the minimax risk in (12.8).*

Another good property of the AIC is that it is *asymptotically efficient* (as defined in (12.1)) in a non-parametric framework. In other words, the predictive performance of its selected model is asymptotically equivalent to the best offered by the candidate models (even though it is highly sensitive to the sample size). However, the BIC is known to be consistent in selecting the data-generating model with the smallest dimension in a parametric framework. For example, suppose that the data are truly generated by a quadratic function corrupted by random noise, and the candidate models are a quadratic polynomial, a cubic polynomial, and an exponential function. Then the quadratic polynomial is selected with probability going to one as the sample size tends to infinity. The exponential function is not selected because it is a wrong model, and the cubic polynomial is not selected because it overfits (even though it nests the true model as a special case).

12.5 The Bridge Criterion – Bridging the Conflicts between the AIC and the BIC

In this section, we review a recent advance in the understanding of the AIC, the BIC, and related criteria. We choose to focus on the AIC and the BIC here because they represent two cornerstones of model-selection principles and theories. We are concerned only with settings where the sample size is larger than the model dimensions. Many details of the following discussion can be found in technical papers such as [8, 13, 15, 39–41] and references therein.

Recall that the AIC is asymptotically efficient for the non-parametric scenario and is also minimax optimal. In contrast, the BIC is consistent and asymptotically efficient for the parametric scenario. Despite the good properties of the AIC and the BIC, they have their own drawbacks. The AIC is known to be inconsistent in a parametric scenario where there are at least two correct candidate models. As a result, the AIC is not asymptotically efficient in such a scenario. For example, if data are truly generated by a quadratic polynomial function corrupted by random noise, and the candidate models include quadratic and cubic polynomials, then the former model cannot be selected with probability going to one as the sample size increases. The asymptotic probability of it being selected can actually be analytically computed in a way similar to that shown in [39]. The BIC, on the other hand, does not exhibit the beneficial properties of minimax-rate optimality and asymptotic efficiency in a non-parametric framework [8, 41].

Why do the AIC and the BIC have the aforementioned differences? In fact, theoretical examinations of those aspects are highly non-trivial and have motivated a vast literature since the arrival of the AIC and the BIC. Here we provide some heuristic explanations. The formulation of the AIC in (12.4) was originally motivated by searching for the candidate model p with the smallest KL divergence (denoted by D) from p to the data-generating model p_*. Since $\min_p D(p_*, p)$ is equivalent to $\min_p \mathbb{E}_*(-\log p)$ for a fixed p_*, the AIC was designed to perform well at minimizing the prediction loss. But the AIC is not consistent for a model class containing a true model and at least one oversized model, because fitting the oversized model would reduce the first term $-2\ell_{n,m}$ in (12.4) only by approximately a chi-square-distributed random variable [42], while the increased penalty on the second item $2d_m$ is at a constant level which is not sufficiently large to suppress the overfitting gain. On the other hand, the selection consistency of the BIC in a parametric framework is not surprising due to its nice Bayesian interpretation. However, its penalty $d_m \log n$ in (12.5) is much larger than the $2d_m$ in the AIC, so it cannot enjoy the predictive optimality in a non-parametric framework (if the AIC already does so).

To briefly summarize, for achieving asymptotic efficiency, the AIC (BIC) is suitable only in non-parametric (parametric) settings. There has been a debate regarding the choice between the AIC and the BIC in model-selection practice, and a key argument is on whether the true model is in a parametric framework or not. The same debate may also appear under other terminologies. In a parametric (non-parametric) framework, the true data-generating model is often said to be well-specified (mis-specified), or finite (infinite)-dimensional. To see a reason for such terminology, consider for instance regression analysis using polynomial basis functions as variables. If the true regression function is indeed a polynomial, then it can be parameterized with a finite number of parameters, or represented by finitely many basis functions. If the true regression function is an exponential function, then it cannot be parameterized with any finite-dimensional parameter and its representation requires infinitely many basis functions. Without prior knowledge on how the observations were generated, determining which method to use becomes very challenging.

The following question is then naturally motivated: "*Is it possible to have a new information criterion that bridges the AIC and the BIC in such a way that the strengths of both of them can be had in combination?*"

12.5.1 Bridging the Fundamental Limits in Inference and Prediction

The above question is of fundamental importance, because, in real applications, data analysts usually do not know whether the data-generating model is correctly specified or not. If there indeed exists an "ideal" model-selection procedure that adaptively achieves the better performance of the AIC and the BIC under all situations, then the risks of misusing the AIC (BIC) in parametric (non-parametric) scenarios will no longer be a matter of concern. Recall the previous discussions on the good properties of the AIC and BIC. An "ideal" procedure could be defined in two possible ways, in terms of what good properties to combine.

First, can the properties of *minimax-rate optimality and consistency* be shared? Unfortunately, it has been theoretically shown that there exists no model-selection method that achieves both types of optimality at the same time [41]. That is, for any model-selection procedure to be consistent in selection, it must behave sub-optimally in terms of the minimax rate of convergence in the prediction loss.

Second, can the properties of *asymptotic efficiency (in both parametric and non-parametric frameworks) and consistency (in the parametric framework)* be shared? Recall that consistency in a parametric framework is typically equivalent to asymptotic efficiency [8, 15]. Clearly, if an ideal method were able to combine asymptotic efficiency and consistency, it would then always achieve asymptotic efficiency irrespective of whether it were operating in a parametric framework or not.

We note that a negative answer to the first question does not imply a negative answer to the second question. Because, in contrast to asymptotic efficiency, minimax-rate optimality allows the true data-generating model to vary for a given sample size (see the definition (12.8)) and is therefore more demanding. Asymptotic efficiency is in a pointwise sense, meaning that the data are generated by some fixed (unknown) data-generating model (i.e., the data-generating process is not of an adversarial nature). Therefore, the minimaxity (uniformity over a model space) is a much stronger requirement than the (pointwise) asymptotic efficiency. That window of opportunity motivated an active line of recent advances in reconciling the two classes of model-selection methods [15, 43–46].

In particular, a new model-selection method called the *bridge criterion* (BC) [10, 15] was recently proposed to simultaneously achieve consistency in a parametric framework and asymptotic efficiency in both (parametric and non-parametric) frameworks. It therefore bridges the advantages of the AIC and the BIC in the asymptotic regime. The idea of the BC, which is similar to the ideas of other types of penalized model selection, is to select a model by minimizing its in-sample loss plus a penalty. Its penalty, which is different from the penalties in the AIC and the BIC, is a nonlinear function of the model dimension. The key idea is to impose a BIC-like heavy penalty for a range of small models, but to alleviate the penalty for larger models if there is more evidence supporting an infinite-dimensional true model. In that way, the selection procedure is automatically adaptive to the appropriate setting (either parametric or non-parametric).

The BC selects the model $\mathcal{M}_m$ that minimizes

$$\mathrm{BC}_m = -2\ell_{n,m} + c_n(1 + 2^{-1} + \cdots + d_m^{-1})$$

(with the default $c_n = n^{2/3}$) over all the candidate models whose dimensions are no larger than $d_{m_{\mathrm{AIC}}}$, defined as the dimension of the model selected by the AIC (d_m in (12.4)). Note that the penalty is approximately $c_n \log d_m$, but it is written as a harmonic number to highlight some of its nice interpretations. Its original derivation was motivated by the recent discovery that the information loss of underfitting a model of dimension d

using dimension $d-1$ is asymptotically χ_1^2/d (where χ_1^2 denotes the chi-squared random variable with one degree of freedom) for large d, assuming that nature generates the model from a non-informative uniform distribution over its model space (in particular the coefficient space of all stationary autoregressions). Its heuristic derivation is reviewed in Section 12.5.2. The BC was later proved to be asymptotically equivalent to the AIC in a non-parametric framework, and equivalent to the BIC otherwise in rather general settings [10, 15]. Some intuitive explanations will be given in Section 12.5.3. A technical explanation of how the AIC, BIC, and BC relate to each other can be found in [15].

12.5.2 Original Motivation

The proposal of the BC was originally motivated from the selection of autoregressive orders, though it can be applied in general settings just like the AIC and the BIC. Suppose that a set of time-series data $\{z_t : t = 1,\ldots,n\}$ is observed, and we specify an autoregressive (AR) model class with the largest size d_n. Each model of size/order d $(d = 1,\ldots,d_n)$ is in the form of

$$z_t = \psi_{d,1}z_{t-1} + \cdots + \psi_{d,d}z_{t-k} + \epsilon_t, \tag{12.9}$$

where $\psi_{d,\ell} \in \mathbb{R}$ $(\ell = 1,\ldots,d)$, $\psi_{d,d} \neq 0$, the roots of the polynomial $z^d + \sum_{\ell=1}^{d} \psi_{d,\ell} z^{d-\ell}$ have modulus less than 1, and the ε_ts are independent random noises with zero mean and variance σ^2. The autoregressive model is referred to as an AR(d) model, and $[\psi_{d,1},\ldots,\psi_{d,d}]^{\mathrm{T}}$ is referred to as the stable autoregressive filter. The best-fitting loss of AR(d) to the data is defined by

$$e_d = \min_{\psi_{d,1},\ldots,\psi_{d,d} \in \mathbb{R}} \mathbb{E}_*(z_t - \psi_{d,1}z_{t-1} - \cdots - \psi_{d,d}z_{t-d})^2,$$

where $\mathbb{E}_*$ is with respect to the stationary distribution of $\{z_t\}$ (see Chapter 3 of [47]). The above quantity can be also regarded as the minimum KL divergence from the space of order-d AR to the true data-generating model (rescaled) under Gaussian noise assumptions. We further define the relative loss $g_d = \log(e_{d-1}/e_d)$ for any positive integer d.

Now suppose that nature generates the data from an AR(d) process, which is in turn randomly generated from the uniform distribution $\mathcal{U}_d$. Here, $\mathcal{U}_d$ is defined over the space of all the stable AR filters of order d whose roots have modulus less than 1:

$$S_d = \Bigg\{ [\psi_{d,1},\ldots,\psi_{d,d}]^{\mathrm{T}} : \\ z^d + \sum_{\ell=1}^{d} \psi_{d,\ell} z^{d-\ell} = \prod_{\ell=1}^{d} (z - a_\ell),\ \psi_{d,\ell} \in \mathbb{R}, |a_\ell| < 1,\ \ell = 1,\ldots,d \Bigg\}.$$

Under this data-generating procedure, g_d is a random variable with distribution described by the following result from [15].

The random variable dg_d converges in distribution to χ_1^2 as d tends to infinity.

It is actually a corollary of the following algorithm [15]. A filter Ψ_d uniformly distributed on S_d can be generated by the following recursive procedure.

Algorithm 12.1 Random generator of uniform distribution on S_d

input d
output $\psi_{d,1},\ldots,\psi_{d,d}$

Randomly draw $\psi_{1,1}$ from the uniform distribution on $(-1,1)$
for $k = 1 \to d$ **do**
  Randomly draw β_k according to the beta distribution

$$\beta_k \sim \mathcal{B}(\lfloor k/2+1 \rfloor, \lfloor (k+1)/2 \rfloor)$$

  Let $\psi_{k,k} = 2\beta_k - 1, \quad \psi_{k,\ell} = \psi_{k-1,\ell} + \psi_{k,k}\psi_{k-1,k-\ell} \quad (\ell = 1,\ldots,k-1)$
end for

This result suggests that the underfitting loss $g_d \approx \chi_1^2/d$ tends to decrease with d. Because the increment of the penalty from dimension $d-1$ to dimension d can be treated as a quantity to compete with the underfitting loss [15], it suggests that we penalize in a way not necessarily linear in model dimension.

12.5.3 Interpretation

In this section, we provide some explanations of how the BC can be related to the AIC and BIC. The explanation does not depend on any specific model assumption, which shows that it can be applied to a wide range of other models.

The penalty curves (normalized by n) for the AIC, BIC, and BC can be respectively denoted by

$$f_{n,d_m}(\text{AIC}) = \frac{2}{n} d_m,$$

$$f_{n,d_m}(\text{BIC}) = \frac{\log(n)}{n} d_m,$$

$$f_{n,d_m}(\text{BC}) = \frac{c_n}{n} \sum_{k=1}^{d_m} \frac{1}{k}.$$

Any of the above penalty curves can be written in the form of $\sum_{k=1}^{d} t_k$, and only the slopes t_k $(k = 1,\ldots,d_n)$ matter to the performance of order selection. For example, suppose that k_2 is selected instead of k_1 $(k_2 > k_1)$ by some criterion. This implies that the gain of prediction loss $\mathcal{L}_{k_1} - \mathcal{L}_{k_2}$ is greater than the sum of slopes $\sum_{k=k_1+1}^{k_2} t_k$. Thus, without loss of generality, we can shift the curves of the above three criteria to be tangent to the bent curve of the BC in order to highlight their differences and connections. Here, two curves are referred to as tangent to each other if one is above the other and they intersect at one point, the tangent point.

Given a sample size n, the tangent point between the $f_{n,d_m}(\text{BC})$ and $f_{n,d_m}(\text{BIC})$ curves is at $T_{\text{BC:BIC}} = c_n/\log n$. Consider the example $c_n = \lfloor n^{1/3} \rfloor$. If the true order d_0 is finite, $T_{\text{BC:BIC}}$ will be larger than d_0 for all sufficiently large n. In other words, there will be an infinitely large region as n tends to infinity, namely $1 \le k \le T_{\text{BC:BIC}}$, where d_0 falls

Table 12.1. Regression variable selection (standard errors are given in the parentheses)

		LOO	AIC	BC	BIC	CV
Case (1)	Efficiency	0.79 (0.03)	0.80 (0.03)	0.96 (0.02)	1.00 (0.00)	0.99 (0.01)
	Dimension	3.94 (0.19)	3.88 (0.18)	3.22 (0.11)	3.00 (0.00)	3.01 (0.01)
	PI			0.94 (0.02)		
Case (2)	Efficiency	0.91 (0.01)	0.92 (0.01)	0.92 (0.01)	0.74 (0.02)	0.65 (0.02)
	Dimension	8.99 (0.11)	9.11 (0.10)	9.11 (0.10)	7.53 (0.13)	6.66 (0.09)
	PI			0.33 (0.05)		

into and where the BC penalizes more than does the BIC. As a result, asymptotically the BC does not overfit. On the other hand, the BC will not underfit because the largest penalty preventing one from selecting dimension $k+1$ versus k is c_n/n, which will be less than any fixed positive constant (close to the KL divergence from a smaller model to the true model) with high probability for large n. This reasoning suggests that the BC is consistent.

Since the BC penalizes less for larger orders and finally becomes similar to the AIC, it is able to share the asymptotic optimality of the AIC under suitable conditions. A full illustration of why the BC is expected to work well in general can be found in [15]. As we shall see, the bent curve of the BC well connects the BIC and the AIC so that a good balance between the underfitting and overfitting risks is achieved.

Moreover, in many applications, the data analyst would like to quantify to what extent the framework under consideration can be virtually treated as parametric, or, in other words, how likely it is that the postulated model class is well specified. This motivated the concept of the "parametricness index" (PI) [15, 45] to evaluate the reliability of model selection. One definition of the PI, which we shall use in the following experiment, is the quantity

$$\mathrm{PI}_n = \frac{|d_{m_{\mathrm{BC}}} - d_{m_{\mathrm{AIC}}}|}{|d_{m_{\mathrm{BC}}} - d_{m_{\mathrm{AIC}}}| + |d_{m_{\mathrm{BC}}} - d_{m_{\mathrm{BIC}}}|}$$

on $[0,1]$ if the denominator is well defined, and $\mathrm{PI}_n = 0$ otherwise. Here, d_{m_δ} is the dimension of the model selected by the method δ. Under some conditions, it can be proved that $\mathrm{PI}_n \to_p 1$ in a parametric framework and $\mathrm{PI}_n \to_p 0$ otherwise.

In an experiment concerning Example 1, we generate data using each of those two true data-generating processes, with sample sizes $n = 500$. We replicate 100 independent trials and then summarize the performances of LOO, GCV, the AIC, the BC, the BIC, and delete-n_v CV (abbreviated as CV) with $n_v = n - n/\log n$ in Table 12.1. Note that CV with such a validation size is proved to be asymptotically close to the BIC, because delete-n_v CV has the same asymptotic behavior as GIC_{λ_n} introduced in Section 12.4 with

$$\lambda_n = n/(n - n_v) + 1 \tag{12.10}$$

in selecting regression models [8]. From Table 12.1, it can be seen that the methods on the right-hand side of the AIC perform better than the others in the parametric setting, while the methods on the left-hand side of the BIC perform better than the others in the non-parametric setting. The PI being close to one (zero) indicates the parametricness (non-parametricness). These are in accord with the existing theory.

12.6 Modeling-Procedure Selection

In this section, we introduce cross-validation as a general tool not only for model selection but also for modeling-procedure selection, and highlight the point that there is no one-size-fits-all data-splitting ratio of cross-validation. In particular, we clarify some widespread folklore on training/validation data size that may lead to improper data analysis.

Before we proceed, it is helpful to first clarify some commonly used terms involved in cross-validation (CV). Recall that a model-selection method would conceptually follow two phases: (A) the estimation and selection phase, where each candidate model is trained using *all* the available data and one of them is selected; and (B) the test phase, where the future data are predicted and the true out-sample performance is checked. For CV methods, the above phase (A) is further split into two phases, namely (A1), the training phase to build up each statistical/machine-learning model on the basis of the training data; and (A2), the validation phase which selects the best-performing model on the basis of the validation data. In the above step (B), analysts are ready to use/implement the obtained model for predicting the future (unseen) data. Nevertheless, in some data practice (e.g., in writing papers), analysts may wonder how that model is going to perform on completely unseen real-world data. In that scenario, part of the original dataset (referred to as the "test set") is taken out before phase (A), in order to approximate the true predictive performance of the finally selected model after phase (B). In a typical application that is based on the given data, the test set is not really available, and thus we need only consider the original dataset being split into two parts: (1) the training set and (2) the validation set.

12.6.1 Cross-Validation for Model Selection

There are widespread general recommendations on how to apply cross-validation (CV) for model selection. For instance, it is stated in the literature that 10-fold CV is the best for model selection. Such guidelines seem to be unwarranted, since they mistakenly disregard the goal of model selection. For prediction purpose, actually LOO may be preferred in tuning parameter selection for traditional non-parametric regression.

In fact, common practices that use 5-fold, 10-fold, or 30%-for-validation do not exhibit asymptotic optimality (neither consistency nor asymptotic efficiency) in simple regression models, and their performance can be very different depending on the goal of applying CV. Recall the equivalence between delete-n_v CV and GIC_{λ_n} in (12.10). It is also known that GIC_{λ_n} achieves asymptotic efficiency in non-parametric scenarios only with $\lambda_n = 2$, and asymptotic efficiency in parametric scenarios only with $\lambda_n \to \infty$ (as $n \to \infty$). In light of that, the optimal splitting ratio n_v/n_t of CV should either converge to zero or diverge to infinity, depending on whether the setting is non-parametric or not, in order to achieve asymptotic efficiency.

In an experiment, we showed how the splitting ratio could affect CV for model selection by using the MNIST database [6]. The MNIST database contains 70000 images, each being a handwritten digit (from 0 to 9) made of 28×28 pixels. Six feed-forward neural network models with different numbers of neurons or layers were considered as candidates. With different splitting ratios, we ran CV and computed the average validation loss of each candidate model based on 10 random partitions. We then selected the model with the smallest average validation loss, and recorded its true predictive performance using the separate test data. The performance in terms of the frequency of correct classifications indicates that a smaller splitting ratio n_v/n_t leads to better accuracy. The results are expected from the existing theory, since the neural network modeling here seems to be of a non-parametric regression nature.

12.6.2 Cross-Validation for Modeling-Procedure Selection

Our discussions in the previous sections have focused on *model selection* in the narrow sense where the candidates are parametric models. In this section, we review the use of CV as a general tool for *modeling-procedure selection*, which aims to select one from a finite set of modeling procedures [48]. For example, one may apply modeling procedures such as the AIC, the BIC, and CV for variable selection. Not knowing the true nature of the data-generating process, one may select one of those procedures (together with the model selected by the procedure), using an appropriately designed CV (which is at the second level). One may also choose among a number of machine-learning methods based on CV. A measure of accuracy is used to evaluate and compare the performances of the modeling procedures, and the "best" procedure is defined in the sense that it outperforms, with high probability, the other procedures in terms of out-sample prediction loss for sufficiently large n (see, for example, Definition 1 of [14]).

There are two main goals of modeling-procedure selection. The first is to *identify* with high probability the best procedure among the candidates. This selection consistency is similar to the selection consistency studied earlier. The second goal of modeling-procedure selection does not intend to pinpoint which candidate procedure is the best. Rather, it tries to achieve the best possible performance offered by the candidates. Note again that, if there are procedures that have similar best performances, we do not need to distinguish their minute differences (so as to single out the best candidate) for the purpose of achieving the asymptotically optimal performance. As in model selection, for the task of modeling-procedure selection, CV randomly splits n data into n_t training data and n_v validation data (so $n = n_t + n_v$). The first n_t data are used to run different modeling procedures, and the remaining n_v data are used to select the better procedure. Judging from the lessons learned from the CV paradox to be given below, for the first goal above, the evaluation portion of CV should be large enough. But, for the second goal, a smaller portion of the evaluation may be enough to achieve optimal predictive performance.

In the literature, much attention has been focused on choosing whether to use the AIC procedure or the BIC procedure for data analysis. For regression-variable selection, it has been proved that the CV method is consistent in choosing between the AIC and the BIC given $n_t \to \infty$, $n_v/n_t \to \infty$, and some other regularity assumptions (Theorem 1 of [48]). In other words, the probability of the BIC being selected goes to 1 in a parametric framework, and the probability of the AIC being selected goes to 1 otherwise. In this way, the modeling-procedure selection using CV naturally leads to a hybrid model-selection criterion that builds upon the strengths of the AIC and the BIC. Such hybrid selection is going to combine some of the theoretical advantages of both the AIC and the BIC, as the BC does. The task of classification is somewhat more relaxed than the task of regression. In order to achieve consistency in selecting the better classifier, the splitting ratio may be allowed to converge to infinity or any positive constant, depending on the situation [14]. In general, it is safe to let $n_t \to \infty$ and $n_v/n_t \to \infty$ for consistency in modeling-procedure selection.

12.6.3 The Cross-Validation Paradox

Closely related to the above discussion is the following paradox. Suppose that a data analyst has used part of the original data for training and the remaining part for validation. Now suppose that a new set of data is given to the analyst. The analyst would naturally add some of the new data in the training phase and some in the validation phase. Clearly, with more data added to the training set, each sensible candidate modeling procedure is improved in accuracy; with more data added to the validation set, the evaluation is also more reliable. It is tempting to think that improving the accuracy on both training and validation would lead to a sharper comparison between procedures. However, this is not the case. The prediction error estimation and procedure comparison are two different targets. In general, we have the following *cross-validation paradox*: "Better estimation of the prediction error by CV does not imply better modeling-procedure selection."

Technically speaking, suppose that an analyst originally splits the data with a large ratio of n_v/n_t; if a new set of data arrives, suppose that half of them were added to the training set and half to the validation set. Then the new ratio can decrease a lot, violating the theoretical requirement $n_v/n_t \to \infty$ (as mentioned above), and leading to worse procedure selection. Intuitively speaking, when one is comparing two procedures that are naturally close to each other, the improved estimation accuracy obtained by adopting more observations in the training part only makes it more difficult to distinguish between the procedures. If the validation size does not diverge fast enough, consistency in identifying the better procedure cannot be achieved.

Example 12.2 A set of real-world data reported by Scheetz *et al.* in [49] (and available from http://bioconductor.org) consists of over 31 000 gene probes represented on an Affymetrix expression microarray, from an experiment using 120 rats. Using domain-specific prescreening procedures as proposed in [49], 18 976 probes were selected that exhibit sufficient signals for reliable analysis. One purpose of the real data was to discover how genetic variation relates to human eye disease. Researchers are interested in finding the genes whose expression is correlated with that of the gene TRIM32, which has recently been found to cause human eye diseases. Its probe is *1389163_at*, one of the 18 976 probes. In other words, we have $n = 120$ data observations and 18975 variables that possibly relate to the observation. From domain knowledge, one expects that only a few genes are related to TRIM32.

In an experiment we demonstrate the cross-validation paradox using the above dataset. We first select the top 200 variables with the largest correlation coefficients as was done in [50], and standardize them. We then compare the following two procedures: the LASSO for the 200 candidate variables, and the LASSO for those 200 variables plus 100 randomly generated independent $\mathcal{N}(0,1)$ variables. We used LASSO here because we have more variables than observations. The first procedure is expected to be better, since the second procedure includes irrelevant variables. We repeat the procedure for 1000 independent replications, and record the frequency of the first procedure being favored (i.e., the first procedure has the smaller quadratic loss on randomly selected $n - n_{\mathrm{t}}$ validation data). We also repeat the experiment for $n = 80, 90, 100, 110$, by sampling from the original 120 data without replacement.

The results are shown in Table 12.2. As the paradox suggests, the accuracy of identifying the better procedure does not necessarily increase when more observations are added to both the estimation phase and the validation phase. With more observations added to both the estimation part and the validation part, the accuracy of identifying the better procedure actually decreases.

Modeling-procedure selection, differently from model selection that aims to select one model according to some principle, focuses more on the selection of a procedure. Such a procedure refers to the whole process of model estimation, model selection, and the approach used for model selection. Thus, the goal of model selection for inference, in its broad sense, includes not only the case of selection of the true/optimal model with confidence but also the case of selection of the optimal modeling procedure among those considered. One example is an emerging online competition platform such as Kaggle that compares new problem-solving techniques/procedures for real-world data analysis. In comparing procedures and awarding a prize, Kaggle usually uses cross-validation (holdout) or its counterpart for time series.

We also note that, in many real applications, the validation size n_{v} itself should be large enough for inference on the prediction accuracy. For example, in a two-label classification competition, the standard error of the estimated success rate (based on n_{v} validation data) is

$$\sqrt{p(1-p)/n_{\mathrm{v}}},$$

where p is the true out-sample success rate of a procedure. Suppose the top-ranked team achieves slightly above $p = 50\%$. Then, in order for the competition to declare it to be the true winner at 95% confidence level from the second-ranked team with only 1% observed difference in classification accuracy, the standard error $1/\sqrt{4n_{\mathrm{v}}}$ is roughly required to be smaller than 0.5%, which demands $n_{\mathrm{v}} \geq 10000$. Thus if the holdout sample size is not enough, the winning team may well be just the lucky one among the

Table 12.2. The cross-validation paradox: more observations do not lead to higher accuracy in selecting the better procedure

Sample size n	80	90	100	110	120
Training size n_t	25	30	35	40	45
Accuracy	70.0%	69.4%	62.7%	63.2%	59.0%

top-ranking teams. It is worth pointing out that the discussion above is based on summary accuracy measures (e.g., classification accuracy on the holdout data), and other hypothesis tests can be employed for formal comparisons of the competing teams (e.g., tests based on differencing of the prediction errors).

To summarize this section, we introduced the use of cross-validation as a general tool for modeling-procedure selection, and we discussed the issue of choosing the test data size so that such selection is indeed reliable for the purpose of consistency. For *modeling-procedure selection*, it is often safe to let the validation size take a large proportion (e.g., half) of the data in order to achieve good selection behavior. In particular, the use of LOO for the goal of comparing procedures is the least trustworthy method. The popular 10-fold CV may leave too few observations in evaluation to be stable. Indeed, the 5-fold CV often produces more stable selection results for high-dimensional regression. Moreover, v-fold CV, regardless of v, in general, is often unstable, and a repeated v-fold approach usually improves performance. A quantitative relation between model stability and selection consistency remains to be established by research. We also introduced the paradox that using more training data and validation data does not necessarily lead to better modeling-procedure selection. That further indicates the importance of choosing an appropriate splitting ratio.

12.7 Conclusion

There has been a debate regarding whether to use the AIC or the BIC in the past decades, centering on whether the true data-generating model is parametric or not with respect to the specified model class, which in turn affects the achievability of fundamental limits of learning the optimal model given a dataset and a model class at hand. Compared with the BIC, the AIC seems to be more widely used in practice, perhaps mainly due to the thought that "all models are wrong" and the minimax-rate optimality of the AIC offers more protection than does the BIC. Nevertheless, the parametric setting is still of vital importance. One reason for this is that being consistent in selecting the true model if it is really among the candidates is certainly mathematically appealing. Also, a non-parametric scenario can be a virtually parametric scenario, where the optimal model (even if it is not the true model) is stable for the current sample size. The war between the AIC and the BIC originates from two fundamentally different goals: one is to minimize the certain loss for prediction purposes, and the other is to select the optimal model for inference purposes. A unified perspective on integrating their fundamental limits is a central issue in model selection, which remains an active line of research.

References

[1] S. Greenland, "Modeling and variable selection in epidemiologic analysis," *Am. J. Public Health*, vol. 79, no. 3, pp. 340–349, 1989.

[2] C. M. Andersen and R. Bro, "Variable selection in regression – a tutorial," *J. Chemometrics*, vol. 24, nos. 11–12, pp. 728–737, 2010.

[3] J. B. Johnson and K. S. Omland, "Model selection in ecology and evolution," *Trends Ecology Evolution*, vol. 19, no. 2, pp. 101–108, 2004.

[4] P. Stoica and Y. Selen, "Model-order selection: A review of information criterion rules," *IEEE Signal Processing Mag.*, vol. 21, no. 4, pp. 36–47, 2004.

[5] J. B. Kadane and N. A. Lazar, "Methods and criteria for model selection," *J. Amer. Statist. Assoc.*, vol. 99, no. 465, pp. 279–290, 2004.

[6] J. Ding, V. Tarokh, and Y. Yang, "Model selection techniques: An overview," *IEEE Signal Processing Mag.*, vol. 35, no. 6, pp. 16–34, 2018.

[7] R. Tibshirani, "Regression shrinkage and selection via the LASSO," *J. Roy. Statist. Soc. Ser. B*, vol. 58, no. 1, pp. 267–288, 1996.

[8] J. Shao, "An asymptotic theory for linear model selection," *Statist. Sinica*, vol. 7, no. 2, pp. 221–242, 1997.

[9] C. R. Rao, "Information and the accuracy attainable in the estimation of statistical parameters," in *Breakthroughs in statistics*. Springer, 1992, pp. 235–247.

[10] J. Ding, V. Tarokh, and Y. Yang, "Optimal variable selection in regression models," http://jding.org/jie-uploads/2017/11/regression.pdf, 2016.

[11] Y. Nan and Y. Yang, "Variable selection diagnostics measures for high-dimensional regression," *J. Comput. Graphical Statist.*, vol. 23, no. 3, pp. 636–656, 2014.

[12] J. P. Ioannidis, "Why most published research findings are false," *PLoS Medicine*, vol. 2, no. 8, p. e124, 2005.

[13] R. Shibata, "Asymptotically efficient selection of the order of the model for estimating parameters of a linear process," *Annals Statist.*, vol. 8, no. 1, pp. 147–164, 1980.

[14] Y. Yang, "Comparing learning methods for classification," *Statist. Sinica*, vol. 16, no. 2, pp. 635–657, 2006.

[15] J. Ding, V. Tarokh, and Y. Yang, "Bridging AIC and BIC: A new criterion for autoregression," *IEEE Trans. Information Theory*, vol. 64, no. 6, pp. 4024–4043, 2018.

[16] H. Akaike, "A new look at the statistical model identification," *IEEE Trans. Automation Control*, vol. 19, no. 6, pp. 716–723, 1974.

[17] H. Akaike, "Information theory and an extension of the maximum likelihood principle," in *Selected papers of Hirotugu Akaike*. Springer, 1998, pp. 199–213.

[18] G. Schwarz, "Estimating the dimension of a model," *Annals Statist.*, vol. 6, no. 2, pp. 461–464, 1978.

[19] A. Gelman, H. S. Stern, J. B. Carlin, D. B. Dunson, A. Vehtari, and D. B. Rubin, *Bayesian data analysis*. Chapman and Hall/CRC, 2013.

[20] A. W. Van der Vaart, *Asymptotic statistics*. Cambridge University Press, 1998, vol. 3.

[21] J. S. Liu, *Monte Carlo strategies in scientific computing*. Springer Science & Business Media, 2008.

[22] D. M. Allen, "The relationship between variable selection and data agumentation and a method for prediction," *Technometrics*, vol. 16, no. 1, pp. 125–127, 1974.

[23] S. Geisser, "The predictive sample reuse method with applications," *J. Amer. Statist. Assoc.*, vol. 70, no. 350, pp. 320–328, 1975.

[24] P. Burman, "A comparative study of ordinary cross-validation, v-fold cross-validation and the repeated learning–testing methods," *Biometrika*, vol. 76, no. 3, pp. 503–514, 1989.

[25] J. Shao, "Linear model selection by cross-validation," *J. Amer. Statist. Assoc.*, vol. 88, no. 422, pp. 486–494, 1993.

[26] P. Zhang, "Model selection via multifold cross validation," *Annals Statist.*, vol. 21, no. 1, pp. 299–313, 1993.

[27] M. Stone, "An asymptotic equivalence of choice of model by cross-validation and Akaike's criterion," *J. Roy. Statist. Soc. Ser. B*, pp. 44–47, 1977.

[28] C. M. Hurvich and C.-L. Tsai, "Regression and time series model selection in small samples," *Biometrika*, vol. 76, no. 2, pp. 297–307, 1989.

[29] P. Craven and G. Wahba, "Smoothing noisy data with spline functions," *Numerische Mathematik*, vol. 31, no. 4, pp. 377–403, 1978.

[30] H. Akaike, "Fitting autoregressive models for prediction," *Ann. Inst. Statist. Math.*, vol. 21, no. 1, pp. 243–247, 1969.

[31] H. Akaike, "Statistical predictor identification," *Ann. Inst. Statist. Math.*, vol. 22, no. 1, pp. 203–217, 1970.

[32] C. L. Mallows, "Some comments on C_P," *Technometrics*, vol. 15, no. 4, pp. 661–675, 1973.

[33] E. J. Hannan and B. G. Quinn, "The determination of the order of an autoregression," *J. Roy. Statist. Soc. Ser. B*, vol. 41, no. 2, pp. 190–195, 1979.

[34] J. Rissanen, "Estimation of structure by minimum description length," *Circuits, Systems Signal Processing*, vol. 1, no. 3, pp. 395–406, 1982.

[35] C.-Z. Wei, "On predictive least squares principles," *Annals Statist.*, vol. 20, no. 1, pp. 1–42, 1992.

[36] R. Nishii *et al.*, "Asymptotic properties of criteria for selection of variables in multiple regression," *Annals Statist.*, vol. 12, no. 2, pp. 758–765, 1984.

[37] R. Rao and Y. Wu, "A strongly consistent procedure for model selection in a regression problem," *Biometrika*, vol. 76, no. 2, pp. 369–374, 1989.

[38] A. Barron, L. Birgé, and P. Massart, "Risk bounds for model selection via penalization," *Probability Theory Related Fields*, vol. 113, no. 3, pp. 301–413, 1999.

[39] R. Shibata, "Selection of the order of an autoregressive model by Akaike's information criterion," *Biometrika*, vol. 63, no. 1, pp. 117–126, 1976.

[40] R. Shibata, "An optimal selection of regression variables," *Biometrika*, vol. 68, no. 1, pp. 45–54, 1981.

[41] Y. Yang, "Can the strengths of AIC and BIC be shared? A conflict between model indentification and regression estimation," *Biometrika*, vol. 92, no. 4, pp. 937–950, 2005.

[42] S. S. Wilks, "The large-sample distribution of the likelihood ratio for testing composite hypotheses," *Annals Math. Statist.*, vol. 9, no. 1, pp. 60–62, 1938.

[43] C.-K. Ing, "Accumulated prediction errors, information criteria and optimal forecasting for autoregressive time series," *Annals Statist.*, vol. 35, no. 3, pp. 1238–1277, 2007.

[44] Y. Yang, "Prediction/estimation with simple linear models: Is it really that simple?" *Economic Theory*, vol. 23, no. 1, pp. 1–36, 2007.

[45] W. Liu and Y. Yang, "Parametric or nonparametric? A parametricness index for model selection," *Annals Statist.*, vol. 39, no. 4, pp. 2074–2102, 2011.

[46] T. van Erven, P. Grünwald, and S. De Rooij, "Catching up faster by switching sooner: A predictive approach to adaptive estimation with an application to the AIC–BIC dilemma," *J. Roy. Statist. Soc. Ser. B.*, vol. 74, no. 3, pp. 361–417, 2012.

[47] G. E. Box, G. M. Jenkins, and G. C. Reinsel, *Time series analysis: Forecasting and control*. John Wiley & Sons, 2011.

[48] Y. Zhang and Y. Yang, "Cross-validation for selecting a model selection procedure," *J. Econometrics*, vol. 187, no. 1, pp. 95–112, 2015.

[49] T. E. Scheetz, K.-Y. A. Kim, R. E. Swiderski, A. R. Philp, T. A. Braun, K. L. Knudtson, A. M. Dorrance, G. F. DiBona, J. Huang, T. L. Casavant, V. C. Sheffield, and E. M. Stone, "Regulation of gene expression in the mammalian eye and its relevance to eye disease," *Proc. Natl. Acad. Sci. USA*, vol. 103, no. 39, pp. 14 429–14 434, 2006.

[50] J. Huang, S. Ma, and C.-H. Zhang, "Adaptive lasso for sparse high-dimensional regression models," *Statist. Sinica*, pp. 1603–1618, 2008.

13 Statistical Problems with Planted Structures: Information-Theoretical and Computational Limits

Yihong Wu and Jiaming Xu

Summary

This chapter provides a survey of the common techniques for determining the sharp statistical and computational limits in high-dimensional statistical problems with planted structures, using community detection and submatrix detection problems as illustrative examples. We discuss tools including the first- and second-moment methods for analyzing the maximum-likelihood estimator, information-theoretic methods for proving impossibility results using mutual information and rate-distortion theory, and methods originating from statistical physics such as the interpolation method. To investigate computational limits, we describe a common recipe to construct a randomized polynomial-time reduction scheme that approximately maps instances of the planted-clique problem to the problem of interest in total variation distance.

13.1 Introduction

The interplay between information theory and statistics is a constant theme in the development of both fields. Since its inception, information theory has been indispensable for understanding the fundamental limits of statistical inference. The classical *information bound* provides fundamental lower bounds for the estimation error, including Cramér–Rao and Hammersley–Chapman–Robbins lower bounds in terms of Fisher information and χ^2-divergence [1, 2]. In the classical "large-sample" regime in parametric statistics, Fisher information also governs the sharp minimax risk in regular statistical models [3]. The prominent role of information-theoretic quantities such as mutual information, metric entropy, and capacity in establishing the minimax rates of estimation has long been recognized since the seminal work of [4–7], etc.

Instead of focusing on the large-sample asymptotics, the attention of contemporary statistics has shifted toward *high dimensions*, where the problem size and the sample size grow simultaneously and the main objective is to obtain a tight characterization of the optimal statistical risk. Certain information-theoretic methods have been remarkably successful for high-dimensional problems. Such methods include those based on metric entropy and Fano's inequality for determining the minimax risk within universal constant factors (minimax rates) [7]. Unfortunately, the aforementioned methods are often too crude for the task of determining the *sharp constant*, which requires more refined analysis and stronger information-theoretic tools.

An additional challenge in dealing with high dimensionality is the need to address the computational aspect of statistical inference. An important element absent from the classical statistical paradigm is the computational complexity of inference procedures, which is becoming increasingly relevant for data scientists dealing with large-scale noisy datasets. Indeed, recent results [8–13] revealed the surprising phenomenon that certain problems concerning large networks and matrices undergo an "easy–hard–impossible" phase transition, and computational constraints can severely penalize the statistical performance. It is worth pointing out that here the notion of complexity differs from the worst-case computational hardness studied in the computer science literature which focused on the time and space complexity of various worst-case problems. In contrast, in a statistical context, the problem is of a stochastic nature and the existing theory on average-case hardness is significantly underdeveloped. Here, the hardness of a statistical problem is often established either within the framework of certain computation models, such as the sums-of-squares relaxation hierarchy, or by means of a reduction argument from another problem, notably the planted-clique problem, which is conjectured to be computationally intractable.

In this chapter, we provide an exposition on some of the methods for determining the information-theoretic as well as the computational limits for high-dimensional statistical problems with a planted structure, with a specific focus on characterizing sharp thresholds. Here the planted structure refers to the true parameter, which is often of a combinatorial nature (e.g., partition) and hidden in the presence of random noise. To characterize the information-theoretic limit, we will discuss tools including the first- and second-moment methods for analyzing the maximum-likelihood estimator, information-theoretic methods for proving impossibility results using mutual information and rate-distortion theory, and methods originating from statistical physics such as the interpolation method. There is no established recipe for determining the computational limit of statistical problems, especially the "easy–hard–impossible" phase transition, and it is usually done on a case-by-case basis; nevertheless, the common element is to construct a randomized polynomial-time reduction scheme that *approximately* maps instances of a given hard problem to one that is close to the problem of interest in total variation distance.

13.2 Basic Setup

To be concrete, in this chapter we consider two representative problems, namely *community detection* and *submatrix detection*, as running examples. Both problems can be cast as the Bernoulli and Gaussian version of the following statistical model with planted community structure.

We first consider a random graph model containing a single hidden community whose size can be sub-linear in data matrix size n.

DEFINITION 13.1 (Single-community model) *Let C^* be drawn uniformly at random from all subsets of $[n]$ of cardinality K. Given probability measures P and Q on a common measurable space, let A be an $n \times n$ symmetric matrix with zero diagonal where, for all $1 \le i < j \le n$, A_{ij} are mutually independent, and $A_{ij} \sim P$ if $i, j \in C^*$ and $A_{ij} \sim Q$ otherwise.*

Here we assume that we have access only to pair-wise information A_{ij} for distinct indices i and j whose distribution is either P or Q depending on the community membership; no direct observation about the individual indices is available (hence the zero diagonal of A). Two choices of P and Q arising in many applications are the following:

- Bernoulli case: $P = \text{Bern}(p)$ and $Q = \text{Bern}(q)$ with $p \neq q$. When $p > q$, this coincides with the *planted dense subgraph model* studied in [10, 14–17], which is also a special case of the general stochastic block model (SBM) [18] with a single community. In this case, the data matrix A corresponds to the adjacency matrix of a graph, where two vertices are connected with probability p if both belong to the community C^*, and with probability q otherwise. Since $p > q$, the subgraph induced by C^* is likely to be denser than the rest of the graph.
- Gaussian case: $P = \mathcal{N}(\mu, 1)$ and $Q = \mathcal{N}(0, 1)$ with $\mu \neq 0$. This corresponds to a symmetric version of the *submatrix detection* problem studied in [9, 16, 19–23]. When $\mu > 0$, the entries of A with row and column indices in C^* have positive mean μ except those on the diagonal, while the rest of the entries have zero mean.

We will also consider a binary symmetric community model with two communities of equal sizes. The Bernoulli case is known as the binary symmetric stochastic block model (SBM).

DEFINITION 13.2 (Binary symmetric community model) *Let (C_1^*, C_2^*) be two communities of equal size that are drawn uniformly at random from all equal-sized partitions of $[n]$. Let A be an $n \times n$ symmetric matrix with empty diagonal where, for all $1 \le i < j \le n$, A_{ij} are mutually independent, and $A_{ij} \sim P$ if i, j are from the same community and $A_{ij} \sim Q$ otherwise.*

Given the data matrix A, the problem of interest is to accurately recover the underlying single community C^* or community partition (C_1^*, C_2^*) up to a permutation of cluster indices. The distributions P and Q as well as the community size K depend on the matrix size n in general. For simplicity we assume that these model parameters are known to the estimator. Common objectives of recovery include the following.

- **Detection**: detect the presence of planted communities versus the absence. This is a hypothesis problem: in the null case the observation consists purely of noise with independently and identically distributed (i.i.d.) entries, while in the alternative case the distribution of the entries is dependent on the hidden communities according to Definition 13.1 or 13.2.
- **Correlated recovery**: recover the hidden communities better than would be achieved by random guessing. For example, for the binary symmetric SBM, the goal is to achieve a misclassification rate strictly less than $1/2$.

- **Almost exact recovery**: the expected number of misclassified vertices is sub-linear in the hidden community sizes.
- **Exact recovery**: all vertices are classified correctly with probability converging to 1 as the dimension $n \to \infty$.

13.3 Information-Theoretic Limits

13.3.1 Detection and Correlated Recovery

In this section, we study detection and correlated recovery under the binary symmetric community model. The community structure under the binary symmetric community model can be represented by a vector $\sigma \in \{\pm 1\}^n$ such that $\sigma_i = 1$ if vertex i is in the first community and $\sigma_i = -1$ otherwise. Let σ^* denote the true community partition and $\widehat{\sigma} \in \{\pm 1\}^n$ denote an estimator of σ. For detection, we assume under the null model, $A_{ii} = 0$ for all $1 \le i \le n$ and $A_{ij} = A_{ij}$ are i.i.d. as $\frac{1}{2}(P+Q)$ for $1 \le i < j \le n$, so that $\mathbb{E}[A]$ is matched between the planted model and the null model.

DEFINITION 13.3 (Detection) *Let $\mathcal{P}$ be the distribution of A in the planted model, and denote by $\mathcal{Q}$ the distribution of A in the null model. A test statistic $\mathcal{T}(A)$ with a threshold τ achieves detection if*[1]

$$\limsup_{n\to\infty}[\mathcal{P}(\mathcal{T}(A) < \tau) + \mathcal{Q}(\mathcal{T}(A) \ge \tau)] = 0,$$

so that the criterion $\mathcal{T}(A) \ge \tau$ determines with high probability whether A is drawn from $\mathcal{P}$ or $\mathcal{Q}$.

DEFINITION 13.4 (Correlated recovery) *The estimator $\widehat{\sigma}$ achieves correlated recovery of σ^* if there exists a fixed constant $\epsilon > 0$ such that $\mathbb{E}[|\langle \sigma, \sigma^* \rangle|] \ge \epsilon n$ for all n.*

The detection problem can be understood as a binary hypothesis-testing problem. Given a test statistic $\mathcal{T}(A)$, we consider its distribution under the planted and null models. If these two distributions are asymptotically disjoint, i.e., their total variation distance tends to 1 in the limit of large datasets, then it is information-theoretically possible to distinguish the two models with high probability by measuring $\mathcal{T}(A)$. A classic choice of statistic for binary hypothesis testing is the likelihood ratio,

$$\frac{\mathcal{P}(A)}{\mathcal{Q}(A)} = \frac{\sum_\sigma \mathcal{P}(A,\sigma)}{\mathcal{Q}(A)} = \frac{\sum_\sigma \mathcal{P}(A|\sigma)\mathcal{P}(\sigma)}{\mathcal{Q}(A)}.$$

This object will figure heavily both in our upper bounds and in our lower bounds of the detection threshold.

Before presenting our proof techniques, we first give the sharp threshold for detection and correlated recovery under the binary symmetric community model.

[1] This criterion is also known as strong detection, in contrast to weak detection which requires only that $\mathcal{P}(\mathcal{T}(A) < \tau) + \mathcal{Q}(\mathcal{T}(A) \ge \tau)$ be bounded away from 1 as $n \to \infty$. Here we focus exclusively on strong detection. See [24, 25] for detailed discussions on weak detection.

THEOREM 13.1 *Consider the binary symmetric community model.*

- *If* $P = \text{Bern}(a/n)$ *and* $Q = \text{Bern}(b/n)$ *for fixed constants* a, b, *then both detection and correlated recovery are information-theoretically possible when* $(a-b)^2 > 2(a+b)$ *and impossible when* $(a-b)^2 < 2(a+b)$.
- *If* $P = \mathcal{N}(\mu/\sqrt{n}, 1)$ *and* $Q = \mathcal{N}(-\mu/\sqrt{n}, 1)$, *then both detection and correlated recovery are information-theoretically possible when* $\mu > 1$ *and impossible when* $\mu < 1$.

We will explain how to prove the converse part of Theorem 13.1 using second-moment analysis of the likelihood ratio $\mathcal{P}(A)/\mathcal{Q}(A)$ and mutual information arguments. For the positive part of Theorem 13.1, we will present a simple first-moment method to derive upper bounds that often concide with the sharp thresholds up to a multiplicative constant.

To achieve the sharp detection upper bound for the SBM, one can use the count of short cycles as test statistics as in [26]. To achieve the sharp detection threshold in the Gaussian model and the correlated recovery threshold in both models, one can resort to spectral methods. For the Gaussian case, this directly follows from a celebrated phase-transition result on the rank-one perturbation of Wigner matrices [27–29]. For the SBM, naive spectral methods fail due to the existence of high-degree vertices [30]. More sophisticated spectral methods based on self-avoiding walks or non-backtracking walks have been shown to achieve the sharp correlated recovery threshold efficiently [26, 31, 32].

First-Moment Method for Detection and Correlated Recovery Upper Bound

Our upper bounds do not use the likelihood ratio directly, since it is hard to furnish lower bounds on the typical value of $\mathcal{P}(A)/\mathcal{Q}(A)$ when A is drawn from $\mathcal{P}$. Instead, we use the generalized likelihood ratio

$$\max_{\sigma} \frac{\mathcal{P}(A|\sigma)}{\mathcal{Q}(A)}$$

as the test statistic. In the planted model where the underlying true community is σ^*, this quantity is trivially bounded below by $\mathcal{P}(A|\sigma^*)/\mathcal{Q}(A)$. Then using a simple first-moment argument (union bound) one can show that, in the null model $\mathcal{Q}$, with high probability, this lower bound is not achieved by any σ and hence the generalized likelihood ratio test succeeds. An easy extension of this argument shows that, in the planted model, the maximum-likelihood estimator (MLE)

$$\widehat{\sigma}_{\text{ML}} = \text{argmax}_{\sigma} \mathcal{P}(A|\sigma)$$

has non-zero correlation with σ^*, achieving the correlated recovery.

Note that the first-moment analysis of the MLE often falls short of proving the sharp detection and correlated recovery upper bound. For instance, as we will explain next, the first-moment calculation of Theorem 2 in [33] shows only that the MLE achieves detection and correlated recovery when $\mu > 2\sqrt{\log 2}$ in the Gaussian model,[2] which is

[2] Throughout this chapter, logarithms are with respect to the natural base.

sub-optimal in view of Theorem 13.1. One reason is that the naive union bound in the first-moment analysis may not be tight; it does not take into the account the correlation between $P(A|\sigma)$ and $P(A|\sigma')$ for two different σ, σ' under the null model.

Next we explain how to carry out the first-moment analysis in the Gaussian case with $P = \mathcal{N}(\mu/\sqrt{n}, 1)$ and $Q = \mathcal{N}(-\mu/\sqrt{n}, 1)$. Specifically, assume $A = (\mu/\sqrt{n})\big(\sigma^*(\sigma^*)^{\mathrm{T}} - \mathbf{I}\big) + W$, where W is a symmetric Gaussian random variable with zero diagonal and $W_{ij} \overset{\text{i.i.d.}}{\sim} \mathcal{N}(0,1)$ for $i < j$. It follows that $\log(\mathcal{P}(A|\sigma)/\mathcal{Q}(A)) = (\mu/\sqrt{n})\sum_{i<j} A_{ij}\sigma_i\sigma_j + \mu^2(n-1)/4$. Therefore, the generalized likelihood test reduces to the test statistic $\max_\sigma \mathcal{T}(\sigma) \triangleq \sum_{i<j} A_{i,j}\sigma_i\sigma_j$. Under the null model $\mathcal{Q}$, $\mathcal{T}(\sigma) \sim \mathcal{N}(0, n(n-1)/2)$. Under the planted model $\mathcal{P}$, $\mathcal{T}(\sigma) = \mu/\sqrt{n}\sum_{i<j}\sigma_i^*\sigma_j^*\sigma_i\sigma_j + \sum_{i<j} W_{i,j}\sigma_i\sigma_j$. Hence the distribution of $\mathcal{T}(\sigma)$ depends on the overlap $|\langle\sigma,\sigma^*\rangle|$ between σ and the planted partition σ^*. Suppose $|\langle\sigma,\sigma^*\rangle| = n\omega$. Then

$$\mathcal{T}(\sigma) \sim \mathcal{N}\left(\frac{\mu n(n\omega^2 - 1)}{2\sqrt{n}}, \frac{n(n-1)}{2}\right).$$

To prove that detection is possible, notice that, in the planted model, $\max_\sigma \mathcal{T}(\sigma) \geq \mathcal{T}(\sigma^*)$. Setting $\omega = 1$, Gaussian tail bounds yield that

$$\mathcal{P}\left[\mathcal{T}(\sigma^*) \leq \frac{\mu n(n-1)}{2\sqrt{n}} - n\sqrt{\log n}\right] \leq n^{-1}.$$

Under the null model, taking the union bound over at most 2^n ways to choose σ, we can bound the probability that *any* partition is as good, according to $\mathcal{T}$, as the planted one, by

$$\mathcal{Q}\left[\max_\sigma \mathcal{T}(\sigma) > \frac{\mu n(n-1)}{2\sqrt{n}} - n\sqrt{\log n}\right] \leq 2^n \exp\left(-n\left(\frac{\mu}{2}\sqrt{\frac{n-1}{n}} - \sqrt{\frac{\log n}{n-1}}\right)^2\right).$$

Thus the probability of this event is $e^{-\Omega(n)}$ whenever $\mu > 2\sqrt{\log 2}$, meaning that above this threshold we can distinguish the null and planted models with the generalized likelihood test.

To prove that correlated recovery is possible, since $\mu > 2\sqrt{\log 2}$, there exists a fixed $\epsilon > 0$ such that $\mu(1-\epsilon^2) > 2\sqrt{\log 2}$. Taking the union bound over every σ with $|\langle\sigma,\sigma^*\rangle| \leq n\epsilon$ gives

$$\mathcal{P}\left[\max_{|\langle\sigma,\sigma^*\rangle| \leq n\epsilon} \mathcal{T}(\sigma) \geq \frac{\mu n(n-1)}{2\sqrt{n}} - n\sqrt{\log n}\right]$$
$$\leq 2^n \exp\left(-n\left(\frac{\mu(1-\epsilon^2)}{2}\sqrt{\frac{n}{n-1}} - \sqrt{\frac{\log n}{n-1}}\right)^2\right).$$

Hence, with probability at least $1 - e^{-\Omega(n)}$,

$$\max_{|\langle\sigma,\sigma^*\rangle| \leq n\epsilon} \mathcal{T}(\sigma) < \frac{n(n-1)\mu}{2\sqrt{n}} - n\sqrt{\log n},$$

and consequently $|\langle\widehat{\sigma}_{\mathrm{ML}}, \sigma^*\rangle| \geq n\epsilon$ with high probability. Thus, $\widehat{\sigma}_{\mathrm{ML}}$ achieves correlated recovery.

Second-Moment Method for Detection Lower Bound

Intuitively, if the planted model $\mathcal{P}$ and the null model $\mathcal{Q}$ are close to being mutually singular, then the likelihood ratio $\mathcal{P}/\mathcal{Q}$ is almost always either very large or close to zero. In particular, its variance under $\mathcal{Q}$, that is, the χ^2-divergence

$$\chi^2(\mathcal{P}\|\mathcal{Q}) \triangleq \mathbb{E}_{A\sim\mathcal{Q}}\left[\left(\frac{\mathcal{P}(A)}{\mathcal{Q}(A)} - 1\right)^2\right],$$

must diverge. This suggests that we can derive lower bounds on the detection threshold by bounding the second moment of $\mathcal{P}(A)/\mathcal{Q}(A)$ under $\mathcal{Q}$, or equivalently its expectation under $\mathcal{P}$. Suppose the second moment is bounded by some constant C, i.e.,

$$\chi^2(\mathcal{P}\|\mathcal{Q}) + 1 = \mathbb{E}_{A\sim\mathcal{Q}}\left[\left(\frac{\mathcal{P}(A)}{\mathcal{Q}(A)}\right)^2\right] = \mathbb{E}_{A\sim\mathcal{P}}\left[\frac{\mathcal{P}(A)}{\mathcal{Q}(A)}\right] \le C. \tag{13.1}$$

A bounded second moment readily implies a bounded Kullback–Leibler divergence between $\mathcal{P}$ and $\mathcal{Q}$, since Jensen's inequality gives

$$D(\mathcal{P}\|\mathcal{Q}) = \mathbb{E}_{A\sim\mathcal{P}}\log\left(\frac{\mathcal{P}(A)}{\mathcal{Q}(A)}\right) \le \log\left(\mathbb{E}_{A\sim\mathcal{P}}\frac{\mathcal{P}(A)}{\mathcal{Q}(A)}\right) \le \log C = O(1). \tag{13.2}$$

Moreover, it also implies non-detectability. To see this, let $E = E_n$ be a sequence of events such that $\mathcal{Q}(E) \to 0$ as $n \to \infty$, and let $\mathbf{1}_E$ denote the indicator random variable for E. Then the Cauchy–Schwarz inequality gives

$$\mathcal{P}(E) = \mathbb{E}_{A\sim\mathcal{Q}}\frac{\mathcal{P}(A)}{\mathcal{Q}(A)}\mathbf{1}_E \le \sqrt{\mathbb{E}_{A\sim\mathcal{Q}}\left(\frac{\mathcal{P}(A)}{\mathcal{Q}(A)}\right)^2 \times \mathbb{E}_{A\sim\mathcal{Q}}\mathbf{1}_E^2} \le \sqrt{C\mathcal{Q}(E)} \to 0. \tag{13.3}$$

In other words, the sequence of distributions $\mathcal{P}$ is *contiguous* to $\mathcal{Q}$ [4]. Therefore, no algorithm can return "yes" with high probability (or even positive probability) in the planted model, and "no" with high probability in the null model. Hence, detection is impossible.

Next we explain how to compute the χ^2-divergence for the binary symmetric SBM. One useful observation due to [34] (see also Lemma 21.1 of [35]) is that, using Fubini's theorem, the χ^2-divergence between a mixture distribution and a simple distribution can be written as

$$\chi^2(\mathcal{P}\|\mathcal{Q}) + 1 = \mathbb{E}_{\sigma,\widetilde{\sigma}}\left[\mathbb{E}_{A\sim\mathcal{Q}}\left[\frac{\mathcal{P}(A|\sigma)\mathcal{P}(A|\widetilde{\sigma})}{\mathcal{Q}(A)}\right]\right],$$

where $\widetilde{\sigma}$ is an independent copy of σ. Note that under the planted model $\mathcal{P}$, the distribution of the ijth entry is given by $P\mathbf{1}_{\{\sigma_i=\sigma_j\}} + Q\mathbf{1}_{\{\sigma_i\neq\sigma_j\}} = (P+Q)/2 + ((P-Q)/2)\sigma_i\sigma_j$. Thus[3]

$$\chi^2(\mathcal{P}\|\mathcal{Q}) + 1 = \mathbb{E}\left[\prod_{i<j}\int \frac{((P+Q)/2 + ((P-Q)/2)\sigma_i\sigma_j)((P+Q)/2 + ((P-Q)/2)\widetilde{\sigma}_i\widetilde{\sigma}_j)}{(P+Q)/2}\right]$$

[3] In fact, the quantity $\rho = \int (P-Q)^2/(2(P+Q))$ is an f-divergence known as the Vincze–Le Cam distance [4, 36].

$$= \mathbb{E}\left[\prod_{i<j}\left(1 + \sigma_i\sigma_j\widetilde{\sigma}_i\widetilde{\sigma}_j \underbrace{\int \frac{(P-Q)^2}{2(P+Q)}}_{\triangleq \rho}\right)\right] \tag{13.4}$$

$$\le \mathbb{E}\left[\exp\left(\rho \sum_{i<j} \sigma_i\widetilde{\sigma}_i\sigma_j\widetilde{\sigma}_j\right)\right]. \tag{13.5}$$

For the Bernoulli setting where $P = \text{Bern}(a/n)$ and $Q = \text{Bern}(b/n)$ for fixed constants a, b, we have $\rho \triangleq \tau/n + O(1/n^2)$, where $\tau \triangleq (a-b)^2/(2(a+b))$. Thus,

$$\chi^2(\mathcal{P}\|\mathcal{Q}) + 1 \le \mathbb{E}\left[\exp\left(\frac{\tau}{2n}\langle\sigma,\widetilde{\sigma}\rangle^2 + O(1)\right)\right].$$

We then write $\sigma = 2\xi - 1$, where $\xi \in \{0,1\}^n$ is the indicator vector for the first community which is drawn uniformly at random from all binary vectors with Hamming weight $n/2$, and $\widetilde{\xi}$ is its independent copy. Then $\langle\sigma,\widetilde{\sigma}\rangle = 4\langle\xi,\widetilde{\xi}\rangle - n$, where $H \triangleq \langle\xi,\widetilde{\xi}\rangle \sim$ Hypergeometric$(n, n/2, n/2)$. Thus

$$\chi^2(\mathcal{P}\|\mathcal{Q}) + 1 \le \mathbb{E}\left[\exp\left(\frac{\tau}{2}\left(\frac{4H-n}{\sqrt{n}}\right)^2 + O(1)\right)\right].$$

Since $(1/\sqrt{n/16})(H - n/4) \to \mathcal{N}(0,1)$ as $n \to \infty$ by the central limit theorem for hypergeometric distributions (see, e.g., p. 194 of [37]), using Theorem 1 of [38] for the convergence of the moment-generating function, we conclude that $\chi^2(\mathcal{P}\|\mathcal{Q})$ is bounded if $\tau < 1$.

Mutual Information-Based Lower Bound for Correlated Recovery

It is tempting to conclude that whenever detection is impossible – that is, whenever we cannot correctly tell with high probability whether the observation was generated from the null or planted model – we cannot infer the planted community structure σ^* better than chance either; this deduction, however, is not true in general (see Section III.D of [33] for a simple counterexample). Instead, we resort to mutual information in proving lower bounds for correlated recovery. In fact, there are two types of mutual information that are relevant in the context of correlated recovery.

Pair-wise mutual information $I(\sigma_1, \sigma_2; A)$. For two communities, it is easy to show that correlated recovery is impossible if and only if

$$I(\sigma_1, \sigma_2; A) = o(1) \tag{13.6}$$

as $n \to \infty$. This in fact also holds for k communities for any constant k. See Appendix A13.1 for a justification in a general setting. Thus (13.6) provides an information-theoretic characterization for correlated recovery.

The intuition is that, since $I(\sigma_1, \sigma_2; A) = I(\mathbf{1}_{\{\sigma_1=\sigma_2\}}; A)$, (13.6) means that the observation A does not provide enough information to distinguish whether any two vertices are in the same community. Alternatively, since $I(\sigma_1; A) = 0$ by symmetry and $I(\sigma_1; \sigma_2) = o(1)$,[4] it follows from the chain rule that $I(\sigma_1, \sigma_2; A) = I(\sigma_1; \sigma_2|A) + o(1)$. Thus (13.6) is

[4] Indeed, since $\mathbb{P}\{\sigma_2 = -|\sigma_1 = +\} = n/(2n-2)$, $I(\sigma_1;\sigma_2) = \log 2 - h(n/(2n-2)) = \Theta(n^{-2})$, where h is the binary entropy function in (13.34).

equivalent to stating that σ_1 and σ_2 are asymptotically independent given the observation A; this is shown in Theorem 2.1 of [39] for the SBM below the recovery threshold $\tau = (a-b)^2/(2(a+b)) < 1$.

Polyanskiy and Wu recently [40] proposed an information-percolation method based on strong data-processing inequalities for mutual information to bound the mutual information in (13.6) in terms of bond percolation probabilities, which yields bounds or a sharp recovery threshold for correlated recovery; a similar program is carried out independently in [41] for a variant of mutual information defined via the χ^2-divergence. For two communities, this method yields the sharp threshold in the Gaussian model but not in the SBM.

Next, we describe another method of proving (13.6) via second-moment analysis that reaches the sharp threshold. Let $\mathcal{P}_+$ and $\mathcal{P}_-$ denote the conditional distribution of A conditioned on $\sigma_1 = \sigma_2$ and on $\sigma_1 \neq \sigma_2$, respectively. The following result can be distilled from [42] (see Appendix A13.2 for a proof): for any probability distribution Q, if

$$\int \frac{(\mathcal{P}_+ - \mathcal{P}_-)^2}{Q} = o(1), \tag{13.7}$$

then (13.6) holds and hence correlated recovery is impossible. The LHS of (13.7) can be computed similarly to the usual second moment (13.4) when Q is chosen to be the distribution of A under the null model. In Appendix A13.2 we verify that (13.7) is satisfied below the correlated recovery threshold $\tau = (a-b)^2/(2(a+b)) < 1$ for the binary symmetric SBM.

Blockwise mutual information $I(\sigma; A)$. Although this quantity is not directly related to correlated recovery *per se*, its derivative with respect to some appropriate signal-to-noise-ratio (SNR) parameter can be related to or coincides with the reconstruction error thanks to the I-MMSE formula [43] or variants. Using this method, we can prove that the Kullback–Leibler divergence $D(\mathcal{P}\|Q) = o(n)$ implies the impossibility of correlated recovery in the Gaussian case. As shown in (13.2), a bounded second moment readily implies a bounded KL divergence. Hence, as a corollary, we prove that a bounded second moment also implies the impossibility of correlated recovery in the Gaussian case. Below, we sketch the proof of the impossibility of correlated recovery in the Gaussian case, by assuming $D(\mathcal{P}\|Q) = o(n)$. The proof makes use of mutual information, the I-MMSE formula, and a type of interpolation argument [44–46].

Assume that $A(\beta) = \sqrt{\beta} M + W$ in the planted model and $A = W$ in the null model, where $\beta \in [0, 1]$ is an SNR parameter, $M = (\mu/\sqrt{n})(\sigma\sigma^{\mathrm{T}} - \mathbf{I})$, W is a symmetric Gaussian random matrix with zero diagonal, and $W_{ij} \overset{\text{i.i.d.}}{\sim} \mathcal{N}(0, 1)$ for all $i < j$. Note that $\beta = 1$ corresponds to the binary symmetric community model in Definition 13.2 with $P = \mathcal{N}(\mu/\sqrt{n}, 1)$ and $Q = \mathcal{N}(-\mu/\sqrt{n}, 1)$. Below we abbreviate $A(\beta)$ as A whenever the context is clear. First, recall that the minimum mean-squared error estimator is given by the posterior mean of M:

$$\widehat{M}_{\mathrm{MMSE}}(A) = \mathbb{E}[M|A],$$

and the resulting (rescaled) minimum mean-squared error is

$$\mathrm{MMSE}(\beta) = \frac{1}{n} \mathbb{E}\|M - \mathbb{E}[M|A]\|_{\mathrm{F}}^2. \tag{13.8}$$

We will start by proving that, if $D(\mathcal{P}\|Q) = o(n)$, then, for all $\beta \in [0,1]$, the MMSE tends to that of the trivial estimator $\widehat{M} = 0$, i.e.,

$$\lim_{n\to\infty} \mathrm{MMSE}(\beta) = \lim_{n\to\infty} \frac{1}{n}\,\mathbb{E}\|M\|_{\mathrm{F}}^2 = \mu^2. \tag{13.9}$$

Note that $\lim_{n\to\infty} \mathrm{MMSE}(\beta)$ exists by virtue of Proposition III.2 of [44]. Let us compute the mutual information between M and A:

$$\begin{aligned} I(\beta) \triangleq I(M;A) &= \mathbb{E}_{M,A} \log\left(\frac{\mathcal{P}(A|M)}{\mathcal{P}(A)}\right) \qquad (13.10)\\ &= \mathbb{E}_A \log\left(\frac{Q(A)}{\mathcal{P}(A)}\right) + \mathbb{E}_{M,A}\log\left(\frac{\mathcal{P}(A|M)}{Q(A)}\right)\\ &= -D(\mathcal{P}\|Q) + \frac{1}{2}\mathbb{E}_{M,A}\left[\sqrt{\beta}\langle M,A\rangle - \frac{\beta\|M\|_{\mathrm{F}}^2}{2}\right]\\ &= -D(\mathcal{P}\|Q) + \frac{\beta}{4}\mathbb{E}\|M\|_{\mathrm{F}}^2. \qquad (13.11) \end{aligned}$$

By assumption, we have that $D(\mathcal{P}\|Q) = o(n)$ holds for $\beta = 1$; by the data-processing inequality for KL divergence [47], this holds for all $\beta < 1$ as well. Thus (13.11) becomes

$$\lim_{n\to\infty} \frac{1}{n} I(\beta) = \frac{\beta}{4} \lim_{n\to\infty} \frac{1}{n}\mathbb{E}\|M\|_{\mathrm{F}}^2 = \frac{\beta\mu^2}{4}. \tag{13.12}$$

Next we compute the MMSE. Recall the I-MMSE formula [43] for Gaussian channels:

$$\frac{\mathrm{d}I(\beta)}{\mathrm{d}\beta} = \frac{1}{2}\sum_{i<j}\left(M_{ij} - \mathbb{E}\left[M_{ij}|A\right]\right)^2 = \frac{n}{4}\mathrm{MMSE}(\beta). \tag{13.13}$$

Note that the MMSE is by definition bounded above by the squared error of the trivial estimator $\widehat{M} = 0$, so that for all β we have

$$\mathrm{MMSE}(\beta) \le \frac{1}{n}\,\mathbb{E}\|M\|_{\mathrm{F}}^2 \le \mu^2. \tag{13.14}$$

On combining these we have

$$\begin{aligned} \frac{\mu^2}{4} &\overset{\text{(a)}}{=} \lim_{n\to\infty} \frac{I(1)}{n} \overset{\text{(b)}}{=} \frac{1}{4}\lim_{n\to\infty}\int_0^1 \mathrm{MMSE}(\beta)\,\mathrm{d}\beta\\ &\overset{\text{(c)}}{\le} \frac{1}{4}\int_0^1 \lim_{n\to\infty} \mathrm{MMSE}(\beta)\,\mathrm{d}\beta\\ &\overset{\text{(d)}}{\le} \frac{1}{4}\int_0^1 \mu^2\,\mathrm{d}\beta = \frac{\mu^2}{4}, \end{aligned}$$

where (a) and (b) hold due to (13.12) and (13.13), (c) follows from Fatou's lemma, and (d) follows from (13.14), i.e., $\mathrm{MMSE}(\beta) \le \mu^2$ pointwise. Since we began and ended with the same expression, these inequalities must all be equalities. In particular, since (d) holds with equality, we have that (13.9) holds for almost all $\beta \in [0,1]$. Since $\mathrm{MMSE}(\beta)$

is a non-increasing function of β, its limit $\lim_{n\to\infty}$ MMSE(β) is also non-increasing in β. Therefore, (13.9) holds for all $\beta \in [0, 1]$. This completes the proof of our claim that the optimal MMSE estimator cannot outperform the trivial one asymptotically.

To show that the optimal estimator actually converges to the trivial one, we expand the definition of MMSE(β) in (13.8) and subtract (13.9) from it. This gives

$$\lim_{n\to\infty} \frac{1}{n}\mathbb{E}\Big[-2\langle M, \mathbb{E}[M|A]\rangle + \|\mathbb{E}[M|A]\|_{\mathrm{F}}^2\Big] = 0. \tag{13.15}$$

From the tower property of conditional expectation and the linearity of the inner product, it follows that

$$\mathbb{E}\langle M, \mathbb{E}[M|A]\rangle = \mathbb{E}\langle \mathbb{E}[M|A], \mathbb{E}[M|A]\rangle = \mathbb{E}\|\mathbb{E}[M|A]\|_{\mathrm{F}}^2,$$

and combining this with (13.15) gives

$$\lim_{n\to\infty} \frac{1}{n}\mathbb{E}\|\mathbb{E}[M|A]\|_{\mathrm{F}}^2 = 0. \tag{13.16}$$

Finally, for any estimator $\widehat{\sigma}(A)$ of the community membership σ, we can define an estimator for M by $\widehat{M} = (\mu/\sqrt{n})(\widehat{\sigma}\widehat{\sigma}^{\mathrm{T}} - \mathbf{I})$. Then using the Cauchy–Schwarz inequality, we have

$$\begin{aligned}\mathbb{E}_{M,A}[\langle M, \widehat{M}\rangle] &= \mathbb{E}_A[\langle \mathbb{E}[M|A], \widehat{M}\rangle]\\ &\le \mathbb{E}_A\Big[\|\mathbb{E}[M|A]\|_{\mathrm{F}}\|\widehat{M}\|_{\mathrm{F}}\Big]\\ &\le \sqrt{\mathbb{E}_A[\|\mathbb{E}[M|A]\|_{\mathrm{F}}^2]} \times \mu\sqrt{n} \overset{(13.16)}{=} o(n).\end{aligned}$$

Since $\langle M, \widehat{M}\rangle = \mu^2(\langle\sigma, \widehat{\sigma}\rangle^2/n - 1)$, it follows that $\mathbb{E}[\langle\sigma, \widehat{\sigma}\rangle^2] = o(n^2)$, which further implies $\mathbb{E}\big[|\langle\sigma, \widehat{\sigma}\rangle|\big] = o(n)$ by Jensen's inequality. Hence, correlated recovery of σ is impossible.

In passing, we remark that, while we focus on the binary symmetric community in this section, the proof techniques are widely applicable for many other high-dimensional inference problems such as detecting a single community [15], sparse PCA, Gaussian mixture clustering [33], synchronization [48], and tensor PCA [24]. In fact, for a more general k-symmetric community model with $P = \mathcal{N}\big((k-1)\mu/\sqrt{n}, 1\big)$ and $Q = \mathcal{N}\big(\mu/\sqrt{n}, 1\big)$, the first moment method shows that both detection and correlated recovery are information-theoretically possible when $\mu > 2\sqrt{\log k/(k-1)}$ and impossible when $\mu < \sqrt{2\log(k-1)/(k-1)}$. The upper and lower bounds differ by a factor of $\sqrt{2}$ when k is asymptotically large. This gap of $\sqrt{2}$ is due to the looseness of the second-moment lower bound. A more refined *conditional* second lower bound can be applied to show that the sharp information-theoretic threshold for detection and correlated recovery is $\mu = 2\sqrt{\log k/k}(1 + o_k(1))$ when $k \to \infty$ [33]. Complete, but not explicit, characterizations of information-theoretic reconstruction thresholds were obtained in [46, 49, 50] for all finite k through the Guerra interpolation technique and cavity method.

13.3.2 Almost Exact and Exact Recovery

In this section, we study almost-exact and exact recovery using the single-community model as an illustrating example. The hidden community can be represented by its indicator vector $\xi \in \{0,1\}^n$ such that $\xi_i = 1$ if vertex i is in the community and $\xi_i = 0$ otherwise. Let ξ^* denote the indicator of the true community and $\widehat{\xi} = \widehat{\xi}(A) \in \{0,1\}^n$ an estimator. The only assumptions on the community size K we impose are that K/n is bounded away from one, and, to avoid triviality, that $K \geq 2$. Of particular interest is the case of $K = o(n)$, where the community size grows sub-linearly with respect to the network size.

DEFINITION 13.5 (Almost-exact recovery) *An estimator* $\widehat{\xi}$ *is said to* almost exactly recover ξ^* *if, as* $n \to \infty$, $d_{\mathrm{H}}(\xi^*, \widehat{\xi})/K \to 0$ *in probability, where* d_{H} *denotes the Hamming distance.*

One can verify that the existence of an estimator satisfying Definition 13.5 is equivalent to the existence of an estimator such that $\mathbb{E}[d_{\mathrm{H}}(\xi^*, \widehat{\xi})] = o(K)$.

DEFINITION 13.6 (Exact recovery) *An estimator* $\widehat{\xi}$ exactly recovers ξ^*, *if, as* $n \to \infty$, $\mathbb{P}[\xi^* \neq \widehat{\xi}] \to 0$, *where the probability is with respect to the randomness of* ξ^* *and* A.

To obtain upper bounds on the thresholds for almost-exact and exact recovery, we turn to the MLE. Specifically,

- to show that the MLE achieves almost-exact recovery, it suffices to prove that there exists $\epsilon_n = o(1)$ such that, with high probability, $\mathcal{P}(A|\xi) < \mathcal{P}(A|\xi^*)$ for all ξ with $d_{\mathrm{H}}(\xi, \xi^*) \geq \epsilon_n K$; and
- to show that the MLE achieves exact recovery, it suffices to prove that, with high probability, $\mathcal{P}(A|\xi) < \mathcal{P}(A|\xi^*)$ for all $\xi \neq \xi^*$.

This type of argument often involves two key steps. First, upper-bound the probability that $\mathcal{P}(A|\xi) \geq \mathcal{P}(A|\xi^*)$ for a fixed ξ using large-deviation techniques. Second, take an appropriate union bound over all possible ξ using a "peeling" argument which takes into account the fact that the further away ξ is from ξ^* the less likely it is for $\mathcal{P}(A|\xi) \geq \mathcal{P}(A|\xi^*)$ to occur. Below we discuss these two key steps in more detail.

Given the data matrix A, a sufficient statistic for estimating the community C^* is the *log likelihood ratio (LLR) matrix* $\boldsymbol{L} \in \mathbb{R}^{n \times n}$, where $L_{ij} = \log(dP/dQ)(A_{ij})$ for $i \neq j$ and $L_{ii} = 0$. For $S, T \subset [n]$, define

$$e(S,T) = \sum_{(i<j):(i,j) \in (S \times T) \cup (T \times S)} L_{ij}. \tag{13.17}$$

Let $\widehat{C}_{\mathrm{ML}}$ denote the MLE of C^*, given by

$$\widehat{C}_{\mathrm{ML}} = \operatorname{argmax}_{C \subset [n]} \{e(C,C) : |C| = K\}, \tag{13.18}$$

which minimizes the error probability $\mathbb{P}\{\widehat{C} \neq C^*\}$ because C^* is equiprobable by assumption. It is worth noting that the optimal estimator that minimizes the misclassification rate (Hamming loss) is the bit-MAP decoder $\widetilde{\xi} = (\widetilde{\xi}_i)$, where $\widetilde{\xi}_i \triangleq \operatorname{argmax}_{j \in \{0,1\}} \mathbb{P}[\xi_i = j | L]$. Therefore, although the MLE is optimal for exact recovery, it need not be optimal for almost-exact recovery; nevertheless, we choose to analyze the MLE due to its simplicity, and it turns out to be asymptotically optimal for almost-exact recovery as well.

To state the main results, we introduce some standard notations associated with binary hypothesis testing based on independent samples. We assume the KL divergences $D(P\|Q)$ and $D(Q\|P)$ are finite. In particular, P and Q are mutually absolutely continuous, and the likelihood ratio, dP/dQ, satisfies $\mathbb{E}_Q[dP/dQ] = \mathbb{E}_P\left[(dP/dQ)^{-1}\right] = 1$. Let $L = \log(dP/dQ)$ denote the LLR. The likelihood-ratio test for n observations and threshold $n\theta$ is to declare P to be the true distribution if $\sum_{k=1}^{n} L_k \geq n\theta$ and to declare Q otherwise. For $\theta \in [-D(Q\|P), D(P\|Q)]$, the standard Chernoff bounds for error probability of this likelihood-ratio test are given by

$$Q\left[\sum_{k=1}^{n} L_k \geq n\theta\right] \leq \exp(-nE_Q(\theta)), \tag{13.19}$$

$$P\left[\sum_{k=1}^{n} L_k \leq n\theta\right] \leq \exp(-nE_P(\theta)), \tag{13.20}$$

where the log moment generating functions of L are denoted by $\psi_Q(\lambda) = \log \mathbb{E}_Q[\exp(\lambda L)]$ and $\psi_P(\lambda) = \log \mathbb{E}_P[\exp(\lambda L)] = \psi_Q(\lambda + 1)$ and the large-deviation exponents are given by Legendre transforms of the log moment generating functions:

$$E_Q(\theta) = \psi_Q^*(\theta) \triangleq \sup_{\lambda \in \mathbb{R}} \lambda\theta - \psi_Q(\lambda), \tag{13.21}$$

$$E_P(\theta) = \psi_P^*(\theta) \triangleq \sup_{\lambda \in \mathbb{R}} \lambda\theta - \psi_P(\lambda) = E_Q(\theta) - \theta. \tag{13.22}$$

In particular, E_P and E_Q are convex functions. Moreover, since $\psi'_Q(0) = -D(Q\|P)$ and $\psi'_Q(1) = D(P\|Q)$, we have $E_Q(-D(Q\|P)) = E_P(D(P\|Q)) = 0$ and hence $E_Q(D(P\|Q)) = D(P\|Q)$ and $E_P(-D(Q\|P)) = D(Q\|P)$.

Under mild assumptions on the distribution (P, Q) (cf. Assumption 1 of [51]) which are satisfied both by the Gaussian distribution and by the Bernoulli distribution, the sharp thresholds for almost exact and exact recovery under the single-community model are given by the following result.

THEOREM 13.2 *Consider the single-community model with $P = \mathcal{N}(\mu, 1)$ and $Q = \mathcal{N}(0, 1)$, or $P = \text{Bern}(p)$ and $Q = \text{Bern}(q)$ with $\log(p/q)$ and $\log((1-p)/(1-q))$ bounded. If*

$$K \cdot D(P\|Q) \to \infty \text{ and } \liminf_{n\to\infty} \frac{(K-1)D(P\|Q)}{\log(n/K)} > 2, \tag{13.23}$$

then almost-exact recovery is information-theoretically possible. If, in addition to (13.23),

$$\liminf_{n\to\infty} \frac{KE_Q((1/K)\log(n/K))}{\log n} > 1 \tag{13.24}$$

holds, then exact recovery is information-theoretically possible.

Conversely, if almost-exact recovery is information-theoretically possible, then

$$K \cdot D(P\|Q) \to \infty \text{ and } \liminf_{n\to\infty} \frac{(K-1)D(P\|Q)}{\log(n/K)} \geq 2. \tag{13.25}$$

If exact recovery is information-theoretically possible, then in addition to (13.25), the following holds:

$$\liminf_{n\to\infty} \frac{KE_Q((1/K)\log(n/K))}{\log n} \geq 1. \tag{13.26}$$

Next we sketch the proof of Theorem 13.2.

Sufficient Conditions
For any $C \subset [n]$ such that $|C| = K$ and $|C \cap C^*| = \ell$, let $S = C^* \setminus C$ and $T = C \setminus C^*$. Then

$$e(C,C) - e(C^*,C^*) = e(T,T) + e(T,C^* \setminus S) - e(S,C^*).$$

Let $m = \binom{K}{2} - \binom{\ell}{2}$. Notice that $e(S,C^*)$ has the same distribution as $\sum_{i=1}^m L_i$ under measure P; $e(T,T) + e(T,C^*\backslash S)$ has the same distribution as $\sum_{i=1}^m L_i$ under measure Q where L_i are i.i.d. copies of $\log(\mathrm{d}P/\mathrm{d}Q)$. It readily follows from large-deviation bounds (13.19) and (13.20) that

$$\mathbb{P}\{e(T,T) + e(T,C^* \setminus S) \geq m\theta\} \geq \exp(-mE_Q(\theta)), \tag{13.27}$$
$$\mathbb{P}\{e(S,C^*) \leq m\theta\} \leq \exp(-mE_P(\theta)).$$

Next we proceed to describe the union bound for the proof of almost-exact recovery. Note that showing that MLE achieves almost exact recovery is equivalent to showing $\mathbb{P}\left\{|\widehat{C}_{\mathrm{ML}} \cap C^*| \leq (1-\epsilon_n)K\right\} = o(1)$. The first layer of the union bound is straightforward:

$$\left\{|\widehat{C}_{\mathrm{ML}} \cap C^*| \leq (1-\epsilon_n)K\right\} = \cup_{\ell=0}^{\lfloor(1-\epsilon_n)K\rfloor}\left\{|\widehat{C}_{\mathrm{ML}} \cap C^*| = \ell\right\}. \tag{13.28}$$

For the second layer of the union bound, one naive way to proceed is

$$\begin{aligned}\left\{|\widehat{C}_{\mathrm{ML}} \cap C^*| = \ell\right\} &\subset \{C \in \mathcal{C}_\ell : e(C,C) \geq e(C^*,C^*)\}\\ &= \cup_{C \in \mathcal{C}_\ell}\{e(C,C) \geq e(C^*,C^*\},\end{aligned}$$

where $\mathcal{C}_\ell = \{C \subset [n] : |C| = K, |C \cap C^*| = \ell\}$. However, this union bound is too loose because of the high correlations among $e(C,C) - e(C^*,C^*)$ for different $C \in \mathcal{C}_\ell$. Instead, we use the following union bound. Let $\mathcal{S}_\ell = \{S \subset C^* : |S| = K - \ell\}$ and $\mathcal{T}_\ell = \{T \subset (C^*)^c : |T| = K - \ell\}$. Then, for any $\theta \in \mathbb{R}$,

$$\begin{aligned}\{|\widehat{C}_{\mathrm{ML}} \cap C^*| = \ell\} &\subset \{\exists S \in \mathcal{S}_\ell, T \in \mathcal{T}_\ell : e(S,C^*) \leq e(T,T) + e(T,C^*\backslash S)\}\\ &\subset \{\exists S \in \mathcal{S}_\ell : e(S,C^*) \leq m\theta\}\\ &\quad \cup \{\exists S \in \mathcal{S}_\ell, T \in \mathcal{T}_\ell : e(T,T) + e(T,C^*\backslash S) \geq m\theta\}\\ &\subset \cup_{S \in \mathcal{S}_\ell}\{e(S,C^*) \leq m\theta\}\\ &\quad \cup_{S \in \mathcal{S}_\ell, T \in \mathcal{T}_\ell}\{e(T,T) + e(T,C^*\backslash S) \geq m\theta\}.\end{aligned} \tag{13.29}$$

Note that we single out $e(S,C^*)$ because the number of different choices of S, $\binom{K}{K-\ell}$, is much smaller than the number of different choices of T, $\binom{n-K}{K-\ell}$, when $K \ll n$. Combining the above union bound with the large-deviation bound (13.27) yields that

$$\mathbb{P}\left\{|\widehat{C}_{\mathrm{ML}} \cap C^*| = \ell\right\} \leq \binom{K}{K-\ell} e^{-mE_P(\theta)} + \binom{n-K}{K-\ell}\binom{K}{K-\ell} e^{-mE_Q(\theta)}. \tag{13.30}$$

Note that, for any $\ell \le (1-\epsilon)K$,

$$\binom{K}{K-\ell} \le \left(\frac{Ke}{K-\ell}\right)^{K-\ell} \le \left(\frac{e}{\epsilon}\right)^{K-\ell},$$
$$\binom{n-K}{K-\ell} \le \left(\frac{(n-K)e}{K-\ell}\right)^{K-\ell} \le \left(\frac{(n-K)e}{K\epsilon}\right)^{K-\ell}.$$

Hence, for any $\ell \le (1-\epsilon)K$,

$$\mathbb{P}\left\{|\widehat{C}_{\mathrm{ML}} \cap C^*| = \ell\right\} \le e^{-(K-\ell)E_1} + e^{-(K-\ell)E_2}, \tag{13.31}$$

where

$$E_1 \triangleq \frac{1}{2}(K-1)E_P(\theta) - \log\left(\frac{e}{\epsilon}\right),$$
$$E_2 \triangleq \frac{1}{2}(K-1)E_Q(\theta) - \log\left(\frac{(n-K)e^2}{K\epsilon^2}\right).$$

Thanks to the second condition in (13.23), we have $(K-1)D(P\|Q)(1-\eta) \ge 2\log(n/K)$ for some $\eta \in (0,1)$. Choose $\theta = (1-\eta)D(P\|Q)$. Under some mild assumption on P and Q which is satisfied in the Gaussian and Bernoulli cases, we have $E_P(\theta) \ge c\eta^2 D(P\|Q)$ for some universal constant $c > 0$. Furthermore, recall from (13.22) that $E_P(\theta) = E_Q(\theta) - \theta$. Hence, since $KD(P\|Q) \to \infty$ by the assumption (13.23), by choosing $\epsilon = 1/\sqrt{KD(P\|Q)}$, we have $\min\{E_1, E_2\} \to \infty$. The proof for almost-exact recovery is completed by taking the first layer of the union bound in (13.28) over ℓ.

For exact recovery, we need to show $\mathbb{P}\left\{|\widehat{C}_{\mathrm{ML}} \cap C^*| \le K-1\right\} = o(1)$. Hence, we need to further bound $\mathbb{P}\left\{|\widehat{C}_{\mathrm{ML}} \cap C^*| = \ell\right\}$ for any $(1-\epsilon)K \le \ell \le K-1$. It turns out that the previous union bound (13.29) is no longer tight. Instead, using $e(T,T) + e(T, C^* \setminus S) = e(T, T \cup C^*) - e(T,S)$, we have the following union bound:

$$\{|\widehat{C}_{\mathrm{ML}} \cap C^*| = \ell\} \subset \cup_{S \in \mathcal{S}_\ell} \{e(S,C^*) \le m_1\theta_1\} \cup_{T \in \mathcal{T}_\ell} \{e(T, T \cup C^*) \ge m_2\theta_2\}$$
$$\cup_{S \in \mathcal{S}_\ell, T \in \mathcal{T}_\ell} \{e(T,S) \le m_2\theta_2 - m_1\theta_1\},$$

where $m_1 = \binom{K}{2} - \binom{\ell}{2}$, $m_2 = \binom{K-\ell}{2} + (K-\ell)K$, and θ_1, θ_2 are to be optimized. Note that we further single out $e(T, T \cup C^*)$ because it depends only on $T \in \mathcal{T}_\ell$ once C^* is fixed. Since $(1-\epsilon)K \le \ell \le K-1$, we have $|T| = |S| = K - \ell \le \epsilon K$ and thus the effect of $e(T,S)$ can be neglected. Therefore, approximately we can set $\theta_1 = \theta_2 = \theta$ and get

$$\mathbb{P}\left\{|\widehat{C}_{\mathrm{ML}} \cap C^*| = \ell\right\} \lesssim \binom{K}{K-\ell} e^{-m_1 E_P(\theta)} + \binom{n-K}{K-\ell} e^{-m_2 E_P(\theta)}. \tag{13.32}$$

Using $\binom{K}{K-\ell} \le K^{K-\ell}$, $\binom{n-K}{K-\ell} \le (n-K)^{K-\ell}$, and $m_2 \ge m_1 \ge (1-\epsilon)(K-\ell)K$, we get that, for any $(1-\epsilon)K \le \ell \le K-1$,

$$\mathbb{P}\left\{|\widehat{C}_{\mathrm{ML}} \cap C^*| = \ell\right\} \le e^{-(K-\ell)E_3} + e^{-(K-\ell)E_4}, \tag{13.33}$$

where

$$E_3 \triangleq (1-\epsilon)KE_P(\theta) - \log K,$$
$$E_4 \triangleq (1-\epsilon)KE_Q(\theta) - \log n.$$

Note that $E_P(\theta) = E_Q(\theta) - \theta$. Hence, we set $\theta = (1/K)\log(n/K)$ so that $E_3 = E_4$, which goes to $+\infty$ under the assumption of (13.24). The proof of exact recovery is completed by taking the union bound over all ℓ.

Necessary Conditions

To derive lower bounds on the almost-exact recovery threshold, we resort to a simple rate-distortion argument. Suppose $\widehat{\xi}$ achieves almost-exact recovery of ξ^*. Then $\mathbb{E}[d_{\mathrm{H}}(\xi,\widehat{\xi})] = \epsilon_n K$ with $\epsilon_n \to 0$. On the one hand, consider the following chain of inequalities, which lower-bounds the amount of information required for a distortion level ϵ_n:

$$
\begin{aligned}
I(A;\xi^*) &\overset{(a)}{\geq} I(\widehat{\xi};\xi^*) \geq \min_{\mathbb{E}[d(\widetilde{\xi},\xi^*)]\leq\epsilon_n K} I(\widetilde{\xi};\xi^*) \\
&\geq H(\xi^*) - \max_{\mathbb{E}[d(\widetilde{\xi},\xi^*)]\leq\epsilon_n K} H(\widetilde{\xi}\oplus\xi^*) \\
&\overset{(b)}{=} \log\binom{n}{K} - nh\left(\frac{\epsilon_n K}{n}\right) \overset{(c)}{\geq} K\log\left(\frac{n}{K}\right)(1+o(1)),
\end{aligned}
$$

where (a) follows from the data-processing inequality for mutual information since $\xi \to A \to \widehat{\xi}$ forms a Markov chain; (b) is due to the fact that $\max_{\mathbb{E}[w(X)]\leq pn} H(X) = nh(p)$ for any $p \leq 1/2$, where

$$
h(p) \triangleq p\log\left(\frac{1}{p}\right) + (1-p)\log\left(\frac{1}{1-p}\right) \tag{13.34}
$$

is the binary entropy function and $w(x) = \sum_i x_i$; and (c) follows from the bound $\binom{n}{K} \geq (n/K)^K$, the assumption K/n is bounded away from one, and the bound $h(p) \leq -p\log p + p$ for $p \in [0,1]$.

On the other hand, consider the following upper bound on the mutual information:

$$
I(A;\xi^*) = \min_{\mathbb{Q}} D(\mathbb{P}_{A|\xi^*}\|\mathbb{Q}|\mathbb{P}_{\xi^*}) \leq D(\mathbb{P}_{A|\xi^*}\|Q^{\otimes\binom{n}{2}}|\mathbb{P}_{\xi^*}) = \binom{K}{2} D(P\|Q),
$$

where the first equality follows from the geometric interpretation of mutual information as an "information radius" (see, e.g., Corollary 3.1 of [52]); the last equality follows from the tensorization property of KL divergence for product distributions. Combining the last two displays, we conclude that the second condition in (13.25) is necessary for almost-exact recovery.

To show the necessity of the first condition in (13.25), we can reduce almost-exact recovery to a local hypothesis testing via a genie-type argument. Given $i, j \in [n]$, let $\xi_{\backslash i,j}$ denote $\{\xi_k : k \neq i, j\}$. Consider the following binary hypothesis-testing problem for determining ξ_i. If $\xi_i = 0$, a node J is randomly and uniformly chosen from $\{j : \xi_j = 1\}$, and we observe $(A, J, \xi_{\backslash i,J})$; if $\xi_i = 1$, a node J is randomly and uniformly chosen from $\{j : \xi_j = 0\}$, and we observe $(A, J, \xi_{\backslash i,J})$. It is straightforward to verify that this hypothesis-testing problem is equivalent to testing $H_0 : Q^{\otimes(K-1)}P^{\otimes(K-1)}$ versus $H_1 : P^{\otimes(K-1)}Q^{\otimes(K-1)}$. Let $\mathcal{E}$ denote the optimal average probability of testing error, $p_{\mathrm{e},0}$ denote the type-I error probability, and $p_{e,1}$ denote the type-II error probability. Then we have the following chain of inequalities:

$$\begin{aligned}\mathbb{E}[d_{\mathrm{H}}(\xi,\widehat{\xi})] &\geq \sum_{i=1}^{n} \min_{\widehat{\xi}_i(A)} \mathbb{P}[\xi_i \neq \widehat{\xi}_i] \\ &\geq \sum_{i=1}^{n} \min_{\widehat{\xi}_i(A,J,\xi_{\backslash i,J})} \mathbb{P}[\xi_i \neq \widehat{\xi}_i] \\ &= n \min_{\widehat{\xi}_1(A,J,\xi_{\backslash 1,J})} \mathbb{P}[\xi_1 \neq \widehat{\xi}_1] = n\mathcal{E}.\end{aligned}$$

By the assumption $\mathbb{E}[d_{\mathrm{H}}(\xi,\widehat{\xi})] = o(K)$, it follows that $\mathcal{E} = o(K/n)$. Under the assumption that K/n is bounded away from one, $\mathcal{E} = o(K/n)$ further implies that the sum of type-I and type-II probabilities of error $p_{e,0} + p_{e,1} = o(1)$, or equivalently, $\mathrm{TV}((P\otimes Q)^{\otimes K-1},(Q\otimes P)^{\otimes K-1}) \to 1$, where $\mathrm{TV}(P,Q) \triangleq \int |dP - dQ|/2$ denotes the total variation distance. Using $D(P\|Q) \geq \log(1/(2(1-\mathrm{TV}(P,Q))))$ (Equation (2.25) of [53]) and the tensorization property of KL divergence for product distributions, we conclude that $(K-1)(D(P\|Q)+D(Q\|P)) \to \infty$ is necessary for almost-exact recovery. It turns out that, both for the Bernoulli distribution and for the Gaussian distribution as specified in the theorem statement, $D(P\|Q) \asymp D(Q\|P)$, and hence $KD(P\|Q) \to \infty$ is necessary for almost-exact recovery.

Clearly, any estimator achieving exact recovery also achieves almost-exact recovery. Hence lower bounds for almost-exact recovery hold automatically for exact recovery. Finally, we show the necessity of (13.26) for exact recovery. Since the MLE minimizes the error probability among all estimators if the true community C^* is uniformly distributed, it follows that, if exact recovery is possible, then, with high probability, C^* has a strictly higher likelihood than any other community $C \neq C^*$, in particular, $C = C^* \setminus \{i\} \cup \{j\}$ for any pair of vertices $i \in C^*$ and $j \notin C^*$. To further illustrate the proof ideas, consider the Bernoulli case of the single-community model. Then C^* has a strictly higher likelihood than $C^* \setminus \{i\} \cup \{j\}$ if and only if $e(i,C^*)$, the number of edges connecting i to vertices in C^*, is larger than $e(j,C^* \setminus \{i\})$, the number of edges connecting j to vertices in $C^* \setminus \{i\}$. Therefore, with high probability, it holds that

$$\min_{i\in C^*} e(i,C^*) > \max_{j\notin C^*} e(j,C^* \setminus \{i_0\}), \qquad (13.35)$$

where i_0 is the random index such that $i_0 \in \arg\min_{i\in C^*} e(i,C^*)$. Note that the $e(j,C^* \setminus \{i_0\})$s are i.i.d. for different $j \notin C^*$ and hence a large-probability lower bound to their maximum can be derived using inverse concentration inequalities. Specifically, for the sake of argument by contradiction, suppose that (13.26) does not hold. Furthermore, for ease of presentation, assume that the large-deviation inequality (13.19) also holds in the reverse direction (cf. Corollary 5 of [51] for a precise statement). Then it follows that

$$\mathbb{P}\left\{e(j,C^* \setminus \{i_0\}) \geq \log\left(\frac{n}{K}\right)\right\} \gtrsim \exp\left(-KE_Q\left(\frac{1}{K}\log\left(\frac{n}{K}\right)\right)\right) \geq n^{-1+\delta}$$

for some small $\delta > 0$. Since the $e(j, C^* \setminus \{i_0\})$s are i.i.d. and there are $n - K$ of them, it further follows that, with large probability,

$$\max_{j \notin C^*} e(j, C^* \setminus \{i_0\}) \geq \log\left(\frac{n}{K}\right).$$

Similarly, by assuming that the large-deviation inequality (13.20) also holds in the opposite direction and using the fact that $E_P(\theta) = E_Q(\theta) - \theta$, we get that

$$\mathbb{P}\left\{e(i, C^*) \leq \log\left(\frac{n}{K}\right)\right\} \gtrsim \exp\left(-K E_P\left(\frac{1}{K}\log\left(\frac{n}{K}\right)\right)\right) \geq K^{-1+\delta}.$$

Although the $e(i, C^*)$s are not independent for different $i \in C^*$, the dependence is weak and can be controlled properly. Hence, following the same argument as above, we get that, with large probability,

$$\min_{i \in C^*} e(i, C^*) \leq \log\left(\frac{n}{K}\right).$$

Combining the large-probability lower and upper bounds and (13.35) yields the contradiction. Hence, (13.26) is necessary for exact recovery.

REMARK 13.1 Note that, instead of using the MLE, one could also apply a two-step procedure to achieve exact recovery: first use an estimator capable of almost-exact recovery and then clean up the residual errors through a local voting procedure for every vertex. Such a two-step procedure has been analyzed in [51]. From the computational perspective, both for the Bernoulli case and for the Gaussian case we have the following results:

- if $K = \Theta(n)$, a linear-time degree-thresholding algorithm achieves the information limit of almost exact recovery (see Appendix A of [54] and Appendix A of [55]);
- if $K = \omega(n/\log n)$, whenever information-theoretically possible, exact recovery can be achieved in polynomial time using semidefinite programming [56];
- if $K \geq (n/\log n)(1/(8e) + o(1))$ for the Gaussian case and $K \geq (n/\log n)(\rho_{\mathrm{BP}}(p/q) + o(1))$ for the Bernoulli case, exact recovery can be attained in nearly linear time via message passing plus clean-up [54, 55] whenever information-theoretically possible. Here $\rho_{\mathrm{BP}}(p/q)$ denotes a constant depending only on p/q.

However, it remains unknown whether any polynomial-time algorithm can achieve the respective information limit of almost exact recovery for $K = o(n)$, or exact recovery for $K \leq (n/\log n)(1/(8e) - \epsilon)$ in the Gaussian case and for $K \leq (n/\log n)(\rho_{\mathrm{BP}}(p/q) - \epsilon)$ in the Bernoulli case, for any fixed $\epsilon > 0$.

Similar techniques can be used to derive the almost-exact and exact recovery thresholds for the binary symmetric community model. For the Bernoulli case, almost-exact recovery is efficiently achieved by a simple spectral method if $n(p-q)^2/(p+q) \to \infty$[57], which turns out to be also information-theoretically necessary [58]. An exact recovery threshold for the binary community model has been derived and further shown to be efficiently achievable by a two-step procedure consisting of a spectral method plus clean-up [58, 59]. For the binary symmetric community model with general discrete distributions P and Q, the information-theoretic limit of exact recovery has been shown to be determined by the Rényi divergence of order 1/2 between P and Q [60]. The analysis of the MLE has been carried out under k-symmetric community models for general k, and the information-theoretic exact recovery threshold has been identified in [16] up to a universal constant. The precise information-theoretic limit of exact recovery has been determined in [61] for $k = \Theta(1)$ with a sharp constant and has further been shown to be efficiently achievable by a polynomial-time two-step procedure.

13.4 Computational Limits

In this section we discuss the computational limits (performance limits of all possible polynomial-time procedures) of detecting the planted structure under the planted-clique hypothesis (to be defined later). To investigate the computational hardness of a given statistical problem, one main approach is to find an *approximate randomized polynomial-time reduction*, which maps certain graph-theoretic problems, in particular, the *planted-clique* problem, to our problem approximately in total variation, thereby showing that these statistical problems are at least as hard as solving the planted-clique problem.

We focus on the single-community model in Definition 13.1 and present results for both the submatrix detection problem (Gaussian) [9] and the community detection problem (Bernoulli) [10]. Surprisingly, under appropriate parameterizations, the two problems share the same "easy–hard–impossible" phase transition. As shown in Fig. 13.1, where the horizontal and vertical axes correspond to the relative community size and the noise level, respectively, the hardness of the detection has a sharp phase transition: optimal detection can be achieved by computationally efficient procedures for a relatively large community, but provably not for a small community. This is one of the first results in high-dimensional statistics where the optimal trade-off between statistical performance and computational efficiency can be precisely quantified. Specifically, consider the submatrix detection problem in the Gaussian case of Definition 13.1, where $P = \mathcal{N}(\mu, 1)$ and $Q = \mathcal{N}(0, 1)$. In other words, the goal is to test the null model, where the observation is an $N \times N$ Gaussian noise matrix, versus the planted model, where there exists a $K \times K$ submatrix of elevated mean μ. Consider the high-dimensional setting of $K = N^{\alpha}$ and $\mu = N^{-\beta}$ with $N \to \infty$, where $\alpha, \beta > 0$ parameterizes the *cluster size* and *signal strength*, respectively. Information-theoretically, it can be shown that there exist detection procedures achieving vanishing error probability if and only if $\beta < \beta^* \triangleq \max(\alpha/2, 2\alpha - 1)$ [21]. In contrast, if only *randomized polynomial-time algorithms* are allowed, then reliable detection is impossible if $\beta > \beta^\sharp \triangleq \max(0, 2\alpha - 1)$; conversely if $\beta < \beta^\sharp$, there exists a *near-linear*-time detection algorithm with vanishing error probability. The plots of β^* and $\beta^\sharp$ in Fig 13.1 correspond to the **statistical and computational limits** of submatrix detection, respectively, revealing the following striking phase transition: for a large community ($\alpha \geq \frac{2}{3}$), optimal detection can be achieved by computationally efficient procedures; however, for a small community ($\alpha < \frac{2}{3}$), computational constraint incurs a severe penalty on the statistical performance and the optimal computationally intensive procedure cannot be mimicked by any efficient algorithms.

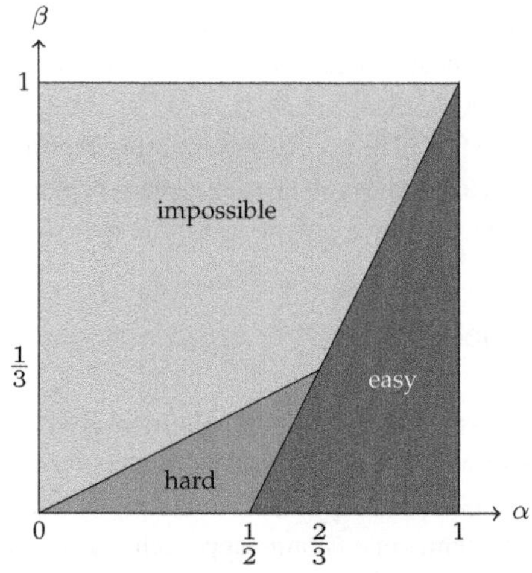

Figure 13.1 Computational versus statistical limits. For the submatrix detection problem, the size of the submatrix is $K = N^\alpha$ and the elevated mean is $\mu = N^{-\beta}$. For the community detection problem, the cluster size is $K = N^\alpha$, and the in-cluster and inter-cluster edge probabilities p and q are both on the order of $N^{-2\beta}$.

For the Bernoulli case, it has been shown that, to detect a planted dense subgraph, when the in-cluster and inter-cluster edge probabilities p and q are on the same order and parameterized as $N^{-2\beta}$ and the cluster size as $K = N^\alpha$, the easy–hard–impossible phase transition obeys the same diagram as that in Fig. 13.1 [10].

Our intractability result is based on the common hardness assumption of the planted-clique problem in the Erdős–Rényi graph when the clique size is of smaller order than the square root of the graph cardinality [62], which has been widely used to establish various hardness results in theoretical computer science [63–68] as well as the hardness of detecting sparse principal components [8]. Recently, the average-case hardness of the planted-clique problem has been established under certain computational models [69, 70] and within the sum-of-squares relaxation hierarchy [71–73].

The rest of the section is organized as follows. Section 13.4.1 gives the precise definition of the planted-clique problem, which forms the basis of reduction both for the submatrix detection problem and for the community detection problem, with the latter requiring a slightly stronger assumption. Section 13.4.2 discusses how to approximately reduce the planted-clique problem to the single-community detection problem in polynomial time in both Bernoulli and Gaussian settings. Finally, Section 13.4.3 presents the key techniques to bound the total variation between the reduced instance and the target hypothesis.

13.4.1 Planted-Clique Problem

Let $\mathcal{G}(n, \gamma)$ denote the Erdős–Rényi graph model with n vertices, where each pair of vertices is connected independently with probability γ. Let $\mathcal{G}(n, k, \gamma)$ denote the planted-clique model in which we add edges to k vertices uniformly chosen from $\mathcal{G}(n, \gamma)$ to form a clique.

DEFINITION 13.7 *The* PC *detection problem with parameters* (n, k, γ), *denoted by* $\mathsf{PC}(n, k, \gamma)$ *henceforth, refers to the problem of testing the following hypotheses:*

$$H_0^{\mathrm{C}}: \quad G \sim \mathcal{G}(n, \gamma), \qquad H_1^{\mathrm{C}}: \quad G \sim \mathcal{G}(n, k, \gamma).$$

The problem of finding the planted clique has been extensively studied for $\gamma = \frac{1}{2}$ and the state-of-the-art polynomial-time algorithms [14, 62, 74–78] work only for $k = \Omega(\sqrt{n})$. There is no known polynomial-time solver for the PC problem for $k = o(\sqrt{n})$ and any constant $\gamma > 0$. It has been conjectured [63, 64, 67, 70, 79] that the PC problem cannot be solved in polynomial time for $k = o(\sqrt{n})$ with $\gamma = \frac{1}{2}$, which we refer to as the PC hypothesis.

PC HYPOTHESIS Fix some constant $0 < \gamma \leq \frac{1}{2}$. For any sequence of randomized polynomial-time tests $\{\psi_{n,k_n}\}$ such that $\limsup_{n\to\infty}(\log k_n / \log n) < 1/2$,

$$\liminf_{n\to\infty} \mathbb{P}_{H_0^{\mathrm{C}}}\{\psi_{n,k}(G) = 1\} + \mathbb{P}_{H_1^{\mathrm{C}}}\{\psi_{n,k}(G) = 0\} \geq 1.$$

The PC hypothesis with $\gamma = \frac{1}{2}$ is similar to Hypothesis 1 of [9] and Hypothesis $\mathrm{B_{PC}}$ of [8]. Our computational lower bounds for submatrix detection require that the PC hypothesis holds for $\gamma = \frac{1}{2}$, and for community detection we need to assume that the PC hypothesis holds for any positive constant γ. The even stronger assumption that the PC hypothesis holds for $\gamma = 2^{-\log^{0.99} n}$ has been used in Theorem 10.3 of [68] for public-key cryptography. Furthermore, Corollary 5.8 of [70] shows that, under a statistical query model, any statistical algorithm requires at least $n^{\Omega\left(\frac{\log n}{\log(1/\gamma)}\right)}$ queries for detecting the planted bi-clique in an Erdős–Rényi random bipartite graph with edge probability γ.

13.4.2 Polynomial-Time Randomized Reduction

We present a polynomial-time randomized reduction scheme for the problem of detecting a single community (Definition 13.1) in both Bernoulli and Gaussian cases. For ease of presentation, we use the Bernoulli case as the main example, and discuss the minor modifications needed for the Gaussian case. The recent work [13] introduces a general reduction recipe for the single-community detection problem under general P, Q distributions, as well as various other detection problems with planted structures.

Let $\mathcal{G}(N, q)$ denote the Erdős–Rényi random graph with N vertices, where each pair of vertices is connected independently with probability q. Let $\mathcal{G}(N, K, p, q)$ denote the planted dense subgraph model with N vertices where (1) each vertex is included in the random set S independently with probability K/N; and (2) any two vertices are connected independently with probability p if both of them are in S and with probability q otherwise, where $p > q$. The planted dense subgraph here has a random size[5] with mean K, instead of a deterministic size K as assumed in [15, 80].

[5] We can also consider a planted dense subgraph with a fixed size K, where K vertices are chosen uniformly at random to plant a dense subgraph with edge probability p. Our reduction scheme extends to this fixed-size model; however, we have not been able to prove that the distributions are approximately matched under the alternative hypothesis. Nevertheless, the recent work [13] showed that the computational limit for detecting fixed-sized community is the same as that in Fig. 13.1, resolving an open problem in [10].

DEFINITION 13.8 *The planted dense subgraph detection problem with parameters* (N,K,p,q)*, henceforth denoted by* $\mathsf{PDS}(N,K,p,q)$*, refers to the problem of distinguishing between the following hypotheses:*

$$H_0: \quad G \sim \mathcal{G}(N,q) \triangleq \mathbb{P}_0, \qquad H_1: \quad G \sim \mathcal{G}(N,K,p,q) \triangleq \mathbb{P}_1.$$

We aim to reduce the $\mathsf{PC}(n,k,\gamma)$ problem to the $\mathsf{PDS}(N,K,cq,q)$ problem. For simplicity, we focus on the case of $c=2$; the general case follows similarly with a change in some numerical constants that come up in the proof. We are given an adjacency matrix $A \in \{0,1\}^{n\times n}$, or, equivalently, a graph G, and, with the help of additional randomness, will map it to an adjacency matrix $\widetilde{A} \in \{0,1\}^{N\times N}$, or, equivalently, a graph $\widetilde{G}$ such that the hypothesis H_0^{C} (H_1^{C}) in Definition 13.7 is mapped to H_0 exactly (H_1 approximately) in Definition 13.8. In other words, if A is drawn from $\mathcal{G}(n,\gamma)$, then $\widetilde{A}$ is distributed according to $\mathbb{P}_0$; if A is drawn from $\mathcal{G}(n,k,1,\gamma)$, then the distribution of $\widetilde{A}$ is close in total variation to $\mathbb{P}_1$.

Our reduction scheme works as follows. Each vertex in $\widetilde{G}$ is randomly assigned a parent vertex in G, with the choice of parent being made independently for different vertices in $\widetilde{G}$, and uniformly over the set $[n]$ of vertices in G. Let V_s denote the set of vertices in $\widetilde{G}$ with parent $s \in [n]$ and let $\ell_s = |V_s|$. Then the set of children nodes $\{V_s : s \in [n]\}$ will form a random partition of $[N]$. For any $1 \le s \le t \le n$, the number of edges, $E(V_s,V_t)$, from vertices in V_s to vertices in V_t in $\widetilde{G}$ will be selected randomly with a conditional probability distribution specified below. Given $E(V_s,V_t)$, the particular set of edges with cardinality $E(V_s,V_t)$ is chosen uniformly at random.

It remains to specify, for $1 \le s \le t \le n$, the conditional distribution of $E(V_s,V_t)$ given ℓ_s, ℓ_t, and $A_{s,t}$. Ideally, conditioned on ℓ_s and ℓ_t, we want to construct a Markov kernel from $A_{s,t}$ to $E(V_s,V_t)$ which maps Bern(1) to the desired edge distribution $\mathrm{Binom}(\ell_s\ell_t,p)$, and Bern(1/2) to $\mathrm{Binom}(\ell_s\ell_t,q)$, depending on whether both s and t are in the clique or not, respectively. Such a kernel, unfortunately, provably does not exist. Nevertheless, this objective can be accomplished approximately in terms of the total variation. For $s=t\in[n]$, let $E(V_s,V_t) \sim \mathrm{Binom}\left(\binom{\ell_t}{2},q\right)$. For $1 \le s < t \le n$, denote $P_{\ell_s\ell_t} \triangleq \mathrm{Binom}(\ell_s\ell_t,p)$ and $Q_{\ell_s\ell_t} \triangleq \mathrm{Binom}(\ell_s\ell_t,q)$. Fix $0<\gamma\le\frac{1}{2}$ and put $m_0 \triangleq \lfloor \log_2(1/\gamma)\rfloor$. Define

$$P'_{\ell_s\ell_t}(m) = \begin{cases} P_{\ell_s\ell_t}(m) + a_{\ell_s\ell_t} & \text{for } m=0, \\ P_{\ell_s\ell_t}(m) & \text{for } 1 \le m \le m_0, \\ (1/\gamma)Q_{\ell_s\ell_t}(m) & \text{for } m_0 < m \le \ell_s\ell_t, \end{cases}$$

where $a_{\ell_s\ell_t} = \sum_{m_0<m\le\ell_s\ell_t}[P_{\ell_s\ell_t}(m) - (1/\gamma)Q_{\ell_s\ell_t}(m)]$. Let $Q'_{\ell_s\ell_t} = (1/(1-\gamma))(Q_{\ell_s\ell_t} - \gamma P'_{\ell_s\ell_t})$. The idea behind our choice of $P'_{\ell_s\ell_t}$ and $Q'_{\ell_s\ell_t}$ is as follows. For a given $P'_{\ell_s\ell_t}$, we choose $Q'_{\ell_s\ell_t}$ to map Bern(γ) to $\mathrm{Binom}(\ell_s\ell_t,q)$ exactly; however, in order for Q' to be a well-defined probability distribution, we need to ensure that $Q_{\ell_s\ell_t}(m) \ge \gamma P'_{\ell_s\ell_t}(m)$, which fails when $m \le m_0$. Thus, we set $P'_{\ell_s\ell_t}(m) = Q_{\ell_s\ell_t}(m)/\gamma$ for $m > m_0$. The remaining probability mass $a_{\ell_s\ell_t}$ is added to $P_{\ell_s\ell_t}(0)$ so that $P'_{\ell_s\ell_t}$ is a well-defined probability distribution.

It is straightforward to verify that $Q'_{\ell_s\ell_t}$ and $P'_{\ell_s\ell_t}$ are well-defined probability distributions, and

$$d_{\mathrm{TV}}(P'_{\ell_s\ell_t}, P_{\ell_s\ell_t}) \le 4(8q\ell^2)^{(m_0+1)} \tag{13.36}$$

as long as $\ell_s, \ell_t \leq 2\ell$ and $16q\ell^2 \leq 1$, where $\ell = N/n$. Then, for $1 \leq s < t \leq n$, the conditional distribution of $E(V_s, V_t)$ given ℓ_s, ℓ_t, and $A_{s,t}$ is given by

$$E(V_s, V_t) \sim \begin{cases} P'_{\ell_s \ell_t} & \text{if } A_{st} = 1, \ell_s, \ell_t \leq 2\ell, \\ Q'_{\ell_s \ell_t} & \text{if } A_{st} = 0, \ell_s, \ell_t \leq 2\ell, \\ Q_{\ell_s \ell_t} & \text{if } \max\{\ell_s, \ell_t\} > 2\ell. \end{cases} \tag{13.37}$$

Next we show that the randomized reduction defined above maps $\mathcal{G}(n,\gamma)$ into $\mathcal{G}(N,q)$ under the null hypothesis and $\mathcal{G}(n,k,\gamma)$ approximately into $\mathcal{G}(N,K,p,q)$ under the alternative hypothesis. By construction, $(1-\gamma)Q'_{\ell_s\ell_t} + \gamma P'_{\ell_s\ell_t} = Q_{\ell_s\ell_t} = \text{Binom}(\ell_s\ell_t, q)$ and therefore the null distribution of the PC problem is exactly matched to that of the PDS problem, i.e., $P_{\widetilde{G}|H_0^C} = \mathbb{P}_0$. The core of the proof lies in establishing that the alternative distributions are approximately matched. The key observation is that, by (13.36), $P'_{\ell_s\ell_t}$ is close to $P_{\ell_s\ell_t} = \text{Binom}(\ell_s\ell_t, p)$ and thus, for nodes with distinct parents $s \neq t$ in the planted clique, the number of edges $E(V_s, V_t)$ is approximately distributed as the desired $\text{Binom}(\ell_s\ell_t, p)$; for nodes with the same parent s in the planted clique, even though $E(V_s, V_s)$ is distributed as $\text{Binom}\left(\binom{\ell_s}{2}, q\right)$ which is not sufficiently close to the desired $\text{Binom}\left(\binom{\ell_s}{2}, p\right)$, after averaging over the random partition $\{V_s\}$, the total variation distance becomes negligible. More formally, we have the following proposition; the proof is postponed to the next section.

PROPOSITION 13.1 Let $\ell, n \in \mathbb{N}$, $k \in [n]$ and $\gamma \in (0, \frac{1}{2}]$. Let $N = \ell n$, $K = k\ell$, $p = 2q$, and $m_0 = \lfloor \log_2(1/\gamma) \rfloor$. Assume that $16q\ell^2 \leq 1$ and $k \geq 6e\ell$. If $G \sim \mathcal{G}(n,\gamma)$, then $\widetilde{G} \sim \mathcal{G}(N,q)$, i.e., $P_{\widetilde{G}|H_0^C} = \mathbb{P}_0$. If $G \sim \mathcal{G}(n,k,1,\gamma)$, then

$$d_{\text{TV}}\left(P_{\widetilde{G}|H_1^C}, \mathbb{P}_1\right) \leq e^{-\frac{K}{12}} + 1.5ke^{-\frac{\ell}{18}} + 2k^2(8q\ell^2)^{m_0+1} + 0.5\sqrt{e^{72e^2q\ell^2} - 1}$$
$$+ \sqrt{0.5k}e^{-\frac{\ell}{36}}. \tag{13.38}$$

Reduction Scheme in the Gaussian Case

The same reduction scheme can be tweaked slightly to work for the Gaussian case, which, in fact, needs only the PC hypothesis for $\gamma = \frac{1}{2}$.[6] In this case, we aim to map an adjacency matrix $A \in \{0,1\}^{n \times n}$ to a symmetric data matrix $\widetilde{A} \in \mathbb{R}^{N \times N}$ with zero diagonal, or, equivalently, a *weighted complete* graph $\widetilde{G}$.

For any $1 \leq s \leq t \leq n$, we let $E(V_s, V_t)$ denote the average weights of edges between V_s and V_t in $\widetilde{G}$. As for the Bernoulli model, we will first generate $E(V_s, V_t)$ randomly with a properly chosen conditional probability distribution. Since $E(V_s, V_t)$ is a sufficient statistic for the set of Gaussian edge weights, the specific weight assignment can be generated from the average weight using the same kernel both for the null and for the alternative.

To see how this works, consider a general setup where $X_1, \ldots, X_n \overset{\text{i.i.d.}}{\sim} \mathcal{N}(\mu, 1)$. Let $\bar{X} = (1/n)\sum_{i=1}^n X_i$. Then we can simulate $X_1, \ldots, X_n$ on the basis of the sufficient statistic $\bar{X}$ as follows. Let $[v_0, v_1, \ldots, v_{n-1}]$ be an orthonormal basis for $\mathbb{R}^n$, with $v_0 = (1/\sqrt{n})\mathbf{1}$ and

[6] The original reduction proof in [9] for the submatrix detection problem crucially relies on the Gaussianity and the reduction maps a bigger planted-clique instance into a smaller instance for submatrix detection by means of averaging.

$\mathbf{1} = (1,\dots,1)^\top$. Generate $Z_1,\dots,Z_{n-1} \overset{\text{i.i.d.}}{\sim} \mathcal{N}(0,1)$. Then $\bar{X}\mathbf{1} + \sum_{i=1}^{n-1} Z_i v_i \sim \mathcal{N}(\mu\mathbf{1}, I_n)$. Using this general procedure, we can generate the weights $\widetilde{A}_{V_s,V_t}$ on the basis of $E(V_s,V_t)$.

It remains to specify, for $1 \le s \le t \le n$, the conditional distribution of $E(V_s,V_t)$ given ℓ_s,ℓ_t, and $A_{s,t}$. As for the Bernoulli case, conditioned on ℓ_s and ℓ_t, ideally we would want to find a Markov kernel from $A_{s,t}$ to $E(V_s,V_t)$ which maps Bern(1) to the desired distribution $\mathcal{N}(\mu, 1/\ell_s\ell_t)$ and Bern(1/2) to $\mathcal{N}(0, 1/\ell_s\ell_t)$, depending on whether both s and t are in the clique or not, respectively. This objective can be accomplished approximately in terms of the total variation. For $s = t \in [n]$, let $E(V_s,V_t) \sim \mathcal{N}(0, 1/\ell_s\ell_t)$. For $1 \le s < t \le n$, denote $P_{\ell_s\ell_t} \triangleq \mathcal{N}(\mu, 1/\ell_s\ell_t)$ and $Q_{\ell_s\ell_t} \triangleq \mathcal{N}(0, 1/\ell_s\ell_t)$, with density functions $p_{\ell_s\ell_t}(x)$ and $q_{\ell_s\ell_t}(x)$, respectively.

Fix $\gamma = \frac{1}{2}$. Note that

$$\frac{q_{\ell_s\ell_t}(x)}{p_{\ell_s\ell_t}(x)} = \exp[\ell_s\ell_t\mu(\mu/2 - x)] \ge \gamma$$

if and only if $x \le x_0 \triangleq \mu/2 + (1/\mu\ell_s\ell_t)\log(1/\gamma)$. Therefore, we define $P'_{\ell_s\ell_t}$ and $Q'_{\ell_s\ell_t}$ with the density $q'_{\ell_s\ell_t} = (1/(1-\gamma))(q_{\ell_s\ell_t} - \gamma p'_{\ell_s\ell_t})$ and

$$p'_{\ell_s\ell_t}(x) = \begin{cases} p_{\ell_s\ell_t}(x) + f_{\ell s \ell_t}(2\mu - x) & \text{for } x < 2\mu - x_0, \\ p_{\ell_s\ell_t}(x) & \text{for } x \le x_0, \\ (1/\gamma) q_{\ell_s\ell_t}(x) & \text{for } x > x_0, \end{cases}$$

where $f_{\ell s \ell_t}(x) = p_{\ell_s\ell_t}(x) - (1/\gamma) q_{\ell_s\ell_t}(x)$. Let

$$a_{\ell_s\ell_t} = \int_{x_0}^{\infty} f_{\ell s \ell_t}(x)dx \le \bar{\Phi}\left(-\frac{\mu}{2}\sqrt{\ell_s\ell_t} + \frac{1}{\mu\sqrt{\ell_s\ell_t}}\log\left(\frac{1}{\gamma}\right)\right).$$

As for the Bernoulli case, it is straightforward to verify that $Q'_{\ell_s\ell_t}$ and $P'_{\ell_s\ell_t}$ are well-defined probability distributions, and

$$d_{\mathrm{TV}}(P'_{\ell_s\ell_t}, P_{\ell_s\ell_t}) = a_{\ell_s\ell_t} \le \bar{\Phi}\left(\frac{1}{2\mu\sqrt{\ell_s\ell_t}}\log\left(\frac{1}{\gamma}\right)\right) \le \exp\left(-\frac{1}{32\mu^2\ell^2}\log^2\left(\frac{1}{\gamma}\right)\right) \tag{13.39}$$

as long as $\ell_s, \ell_t \le 2\ell$ and $4\mu^2\ell^2 \le \log(1/\gamma)$, where $\ell = N/n$. Following the same argument as in the Bernoulli case, we can obtain a counterpart to Proposition 13.1.

PROPOSITION 13.2 Let $\ell, n \in \mathbb{N}$, $k \in [n]$ and $\gamma = 1/2$. Let $N = \ell n$ and $K = k\ell$. Assume that $16\mu^2\ell \le 1$ and $k \ge 6e\ell$. Let $\mathbb{P}_0$ and $\mathbb{P}_1$ denote the desired null and alternative distributions of the submatrix detection problem (N, K, μ). If $G \sim \mathcal{G}(n,\gamma)$, then $P_{\widetilde{G}|H_0^C} = \mathbb{P}_0$. If $G \sim \mathcal{G}(n,k,1,\gamma)$, then

$$d_{\mathrm{TV}}\left(P_{\widetilde{G}|H_1^C}, \mathbb{P}_1\right) \le e^{-\frac{K}{12}} + 1.5ke^{-\frac{\ell}{18}} + \frac{k^2}{2}\exp\left(-\frac{\log^2 2}{32\mu^2\ell^2}\right) + 0.5\sqrt{e^{72e^2\mu^2\ell^2} - 1} + \sqrt{0.5k}e^{-\frac{\ell}{36}}. \tag{13.40}$$

Let us close this section with two remarks. First, to investigate the computational aspect of inference in the Gaussian model, an immediate hurdle is that the computational complexity is not well defined for tests dealing with samples drawn from non-discrete distributions, which cannot be represented by finitely many bits almost surely. To overcome this difficulty, we consider a sequence of *discretized* Gaussian models that is *asymptotically equivalent* to the original model in the sense of Le Cam [4] and hence preserves the statistical difficulty of the problem. In other words, the continuous model and its appropriately discretized counterpart are statistically indistinguishable and, more importantly, the computational complexity of tests on the latter is well defined. More precisely, for the submatrix detection model, provided that each entry of the $n \times n$ matrix A is quantized by $\Theta(\log n)$ bits, the discretized model is asymptotically equivalent to the previous model (cf. Section 3 and Theorem 1 of [9] for a precise bound on the Le Cam distance). With a slight modification, the above reduction scheme can be applied to the discretized model (cf. Section 4.2 of [9]).

Second, we comment on the distinctions between the reduction scheme here and the prior work that relies on the planted clique as the hardness assumption. Most previous work [63, 64, 68, 81] in the theoretical computer science literature uses the reduction from the PC problem to generate computationally hard instances of other problems and establish *worst-case* hardness results; the underlying distributions of the instances could be arbitrary. The idea of proving the hardness of a hypothesis-testing problem by means of approximate reduction from the planted-clique problem such that the reduced instance is close to the target hypothesis in total variation originates from the seminal work by Berthet and Rigollet [8] and the subsequent paper by Ma and Wu [9]. The main distinction between these works and the results presented here, which are based on the techniques in [10], is that Berthet and Rigollet [8] studied a composite-versus-composite testing problem and Ma and Wu [9] studied a simple-versus-composite testing problem, both in the minimax sense, as opposed to the simple-versus-simple hypothesis considered here and in [10], which constitutes a stronger hardness result. For the composite hypothesis, a reduction scheme works as long as the distribution of the reduced instance is close to *some* mixture distribution under the hypothesis. This freedom is absent in constructing reduction for the simple hypothesis, which renders the reduction scheme as well as the corresponding calculation of the total variation considerably more difficult. In contrast, for the simple-versus-simple hypothesis, the underlying distributions of the problem instances generated from the reduction must be close to the desired distributions in total variation both under the null hypothesis and under the alternative hypothesis.

13.4.3 Bounding the Total Variation Distance

Below we prove Proposition 13.1 and obtain the desired computational limits given by Fig. 13.1. We consider only the Bernoulli case as the derivations for the Gaussian case are analogous. The main technical challenge is bounding the total variation distance in (13.38).

Proof of Proposition 13.1 Let $[i,j]$ denote the unordered pair of i and j. For any set $I \subset [N]$, let $\mathcal{E}(I)$ denote the set of unordered pairs of distinct elements in I, i.e., $\mathcal{E}(I) = \{[i,j] : i,j \in S, i \neq j\}$, and let $\mathcal{E}(I)^c = \mathcal{E}([N]) \setminus \mathcal{E}(I)$. For $s,t \in [n]$ with $s \neq t$, let $\widetilde{G}_{V_sV_t}$ denote the bipartite graph where the set of left (right) vertices is V_s (V_t) and the set of edges is the set of edges in $\widetilde{G}$ from vertices in V_s to vertices in V_t. For $s \in [n]$, let $\widetilde{G}_{V_sV_s}$ denote the subgraph of $\widetilde{G}$ induced by V_s. Let $\widetilde{P}_{V_sV_t}$ denote the edge distribution of $\widetilde{G}_{V_sV_t}$ for $s,t \in [n]$.

It is straightforward to verify that the null distributions are exactly matched by the reduction scheme. Henceforth, we consider the alternative hypothesis, under which G is drawn from the planted-clique model $\mathcal{G}(n,k,\gamma)$. Let $C \subset [n]$ denote the planted clique. Define $S = \cup_{t\in C} V_t$ and recall that $K = k\ell$. Then $|S| \sim \mathrm{Binom}(N, K/N)$ and, conditional on $|S|$, S is uniformly distributed over all possible subsets of size $|S|$ in $[N]$. By the symmetry of the vertices of G, the distribution of $\widetilde{A}$ conditional on C does not depend on C. Hence, without loss of generality, we shall assume that $C = [k]$ henceforth. The distribution of $\widetilde{A}$ can be written as a mixture distribution indexed by the random set S as

$$\widetilde{A} \sim \widetilde{\mathbb{P}}_1 \triangleq \mathbb{E}_S\left[\widetilde{P}_{SS} \times \prod_{[i,j]\in\mathcal{E}(S)^c} \mathrm{Bern}(q)\right].$$

By the definition of $\mathbb{P}_1$,

$$\begin{aligned}
& d_{\mathrm{TV}}(\widetilde{\mathbb{P}}_1, \mathbb{P}_1) \\
&= d_{\mathrm{TV}}\left(\mathbb{E}_S\left[\widetilde{P}_{SS} \times \prod_{[i,j]\in\mathcal{E}(S)^c} \mathrm{Bern}(q)\right], \mathbb{E}_S\left[\prod_{[i,j]\in\mathcal{E}(S)} \mathrm{Bern}(p) \prod_{[i,j]\in\mathcal{E}(S)^c} \mathrm{Bern}(q)\right]\right) \\
&\le \mathbb{E}_S\left[d_{\mathrm{TV}}\left(\widetilde{P}_{SS} \times \prod_{[i,j]\in\mathcal{E}(S)^c} \mathrm{Bern}(q), \prod_{[i,j]\in\mathcal{E}(S)} \mathrm{Bern}(p) \prod_{[i,j]\in\mathcal{E}(S)^c} \mathrm{Bern}(q)\right)\right] \\
&= \mathbb{E}_S\left[d_{\mathrm{TV}}\left(\widetilde{P}_{SS}, \prod_{[i,j]\in\mathcal{E}(S)} \mathrm{Bern}(p)\right)\right] \\
&\le \mathbb{E}_S\left[d_{\mathrm{TV}}\left(\widetilde{P}_{SS}, \prod_{[i,j]\in\mathcal{E}(S)} \mathrm{Bern}(p)\right)\mathbf{1}_{\{|S|\le 1.5K\}}\right] + \exp(-K/12), \qquad (13.41)
\end{aligned}$$

where the first inequality follows from the convexity of $(P,Q) \mapsto d_{\mathrm{TV}}(P,Q)$, and the last inequality follows from applying the Chernoff bound to $|S|$. Fix an $S \subset [N]$ such that $|S| \le 1.5K$. Define $P_{V_tV_t} = \prod_{[i,j]\in\mathcal{E}(V_t)} \mathrm{Bern}(q)$ for $t \in [k]$ and $P_{V_sV_t} = \prod_{(i,j)\in V_s\times V_t} \mathrm{Bern}(p)$ for $1 \le s < t \le k$. By the triangle inequality,

$$\begin{aligned}
d_{\mathrm{TV}}\left(\widetilde{P}_{SS}, \prod_{[i,j]\in\mathcal{E}(S)} \mathrm{Bern}(p)\right) &\le d_{\mathrm{TV}}\left(\widetilde{P}_{SS}, \mathbb{E}_{V_1^k}\left[\prod_{1\le s\le t\le k} P_{V_sV_t} \,\middle|\, S\right]\right) \\
&\quad + d_{\mathrm{TV}}\left(\mathbb{E}_{V_1^k}\left[\prod_{1\le s\le t\le k} P_{V_sV_t} \,\middle|\, S\right], \prod_{[i,j]\in\mathcal{E}(S)} \mathrm{Bern}(p)\right). \qquad (13.42)
\end{aligned}$$

To bound the term on the first line on the right-hand side of (13.42), first note that, conditioned on the set S, $\{V_1^k\}$ can be generated as follows. Throw balls indexed by S into bins indexed by $[k]$ independently and uniformly at random; let V_t is the set of balls in the tth bin. Define the event $E = \{V_1^k : |V_t| \le 2\ell, t \in [k]\}$. Since $|V_t| \sim \mathrm{Binom}(|S|, 1/k)$ is stochastically dominated by $\mathrm{Binom}(1.5K, 1/k)$ for each fixed $1 \le t \le k$, it follows from the Chernoff bound and the union bound that $\mathbb{P}\{E^c\} \le k\exp(-\ell/18)$. Then we have

$$
\begin{aligned}
&d_{\mathrm{TV}}\left(\widetilde{P}_{SS}, \mathbb{E}_{V_1^k}\left[\prod_{1\le s\le t\le k} P_{V_sV_t} \,\middle|\, S\right]\right)\\
&\overset{(a)}{=} d_{\mathrm{TV}}\left(\mathbb{E}_{V_1^k}\left[\prod_{1\le s\le t\le k} \widetilde{P}_{V_sV_t} \,\middle|\, S\right], \mathbb{E}_{V_1^k}\left[\prod_{1\le s\le t\le k} P_{V_sV_t} \,\middle|\, S\right]\right)\\
&\le \mathbb{E}_{V_1^k}\left[d_{\mathrm{TV}}\left(\prod_{1\le s\le t\le k} \widetilde{P}_{V_sV_t}, \prod_{1\le s\le t\le k} P_{V_sV_t}\right) \,\middle|\, S\right]\\
&\le \mathbb{E}_{V_1^k}\left[d_{\mathrm{TV}}\left(\prod_{1\le s\le t\le k} \widetilde{P}_{V_sV_t}, \prod_{1\le s\le t\le k} P_{V_sV_t}\right)\mathbf{1}_{\{V_1^k\in E\}} \,\middle|\, S\right] + k\exp(-\ell/18),
\end{aligned}
$$

where (a) holds because, conditional on V_1^k, $\{\widetilde{A}_{V_sV_t} : s,t\in[k]\}$ are independent. Recall that $\ell_t = |V_t|$. For any fixed $V_1^k \in E$, we have

$$
\begin{aligned}
&d_{\mathrm{TV}}\left(\prod_{1\le s\le t\le k} \widetilde{P}_{V_sV_t}, \prod_{1\le s\le t\le k} P_{V_sV_t}\right)\\
&\overset{(a)}{=} d_{\mathrm{TV}}\left(\prod_{1\le s<t\le k} \widetilde{P}_{V_sV_t}, \prod_{1\le s<t\le k} P_{V_sV_t}\right)\\
&\overset{(b)}{=} d_{\mathrm{TV}}\left(\prod_{1\le s<t\le k} P'_{\ell_s\ell_t}, \prod_{1\le s<t\le k} P_{\ell_s\ell_t}\right)\\
&\le d_{\mathrm{TV}}\left(\prod_{1\le s<t\le k} P'_{\ell_s\ell_t}, \prod_{1\le s<t\le k} P_{\ell_s\ell_t}\right)\\
&\le \sum_{1\le s<t\le k} d_{\mathrm{TV}}\left(P'_{\ell_s\ell_t}, P_{\ell_s\ell_t}\right) \overset{(c)}{\le} 2k^2(8q\ell^2)^{(m_0+1)},
\end{aligned}
$$

where (a) follows since $\widetilde{P}_{V_tV_t} = P_{V_tV_t}$ for all $t\in[k]$; (b) is because the number of edges $E(V_s, V_t)$ is a sufficient statistic for testing $\widetilde{P}_{V_sV_t}$ versus $P_{V_sV_t}$ on the submatrix $A_{V_sV_t}$ of the adjacency matrix; and (c) follows from the total variation bound (13.36). Therefore,

$$
d_{\mathrm{TV}}\left(\widetilde{P}_{SS}, \mathbb{E}_{V_1^k}\left[\prod_{1\le s\le t\le k} P_{V_sV_t} \,\middle|\, S\right]\right) \le 2k^2(8q\ell^2)^{(m_0+1)} + k\exp(-\ell/18). \tag{13.43}
$$

To bound the term on the second line on the right-hand side of (13.42), applying Lemma 9 of [10], which is a conditional version of the second-moment method, yields

$$
\begin{aligned}
&d_{\mathrm{TV}}\left(\mathbb{E}_{V_1^k}\left[\prod_{1\le s\le t\le k} P_{V_sV_t} \,\middle|\, S\right], \prod_{[i,j]\in\mathcal{E}(S)} \mathrm{Bern}(p)\right)\\
&\le \frac{1}{2}\mathbb{P}\{E^{\mathrm{c}}\} + \frac{1}{2}\sqrt{\mathbb{E}_{V_1^k;\widetilde{V}_1^k}\left[g(V_1^k,\widetilde{V}_1^k)\mathbf{1}_{\{V_1^k\in E\}}\mathbf{1}_{\{\widetilde{V}_1^k\in E\}} \,\middle|\, S\right] - 1 + 2\mathbb{P}\{E^{\mathrm{c}}\}},
\end{aligned} \tag{13.44}
$$

where

$$
\begin{aligned}
g(V_1^k, \widetilde{V}_1^k) &= \int \frac{\prod_{1\le s\le t\le k} P_{V_sV_t} \prod_{1\le s\le t\le k} P_{\widetilde{V}_s\widetilde{V}_t}}{\prod_{[i,j]\in\mathcal{E}(S)} \mathrm{Bern}(p)} \\
&= \prod_{s,t=1}^{k} \left(\frac{q^2}{p} + \frac{(1-q)^2}{1-p}\right)^{\binom{|V_s\cap\widetilde{V}_t|}{2}} = \prod_{s,t=1}^{k} \left(\frac{1-\frac{3}{2}q}{1-2q}\right)^{\binom{|V_s\cap\widetilde{V}_t|}{2}}. \qquad (13.45)
\end{aligned}
$$

Let $X \sim \mathrm{Bin}(1.5K, 1/k^2)$ and $Y \sim \mathrm{Bin}(3\ell, e/k)$. It follows that

$$
\begin{aligned}
&\mathbb{E}_{V_1^k;\widetilde{V}_1^k}\left[\prod_{s,t=1}^{k}\left(\frac{1-\frac{3}{2}q}{1-2q}\right)^{\binom{|V_s\cap\widetilde{V}_t|}{2}} \prod_{s,t=1}^{k} \mathbf{1}_{\{|V_s|\le 2\ell, |\widetilde{V}_t|\le 2\ell\}} \,\middle|\, S\right] \\
&\overset{(a)}{\le} \mathbb{E}_{V_1^k;\widetilde{V}_1^k}\left[\prod_{s,t=1}^{k} e^{q\binom{|V_s\cap\widetilde{V}_t|\wedge 2\ell}{2}} \,\middle|\, S\right] \\
&\overset{(b)}{\le} \prod_{s,t=1}^{k} \mathbb{E}\left[e^{q\binom{|V_s\cap\widetilde{V}_t|\wedge 2\ell}{2}} \,\middle|\, S\right] \\
&\overset{(c)}{\le} \left(\mathbb{E}\left[e^{q\binom{X\wedge 2\ell}{2}}\right]\right)^{k^2} \overset{(d)}{\le} \mathbb{E}\left[e^{q\binom{Y}{2}}\right]^{k^2} \overset{(e)}{\le} \exp(72e^2q\ell^2), \qquad (13.46)
\end{aligned}
$$

where (a) follows from $1+x \le e^x$ for all $x \ge 0$ and $q < 1/4$; (b) follows from the negative association property of $\{|V_s \cap \widetilde{V}_t| : s,t \in [k]\}$ proved in Lemma 10 of [10] in view of the monotonicity of $x \mapsto e^{q\binom{x\wedge 2\ell}{2}}$ on $\mathbb{R}_+$; (c) follows because $|V_s \cap \widetilde{V}_t|$ is stochastically dominated by $\mathrm{Binom}(1.5K, 1/k^2)$ for all $(s,t) \in [k]^2$; (d) follows from Lemma 11 of [10]; (e) follows from Lemma 12 of [10] with $\lambda = q/2$ and $q\ell \le 1/8$. Therefore, by (13.44)

$$
\begin{aligned}
d_{\mathrm{TV}}\left(\widetilde{P}_{SS}, \prod_{[i,j]\in\mathcal{E}(S)} \mathrm{Bern}(p)\right) &\le 0.5ke^{-\frac{\ell}{18}} + 0.5\sqrt{e^{72e^2q\ell^2} - 1 + 2ke^{-\frac{\ell}{18}}} \\
&\le 0.5ke^{-\frac{\ell}{18}} + 0.5\sqrt{e^{72e^2q\ell^2} - 1} + \sqrt{0.5k}e^{-\frac{\ell}{36}}. \qquad (13.47)
\end{aligned}
$$

Proposition 13.1 follows by combining (13.41), (13.42), (13.44), and (13.47).

The following theorem establishes the computational hardness of the PDS problem in the interior of the hard region in Fig. 13.1.

THEOREM 13.3 *Assume the* PC *hypothesis holds for all* $0 < \gamma \le 1/2$. *Let* $\alpha > 0$ *and* $0 < \beta < 1$ *be such that*

$$
\max\{0, 2\alpha - 1\} \triangleq \beta^\sharp < \beta < \frac{\alpha}{2}. \qquad (13.48)
$$

Then there exists a sequence $\{(N_\ell, K_\ell, q_\ell)\}_{\ell\in\mathbb{N}}$ *satisfying*

$$
\lim_{\ell\to\infty} \frac{\log(1/q_\ell)}{\log N_\ell} = 2\beta, \quad \lim_{\ell\to\infty} \frac{\log K_\ell}{\log N_\ell} = \alpha
$$

such that, for any sequence of randomized polynomial-time tests $\phi_\ell : \{0,1\}^{\binom{N_\ell}{2}} \to \{0,1\}$ *for the* $\mathsf{PDS}(N_\ell, K_\ell, 2q_\ell, q_\ell)$ *problem, the type-I-plus-type-II error probability is lower-bounded by*

$$\liminf_{\ell\to\infty} \mathbb{P}_0\{\phi_\ell(G') = 1\} + \mathbb{P}_1\{\phi_\ell(G') = 0\} \geq 1,$$

where $G' \sim \mathcal{G}(N,q)$ *under* H_0 *and* $G' \sim \mathcal{G}(N,K,p,q)$ *under* H_1.

Proof Let $m_0 = \lfloor \log_2(1/\gamma) \rfloor$. By (13.48), there exist $0 < \gamma \leq 1/2$ and thus m_0 such that

$$2\beta < \alpha < \frac{1}{2} + \frac{m_0\beta + 2}{2m_0\beta + 1}\beta - \frac{1}{m_0\beta}. \tag{13.49}$$

Fix $\beta > 0$ and $0 < \alpha < 1$ that satisfy (13.49). Let $\delta = 1/(m_0\beta)$. Then it is straightforward to verify that $((2+m_0\delta)/(2+\delta))\beta \geq \frac{1}{2} - \delta + ((1+2\delta)/(2+\delta))\beta$. It follows from the assumption (13.49) that

$$2\beta < \alpha < \min\left\{\frac{2+m_0\delta}{2+\delta}\beta,\ \frac{1}{2} - \delta + \frac{1+2\delta}{2+\delta}\beta\right\}. \tag{13.50}$$

Let $\ell \in \mathbb{N}$ and $q_\ell = \ell^{-(2+\delta)}$. Define

$$n_\ell = \lfloor \ell^{\frac{2+\delta}{2\beta}-1} \rfloor,\ k_\ell = \lfloor \ell^{\frac{(2+\delta)\alpha}{2\beta}-1} \rfloor,\ N_\ell = n_\ell \ell,\ K_\ell = k_\ell \ell. \tag{13.51}$$

Then

$$\begin{aligned} \lim_{\ell\to\infty} \frac{\log(1/q_\ell)}{\log N_\ell} &= \frac{2+\delta}{(2+\delta)/(2\beta) - 1 + 1} = 2\beta, \\ \lim_{\ell\to\infty} \frac{\log K_\ell}{\log N_\ell} &= \frac{(2+\delta)\alpha/(2\beta) - 1 + 1}{(2+\delta)/(2\beta) - 1 + 1} = \alpha. \end{aligned} \tag{13.52}$$

Suppose that for the sake of contradiction there exists a small $\epsilon > 0$ and a sequence of randomized polynomial-time tests $\{\phi_\ell\}$ for $\mathsf{PDS}(N_\ell, K_\ell, 2q_\ell, q_\ell)$, such that

$$\mathbb{P}_0\{\phi_{N_\ell,K_\ell}(G') = 1\} + \mathbb{P}_1\{\phi_{N_\ell,K_\ell}(G') = 0\} \leq 1 - \epsilon$$

holds for arbitrarily large ℓ, where G' is the graph in the $\mathsf{PDS}(N_\ell, K_\ell, 2q_\ell, q_\ell)$. Since $\alpha > 2\beta$, we have $k_\ell \geq \ell^{1+\delta}$. Therefore, $16q_\ell\ell^2 \leq 1$ and $k_\ell \geq 6e\ell$ for all sufficiently large ℓ. Applying Proposition 13.1, we conclude that $G \mapsto \phi(\widetilde{G})$ is a randomized polynomial-time test for $\mathsf{PC}(n_\ell, k_\ell, \gamma)$ whose type-I-plus-type-II error probability satisfies

$$\mathbb{P}_{H_0^C}\{\phi_\ell(\widetilde{G}) = 1\} + \mathbb{P}_{H_1^C}\{\phi_\ell(\widetilde{G}) = 0\} \leq 1 - \epsilon + \xi, \tag{13.53}$$

where ξ is given by the right-hand side of (13.38). By the definition of q_ℓ, we have $q_\ell\ell^2 = \ell^{-\delta}$ and thus

$$k_\ell^2(q_\ell\ell^2)^{m_0+1} \leq \ell^{(2+\delta)\alpha/\beta - 2 - (m_0+1)\delta} \leq \ell^{-\delta},$$

where the last inequality follows from (13.50). Therefore $\xi \to 0$ as $\ell \to \infty$. Moreover, by the definition in (13.51),

$$\lim_{\ell\to\infty} \frac{\log k_\ell}{\log n_\ell} = \frac{(2+\delta)\alpha/(2\beta) - 1}{(2+\delta)/(2\beta) - 1} \leq \frac{1}{2} - \delta,$$

where the above inequality follows from (13.50). Therefore, (13.53) contradicts the assumption that the PC hypothesis holds for γ.

13.5 Discussion and Open Problems

Recent years have witnessed a great deal of progress on understanding the information-theoretical and computational limits of various statistical problems with planted structures. As outlined in this survey, various techniques to identify the information-theoretic limits are available. In some cases, polynomial-time procedures have been shown to achieve the information-theoretic limits. However, in many other cases, it is believed that there exists a wide gap between the information-theoretic limits and the computational limits. For the planted-clique problem, a recent exciting line of research has identified the performance limits of a sum-of-squares hierarchy [71–73, 82, 83]. Under the PC hypothesis, complexity-theoretic computational lower bounds have been derived for sparse PCA [8], submatrix location [9], single-community detection [10], and various other detection problems with planted structures [13]. Despite these encouraging results, a variety of interesting questions remain open. Below we list a few representative problems. Closing the observed computational gap, or, equally importantly, disproving the possibility thereof on rigorous complexity-theoretic grounds, is an exciting new topic at the intersection of high-dimensional statistics, information theory, and computer science.

Computational Lower Bounds for Recovering the Planted Dense Subgraph
Closely related to the PDS detection problem is the recovery problem, where, given a graph generated from $\mathcal{G}(N,K,p,q)$, the task is to recover the planted dense subgraph. Consider the asymptotic regime depicted in Fig. 13.1. It has been shown in [16, 84] that exact recovery is information-theoretically possible if and only if $\beta < \alpha/2$ and can be achieved in polynomial time if $\beta < \alpha - \frac{1}{2}$. Our computational lower bounds for the PDS detection problem imply that the planted dense subgraph is hard to approximate to any constant factor if $\max(0, 2\alpha - 1) < \beta < \alpha/2$ (the hard regime in Fig. 13.1). Whether the planted dense subgraph is hard to approximate with any constant factor in the regime of $\alpha - \frac{1}{2} \le \beta \le \min\{2\alpha - 1, \alpha/2\}$ is an interesting open problem. For the Gaussian case, Cai *et al.* [23] showed that exact recovery is computationally hard, $\beta > \alpha - \frac{1}{2}$, by assuming a variant of the standard PC hypothesis (see p. 1425 of [23]).

Finally, we note that in order to prove our computational lower bounds for the planted dense subgraph detection problem in Theorem 13.3, we have assumed that the PC detection problem is hard for any constant $\gamma > 0$. An important open problem is to show by means of reduction that, if the PC detection problem is hard with $\gamma = 0.5$, then it is also hard with $\gamma = 0.49$.

Computational Lower Bounds within the Sum-of-Squares Hierarchy
For the single-community model, Hajek *et al.* [56] obtained a tight characterization of the performance limits of semidefinite programming (SDP) relaxations, corresponding to the sum-of-squares hierarchy with degree 2. In particular, (1) if $K = \omega(n/\log n)$, SDP attains the information-theoretic threshold with sharp constants; (2) if $K = \Theta(n/\log n)$,

SDP is sub-optimal by a constant factor; and (3) if $K = o(n/\log n)$ and $K \to \infty$, SDP is order-wise sub-optimal. An interesting future direction would be to generalize this result to the sum-of-squares hierarchy, showing that sum-of-squares results with any constant degree are sub-optimal when $K = o(n \log n)$.

Furthermore, if $K \geq (n/\log n)(1/(8e) + o(1))$ for the Gaussian case and $K \geq (n/\log n)(\rho_{\text{BP}}(p/q) + o(1))$ for the Bernoulli case, exact recovery can be attained in nearly linear time via message passing plus clean-up [54, 55] whenever information-theoretically possible. An interesting question is whether exact recovery beyond the aforementioned two limits is possible in polynomial time.

Recovering Multiple Communities

Consider the stochastic block model under which n vertices are partitioned into k equal-sized communities, and two vertices are connected by an edge with probability p if they are from the same community and q otherwise.

First let us focus on correlated recovery in the sparse regime where $p = a/n$ and $q = b/n$ for two fixed constants $a > b$ in the assortative case. For $k = 2$, it has been shown [26, 31, 39] that the information-theoretic and computational thresholds coincide at $(a-b)^2 = 2(a+b)$. Employing statistical physics heuristics, it is further conjectured that the information-theoretic and computational thresholds continue to coincide for $k = 3, 4$, but depart from each other for $k \geq 5$; however, a rigorous proof remains to be derived.

Next let us turn to exact recovery in the relatively sparse regime where $p = a\log n/n$ and $q = b\log n/n$ for two fixed constants $a > b$. For $k = \Theta(1)$, it has been shown that the SDP relaxations achieve the information-theoretic limit $\sqrt{a} - \sqrt{b} > \sqrt{k}$. Furthermore, it has been shown that SDP continues to be optimal for $k = o(\log n)$, but ceases to be optimal for $k = \Theta(\log n)$. It is conjectured in [16] that no polynomial-time procedure can be optimal for $k = \Theta(\log n)$.

Estimating Graphons

Graphon is a powerful network model for studying large networks [85]. Concretely, given n vertices, the edges are generated independently, connecting each pair of two distinct vertices i and j with a probability $M_{ij} = f(x_i, x_j)$, where $x_i \in [0,1]$ is the latent feature vector of vertex i that captures various characteristics of vertex i; $f : [0,1] \times [0,1] \to [0,1]$ is a symmetric function called a graphon. The problem of interest is to estimate either the edge probability matrix M or the graphon f on the basis of the observed graph.

- When f is a step function which corresponds to the stochastic block model with k blocks for some k, the minimax optimal estimation error rate is shown to be on the order of $k^2/n^2 + \log k/n$ [86], while the currently best error rate achievable in polynomial time is k/n [87].
- When f belongs to Hölder or Sobolev space with smoothness index α, the minimax optimal rate is shown to be $n^{-2\alpha/(\alpha+1)}$ for $\alpha < 1$ and $\log n/n$ for $\alpha > 1$ [86], while the best error rate achievable in polynomial time that is known in the literature is $n^{-2\alpha/(2\alpha+1)}$ [88].

For both cases, it remains to be determined whether the minimax optimal rate can be achieved in polynomial time.

Sparse PCA

Consider the following *spiked Wigner* model, where the underlying signal is a rank-one matrix:

$$X = \frac{\lambda}{\sqrt{n}} v v^{\mathrm{T}} + W. \tag{13.54}$$

Here, $v \in \mathbb{R}^n$, $\lambda > 0$ and $W \in \mathbb{R}^{n \times n}$ is a Wigner random matrix with $W_{ii} \overset{\text{i.i.d.}}{\sim} \mathcal{N}(0,2)$ and $W_{ij} = W_{ij} \overset{\text{i.i.d.}}{\sim} \mathcal{N}(0,1)$ for $i < j$. We assume that for some $\gamma \in [0,1]$ the support of v is drawn uniformly from all $\binom{n}{\gamma n}$ subsets $S \subset [n]$ with $|S| = \gamma n$. Once the support has been chosen, each non-zero component v_i is drawn independently and uniformly from $\{\pm \gamma^{-1/2}\}$, so that $\|v\|_2^2 = n$. When γ is small, the data matrix X is a sparse, rank-one matrix contaminated by Gaussian noise. For detection, we also consider a null model of $\lambda = 0$ where $X = W$.

One natural approach for this problem is PCA: that is, diagonalize X and use its leading eigenvector $\widehat{v}$ as an estimate of v. Using the theory of random matrices with rank-one perturbations [27–29], both detection and correlated recovery of v are possible if and only if $\lambda > 1$. Intuitively, PCA exploits only the low-rank structure of the underlying signal, and not the sparsity of v; it is natural to ask whether one can succeed in detection or reconstruction for some $\lambda < 1$ by taking advantage of this additional structure. Through analysis of an approximate message-passing algorithm and the free energy, it has been conjectured [46, 89] that there exists a critical sparsity threshold $\gamma^* \in (0,1)$ such that, if $\gamma \geq \gamma^*$, then both the information-theoretic threshold and the computational threshold are given by $\lambda = 1$; if $\gamma < \gamma^*$, then the computational threshold is given by $\lambda = 1$, but the information-theoretic threshold for λ is strictly smaller. A recent series of papers has identified the sharp information-theoretic threshold for correlated recovery through the Guerra interpolation technique and the cavity method [46, 49, 50, 90]. Also, the sharp information-theoretic threshold for detection has recently been determined in [25]. However, there is no rigorous evidence justifying the clain that $\lambda = 1$ is the computational threshold.

Tensor PCA

We can also consider a planted tensor model, in which we observe an order-k tensor

$$X = \lambda v^{\otimes k} + W, \tag{13.55}$$

where v is uniformly distributed over the unit sphere in $\mathbb{R}^n$ and $W \in (\mathbb{R}^n)^{\otimes k}$ is a totally symmetric noise tensor with Gaussian entries $\mathcal{N}(0, 1/n)$ (see Section 3.1 of [91] for a precise definition). This model is known as the *p-spin model* in statistical physics, and is widely used in machine learning and data analysis to model high-order correlations in a dataset. A natural approach is tensor PCA, which coincides with the MLE: $\min_{\|u\|_2=1} \langle X, u^{\otimes k} \rangle$. When $k = 2$, this reduces to standard PCA, which can be efficiently computed by singular value decomposition; however, as soon as $k \geq 3$, tensor PCA becomes NP-hard in the worst case [92].

Previous work [24, 91, 93] has shown that tensor PCA achieves consistent estimation of v if $\lambda \gtrsim \sqrt{k \log k}$, while this is information-theoretically impossible if $\lambda \lesssim \sqrt{k \log k}$. The exact location of the information-theoretic threshold for any k was determined recently in [94], but all known polynomial-time algorithms fail far from this threshold. A "tensor unfolding" algorithm is shown in [93] to succeed if $\lambda \gtrsim n^{(\lceil k/2 \rceil - 1)/2}$. In the special case $k = 3$, it is further shown in [95] that a degree-4 sum-of-squares relaxation succeeds if $\lambda = \omega(n \log n)^{1/4}$ and fails if $\lambda = O(n/\log n)^{1/4}$. More recent work [96] shows that a spectral method achieves consistent estimation provided that $\lambda = \Omega(n^{1/4})$, improving the positive result in [95] by a poly-logarithmic factor. It remains to be determined whether any polynomial-time algorithm succeeds in the regime of $1 \lesssim \lambda \lesssim n^{1/4}$. Under a hypergraph version of the planted-clique detection hypothesis, it is shown in [96] that no polynomial-time algorithm can succeed when $\lambda \le n^{1/4-\epsilon}$ for an arbitrarily small constant $\epsilon > 0$. It remains to be determined whether the usual planted-clique problem can be reduced to the hypergraph version.

Gaussian Mixture Clustering
Consider the following model of clustering in high dimensions. Let $v_1, \dots, v_k$ be i.i.d. as $\mathcal{N}(0, k/(k-1)\mathbf{I}_n)$, and define $\overline{v} = (1/k)\sum_s v_s$ to be their mean. The scaling of the expected norm of each v_s with k ensures that $\mathbb{E}\|v_s - \overline{v}\|_2^2 = n$ for all $1 \le s \le k$. For a fixed parameter $\alpha > 0$, we then generate $m = \alpha n$ points $x_i \in \mathbb{R}^n$ which are partitioned into k clusters of equal size by a balanced partition $\sigma : [n] \to [k]$, again chosen uniformly at random from all such partitions. For each data point i, let $\sigma_i \in [k]$ denote its cluster index, and generate x_i independently according to a Gaussian distribution with mean $\sqrt{\rho/n}(v_{\sigma_i} - \overline{v})$ and identity covariance matrix, where $\rho > 0$ is a fixed parameter characterizing the separation between clusters. Equivalently, this model can be described by the following matrix form:

$$X = \sqrt{\frac{\rho}{n}}\left(S - \frac{1}{k}\mathbf{J}_{m,k}\right)V^{\mathrm{T}} + W, \tag{13.56}$$

where $X = [x_1, \dots, x_m]^{\mathrm{T}}$, $V = [v_1, \dots, v_k]$, S is an $m \times k$ matrix with $S_{i,t} = \mathbf{1}_{\sigma_i = t}$, $\mathbf{J}_{m,k}$ is the $m \times k$ all-one matrix, and $W_{i,j} \overset{\text{i.i.d.}}{\sim} \mathcal{N}(0,1)$. In the null model, there is no cluster structure and $X = W$. The subtraction of $\mathbf{J}_{m,k}/k$ centers the signal matrix so that $\mathbb{E}X = 0$ in both models. It follows from the celebrated BBP phase transition [27, 97] that detection and correlated recovery using spectral methods is possible if and only if $\rho\sqrt{\alpha} > (k-1)$. In contrast, detection and correlated recovery are shown to be information-theoretically possible if $\rho > 2\sqrt{k \log k/\alpha} + 2\log k$. The sharp characterization of the information-theoretic limit still remains to be done, and it has been conjectured [98] that the computational threshold coincides with the spectral detection threshold.

A13.1 Mutual Information Characterization of Correlated Recovery

We consider a general setup. Let the number of communities k be a constant. Denote the membership vector by $\sigma = (\sigma_1, \dots, \sigma_n) \in [k]^n$ and the observation is $A = (A_{ij} : 1 \le i < j \le n)$. Assume the following conditions.

A1 For any permutation $\pi \in S_k$, (σ, A) and $(\pi(\sigma), A)$ are equal in law, where $\pi(\sigma) \triangleq (\pi(\sigma_1), \ldots, \pi(\sigma_n))$.

A2 For any $i \neq j \in [n]$, $I(\sigma_i, \sigma_j; A) = I(\sigma_1, \sigma_2; A)$.

A3 For any $z_1, z_2 \in [k]$, $\mathbb{P}\{\sigma_1 = z_1, \sigma_2 = z_2\} = 1/k^2 + o(1)$ as $n \to \infty$.

These assumptions are satisfied for example for k-community SBM (where each pair of vertices i and j are connected independently with probability p if $\sigma_i = \sigma_j$ and q otherwise), and the membership vector σ can be uniformly distributed either on $[k]^n$ or on the set of equal-sized k-partition of $[n]$.

Recall that correlated recovery entails the following. For any $\sigma, \widehat{\sigma} \in [k]^n$, define the overlap:

$$o(\sigma, \widehat{\sigma}) = \frac{1}{n} \max_{\pi \in S_k} \sum_{i \in [n]} \left(\mathbf{1}_{\{\pi(\sigma_i) = \widehat{\sigma}_i\}} - \frac{1}{k} \right). \tag{13.57}$$

We say an estimator $\widehat{\sigma} = \widehat{\sigma}(A)$ achieves correlated recovery if[7]

$$\mathbb{E}[o(\sigma, \widehat{\sigma})] = \Omega(1), \tag{13.58}$$

that is, the misclassification rate, up to a global permutation, outperforms random guessing. Under the above three assumptions, we have the following characterization of correlated recovery.

LEMMA A13.1 *Correlated recovery is possible if and only if* $I(\sigma_1, \sigma_2; A) = \Omega(1)$.

Proof We start by recalling the relation between the mutual information and the total variation. For any pair of random variables (X, Y), define the so-called T-information [99]: $T(X;Y) \triangleq d_{\mathrm{TV}}(P_{XY}, P_X P_Y) = \mathbb{E}[d_{\mathrm{TV}}(P_{Y|X}, P_Y)]$. For $X \sim \mathrm{Bern}(p)$, this simply reduces to

$$T(X;Y) = 2p(1-p) d_{\mathrm{TV}}(P_{Y|X=0}, P_{Y|X=1}). \tag{13.59}$$

Furthermore, the mutual information can be bounded by the T-information, by Pinsker's and Fano's inequality, as follows (from Equation (84) and Proposition 12 of [100]):

$$2T(X;Y)^2 \leq I(X;Y) \leq \log(M-1) T(X;Y) + h(T(X;Y)), \tag{13.60}$$

where in the upper bound M is the number of possible values of X, and h is the binary entropy function in (13.34).

We prove the "if" part. Suppose $I(\sigma_1, \sigma_2; A) = \Omega(1)$. We first claim that assumption A1 implies that

$$I(\mathbf{1}_{\{\sigma_1 = \sigma_2\}}; A) = I(\sigma_1, \sigma_2; A), \tag{13.61}$$

that is, A is independent of σ_1, σ_2 conditional on $\mathbf{1}_{\{\sigma_1 = \sigma_2\}}$. Indeed, for any $z \neq z' \in [k]$, let π be any permutation such that $\pi(z') = z$. Since $P_{\sigma, A} = P_{\pi(\sigma), A}$, we have $P_{A|\sigma_1 = z, \sigma_2 = z} = P_{A|\pi(\sigma_1) = z, \pi(\sigma_2) = z}$, i.e., $P_{A|\sigma_1 = z, \sigma_2 = z} = P_{A|\sigma_1 = z', \sigma_2 = z'}$. Similarly, one can

[7] For the special case of $k = 2$, (13.58) is equivalent to $(1/n)\mathbb{E}[|\langle \sigma, \widehat{\sigma} \rangle|] = \Omega(1)$, where $\sigma, \widehat{\sigma}$ are assumed to be $\{\pm\}^n$-valued.

show that $P_{A|\sigma_1=z_1,\sigma_2=z_2} = P_{A|\sigma_1=z'_1,\sigma_2=z'_2}$, for any $z_1 \neq z_2$ and $z'_1 \neq z'_2$, and this proves the claim.

Let $x_j = \mathbf{1}_{\{\sigma_1=\sigma_j\}}$. By the symmetry assumption A2, $I(x_j;A) = I(x_2;A) = \Omega(1)$ for all $j \neq 1$. Since $\mathbb{P}\{x_j = 1\} = 1/k + o(1)$ by assumption A3, applying (13.60) with $M = 2$ and in view of (13.59), we have $d_{\mathrm{TV}}(P_{A|x_j=0}, P_{A|x_j=1}) = \Omega(1)$. Thus, there exists an estimator $\widehat{x}_j \in \{0,1\}$ as a function of A, such that

$$\mathbb{P}\{\widehat{x}_j = 1 \mid x_j = 1\} + \mathbb{P}\{\widehat{x}_j = 0 \mid x_j = 0\} \geq 1 + d_{\mathrm{TV}}(P_{A|x_j=0}, P_{A|x_j=1}) = 1 + \Omega(1). \quad (13.62)$$

Define $\widehat{\sigma}$ as follows: set $\widehat{\sigma}_1 = 1$; for $j \neq 1$, set $\widehat{\sigma}_j = 1$ if $\widehat{x}_j = 1$ and draw $\widehat{\sigma}_j$ from $\{2,\ldots,k\}$ uniformly at random if $\widehat{x}_j = 0$. Next, we show that $\widehat{\sigma}$ achieves correlated recovery. Indeed, fix a permutation $\pi \in S_k$ such that $\pi(\sigma_1) = 1$. It follows from the definition of overlap that

$$\mathbb{E}[o(\sigma,\widehat{\sigma})] \geq \frac{1}{n} \sum_{j \neq 2} \mathbb{P}\{\pi(\sigma_j) = \widehat{\sigma}_j\} - \frac{1}{k}. \quad (13.63)$$

Furthermore, since $\pi(\sigma_1) = 1$, we have, for any $j \neq 1$,

$$\mathbb{P}\{\pi(\sigma_j) = \widehat{\sigma}_j, x_j = 1\} = \mathbb{P}\{\widehat{x}_j = 1, x_j = 1\}$$

and

$$\mathbb{P}\{\pi(\sigma_j) = \widehat{\sigma}_j, x_j = 0\} = \mathbb{P}\{\pi(\sigma_j) = \widehat{\sigma}_j, \widehat{x}_j = 0, x_j = 0\} = \frac{1}{k-1}\mathbb{P}\{\widehat{x}_j = 0, x_j = 0\},$$

where the last step is because, conditional on $\widehat{x}_j = 0$, $\widehat{\sigma}_j$ is chosen from $\{2,\ldots,k\}$ uniformly and independently of everything else. Since $\mathbb{P}\{x_j = 1\} = 1/k + o(1)$, we have

$$\mathbb{P}\{\pi(\sigma_j) = \widehat{\sigma}_j\} = \frac{1}{k}\left(\mathbb{P}\{\widehat{x}_j = 1 \mid x_j = 1\} + \mathbb{P}\{\widehat{x}_j = 0 \mid x_j = 0\}\right) + o(1) \overset{(13.62)}{\geq} \frac{1}{k} + \Omega(1).$$

By (13.63), we conclude that $\widehat{\sigma}$ achieves correlated recovery of σ.

Next we prove the "only if" part. Suppose $I(\sigma_1,\sigma_2;A) = o(1)$ and we aim to show that $\mathbb{E}[o(\sigma,\widehat{\sigma})] = o(1)$ for any estimator $\widehat{\sigma}$. By the definition of overlap, we have

$$o(\sigma,\widehat{\sigma}) \leq \frac{1}{n} \sum_{\pi \in S_k} \left| \sum_{i \in [n]} \left(\mathbf{1}_{\{\pi(\sigma_i)=\widehat{\sigma}_i\}} - \frac{1}{k} \right) \right|.$$

Since there are $k! = \Omega(1)$ permutations in S_k, it suffices to show that, for any fixed permutation π,

$$\mathbb{E}\left[\left| \sum_{i \in [n]} \left(\mathbf{1}_{\{\pi(\sigma_i)=\widehat{\sigma}_i\}} - \frac{1}{k} \right) \right| \right] = o(n).$$

Since $I(\pi(\sigma_i),\pi(\sigma_j);A) = I(\sigma_i,\sigma_j;A)$, without loss of generality, we assume $\pi = \mathrm{id}$ in the following. By the Cauchy–Schwarz inequality, it further suffices to show

$$\mathbb{E}\left[\left(\sum_{i \in [n]} \left(\mathbf{1}_{\{\sigma_i=\widehat{\sigma}_i\}} - \frac{1}{k} \right) \right)^2 \right] = o(n^2). \quad (13.64)$$

Note that

$$\mathbb{E}\left[\left(\sum_{i\in[n]}\left(\mathbf{1}_{\{\sigma_i=\widehat{\sigma}_i\}}-\frac{1}{k}\right)\right)^2\right]$$
$$=\sum_{i,j\in[n]}\mathbb{E}\left[\left(\mathbf{1}_{\{\sigma_i=\widehat{\sigma}_i\}}-\frac{1}{k}\right)\left(\mathbf{1}_{\{\sigma_j=\widehat{\sigma}_j\}}-\frac{1}{k}\right)\right]$$
$$=\sum_{i,j\in[n]}\mathbb{P}\left\{\sigma_i=\widehat{\sigma}_i,\sigma_j=\widehat{\sigma}_j\right\}-\frac{2n}{k}\sum_{i\in[n]}\mathbb{P}\{\sigma_i=\widehat{\sigma}_i\}+\frac{n^2}{k^2}.$$

For the first term in the last displayed equation, let σ' be identically distributed as $\widehat{\sigma}$ but independent of σ. Since $I(\sigma_i,\sigma_j;\widehat{\sigma}_i,\widehat{\sigma}_j)\le I(\sigma_i,\sigma_j;A)=o(1)$ by the data-processing inequality, it follows from the lower bound in (13.60) that $d_{\mathrm{TV}}(P_{\sigma_i,\sigma_j,\widehat{\sigma}_i,\widehat{\sigma}_j},P_{\sigma_i,\sigma_j,\sigma'_i,\sigma'_j})=o(1)$. Since $\mathbb{P}\{\sigma_i=\sigma'_i,\sigma_j=\sigma'_j\}\le\max_{a,b\in[k]}\mathbb{P}\left\{\sigma_i=a,\sigma_j=b\right\}\le 1/k^2+o(1)$ by assumption A3, we have

$$\mathbb{P}\left\{\sigma_i=\widehat{\sigma}_i,\sigma_j=\widehat{\sigma}_j\right\}\le\frac{1}{k^2}+o(1).$$

Similarly, for the second term, we have

$$\mathbb{P}\{\sigma_i=\widehat{\sigma}_i\}=\frac{1}{k}+o(1),$$

where the last equality holds due to $I(\sigma_i;A)=o(1)$. Combining the last three displayed equations gives (13.64) and completes the proof.

A13.2 Proof of (13.7) ⇒ (13.6) and Verification of (13.7) in the Binary Symmetric SBM

Combining (13.61) with (13.60) and (13.59), we have $I(\sigma_1,\sigma_2;A)=o(1)$ if and only if $d_{\mathrm{TV}}(\mathcal{P}_+,\mathcal{P}_-)=o(1)$, where $\mathcal{P}_+=P_{A|\sigma_1=\sigma_2}$ and $\mathcal{P}_-=P_{A|\sigma_1\neq\sigma_2}$. Note the following characterization about the total variation distance, which simply follows from the Cauchy–Schwartz inequality:

$$d_{\mathrm{TV}}(\mathcal{P}_+,\mathcal{P}_-)=\frac{1}{2}\sqrt{\inf_{Q}\int\frac{(\mathcal{P}_+-\mathcal{P}_-)^2}{Q}},\tag{13.65}$$

where the infimum is taken over all probability distributions Q. Therefore (13.7) implies (13.6).

Finally, we consider the binary symmetric SBM and show that, below the correlated recovery threshold $\tau=(a-b)^2/(2(a+b))<1$, (13.7) is satisfied if the reference distribution Q is the distribution of A in the null (Erdős–Rényi) model. Note that

$$\int\frac{(\mathcal{P}_+-\mathcal{P}_-)^2}{Q}=\int\frac{\mathcal{P}_+^2}{Q}+\int\frac{\mathcal{P}_-^2}{Q}-2\int\frac{\mathcal{P}_+\mathcal{P}_-}{Q}.$$

Hence, it is sufficient to show

$$\int\frac{\mathcal{P}_z\mathcal{P}_{\widetilde{z}}}{Q}=C+o(1),\quad\forall z,\widetilde{z}\in\{\pm\}$$

for some constant C that is independent of z and $\widetilde{z}$. Specifically, following the derivations in (13.4), we have

$$\int \frac{\mathcal{P}_z \mathcal{P}_{\widetilde{z}}}{Q} = \mathbb{E}\left[\prod_{i<j}\left(1+\sigma_i\sigma_j\widetilde{\sigma}_i\widetilde{\sigma}_j\rho\right) \,\middle|\, \sigma_1\sigma_2 = z, \widetilde{\sigma}_1\widetilde{\sigma}_2 = \widetilde{z}\right]$$
$$= (1+o(1))e^{-\tau^2/4-\tau/2} \times \mathbb{E}\left[\exp\left(\frac{\rho}{2}\langle\sigma,\widetilde{\sigma}\rangle^2\right) \,\middle|\, \sigma_1\sigma_2 = z, \widetilde{\sigma}_1\widetilde{\sigma}_2 = \widetilde{z}\right], \quad (13.66)$$

where the last equality holds for $\rho = \tau/n + O(1/n^2)$ and $\log(1+x) = x - x^2/2 + O(x^3)$.

Write $\sigma = 2\xi - 1$ for $\xi \in \{0,1\}^n$ and let

$$H_1 \triangleq \xi_1\widetilde{\xi}_1 + \xi_2\widetilde{\xi}_2 \quad \text{and} \quad H_2 \triangleq \sum_{j\geq 3}^{n} \xi_j\widetilde{\xi}_j.$$

Then $\langle\sigma,\widetilde{\sigma}\rangle = 4(H_1+H_2) - n$. Moreover, conditional on σ_1, σ_2 and $\widetilde{\sigma}_1, \widetilde{\sigma}_2$,

$$H_2 \sim \text{Hypergeometric}\left(n-2, n/2 - \xi_1 - \xi_2, n/2 - \widetilde{\xi}_1 - \widetilde{\xi}_2\right).$$

Since $|H_1| \leq 2$, $\xi_1 + \xi_2 \leq 2$, and $\widetilde{\xi}_1 + \widetilde{\xi}_2 \leq 2$, it follows that, conditional on $\sigma_1\sigma_2 = z$, $\widetilde{\sigma}_1\widetilde{\sigma}_2 = \widetilde{z}$, $(1/\sqrt{n})(4H_1 + 4H_2 - n)$ converges to $\mathcal{N}(0,1)$ in distribution as $n \to \infty$ by the central limit theorem for a hypergeometric distribution. Therefore

$$\mathbb{E}\left[\exp\left(\frac{\rho}{2}\langle\sigma,\widetilde{\sigma}\rangle^2\right) \,\middle|\, \sigma_S = z, \widetilde{\sigma}_S = \widetilde{z}\right]$$
$$= \mathbb{E}\left[\exp\left(\frac{n\rho}{2}\left(\frac{4H_1+4H_2-n}{\sqrt{n}}\right)^2\right) \,\middle|\, \sigma_1\sigma_2 = z, \widetilde{\sigma}_1\widetilde{\sigma}_2 = \widetilde{z}\right]$$
$$= \frac{1+o(1)}{\sqrt{1-\tau}},$$

where the last equality holds due to $n\rho = \tau + o(1/n)$, $\tau < 1$, and the convergence of the moment-generating function.

References

[1] E. L. Lehmann and G. Casella, *Theory of point estimation*, 2nd edn. Springer, 1998.

[2] L. D. Brown and M. G. Low, "Information inequality bounds on the minimax risk (with an application to nonparametric regression)," *Annals Statist.*, vol. 19, no. 1, pp. 329–337, 1991.

[3] A. W. Van der Vaart, *Asymptotic statistics*. Cambridge University Press, 2000.

[4] L. Le Cam, *Asymptotic methods in statistical decision theory*. Springer, 1986.

[5] I. A. Ibragimov and R. Z. Khas'minskĭ, *Statistical estimation: Asymptotic theory*. Springer, 1981.

[6] L. Birgé, "Approximation dans les espaces métriques et théorie de l'estimation," *Z. Wahrscheinlichkeitstheorie verwandte Gebiete*, vol. 65, no. 2, pp. 181–237, 1983.

[7] Y. Yang and A. R. Barron, "Information-theoretic determination of minimax rates of convergence," *Annals Statist.*, vol. 27, no. 5, pp. 1564–1599, 1999.

[8] Q. Berthet and P. Rigollet, "Complexity theoretic lower bounds for sparse principal component detection," *Journal of Machine Learning Research: Workshop and Conference Proceedings*, vol. 30, pp. 1046–1066, 2013.

[9] Z. Ma and Y. Wu, "Computational barriers in minimax submatrix detection," *Annals Statist.*, vol. 43, no. 3, pp. 1089–1116, 2015.

[10] B. Hajek, Y. Wu, and J. Xu, "Computational lower bounds for community detection on random graphs," in *Proc. COLT 2015*, 2015, pp. 899–928.

[11] T. Wang, Q. Berthet, and R. J. Samworth, "Statistical and computational trade-offs in estimation of sparse principal components," *Annals Statist.*, vol. 44, no. 5, pp. 1896–1930, 2016.

[12] C. Gao, Z. Ma, and H. H. Zhou, "Sparse CCA: Adaptive estimation and computational barriers," *Annals Statist.*, vol. 45, no. 5, pp. 2074–2101, 2017.

[13] M. Brennan, G. Bresler, and W. Huleihel, "Reducibility and computational lower bounds for problems with planted sparse structure," in *Proc. COLT 2018*, 2018, pp. 48–166.

[14] F. McSherry, "Spectral partitioning of random graphs," in *42nd IEEE Symposium on Foundations of Computer Science*, 2001, pp. 529–537.

[15] E. Arias-Castro and N. Verzelen, "Community detection in dense random networks," *Annals Statist.*, vol. 42, no. 3, pp. 940–969, 2014.

[16] Y. Chen and J. Xu, "Statistical–computational tradeoffs in planted problems and submatrix localization with a growing number of clusters and submatrices," in *Proc. ICML 2014*, 2014, *arXiv:1402.1267*.

[17] A. Montanari, "Finding one community in a sparse random graph," *J. Statist. Phys.*, vol. 161, no. 2, pp. 273–299, *arXiv:1502.05680*, 2015.

[18] P. W. Holland, K. B. Laskey, and S. Leinhardt, "Stochastic blockmodels: First steps," *Social Networks*, vol. 5, no. 2, pp. 109–137, 1983.

[19] A. A. Shabalin, V. J. Weigman, C. M. Perou, and A. B. Nobel, "Finding large average submatrices in high dimensional data," *Annals Appl. Statist.*, vol. 3, no. 3, pp. 985–1012, 2009.

[20] M. Kolar, S. Balakrishnan, A. Rinaldo, and A. Singh, "Minimax localization of structural information in large noisy matrices," in *Advances in Neural Information Processing Systems*, 2011.

[21] C. Butucea and Y. I. Ingster, "Detection of a sparse submatrix of a high-dimensional noisy matrix," *Bernoulli*, vol. 19, no. 5B, pp. 2652–2688, 2013.

[22] C. Butucea, Y. Ingster, and I. Suslina, "Sharp variable selection of a sparse submatrix in a high-dimensional noisy matrix," *ESAIM: Probability and Statistics*, vol. 19, pp. 115–134, 2015.

[23] T. T. Cai, T. Liang, and A. Rakhlin, "Computational and statistical boundaries for submatrix localization in a large noisy matrix," *Annals Statist.*, vol. 45, no. 4, pp. 1403–1430, 2017.

[24] A. Perry, A. S. Wein, and A. S. Bandeira, "Statistical limits of spiked tensor models," *arXiv:1612.07728*, 2016.

[25] A. E. Alaoui, F. Krzakala, and M. I. Jordan, "Finite size corrections and likelihood ratio fluctuations in the spiked Wigner model," *arXiv:1710.02903*, 2017.

[26] E. Mossel, J. Neeman, and A. Sly, "A proof of the block model threshold conjecture," *Combinatorica*, vol. 38, no. 3, pp. 665–708, 2013.

[27] J. Baik, G. Ben Arous, and S. Péché, "Phase transition of the largest eigenvalue for nonnull complex sample covariance matrices," *Annals Probability*, vol. 33, no. 5, pp. 1643–1697, 2005.

[28] S. Péché, "The largest eigenvalue of small rank perturbations of hermitian random matrices," *Probability Theory Related Fields*, vol. 134, no. 1, pp. 127–173, 2006.

[29] F. Benaych-Georges and R. R. Nadakuditi, "The eigenvalues and eigenvectors of finite, low rank perturbations of large random matrices," *Adv. Math.*, vol. 227, no. 1, pp. 494–521, 2011.

[30] F. Krzakala, C. Moore, E. Mossel, J. Neeman, A. Sly, L. Zdeborová, and P. Zhang, "Spectral redemption in clustering sparse networks," *Proc. Natl. Acad. Sci. USA*, vol. 110, no. 52, pp. 20 935–20 940, 2013.

[31] L. Massoulié, "Community detection thresholds and the weak Ramanujan property," in *Proceedings of the Forty-Sixth Annual ACM Symposium on Theory of Computing*, 2014, pp. 694–703, *arXiv:1109.3318*.

[32] C. Bordenave, M. Lelarge, and L. Massoulié, "Non-backtracking spectrum of random graphs: Community detection and non-regular Ramanujan graphs," in *2015 IEEE 56th Annual Symposium on Foundations of Computer Science (FOCS)*, 2015, pp. 1347–1357, *arXiv: 1501.06087*.

[33] J. Banks, C. Moore, R. Vershynin, N. Verzelen, and J. Xu, "Information-theoretic bounds and phase transitions in clustering, sparse PCA, and submatrix localization," *IEEE Trans. Information Theory*, vol. 64, no. 7, pp. 4872–4894, 2018.

[34] Y. I. Ingster and I. A. Suslina, *Nonparametric goodness-of-fit testing under Gaussian models*. Springer, 2003.

[35] Y. Wu, "Lecture notes on information-theoretic methods for high-dimensional statistics," 2017, www.stat.yale.edu/~yw562/teaching/598/it-stats.pdf.

[36] I. Vajda, "On metric divergences of probability measures," *Kybernetika*, vol. 45, no. 6, pp. 885–900, 2009.

[37] W. Feller, *An introduction to probability theory and its applications*, 3rd edn. Wiley, 1970, vol. I.

[38] W. Kozakiewicz, "On the convergence of sequences of moment generating functions," *Annals Math. Statist.*, vol. 18, no. 1, pp. 61–69, 1947.

[39] E. Mossel, J. Neeman, and A. Sly, "Reconstruction and estimation in the planted partition model," *Probability Theory Related Fields*, vol. 162, nos. 3–4, pp. 431–461, 2015.

[40] Y. Polyanskiy and Y. Wu, "Application of information-percolation method to reconstruction problems on graphs," *arXiv:1804.05436*, 2018.

[41] E. Abbe and E. Boix, "An information-percolation bound for spin synchronization on general graphs," *arXiv:1806.03227*, 2018.

[42] J. Banks, C. Moore, J. Neeman, and P. Netrapalli, "Information-theoretic thresholds for community detection in sparse networks," in *Proc. 29th Conference on Learning Theory, COLT 2016*, 2016, pp. 383–416.

[43] D. Guo, S. Shamai, and S. Verdú, "Mutual information and minimum mean-square error in Gaussian channels," *IEEE Trans. Information Theory*, vol. 51, no. 4, pp. 1261–1282, 2005.

[44] Y. Deshpande and A. Montanari, "Information-theoretically optimal sparse PCA," in *IEEE International Symposium on Information Theory*, 2014, pp. 2197–2201.

[45] Y. Deshpande, E. Abbe, and A. Montanari, "Asymptotic mutual information for the two-groups stochastic block model," *arXiv:1507.08685*, 2015.

[46] F. Krzakala, J. Xu, and L. Zdeborová, "Mutual information in rank-one matrix estimation," in *2016 IEEE Information Theory Workshop (ITW)*, 2016, pp. 71–75, *arXiv: 1603.08447*.

[47] I. Csiszár, "Information-type measures of difference of probability distributions and indirect observations," *Studia Scientiarum Mathematicarum Hungarica*, vol. 2, pp. 299–318, 1967.

[48] A. Perry, A. S. Wein, A. S. Bandeira, and A. Moitra, "Optimality and sub-optimality of PCA for spiked random matrices and synchronization," *arXiv:1609.05573*, 2016.

[49] J. Barbier, M. Dia, N. Macris, F. Krzakala, T. Lesieur, and L. Zdeborová, "Mutual information for symmetric rank-one matrix estimation: A proof of the replica formula," in *Advances in Neural Information Processing Systems*, 2016, pp. 424–432, *arXiv: 1606.04142*.

[50] M. Lelarge and L. Miolane, "Fundamental limits of symmetric low-rank matrix estimation," in *Proceedings of the 2017 Conference on Learning Theory*, 2017, pp. 1297–1301.

[51] B. Hajek, Y. Wu, and J. Xu, "Information limits for recovering a hidden community," *IEEE Trans. Information Theory*, vol. 63, no. 8, pp. 4729–4745, 2017.

[52] Y. Polyanskiy and Y. Wu, "Lecture notes on information theory," 2015, http://people.lids.mit.edu/yp/homepage/data/itlectures_v4.pdf.

[53] A. B. Tsybakov, *Introduction to nonparametric estimation*. Springer, 2009.

[54] B. Hajek, Y. Wu, and J. Xu, "Recovering a hidden community beyond the Kesten–Stigum threshold in $O(|E|\log^*|V|)$ time," *J. Appl. Probability*, vol. 55, no. 2, pp. 325–352, 2018.

[55] B. Hajek, Y. Wu, and J. Xu "Submatrix localization via message passing," *J. Machine Learning Res.*, vol. 18, no. 186, pp. 1–52, 2018.

[56] B. Hajek, Y. Wu, and J. Xu "Semidefinite programs for exact recovery of a hidden community," in *Proc. Conference on Learning Theory (COLT)*, 2016, pp. 1051–1095, *arXiv:1602.06410*.

[57] S.Y. Yun and A. Proutiere, "Community detection via random and adaptive sampling," in *Proc. 27th Conference on Learning Theory*, 2014, pp. 138–175.

[58] E. Mossel, J. Neeman, and A. Sly, "Consistency thresholds for the planted bisection model," in *Proc. Forty-Seventh Annual ACM on Symposium on Theory of Computing*, 2015, pp. 69–75.

[59] E. Abbe, A. S. Bandeira, and G. Hall, "Exact recovery in the stochastic block model," *IEEE Trans. Information Theory*, vol. 62, no. 1, pp. 471–487, 2016.

[60] V. Jog and P.-L. Loh, "Information-theoretic bounds for exact recovery in weighted stochastic block models using the Rényi divergence," *arXiv:1509.06418*, 2015.

[61] E. Abbe and C. Sandon, "Community detection in general stochastic block models: Fundamental limits and efficient recovery algorithms," in *2015 IEEE 56th Annual Symposium on Foundations of Computer Science (FOCS)*, 2015, pp. 670–688, *arXiv:1503.00609*.

[62] N. Alon, M. Krivelevich, and B. Sudakov, "Finding a large hidden clique in a random graph," *Random Structures and Algorithms*, vol. 13, nos. 3–4, pp. 457–466, 1998.

[63] E. Hazan and R. Krauthgamer, "How hard is it to approximate the best Nash equilibrium?" *SIAM J. Computing*, vol. 40, no. 1, pp. 79–91, 2011.

[64] N. Alon, A. Andoni, T. Kaufman, K. Matulef, R. Rubinfeld, and N. Xie, "Testing k-wise and almost k-wise independence," in *Proc. Thirty-Ninth Annual ACM Symposium on Theory of Computing*, 2007, pp. 496–505.

[65] P. Koiran and A. Zouzias, "On the certification of the restricted isometry property," *arXiv:1103.4984*, 2011.

[66] L. Kučera, "A generalized encryption scheme based on random graphs," in *Graph-Theoretic Concepts in Computer Science*, 1992, pp. 180–186.

[67] A. Juels and M. Peinado, "Hiding cliques for cryptographic security," *Designs, Codes and Cryptography*, vol. 20, no. 3, pp. 269–280, 2000.

[68] B. Applebaum, B. Barak, and A. Wigderson, "Public-key cryptography from different assumptions," in *Proc. 42nd ACM Symposium on Theory of Computing*, 2010, pp. 171–180.

[69] B. Rossman, "Average-case complexity of detecting cliques," Ph.D. dissertation, Massachusetts Institute of Technology, 2010.

[70] V. Feldman, E. Grigorescu, L. Reyzin, S. Vempala, and Y. Xiao, "Statistical algorithms and a lower bound for detecting planted cliques," in *Proc. 45th Annual ACM Symposium on Theory of Computing*, 2013, pp. 655–664.

[71] R. Meka, A. Potechin, and A. Wigderson, "Sum-of-squares lower bounds for planted clique," in *Proc. Forty-Seventh Annual ACM Symposium on Theory of Computing*, 2015, pp. 87–96.

[72] Y. Deshpande and A. Montanari, "Improved sum-of-squares lower bounds for hidden clique and hidden submatrix problems," in *Proc. COLT 2015*, 2015, pp. 523–562.

[73] B. Barak, S. B. Hopkins, J. A. Kelner, P. Kothari, A. Moitra, and A. Potechin, "A nearly tight sum-of-squares lower bound for the planted clique problem," in *IEEE 57th Annual Symposium on Foundations of Computer Science, FOCS*, 2016, pp. 428–437.

[74] U. Feige and R. Krauthgamer, "Finding and certifying a large hidden clique in a semirandom graph," *Random Structures and Algorithms*, vol. 16, no. 2, pp. 195–208, 2000.

[75] U. Feige and D. Ron, "Finding hidden cliques in linear time," in *Proc. DMTCS*, 2010, pp. 189–204.

[76] Y. Dekel, O. Gurel-Gurevich, and Y. Peres, "Finding hidden cliques in linear time with high probability," *Combinatorics, Probability and Computing*, vol. 23, no. 01, pp. 29–49, 2014.

[77] B. P. Ames and S. A. Vavasis, "Nuclear norm minimization for the planted clique and biclique problems," *Mathematical Programming*, vol. 129, no. 1, pp. 69–89, 2011.

[78] Y. Deshpande and A. Montanari, "Finding hidden cliques of size $\sqrt{N/e}$ in nearly linear time," *Foundations Comput. Math.*, vol. 15, no. 4, pp. 1069–1128, 2015.

[79] M. Jerrum, "Large cliques elude the Metropolis process," *Random Structures and Algorithms*, vol. 3, no. 4, pp. 347–359, 1992.

[80] N. Verzelen and E. Arias-Castro, "Community detection in sparse random networks," *Annals Appl. Probability*, vol. 25, no. 6, pp. 3465–3510, 2015.

[81] N. Alon, S. Arora, R. Manokaran, D. Moshkovitz, and O. Weinstein., "Inapproximability of densest κ-subgraph from average case hardness," 2011, www.nada.kth.se/~rajsekar/papers/dks.pdf.

[82] S. B. Hopkins, P. K. Kothari, and A. Potechin, "SoS and planted clique: Tight analysis of MPW moments at all degrees and an optimal lower bound at degree four," *arXiv:1507.05230*, 2015.

[83] P. Raghavendra and T. Schramm, "Tight lower bounds for planted clique in the degree-4 SOS program," *arXiv:1507.05136*, 2015.

[84] B. Ames, "Robust convex relaxation for the planted clique and densest k-subgraph problems," *arXiv:1305.4891*, 2013.

[85] L. Lovász, *Large networks and graph limits*. American Mathematical Society, 2012.

[86] C. Gao, Y. Lu, and H. H. Zhou, "Rate-optimal graphon estimation," *Annals Statist.*, vol. 43, no. 6, pp. 2624–2652, 2015.

[87] O. Klopp and N. Verzelen, "Optimal graphon estimation in cut distance," *arXiv:1703.05101*, 2017.

[88] J. Xu, "Rates of convergence of spectral methods for graphon estimation," in *Proc. 35th International Conference on Machine Learning*, 2018, *arXiv:1709.03183*.

[89] T. Lesieur, F. Krzakala, and L. Zdeborová, "Phase transitions in sparse PCA," in *IEEE International Symposium on Information Theory*, 2015, pp. 1635–1639.

[90] A. E. Alaoui and F. Krzakala, "Estimation in the spiked Wigner model: A short proof of the replica formula," *arXiv:1801.01593*, 2018.

[91] A. Montanari, D. Reichman, and O. Zeitouni, "On the limitation of spectral methods: From the Gaussian hidden clique problem to rank one perturbations of Gaussian tensors," in *Advances in Neural Information Processing Systems*, 2015, pp. 217–225, *arXiv: 1411.6149*.

[92] C. J. Hillar and L.-H. Lim, "Most tensor problems are NP-hard," *J. ACM*, vol. 60, no. 6, pp. 45:1–45:39, 2013.

[93] A. Montanari and E. Richard, "A statistical model for tensor PCA," in *Proc. 27th International Conference on Neural Information Processing Systems*, 2014, pp. 2897–2905.

[94] T. Lesieur, L. Miolane, M. Lelarge, F. Krzakala, and L. Zdeborová, "Statistical and computational phase transitions in spiked tensor estimation," *arXiv:1701.08010*, 2017.

[95] S. B. Hopkins, J. Shi, and D. Steurer, "Tensor principal component analysis via sum-of-square proofs," in *COLT*, 2015, pp. 956–1006.

[96] A. Zhang and D. Xia, "Tensor SVD: Statistical and computational limits," *arXiv:1703.02724*, 2017.

[97] D. Paul, "Asymptotics of sample eigenstruture for a large dimensional spiked covariance model," *Statistica Sinica*, vol. 17, no. 4, pp. 1617–1642, 2007.

[98] T. Lesieur, C. D. Bacco, J. Banks, F. Krzakala, C. Moore, and L. Zdeborová, "Phase transitions and optimal algorithms in high-dimensional Gaussian mixture clustering," *arXiv:1610.02918*, 2016.

[99] I. Csiszár, "Almost independence and secrecy capacity," *Problemy peredachi informatsii*, vol. 32, no. 1, pp. 48–57, 1996.

[100] Y. Polyanskiy and Y. Wu, "Dissipation of information in channels with input constraints," *IEEE Trans. Information Theory*, vol. 62, no. 1, pp. 35–55, 2016.

14 Distributed Statistical Inference with Compressed Data

Wenwen Zhao and Lifeng Lai

Summary

This chapter introduces basic ideas of information-theoretic models for distributed statistical inference problems with compressed data, discusses current and future research directions and challenges in applying these models to various statistical learning problems. In these applications, data are distributed in multiple terminals, which can communicate with each other via limited-capacity channels. Instead of recovering data at a centralized location first and then performing inference, this chapter describes schemes that can perform statistical inference without recovering the underlying data. Information-theoretic tools are borrowed to characterize the fundamental limits of the classical statistical inference problems using compressed data directly. In this chapter, distributed statistical learning problems are first introduced. Then, models and results of distributed inference are discussed. Finally, new directions that generalize and improve the basic scenarios are described.

14.1 Introduction

Nowadays, large amounts of data are collected by devices and sensors in multiple terminals. In many scenarios, it is risky and expensive to store all of the data in a centralized location, due to the data size, privacy concerns, etc. Thus, distributed inference, a class of problems aiming to infer useful information from these distributed data without collecting all of the data in a centralized location, has attracted significant attention. In these problems, communication among different terminals is usually beneficial but the communication channels between terminals are typically of limited capacity. Compared with the relatively well-studied centralized setting, where data are stored in one terminal, the distributed problems with a limited communication budget are more challenging.

Two main scenarios are considered in the existing works on distributed inference. In the first scenario, named *sample partitioning*, each terminal has data samples related to all random variables [1, 2], as shown in Fig. 14.1. In this figure, we use a matrix to represent the available data. Here, different columns in the matrix denote corresponding random variables to which the samples are related. The data matrix is partitioned in a

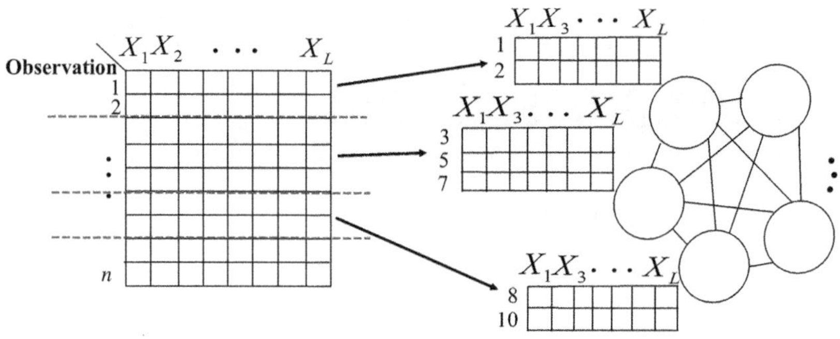

Figure 14.1 Sample partitioning.

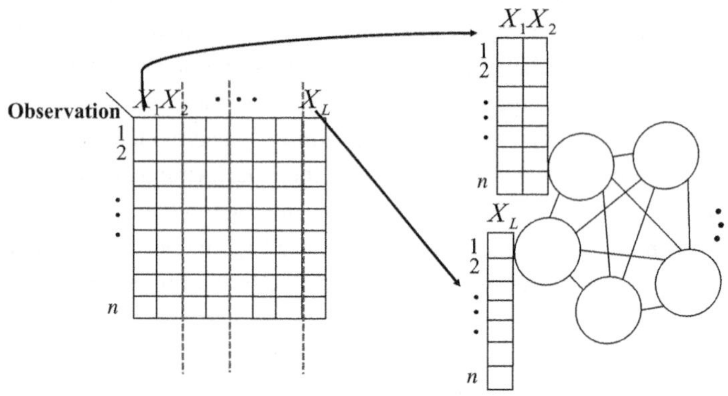

Figure 14.2 Feature partitioning.

row-wise manner, and each terminal observes a subset of the samples, which relates to all random variables $(X_1, X_2, \ldots, X_L)$. This scenario is quite common in real life. For example, there are large quantities of voice and image data stored in personal smart devices but, due to the sensitive nature of the data, we cannot ask all users to send their voice messages or photos to those in the location. Generally, in this scenario, even though each terminal has fewer data than those the centralized setting, terminals can still apply learning methods to their local data. Certainly, communicating and combining learning results from distributed terminals may improve the performance.

In the second scenario, named as *feature partitioning*, the data stored in each terminal are related to only a subset, not all, of the random variables. Fig. 14.2 illustrates the feature partitioning scenario, in which the data matrix is partitioned in a column-wise manner and each terminal observes the data related to a subset of random variables. For example, terminal X_1 has all observations related to random variable X_1. This scenario is also quite common in practice. For example, different information about each patient is typically stored in different locations as patients may go to different departments or different hospitals for different tests. In general, this scenario is more challenging than sample partitioning as each terminal in the feature partitioning scenario is not able to obtain meaningful information from local data alone. Moreover, due to the limited communication budget (it can be as low as a diminishing value), recovering data first and then conducting inference is neither optimal nor necessary. Thus, we need to design inference algorithms that can deal with compressed data, which is a much more complicated and challenging problem than the problems discussed above.

This chapter will focus on the inference problems for the feature partitioning scenario, which was first proposed by Berger [3] in 1979. Many existing works on the classic distributed inference problem focus on the following three branches: distributed hypothesis testing or distributed detection, distributed pattern classification, and distributed estimation [4–14]. Using powerful information-theoretic tools such as typical sequences, the covering lemma, etc., good upper and lower bounds on the inference performance are derived for the basic model and many special cases. Moreover, some results have been extended to more general cases. In this chapter, we focus on the distributed hypothesis-testing problem. In Section 14.2, we study the basic distributed hypothesis-testing model with non-interactive communication in the sense that each terminal sends only one message that is a function of its local data to the decision-maker [4–9, 11, 15]. More details are introduced in Section 14.2.

In Section 14.3, we consider more sophisticated models that allow interactive communications. In the interactive communication cases, there can be multiple rounds of communication among different terminals. In each round, each terminal can utilize messages received from other terminals in previous rounds along with its own data to determine the transmitted message. We start with a special form of interaction, i.e., cascaded communication among terminals [16, 17], in which terminals broadcast their messages in a sequential order and each terminal uses all messages received so far along with its own observations for encoding. Then we discuss the full interaction between two terminals and analyze their performance [18–20]. More details will be introduced in Section 14.3.

In Sections 14.2 and 14.3, the probability mass functions (PMFs) are fully specified. In Section 14.4, we generalize the discussion to the scenario with model uncertainty. In particular, we will discuss the identity-testing problem, in which the goal is to determine whether given samples are generated from a certain distribution or not. By interpreting the distributed identity-testing problem as composite hypothesis-testing problems, the type-2 error exponent can be characterized using information-theoretic tools. We will introduce more details in Section 14.4.

14.2 Basic Model

In this section, we review basic theoretic models for distributed hypothesis-testing problems. We first introduce the general model, then discuss basic ideas and results for two special cases: (1) hypothesis testing against independence; and (2) zero-rate data compression. In this chapter, we will present some results and the main ideas behind these results. For detailed proofs, interested readers can refer to [4–11].

14.2.1 Model

Consider a system with L terminals: $\mathcal{X}_i$, $i = 1,\ldots,L$ and a decision-maker $\mathcal{Y}$. Each terminal and the decision-maker observe a component of the random vector $(X_1,\ldots,X_L,Y)$ that takes values in a finite set $\mathcal{X}_1 \times\cdots\times \mathcal{X}_L \times \mathcal{Y}$ and admits a PMF with two possible forms:

$$H_0 : P_{X_1\cdots X_L Y}, \qquad H_1 : Q_{X_1\cdots X_L Y}. \tag{14.1}$$

With a slight abuse of notation, $\mathcal{X}_i$ is used to denote both the terminal and the alphabet set from which the random variable X_i takes values. $(X_1^n,\ldots,X_L^n,Y^n)$ are independently and identically (i.i.d.) generated according to one of the above joint PMFs. In other words, $(X_1^n,\ldots,X_L^n,Y^n)$ is generated by either $P^n_{X_1\cdots X_L Y}$ or $Q^n_{X_1\cdots X_L Y}$. In a typical hypothesis-testing problem, one determines which hypothesis is true under the assumption that $(X_1^n,\ldots,X_L^n,Y^n)$ are fully available at the decision-maker. In the distributed setting, X_i^n, $i = 1,\ldots,L$ and Y^n are at different locations. In particular, terminal $\mathcal{X}_i$ observes only X_i^n and terminal $\mathcal{Y}$ observes only Y^n. Terminals $\mathcal{X}_i$ are allowed to send messages to the decision-maker $\mathcal{Y}$. Using Y^n and the received messages, $\mathcal{Y}$ determines which hypothesis is true. We denote this system as $S_{X_1\cdots X_L|Y}$. Fig. 14.3 illustrates the system model. In the following, we will use the term "decision-maker" and terminal $\mathcal{Y}$ interchangeably. Here, Y^n is used to model any side information available at the decision-maker. If $\mathcal{Y}$ is defined to be an empty set, then the decision-maker does not have side information.

After observing the data sequence $x_i^n \in \mathcal{X}_i^n$, terminal $\mathcal{X}_i$ will use an encoder f_i to transform the sequence x_i^n into a message $f_i(x_i^n)$, which takes values from the message set $\mathcal{M}_i$,

$$f_i : \mathcal{X}_i^n \to \mathcal{M}_i = \{1,2,\ldots,M_i\}, \tag{14.2}$$

with rate constraint

$$\limsup_{n\to\infty} \frac{1}{n} \log M_i \le R_i, \quad i = 1,\ldots,L. \tag{14.3}$$

Using messages M_i, $i = 1,\ldots,L$ and its side information Y^n, the decision-maker will employ a decision function ψ to determine which hypothesis is true:

$$\psi : \mathcal{M}_1 \times\cdots\times \mathcal{M}_L \times \mathcal{Y}^n \to \{H_0,H_1\}. \tag{14.4}$$

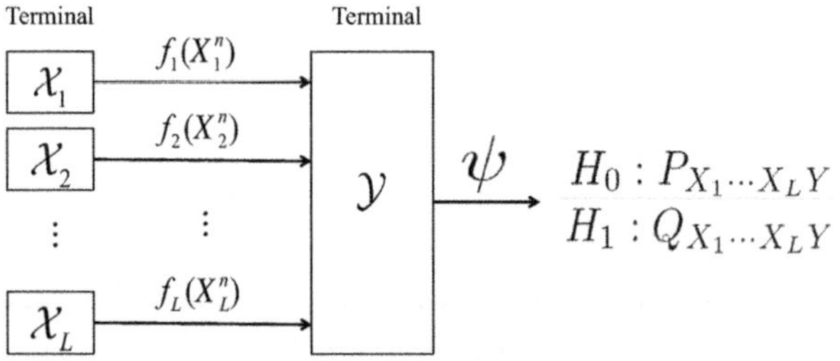

Figure 14.3 The basic model.

For any given encoding functions f_i, $i = 1,\dots,L$ and decision function ψ, one can define the acceptance region as

$$\begin{aligned}\mathcal{A}_n = \{(x_1^n,\dots,x_L^n,y^n) \in \mathcal{X}_1^n \times \cdots \times \mathcal{X}_L^n \times \mathcal{Y}^n : \\ \psi(f_1(x_1^n)\cdots f_L(x_L^n)y^n) = H_0\}.\end{aligned} \tag{14.5}$$

Correspondingly, the type-1 error probability is defined as

$$\alpha_n = P^n_{X_1\cdots X_L Y}(\mathcal{A}_n^c), \tag{14.6}$$

in which $\mathcal{A}_n^c$ denotes the complement of $\mathcal{A}_n$, and the type-2 error probability is defined as

$$\beta_n = Q^n_{X_1\cdots X_L Y}(\mathcal{A}_n). \tag{14.7}$$

The goal is to design the encoding functions f_i, $i = 1,\dots,L$ and the decision function ψ to maximize the type-2 error exponent under certain type-1 error and communication-rate constraints (14.3).

More specifically, one can consider two kinds of type-1 error constraint, namely the following.

- Constant-type constraint

$$\alpha_n \leq \epsilon \tag{14.8}$$

  for a prefixed $\epsilon > 0$, which implies that the type-1 error probability must be smaller than a given threshold.

- Exponential-type constraint

$$\alpha_n \leq \exp(-nr) \tag{14.9}$$

  for a given $r > 0$, which implies that the type-1 error probability must decrease exponentially fast with an exponent no less than r. Hence the exponential-type constraint is stricter than the constant-type constraint.

To distinguish these two different type-1 error constraints, we use different notations to denote the corresponding type-2 error exponent, and we use the subscript 'b' to denote that the error exponent is under the basic model.

- Under the constant-type constraint, we define the type-2 error exponent as

$$\theta_b(R_1,\dots,R_L,\epsilon) = \liminf_{n\to\infty}\left(-\frac{1}{n}\log\left(\min_{f_1,\dots,f_L,\psi}\beta_n\right)\right),$$

  in which the minimization is over all f_is and ψ satisfying condition (14.3) and (14.8).
- Under the exponential-type constraint, we define the type-2 error exponent as

$$\sigma_b(R_1,\dots,R_L,r) = \liminf_{n\to\infty}\left(-\frac{1}{n}\log\left(\min_{f_1,\dots,f_L,\psi}\beta_n\right)\right),$$

  in which the minimization is over all f_is and ψ satisfying condition (14.3) and (14.9).

14.2.2 Connections and Differences with Distributed Source Coding Problems

From the introduction of the model for the distributed testing problem, we can see that it is different from the classic distributed source coding problems [21]. In the source coding problem, which is illustrated in Fig. 14.4, the goal of the decoder is to recover the source sequences after it has received the compressed messages from terminal $\{\mathcal{X}_l\}_{l=1}^L$. According to the Slepian–Wolf theorem [21], the decoder can recover the original sequences with diminishing error probability when the compression rates are larger than certain values. Hence, when the compression rates are sufficiently large, we can adopt the source coding method to first recover the original source sequences and then perform inferences. However, in general, the rate constraints are typically too strict for the decision-maker to fully recover $\{X_l^n\}_{l=1}^L$ in the inference problem. Moreover, in the inference problem, recovery of source sequences is not its goal and typically is not necessary.

On the other hand, this inference problem is closely connected to distributed source coding problems. In particular, the general idea of the existing schemes in distributed inference problems is to mimic the schemes used in distributed source coding problems. In the existing studies [4–7, 22], each terminal $\mathcal{X}_l$ compresses its sequence X_l^n into U_l^n. Then these terminals send the auxiliary sequences $\{U_l^n\}_{l=1}^L$ to the decision-maker using source coding ideas so that the decision-maker can obtain $\{\widehat{U}_l^n\}_{l=1}^L$, which has a high probability of being the same as $\{U_l^n\}_{l=1}^L$. The compression step is to make sure each terminal sends enough information for one to recover U_l^n but does not exceed the rate constraint. Finally, the decision-maker will decide between the two hypotheses using $\{\widehat{U}_l^n\}_{l=1}^L$. Hence, we can see that, even though the decision-maker does not need to recover the sequences $\{X_l^n\}_{l=1}^L$, it does need to recover $\{U_l^n\}_{l=1}^L$ from the compressed messages.

14.2.3 Typical Sequences

The notation and properties of typical sequences are heavily used in the schemes and proofs in the existing work. Here, for reference, we provide a brief overview of the definition and key properties. Following [6], for any sequence $x^n = (x(1), \ldots, x(n)) \in \mathcal{X}^n$, we use $n(a|x^n)$ to denote the total number of indices t at which $x(t) = a$. Then, the relative frequency or empirical PMF, $\pi(a|x^n) \triangleq n(a|x^n)/n, \forall a \in \mathcal{X}$ of the components of x^n, is called the type of x^n and is denoted by $\text{tp}(x^n)$. The set of all types of sequences in $\mathcal{X}^n$

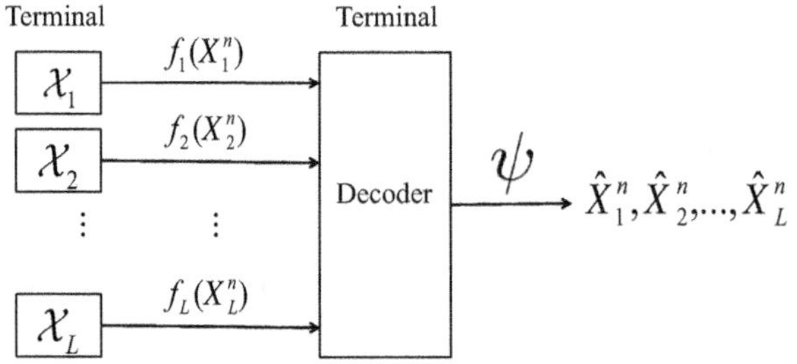

Figure 14.4 A canonical example for the source coding problem.

is denoted by $\mathcal{P}^n(\mathcal{X})$. Furthermore, we call a random variable $X^{(n)}$ that has the same distribution as $\text{tp}(x^n)$ the type variable of x^n.

For any given sequence x^n, we use a typical sequence to measure how likely it is that this sequence is generated from a PMF P_X.

DEFINITION 14.1 *(Definition 1 of [6]) For a given a type $P_X \in \mathcal{P}^n(\mathcal{X})$ and a constant η, we denote by $T^n_\eta(X)$ the set of (P_X,η)-typical sequences in $\mathcal{X}^n$:*

$$T^n_\eta(X) \triangleq \{x^n \in \mathcal{X}^n : |\pi(a|x^n) - P_X(a)| \leq \eta P_X(a), \forall a \in \mathcal{X}\}.$$

In the same manner, we use $\widetilde{T}^n_\eta(X)$ to denote the set of $(\widetilde{P}_X,\eta)$-typical sequences. Note that, when $\eta = 0$, $T^n_0(X)$ denotes the set of sequences $x^n \in \mathcal{X}^n$ of type P_X, and we use $T^n(X)$ for simplicity.

We use the following lemma to summarize key properties of typical sequences.

LEMMA 14.1 *(Lemma 1 of [6]) Let $\lambda > 0$ be arbitrary.*

1. $P^n_X(T^n_\eta(X)) \geq 1 - \lambda$.
2. *Let $X^{(n)}$ be a type variable for a sequence in $\mathcal{X}^n$, then*

$$(n+1)^{-|\mathcal{X}|}\exp[nH(X^{(n)})] \leq |T^n_0(X^{(n)})| \leq \exp[n(H(X^{(n)}))]. \tag{14.10}$$

3. *Let $x^n \in \mathcal{X}^n$ and X be a random variable in $\mathcal{X}$, then*

$$\Pr(X^n = x^n) = \exp[-n(H(X^{(n)}) + D(X^{(n)}\|X))]. \tag{14.11}$$

Let $\Lambda_n(\mathcal{X})$ be the set of all the different types of x^n over $\mathcal{X}^n$. Then, the data space $\mathcal{X}^n$ is partitioned into classes each with the same type

$$\mathcal{X}^n = \bigcup_{X^{(n)} \in \Lambda_n(\mathcal{X})} T^n_0(X^{(n)}). \tag{14.12}$$

Similarly, for multiple random variables, define their joint empirical PMF as

$$\pi(a,b|x^n,y^n) \triangleq \frac{n(a,b|x^n,y^n)}{n}, \ \forall (a,b) \in \mathcal{X} \times \mathcal{Y}. \tag{14.13}$$

For a given a type $P_{XY} \in \mathcal{P}^n(\mathcal{X} \times \mathcal{Y})$ and a constant η, we denote by $T^n_\eta(XY)$ the set of jointly (P_{XY},η)-typical sequences in $\mathcal{X}^n \times \mathcal{Y}^n$:

$$\begin{aligned} T^n_\eta(XY) \triangleq \{&(x^n,y^n) \in \mathcal{X}^n \times \mathcal{Y}^n : \\ &|\pi(a,b|x^n,y^n) - P_{XY}(a,b)| \leq \eta P_{XY}(a,b), \forall (a,b) \in \mathcal{X} \times \mathcal{Y}\}. \end{aligned} \tag{14.14}$$

Furthermore, for $y^n \in \mathcal{Y}^n$, we define $T^n_\eta(X|y^n)$ as the set of all x^ns that are jointly typical with y^n:

$$T^n_\eta(X|y^n) = \{x^n \in \mathcal{X}^n : (x^n,y^n) \in T^n_\eta(XY)\}. \tag{14.15}$$

More details and properties can be found in [6, 21].

14.2.4 Testing against Independence with Positive-Rate Data Compression

First, we focus on a special case: testing against independence, in which we are interested in determining whether $(X_1,\ldots,X_L)$ and Y are independent or not. In this case,

$Q_{X_1 \cdots X_L Y}$ in (14.1) has the special form

$$Q_{X_1 \cdots X_L Y} = P_{X_1 \cdots X_L} P_Y$$

and the two hypotheses in (14.1) become

$$H_0 : P_{X_1 \cdots X_L Y} \quad \text{versus} \quad H_1 : P_{X_1 \cdots X_L} P_Y. \tag{14.16}$$

Note that the marginal distribution of $(X_1, \ldots, X_L)$ and Y are the same under both hypotheses in the case of testing against independence. The problem has been studied in [4, 9].

When $L = 1$ (the system is then denoted as $S_{X_1|Y}$), this problem can be shown to have a close connection with the problem of source coding with a single helper problem of [22]. Building on this connection and the results in [22], Ahlswede and Csiszár [4] provided a single-letter characterization of the optimal type-2 error exponent.

THEOREM 14.1 *(Theorem 2 of [4]) In the system $S_{X_1|Y}$ with $R_1 \geq 0$, when the constraint on the type-1 error probability* (14.8) *and communication constraints* (14.3) *are satisfied, the best error exponent for the type-2 error probability satisfies*

$$\theta_b(R_1, \epsilon) = \max_{U_1 \in \varphi_1} I(U_1; Y), \tag{14.17}$$

where

$$\varphi_1 = \{U_1 : R_1 \geq I(U_1; X_1), \quad U_1 \leftrightarrow X_1 \leftrightarrow Y, \quad |\mathcal{U}_1| \leq |\mathcal{X}_1| + 1,\}. \tag{14.18}$$

When $L \geq 2$, one can follow the similar approach in [4] to connect the testing against independence problem to the problem of source coding with multiple helpers. However, unlike the problem of source coding with a single helper, the general problem of source coding with multiple helpers is still open. Hence, a different approach to exploit the more complicated problem is needed. First, we provide a lower bound on the type-2 error exponent by generalizing Theorem 6 in [9] to the case of L terminals.

THEOREM 14.2 *(Theorem 6 of [9]) In the system $S_{X_1 \cdots X_L|Y}$ with $R_i > 0$, $i = 1, \ldots, L$, when the constraint on the type-1 error probability* (14.8) *and communication constraints* (14.3) *are satisfied, the error exponent of the type-2 error probability is lower-bounded by*

$$\theta_b(R_1, \ldots, R_L, \epsilon) \geq \max_{P_{U_1|X_1} \cdots P_{U_L|X_L}} I(U_1 \cdots U_L; Y), \tag{14.19}$$

in which the maximization is over the $P_{U_i|X_i}$s such that $I(U_i; X_i) \leq R_i$ and $|\mathcal{U}_i| \leq |\mathcal{X}_i| + 1$ for $i = 1, \ldots, L$.

The lower bound in Theorem 14.2 can be viewed as a generalization of the bound in Theorem 14.1. The constraints in Theorem 14.2 can be interpreted as the following Markov-chains condition on the auxiliary random variables:

$$U_i \leftrightarrow X_i \leftrightarrow (U_1, \ldots, U_{i-1}, X_1, \ldots, X_{i-1}, X_{i+1}, \ldots, X_L, Y). \tag{14.20}$$

To achieve this lower bound, we design the following encoding/decoding scheme. For a given rate constraint R_i, terminal X_i first generates a quantization codebook containing 2^{nR_i} quantization sequences u_i^n. After observing x_i^n, terminal X_i picks one sequence u_i^n from the quantization codebook to describe x_i^n and sends this sequence to the decision-maker. After receiving the descriptions from terminals, the decision-maker will declare that the hypothesis H_0 is true if the descriptions from these terminals and the side information at the decision-maker are correlated. Otherwise, the decision-maker will declare H_1 true.

Then, we establish an upper bound on the type-2 error exponent that any scheme can achieve by generalizing Theorem 7 in [9] to the case of L terminals.

THEOREM 14.3 *(Theorem 7 of [9]) In the system $S_{X_1 \cdots X_L|Y}$ with $R_i \geq 0$, $i = 1,\ldots,L$, when the constraint on the type-1 error probability* (14.8) *and communication constraints* (14.3) *are satisfied, the best error exponent for the type-2 error probability is upper-bounded by*

$$\lim_{\epsilon \to 0} \theta_b(R_1,\ldots R_L,\epsilon) \leq \max_{U_1 \cdots U_L} I(U_1 \cdots U_L;Y) \tag{14.21}$$

in which the maximization is over the U_is such that $R_i \geq I(U_i;X_i)$, $|\mathcal{U}_i| \leq |\mathcal{X}_i|+1$, $U_i \leftrightarrow X_i \leftrightarrow (X_1,\ldots,X_{i-1},X_{i+1},\ldots,X_L,Y)$ for $i = 1,\ldots,L$.

We note that the constraints on auxiliary random variables in Theorems 14.2 and 14.3 are different. In particular, the Markov constraints in Theorem 14.3 are less strict than those in Theorem 14.2. This implies that the lower bound and the upper bound do not match with each other. Hence, more exploration of this problem is needed.

For the case of testing against independence under an exponential-type constraint on the type-1 error probability, only a lower bound is established in [7], which is stated in the following theorem.

First, let Φ denote the set of all continuous mappings from $\mathcal{P}(\mathcal{X}_1)$ to $\mathcal{P}(\mathcal{U}_1|\mathcal{X}_1)$. $\omega(\widehat{X}_1)$ is an element in Φ for one particular $\widehat{X}_1$. $\widehat{U}_1$ is an auxiliary random variable that satisfies $\widehat{P}_{U_1|X_1} = \omega(\widehat{X}_1)$. Then, define

$$\phi_{X_1}(R_1,r) = \left\{ \omega \in \Phi : \max_{\substack{\widehat{X}: D(\widehat{X}_1\|X_1) \leq r \\ \widehat{P}_{U_1|X_1} = \omega(\widehat{X}_1)}} I(\widehat{U}_1;\widehat{X}_1) \leq R_1 \right\}. \tag{14.22}$$

THEOREM 14.4 *(Corollary 2 of [7]) In the system $S_{X_1|Y}$ with $R_1 \geq 0$, when the constraint on the type 1 error probability* (14.9) *and communication constraints* (14.3) *are satisfied, the best error exponent for the type-2 error probability is lower-bounded by*

$$\sigma_b(R_1,r) \geq \max_{\substack{\omega \in \phi_{P_{X_1 Y}}(R_1,r) \\ \|\widehat{U}_1\| \leq \|\mathcal{X}_1\|+2}} \min_{\substack{\widetilde{U}_1\widetilde{X}_1\widetilde{Y} \\ D(\widetilde{U}_1\widetilde{X}_1\widetilde{Y}\|U_1X_1Y) \leq r \\ \widehat{P}_{U_1|X_1} = P_{U_1|X_1} = \omega(\widetilde{X}_1) \\ U \leftrightarrow X_1 \leftrightarrow Y}} \left(D(\widehat{X}_1\|X_1) + I(\widehat{U}_1;\widehat{Y}) \right). \tag{14.23}$$

If we let $r = 0$, then $D(\widehat{X}_1\|X_1) + I(\widehat{U}_1;\widehat{Y})$ reduces to $I(\widehat{U}_1;\widehat{Y})$, which is the same as in (14.19). Unlike Theorem 14.1, a matching upper bound is hard to establish due to the stronger constraint on the type-1 error probability.

14.2.5 General PMF with Zero-Rate Data Compression

In this section, we focus on the "zero-rate" data compression, i.e.,

$$M_i \geq 2, \quad \lim_{n\to\infty} \frac{1}{n} \log M_i = 0, \quad i = 1,\ldots,L. \tag{14.24}$$

In this case, $\sigma_b(R_1,\ldots,R_L,r)$ will be denoted as $\sigma_b(0,\ldots,0,r)$. This zero-rate compression is of practical interest, as the normalized (by the length of the data) communication cost is minimal. It is well known that this kind of zero-rate information is not useful in the traditional distributed source coding with side-information problems [21, 23], whose goal is to recover $(X_1^n,\ldots,X_L^n)$ at terminal $\mathcal{Y}$. However, in the distributed inference setup, the goal is only to determine which hypothesis is true. The limited information from zero-rate compressed messages will be very useful. A clear benefit of this zero-rate compression approach is that the terminals need to consume only a limited amount of communication resources.

Under a Constant-Type Constraint

We first discuss the zero-rate date compression under a constant-type constraint on the type-1 error probability.

A special case of general zero-rate compression was first studied in Theorem 5 of [6], in which Han assumed that $M_i = 2$, i.e., each terminal can send only one bit of information. Using the properties of typical sequences, a matching upper and lower bound were derived, as shown in Theorem 14.5.

THEOREM 14.5 *(Theorem 5 of [6]) Suppose that* $D(P_{X_1Y}\|Q_{X_1Y}) < +\infty$. *In the system* $S_{X_1|Y}$ *with one-bit data compression, denoted as* $R_1 = 0_2$, *when the constraint on the type-1 error probability* (14.8) *is satisfied for some* $0 < \epsilon_0 \leq 1$ *and for all* $0 < \epsilon < \epsilon_0$, *the best error exponent for the type-2 error probability is*

$$\theta_b(0_2,\epsilon) = \min_{\substack{\tilde{P}_{X_1Y}:\\ \tilde{P}_{X_1}=P_{X_1},\tilde{P}_Y=P_Y}} D(\tilde{P}_{X_1Y}\|Q_{X_1Y}). \tag{14.25}$$

To achieve this bound, one can use the following simple encoding/coding scheme: if the observed sequence $x_1^n \in T_\epsilon^n(P_{X_1})$, i.e., it is a typical sequence of P_{X_1}, then we send 1 to the decision-maker $\mathcal{Y}$; otherwise, we send 0. If terminal $\mathcal{Y}$ receives 1, then it decides H_0 is true; otherwise, it decides H_1 is true. Using the properties of typical sequences, we can easily get the lower bound on the type-2 error exponent. To get the matching upper bound, the condition $D(P_{X_1Y}\|Q_{X_1Y}) < +\infty$ is required. Readers can refer to [6] for more details.

For the general zero-rate data compression, as we have $M_i \geq 2$, more information is sent to the decision-maker, which may lead to a better performance. Hence, the following inequality holds:

$$\theta_b(0_2,\ldots,0_2,\epsilon) \leq \theta_b(0,\ldots,0,\epsilon) \leq \theta_b(R_1,\ldots,R_L,\epsilon). \tag{14.26}$$

The scenario with general zero-rate data compression under a constant-type constraint has been considered in [8]. Shalaby adopted the blowing-up lemma [24] to give a tight upper bound on the type-2 error exponent.

THEOREM 14.6 *(Theorem 1 of [8]) Let $Q_{X_1Y} > 0$. In the system $S_{X_1|Y}$ with $R_1 \geq 0$, when the constraint on the type-1 error probability* (14.8) *is satisfied for all* $\epsilon \in (0,1)$, *the best error exponent for the type-2 error probability is upper-bounded by*

$$\theta_b(0,\epsilon) \leq \min_{\substack{\tilde{P}_{X_1Y}: \\ \tilde{P}_{X_1} = P_{X_1}, \tilde{P}_Y = P_Y}} D(\tilde{P}_{X_1Y} \| Q_{X_1Y}). \tag{14.27}$$

Here, the positive condition $Q_{X_1Y}(x_1,y) > 0, \forall (x_1,y) \in \mathcal{X}_1 \times \mathcal{Y}$ is required by the blowing-up lemma.

Using the inequality (14.26), the combination of Theorems 14.5 and 14.6 yields the following theorem.

THEOREM 14.7 *(Theorem 2 of [8]) Let $Q_{X_1Y} > 0$. In the system $S_{X_1|Y}$ with $R_1 \geq 0$, when the constraint on the type-1 error probability* (14.8) *is satisfied for all* $\epsilon \in (0,1)$, *the best error exponent for the type-2 error probability is*

$$\theta_b(0,\epsilon) = \min_{\substack{\tilde{P}_{X_1Y}: \\ \tilde{P}_{X_1} = P_{X_1}, \tilde{P}_Y = P_Y}} D(\tilde{P}_{X_1Y} \| Q_{X_1Y}). \tag{14.28}$$

The above results are given for the case with $L = 1$, in [8] Shalaby and Papamarcou also discussed the results in the case of general L.

Under an Exponential-Type Constraint

Compared with the constant-type constraint, the exponential-type constraint is much stricter, which requires the type-1 error probability to decrease exponentially fast with an exponent no less than r. Hence, the encoding/decoding scheme is more complexed.

Zero-rate data compression under an exponential-type constraint was first studied by Han and Kobayashi [7]. They studied the case S_{X_1Y}, which means $L = 1$, terminal $\mathcal{Y}$ also sends encoded messages to the decision-maker, and the decision-maker does not have any side information. They provided a lower bound on the type-2 error exponent, which is stated in the following theorem.

THEOREM 14.8 *(Theorem 6 of [7]) For zero-rate data compression in S_{X_1Y} with $R_1 = R_Y = 0$, the error exponent satisfies*

$$\sigma_b(0,0,r) \geq \sigma_{\text{opt}}, \tag{14.29}$$

in which

$$\sigma_{\text{opt}} \triangleq \min_{\tilde{P}_{X_1Y} \in \mathcal{H}_r} D(\tilde{P}_{X_1Y} \| Q_{X_1Y}) \tag{14.30}$$

with

$$\mathcal{H}_r = \left\{ \tilde{P}_{X_1Y} : \tilde{P}_{X_1} = \hat{P}_{X_1}, \tilde{P}_Y = \hat{P}_Y \; for \; some \; \hat{P}_{X_1Y} \in \varphi_r \right\}, \tag{14.31}$$

where

$$\varphi_r = \{ \hat{P}_{X_1Y} : D(\hat{P}_{X_1Y} \| P_{X_1Y}) \leq r \}. \tag{14.32}$$

To show this bound, the following coding scheme is adopted. After observing x_1^n, terminal $\mathcal{X}_1$ knows the type $\text{tp}(x_1^n)$ and sends $\text{tp}(x_1^n)$ (or an approximation of it, see below) to the decision-maker. Terminal $\mathcal{Y}$ does the same. As there are at most $n^{|\mathcal{X}_1|}$ types [25], the rate required for sending the type from terminal $\mathcal{X}_1$ is $(|\mathcal{X}_1|\log n)/n$, which goes to zero as n increases. After receiving all type information from the terminals, the decision-maker will check whether there is a joint type $\widetilde{P}_{X_1Y} \in \mathcal{H}_r$ such that its marginal types are the same as the information received from the terminals. If so, the decision-maker declares H_0 to be true, otherwise it declares H_1 to be true. If the message size M_1 is less than $n^{|\mathcal{X}_1|}$, then, instead of the exact type information $\text{tp}(x_1^n)$, each terminal will send an approximated version. For more details, please refer to [7].

Later, Han and Amari [5] proved an upper bound that matched the lower bound in Theorem 14.8 by converting the problem under an exponential-type constraint to the problem under a constant-type constraint.

THEOREM 14.9 *(Theorem 5.5 of [5]) Let P_{X_1Y} be arbitrary and $Q_{X_1Y} > 0$. For zero-rate compression in S_{X_1Y} with $R_1 = R_Y = 0$, the error exponent is upper-bounded by*

$$\sigma_b(0,0,r) \leq \sigma_{\text{opt}}, \tag{14.33}$$

with σ_{opt} defined in (14.30).

A consequence of Theorems 14.8 and 14.9 is the following theorem.

THEOREM 14.10 *(Corollary 5.3 of [5]) Let P_{X_1Y} be arbitrary and $Q_{X_1Y} > 0$. For zero-rate compression in S_{X_1Y} with $R_1 = R_Y = 0$, the error exponent is upper-bounded by*

$$\sigma_b(0,0,r) = \sigma_{\text{opt}}, \tag{14.34}$$

with σ_{opt} defined in (14.30).

For the case with $L \geq 2$, the following results hold.

THEOREM 14.11 *(Theorem 4 of [11]) Let $P_{X_1,\ldots,X_LY}$ be arbitrary and $Q_{X_1,\ldots,X_LY} > 0$. For zero-rate compression in $S_{X_1\ldots X_L|Y}$ with $R_i = 0$, $i = 1,\ldots,L$ and type-1 error constraint* (14.9), *the best type-2 error exponent is*

$$\sigma_b(0,\ldots,0,r) = \min_{\widetilde{P}_{X_1\cdots X_LY} \in \mathcal{H}_r} D(\widetilde{P}_{X_1\cdots X_LY} \| Q_{X_1\cdots X_LY}), \tag{14.35}$$

with

$$\mathcal{H}_r = \left\{ \widetilde{P}_{X_1\cdots X_LY} : \widetilde{P}_{X_i} = \widehat{P}_{X_i}, \widetilde{P}_Y = \widehat{P}_Y,\ i = 1,\ldots,L \text{ for some } \widehat{P}_{X_1\cdots X_LY} \in \varphi_r \right\}, \tag{14.36}$$

where

$$\varphi_r = \{\widehat{P}_{X_1\cdots X_LY} : D(\widehat{P}_{X_1\cdots X_LY} \| P_{X_1\cdots X_LY}) \leq r\}. \tag{14.37}$$

Thus we have single-letter characterized the distributed testing problem under zero-rate data compression with an exponential-type constraint on the type-1 error probability. Furthermore, the minimization problems in (14.30) and (14.35) are convex optimization problems, which can be solved efficiently.

COROLLARY 14.1 *Given $P_{X_1 \cdots X_L Y}$ and $Q_{X_1 \cdots X_L Y}$, the problem of finding σ_{opt} defined in* (14.35) *is a convex optimization problem.*

14.2.6 General PMF with Positive-Rate Data Compression

In this section, we discuss the general case of hypothesis testing with positive data compression. Unfortunately, establishing a tight upper bound on the type-2 error exponent is still an open problem, so here we give only a classic lower bound.

The result under a constant-type constraint on the type-1 error probability is stated in the following theorem.

THEOREM 14.12 *(Theorem 1 of [6]) In the system $S_{X_1|Y}$ with $R_1 \geq 0$, when the constraint on the type-1 error probability* (14.8) *and communication constraints* (14.3) *are satisfied, the best error exponent for the type-2 error probability satisfies*

$$\theta_{\text{b}}(R_1, \epsilon) \geq \max_{U_1 \in \varphi_0} \min_{\widetilde{P}_{U_1 X_1 Y} \in \xi_0} D(\widetilde{P}_{U_1 X_1 Y} \| Q_{U_1 X_1 Y}), \tag{14.38}$$

where

$$\varphi_0 = \{U_1 : R_1 \geq I(U_1; X_1), U_1 \leftrightarrow X_1 \leftrightarrow Y, |\mathcal{U}_1| \leq |\mathcal{X}_1| + 1\},$$
$$\xi_0 = \{\widetilde{P}_{U_1 X_1 Y} : \widetilde{P}_{U_1 X_1} = P_{U_1 X_1}, \widetilde{P}_{U_1 Y} = P_{U_1 Y}\},$$

and $Q_{U_1|X_1} = P_{U_1|X_1}$.

For the case of an exponential-type constraint on the type-1 error probability, we first define

$$\widehat{\phi}(\omega) = \left\{ \widehat{U}\widehat{X}_1\widehat{Y} : \begin{array}{c} D(\widehat{U}\widehat{X}_1\widehat{Y} \| UX_1Y) \leq r \\ P_{U|X_1} = \widehat{P}_{U|X_1} = \omega(\widehat{X}_1) \\ U \leftrightarrow X_1 \leftrightarrow Y \end{array} \right\} \tag{14.39}$$

and

$$\phi(\omega) = \left\{ \widetilde{U}\widetilde{X}_1\widetilde{Y} : \begin{array}{c} \widetilde{P}_{UX_1} = \widehat{P}_{UX_1} \\ \widetilde{P}_{UY} = \widehat{P}_{UY} \\ \text{for some } \widehat{U}\widehat{X}_1\widehat{Y} \in \widehat{\phi}(\omega) \end{array} \right\}. \tag{14.40}$$

THEOREM 14.13 *(Theorem 1 of [7]) In the system $S_{X_1|Y}$ with $R_1 \geq 0$, when the constraint on the type-1 error probability* (14.9) *and communication constraints* (14.3) *are satisfied, the best error exponent for the type-2 error probability satisfies*

$$\sigma_{\text{b}}(R_1, r) \geq \max_{\substack{\omega \in \phi_{X_1}(R_1, r) \\ \|\widetilde{U}\| \leq \|\mathcal{X}_1\| + 2}} \min_{\widetilde{U}\widetilde{X}_1\widetilde{Y} \in \phi(\omega)} D(\widetilde{P}_{UX_1Y} \| Q_{UX_1Y}), \tag{14.41}$$

with $Q_{U|X_1} = \widetilde{P}_{U|X_1}$ and $\phi_{X_1}(R_1, r)$ defined in (14.22).

14.3 Interactive Communication

In Section 14.2, we discussed basic models with non-interactive communications. In this section, we discuss results for more sophisticated models with interactive communication among different terminals. The basic idea is that, by allowing interactive communications among terminals, the decision-maker could receive more information that may lead to a smaller error probability. We discuss two forms of interaction. First, we study cascaded communication [16, 17], in which these terminals broadcast their messages in a sequential order from terminal $\mathcal{X}_1$ to terminal $\mathcal{X}_L$ and each terminal uses all previously received messages along with its own observations for encoding. Then, we discuss the scenario with multiple rounds of communication between two terminals $\mathcal{X}_1$ and $\mathcal{Y}$ [18–20].

14.3.1 Cascaded Communication

Model

As in Section 14.2, we consider the following hypothesis-testing problem:

$$H_0 : P_{X_1 \cdots X_L Y}, \quad H_1 : Q_{X_1 \cdots X_L Y}. \tag{14.42}$$

$(X_1^n, \ldots, X_L^n, Y^n)$ are i.i.d. generated according to one of the above joint PMFs, and are observed at different terminals. These terminals broadcast messages in a sequential order from terminal 1 until terminal L, and each terminal will use all messages received so far along with its own observations for encoding. More specifically, terminal $\mathcal{X}_1$ will first broadcast its encoded message, which depends only on X_1^n, and then terminal $\mathcal{X}_2$ will broadcast its encoded message, which now depends not only on its own observations X_2^n but also on the message received from terminal $\mathcal{X}_1$. The process continues until terminal $\mathcal{X}_L$, which will use messages received from $\mathcal{X}_1$ until $\mathcal{X}_{L-1}$ and its own observations X_L^n for encoding. Finally, terminal $\mathcal{Y}$ decides which hypothesis is true on the basis of its own information and the messages received from terminals $\mathcal{X}_1$, …, $\mathcal{X}_L$. The system model is illustrated in Fig. 14.5.

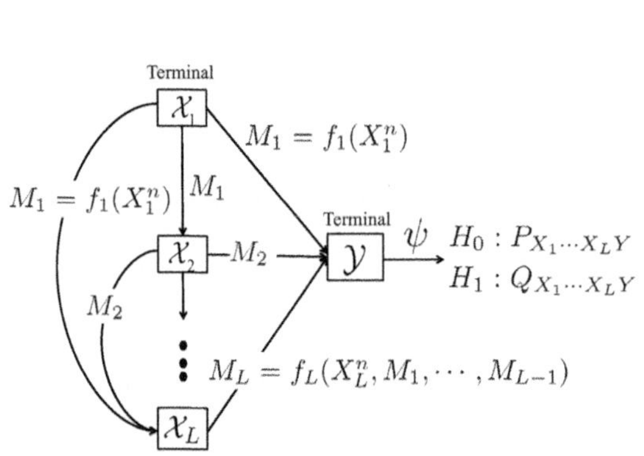

Figure 14.5 Model with cascaded encoders.

More specifically, terminal $\mathcal{X}_1$ uses an encoder

$$f_1 : \mathcal{X}_1^n \to \mathcal{M}_1 = \{1, 2, \ldots, M_1\}, \tag{14.43}$$

which is a map from $\mathcal{X}_1^n$ to $\mathcal{M}_1$. Terminal $\mathcal{X}_l$, $l = 2, \ldots, L$, uses an encoder

$$f_l : (\mathcal{X}_l^n, \mathcal{M}_1, \ldots, \mathcal{M}_{l-1}) \to \mathcal{M}_l = \{1, 2, \ldots, M_l\}, \tag{14.44}$$

with rates R_l such that

$$\limsup_{n\to\infty} \frac{1}{n} \log M_l \le R_l, \quad l = 1, \ldots, L. \tag{14.45}$$

Using its own observations and messages received from encoders, terminal $\mathcal{Y}$ will use a decoding function ψ to decide which hypothesis is true:

$$\psi : (\mathcal{M}_1, \ldots, \mathcal{M}_L, \mathcal{Y}^n) \to \{H_0, H_1\}. \tag{14.46}$$

Given the encoding and decoding functions, we can define the acceptance region and corresponding type-1 error probability, type-2 error probability, and type-2 error exponents under different types of constraints on the type-1 error probability in a similar manner to those in Section 14.2. To distinguish this case from the basic model, we use θ_c and σ_c to denote the type-2 error exponent under a constant-type constraint and the type-2 error exponent under an exponential-type constraint, respectively.

Testing against Independence with Positive-Rate Data Compression

For testing against independence, we fully characterize the type-2 error exponent for the case of general L. To simplify our presentation, we use $L = 2$ as an example.

THEOREM 14.14 *(Theorem 3 of [17]) For testing against independence with $L = 2$ cascaded encoders, the best error exponent for the type-2 error probability satisfies*

$$\lim_{\epsilon\to 0} \theta_c(R_1, R_2, \epsilon) = \max_{U_1U_2 \in \varphi_0} I(U_1U_2; Y), \tag{14.47}$$

where

$$\begin{aligned} \varphi_0 = \{U_1U_2 : R_1 \ge I(U_1; X_1),\ & R_2 \ge I(U_2; X_2|U_1), \\ & U_1 \leftrightarrow X_1 \leftrightarrow (X_2, Y), \\ & U_2 \leftrightarrow (X_2, U_1) \leftrightarrow (X_1, Y), \\ & |\mathcal{U}_1| \le |\mathcal{X}_1| + 1, |\mathcal{U}_2| \le |\mathcal{X}_2| \cdot |\mathcal{U}_1| + 1\}. \end{aligned} \tag{14.48}$$

To achieve this bound, we employ the following coding scheme. For a given rate constraint R_1, terminal $\mathcal{X}_1$ first generates a quantization codebook containing 2^{nR_1} quantization sequences u_1^n that are based on x_1^n. After observing x_1^n, terminal $\mathcal{X}_1$ picks one sequence u_1^n from the quantization codebook which is jointly typical with x_1^n and broadcasts this sequence. After receiving u_1^n from terminal $\mathcal{X}_1$, terminal $\mathcal{X}_2$ generates a quantization codebook containing 2^{nR_2} quantization sequences u_2^n that are based on both x_2^n and u_1^n. Then after observing x_2^n, terminal $\mathcal{X}_2$ picks one sequence u_2^n such that it is jointly typical with u_1^n and x_2^n and broadcasts this sequence. Upon receiving both u_1^n

Table 14.1. The joint PMF $P_{X_1X_2Y}$

X_1X_2Y	000	010	100	110
$P_{X_1X_2Y}$	0.0704	0.2108	0.0015	0.3233
X_1X_2Y	001	011	101	111
$P_{X_1X_2Y}$	0.2206	0.0667	0.0046	0.1021

Table 14.2. $P_{U_1|X_1}$ and $P_{U_2|X_2}$ for the non-interactive case when $R = 0.48$

$U_1|X_1$	0|0	1|0	0|1	1|1
$P_{U_1|X_1}$	0.9991	0.0009	0.1564	0.8436
$U_2|X_2$	0|0	1|0	0|1	1|1
$P_{U_2|X_2}$	0.9686	0.0314	0.0357	0.9643

Table 14.3. $P_{U_1|X_1}$ and $P_{U_2|X_2U_1}$ for the cascaded case when $R = 0.48$

$U_1|X_1$	0|0	1|0	0|1	1|1
$P_{U_1|X_1}$	0.0155	0.9845	0.5829	0.4171
$U_2|X_2U_1$	0|00	1|00	0|01	1|01
$P_{U_2|X_2U_1}$	0.0636	0.9364	0.9727	0.0273
$U_2|X_2U_1$	0|10	1|10	0|11	1|11
$P_{U_2|X_2U_1}$	0.9898	0.0102	0.0005	0.9995

and u_2^n, the decision-maker will declare that the hypothesis H_0 is true if the descriptions from these terminals and the side information at the decision-maker are correlated. Otherwise, the decision-maker will declare H_1 true. To prove the converse part, please refer to [17] for more details.

From the description above, we can get an intuitive idea that the decision-maker in the interactive case receives more information than does the decision-maker in the non-interactive case; thus a better performance is expected. In the following, we provide a numerical example to illustrate the gain obtained from interactive communications.

In the example, we let X_1, X_2, and Y be binary random variables with joint PMF $P_{X_1X_2Y}$, which is shown in Table 14.1. With $Q_{X_1X_2Y} = P_{X_1X_2}P_Y$ and increasing communication constraints $R = R_1 = R_2$, we use Theorem 14.14 to find the best value of the type-2 error exponent that we can achieve using our cascaded scheme. For comparison, we also use Theorem 14.2 to find an upper bound on the type-2 error exponent of the non-interactive case. By applying a grid search, we find the optimal conditional distributions $P_{U_1|X_1}$ and $P_{U_2|X_2}$ for the non-interactive case and the optimal conditional distributions $P_{U_1|X_1}$ and $P_{U_2|X_2U_1}$ for the cascaded case. We then calculate the bound on the type-2 error exponent for both cases. For $R = 0.48$, we list the conditional distributions $P_{U_1|X_1}$ and $P_{U_2|X_2}$ for the non-interactive case in Table 14.2 and the conditional distributions $P_{U_1|X_1}$ and $P_{U_2|X_2U_1}$ for the cascaded case in Table 14.3.

The simulation results for different Rs are shown in Fig. 14.6. From Fig. 14.6, we can see that the type-2 error exponents in both cases increase with the increasing R, which makes sense as the more information we can send, the fewer errors we will make.

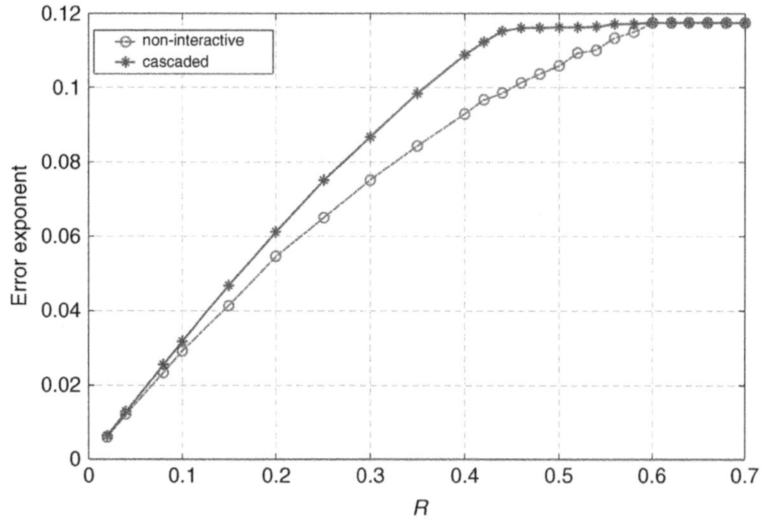

Figure 14.6 Simulation results.

We also observe that the type-2 error exponent achieved using our cascaded communication scheme is even larger than an upper bound on the type-2 error exponent of any non-interactive schemes. Hence, we confirm the intuitive idea that the greater amount of information offered by cascaded communication facilitates better decision-making for certain forms of testing against independence cases with positive communication rates.

General PMF with Zero-Rate Data Compression

For hypothesis testing with zero-rate data compression, we show that the cascaded scheme has the same performance as that of the non-interactive communication scenario. In the non-interactive communication scenario with zero-rate compression, a matching upper bound and lower bound on the type-2 error exponent were provided in Theorem 14.5. If we can prove an upper bound on the type-2 error exponent for the cascaded communication case that is no larger than the error exponent shown in Theorem 14.5, then we can arrive at the conclusion that cascaded communication won't help in the zero-rate compression case.

In the following, we provide an upper bound on the type-2 error exponent for the cascaded case.

THEOREM 14.15 *(Theorem 7 of [17]) Letting $P_{X_1 \cdots X_l Y}$ be arbitrary and $Q_{X_1 \cdots X_l Y} > 0$, for all $\epsilon \in [0, 1)$, the best type-2 error exponent for zero-rate compression under the type-1 error constraint* (14.8), *with L cascaded encoders, satisfies*

$$\theta_c(0, \ldots, 0, \epsilon) \leq \min_{\widetilde{P}_{X_1 \cdots X_L Y} \in \mathcal{L}} D(\widetilde{P}_{X_1 \cdots X_L Y} \| Q_{X_1 \cdots X_L Y}), \tag{14.49}$$

where

$$\mathcal{L} = \left\{ \widetilde{P}_{X_1 \cdots X_L Y} : \widetilde{P}_{X_i} = P_{X_i}, \ \widetilde{P}_Y = P_Y, \ i = 1, \ldots, L \right\}. \tag{14.50}$$

On comparing Theorem 14.15 with Theorem 14.5, we can see that the upper bound on the type-2 error exponent for the cascaded communication scheme is the same as the type-2 error exponent achievable by the non-interactive communication scheme. This implies that the performance of the cascaded communication scheme is the same as that of the non-interactive communication scheme in the zero-rate data compression case.

The conclusion that cascaded communication does not improve the type-2 error exponent in the zero-rate data compression case also holds for the scenario with an exponential-type constraint on the type-1 error probability. In the cascaded communication case, according to the results in Theorem 14.15, we can use a similar strategy to that in [5] to convert the problem under the exponential-type constraint (14.9) to the corresponding problem under the constraint in (14.8). As the converting strategy is independent of the communication style, it will be the same as that in Theorem 14.11. Then an upper bound on the type-2 error exponent under the exponential-type constraint can be easily derived without going into details, which will be shown in what follows.

THEOREM 14.16 *(Theorem 8 of [17]) Letting $P_{X_1 \cdots X_L Y}$ be arbitrary and $Q_{X_1 \cdots X_L Y} > 0$, the best type-2 error exponent for zero-rate compression case under the type-1 error constraint* (14.9), *with L cascaded encoders, satisfies*

$$\sigma_{\mathrm{c}}(0,\ldots,0,r) \le \min_{\widetilde{P}_{X_1 \cdots X_L Y} \in \mathcal{H}_r} D(\widetilde{P}_{X_1 \cdots X_L Y} \| Q_{X_1 \cdots X_L Y}), \tag{14.51}$$

with

$$\mathcal{H}_r = \Big\{\widetilde{P}_{X_1 \cdots X_L Y} : \widetilde{P}_{X_l} = \widehat{P}_{X_l}, \widetilde{P}_Y = \widehat{P}_Y,\ l = 1,\ldots,L \quad \text{for some } \widehat{P}_{X_1 \cdots X_L Y} \in \varphi_r\Big\}, \tag{14.52}$$

where

$$\varphi_r = \{\widehat{P}_{X_1 \cdots X_L Y} : D(\widehat{P}_{X_1 \cdots X_L Y} \| P_{X_1 \cdots X_L Y}) \le r\}. \tag{14.53}$$

On comparing Theorem 14.16 with Theorem 14.11, where a matching upper and lower bound are provided for the non-interactive scheme, we can conclude that there is no gain in performance on the type-2 error exponent under zero-rate compression with the exponential-type constraint on the type-1 error probability.

Genral PMF with Positive-Rate Data Compression

We can also get a lower bound on the type-2 error exponent for the case of a general PMF with positive-rate data compression.

THEOREM 14.17 *(Theorem 5 of [17]) For the case with general hypothesis $P_{X_1 X_2 Y}$ versu $Q_{X_1 X_2 Y}$ with $L = 2$ cascaded encoders, the best error exponent of the type-2 error probability satisfies*

$$\theta_{\mathrm{c}}(R_1, R_2, \epsilon) \ge \max_{U_1 U_2 \in \varphi_0} \min_{\widetilde{P}_{U_1 U_2 X_1 X_2 Y} \in \xi_0} D(\widetilde{P}_{U_1 U_2 X_1 X_2 Y} \| Q_{U_1 U_2 X_1 X_2 Y}), \tag{14.54}$$

where φ_0 *is defined in Theorem 14.14,*

$$\xi_0 = \{\widetilde{P}_{U_1U_2X_1X_2Y} : \widetilde{P}_{U_1X_1} = P_{U_1X_1}, \widetilde{P}_{U_1U_2X_2} = P_{U_1U_2X_2}, \widetilde{P}_{U_1U_2Y} = P_{U_1U_2Y}\}, \quad (14.55)$$

and $Q_{U_1|X_1} = P_{U_1|X_1}, Q_{U_2|U_1X_2} = P_{U_2|U_1X_2}$.

14.3.2 Fully Interactive Communication

For fully interactive communication, only the $L = 1$ case is studied under constant-type constraint on the type-1 error probability [20].

Model

As in Section 14.2, we consider the following hypothesis-testing problem:

$$H_0 : P_{X_1Y}, \quad H_1 : Q_{X_1Y}. \quad (14.56)$$

(X_1^n, Y^n) are i.i.d. generated according to one of the above joint PMFs, and are observed at two different terminals. Terminal $\mathcal{X}_1$ first encodes its local information to messages M_{11} and broadcasts it. Terminal $\mathcal{Y}$ utilizes the message M_{11} to encode its own information as M_{21} and broadcasts it. This is called one round of interactive communication of $\mathcal{X}_1$ and $\mathcal{Y}$. After receiving M_{21}, terminal $\mathcal{X}_1$ can further encode its local information as M_{21} and broadcast it. This process goes on until N rounds of interactive communication have been carried out. After receiving all messages from terminal $\mathcal{X}_1$ and $\mathcal{Y}$, terminal $\mathcal{X}_1$ will act as the decision-maker $\mathcal{Y}$ and makes a decision about the joint PMF of (X_1, Y). The system model is illustrated in Fig. 14.7.

More specifically, terminal $\mathcal{X}_1$ uses the encoding function

$$f_{1i} : \{\mathcal{X}_1^n, \mathcal{M}_{2(i-1)}, \ldots, \mathcal{M}_{21}\} \to \mathcal{M}_{1i} = \{1, \ldots, M_{1i}\}, i = 1, \ldots, N, \quad (14.57)$$

which is a map from $\mathcal{X}_1^n$ to $\mathcal{M}_{1i}$. Terminal $\mathcal{Y}$ uses an encoder

$$f_{2i} : \{\mathcal{Y}^n, \mathcal{M}_{1i}, \ldots, \mathcal{M}_{11}\} \to \mathcal{M}_{2i} = \{1, \ldots, M_{2i}\}, i = 1, \ldots, N, \quad (14.58)$$

with rate R such that

$$\limsup_{n\to\infty} \frac{1}{n} \sum_{i=1}^{N} \log(M_{1i}M_{2i}) \leq R, \quad i = 1, \ldots, N. \quad (14.59)$$

Terminal $\mathcal{X}_1$ uses a decoding function to decide which hypothesis is true:

$$\psi : \{\mathcal{M}_{11}, \mathcal{M}_{21}, \ldots, \mathcal{M}_{1N}, \mathcal{M}_{2N}\} \to \{H_0, H_1\}. \quad (14.60)$$

Given the encoding and decoding function, we can define the acceptance region and corresponding type-1 error probability, type-2 error probability, and type-2 error exponents under different types of constraints on type-1 error probability in a similar way

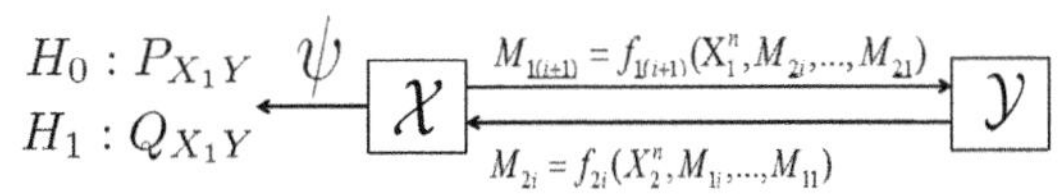

Figure 14.7 Model with interactive encoders.

to what we did in Section 14.2. To distinguish this case from the basic model and the cascaded model, we use θ_{i} and σ_{i} to denote the type-2 error exponent under a constant-type constraint and the type-2 error exponent under an exponential-type constraint, respectively.

Testing against Independence with Positive-Rate Data Compression

THEOREM 14.18 *(Proposition 2 of [20]) For the system S_{X_1Y} with N rounds of communication, the type-2 error exponent satisfies*

$$\lim_{\epsilon\to 0}\theta_{\mathrm{i}} \geq \max_{U_{[1:N]}V_{[1:N]}\in\varphi(R)} \sum_{k=1}^{N}[I(U_{[k]};Y|U_{[1:k-1]}V_{[1:k-1]})+I(V_{[k]};Y|U_{[1:k]}V_{[1:k-1]})], \tag{14.61}$$

where

$$\varphi(R) \triangleq \Bigg\{ U_{[1:N]}V_{[1:N]} : R \geq \sum_{k=1}^{N}[I(U_{[k]};Y|U_{[1:k-1]}V_{[1:k-1]}) + I(V_{[k]};Y|U_{[1:k]}V_{[1:k-1]})], \\ U_{[k]} \leftrightarrow (X, U_{1:k-1}, V_{1:k-1}) \leftrightarrow Y, |U_{[k]}| < \infty, \\ V_{[k]} \leftrightarrow (Y, U_{1:k}, V_{1:k-1}) \leftrightarrow X, |V_{[k]}| < \infty, \\ k = 1,\ldots,N \Bigg\}. \tag{14.62}$$

For $N = 1$, which means there is only one round of communication between terminal $\mathcal{X}_1$ and $\mathcal{Y}$, one can also prove a matching upper bound on the type-2 error exponent. Hence one has the following theorem.

THEOREM 14.19 *(Theorem 3 of [20]) For the system S_{X_1Y} with N rounds of communication, the type-2 error exponent satisfies*

$$\lim_{\epsilon\to 0}\theta_{\mathrm{i}} = \max_{\substack{P_{U|X}P_{V|UY} \\ I(U;Y)+I(V;Y|U)\leq R}} I(U;Y)+I(V;Y|U). \tag{14.63}$$

To achieve this bound, one can employ the following coding scheme. Terminal $\mathcal{X}_1$ first generates a quantization codebook containing $2^{n(I(U;X)+\eta)}$ quantization sequences u^n that are based on x_1^n. After observing x_1^n, terminal $\mathcal{X}_1$ picks one sequence u^n from the quantization codebook which is jointly typical with x_1^n and broadcasts this sequence. After receiving u_1^n from terminal $\mathcal{X}_1$, terminal $\mathcal{Y}$ generates a quantization codebook containing $2^{n(I(V;Y)+\epsilon)}$ quantization sequences v^n that are based on both y^n and u_1^n. Then, after observing y^n, terminal $\mathcal{Y}$ picks one sequence v^n such that it is jointly typical with u^n and y^n and broadcasts this sequence. Upon receiving v^n, the decision-maker will declare that the hypothesis H_0 is true if the descriptions from terminal $\mathcal{Y}$ and the information at terminal $\mathcal{X}$ are correlated. Otherwise, the decision-maker will declare H_1 true. To prove the converse part, please refer to [20] for more details.

General PMF with Zero-Rate Data Compression

One can also prove that interactive communication does not help to improve the performance with zero-rate data compression. In particular, [20] provides a matching upper bound and lower bound on the type-2 error exponent for the fully interactive case.

THEOREM 14.20 *(Theorem 4 of [20]) Letting P_{X_1Y} be arbitrary and $Q_{X_1Y} > 0$, for all $\epsilon \in [0,1)$, the best type-2 error exponent for zero-rate compression under the type-1 error constraint* (14.8) *satisfies*

$$\theta_{\text{i}}(0,\epsilon) = \min_{\widetilde{P}_{X_1Y} \in \mathcal{L}_0} D(\widetilde{P}_{X_1Y} \| Q_{X_1Y}), \tag{14.64}$$

where

$$\mathcal{L}_0 = \left\{ \widetilde{P}_{X_1Y} : \widetilde{P}_{X_1} = P_{X_1},\ \widetilde{P}_Y = P_Y \right\}.$$

General PMF with Positive-Rate Data Compression

One can also get a lower bound on the type-2 error exponent for the case of a general PMF with positive-rate data compression.

THEOREM 14.21 *(Proposition 2 [20]) In the system S_{X_1Y}, the best error exponent of the type-2 error probability satisfies*

$$\theta_{\text{i}}(R,\epsilon) \geq \max_{\varphi(R)} \min_{\xi(U_{[1:N]}V_{[1:N]})} D(\widetilde{P}_{U_{[1:N]}V_{[1:N]}X_1Y} \| Q_{U_{[1:N]}V_{[1:N]}X_1Y}), \tag{14.65}$$

where φ_0 is defined in Theorem 14.18, and

$$\xi(U_{[1:N]}V_{[1:N]}) = \Big\{ \widetilde{P}_{U_{[1:N]}V_{[1:N]}X_1Y} : \widetilde{P}_{U_{[1:N]}V_{[1:N]}X_1} = P_{U_{[1:N]}V_{[1:N]}X_1},$$
$$\widetilde{P}_{U_{[1:N]}V_{[1:N]}Y} = P_{U_{[1:N]}V_{[1:N]}Y}, \Big\}. \tag{14.66}$$

14.4 Identity Testing

In Sections 14.2 and 14.3, we dealt with the scenarios where the PMF under each hypothesis is fully specified. However, there are practical scenarios where the probabilistic models are not fully specified. One of these problems is the identity-testing problem, in which the goal is to determine whether given samples are generated from a certain distribution or not.

In this section, we discuss the identity-testing problem in the feature partitioning scenario with non-interactive communication of the encoders. As in Section 14.2, we consider a setup with L terminals (encoders) $\mathcal{X}_l$, $l = 1,\ldots,L$ and a decision-making terminal $\mathcal{Y}$. $(X_1^n,\ldots,X_L^n,Y^n)$ are generated according to some unknown PMF $P_{X_1\cdots X_LY}$. Terminals $\{\mathcal{X}_l\}_{l=1}^L$ can send compressed messages related to their own data with limited rates to the decision-maker, then the decision-maker performs statistical inference on the basis of the messages received from terminals $\{\mathcal{X}_l\}_{l=1}^L$ and its local information related to Y. In particular, we focus on the problem in which that the decision-maker tries to decide whether $P_{X_1\cdots X_LY}$ is the same as a given distribution $Q_{X_1\cdots X_LY}$, i.e., $P_{X_1\cdots X_LY} = Q_{X_1\cdots X_LY}$, or they are λ-far away, i.e., $\|P_{X_1\cdots X_LY} - Q_{X_1\cdots X_LY}\|_1 \geq \lambda (\lambda > 0)$, where $\|\cdot\|_1$ denotes the ℓ_1

norm of its argument. This identity-testing problem can be interpreted as the following two hypothesis-testing problems.

- *Problem 1*:

$$H_0 : \|P_{X_1 \cdots X_L Y} - Q_{X_1 \cdots X_L Y}\|_1 \geq \lambda \text{ versus } H_1 : P_{X_1 \cdots X_L Y} = Q_{X_1 \cdots X_L Y}. \tag{14.67}$$

- *Problem 2*:

$$H_0 : P_{X_1 \cdots X_L Y} = Q_{X_1 \cdots X_L Y} \text{ versus } H_1 : \|P_{X_1 \cdots X_L Y} - Q_{X_1 \cdots X_L Y}\|_1 \geq \lambda. \tag{14.68}$$

In both problems, our goal is to characterize the type-2 error exponent under the constraints on the communication rates and type-1 error probability.

This distributed identity-testing problem with composite hypotheses can be viewed as a generalization of the problems considered in Section 14.2. Between the two possible problems defined in (14.67) and (14.68), Problem 2 is relatively simple and it can be solved using similar schemes to those proposed in Section 14.2. In particular, the encoding schemes and the definition of the acceptance regions at the decision-maker in Section 14.2 depend only on the form of the PMF under H_0. Since the form of the PMF under H_0 in Problem 2 is known, we can apply the existing coding/decoding schemes such as that in Section 14.2 and take the type-2 error probability as the supremum of the type-2 error probabilities under each $P_{X_1 \cdots X_L Y}$ that satisfies $\|P_{X_1 \cdots X_L Y} - Q_{X_1 \cdots X_L Y}\|_1 \geq \lambda$. As well, it can be shown that these schemes are optimal for Problem 2. However, in Problem 1, as H_0 is composite, we need to design universal encoding/decoding schemes so that our schemes can provide a guaranteed performance regardless of the true PMF under H_0. In this section, we will focus on the more challenging Problem 1 with $L = 1$, and, to simplify our presentation, we use terminal $\mathcal{X}$ and terminal $\mathcal{Y}$ to denote the two terminals.

14.4.1 Model

To simplify the presentation, we assume that there are only terminal $\mathcal{X}$ and terminal $\mathcal{Y}$. Our goal is to determine whether the true joint distribution is the same as the given distribution Q_{XY} or far away from it. We interpret this problem as a hypothesis-testing problem with a composite null hypothesis and a simple alternative hypothesis:

$$H_0 : P_{XY} \in \Pi \quad \text{versus} \quad H_1 : Q_{XY}, \tag{14.69}$$

where $\Pi = \{P_{XY} \in \mathcal{P}_{XY} : \|P_{XY} - Q_{XY}\|_1 \geq \lambda\}$ and λ is some fixed positive number. The model is shown in Fig. 14.8. In a typical identity-testing problem, one determines which

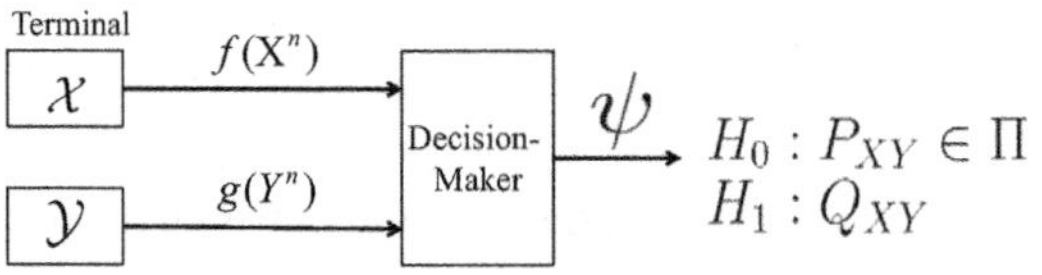

Figure 14.8 Model.

hypothesis is true under the assumption that (X^n, Y^n) are fully available at the decision-maker. We assume terminal $\mathcal{X}$ observes only X^n and terminal $\mathcal{Y}$ observes only Y^n. Terminals $\mathcal{X}$ and $\mathcal{Y}$ are allowed to send encoded messages to the decision-maker, which then decides which hypothesis is true using the encoded messages directly. We denote the system as S_{XY} in what follows.

More formally, the system consists of two encoders, f_1, f_2, and one decision function, ψ. Terminal $\mathcal{X}$ has the encoder f_1, terminal $\mathcal{Y}$ has the encoder f_2, and the decision-maker has the decision function ψ. After observing the data sequences $x^n \in \mathcal{X}^n$ and $y^n \in \mathcal{Y}^n$, encoders f_1 and f_2 transform sequences x^n and y^n into messages $f_1(x^n)$ and $f_2(y^n)$, respectively, which take values from the message sets $\mathcal{M}_n$ and $\mathcal{N}_n$,

$$f_1 : \mathcal{X}^n \to \mathcal{M}_1 = \{1,2,\dots,M_1\}, \tag{14.70}$$

$$f_2 : \mathcal{Y}^n \to \mathcal{M}_2 = \{1,2,\dots,M_2\}, \tag{14.71}$$

with rate constraints

$$\limsup_{n\to\infty} \frac{1}{n} \log M_1 \le R_1, \tag{14.72}$$

$$\limsup_{n\to\infty} \frac{1}{n} \log M_2 \le R_2. \tag{14.73}$$

Using the messages $f_1(X^n)$ and $f_2(Y^n)$, the decision-maker will use the decision function ψ to determine which hypothesis is true:

$$\psi : (\mathcal{M}_n, \mathcal{N}_n) \to \{H_0, H_1\}. \tag{14.74}$$

For any given decision function ψ, one can define the acceptance region as

$$\mathcal{A}_n = \{(x^n, y^n) \in \mathcal{X}^n \times \mathcal{Y}^n : \psi(f_1(x^n), f_2(y^n)) = H_0\}.$$

For any given f_i, $i = 1,2$, and ψ, the type-1 error probability α_n and the type-2 error probability β_n are defined as follows:

$$\alpha_n = \sup_{P_{XY}\in\Pi} P^n_{XY}(\mathcal{A}^c_n) \text{ and } \beta_n = Q^n_{XY}(\mathcal{A}_n). \tag{14.75}$$

Given the type-1 error probability and type-2 error probability, we define the type-2 error exponents under different types of constraints on the type-1 error probability in a similar way to that in Section 14.2. To distinguish this case from the basic model, we use θ_{id} and σ_{id} to denote the type-2 error exponent under a constant-type constraint and the type-2 error exponent under an exponential-type constraint, respectively.

14.4.2 Testing against Independence with Positive-Rate Data Compression

We now focus on a special case, namely testing against independence, for which the hypotheses are

$$H_0 : P_{XY} \in \Pi^{\perp} \qquad \text{versus} \qquad H_1 : Q_X Q_Y, \tag{14.76}$$

where $\Pi^{\perp} = \Pi \cap \{P_{XY} : P_X = Q_X, P_Y = Q_Y\}$. This special case has a two-fold meaning: testing whether (X,Y) are independent or not, and testing whether the joint distribution of (X,Y) is $Q_X Q_Y$ or not. Owing to the fact that the marginal distributions for X and Y in this special case are the same under both hypotheses, we can get a simple encoding/decoding scheme and derive matching lower and upper bounds on the type-2 error exponent, which allows us to fully characterize the optimal type-2 error exponent.

THEOREM 14.22 *(Theorem 3 of [26]) For $R_1 \geq 0$, $R_2 \geq 0$, we have*

$$\theta_{\mathrm{id}}(R_1, R_2, \epsilon) = \inf_{P_{XY} \in \Pi^{\perp}} \max_{UV \in \varphi_{P_{XY}}} I(U;V), \tag{14.77}$$

where

$$\varphi_{P_{XY}} = \{UV : I(U;X) \leq R_1, I(V;Y) \leq R_2, U \leftrightarrow X \leftrightarrow Y \leftrightarrow V\}. \tag{14.78}$$

To achieve this lower bound, we design the following encoding/decoding scheme. For a given rate constraint R_1, terminal $\mathcal{X}_1$ first generates a quantization codebook containing 2^{nR_1} quantization sequences u_1^n. After observing x_1^n, terminal $\mathcal{X}_1$ picks one sequence u_1^n from the quantization codebook to describe x_1^n and sends this sequence to the decision-maker. Then terminal $\mathcal{Y}$ uses a similar scheme and sends its quantization codebook v^n. After receiving the descriptions from both terminals, the decision-maker will employ a universal decoding method to declare either H_0 or H_1 true. More details about the universal decoding method can be found in [26].

14.4.3 General PMF with Zero-Rate Data Compression

Under a Constant-Type Constraint

Shalaby and Papamarcou [8] studied the problem with zero-rate data compression under a constant-type constraint for a more general case:

$$H_0 : P_{XY} \in \Pi \quad \text{versus} \quad H_1 : Q_{XY} \in \Xi, \tag{14.79}$$

where Π and Ξ are two disjoint subsets of $\mathcal{P}(\mathcal{X} \times \mathcal{Y})$. In this case, the type-2 error probability is defined as

$$\beta_n = \sup_{Q_{XY} \in \Xi} Q_{XY}^n(\mathcal{A}_n). \tag{14.80}$$

To show a matching upper bound and lower bond, we assume the uniform positivity constraint

$$\rho_{\inf} \triangleq \inf_{Q_{XY} \in \Xi} \min_{(x,y) \in \mathcal{X} \times \mathcal{Y}} Q_{XY}(x,y) > 0. \tag{14.81}$$

The result is shown in the following theorem.

THEOREM 14.23 *(Theorem 4 of [8]) Let $P_{XY} \in \Pi$ be arbitrary and (14.81) be satisfied. For zero-rate compression in S_{XY} and the type-1 error constraint* (14.8), *the error exponent satisfies*

$$\theta_{\mathrm{id}}(0,0,\epsilon) = \inf_{P_{XY} \in \Pi, Q_{XY} \in \Xi} \min_{\substack{\widetilde{P}_{XY}: \\ \widetilde{P}_X = P_X, \widetilde{P}_Y = P_Y}} D(\widetilde{P}_{XY} \| Q_{XY}). \tag{14.82}$$

Under an Exponential-Type Constraint

As H_0 is a composite hypothesis, we provide a universal encoding and decoding scheme to establish a lower bound on the error exponent of the type-2 error. We further establish a matching upper bound and hence fully characterize the error exponent of the type-2 error for this scenario.

THEOREM 14.24 *(Theorem 1 of [27]) Let $P_{XY} \in \Pi$ be arbitrary and $Q_{XY} > 0$. For zero-rate compression in S_{XY} and the type-1 error constraint* (14.9), *the error exponent satisfies*

$$\sigma_{\text{id}}(0,0,r) = \inf_{P_{XY}\in\Pi} \min_{\widetilde{P}_{XY}\in\mathcal{H}_r^{P_{XY}}} D(\widetilde{P}_{XY}\|Q_{XY}) \tag{14.83}$$

with

$$\mathcal{H}_r^{P_{XY}} = \{\widetilde{P}_{XY} : \widetilde{P}_X = \widehat{P}_X, \widetilde{P}_Y = \widehat{P}_Y \textit{ for some } \widehat{P}_{XY} \in \varphi_r^{P_{XY}}\},$$

where

$$\varphi_r^{P_{XY}} = \{\widehat{P}_{XY} : D(\widehat{P}_{XY}\|P_{XY}) \le r\}.$$

To achieve this bound, we use the same encoding scheme as in Section 14.2, but employ a universal decoding scheme at the decision-maker so that the type-1 error constraint is satisfied regardless of what the true value of P_{XY} is. One can certainly design an individual acceptance region that satisfies the type-1 error constraint for each possible value of $P_{XY} \in \Pi$ using the approach in the simple hypothesis case, and then take the union of these individual regions as the final acceptance region. This will clearly satisfy the type-1 error constraint regardless of the true value of P_{XY}. This approach might work if the number of possible P_{XY}s is finite or grows polynomially in n. However, in our case, there are uncountably infinitely many possible P_{XY}s in Π. This approach will lead to a very loose performance bound. Hence, we need to design a new approach that will lead to a performance bound matching with the converse bound. More details can be found in [26].

14.4.4 General PMF with Positive-Rate Data Compression

In this section, we investigate the identity-testing problem under the positive rate compression constraints (14.72) and (14.73). We first establish a lower bound on the type-2 error exponent under the constant type constraint (14.8) for a general PMF. Then we provide a lower bound on the type-2 error exponent under the exponential-type constraint (14.9).

Under a Constant-Type Constraint

Let $\mathcal{U}$ and $\mathcal{V}$ be arbitrary finite sets. For each distribution P_X on $\mathcal{X}$, let $\omega(\cdot|\cdot;P_X)$ be any stochastic mapping from $\mathcal{X}$ to $\mathcal{U}$, i.e., let $\omega(u|x;P_X)$ be the conditional probability of $u \in \mathcal{U}$ given $x \in \mathcal{X}$. Similarly, for each distribution P_Y on $\mathcal{Y}$, let $\varrho(\cdot|\cdot;P_Y)$ be any stochastic mapping from $\mathcal{Y}$ to $\mathcal{V}$, i.e., let $\varrho(v|y;P_Y)$ be the probability of $v \in \mathcal{V}$ given $y \in \mathcal{Y}$.

THEOREM 14.25 *(Theorem 2 of [26]) For $R_1 \geq 0$, $R_2 \geq 0$, we have*

$$\theta_{\text{id}}(R_1, R_2, \epsilon) \geq \inf_{P_{XY} \in \Pi} \max_{(\omega,\varrho) \in \varphi_{P_{XY}}} \min_{\widetilde{P}_{UVXY} \in \xi_{P_{XY}}} D(\widetilde{P}_{UVXY} \| Q_{UVXY}), \tag{14.84}$$

where

$$\begin{aligned}\varphi_{P_{XY}} = \{(\omega, \varrho) : & R_1' \geq I(X;U), R_2' \geq I(Y;V),\\ & R_1' - R_1 \leq I(U;V),\\ & R_2' - R_2 \leq I(U;V),\\ & R_1' - R_1 + R_2' - R_2 \leq I(U;V)\\ & P_{U|X} = \omega(u|x; P_X), P_{V|Y} = \varrho(v|y; P_Y),\\ & U \leftrightarrow X \leftrightarrow Y \leftrightarrow V\}\end{aligned} \tag{14.85}$$

and

$$\xi_{P_{XY}} = \{\widetilde{P}_{UVXY} : \widetilde{P}_{UX} = P_{UX}, \widetilde{P}_{VY} = P_{VY}, \widetilde{P}_{UV} = P_{UV}\}. \tag{14.86}$$

Here $\varphi_{P_{XY}}$ denotes the set of (ω, ϱ) when the distribution of (X, Y) is P_{XY}. The notation $\xi_{P_{XY}}$ has a similar interpretation.

To achieve this bound, we employ a universal encoding/decoding scheme with a binning method. More details can be found in [26].

Under an Exponential-Type Constraint

In this section, we consider the case with an exponential-type constraint, i.e., we require that the exponent of the type-1 error probability should be larger than r. Owing to the complexity of the problem for the general case S_{XY}, we here give the result assuming terminal $\mathcal{Y}$ can communicate with a large rate $R_2 > \log|\mathcal{Y}|$ so that the decision-maker has full information about Y^n. We will use $\sigma(R_1, r)$ to denote the corresponding type-2 error exponent.

Let $\mathcal{U}$ be an arbitrary finite set and $\mathcal{P}(\mathcal{U}|\mathcal{X})$ be the set of all possible conditional probability distributions $(P_{U|X}(u|x))_{(u,x)\in\mathcal{U}\times\mathcal{X}}$ on $\mathcal{U}$ given values in $\mathcal{X}$. Let ω denote a continuous mapping from $\mathcal{P}(\mathcal{X})$ to $\mathcal{P}(\mathcal{U}|\mathcal{X})$ and Φ be the set of all possible mappings.

THEOREM 14.26 *(Theorem 4 of [26]) For $R_1 \geq 0$, $r \geq 0$, we have*

$$\sigma_{\text{id}}(R_1, r) \geq \inf_{P_{XY} \in \Pi} \sup_{\omega \in \phi_{P_{XY}}(R_1, r)} \min_{\widetilde{P}_{UXY} \in \Xi_{P_{XY}}(\omega)} D(\widetilde{P}_{UXY} \| Q_{UXY}), \tag{14.87}$$

where

$$\phi_{P_{XY}}(R_1, r) = \left\{\omega \in \Phi : \max_{\substack{\widehat{X} : D(\widehat{X}\|X) \leq r \\ P_{U|X} = \omega(\widehat{X})}} I(\widehat{U}; \widehat{X}) \leq R_1\right\},$$

$$\widehat{\Xi}_{P_{XY}}(\omega) = \left\{\widehat{P}_{UXY} : \begin{array}{c} D(\widehat{P}_{UXY} \| P_{UXY}) \leq r \\ P_{U|X} = \widehat{P}_{U|X} = \omega(\widehat{X}) \\ U \leftrightarrow X \leftrightarrow Y \end{array}\right\},$$

$$\Xi_{P_{XY}}(\omega) = \Big\{\widetilde{P}_{UXY} : \widetilde{P}_{UX} = \widehat{P}_{UX}, \widetilde{P}_{UY} = \widehat{P}_{UY},$$
$$\text{for some } \widehat{P}_{UXY} \in \widehat{\Xi}_{t_{XY}}(\omega)\Big\},$$

and $Q_{U|X} = \widetilde{P}_{U|X}, Q_{UXY} = Q_{U|X} Q_{XY}$.

14.5 Conclusion and Extensions

14.5.1 Conclusion

This chapter has explored distributed inference problems from an information-theoretic perspective. First, we discussed distributed inference problems with non-interactive encoders. Second, we considered distributed testing problems with interactive encoders. We investigated the case of cascaded communication among multiple terminals and then discussed the fully interactive communication between two terminals. Finally, we studied the distributed identity-testing problem, in which the decision-maker decides whether the distribution indirectly revealed from the compressed data from multiple distributed terminals is the same as or λ-far from a given distribution.

14.5.2 Future Directions

Despite the many interesting works reviewed in this chapter, there are still many open problems in distributed inference with multiterminal data compression that require further research efforts. In addition, one can further generalize the models discussed in this chapter to deal with more realistic scenarios. We list here some of the open problems and possible generalizations.

- For testing against independence with multiple non-interactive encoders, it remains to find a matching upper bound on the type-2 error exponent under a constant-type constraint on the type-1 error probability.
- Deriving an upper bound on the type-2 error exponent for the general PMF case for all three cases in this chapter remains to be done.
- Under the exponential-type constraint on the type-1 error probability, it remains to prove the converse part for testing against independence for all three cases in this chapter.
- In Section 14.3, we discussed cascaded communication among L terminals and fully interactive communication between two terminals. One possible generalization of this problem is to allow fully interactive communication among L terminals. This model is illustrated in Fig. 14.9.
- The models discussed in this chapter can be viewed as models with multiple encoders and one receiver. Given these results, another interesting topic to consider is models with multiple encoders and multiple receivers. Different receivers might have different learning goals. Hence, it is of interest to investigate how to encode messages to strike a desirable trade-off among the performances of multiple receivers. There have been some interesting recent papers that address problems in this direction [28, 29],

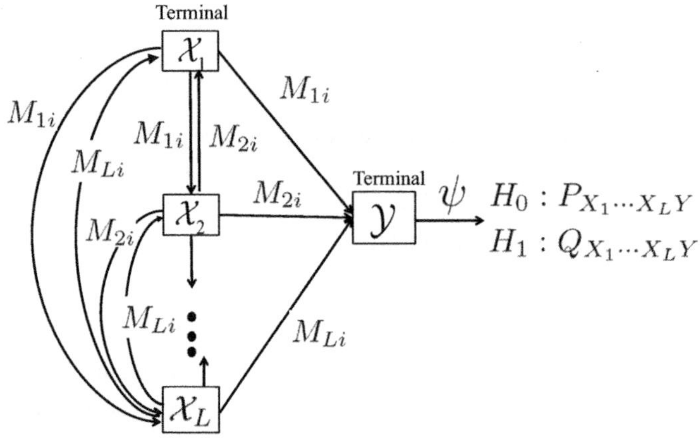

Figure 14.9 Model with L fully interacting encoders.

focusing on the special case of testing against (conditional) independence with multiple receivers.

- As discussed in the introduction, the distributed scenario is very common in real life. Hence, besides the basic inference problems discussed in this chapter, one can study more sophisticated statistical learning tasks such as linear regression, clustering, etc. While there are many interesting works that address these problems for the sample partitioning scenario [2, 30–33], these problems remain largely unexplored for the feature partitioning scenario.

Acknowledgments

The work of W. Zhao and L. Lai was supported by the National Science Foundation under grants CNS-1660128, ECCS-1711468, and CCF-1717943.

References

[1] Y. Zhang, J. C. Duchi, M. I. Jordan, and M. J. Wainwright, "Information-theoretic lower bounds for distributed statistical estimation with communication constraints," in *Advances in Neural Information Processing Systems 26*, 2013, pp. 2328–2336.

[2] O. Shamir, N. Srebro, and T. Zhang, "Communication-efficient distributed optimization using an approximate Newton-type method," in *Proc. International Conference on Machine Learning*, 2014.

[3] T. Berger, "Decentralized estimation and decision theory," in *Proc. IEEE Information Theory Workshop*, 1979.

[4] R. Ahlswede and I. Csiszár, "Hypothesis testing with communication constraints," *IEEE Trans. Information Theory*, vol. 32, no. 4, pp. 533–542, July 1986.

[5] T. S. Han and S. Amari, "Statistical inference under multiterminal data compression," *IEEE Trans. Information Theory*, vol. 44, no. 6, pp. 2300–2324, 1998.

[6] T. S. Han, "Hypothesis testing with multiterminal data compression," *IEEE Trans. Information Theory*, vol. 33, no. 6, pp. 759–772, 1987.

[7] T. S. Han and K. Kobayashi, "Exponential-type error probabilities for multiterminal hypothesis testing," *IEEE Trans. Information Theory*, vol. 35, no. 1, pp. 2–14, 1989.

[8] H. M. H. Shalaby and A. Papamarcou, "Multiterminal detection with zero-rate data compression," *IEEE Trans. Information Theory*, vol. 38, no. 2, pp. 254–267, 1992.

[9] W. Zhao and L. Lai, "Distributed detection with vector quantizer," *IEEE Trans. Signal and Information Processing over Networks*, vol. 2, no. 2, pp. 105–119, 2016.

[10] M. S. Rahman and A. B. Wagner, "The optimality of binning for distributed hypothesis testing," *IEEE Trans. Information Theory*, vol. 58, no. 10, pp. 6282–6303, 2012.

[11] W. Zhao and L. Lai, "Distributed testing with zero-rate compression," in *Proc. IEEE International Symposium on Information Theory*, 2015.

[12] C. Tian and J. Chen, "Successive refinement for hypothesis testing and lossless one-helper problem," *IEEE Trans. Information Theory*, vol. 54, no. 10, pp. 4666–4681, 2008.

[13] G. Katz, P. Piantanida, R. Couillet, and M. Debbah, "On the necessity of binning for the distributed hypothesis testing problem," in *Proc. IEEE International Symposium on Information Theory*, 2015.

[14] M. Mhanna and P. Piantanida, "On secure distributed hypothesis testing," in *Proc. IEEE International Symposium on Information Theory*, 2015.

[15] W. Zhao and L. Lai, "Distributed testing against independence with multiple terminals," in *Proc. Allerton Conference on Communication, Control, and Computing*, 2014, pp. 1246–1251.

[16] W. Zhao and L. Lai, "Distributed testing against independence with conferencing encoders," in *Proc. IEEE Information Theory Workshop*, 2015.

[17] W. Zhao and L. Lai, "Distributed testing with cascaded encoders," *IEEE Trans. Information Theory*, vol. 64, no. 11, pp. 7339–7348, 2018.

[18] Y. Xiang and Y. Kim, "Interactive hypothesis testing with communication constraints," in *Proc. Allerton Conference on Communication, Control, and Computing*, 2012, pp. 1065–1072.

[19] Y. Xiang and Y. Kim, "Interactive hypothesis testing against independence," in *Proc. IEEE International Symposium on Information Theory*, 2013, pp. 2840–2844.

[20] G. Katz, P. Piantanida, and M. Debbah, "Collaborative distributed hypothesis testing," *arXiv:1604.01292*, 2016.

[21] A. El Gamal and Y. Kim, *Network information theory*. Cambridge Unversity Press, 2011.

[22] R. Ahlswede and J. Körner, "Source coding with side information and a converse for degraded broadcast channels," *IEEE Trans. Information Theory*, vol. 21, no. 6, pp. 629–637, 1975.

[23] D. Slepian and J. Wolf, "Noiseless coding of correlated information sources," *IEEE Trans. Information Theory*, vol. 19, no. 4, pp. 471–480, 1973.

[24] R. Ahlswede, P. Gács, and J. Körner, "Bounds on conditional probabilities with applications in multi-user communication," *Z. Wahrscheinlichkeitstheorie verwandte Gebiete*, vol. 34, no. 2, pp. 157–177, 1976.

[25] T. M. Cover and J. A. Thomas, *Elements of information theory*. Wiley, 2005.

[26] W. Zhao and L. Lai, "Distributed identity testing with data compression," submitted to *IEEE Trans. Information Theory*, 2017.

[27] W. Zhao and L. Lai, "Distributed identity testing with zero-rate compression," in *Proc. IEEE International Symposium on Information Theory*, 2017, pp. 3135–3139.

[28] M. Wigger and R. Timo, "Testing against independence with multiple decision centers," in *Proc. IEEE International Conference on Signal Processing and Communications*, 2016, pp. 1–5.

[29] S. Salehkalaibar, M. Wigger, and R. Timo, "On hypothesis testing against conditional independence with multiple decision centers," *IEEE Trans. Communications*, vol. 66, no. 6, pp. 2409–2420, 2018.

[30] J. D. Lee, Y. Sun, Q. Liu, and J. E. Taylor, "Communication-efficient sparse regression: A one-shot approach," *arXiv:1503.04337*, 2015.

[31] H. B. McMahan, E. Moore, D. Ramage, S. Hampson, and B. Agüera y Arcas, "Communication-efficient learning of deep networks from decentralized data," *arXiv:1602.05629*, 2016.

[32] J. Konečný, "Stochastic, distributed and federated optimization for machine learning," *arXiv:1707.01155*, 2017.

[33] V. M. A. Martin, K. David, and B. Merlinsuganthi, "Distributed data clustering: A comparative analysis," *Int. J. Sci. Res. Computer Science, Engineering and Information Technol.*, vol. 3, no. 3, article CSEITI83376, 2018.

15 Network Functional Compression

Soheil Feizi and Muriel Médard

Summary

In this chapter,[1] we study the problem of compressing for function computation across a network from an information-theoretic point of view. We refer to this problem as *network functional compression*. In network functional compression, computation of a function (or some functions) of sources located at certain nodes in a network is desired at receiver(s). The rate region of this problem has been considered in the literature under certain restrictive assumptions, particularly in terms of the network topology, the functions, and the characteristics of the sources. In this chapter, we present results that significantly relax these assumptions. For a one-stage tree network, we characterize a rate region by introducing a necessary and sufficient condition for any achievable coloring-based coding scheme called the *coloring connectivity condition* (CCC). We also propose a modularized coding scheme based on graph colorings to perform arbitrarily closely to derived rate lower bounds. For a general tree network, we provide a rate lower bound based on graph entropies and show that this bound is tight in the case of having independent sources. In particular, we show that, in a general tree network case with independent sources, to achieve the rate lower bound, intermediate nodes should perform computations. However, for a family of functions and random variables, which we call *chain-rule proper sets*, it is sufficient to have no computations at intermediate nodes in order for the system to perform arbitrarily closely to the rate lower bound. Moreover, we consider practical issues of coloring-based coding schemes and propose an efficient algorithm to compute a minimum-entropy coloring of a characteristic graph under some conditions on source distributions and/or the desired function. Finally, extensions of these results for cases with feedback and lossy function computations are discussed.

15.1 Introduction

In modern applications, data are often stored in clouds in a distributed way (i.e., different portions of data are located in different nodes in the cloud/network). Therefore, to compute certain functions of the data, nodes in the network need to communicate with each other. Depending on the desired computation, however, nodes can first compress

[1] The content of this chapter is based on [1].

the data before transmitting them in the network, providing a gain in communication costs. We refer to this problem as *network functional compression.*

In this chapter, we consider different aspects of this problem from an information-theoretic point of view. In the network functional compression problem, we would like to compress source random variables for the purpose of computing a deterministic function (or some deterministic functions) at the receiver(s), when these sources and receivers are nodes in a network. Traditional data-compression schemes are special cases of functional compression, where the desired function is the identity function. However, if the receiver is interested in computing a function (or some functions) of sources, further compression is possible.

Several approaches have been applied to investigate different aspects of this problem. One class of works considered the functional computation problem for specific functions. For example, Kowshik and Kumar [2] investigated computation of symmetric Boolean functions in tree networks, and Shenvi and Dey [3] and Ramamoorthy [4] studied the sum network with three sources and three terminals. Other authors investigated the asymptotic analysis of the transmission rate in noisy broadcast networks [5], and also in random geometric graph models (e.g., [6, 7]). Also, Ma *et al.* [8] investigated information-theoretic bounds for multiround function computation in collocated networks. Network flow techniques (also known as multi-commodity methods) have been used to study multiple unicast problems [9, 10]. Shah *et al.* [11] used this framework, with some modifications, for function computation considering communication constraints.

A major body of work on in-network computation investigates information-theoretic rate bounds, when a function of sources is desired to be computed at the receiver. These works can be categorized into the study of lossless functional compression and that of functional compression with distortion. By lossless computation, we mean asymptotically lossless computation of a function: the error probability goes to zero as the blocklength goes to infinity. However, there are several works investigating zero-error computation of functions (e.g., [12, 13]).

Shannon was the first to consider the function computation problem in [14] for a special case when $f(X_1,X_2)=(X_1,X_2)$ (the identity function) and for the network topology depicted in Fig. 15.1(a) (the side-information problem). For a general function, Orlitsky and Roche provided a single-letter characterization of the rate region in [15]. In [16], Doshi *et al.* proposed an optimal coding scheme for this problem.

For the network topology depicted in Fig. 15.1(b), and for the case in which the desired function at the receiver is the identity function (i.e., $f(X_1,X_2)=(X_1,X_2)$), Slepian

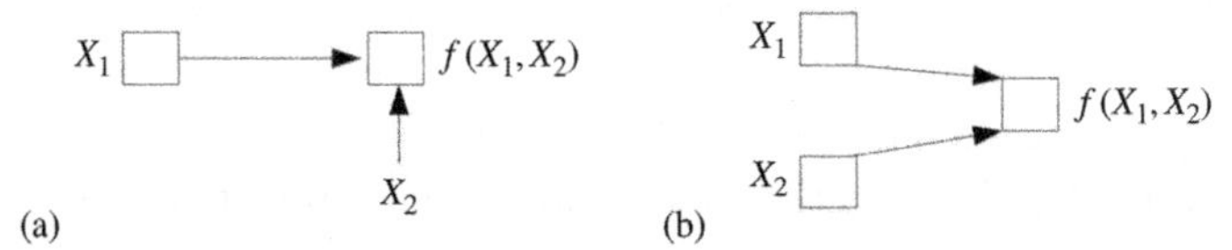

Figure 15.1 (a) Functional compression with side information. (b) A distributed functional compression problem with two transmitters and a receiver.

and Wolf [17] provided a characterization of the rate region and an optimal achievable coding scheme. Some other practical but sub-optimal coding schemes were proposed by Pradhan and Ramchandran [18]. Also, a rate-splitting technique for this problem was developed in [19, 20]. Special cases when $f(X_1,X_2) = X_1$ and $f(X_1,X_2) = (X_1 + X_2)$ mod 2 have been investigated by Ahlswede and Körner [21] and by Körner and Marton [22], respectively. Under some special conditions on source distributions, Doshi *et al.* [16] investigated this problem for a general function and proposed some achievable coding schemes.

There have been several prior works that studied lossy functional compression where the function at the receiver is desired to be computed within a distortion level. Wyner and Ziv [23] considered the side-information problem for computing the identity function at the receiver within some distortion. Yamamoto [24] solved this problem for a general function $f(X_1,X_2)$. Doshi *et al.* [16] gave another characterization of the rate-distortion function given by Yamamoto. Feng *et al.* [25] considered the side-information problem for a general function at the receiver in the case in which the encoder and decoder have some noisy information. For the distributed function computation problem and for a general function, the rate-distortion region remains unknown, but some bounds have been given by Berger and Yeung [26], Barros and Servetto [27], and Wagner *et al.* [28], who considered a specific quadratic distortion function.

In this chapter, we present results that significantly relax previously considered restrictive assumptions, particularly in terms of the network topology, the functions, and the characteristics of the sources. For a one-stage tree network, we introduce a necessary and sufficient condition for any achievable coloring-based coding scheme called the *coloring connectivity condition* (CCC), thus relaxing the previous sufficient zigzag condition of Doshi *et al.* [16]. By using the CCC, we characterize a rate region for distributed functional compression and propose a modularized coding scheme based on graph colorings in order for the system to perform arbitrarily closely to rate lower bounds. These results are presented in Section 15.3.1.

In Section 15.3.2, we consider a general tree network and provide a rate lower bound based on graph entropies. We show that this bound is tight in the case with independent sources. In particular, we show that, to achieve the rate lower bound, intermediate nodes should perform computations. However, for a family of functions and random variables, which we call *chain-rule proper sets*, it is sufficient to have intermediate nodes act like relays (i.e., no computations are performed at intermediate nodes) in order for the system to perform arbitrarily closely to the rate lower bound.

In Section 15.3.3, we discuss practical issues of coloring-based coding schemes and propose an efficient algorithm to compute a minimum-entropy coloring of a characteristic graph under some conditions on source distributions and/or the desired function. Finally, extensions of proposed results for cases with feedback and lossy function computations are discussed in Section 15.4. In particular, we show that, in functional compression, unlike in the Slepian–Wolf case, by having feedback, one may outperform the rate bounds of the case without feedback. These results extend those of Bakshi *et al.* We also present a practical coding scheme for the distributed lossy functional compression problem with a non-trivial performance guarantee.

15.2 Problem Setup and Prior Work

In this section, we set up the functional compression problem and review some prior work.

15.2.1 Problem Setup

Consider k discrete memoryless random processes, $\{X_1^i\}_{i=1}^{\infty}, \ldots, \{X_k^i\}_{i=1}^{\infty}$, as source processes. Memorylessness is not necessary, and one can approximate a source by a memoryless one with an arbitrary precision [29]. Suppose these sources are drawn from finite sets $\mathcal{X}_1 = \{x_1^1, x_1^2, \ldots, x_1^{|\mathcal{X}_1|}\}, \ldots, \mathcal{X}_k = \{x_k^1, x_k^2, \ldots, x_k^{|\mathcal{X}_k|}\}$. These sources have a joint probability distribution $p(x_1, \ldots, x_k)$. We express n-sequences of these random variables as $\mathbf{X}_1, \ldots, \mathbf{X}_k$ with the joint probability distribution $p(\mathbf{x}_1, \ldots, \mathbf{x}_k)$. To simplify the notation, n will be implied by the context if no confusion arises. We refer to the ith element of $\mathbf{x}_j$ as x_{ji}. We use $\mathbf{x}_j^1, \mathbf{x}_j^2, \ldots$ as different n-sequences of $\mathbf{X}_j$. We shall omit the superscript when no confusion arises. Since the sequence $(\mathbf{x}_1, \ldots, \mathbf{x}_k)$ is drawn i.i.d. according to $p(x_1, \ldots, x_k)$, one can write $p(\mathbf{x}_1, \ldots, \mathbf{x}_k) = \prod_{i=1}^{n} p(x_{1i}, \ldots, x_{ki})$.

Consider a tree network with k source nodes in its leaves and a receiver in its root. Other nodes of this tree are referred to as intermediate nodes. Source node j has an input random process $\{X_j^i\}_{i=1}^{\infty}$. The receiver wishes to compute a deterministic function $f : \mathcal{X}_1 \times \cdots \times \mathcal{X}_k \to \mathcal{Z}$, or $f : \mathcal{X}_1^n \times \cdots \times \mathcal{X}_k^n \to \mathcal{Z}^n$, its vector extension.

Note that sources can be at any nodes of the network. However, without loss of generality, we can modify the network by adding some fake leaves to source nodes which are not located in leaves of the network. So, in the achieved network, sources are always located in leaves. Also, by adding some auxiliary nodes, one can make sources be at the same distance from the receiver. Nodes of this tree are labeled as i for different is, where source nodes are denoted by $\{1, \ldots, k\}$ and the outgoing link of node i is denoted by e_i. Node i sends M_i over its outgoing edge e_i with a rate R_i (it maps length-n blocks of M_i, referred to as $\mathbf{M}_i$, to $\{1, 2, \ldots, 2^{nR_i}\}$).

For a source node, $\mathbf{M}_i = \mathrm{en}_{X_i}(\mathbf{X}_i)$, where en_{X_i} is the encoding function of the source node i. For an intermediate node i, $i \notin \{1, \ldots, k\}$, with incoming edges $e_j, \ldots, e_q$, $\mathbf{M}_i = g_i(\mathbf{M}_j, \ldots, \mathbf{M}_q)$, where $g_i(\cdot)$ is a function to be computed in that node.

The receiver maps incoming messages $\{\mathbf{M}_i, \ldots, \mathbf{M}_j\}$ to $\mathcal{Z}^n$ by using a function $r(\cdot)$; i.e., $r : \prod_{l=i}^{j} \{1, \ldots, 2^{nR_l}\} \to \mathcal{Z}^n$. Thus, the receiver computes $r(\mathbf{M}_i, \ldots, \mathbf{M}_j) = r'(\mathrm{en}_{X_1}(\mathbf{X}_1), \ldots, \mathrm{en}_{X_k}(\mathbf{X}_k))$. We refer to this encoding/decoding scheme as an n-distributed functional code. Intermediate nodes are allowed to compute functions, but have no demand of their own. The desired function $f(\mathbf{X}_1, \ldots, \mathbf{X}_k)$ at the receiver is the only demand in the network. For any encoding/decoding scheme, the probability of error is defined as

$$P_e^n = \Pr[(\mathbf{x}_1, \ldots, \mathbf{x}_k) : f(\mathbf{x}_1, \ldots, \mathbf{x}_k) \neq r'(\mathrm{en}_{X_1}(\mathbf{x}_1), \ldots, \mathrm{en}_{X_k}(\mathbf{x}_k))]. \quad (15.1)$$

A rate tuple of the network is the set of rates of its edges (i.e., $\{R_i\}$ for valid is). We say a rate tuple is achievable iff there exists a coding scheme operating at these rates so that $P_e^n \to 0$ as $n \to \infty$. The achievable rate region is the set closure of the set of all achievable rates.

15.2.2 Definitions and Prior Results

In this part, we present some definitions and review prior results.

DEFINITION 15.1 *The characteristic (conflict) graph* $G_{X_1} = (V_{X_1}, E_{X_1})$ *of* X_1 *with respect to* X_2, $p(x_1, x_2)$, *and the function* $f(X_1, X_2)$ *is defined as follows:* $V_{X_1} = \mathcal{X}_1$ *and an edge* $(x_1^1, x_1^2) \in \mathcal{X}_1^2$ *is in* E_{X_1} *iff there exists an* $x_2^1 \in \mathcal{X}_2$ *such that* $p(x_1^1, x_2^1)p(x_1^2, x_2^1) > 0$ *and* $f(x_1^1, x_2^1) \neq f(x_1^2, x_2^1)$.

In other words, in order to avoid confusion about the function $f(X_1, X_2)$ at the receiver, if $(x_1^1, x_1^2) \in E_{X_1}$, the descriptions of x_1^1 and x_1^2 must be different. Shannon first defined this when studying the zero-error capacity of noisy channels [14]. Witsenhausen [30] used this concept to study a simplified version of the functional compression problem where one encodes X_1 to compute $f(X_1) = X_1$ with zero distortion and showed that the chromatic number of the strong graph-product characterizes the rate. The characteristic graph of X_2 with respect to X_1, $p(x_1, x_2)$, and $f(X_1, X_2)$ is defined analogously and denoted by G_{X_2}. One can extend the definition of the characteristic graph to the case with more than two random variables as follows. Suppose $X_1, \ldots, X_k$ are k random variables as defined in Section 15.2.1.

DEFINITION 15.2 *The characteristic graph* $G_{X_1} = (V_{X_1}, E_{X_1})$ *of* X_1 *with respect to random variables* $X_2, \ldots, X_k$, $p(x_1, \ldots, x_k)$, *and* $f(X_1, \ldots, X_k)$ *is defined as follows:* $V_{X_1} = \mathcal{X}_1$ *and an edge* $(x_1^1, x_1^2) \in \mathcal{X}_1^2$ *is in* E_{X_1} *if there exist* $x_j^1 \in \mathcal{X}_j$ *for* $2 \leq j \leq k$ *such that* $p(x_1^1, x_2^1, \ldots, x_k^1)p(x_1^2, x_2^1, \ldots, x_k^1) > 0$ *and* $f(x_1^1, x_2^1, \ldots, x_k^1) \neq f(x_1^2, x_2^1, \ldots, x_k^1)$.

Example 15.1 To illustrate the idea of confusability and the characteristic graph, consider two random variables X_1 and X_2 such that $\mathcal{X}_1 = \{0, 1, 2, 3\}$ and $\mathcal{X}_2 = \{0, 1\}$, where they are uniformly and independently distributed on their own supports. Suppose $f(X_1, X_2) = (X_1 + X_2) \bmod 2$ is to be perfectly reconstructed at the receiver. Then, the characteristic graph of X_1 with respect to X_2, $p(x_1, x_2) = \frac{1}{8}$, and f is as shown in Fig. 15.2(a). Similarly, G_{X_2} is depicted in Fig. 15.2(b).

The following definition can be found in [31].

DEFINITION 15.3 *Given a graph* $G_{X_1} = (V_{X_1}, E_{X_1})$ *and a distribution on its vertices* V_{X_1}, *the graph entropy is*

$$H_{G_{X_1}}(X_1) = \min_{X_1 \in W_1 \in \Gamma(G_{X_1})} I(X_1; W_1), \tag{15.2}$$

where $\Gamma(G_{X_1})$ *is the set of all maximal independent sets of* G_{X_1}.

The notation $X_1 \in W_1 \in \Gamma(G_{X_1})$ means that we are minimizing over all distributions $p(w_1, x_1)$ such that $p(w_1, x_1) > 0$ implies $x_1 \in w_1$, where w_1 is a maximal independent set of the graph G_{X_1}.

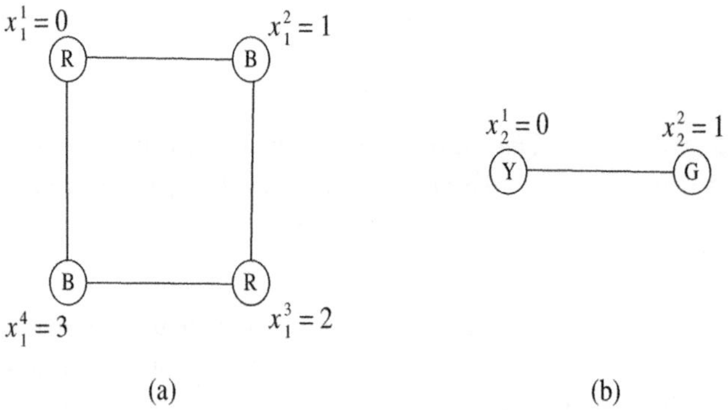

Figure 15.2 Characteristic graphs (a) G_{X_1} and (b) G_{X_2} for the setup of Example 15.1. (Different letters written over graph vertices indicate different colors.)

Example 15.2 Consider the scenario described in Example 15.1. For the characteristic graph of X_1 shown in Fig. 15.2(a), the set of maximal independent sets is $W_1 = \{\{0,2\},\{1,3\}\}$. To minimize $I(X_1;W_1) = H(X_1) - H(X_1|W_1) = \log(4) - H(X_1|W_1)$, one should maximize $H(X_1|W_1)$. Because of the symmetry of the problem, to maximize $H(X_1|W_1)$, $p(w_1)$ must be uniform over two possible maximal independent sets of G_{X_1}. Since each maximal independent set $w_1 \in W_1$ has two X_1 values, $H(X_1|w_1) = \log(2)$ bit, and since $p(w_1)$ is uniform, $H(X_1|W_1) = \log(2)$ bit. Therefore, $H_{G_{X_1}}(X_1) = \log(4) - \log(2) = 1$ bit. One can see that, if we want to encode X_1 ignoring the effect of the function f, we need $H(X_1) = \log(4) = 2$ bits. We will show that, for this example, functional compression saves us 1 bit in every 2 bits compared with the traditional data compression.

Witsenhausen [30] showed that the chromatic number of the strong graph-product characterizes the minimum rate at which a single source can be encoded so that the identity function of that source can be computed with zero distortion. Orlitsky and Roche [15] defined an extension of Körner's graph entropy, the *conditional graph entropy*.

DEFINITION 15.4 *The conditional graph entropy is*

$$H_{G_{X_1}}(X_1|X_2) = \min_{\substack{X_1 \in W_1 \in \Gamma(G_{X_1}) \\ W_1 - X_1 - X_2}} I(W_1;X_1|X_2). \tag{15.3}$$

The notation $W_1 - X_1 - X_2$ indicates a Markov chain. If X_1 and X_2 are independent, $H_{G_{X_1}}(X_1|X_2) = H_{G_{X_1}}(X_1)$. To illustrate this concept, let us consider an example borrowed from [15].

Example 15.3 When $f(X_1,X_2) = X_1$, $H_{G_{X_1}}(X_1|X_2) = H(X_1|X_2)$.

To show this, consider the characteristic graph of X_1, denoted as G_{X_1}. Since $f(X_1,X_2) = X_1$, then, for every $x_2^1 \in \mathcal{X}_2$, the set $\{x_1^i : p(x_1^i, x_2^1) > 0\}$ of possible x_1^i are connected to each other (i.e., this set is a clique of G_{X_1}). Since the intersection of a clique and a maximal independent set is a singleton, X_2 and the maximal independent set W_1 containing X_1 determine X_1. So,

$$\begin{aligned} H_{G_{X_1}}(X_1|X_2) &= \min_{\substack{X_1 \in W_1 \in \Gamma(G_{X_1}) \\ W_1 - X_1 - X_2}} I(W_1; X_1|X_2) \\ &= H(X_1|X_2) - \max_{X_1 \in W_1 \in \Gamma(G_{X_1})} H(X_1|W_1, X_2) \\ &= H(X_1|X_2). \end{aligned} \tag{15.4}$$

DEFINITION 15.5 *A vertex coloring of a graph is a function $c_{G_{X_1}}(X_1) : V_{X_1} \to \mathbb{N}$ of a graph $G_{X_1} = (V_{X_1}, E_{X_1})$ such that $(x_1^1, x_1^2) \in E_{X_1}$ implies $c_{G_{X_1}}(x_1^1) \neq c_{G_{X_1}}(x_1^2)$. The entropy of a coloring is the entropy of the induced distribution on colors. Here, $p(c_{G_{X_1}}(x_1^i)) = p(c^{-1}_{G_{X_1}}(c_{G_{X_1}}(x_1^i)))$, where $c^{-1}_{G_{X_1}}(c_{G_{X_1}}(x_1^i)) = \{x_1^j : c_{G_{X_1}}(x_1^j) = c_{G_{X_1}}(x_1^i)\}$ for all valid j. This subset of vertices with the same color is called a color class. We refer to a coloring which minimizes the entropy as a minimum-entropy coloring. We use $C_{G_{X_1}}$ as the set of all valid colorings of a graph G_{X_1}.*

Example 15.4 Consider again the random variable X_1 described in Example 15.1, whose characteristic graph G_{X_1} and its valid coloring are shown in Fig. 15.2a. One can see that, in this coloring, two connected vertices are assigned to different colors. Specifically, $c_{G_{X_1}}(X_1) = \{r, b\}$. Therefore, $p(c_{G_{X_1}}(x_1^i) = r) = p(x_1^i = 0) + p(x_1^i = 2)$ and $p(c_{G_{X_1}}(x_1^i) = b) = p(x_1^i = 1) + p(x_1^i = 3)$.

We define a power graph of a characteristic graph as its co-normal products.

DEFINITION 15.6 *The nth power of a graph G_{X_1} is a graph $G^n_{\mathbf{X}_1} = (V^n_{X_1}, E^n_{X_1})$ such that $V^n_{X_1} = \mathcal{X}_1^n$ and $(\mathbf{x}_1^1, \mathbf{x}_1^2) \in E^n_{X_1}$ when there exists at least one i such that $(x^1_{1i}, x^2_{1i}) \in E_{X_1}$. We denote a valid coloring of $G^n_{\mathbf{X}_1}$ by $c_{G^n_{\mathbf{X}_1}}(\mathbf{X}_1)$.*

One may ignore atypical sequences in a sufficiently large power graph of a conflict graph and then color that graph. This coloring is called an ϵ-coloring of a graph and is defined as follows.

DEFINITION 15.7 *Given a non-empty set $\mathcal{A} \subset \mathcal{X}_1 \times \mathcal{X}_2$, define $\widehat{p}(x_1, x_2) = p(x_1, x_2)/p(\mathcal{A})$ when $(x_1, x_2) \in \mathcal{A}$, and $\widehat{p}(x, y) = 0$ otherwise. $\widehat{p}$ is the distribution over (x_1, x_2) conditioned on $(x_1, x_2) \in \mathcal{A}$. Denote the characteristic graph of X_1 with respect to X_2, $\widehat{p}(x_1, x_2)$, and $f(X_1, X_2)$ as $\widehat{G}_{X_1} = (\widehat{V}_{X_1}, \widehat{E}_{X_1})$ and the characteristic graph of X_2 with respect to X_1, $\widehat{p}(x_1, x_2)$, and $f(X_1, X_2)$ as $\widehat{G}_{X_2} = (\widehat{V}_{X_2}, \widehat{E}_{X_2})$. Note that $\widehat{E}_{X_1} \subseteq E_{X_1}$ and $\widehat{E}_{X_2} \subseteq E_{X_2}$. Suppose $p(\mathcal{A}) \geq 1 - \epsilon$. We say that $c_{G_{X_1}}(X_1)$ and $c_{G_{X_2}}(X_2)$ are ϵ-colorings of G_{X_1} and G_{X_2} if they are valid colorings of $\widehat{G}_{X_1}$ and $\widehat{G}_{X_2}$.*

In [32], the *chromatic entropy* of a graph G_{X_1} is defined as follows.

DEFINITION 15.8

$$H^{\mathcal{X}}_{G_{X_1}}(X_1) = \min_{c_{G_{X_1}} \text{ is an } \epsilon\text{-coloring of } G_{X_1}} H(c_{G_{X_1}}(X_1)).$$

The chromatic entropy is a representation of the chromatic number of high-probability subgraphs of the characteristic graph. In [16], the conditional chromatic entropy is defined as follows.

DEFINITION 15.9

$$H^{\mathcal{X}}_{G_{X_1}}(X_1|X_2) = \min_{c_{G_{X_1}} \text{ is an } \epsilon\text{-coloring of } G_{X_1}} H(c_{G_{X_1}}(X_1)|X_2).$$

Regardless of ϵ, the above optimizations are minima, rather than infima, because there are finitely many subgraphs of any fixed graph G_{X_1}, and therefore there are only finitely many ϵ-colorings, regardless of ϵ.

In general, these optimizations are NP-hard [33]. But, depending on the desired function f, there are some interesting cases for which optimal solutions can be computed efficiently. We discuss these cases in Section 15.3.3.

Körner showed in [31] that, in the limit of large n, there is a relation between the chromatic entropy and the graph entropy.

THEOREM 15.1

$$\lim_{n\to\infty} \frac{1}{n} H^{\mathcal{X}}_{G^n_{\mathbf{X}_1}}(\mathbf{X}_1) = H_{G_{X_1}}(X_1). \tag{15.5}$$

This theorem implies that the receiver can asymptotically compute a deterministic function of a discrete memoryless source. The source first colors a sufficiently large power of the characteristic graph of the random variable with respect to the function, and then encodes achieved colors using any encoding scheme that achieves the entropy bound of the coloring random variable. In the previous approach, to achieve the encoding rate close to graph entropy of X_1, one should find the optimal distribution over the set of maximal independent sets of G_{X_1}. However, this theorem allows us to find the optimal coloring of $G^n_{\mathbf{X}_1}$, instead of the optimal distribution on maximal independent sets. One can see that this approach modularizes the encoding scheme into two parts, a graph-coloring module, followed by a Slepian–Wolf compression module.

The conditional version of the above theorem is proven in [16].

THEOREM 15.2

$$\lim_{n\to\infty} \frac{1}{n} H^{\mathcal{X}}_{G^n_{\mathbf{X}_1}}(\mathbf{X}_1|\mathbf{X}_2) = H_{G_{X_1}}(X_1|X_2). \tag{15.6}$$

This theorem implies a practical encoding scheme for the problem of functional compression with side information where the receiver wishes to compute $f(X_1,X_2)$, when X_2 is available at the receiver as the side information. Orlitsky and Roche showed in [15] that $H_{G_{X_1}}(X_1|X_2)$ is the minimum achievable rate for this problem. Their proof uses random coding arguments and shows the existence of an optimal coding scheme. This theorem presents a modularized encoding scheme where one first finds the minimum-entropy coloring of $G^n_{\mathbf{X}_1}$ for large enough n, and then uses a compression scheme on the coloring random variable (such as the Slepian–Wolf Scheme in [17]) to achieve a rate arbitrarily close to $H(c_{G^n_{\mathbf{X}_1}}(\mathbf{X}_1)|\mathbf{X}_2)$. This encoding scheme guarantees computation of the function at the receiver with a vanishing probability of error.

All these results considered only functional compression with side information at the receiver (Fig. 15.1(a)). In general, the rate region of the distributed functional compression problem (Fig. 15.1(b)) has not been determined. However, Doshi *et al.* [16] characterized a rate region of this network when source random variables satisfy a condition called the zigzag condition, defined below.

We refer to the ϵ-joint-typical set of sequences of random variables $\mathbf{X}_1,\ldots,\mathbf{X}_k$ as T^n_ϵ. k is implied in this notation for simplicity. T^n_ϵ can be considered as a strong or weak typical set [29].

DEFINITION 15.10 *A discrete memoryless source* $\{(X^i_1,X^i_2)\}_{i\in\mathbb{N}}$ *with a distribution* $p(x_1,x_2)$ *satisfies the zigzag condition if for any* ϵ *and some* n, $(\mathbf{x}^1_1,\mathbf{x}^1_2)$, $(\mathbf{x}^2_1,\mathbf{x}^2_2)\in T^n_\epsilon$, *there exists some* $(\mathbf{x}^3_1,\mathbf{x}^3_2)\in T^n_\epsilon$ *such that* $(\mathbf{x}^3_1,\mathbf{x}^i_2),(\mathbf{x}^i_1,\mathbf{x}^3_2)\in T^n_{\frac{\epsilon}{2}}$ *for each* $i\in\{1,2\}$, *and* $(x^3_{1j},x^3_{2j})=(x^i_{1j},x^{3-i}_{2j})$ *for some* $i\in\{1,2\}$ *for each* j.

In fact, the zigzag condition forces many source sequences to be typical. Doshi *et al.* [16] show that, if the source random variables satisfy the zigzag condition, an achievable rate region for this network is the set of all rates that can be achieved through graph colorings. The zigzag condition is a restrictive condition which does not depend on the desired function at the receiver. This condition is not necessary, but it is sufficient. In the next section, we relax this condition by introducing a necessary and sufficient condition for any achievable coloring-based coding scheme and characterize a rate region for the distributed functional compression problem.

15.3 Network Functional Compression

In this section, we present the main results for network functional compression.

15.3.1 A Rate Region for One-Stage Tree Networks

In this section, we compute a rate region for a general one-stage tree network without any restrictive conditions such as the zigzag condition.

Consider a one-stage tree network with k sources.

DEFINITION 15.11 *A path with length* m *between two points* $Z_1=(x^1_1,x^1_2,\ldots,x^1_k)$ *and* $Z_m=(x^2_1,x^2_2,\ldots,x^2_k)$ *is determined by* $m-1$ *points* Z_i, $1\le i\le m$, *such that*

(i) $\Pr(Z_i)>0$, *for all* $1\le i\le m$; *and*
(ii) Z_i *and* Z_{i+1} *differ in only one of their coordinates.*

Definition 15.11 can be generalized to two length-n vectors as follows.

DEFINITION 15.12 *A path with length m between two points* $\mathbf{Z}_1 = (\mathbf{x}_1^1, \mathbf{x}_2^1, \ldots, \mathbf{x}_k^1) \in T_\epsilon^n$ *and* $\mathbf{Z}_m = (\mathbf{x}_1^2, \mathbf{x}_2^2, \ldots, \mathbf{x}_k^2) \in T_\epsilon^n$ *is determined by* $m-1$ *points* $\mathbf{Z}_i$, $1 \leq i \leq m$ *such that*

(i) $\mathbf{Z}_i \in T_\epsilon^n$, *for all* $1 \leq i \leq m$; *and*
(ii) $\mathbf{Z}_i$ *and* $\mathbf{Z}_{i+1}$ *differ in only one of their coordinates.*

Note that each coordinate of $\mathbf{Z}_i$ is a vector with length n.

DEFINITION 15.13 *A joint coloring family* J_C *for random variables* $X_1, \ldots, X_k$ *with characteristic graphs* $G_{X_1}, \ldots, G_{X_k}$, *and any valid respective colorings* $c_{G_{X_1}}, \ldots, c_{G_{X_k}}$, *is defined as* $J_C = \{j_c^1, \ldots, j_c^l\}$, *where* j_c^i *is the collection of points* $(x_1^{i_1}, x_2^{i_2}, \ldots, x_k^{i_k})$ *whose coordinates have the same color. Each* j_c^i *is called a joint coloring class.*

We say a joint coloring class j_c^i is connected if, between any two points in j_c^i, there exists a path that lies in j_c^i. Otherwise, it is disconnected. Definition 15.11 can be expressed for random vectors $\mathbf{X}_1, \ldots, \mathbf{X}_k$ with characteristic graphs $G_{\mathbf{X}_1}^n, \ldots, G_{\mathbf{X}_k}^n$ and any valid respective ϵ-colorings $c_{G_{\mathbf{X}_1}^n}, \ldots, c_{G_{\mathbf{X}_k}^n}$.

In the following, we present the *coloring connectivity condition* (CCC) which is a necessary and sufficient condition for any coloring-based coding scheme.

DEFINITION 15.14 *Consider random variables* $X_1, \ldots, X_k$ *with characteristic graphs* $G_{X_1}, \ldots, G_{X_k}$, *and any valid colorings* $c_{G_{X_1}}, \ldots, c_{G_{X_k}}$. *We say a joint coloring class* $j_c^i \in J_C$ *satisfies the coloring connectivity condition (CCC) when it is connected, or its disconnected parts have the same function values. We say colorings* $c_{G_{X_1}}, \ldots, c_{G_{X_k}}$ *satisfy the CCC when all joint coloring classes satisfy the CCC.*

The CCC can be expressed for random vectors $\mathbf{X}_1, \ldots, \mathbf{X}_k$ with characteristic graphs $G_{\mathbf{X}_1}^n, \ldots, G_{\mathbf{X}_k}^n$, and any valid respective ϵ-colorings $c_{G_{\mathbf{X}_1}^n}, \ldots, c_{G_{\mathbf{X}_k}^n}$.

Example 15.5 Suppose we have two random variables X_1 and X_2 with characteristic graphs G_{X_1} and G_{X_2}. Let us assume $c_{G_{X_1}}$ and $c_{G_{X_2}}$ are two valid colorings of G_{X_1} and G_{X_2}, respectively. Assume $c_{G_{X_1}}(x_1^1) = c_{G_{X_1}}(x_1^2)$ and $c_{G_{X_2}}(x_2^1) = c_{G_{X_2}}(x_2^2)$. Suppose j_c^1 represents this joint coloring class. In other words, $j_c^1 = \{(x_1^i, x_2^j)\}$, for all $1 \leq i, j \leq 2$ when $p(x_1^i, x_2^j) > 0$. Figure 15.3 considers two different cases. The first case is when $p(x_1^1, x_2^2) = 0$, and other points have a non-zero probability. It is illustrated in Fig. 15.3(a). One can see that there exists a path between any two points in this joint coloring class. Therefore, this joint coloring class satisfies the CCC. If other joint coloring classes of $c_{G_{X_1}}$ and $c_{G_{X_2}}$ satisfy the CCC, we say $c_{G_{X_1}}$ and $c_{G_{X_2}}$ satisfy the CCC. Now, consider the second case depicted in Fig. 15.3(b). In this case, we have $p(x_1^1, x_2^2) = 0$, $p(x_1^2, x_2^1) = 0$, and other points have a non-zero probability. One can see that there is no path between (x_1^1, x_2^1) and (x_1^2, x_2^2) in j_c^1. So, though these two points belong to a same joint coloring class, their corresponding function values can be different from each other. Thus, j_c^1 does not satisfy the CCC for this example. Therefore, $c_{G_{X_1}}$ and $c_{G_{X_2}}$ do not satisfy the CCC.

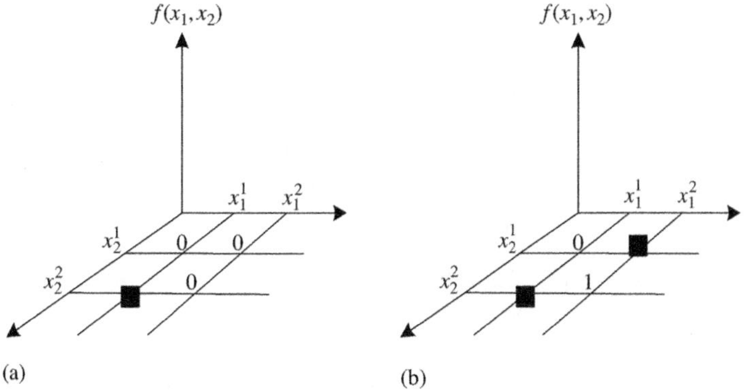

Figure 15.3 Two examples of a joint coloring class: (a) satisfying the CCC and (b) not satisfying the CCC. Dark squares indicate points with zero probability. Function values are depicted in the picture.

There are several examples of source distributions and functions that satisfy the CCC.

LEMMA 15.1 *Consider two random variables X_1 and X_2 with characteristic graphs G_{X_1} and G_{X_2} and any valid colorings $c_{G_{X_1}}(X_1)$ and $c_{G_{X_2}}(X_2)$, respectively, where $c_{G_{X_2}}(X_2)$ is a trivial coloring, assigning different colors to different vertices (to simplify the notation, we use $c_{G_{X_2}}(X_2) = X_2$ to refer to this coloring). These colorings satisfy the CCC. Also, $c_{G^n_{\mathbf{X}_1}}(\mathbf{X}_1)$ and $c_{G^n_{\mathbf{X}_2}}(\mathbf{x}_2) = \mathbf{X}_2$ satisfy the CCC, for any n.*

Proof of this lemma is presented in Section 15.5.1.

The following lemmas demonstrate why the CCC is a necessary and sufficient condition for any achievable coloring-based coding scheme.

LEMMA 15.2 *Consider random variables $X_1, \ldots, X_k$ with characteristic graphs $G_{X_1}, \ldots, G_{X_k}$, and any valid colorings $c_{G_{X_1}}, \ldots, c_{G_{X_k}}$ with joint coloring class $J_C = \{j_c^i : i\}$. For any two points $(x_1^1, \ldots, x_k^1)$ and $(x_1^2, \ldots, x_k^2)$ in j_c^i, $f(x_1^1, \ldots, x_k^1) = f(x_1^2, \ldots, x_k^2)$ if and only if j_c^i satisfies the CCC.*

Proof of this lemma is presented in Section 15.5.2.

LEMMA 15.3 *Consider random variables $\mathbf{X}_1, \ldots, \mathbf{X}_k$ with characteristic graphs $G^n_{\mathbf{X}_1}, \ldots, G^n_{\mathbf{X}_k}$, and any valid ϵ-colorings $c_{G^n_{\mathbf{X}_1}}, \ldots, c_{G^n_{\mathbf{X}_k}}$ with joint coloring class $J_C = \{j_c^i : i\}$. For any two points $(\mathbf{x}_1^1, \ldots, \mathbf{x}_k^1)$ and $(\mathbf{x}_1^2, \ldots, \mathbf{x}_k^2)$ in j_c^i, $f(\mathbf{x}_1^1, \ldots, \mathbf{x}_k^1) = f(\mathbf{x}_1^2, \ldots, \mathbf{x}_k^2)$ if and only if j_c^i satisfies the CCC.*

Proof of this lemma is presented in Section 15.5.3.

Next, we show that, if X_1 and X_2 satisfy the zigzag condition given in Definition 15.10, any valid colorings of their characteristic graphs satisfy the CCC, but not vice versa. In other words, we show that the zigzag condition used in [16] is sufficient but not necessary.

LEMMA 15.4 *If two random variables X_1 and X_2 with characteristic graphs G_{X_1} and G_{X_2} satisfy the zigzag condition, any valid colorings $c_{G_{X_1}}$ and $c_{G_{X_2}}$ of G_{X_1} and G_{X_2} satisfy the CCC, but not vice versa.*

Proof of this lemma is presented in Section 15.5.4.

We use the CCC to characterize a rate region of functional compression for a one-stage tree network as follows.

DEFINITION 15.15 *For random variables $X_1, \dots, X_k$ with characteristic graphs $G_{X_1}, \dots, G_{X_k}$, the joint graph entropy is defined as follows:*

$$H_{G_{X_1},\dots,G_{X_k}}(X_1,\dots,X_k) \triangleq \lim_{n\to\infty} \min_{c_{G^n_{\mathbf{X}_1}},\dots,c_{G^n_{\mathbf{X}_k}}} \frac{1}{n} H(c_{G^n_{\mathbf{X}_1}}(\mathbf{X}_1),\dots,c_{G^n_{\mathbf{X}_k}}(\mathbf{X}_k)), \tag{15.7}$$

in which $c_{G^n_{\mathbf{X}_1}}(\mathbf{X}_1), \dots, c_{G^n_{\mathbf{X}_k}}(\mathbf{X}_k)$ are ϵ-colorings of $G^n_{\mathbf{X}_1}, \dots, G^n_{\mathbf{X}_k}$ satisfying the CCC. We refer to this joint graph entropy as $H_{[G_{X_i}]_{i\in S}}$, where $S = \{1,2,\dots,k\}$. Note that this limit exists because we have a monotonically decreasing sequence bounded from below. Similarly, we can define the conditional graph entropy.

DEFINITION 15.16 *For random variables $X_1, \dots, X_k$ with characteristic graphs $G_{X_1}, \dots, G_{X_k}$, the conditional graph entropy can be defined as follows:*

$$\begin{aligned} &H_{G_{X_1},\dots,G_{X_i}}(X_1,\dots,X_i|X_{i+1},\dots,X_k) \\ &\triangleq \lim_{n\to\infty} \min \frac{1}{n} H(c_{G^n_{\mathbf{X}_1}}(\mathbf{X}_1),\dots,c_{G^n_{\mathbf{X}_i}}(\mathbf{X}_i)|c_{G^n_{\mathbf{X}_{i+1}}}(\mathbf{X}_{i+1}),\dots,c_{G^n_{\mathbf{X}_k}}(\mathbf{X}_k)), \end{aligned} \tag{15.8}$$

where the minimization is over $c_{G^n_{\mathbf{X}_1}}(\mathbf{X}_1), \dots, c_{G^n_{\mathbf{X}_k}}(\mathbf{X}_k)$, which are ϵ-colorings of $G^n_{\mathbf{X}_1}, \dots, G^n_{\mathbf{X}_k}$ satisfying the CCC.

LEMMA 15.5 *For $k = 2$, Definitions 15.4 and 15.14 are the same.*

Proof of this lemma is presented in Section 15.5.5.

Note that, by this definition, *the graph entropy does not satisfy the chain rule.*

Suppose $\mathcal{S}(k)$ denotes the power set of the set $\{1,2,\dots,k\}$ excluding the empty subset. Then, for any $S \in \mathcal{S}(k)$,

$$X_S \triangleq \{X_i : i \in S\}.$$

Let S^c denote the complement of S in $\mathcal{S}(k)$. For $S = \{1,2,\dots,k\}$, denote S^c as the empty set. To simplify the notation, we refer to a subset of sources by X_S. For instance, $\mathcal{S}(2) = \{\{1\},\{2\},\{1,2\}\}$, and for $S = \{1,2\}$, we write $H_{[G_{X_i}]_{i\in S}}(X_S)$ instead of $H_{G_{X_1},G_{X_2}}(X_1,X_2)$.

THEOREM 15.3 *A rate region of a one-stage tree network is characterized by the following conditions:*

$$\forall S \in \mathcal{S}(k) \Longrightarrow \sum_{i\in S} R_i \geq H_{[G_{X_i}]_{i\in S}}(X_S|X_{S^c}). \tag{15.9}$$

Proof of this theorem is presented in Section 15.5.6.

If we have two transmitters ($k = 2$), Theorem 15.3 can be simplified as follows.

COROLLARY 15.1 *A rate region of the network shown in Fig. 15.1(b) is determined by the following three conditions:*

$$\begin{aligned} R_{11} &\geq H_{G_{X_1}}(X_1|X_2), \\ R_{12} &\geq H_{G_{X_2}}(X_2|X_1), \\ R_{11}+R_{12} &\geq H_{G_{X_1},G_{X_2}}(X_1,X_2). \end{aligned} \tag{15.10}$$

Algorithm 15.1

The following algorithm proposes a modularized coding scheme which performs arbitrarily closely to the rate bounds of Theorem 15.1.

- Source nodes compute ϵ-colorings of sufficiently large power of their characteristic graphs satisfying the CCC, followed by Slepian–Wolf compression.
- The receiver first uses a Slepian–Wolf decoder to decode transmitted coloring variables. Then, it uses a look-up table to compute the function values.

The achievablity proof of this algorithm directly follows from the proof of Theorem 15.3.

15.3.2 A Rate Lower Bound for a General Tree Network

In this section, we consider a general tree structure with intermediate nodes that are allowed to perform computations. However, to simplify the notation, we limit the arguments to the tree structure of Fig. 15.4. Note that all discussions can be extended to a general tree network.

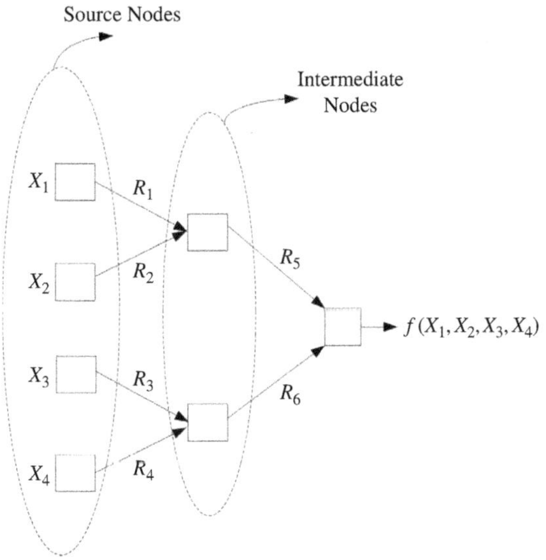

Figure 15.4 An example of a two-stage tree network.

The problem of function computations for a general tree network has been considered in [13, 34]. Kowshik and Kumar [13] derive a necessary and sufficient condition for the encoders on each edge of the tree for a zero-error computation of the desired function. Appuswamy *et al.* [34] show that, for a tree network with independent sources, a min-cut rate is a tight upper bound. Here, we consider an asymptotically lossless functional compression problem. For a general tree network with correlated sources, we derive rate bounds using graph entropies. We show that these rates are achievable for the case of independent sources and propose a modularized coding scheme based on graph colorings that performs arbitrarily closely to rate bounds. We also show that, for a family of functions and random variables, which we call *chain-rule proper sets*, it is sufficient to have no computations at intermediate nodes in order for the system to perform arbitrarily closely to the rate lower bound.

In the tree network depicted in Fig. 15.4, nodes $\{1,\ldots,4\}$ represent source nodes, nodes $\{5,6\}$ are intermediate nodes, and node 7 is the receiver. The receiver wishes to compute a deterministic function of source random variables. Intermediate nodes have no demand of their own, but they are allowed to perform computation. Computing the desired function f at the receiver is the only demand of the network. For this network, we compute a rate lower bound and show that this bound is tight in the case of independent sources. We also propose a modularized coding scheme to perform arbitrarily closely to derived rate lower bounds in this case.

Sources transmit variables $M_1,\ldots,M_4$ through links $e_1,\ldots,e_4$, respectively. Intermediate nodes transmit variables M_5 and M_6 over e_5 and e_6, respectively, where $M_5 = g_5(M_1,M_2)$ and $M_6 = g_6(M_3,M_4)$.

Let $\mathcal{S}(4)$ and $\mathcal{S}(5,6)$ be the power sets of the set $\{1,\ldots,4\}$ and the set $\{5,6\}$ except the empty set, respectively.

THEOREM 15.4 *A rate lower bound for the tree network of Fig. 15.4 can be characterized as follows:*

$$\begin{aligned}\forall S \in \mathcal{S}(4) &\Longrightarrow \sum_{i\in S} R_i \geq H_{[G_{X_i}]_{i\in S}}(X_S|X_{S^c}),\\ \forall S \in \mathcal{S}(5,6) &\Longrightarrow \sum_{i\in S} R_i \geq H_{[G_{X_i}]_{i\in S}}(X_S|X_{S^c}).\end{aligned} \tag{15.11}$$

Proof of this theorem is presented in Section 15.5.7. Note that the result of Theorem 15.4 can be extended to a general tree network topology.

In the following, we show that, for independent source variables, the rate bounds of Theorem 15.4 are tight, and we propose a coding scheme that performs arbitrarily closely to these bounds.

Tightness of the Rate Lower Bound for Independent Sources

Suppose random variables $\mathbf{X}_1,\ldots,\mathbf{X}_4$ with characteristic graphs $G^n_{\mathbf{X}_1},\ldots,G^n_{\mathbf{X}_4}$ are independent. Assume $c_{G^n_{\mathbf{X}_1}},\ldots,c_{G^n_{\mathbf{X}_4}}$ are valid ϵ-colorings of these characteristic graphs satisfying the CCC. The following coding scheme performs arbitrarily closely to rate bounds of Theorem 15.4.

Source nodes first compute colorings of high-probability subgraphs of their characteristic graphs satisfying CCC, and then perform source coding on these coloring random variables. Intermediate nodes first compute their parents' coloring random variables, and then, by using a look-up table, find corresponding source values of their received colorings. Then, they compute ϵ-colorings of their own characteristic graphs. The corresponding source values of their received colorings form an independent set in the graph. If all are assigned to a single color in the minimum-entropy coloring, intermediate nodes send this coloring random variable followed by a source coding. However, if vertices of this independent set are assigned to different colors, intermediate nodes send the coloring with the lowest entropy followed by source coding (Slepian–Wolf). The receiver first performs a minimum-entropy decoding [29] on its received information and achieves coloring random variables. Then, it uses a look-up table to compute its desired function by using the achieved colorings.

In the following, we summarize this proposed algorithm.

Algorithm 15.2

The following algorithm proposes a modularized coding scheme which performs arbitrarily closely to the rate bounds of Theorem 15.4 when the sources are independent.

- Source nodes compute ϵ-colorings of sufficiently large power of their characteristic graphs satisfying the CCC, followed by Slepian–Wolf compression.
- Intermediate nodes compute ϵ-colorings of sufficiently large power of their characteristic graphs by using their parents' colorings.
- The receiver first uses a Slepian–Wolf decoder to decode transmitted coloring variables. Then, it uses a look-up table to compute the function values.

The achievablity proof of this algorithm is presented in Section 15.5.8. Also, in Section 15.3.3, we show that minimum-entropy colorings of independent random variables can be computed efficiently.

A Case When Intermediate Nodes Do Not Need to Compute

Though the proposed coding scheme in Algorithm 15.1 can perform arbitrarily closely to the rate lower bound, it may require computation at intermediate nodes. Here, we show that, for a family of functions and random variables, intermediate nodes do not need to perform computations.

DEFINITION 15.17 *Suppose $f(X_1,\ldots,X_k)$ is a deterministic function of random variables $X_1,\ldots,X_k$. $(f,X_1,\ldots,X_k)$ is called a chain-rule proper set when, for any $s \in \mathcal{S}(k)$, $H_{[G_{X_i}]_{i\in s}} = H_{G_{X_s}}(X_s)$.*

THEOREM 15.5 *In a general tree network, if sources $X_1,\ldots,X_k$ are independent random variables and $(f,X_1,\ldots,X_k)$ is a chain-rule proper set, it is sufficient to have intermediate nodes as relays (with no computations) in order for the system to perform arbitrarily closely to the rate lower bound mentioned in Theorem 15.4.*

Proof of this theorem is presented in Section 15.5.9.

In the following lemma, we provide a sufficient condition to guarantee that a set is a chain-rule proper set.

LEMMA 15.6 *Suppose X_1 and X_2 are independent and $f(X_1,X_2)$ is a deterministic function. If, for any x_2^1 and x_2^2 in $\mathcal{X}_2$, we have $f(x_1^i,x_2^1) \neq f(x_1^j,x_2^2)$ for any possible i and j, then (f,X_1,X_2) is a chain-rule proper set.*

Proof of this lemma is presented in Section 15.5.10.

15.3.3 Polynomial Time Cases for Finding the Minimum-Entropy Coloring of a Characteristic Graph

In the proposed coding schemes of Sections 15.3.1 and 15.3.2, one needs to compute a minimum-entropy coloring (a coloring random variable which minimizes the entropy) of a characteristic graph. In general, finding this coloring is an NP-hard problem (as shown by Cardinal *et al.* [33]). However, in this section we show that, depending on the characteristic graph's structure, there are some interesting cases where finding the minimum-entropy coloring is not NP-hard, but tractable and practical. In one of these cases, we show that having a non-zero joint probability condition on random variables' distributions, for any desired function f, means that the characteristic graphs are formed of some non-overlapping fully connected maximal independent sets. We will show that, in this case, the minimum-entropy coloring can be computed in polynomial time, according to Algorithm 15.3. In another case, we show that, if the function we seek to compute is a type of quantization function, this problem is also tractable.

For simplicity, we consider functions with two input random variables, but one can extend all of the discussions to functions with more input random variables than two. Suppose $c_{G_{X_1}}^{\min}$ represents a minimum-entropy coloring of a characteristic graph G_{X_1}.

Non-Zero Joint Probability Distribution Condition

Consider the network shown in Fig. 15.1(b). Source random variables have a joint probability distribution $p(x_1,x_2)$, and the receiver wishes to compute a deterministic function of sources (i.e., $f(X_1,X_2)$). In this section, we consider the effect of the probability distribution of sources in computations of minimum-entropy colorings.

THEOREM 15.6 *Suppose that, for all $(x_1,x_2) \in \mathcal{X}_1 \times \mathcal{X}_2$, $p(x_1,x_2) > 0$. Then, the maximally independent sets of the characteristic graph G_{X_1} (and its nth power $G_{\mathbf{X}_1}^n$, for any n) are some non-overlapping fully connected sets. Under this condition, the minimum-entropy coloring can be achieved by assigning different colors to its different maximally independent sets as described in Algorithm 15.3.*

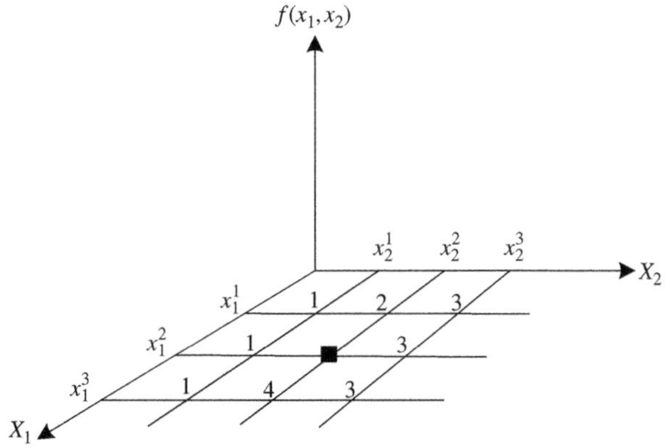

Figure 15.5 Having non-zero joint probability condition is necessary for Theorem 15.6. A dark square represents a zero-probability point.

Here are some remarks about Theorem 15.6.

- The condition $p(x_1,x_2) > 0$, for all $(x_1,x_2) \in \mathcal{X}_1 \times \mathcal{X}_2$, is a necessary condition for Theorem 15.6. In order to illustrate this, consider Fig. 15.5. In this example, x_1^1, x_1^2, and x_1^3 are in $\mathcal{X}_1$, and x_2^1, x_2^2, and x_2^3 are in $\mathcal{X}_2$. Suppose $p(x_1^2,x_2^2) = 0$. By considering the value of the function f at these points depicted in the figure, one can see that, in G_{X_1}, x_1^2 is not connected to x_1^1 and x_1^3. However, x_1^1 and x_1^3 are connected to each other. Thus, Theorem 15.6 does not hold here.
- The condition used in Theorem 15.6 merely restricts the probability distribution and it does not depend on the function f. Thus, for any function f at the receiver, if we have a non-zero joint probability distribution of source random variables (for example, when the source random variables are independent), finding the minimum-entropy coloring is easy and tractable.
- Orlitsky and Roche [15] showed that, for the side-information problem, having a non-zero joint probability condition yields a simplified graph entropy calculation. Here, we show that, under this condition, characteristic graphs have certain structures so that optimal coloring-based coding schemes that perform arbitrarily closely to rate lower bounds can be designed efficiently.

Quantization Functions

In this section, we consider some special functions which lead to practical minimum-entropy coloring computation.

An interesting function in this context is a quantization function. A natural quantization function is a function which separates the X_1–X_2 plane into some rectangles so that each rectangle corresponds to a different value of that function. The sides of these rectangles are parallel to the plane axes. Figure 15.6(a) depicts such a quantization function.

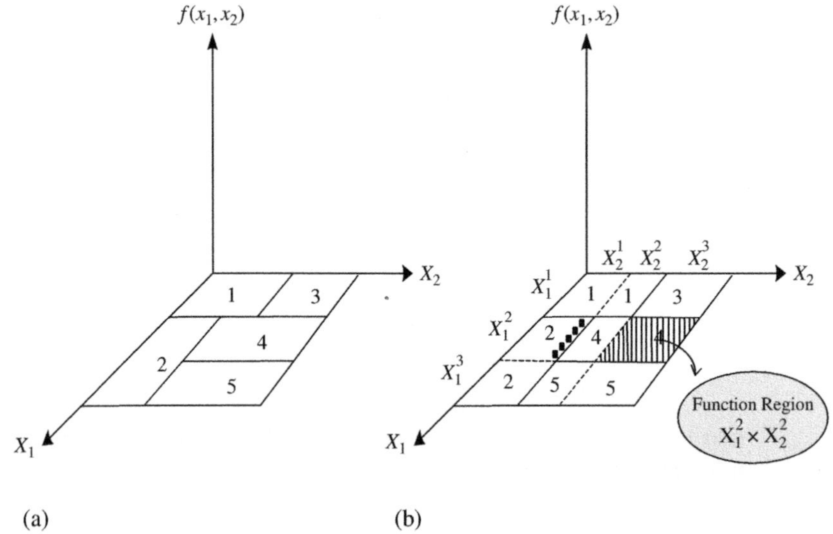

Figure 15.6 (a) A quantization function. Function values are depicted in the figure on each rectangle. (b) By extending the sides of rectangles, the plane is covered by some function regions.

Given a quantization function, one can extend different sides of each rectangle in the X_1–X_2 plane. This may make some new rectangles. We call each of them *a function region*. Each function region can be determined by two subsets of $\mathcal{X}_1$ and $\mathcal{X}_2$. For example, in Fig. 15.6(b), one of the function regions is distinguished by the shaded area.

DEFINITION 15.18 *Consider two function regions $\mathcal{X}_1^1 \times \mathcal{X}_2^1$ and $\mathcal{X}_1^2 \times \mathcal{X}_2^2$. If, for any $x_1^1 \in \mathcal{X}_1^1$ and $x_1^2 \in \mathcal{X}_1^2$, there exists x_2^1 such that $p(x_1^1, x_2^1)p(x_1^2, x_2^1) > 0$ and $f(x_1^1, x_2^1) \neq f(x_1^2, x_2^1)$, we say these two function regions are pair-wise X_1-proper.*

THEOREM 15.7 *Consider a quantization function f such that its function regions are pair-wise X_1-proper. Then, G_{X_1} (and $G^n_{\mathbf{X}_1}$, for any n) is formed of some non-overlapping fully connected maximally independent sets, and its minimum-entropy coloring can be achieved by assigning different colors to different maximally independent sets.*

Proof of this theorem is presented in Section 15.5.12.

Note that, without X_1-proper condition of Theorem 15.7, assigning different colors to different partitions still leads to an achievable coloring scheme. However, it is not necessarily a minimum-entropy coloring. In other words, without this condition, maximally independent sets may overlap.

COROLLARY 15.2 *If a function f is strictly monotonic with respect to X_1, and $p(x_1, x_2) \neq 0$, for all $x_1 \in \mathcal{X}_1$ and $x_2 \in \mathcal{X}_2$, then G_{X_1} (and $G^n_{\mathbf{X}_1}$, for any n) is a complete graph.*

Under the conditions of Corollary 15.2, functional compression does not give us any gain, because, in a complete graph, one should assign different colors to different vertices. Traditional compression where f is the identity function is a special case of Corollary 15.2.

Section 15.3.3 presents conditions on either source probability distributions and/or the desired function such that characteristic graphs of random variables are composed of fully connected non-overlapping maximally independent sets. The following algorithm shows how a minimum-entropy coloring can be computed for these graphs in polynomial time complexity.

Algorithm 15.3

Suppose $G_{X_1} = (V, E)$ is a graph composed of fully connected non-overlapping maximally independent sets and $\bar{G}_{X_1} = (V, \bar{E})$ represents its complement, where E and $\bar{E}$ are partitions of complete graph edges. Say C is the set of used colors formed as follows.

- Choose a node $v \in V$.
- Color node v and its neighbors in the graph $\bar{G}_{X_1}$ by a color c_v such that $c_v \notin C$.
- Add c_v to C. Repeat until all nodes are colored.

This algorithm finds minimum colorings of G_{X_1} in polynomial time with respect to the number of vertices of G_{X_1}.

The achievablity proof of this algorithm is presented in Section 15.5.13.

15.4 Discussion and Future Work

In this section, we discuss other aspects of the functional compression problem such as the effect of having feedback and lossy computations. First, by presenting an example, we show that, unlike in the Slepian–Wolf case, by having feedback in functional compression, one can outperform the rate bounds of the case without feedback. Then, we investigate the problem of distributed functional compression with distortion, where computation of a function within a distortion level is desired at the receiver. Here, we propose a simple sub-optimal coding scheme with a non-trivial performance guarantee. Finally, we explain some future research directions.

15.4.1 Feedback in Functional Compression

If the function at the receiver is the identity function, the functional compression problem is Slepian–Wolf compression with feedback. For this case, having feedback does not improve the rate bounds. For example, see [12], which considers both zero-error and asymptotically zero-error Slepian–Wolf compression with feedback. However, here by presenting an example we show that, for a general desired function at the receiver, having feedback can improve the rate bounds of the case without feedback.

Example 15.6 Consider a distributed functional compression problem with two sources and a receiver as depicted in Fig. 15.7(a). Suppose each source has one byte (8 bits) to transmit to the receiver. Bits are sorted from the (most significant bit MSB) to the least significant bit (LSB). Bits can be 0 or 1 with the same probability. The desired function at the receiver is $f(\mathbf{X}_1, \mathbf{X}_2) = \max(\mathbf{X}_1, \mathbf{X}_2)$.

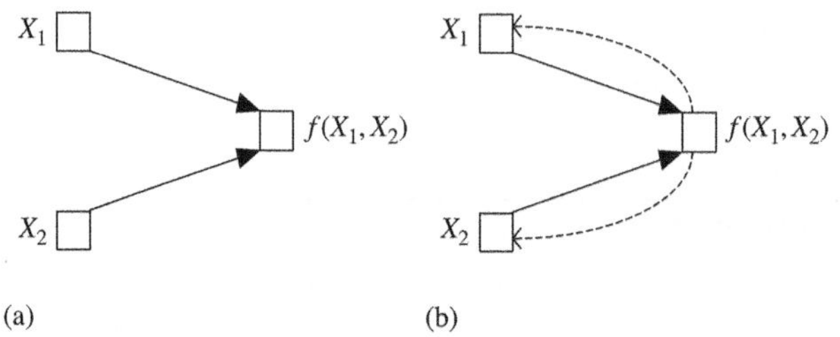

Figure 15.7 A distributed functional compression network (a) without feedback and (b) with feedback.

In the case without feedback, the characteristic graphs of sources are trivially complete graphs. Therefore, each source should transmit all bits to the receiver (i.e., the un-scaled rates are $R_1 = 8$ and $R_2 = 8$).

Now, suppose the receiver can broadcast some feedback bits to sources. In the following, we propose a communication scheme that has a reduced sum transmission rate compared with the case without feedback:

First, each source transmits its MSB. The receiver compares two received bits. If they are the same, the receiver broadcasts 0 to sources; otherwise it broadcasts 1. If sources receive 1 from feedback, they stop transmitting. Otherwise, they transmit their next significant bits. For this communication scheme, the un-scaled sum rate of the forward links can be calculated as follows:

$$\begin{aligned} R_1^{\mathrm{f}} + R_2^{\mathrm{f}} &= \underbrace{\frac{1}{2}(2 \times 1)}_{\text{(a)}} + \underbrace{\frac{1}{4}(2 \times 2)}_{\text{(b)}} + \cdots + \frac{1}{2^n}(2 \times n) \\ &\overset{\text{(c)}}{=} 2\left(2 - \frac{n+2}{2^n}\right), \end{aligned} \tag{15.12}$$

where n is the blocklength (in this example, $n = 8$), and R_1^{f} and R_2^{f} are the transmission rates of sources X_1 and X_2, respectively. Sources stop transmitting after the first bit transmission if these bits are not equal. The probability of this event is $1/2$ (term (a) in equation (15.12)). Similarly, forward transmissions stop in the second round with probability $1/4$ (term (b) in equation (15.12)) and so on. Equality (c) follows from a closed-form solution for the series $\sum_i (i/2^i)$. For $n = 8$, this rate is around 3.92, which is less than the sum rate in the case without feedback. With similar calculations, the feedback rate is around 1.96. Hence, the total forward and feedback transmission rate is around 5.88, which is less than that in the case without feedback.

15.4.2 A Practical Rate-Distortion Scheme for Distributed Functional Compression

In this section, we consider the problem of distributed functional compression with distortion. The objective is to compress correlated discrete sources so that an arbitrary deterministic function of sources can be computed up to a distortion level at the receiver.

Here, we present a practical coding scheme for this problem with a non-trivial performance guarantee. All discussions can be extended to more general networks similar to results of Section 15.3.

Consider two sources as described in Section 15.2.1. Here, we assume that the receiver wants to compute a deterministic function $f : \mathcal{X}_1 \times \mathcal{X}_2 \to \mathcal{Z}$ or $f : \mathcal{X}_1^n \times \mathcal{X}_2^n \to \mathcal{Z}^n$, its vector extension up to distortion D with respect to a given distortion function $d : \mathcal{Z} \times \mathcal{Z} \to [0,\infty)$. A vector extension of the distortion function is defined as follows:

$$d(\mathbf{z}_1,\mathbf{z}_2) = \frac{1}{n}\sum_{i=1}^{n} d(z_{1i},z_{2i}), \tag{15.13}$$

where $\mathbf{z}_1,\mathbf{z}_2 \in \mathcal{Z}^n$. We assume that $d(z_1,z_2) = 0$ if and only if $z_1 = z_2$. This assumption causes vector extension to satisfy the same property (i.e., $d(\mathbf{z}_1,\mathbf{z}_2) = 0$ if and only if $\mathbf{z}_1 = \mathbf{z}_2$).

The probability of error in this case is

$$P_{\mathrm{e}}^n = \Pr[\{(\mathbf{x}_1,\mathbf{x}_2) : d(f(\mathbf{x}_1,\mathbf{x}_2), r(\mathrm{en}_{X_1}(\mathbf{x}_1),\mathrm{en}_{X_2}(\mathbf{x}_2))) > D\}].$$

We say a rate pair (R_1,R_2) is achievable up to distortion D if there exist en_{X_1}, en_{X_2}, and r such that $P_{\mathrm{e}}^n \to 0$ when $n \to \infty$.

Yamamoto gives a characterization of a rate-distortion function for the side-information functional compression problem (i.e., X_2 is available at the receiver) in [24]. The rate-distortion function proposed in [24] is a generalization of the Wyner–Ziv side-information rate-distortion function [23]. Another multi-letter characterization of the rate-distortion function for the side-information problem given by Yamamoto was discussed in [16]. The multi-letter characterization of [16] can be extended naturally to a distributed functional compression case by using results of Section 15.3.

Here, we present a practical coding scheme with a non-trivial performance gaurantee for a given distributed lossy functional compression setup.

Define the D-characteristic graph of X_1 with respect to X_2, $p(x_1,x_2)$, and $f(X_1,X_2)$ as having vertices $V = \mathcal{X}_1$ and the pair (x_1^1,x_1^2) is an edge if there exists some $x_2^1 \in \mathcal{X}_2$ such that $p(x_1^1,x_2^1)p(x_1^2,x_2^1) > 0$ and $d(f(x_1^1,x_2^1), f(x_1^2,x_2^1)) > D$ as in [16]. Denote this graph as $G_{X_1}(D)$. Similarly, we define $G_{X_2}(D)$.

THEOREM 15.8 *For the network depicted in Fig. 15.1(b) with independent sources, if the distortion function is a metric, then the following rate pair (R_1,R_2) is achievable for the distributed lossy functional compression problem with distortion D:*

$$\begin{aligned} R_1 &\geq H_{G_{X_1}(D/2)}(X_1),\\ R_2 &\geq H_{G_{X_2}(D/2)}(X_2),\\ R_1 + R_2 &\geq H_{G_{X_1}(D/2),G_{X_2}(D/2)}(X_1,X_2). \end{aligned} \tag{15.14}$$

The proof of this theorem is presented in Section 15.5.14. A modularized coding scheme similar to Algorithm 15.2 can be used on D-characteristic graphs to perform arbitrarily closely to rate bounds of Theorem 15.8.

15.4.3 Future Work in Functional Compression

Throughout this chapter, we considered the case of having only one desired function at the receiver. However, all results can be extended naturally to the case of having several desired functions at the receiver by considering a vector extension of functions, and computing characteristic graphs of variables with respect to that vector. In fact, one can show that the characteristic graph of a random variable with respect to several functions is equal to the union of individual characteristic graphs (the union of two graphs with the same vertices is a graph with the same vertex set whose edges are the union of the edges of the individual graphs).

For possible future work, one may consider a general network topology rather than tree networks. For instance, one can consider a general multi-source multicast network in which receivers desire to have a deterministic function of source random variables. For the case of having the identity function at the receivers, this problem has been well studied in [35–37] under the name of network coding for multi-source multicast networks. Ho *et al.* [37] show that random linear network coding can perform arbitrarily closely to min-cut max-flow bounds. To have an achievable scheme for the functional version of this problem, one may perform random network coding on coloring random variables satisfying the CCC. If receivers desire different functions, one can use colorings of multi-functional characteristic graphs satisfying the CCC, and then use random network coding for these coloring random variables. This achievable scheme can be extended to the disjoint multicast and disjoint multicast plus multicast cases described in [36]. This scheme is an achievable scheme; however, it is not optimal in general. If the sources are independent, one may use encoding/decoding functions derived for tree networks at intermediate nodes, along with network coding.

Throughout this chapter, we considered asymptotically lossless or lossy computation of a function. For possible future work, one may consider this problem for the zero-error computation of a function, which leads to a communication complexity problem. One can use the tools and schemes we have developed in this chapter to attain some achievable schemes in the zero-error computation case as well.

15.5 Proofs

15.5.1 Proof of Lemma 15.1

Proof First, we know that any random variable X_2 by itself is a trivial coloring of G_{X_2} such that each vertex of G_{X_2} is assigned to a different color. So, J_C for $c_{G_{X_1}}(X_1)$ and $c_{G_{X_2}}(X_2) = X_2$ can be written as $J_C = \{j_c^1, \ldots, j_c^{n_{j_c}}\}$ such that $j_c^1 = \{(x_1^i, x_2^1) : c_{G_{X_1}}(x_1^i) = \sigma_i\}$, where σ_i is a generic color. Any two points in j_c^1 are connected to each other by a path with length one. So, j_c^1 satisfies the CCC. This arguments hold for any j_c^i for any valid i. Thus, all joint coloring classes and, therefore, $c_{G_{X_1}}(X_1)$ and $c_{G_{X_2}}(X_2) = X_2$ satisfy the CCC. The argument for $c_{G^n_{\mathbf{X}_1}}(\mathbf{X}_1)$ and $c_{G^n_{\mathbf{X}_2}}(\mathbf{X}_2) = \mathbf{X}_2$ is similar.

15.5.2 Proof of Lemma 15.2

Proof We first show that if j_c^i satisfies the CCC, then, for any two points $(x_1^1,\ldots,x_k^1)$ and $(x_1^2,\ldots,x_k^2)$ in j_c^i, $f(x_1^1,\ldots,x_k^1) = f(x_1^2,\ldots,x_k^2)$. Since j_c^i satisfies the CCC, by definition, either $f(x_1^1,\ldots,x_k^1) = f(x_1^2,\ldots,x_k^2)$ or there exists a path with length $m-1$ between these two points $Z_1 = (x_1^1,\ldots,x_k^1)$ and $Z_m = (x_1^2,\ldots,x_k^2)$, for some m, where two consecutive points Z_j and Z_{j+1} on this path differ in exactly one of their coordinates. Without loss of generality, suppose Z_j and Z_{j+1} differ in their first coordinate, i.e., $Z_j = (x_1^{j_1}, x_2^{j_2} \ldots, x_k^{j_k})$ and $Z_{j+1} = (x_1^{j_0}, x_2^{j_2} \ldots, x_k^{j_k})$. Since these two points belong to j_c^i, $c_{G_{X_1}}(x_1^{j_1}) = c_{G_{X_1}}(x_1^{j_0})$. If $f(Z_j) \neq f(Z_{j+1})$, there would exist an edge between $x_1^{j_1}$ and $x_1^{j_0}$ in G_{X_1}, and they could not have the same color. Therefore, $f(Z_j) = f(Z_{j+1})$. By applying the same argument inductively for all consecutive points on the path between Z_1 and Z_m, we have $f(Z_1) = f(Z_2) = \cdots = f(Z_m)$.

If j_c^i does not satisfy the CCC, by definition there exist at least two points Z_1 and Z_2 in j_c^i with different function values.

15.5.3 Proof of Lemma 15.3

Proof The proof is similar to Lemma 15.2. The only difference is that we use the definition of the CCC for $c_{G^n_{\mathbf{X}_1}}, \ldots, c_{G^n_{\mathbf{X}_k}}$. Since j_c^i satisfies the CCC, either $f(\mathbf{x}_1^1,\ldots,\mathbf{x}_k^1) = f(\mathbf{x}_1^2,\ldots,\mathbf{x}_k^2)$ or there exists a path with length $m-1$ between any two points $\mathbf{Z}_1 = (\mathbf{x}_1^1,\ldots,\mathbf{x}_k^1) \in T_\epsilon^n$ and $\mathbf{Z}_m = (\mathbf{x}_1^2,\ldots,\mathbf{x}_k^2) \in T_\epsilon^n$ in j_c^i, for some m. Consider two consecutive points $\mathbf{Z}_j$ and $\mathbf{Z}_{j+1}$ in this path. They differ in one of their coordinates (suppose they differ in their first coordinate). In other words, suppose $\mathbf{Z}_j = (\mathbf{x}_1^{j_1}, \mathbf{x}_2^{j_2}, \ldots, \mathbf{x}_k^{j_k}) \in T_\epsilon^n$ and $\mathbf{Z}_{j+1} = (\mathbf{x}_1^{j_0}, \mathbf{x}_2^{j_2}, \ldots, \mathbf{x}_k^{j_k}) \in T_\epsilon^n$. Since these two points belong to j_c^i, $c_{G_{X_1}}(\mathbf{x}_1^{j_1}) = c_{G_{X_1}}(\mathbf{x}_1^{j_0})$. If $f(\mathbf{Z}_j) \neq f(\mathbf{Z}_{j+1})$, there would exist an edge between $\mathbf{x}_1^{j_1}$ and $\mathbf{x}_1^{j_0}$ in $G^n_{\mathbf{X}_1}$ and they could not acquire the same color. Thus, $f(\mathbf{Z}_j) = f(\mathbf{Z}_{j+1})$. By applying the same argument for all consecutive points on the path between $\mathbf{Z}_1$ and $\mathbf{Z}_m$, one can get $f(\mathbf{Z}_1) = f(\mathbf{Z}_2) = \cdots = f(\mathbf{Z}_m)$. The converse part is similar to Lemma 15.2.

15.5.4 Proof of Lemma 15.4

Proof Suppose X_1 and X_2 satisfy the zigzag condition, and $c_{G_{X_1}}$ and $c_{G_{X_2}}$ are two valid colorings of G_{X_1} and G_{X_2}, respectively. We want to show that these colorings satisfy the CCC. To do this, consider two points (x_1^1, x_2^1) and (x_1^2, x_2^2) in a joint coloring class j_c^i. The definition of the zigzag condition guarantees the existence of a path with length two between these two points. Thus, $c_{G_{X_1}}$ and $c_{G_{X_2}}$ satisfy the CCC.

The second part of this lemma says that the converse part is not true. For example, one can see that in a special case considered in Lemma 15.1, those colorings always satisfy the CCC in the absence of any condition such as the zigzag condition.

15.5.5 Proof of Lemma 15.5

Proof By using the data-processing inequality, we have

$$H_{G_{X_1}}(X_1|X_2) = \lim_{n\to\infty} \min_{c_{G^n_{\mathbf{X}_1}}, c_{G^n_{\mathbf{X}_2}}} \frac{1}{n} H(c_{G^n_{\mathbf{X}_1}}(\mathbf{X}_1)|c_{G^n_{\mathbf{X}_2}}(\mathbf{X}_2))$$
$$= \lim_{n\to\infty} \min_{c_{G^n_{\mathbf{X}_1}}} \frac{1}{n} H(c_{G^n_{\mathbf{X}_1}}(\mathbf{X}_1)|\mathbf{X}_2).$$

Then, Lemma 15.1 implies that $c_{G^n_{\mathbf{X}_1}}(\mathbf{X}_1)$ and $c_{G^n_{\mathbf{X}_2}}(\mathbf{x}_2) = \mathbf{X}_2$ satisfy the CCC. A direct application of Theorem 15.2 completes the proof.

15.5.6 Proof of Theorem 15.3

Proof We first show the achievability of this rate region. We also propose a modularized encoding/decoding scheme in this part. Then, for the converse, we show that no encoding/decoding scheme can outperform this rate region.

(1) *Achievability.*

LEMMA 15.7 *Consider random variables* $\mathbf{X}_1, \ldots, \mathbf{X}_k$ *with characteristic graphs* $G^n_{\mathbf{X}_1}, \ldots, G^n_{\mathbf{X}_k}$, *and any valid* ϵ*-colorings* $c_{G^n_{\mathbf{X}_1}}, \ldots, c_{G^n_{\mathbf{X}_k}}$ *satisfying the CCC over typical points* T^n_ϵ, *for sufficiently large n. There exists*

$$\widehat{f} : c_{G^n_{\mathbf{X}_1}}(X_1) \times \cdots \times c_{G^n_{\mathbf{X}_k}}(X_k) \to \mathcal{Z}^n \tag{15.15}$$

such that $\widehat{f}(c_{G^n_{\mathbf{X}_1}}(\mathbf{x}_1), \ldots, c_{G^n_{\mathbf{X}_k}}(\mathbf{x}_k)) = f(\mathbf{x}_1, \ldots, \mathbf{x}_k)$, *for all* $(\mathbf{x}_1, \ldots, \mathbf{x}_k) \in T^n_\epsilon$.

Proof Suppose the joint coloring family for these colorings is $J_C = \{j^i_c : i\}$. We proceed by constructing $\widehat{f}$. Assume $(\mathbf{x}^1_1, \ldots, \mathbf{x}^1_k) \in j^i_c$ and $c_{G^n_{\mathbf{X}_1}}(\mathbf{x}^1_1) = \sigma_1, \ldots, c_{G^n_{\mathbf{X}_1}}(\mathbf{x}^1_k) = \sigma_k$. Define $\widehat{f}(\sigma_1, \ldots \sigma_k) = f(\mathbf{x}^1_1, \ldots, \mathbf{x}^1_k)$.

To show that this function is well defined on elements in its support, we should show that, for any two points $(\mathbf{x}^1_1, \ldots, \mathbf{x}^1_k)$ and $(\mathbf{x}^2_1, \ldots, \mathbf{x}^2_k)$ in T^n_ϵ, if $c_{G^n_{\mathbf{X}_1}}(\mathbf{x}^1_1) = c_{G^n_{\mathbf{X}_1}}(\mathbf{x}^2_1), \ldots, c_{G^n_{\mathbf{X}_k}}(\mathbf{x}^1_k) = c_{G^n_{\mathbf{X}_k}}(\mathbf{x}^2_k)$, then $f(\mathbf{x}^1_1, \ldots, \mathbf{x}^1_k) = f(\mathbf{x}^2_1, \ldots, \mathbf{x}^2_k)$.

Since $c_{G^n_{\mathbf{X}_1}}(\mathbf{x}^1_1) = c_{G^n_{\mathbf{X}_1}}(\mathbf{x}^2_1), \ldots, c_{G^n_{\mathbf{X}_k}}(\mathbf{x}^1_k) = c_{G^n_{\mathbf{X}_k}}(\mathbf{x}^2_k)$, these two points belong to a joint coloring class such as j^i_c. Since $c_{G^n_{\mathbf{X}_1}}, \ldots, c_{G^n_{\mathbf{X}_k}}$ satisfy the CCC, we have, by using Lemma 15.3, $f(\mathbf{x}^1_1, \ldots, \mathbf{x}^1_k) = f(\mathbf{x}^2_1, \ldots, \mathbf{x}^2_k)$. Therefore, our function $\widehat{f}$ is well defined and has the desired property.

Lemma 15.7 implies that, given ϵ-colorings of characteristic graphs of random variables satisfying the CCC, at the receiver, we can successfully compute the desired function f with a vanishing probability of error as n goes to infinity. Thus, if the decoder at the receiver is given colors, it can look up f from its table of $\widehat{f}$. It remains to be ascertained at which rates encoders can transmit these colors to the receiver faithfully (with a probability of error less than ϵ).

LEMMA 15.8 *(Slepian–Wolf theorem) A rate region of a one-stage tree network with the desired identity function at the receiver is characterized by the following conditions:*

$$\forall S \in \mathcal{S}(k) \Longrightarrow \sum_{i \in S} R_i \geq H(X_S | X_{S^c}). \tag{15.16}$$

Proof See [17].

We now use the Slepian–Wolf (SW) encoding/decoding scheme on the achieved coloring random variables. Suppose the probability of error in each decoder of SW is less than ϵ/k. Then, the total error in the decoding of colorings at the receiver is less than ϵ. Therefore, the total error in the coding scheme of first coloring $G^n_{\mathbf{X}_1}, \ldots, G^n_{\mathbf{X}_k}$ and then encoding those colors by using the SW encoding/decoding scheme is upper-bounded by the sum of errors in each stage. By using Lemmas 15.7 and 15.8, we find that the total error is less than ϵ, and goes to zero as n goes to infinity. By applying Lemma 15.8 on the achieved coloring random variables, we have

$$\forall S \in \mathcal{S}(k) \Longrightarrow \sum_{i \in S} R_i \geq \frac{1}{n} H(c_{G^n_{\mathbf{X}_S}} | c_{G^n_{\mathbf{X}_{S^c}}}), \tag{15.17}$$

where $c_{G^n_{\mathbf{X}_S}}$ and $c_{G^n_{\mathbf{X}_{S^c}}}$ are ϵ-colorings of characteristic graphs satisfying the CCC. Thus, using Definition 15.6 completes the achievability part.

(2) Converse. Here, we show that any distributed functional source coding scheme with a small probability of error induces ϵ-colorings on characteristic graphs of random variables satisfying the CCC. Suppose $\epsilon > 0$. Define $\mathcal{F}^n_\epsilon$ for all (n, ϵ) as follows:

$$\mathcal{F}^n_\epsilon = \{\widehat{f} : \Pr[\widehat{f}(\mathbf{X}_1, \ldots, \mathbf{X}_k) \neq f(\mathbf{X}_1, \ldots, \mathbf{X}_k)] < \epsilon\}. \tag{15.18}$$

In other words, $\mathcal{F}^n_\epsilon$ is the set of all functions that differ from f with probability ϵ. Suppose $\widehat{f}$ is an achievable code with vanishing error probability, where

$$\widehat{f}(\mathbf{x}_1, \ldots, \mathbf{x}_k) = r_n(\text{en}_{X_1,n}(\mathbf{x}_1), \ldots, \text{en}_{X_k,n}(\mathbf{x}_k)), \tag{15.19}$$

where n is the blocklength. Then there exists n_0 such that for all $n > n_0$, $\Pr(\widehat{f} \neq f) < \epsilon$. In other words, $\widehat{f} \in \mathcal{F}^n_\epsilon$. We call these codes ϵ-error functional codes.

LEMMA 15.9 *Consider some function $f : \mathcal{X}_1 \times \cdots \times \mathcal{X}_k \to \mathcal{Z}$. Any distributed functional code which reconstructs this function with zero-error probability induces colorings on $G_{X_1}, \ldots, G_{X_k}$ satisfying the CCC with respect to this function.*

Proof Say we have a zero-error distributed functional code represented by encoders $\text{en}_{X_1}, \ldots, \text{en}_{X_k}$ and a decoder r. For any two points $(x_1^1, \ldots, x_k^1)$ and $(x_1^2, \ldots, x_k^2)$ with positive probabilities, if their encoded values are the same (i.e., $\text{en}_{X_1}(x_1^1) = \text{en}_{X_1}(x_1^2), \ldots, \text{en}_{X_k}(x_k^1) = \text{en}_{X_k}(x_k^2)$), their function values will be the same as well since it is an error-free scheme:

$$f(x_1^1, \ldots, x_k^1) = f(x_1^2, \ldots, x_k^2). \tag{15.20}$$

We show that $\mathrm{en}_{X_1}, \ldots, \mathrm{en}_{X_k}$ are in fact some valid colorings of $G_{X_1}, \ldots, G_{X_k}$ satisfying the CCC. We demonstrate this argument for X_1. The argument for other random variables is analogous. First, we show that en_{X_1} induces a valid coloring on G_{X_1}, and then we show that this coloring satisfies the CCC.

Let us proceed by contradiction. If en_{X_1} did not induce a coloring on G_{X_1}, there must be some edge in G_{X_1} connecting two vertices with the same color. Let us call these vertices x_1^1 and x_1^2. Since these vertices are connected in G_{X_1}, there must exist an $(x_2^1, \ldots, x_k^1)$ such that $p(x_1^1, x_2^1, \ldots, x_k^1)p(x_1^2, x_2^1, \ldots, x_k^1) > 0$, $\mathrm{en}_{X_1}(x_1^1) = \mathrm{en}_{X_1}(x_1^2)$, and $f(x_1^1, x_2^1, \ldots, x_k^1) \neq f(x_1^2, x_2^1, \ldots, x_k^1)$. Taking $x_2^1 = x_2^2, \ldots, x_k^1 = x_k^2$ as in equation (15.20) leads to a contradiction. Therefore, the contradiction assumption is wrong and en_{X_1} induces a valid coloring on G_{X_1}.

Now, we show that these induced colorings satisfy the CCC. If this were not true, it would mean that there must exist two points $(x_1^1, \ldots, x_k^1)$ and $(x_1^2, \ldots, x_k^2)$ in a joint coloring class j_c^i so that there is no path between them in j_c^i. So, Lemma 15.2 says that the function f can acquire different values at these two points. In other words, it is possible to have $f(x_1^1, \ldots, x_k^1) \neq f(x_1^2, \ldots, x_k^2)$, where $c_{G_{X_1}}(x_1^1) = c_{G_{X_1}}(x_1^2), \ldots, c_{G_{X_k}}(x_k^1) = c_{G_{X_k}}(x_k^2)$, which is in contradiction with equation (15.20). Thus, these colorings satisfy the CCC.

In the last step, we must show that any achievable functional code represented by $\mathcal{F}_\epsilon^n$ induces ϵ-colorings on characteristic graphs satisfying the CCC.

LEMMA 15.10 *Consider random variables $\mathbf{X}_1, \ldots, \mathbf{X}_k$. All ϵ-error functional codes of these random variables induce ϵ-colorings on characteristic graphs satisfying the CCC.*

Proof Suppose $\widehat{f}(\mathbf{x}_1, \ldots, \mathbf{x}_k) = r(\mathrm{en}_{X_1}(\mathbf{x}_1), \ldots, \mathrm{en}_{X_k}(\mathbf{x}_k)) \in \mathcal{F}_\epsilon^n$ is such a code. If the function it is desired to compute is $\widehat{f}$, then, according to Lemma 15.9, a zero-error reconstruction of $\widehat{f}$ induces colorings on characteristic graphs satisfying the CCC with respect to $\widehat{f}$. Let the set of all points $(\mathbf{x}_1, \ldots, \mathbf{x}_k)$ such that $\widehat{f}(\mathbf{x}_1, \ldots, \mathbf{x}_k) \neq f(\mathbf{x}_1, \ldots, \mathbf{x}_k)$ be denoted by C. Since $\Pr(\widehat{f} \neq f) < \epsilon$, $\Pr[C] < \epsilon$. Therefore, functions $\mathrm{en}_{X_1}, \ldots, \mathrm{en}_{X_k}$ restricted to C are ϵ-colorings of characteristic graphs satisfying the CCC. with respect to f.

According to Lemmas 15.9 and 15.10, any distributed functional source code with vanishing error probability induces ϵ-colorings on characteristic graphs of source variables satisfying the CCC with respect to the desired function f.

Then, according to the Slepian–Wolf theorem, Theorem 15.8, we have

$$\forall S \in \mathcal{S}(k) \Longrightarrow \sum_{i \in S} R_i \geq \frac{1}{n} H(c_{G^n_{\mathbf{X}_S}} | c_{G^n_{\mathbf{X}_{S^c}}}), \tag{15.21}$$

where $c_{G^n_{\mathbf{X}_S}}$ and $c_{G^n_{\mathbf{X}_{S^c}}}$ are ϵ-colorings of characteristic graphs satisfying the CCC with respect to f. Using Definition 15.6 completes the converse part.

15.5.7 Proof of Theorem 15.4

Proof Here, we show that no coding scheme can outperform this rate region. Suppose source nodes $\{1,\ldots,4\}$ are directly connected to the receiver. By direct application of Theorem 15.3, the first set of conditions of Theorem 15.4 can be derived. Repeating the argument for intermediate nodes $\{5,6\}$ completes the proof.

15.5.8 Achievablity Proof of Algorithm 15.1

Proof To show the achievability, we show that, if the nodes of each stage were directly connected to the receiver, the receiver could compute its desired function. For nodes $\{1,\ldots,4\}$ in the first stage, the argument follows directly from Theorem 15.3. Now we show that the argument holds for intermediate nodes $\{5,6\}$ as well. Consider node 5 in the second stage of the network. Since the corresponding source values of its received colorings form an independent set on its characteristic graph and since this node computes the minimum-entropy coloring of this graph, it is equivalent to the case where it would receive the exact source information, because both of them lead to the same coloring random variable. Therefore, by having nodes 5 and 6 directly connected to the receiver, and a direct application of Theorem 15.3, the receiver is able to compute its desired function by using colorings of characteristic graphs of nodes 5 and 6.

15.5.9 Proof of Theorem 15.5

Proof Here, we present the proof for the tree network structure depicted in Fig. 15.4. However, all of the arguments can be extended to a general case. Suppose intermediate nodes 5 and 6 perform no computations and act as relays. Therefore, we have

$$R_5 = R_1 + R_2 = H_{G_{X_1},G_{X_2}}(X_1,X_2|X_3,X_4).$$

By using the chain-rule proper set condition, we can rewrite this as

$$R_5 = H_{G_{X_1,X_2}}(X_1,X_2|X_3,X_4),$$

which is the same condition as that of Theorem 15.4. Repeating this argument for R_6 and $R_5 + R_6$ establishes the proof.

15.5.10 Proof of Lemma 15.6

Proof To prove this lemma, it is sufficient to show that, under the conditions of this lemma, any colorings of the graph G_{X_1,X_2} can be expressed as colorings of G_{X_1} and G_{X_2}, and vice versa. The converse part is straightforward because any colorings of G_{X_1} and G_{X_2} can be viewed as a coloring of G_{X_1,X_2}.

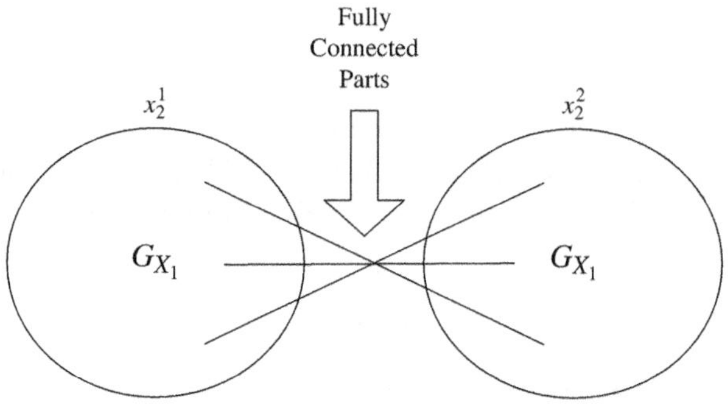

Figure 15.8 An example of G_{X_1,X_2} satisfying conditions of Lemma 15.6, when $\mathcal{X}_2$ has two members.

Consider Fig. 15.8, which illustrates the conditions of this lemma. Under these conditions, since all x_2 in $\mathcal{X}_2$ have different function values, the graph G_{X_1,X_2} can be decomposed into subgraphs which have the same topology as G_{X_1} (i.e., isomorphism to G_{X_1}), corresponding to each x_2 in $\mathcal{X}_2$. These subgraphs are fully connected to each other under the conditions of this lemma. Thus, any coloring of this graph can be represented as two colorings of G_{X_1} (within each subgraph) and G_{X_2} (across subgraphs). Therefore, the minimum-entropy coloring of G_{X_1,X_2} is equal to the minimum-entropy coloring of (G_{X_1}, G_{X_2}), i.e., $H_{G_{X_1},G_{X_2}}(X_1,X_2) = H_{G_{X_1,X_2}}(X_1,X_2)$.

15.5.11 Proof of Theorem 15.6

Proof Suppose $\Gamma(G_{X_1})$ is the set of all maximal independent sets of G_{X_1}. Let us proceed by contradiction. Consider Fig. 15.9(a). Suppose w_1 and w_2 are two different non-empty maximally independent sets. Without loss of generality, assume x_1^1 and x_1^2 are in w_1, and x_1^2 and x_1^3 are in w_2. These sets have a common element x_1^2. Since w_1 and w_2 are two different maximally independent sets, $x_1^1 \notin w_2$ and $x_1^3 \notin w_1$. Since x_1^1 and x_1^2 are in w_1, there is no edge between them in G_{X_1}. The same argument holds for x_1^2 and x_1^3. However, we have an edge between x_1^1 and x_1^3, because w_1 and w_2 are two different maximally independent sets, and at least there should exist such an edge between them. Now, we want to show that this is not possible.

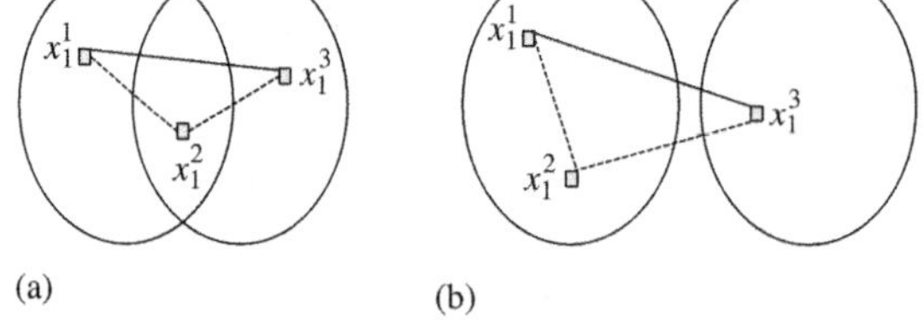

Figure 15.9 Having non-zero joint probability distribution, (a) maximally independent sets cannot overlap with each other (this figure depicts the contradiction); and (b) maximally independent sets should be fully connected to each other. In this figure, a solid line represents a connection, and a dashed line means that no connection exists.

Since there is no edge between x_1^1 and x_1^2, for any $x_2^1 \in \mathcal{X}_2$, $p(x_1^1, x_2^1)p(x_1^2, x_2^1) > 0$ and $f(x_1^1, x_2^1) = f(x_1^2, x_2^1)$. A similar argument can be expressed for x_1^2 and x_1^3. In other words, for any $x_2^1 \in \mathcal{X}_2$, $p(x_1^2, x_2^1)p(x_1^3, x_2^1) > 0$ and $f(x_1^2, x_2^1) = f(x_1^3, x_2^1)$. Thus, for all $x_2^1 \in \mathcal{X}_2$, $p(x_1^1, x_2^1)p(x_1^3, x_2^1) > 0$ and $f(x_1^1, x_2^1) = f(x_1^3, x_2^1)$. However, since x_1^1 and x_1^3 are connected to each other, there should exist an $x_2^1 \in \mathcal{X}_2$ such that $f(x_1^1, x_2^1) \neq f(x_1^3, x_2^1)$, which is not possible. So, the contradiction assumption is not correct and these two maximally independent sets do not overlap with each other.

We showed that maximally independent sets cannot have overlaps with each other. Now, we want to show that they are also fully connected to each other. Again, let us proceed by contradiction. Consider Fig. 15.9(b). Suppose w_1 and w_2 are two different non-overlapping maximally independent sets. Suppose there exists an element in w_2 (call it x_1^3) which is connected to one of the elements in w_1 (call it x_1^1) and is not connected to another element of w_1 (call it x_1^2). By using a similar argument to the one in the previous paragraph, we may show that it is not possible. Thus, x_1^3 should be connected to x_1^1. Therefore, if, for all $(x_1, x_2) \in \mathcal{X}_1 \times \mathcal{X}_2$, $p(x_1, x_2) > 0$, then the maximally independent sets of G_{X_1} are some separate fully connected sets. In other words, the complement of G_{X_1} is formed by some non-overlapping cliques. Finding the minimum-entropy coloring of this graph is trivial and can be achieved by assigning different colors to these non-overlapping fully connected maximally independent sets.

This argument also holds for any power of G_{X_1}. Suppose $\mathbf{x}_1^1$, $\mathbf{x}_1^2$, and $\mathbf{x}_1^3$ are some typical sequences in $\mathcal{X}_1^n$. If $\mathbf{x}_1^1$ is not connected to $\mathbf{x}_1^2$ and $\mathbf{x}_1^3$, it is not possible to have $\mathbf{x}_1^2$ and $\mathbf{x}_1^3$ connected. Therefore, one can apply a similar argument to prove the theorem for $G_{\mathbf{X}_1}^n$, for some n. This completes the proof.

15.5.12 Proof of Theorem 15.7

Proof We first prove it for G_{X_1}. Suppose $\mathcal{X}_1^1 \times \mathcal{X}_2^1$ and $\mathcal{X}_1^2 \times \mathcal{X}_2^2$ are two X_1-proper function regions of a quantization function f, where $\mathcal{X}_1^1 \neq \mathcal{X}_1^2$. We show that $\mathcal{X}_1^1$ and $\mathcal{X}_1^2$ are two non-overlapping fully connected maximally independent sets. By definition, $\mathcal{X}_1^1$ and $\mathcal{X}_1^2$ are two non-equal partition sets of $\mathcal{X}_1$. Thus, they do not have any element in common.

Now, we want to show that the vertices of each of these partition sets are not connected to each other. Without loss of generality, we show it for $\mathcal{X}_1^1$. If this partition set of $\mathcal{X}_1$ has only one element, this is a trivial case. So, suppose x_1^1 and x_1^2 are two elements in $\mathcal{X}_1^1$. From the definition of function regions, one can see that, for any $x_2^1 \in \mathcal{X}_2$ such that $p(x_1^1, x_2^1)p(x_1^2, x_2^1) > 0$, we have $f(x_1^1, x_2^1) = f(x_1^2, x_2^1)$. Thus, these two vertices are not connected to each other. Now, suppose x_1^3 is an element in $\mathcal{X}_1^2$. Since these function regions are X_1-proper, there should exist at least one $x_2^1 \in \mathcal{X}_2$ such that $p(x_1^1, x_2^1)p(x_1^3, x_2^1) > 0$ and $f(x_1^1, x_2^1) \neq f(x_1^3, x_2^1)$. Thus, x_1^1 and x_1^3 are connected to each other. Therefore, $\mathcal{X}_1^1$ and $\mathcal{X}_1^2$ are two non-overlapping fully connected maximally independent sets. One can easily apply this argument to other partition sets. Thus, the minimum-entropy coloring can be achieved by assigning different colors to different maximally independent sets (partition sets). The proof for $G_{\mathbf{X}_1}^n$, for any n, is similar to the one mentioned in Theorem 15.6. This completes the proof.

15.5.13 Achievablity Proof of Algorithm 15.3

Proof Suppose G_{X_1} has P vertices labeled as $\{1,\ldots,P\}$ and sorted in a list. Say that, if a vertex v has d_v neighbors in $\bar{G}_{X_1}$, the complexity of finding them is on the order of $O(d_v)$. Therefore, for a vertex v, the complexity of the first two steps of the algorithm is on the order of $O(d_v \log(P))$, where $O(\log(P))$ is the complexity of updating the list of un-colored vertices. Therefore, the overall worst-case complexity of the algorithm is on the order of $O(P^2 \log(P))$. Since the maximally independent sets of graph G_{X_1} are non-overlapping and also fully connected, any valid coloring scheme should assign them to different colors. Therefore, the minimum number of required colors is equal to the number of non-overlapping maximally independent sets of G_{X_1}, which in fact is the number of colors used in Algorithm 15.2. This completes the proof.

15.5.14 Proof of Theorem 15.8

Proof From Theorem 15.2, we have that, by sending colorings of high-probability subgraphs of sources $D/2$-characteristic graphs satisfying the CCC, one can achieve the rate region described in (15.14). For simplicity, we assume that the power of the graphs is one. Extensions to an arbitrary power are analogous. Suppose the receiver gets two colors from sources (say c_1 from source 1, and c_2 from source 2). To show that the receiver is able to compute its desired function up to distortion level D, we need to show that, for every (x_1^1, x_2^1) and (x_1^2, x_2^2) such that $C_{G_{X_1}(D/2)}(x_1^1) = C_{G_{X_1}(D/2)}(x_1^2)$ and $C_{G_{X_2}(D/2)}(x_2^1) = C_{G_{X_2}(D/2)}(x_2^2)$, we have $d(f(x_1^1,x_2^1), f(x_1^2,x_2^2)) \leq D$. Since the distortion function d is a metric, we have

$$
\begin{aligned}
d(f(x_1^1,x_2^1), f(x_1^2,x_2^2)) &\leq d(f(x_1^1,x_2^1), f(x_1^2,x_2^1)) + d(f(x_1^2,x_2^1), f(x_1^2,x_2^2)) \\
&\leq D/2 + D/2 = D. \qquad (15.22)
\end{aligned}
$$

This completes the proof.

References

[1] S. Feizi and M. Médard, "On network functional compression," *IEEE Trans. Information Theory*, vol. 60, no. 9, pp. 5387–5401, 2014.

[2] H. Kowshik and P. R. Kumar, "Optimal computation of symmetric Boolean functions in tree networks," in *Proc. 2010 IEEE International Symposium on Information Theory, ISIT 2010*, 2010, pp. 1873–1877.

[3] S. Shenvi and B. K. Dey, "A necessary and sufficient condition for solvability of a 3s/3t sum-network," in *Proc. 2010 IEEE International Symposium on Information Theory, ISIT 2010*, 2010, pp. 1858–1862.

[4] A. Ramamoorthy, "Communicating the sum of sources over a network," in *Proc. 2008 IEEE International Symposium on Information Theory, ISIT 2008*, 2008, pp. 1646–1650.

[5] R. Gallager, "Finding parity in a simple broadcast network," *IEEE Trans. Information Theory*, vol. 34, no. 2, pp. 176–180, 1988.

[6] A. Giridhar and P. Kumar, "Computing and communicating functions over sensor networks," *IEEE J. Selected Areas in Communications*, vol. 23, no. 4, pp. 755–764, 2005.

[7] S. Kamath and D. Manjunath, "On distributed function computation in structure-free random networks," in *Proc. 2008 IEEE International Symposium on Information Theory, ISIT 2008*, 2008, pp. 647–651.

[8] N. Ma, P. Ishwar, and P. Gupta, "Information-theoretic bounds for multiround function computation in collocated networks," in *Proc. 2009 IEEE International Symposium on Information Theory, ISIT 2009*, 2009, pp. 2306–2310.

[9] R. Ahuja, T. Magnanti, and J. Orlin, *Network flows: Theory, algorithms, and applications*. Prentice Hall, 1993.

[10] F. Shahrokhi and D. Matula, "The maximum concurrent flow problem," *J. ACM*, vol. 37, no. 2, pp. 318–334, 1990.

[11] V. Shah, B. Dey, and D. Manjunath, "Network flows for functions," in *Proc. 2011 IEEE International Symposium on Information Theory*, 2011, pp. 234–238.

[12] M. Bakshi and M. Effros, "On zero-error source coding with feedback," in *Proc. 2010 IEEE International Symposium on Information Theory, ISIT 2010*, 2010.

[13] H. Kowshik and P. Kumar, "Zero-error function computation in sensor networks," in *Proc. 48th IEEE Conference on Decision and Control, 2009 held jointly with the 2009 28th Chinese Control Conference, CDC/CCC 2009*, 2009, pp. 3787–3792.

[14] C. E. Shannon, "The zero error capacity of a noisy channel," *IRE Trans. Information Theory*, vol. 2, no. 3, pp. 8–19, 1956.

[15] A. Orlitsky and J. R. Roche, "Coding for computing," *IEEE Trans. Information Theory*, vol. 47, no. 3, pp. 903–917, 2001.

[16] V. Doshi, D. Shah, M. Médard, and M. Effros, "Functional compression through graph coloring," *IEEE Trans. Information Theory*, vol. 56, no. 8, pp. 3901–3917, 2010.

[17] D. Slepian and J. K. Wolf, "Noiseless coding of correlated information sources," *IEEE Trans. Information Theory*, vol. 19, no. 4, pp. 471–480, 1973.

[18] S. S. Pradhan and K. Ramchandran, "Distributed source coding using syndromes (DISCUS): Design and construction," *IEEE Trans. Information Theory*, vol. 49, no. 3, pp. 626–643, 2003.

[19] B. Rimoldi and R. Urbanke, "Asynchronous Slepian–Wolf coding via source-splitting," in *Proc. 1997 IEEE International Symposium on Information Theory*, 1997, p. 271.

[20] T. P. Coleman, A. H. Lee, M. Médard, and M. Effros, "Low-complexity approaches to Slepian–Wolf near-lossless distributed data compression," *IEEE Trans. Information Theory*, vol. 52, no. 8, pp. 3546–3561, 2006.

[21] R. F. Ahlswede and J. Körner, "Source coding with side information and a converse for degraded broadcast channels," *IEEE Trans. Information Theory*, vol. 21, no. 6, pp. 629–637, 1975.

[22] J. Körner and K. Marton, "How to encode the modulo-two sum of binary sources," *IEEE Trans. Information Theory*, vol. 25, no. 2, pp. 219–221, 1979.

[23] A. Wyner and J. Ziv, "The rate-distortion function for source coding with side information at the decoder," *IEEE Trans. Information Theory*, vol. 22, no. 1, pp. 1–10, 1976.

[24] H. Yamamoto, "Wyner–Ziv theory for a general function of the correlated sources," *IEEE Trans. Information Theory*, vol. 28, no. 5, pp. 803–807, 1982.

[25] H. Feng, M. Effros, and S. Savari, "Functional source coding for networks with receiver side information," in *Proc. Allerton Conference on Communication, Control, and Computing*, 2004, pp. 1419–1427.

[26] T. Berger and R. W. Yeung, "Multiterminal source encoding with one distortion criterion," *IEEE Trans. Information Theory*, vol. 35, no. 2, pp. 228–236, 1989.

[27] J. Barros and S. Servetto, "On the rate-distortion region for separate encoding of correlated sources," in *IEEE Trans. Information Theory (ISIT)*, 2003, p. 171.

[28] A. B. Wagner, S. Tavildar, and P. Viswanath, "Rate region of the quadratic Gaussian two-terminal source-coding problem," in *Proc. 2006 IEEE International Symposium on Information Theory*, 2006.

[29] I. Csiszár and J. Körner, in *Information theory: Coding theorems for discrete memoryless systems*. New York, 1981.

[30] H. S. Witsenhausen, "The zero-error side information problem and chromatic numbers," *IEEE Trans. Information Theory*, vol. 22, no. 5, pp. 592–593, 1976.

[31] J. Körner, "Coding of an information source having ambiguous alphabet and the entropy of graphs," in *Proc. 6th Prague Conference on Information Theory*, 1973, pp. 411–425.

[32] N. Alon and A. Orlitsky, "Source coding and graph entropies," *IEEE Trans. Information Theory*, vol. 42, no. 5, pp. 1329–1339, 1996.

[33] J. Cardinal, S. Fiorini, and G. Joret, "Tight results on minimum entropy set cover," *Algorithmica*, vol. 51, no. 1, pp. 49–60, 2008.

[34] R. Appuswamy, M. Franceschetti, N. Karamchandani, and K. Zeger, "Network coding for computing: Cut-set bounds," *IEEE Trans. Information Theory*, vol. 57, no. 2, pp. 1015–1030, 2011.

[35] R. Ahlswede, N. Cai, S.-Y. R. Li, and R. W. Yeung, "Network information flow," *IEEE Trans. Information Theory*, vol. 46, pp. 1204–1216, 2000.

[36] R. Koetter and M. Médard, "An algebraic approach to network coding," *IEEE/ACM Trans. Networking*, vol. 11, no. 5, pp. 782–795, 2003.

[37] T. Ho, M. Médard, R. Koetter, D. R. Karger, M. Effros, J. Shi, and B. Leong, "A random linear network coding approach to multicast," *IEEE Trans. Information Theory*, vol. 52, no. 10, pp. 4413–4430, 2006.

16 An Introductory Guide to Fano's Inequality with Applications in Statistical Estimation

Jonathan Scarlett and Volkan Cevher

Summary

Information theory plays an indispensable role in the development of algorithm-independent impossibility results, both for communication problems and for seemingly distinct areas such as statistics and machine learning. While numerous information-theoretic tools have been proposed for this purpose, the oldest one remains arguably the most versatile and widespread: Fano's inequality. In this chapter, we provide a survey of Fano's inequality and its variants in the context of statistical estimation, adopting a versatile framework that covers a wide range of specific problems. We present a variety of key tools and techniques used for establishing impossibility results via this approach, and provide representative examples covering group testing, graphical model selection, sparse linear regression, density estimation, and convex optimization.

16.1 Introduction

The tremendous progress in large-scale statistical inference and learning in recent years has been spurred by both practical and theoretical advances, with strong interactions between the two: algorithms that come with *a priori* performance guarantees are clearly desirable, if not crucial, in practical applications, and practical issues are indispensable in guiding the theoretical studies.

A key role, complementary to that of performance bounds for specific algorithms, is played by algorithm-independent impossibility results, stating conditions under which one cannot hope to achieve a certain goal. Such results provide definitive benchmarks for practical methods, serve as certificates for near-optimality, and help guide practical developments toward directions where the greatest improvements are possible.

Since its introduction in 1948, the field of information theory has continually provided such benefits for the problems of storing and transmitting data, and has accordingly shaped the design of practical communication systems. In addition, recent years have seen mounting evidence that the tools and methodology of information theory reach far beyond communication problems, and can provide similar benefits *within the entire data-processing pipeline*.

Table 16.1. Examples of applications for which impossibility results have been derived using Fano's inequality

Sparse and low-rank problems		**Other estimation problems**	
Problem	References	Problem	References
Group testing	[2, 3]	Regression	[12, 13]
Compressive sensing	[4, 5]	Density estimation	[13, 14]
Sparse Fourier transform	[6, 7]	Kernel methods	[15, 16]
Principal component analysis	[8, 9]	Distributed estimation	[17, 18]
Matrix completion	[10, 11]	Local privacy	[19]
Sequential decision problems		**Other learning problems**	
Problem	References	Problem	References
Convex optimization	[20, 21]	Graph learning	[26, 27]
Active learning	[22]	Ranking	[28, 29]
Multi-armed bandits	[23]	Classification	[30, 31]
Bayesian optimization	[24]	Clustering	[32]
Communication complexity	[25]	Phylogeny	[33]

While many information-theoretic tools have been proposed for establishing impossibility results, the oldest one remains arguably the most versatile and widespread: Fano's inequality [1]. This fundamental inequality is not only ubiquitous in studies of communication, but has also been applied extensively in statistical inference and learning problems; several examples are given in Table 16.1.

In applying Fano's inequality to such problems, one typically encounters a number of distinct challenges different from those found in communication problems. The goal of this chapter is to introduce the reader to some of the key tools and techniques, explain their interactions and connections, and provide several representative examples.

16.1.1 Overview of Techniques

Throughout the chapter, we consider the following statistical estimation framework, which captures a broad range of problems including the majority of those listed in Table 16.1.

- There exists an unknown parameter θ, known to lie in some set Θ (e.g., a subset of $\mathbb{R}^p$), that we would like to estimate.
- In the simplest case, the estimation algorithm has access to a set of *samples* $\mathbf{Y} = (Y_1, \dots, Y_n)$ drawn from some joint distribution $P_\theta^n(\mathbf{y})$ parameterized by θ. More generally, the samples may be drawn from some joint distribution $P_{\theta,\mathbf{X}}^n(\mathbf{y})$ parameterized by $(\theta, \mathbf{X})$, where $\mathbf{X} = (X_1, \dots, X_n)$ are *inputs* that are either known in advance or selected by the algorithm itself.
- Given knowledge of $\mathbf{Y}$, as well as $\mathbf{X}$ if inputs are present, the algorithm forms an estimate $\widehat{\theta}$ of θ, with the goal of the two being "close" in the sense that some *loss function* $\ell(\theta, \widehat{\theta})$ is small. When referring to this step of the estimation algorithm, we will use the terms *algorithm* and *decoder* interchangeably.

We will initially use the following simple running example to exemplify some of the key concepts, and then turn to detailed applications in Sections 16.4 and 16.6.

Example 16.1 (1-sparse linear regression) A vector parameter $\theta \in \mathbb{R}^p$ is known to have at most one non-zero entry, and we are given n linear samples of the form $\mathbf{Y} = \mathbf{X}\theta + \mathbf{Z}$,[1] where $\mathbf{X} \in \mathbb{R}^{n\times p}$ is a known input matrix, and $\mathbf{Z} \sim \mathcal{N}(\mathbf{0}, \sigma^2\mathbf{I})$ is additive Gaussian noise. In other words, the ith sample Y_i is a noisy sample of $\langle X_i, \theta\rangle$, where $X_i \in \mathbb{R}^p$ is the transpose of the ith row of $\mathbf{X}$. The goal is to construct an estimate $\widehat{\theta}$ such that the squared distance $\ell(\theta, \widehat{\theta}) = \|\theta - \widehat{\theta}\|_2^2$ is small.

This example is an extreme case of *k-sparse linear regression*, in which θ has at most $k \ll p$ non-zero entries, i.e., at most k columns of $\mathbf{X}$ impact the output. The more general k-sparse recovery problem will be considered in Section 16.6.1.

We seek to establish algorithm-independent impossibility results, henceforth referred to as *converse bounds*, in the form of lower bounds on the *sample complexity*, i.e., the number of samples n required to achieve a certain average target loss. The following aspects of the problem significantly impact this goal, and their differences are highlighted throughout the chapter.

- *Discrete versus continuous*. Depending on the application, the parameter set Θ may be discrete or continuous. For instance, in the 1-sparse linear regression example, one may consider the case in which θ is known to lie in a finite set $\Theta \subseteq \mathbb{R}^p$, or one may consider the general estimation of a vector in the set

$$\Theta = \{\theta \in \mathbb{R}^p : \|\theta\|_0 \le 1\}, \tag{16.1}$$

  where $\|\theta\|_0$ is the number of non-zero entries in θ.
- *Minimax versus Bayesian*. In the minimax setting, one seeks a decoder that attains a small loss for any given $\theta \in \Theta$, whereas in the Bayesian setting, one considers the average performance under some prior distribution on θ. Hence, these two variations respectively consider the worst-case and average-case performance with respect to θ. We focus primarily on the minimax setting throughout the chapter, and further discuss Bayesian settings in Section 16.7.2.
- *Choice of target goal*. Naturally, the target goal can considerably impact the fundamental performance limits of an estimation problem. For instance, in discrete settings, it is common to consider exact recovery, requiring that $\widehat{\theta} = \theta$ (i.e., the 0-1 loss $\ell(\theta, \widehat{\theta}) = \mathbb{1}\{\widehat{\theta} \neq \theta\}$), but it is also of interest to understand to what extent approximate recovery criteria make the problem easier.
- *Non-adaptive versus adaptive sampling*. In settings consisting of an input $\mathbf{X} = (X_1, \ldots, X_n)$ as introduced above, one often distinguishes between the *non-adaptive* setting, in which $\mathbf{X}$ is specified prior to observing any samples, and the *adaptive* setting, in which a given input X_i can be designed starting from the past inputs $(X_1, \ldots, X_{i-1})$ and samples $(Y_1, \ldots, Y_{i-1})$. It is of significant interest to understand to what extent the additional freedom of adaptivity impacts the performance.

With these variations in mind, we proceed by outlining the main steps in obtaining converse bounds for statistical estimation via Fano's inequality.

[1] Throughout the chapter, we interchange tuple-based notations such as $\mathbf{X} = (X_1, \ldots, X_n)$, $\mathbf{Y} = (Y_1, \ldots, Y_n)$ with vector/matrix notation such as $\mathbf{X} \in \mathbb{R}^{n\times p}$, $\mathbf{Y} \in \mathbb{R}^n$.

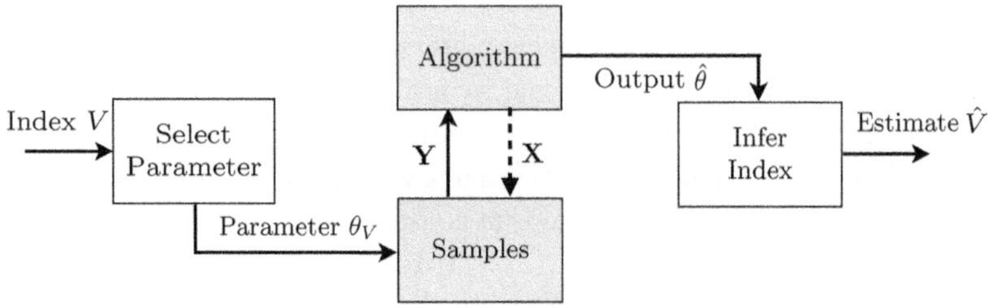

Figure 16.1 Reduction of minimax estimation to multiple hypothesis testing. The gray boxes are fixed as part of the problem statement, whereas the white boxes are constructed to our liking for the purpose of proving a converse bound. The dashed line marked with **X** is optional, depending on whether inputs are present.

Step 1: Reduction to Multiple Hypothesis Testing

The *multiple-hypothesis-testing* problem is defined as follows. An index $V \in \{1,\dots,M\}$ is drawn from a prior distribution P_V, and a sequence of samples $\mathbf{Y} = (Y_1,\dots,Y_n)$ is drawn from a probability distribution $P_{\mathbf{Y}|V}$ parameterized by V. The M possible conditional distributions are known in advance, and the goal is to identify the index V with high probability given the samples.

In Fig. 16.1, we provide a general illustration of how an estimation problem can be reduced to multiple hypothesis testing, possibly with the added twist of including inputs $\mathbf{X} = (X_1,\dots,X_n)$. Supposing for the time being that we are in the minimax setting, the idea is to construct a hard subset of parameters $\{\theta_1,\dots,\theta_M\}$ that are difficult to distinguish given the samples. We then lower-bound the worst-case performance by the average over this hard subset. As a concrete example, a good choice for the 1-sparse linear regression problem is to set $M = 2p$ and consider the set of vectors of the form

$$\theta = (0,\dots,0,\pm\epsilon,0,\dots,0), \tag{16.2}$$

where $\epsilon > 0$ is a constant. Hence, the non-zero entry of θ has a given magnitude, which can be selected to our liking for the purpose of proving a converse.

We envision an index $V \in \{1,\dots,M\}$ being drawn uniformly at random and used to select the corresponding parameter θ_V, and the estimation algorithm being run to produce an estimate $\widehat{\theta}$. If the parameters $\{\theta_1,\dots,\theta_M\}$ are not too close and the algorithm successfully produces $\widehat{\theta} \approx \theta_V$, then we should be able to infer the index V from $\widehat{\theta}$. This entire process can be viewed as a problem of multiple hypothesis testing, where the vth hypothesis is that the underlying parameter is θ_v ($v = 1,\dots,M$). With this reduction, we can deduce that if the algorithm performs well then the hypothesis test is successful; the contrapositive statement is then that *if the hypothesis test cannot be successful, then the algorithm cannot perform well.*

In the 1-sparse linear regression example, we find from (16.2) that distinct $\theta_j,\theta_{j'}$ must satisfy $\|\theta_j - \theta_{j'}\|_2 \geq \sqrt{2}\cdot\epsilon$. As a result, we immediately obtain from the triangle inequality that the following holds:

$$\text{If} \quad \|\widehat{\theta} - \theta_v\|_2 < \frac{\sqrt{2}}{2}\cdot\epsilon, \quad \text{then} \quad \operatorname*{arg\,min}_{v'=1,\dots,M} \|\widehat{\theta} - \theta_{v'}\| = v. \tag{16.3}$$

In other words, if the algorithm yields $\|\widehat{\theta}-\theta_v\|_2^2 < (\sqrt{2}/2)\epsilon$, then V can be identified as the index corresponding to the closest vector to $\widehat{\theta}$. Thus, sufficiently accurate estimation implies success in identifying V.

Discussion. Selecting the hard subset $\{\theta_1,\ldots,\theta_M\}$ of parameters is often considered something of an art. While the proofs of existing converse bounds may seem easy in hindsight when the hard subset is known, coming up with a suitable choice for a new problem usually requires some creativity and/or exploration. Despite this, there exist general approaches that have proved to been effective in a wide range of problems, which we exemplify in Sections 16.4 and 16.6.

In general, selecting the hard subset requires balancing conflicting goals: increasing M so that the hypothesis test is more difficult, keeping the elements "close" so that they are difficult to distinguish, and keeping the elements "sufficiently distant" so that one can recover V from $\widehat{\theta}$. Typically, one of the following three approaches is adopted: (i) explicitly construct a set whose elements are known or believed to be difficult to distinguish; (ii) prove the existence of such a set using probabilistic arguments; or (iii) consider packing as many elements as possible into the entire space. We will provide examples of all three kinds.

In the Bayesian setting, θ is already random, so we cannot use the above-mentioned method of lower-bounding the worst-case performance by the average. Nevertheless, if Θ is discrete, we can still use the trivial reduction $V=\theta$ to form a multiple-hypothesis-testing problem with a possibly non-uniform prior. In the continuous Bayesian setting, one typically requires more advanced methods that are not covered in this chapter; we provide further discussion in Section 16.7.2.

Step 2: Application of Fano's Inequality

Once a multiple hypothesis test has been set up, Fano's inequality provides a lower bound on its error probability in terms of the mutual information, which is one of the most fundamental information measures in information theory. The mutual information can often be explicitly characterized given the problem formulation, and a variety of useful properties are known for doing so, as outlined below.

We briefly state the standard form of Fano's inequality for the case in which V is uniform on $(1,\ldots,M)$ and $\widehat{V}$ is some estimate of V:

$$\mathbb{P}[\widehat{V}\neq V] \geq 1 - \frac{I(V;\widehat{V})+\log 2}{\log M}. \tag{16.4}$$

The intuition is as follows. The term $\log M$ represents the prior uncertainty (i.e., entropy) of V, and the mutual information $I(V;\widehat{V})$ represents how much information $\widehat{V}$ reveals about V. In order to have a small probability of error, we require that the information revealed is close to the prior uncertainty.

Beyond the standard form of Fano's inequality (16.4), it is useful to consider other variants, including approximate recovery and conditional versions. These are the topic of Section 16.2, and we discuss other alternatives in Section 16.7.2.

Step 3: Bounding the Mutual Information

In order to make lower bounds such as (16.4) explicit, we need to upper-bound the mutual information therein. This often consists of tedious yet routine calculations, but there are cases where it is highly non-trivial. The mutual information depends crucially on the choice of reduction in the first step.

The joint distribution of $(V, \widehat{V})$ is decoder-dependent and usually very complicated, so, to simplify matters, the typical first step is to apply an upper bound known as the data-processing inequality. In the simplest case in which there is no extra input to the sampling mechanism (i.e., $\mathbf{X}$ is absent in Fig. 16.1), this inequality takes the form $I(V;\widehat{V}) \le I(V;\mathbf{Y})$ under the Markov chain $V \to \mathbf{Y} \to \widehat{V}$. Thus, we are left to answer the question of how much information the samples reveal about the index V.

In Section 16.3, we introduce several useful tools for this purpose, including the following.

- *Tensorization.* If the samples $\mathbf{Y} = (Y_1, \ldots, Y_n)$ are conditionally independent given V, we have $I(V;\mathbf{Y}) \le \sum_{i=1}^{n} I(V;Y_i)$. Bounds of this type simplify the mutual information containing a set of observations to simpler terms containing only a single observation.
- *Kullback–Leibler (KL)-divergence-based bounds.* Straightforward bounds on the mutual information reveal that, if $\{P^n_{\theta_v}\}_{v=1,\ldots,M}$ are close in terms of KL divergence, then the mutual information is small. Results of this type are useful, as the relevant KL divergences can often be evaluated exactly or tightly bounded.

In addition to these, we introduce variations for cases in which the input $\mathbf{X}$ is present in Fig. 16.1, distinguishing between non-adaptive and adaptive sampling.

Toy example. To give a simple example of how this step is combined with the previous one, consider the case in which we wish to identify one of M hypotheses, with the vth hypothesis being that $\mathbf{Y} \sim P_v(\mathbf{y})$ for some distribution P_v on $\{0,1\}^n$. That is, the n observations $(Y_1, \ldots, Y_n)$ are binary-valued. Starting with the above-mentioned bound $I(V;\widehat{V}) \le I(V;\mathbf{Y})$, we simply write $I(V;\mathbf{Y}) \le H(\mathbf{Y}) \le n \log 2$, which follows since $\mathbf{Y}$ takes one of at most 2^n values. Substitution into (16.4) yields $P_e \ge 1 - (n+1)/\log_2 M$, which means that achieving $P_e \le \delta$ requires $n \ge (1-\delta)\log_2 M - 1$. This formalizes the intuitive fact that reliably identifying one of $M \gg 1$ hypotheses requires roughly $\log_2 M$ binary observations.

16.2 Fano's Inequality and Its Variants

In this section, we state various forms of Fano's inequality that will form the basis for the results in the remainder of the chapter.

16.2.1 Standard Version

We begin with the most simple and widely used form of Fano's inequality. We use the generic notation V for the discrete random variable in a multiple hypothesis test, and we write its estimate as $\widehat{V}$. In typical applications, one has a Markov-chain relation such as

$V \to \mathbf{Y} \to \widehat{V}$, where $\mathbf{Y}$ is the collection of samples; we will exploit this fact in Section 16.3, but, for now, one can think of $\widehat{V}$ being randomly generated by any means given V.

The two fundamental quantities appearing in Fano's inequality are the conditional entropy $H(V|\widehat{V})$, representing the uncertainty of V given its estimate, and the *error probability*:

$$P_e = \mathbb{P}[\widehat{V} \neq V]. \tag{16.5}$$

Since $H(V|\widehat{V}) = H(V) - I(V;\widehat{V})$, the conditional entropy is closely related to the mutual information, representing how much information $\widehat{V}$ reveals about V.

THEOREM 16.1 (Fano's inequality) *For any discrete random variables V and $\widehat{V}$ on a common finite alphabet $\mathcal{V}$, we have*

$$H(V|\widehat{V}) \leq H_2(P_e) + P_e \log(|\mathcal{V}| - 1), \tag{16.6}$$

where $H_2(\alpha) = \alpha \log(1/\alpha) + (1-\alpha)\log(1/(1-\alpha))$ is the binary entropy function. In particular, if V is uniform on $\mathcal{V}$, we have

$$I(V;\widehat{V}) \geq (1 - P_e)\log|\mathcal{V}| - \log 2, \tag{16.7}$$

or equivalently,

$$P_e \geq 1 - \frac{I(V;\widehat{V}) + \log 2}{\log|\mathcal{V}|}. \tag{16.8}$$

Since the proof of Theorem 16.1 is widely accessible in standard references such as [34], we provide only an intuitive explanation of (16.6). To resolve the uncertainty in V given $\widehat{V}$, we can first ask whether the two are equal, which bears uncertainty $H_2(P_e)$. If they differ, which only occurs a fraction P_e of the time, the remaining uncertainty is at most $\log(|\mathcal{V}| - 1)$.

REMARK 16.1 For uniform V, we obtain (16.7) by upper-bounding $|\mathcal{V}| - 1 \leq |\mathcal{V}|$ and $H_2(P_e) \leq \log 2$ in (16.6), and subtracting $H(V) = \log|\mathcal{V}|$ on both sides. While these additional bounds have a minimal impact for moderate to large values of $|\mathcal{V}|$, a notable case where one should use (16.6) is the binary setting, i.e., $|\mathcal{V}| = 2$. In this case, (16.7) is meaningless due to the right-hand side being negative, whereas (16.6) yields the following for uniform V:

$$I(V;\widehat{V}) \geq \log 2 - H_2(P_e). \tag{16.9}$$

It follows that the error probability is lower-bounded as

$$P_e \geq H_2^{-1}(\log 2 - I(V;\widehat{V})), \tag{16.10}$$

where $H_2^{-1}(\cdot) \in [0, \frac{1}{2}]$ is the inverse of $H_2(\cdot) \in [0, \log 2]$ on the domain $[0, \frac{1}{2}]$.

16.2.2 Approximate Recovery

The notion of error probability considered in Theorem 16.1 is that of exact recovery, insisting that $\widehat{V} = V$. More generally, one can consider notions of *approximate recovery*, where one only requires $\widehat{V}$ to be "close" to V in some sense. This is useful for at least two reasons.

- Exact recovery is often a highly stringent criterion in discrete statistical estimation problems, and it is of considerable interest to understand to what extent moving to approximate recovery makes the problem easier.
- When we reduce continuous estimation problems to the discrete setting (Section 16.5), permitting approximate recovery will provide a useful additional degree of freedom.

We consider a general setup with a random variable V, an estimate $\widehat{V}$, and an error probability of the form

$$P_{\mathrm{e}}(t) = \mathbb{P}[d(V,\widehat{V}) > t] \tag{16.11}$$

for some real-valued function $d(v,\widehat{v})$ and threshold $t \in \mathbb{R}$. In contrast to the exact recovery setting, there are interesting cases where V and $\widehat{V}$ are defined on different alphabets, so we denote these by $\mathcal{V}$ and $\widehat{\mathcal{V}}$, respectively.

One can interpret (16.11) as requiring $\widehat{V}$ to be within a "distance" t of V. However, d need not be a true distance function, and it need not even be symmetric or take non-negative values. This definition of the error probability in fact entails no loss of generality, since one can set $t = 0$ and $d(V,\widehat{V}) = \mathbb{1}\{(V,\widehat{V}) \in \mathcal{E}\}$ for an arbitrary set $\mathcal{E}$ containing the pairs that are considered errors.

In the following, we make use of the quantities

$$N_{\max}(t) = \max_{\widehat{v}\in\widehat{\mathcal{V}}} N_{\widehat{v}}(t), \qquad N_{\min}(t) = \min_{\widehat{v}\in\widehat{\mathcal{V}}} N_{\widehat{v}}(t), \tag{16.12}$$

where

$$N_{\widehat{v}}(t) = \sum_{v\in\mathcal{V}} \mathbb{1}\{d(v,\widehat{v}) \le t\} \tag{16.13}$$

counts the number of $v \in \mathcal{V}$ within a "distance" t of $\widehat{v} \in \widehat{\mathcal{V}}$.

THEOREM 16.2 (Fano's inequality with approximate recovery) *For any random variables $V,\widehat{V}$ on the finite alphabets $\mathcal{V},\widehat{\mathcal{V}}$, we have*

$$H(V|\widehat{V}) \le H_2(P_{\mathrm{e}}(t)) + P_{\mathrm{e}}(t)\log\left(\frac{|\mathcal{V}| - N_{\min}(t)}{N_{\max}(t)}\right) + \log N_{\max}(t). \tag{16.14}$$

In particular, if V is uniform on $\mathcal{V}$, then

$$I(V;\widehat{V}) \ge (1 - P_{\mathrm{e}}(t))\log\left(\frac{|\mathcal{V}|}{N_{\max}(t)}\right) - \log 2, \tag{16.15}$$

or equivalently

$$P_{\mathrm{e}}(t) \ge 1 - \frac{I(V;\widehat{V}) + \log 2}{\log(|\mathcal{V}|/N_{\max}(t))}. \tag{16.16}$$

The proof is similar to that of Theorem 16.1, and can be found in [35].

By setting $d(v,\widehat{v}) = \mathbb{1}\{v \neq \widehat{v}\}$ and $t = 0$, we find that Theorem 16.2 recovers Theorem 16.1 as a special case. More generally, the bounds (16.15) and (16.16) resemble those for exact recovery in (16.7) and (16.8), except that $\log |\mathcal{V}|$ is replaced by $\log(|\mathcal{V}|/N_{\max}(t))$. When $\mathcal{V} = \widehat{\mathcal{V}}$, one can intuitively think of the approximate recovery setting as dividing the space into regions of size $N_{\max}(t)$, and only requiring the correct region to be identified, thereby reducing the effective alphabet size to $|\mathcal{V}|/N_{\max}(t)$.

16.2.3 Conditional Version

When applying Fano's inequality, it is often useful to condition on certain random events and random variables. The following theorem states a general variant of Theorem 16.1 with such conditioning. Conditional forms for the case of approximate recovery (Theorem 16.2) follow in an identical manner.

THEOREM 16.3 (Conditional Fano inequality) *For any discrete random variables V and $\widehat{V}$ on a common alphabet $\mathcal{V}$, any discrete random variable A on an alphabet $\mathcal{A}$, and any subset $\mathcal{A}' \subseteq \mathcal{A}$, the error probability $P_e = \mathbb{P}[\widehat{V} \neq V]$ satisfies*

$$P_e \geq \sum_{a \in \mathcal{A}'} \mathbb{P}[A = a] \frac{H(V|\widehat{V}, A = a) - \log 2}{\log(|\mathcal{V}_a| - 1)}, \tag{16.17}$$

where $\mathcal{V}_a = \{v \in \mathcal{V} : \mathbb{P}[V = v | A = a] > 0\}$. For possibly continuous A, the same holds true with $\sum_{a\in\mathcal{A}'} \mathbb{P}[A = a](\cdots)$ replaced by $\mathbb{E}[\mathbb{1}\{A \in \mathcal{A}'\}(\cdots)]$.

Proof We write $P_e \geq \sum_{a\in\mathcal{A}'} \mathbb{P}[A = a]\mathbb{P}[\widehat{V} \neq V | A = a]$, and lower-bound the conditional error probability using Fano's inequality (Theorem 16.1) under the joint distribution of $(V, \widehat{V})$ conditioned on $A = a$.

REMARK 16.2 Our main use of Theorem 16.3 will be to average over the input $\mathbf{X}$ (Fig. 16.1) in the case in which it is random and independent of V. In such cases, by setting $A = \mathbf{X}$ in (16.17) and letting $\mathcal{A}'$ contain all possible outcomes, we simply recover Theorem 16.1 with conditioning on $\mathbf{X}$ in the conditional entropy and mutual information terms. The approximate recovery version, Theorem 16.2, extends in the same way. In Section 16.4, we will discuss more advanced applications of Theorem 16.3, including (i) genie arguments, in which some information about V is revealed to the decoder, and (ii) typicality arguments, where we condition on V falling in some high-probability set.

16.3 Mutual Information Bounds

We saw in Section 16.2 that the mutual information $I(V; \widehat{V})$ naturally arises from Fano's inequality when V is uniform. More generally, we have $H(V|\widehat{V}) = H(V) - I(V; \widehat{V})$, so we can characterize the conditional entropy by characterizing both the entropy and the mutual information. In this section, we provide some of the main useful tools for upper-bounding the mutual information. For brevity, we omit the proofs of standard results commonly found in information-theory textbooks, or simple variations thereof.

Throughout the section, the random variables V and $\widehat{V}$ are assumed to be discrete, whereas the other random variables involved, including the inputs $\mathbf{X} = (X_1, \ldots, X_n)$ and samples $\mathbf{Y} = (Y_1, \ldots, Y_n)$, may be continuous. Hence, notation such as $P_Y(y)$ may represent either a probability mass function (PMF) or a probability density function (PDF).

16.3.1 Data-Processing Inequality

Recall the random variables V, $\mathbf{X}$, $\mathbf{Y}$, and $\widehat{V}$ in the multiple-hypothesis-testing reduction depicted in Fig. 16.1. In nearly all cases, the first step in bounding a mutual information term such as $I(V;\widehat{V})$ is to upper-bound it in terms of the samples $\mathbf{Y}$, and possibly the inputs $\mathbf{X}$. By doing so, we remove the dependence on $\widehat{V}$, and form a bound that is algorithm-independent.

The following lemma provides three variations along these lines. The three are all essentially equivalent, but are written separately since each will be more naturally suited to certain settings, as described below. Recall the terminology that $X \to Y \to Z$ *forms a Markov chain* if X and Z are conditionally independent given Y, or equivalently, Z depends on (X, Y) only through Y.

LEMMA 16.1 (Data-processing inequality)

(i) *If* $V \to \mathbf{Y} \to \widehat{V}$ *forms a Markov chain, then* $I(V;\widehat{V}) \leq I(V;\mathbf{Y})$.
(ii) *If* $V \to \mathbf{Y} \to \widehat{V}$ *forms a Markov chain conditioned on* $\mathbf{X}$, *then* $I(V;\widehat{V}|\mathbf{X}) \leq I(V;\mathbf{Y}|\mathbf{X})$.
(iii) *If* $V \to (\mathbf{X}, \mathbf{Y}) \to \widehat{V}$ *forms a Markov chain, then* $I(V;\widehat{V}) \leq I(V;\mathbf{X}, \mathbf{Y})$.

We will use the first part when $\mathbf{X}$ is absent or deterministic, the second part for random non-adaptive $\mathbf{X}$, and the third when the elements of $\mathbf{X}$ can be chosen adaptively on the basis of the past samples (Section 16.1.1).

16.3.2 Tensorization

One of the most useful properties of mutual information is *tensorization*. Under suitable conditional independence assumptions, mutual information terms containing length-n sequences (e.g., $\mathbf{Y} = (Y_1, \ldots, Y_n)$) can be upper-bounded by a sum of n mutual information terms, the ith of which contains the corresponding entry of each associated vector (e.g., Y_i). Thus, we can reduce a complicated mutual information term containing sequences to a sum of simpler terms containing individual elements. The following lemma provides some of the most common scenarios in which such tensorization can be performed.

LEMMA 16.2 (Tensorization of mutual information) *(i) If the entries of* $\mathbf{Y} = (Y_1, \ldots, Y_n)$ *are conditionally independent given* V, *then*

$$I(V;\mathbf{Y}) \leq \sum_{i=1}^{n} I(V;Y_i). \tag{16.18}$$

(ii) If the entries of $\mathbf{Y}$ *are conditionally independent given* $(V,\mathbf{X})$*, and* Y_i *depends on* $(V,\mathbf{X})$ *only through* (V,X_i)*, then*

$$I(V;\mathbf{Y}|\mathbf{X}) \le \sum_{i=1}^{n} I(V;Y_i|X_i). \tag{16.19}$$

(iii) If, in addition to the assumptions in part (ii), Y_i *depends on* (V,X_i) *only through* $U_i = \psi_i(V,X_i)$ *for some deterministic function* ψ_i*, then*

$$I(V;\mathbf{Y}|\mathbf{X}) \le \sum_{i=1}^{n} I(U_i;Y_i). \tag{16.20}$$

The proof is based on the sub-additivity of entropy, along with the conditional independence assumptions given. We will use the first part of the lemma when $\mathbf{X}$ is absent or deterministic, and the second and third parts for random non-adaptive $\mathbf{X}$. When $\mathbf{X}$ can be chosen adaptively on the basis of the past samples (Section 16.1.1), the following variant is used.

LEMMA 16.3 (Tensorization of mutual information for adaptive settings) *(i) If* X_i *is a function of* (X_1^{i-1},Y_1^{i-1})*, and* Y_i *is conditionally independent of* (X_1^{i-1},Y_1^{i-1}) *given* (V,X_i)*, then*

$$I(V;\mathbf{X},\mathbf{Y}) \le \sum_{i=1}^{n} I(V;Y_i|X_i). \tag{16.21}$$

(ii) If, in addition to the assumptions in part (i), Y_i *depends on* (V,X_i) *only through* $U_i = \psi_i(V,X_i)$ *for some deterministic function* ψ_i*, then*

$$I(V;\mathbf{X},\mathbf{Y}) \le \sum_{i=1}^{n} I(U_i;Y_i). \tag{16.22}$$

The proof is based on the chain rule for mutual information, i.e., $I(V;\mathbf{X},\mathbf{Y}) = \sum_{i=1}^{n} I(X_i,Y_i;V\,|\,X_1^{i-1},Y_1^{i-1})$, as well as suitable simplifications via the conditional independence assumptions.

REMARK 16.3 The mutual information bounds in Lemma 16.3 are analogous to those used in the problem of communication with feedback (see Section 7.12 of [34]). A key difference is that, in the latter setting, the channel input X_i is a function of (V,X_1^{i-1},Y_1^{i-1}), with V representing the message. In statistical estimation problems, the quantity V being estimated is typically unknown to the decision-maker, so the input X_i is only a function of (X_1^{i-1},Y_1^{i-1}).

REMARK 16.4 Lemma 16.3 should be applied with care, since, even if V is uniform on some set *a priori*, it may not be uniform conditioned on X_i. This is because, in the adaptive setting, X_i depends on Y_1^{i-1}, which in turn depends on V.

16.3.3 KL-Divergence-Based Bounds

By definition, the mutual information is the KL divergence between the joint distribution and the product of marginals, $I(V;Y) = D(P_{VY} \| P_V \times P_Y)$, and can equivalently be viewed as a conditional divergence $I(V;Y) = D(P_{Y|V} \| P_Y | P_V)$. Viewing the mutual information in this way leads to a variety of useful bounds in terms of related KL-divergence quantities, as the following lemma shows.

LEMMA 16.4 (KL-divergence-based bounds) *Let P_V, P_Y, and $P_{Y|V}$ be the marginal distributions corresponding to a pair (V,Y), where V is discrete. For any auxiliary distribution Q_Y, we have*

$$I(V;Y) = \sum_v P_V(v) D\Big(P_{Y|V}(\cdot|v) \Big\| P_Y\Big) \tag{16.23}$$

$$\leq \sum_v P_V(v) D\Big(P_{Y|V}(\cdot|v) \Big\| Q_Y\Big) \tag{16.24}$$

$$\leq \max_v D\Big(P_{Y|V}(\cdot|v) \Big\| Q_Y\Big), \tag{16.25}$$

and, in addition,

$$I(V;Y) \leq \sum_{v,v'} P_V(v) P_V(v') D\Big(P_{Y|V}(\cdot|v) \Big\| P_Y(\cdot|v')\Big) \tag{16.26}$$

$$\leq \max_{v,v'} D\Big(P_{Y|V}(\cdot|v) \Big\| P_{Y|V}(\cdot|v')\Big). \tag{16.27}$$

Proof We obtain (16.23) from the definition of mutual information, and (16.24) from the fact that $\mathbb{E}[\log(P_{Y|V}(Y|V)/P_Y(Y))] = \mathbb{E}[\log(P_{Y|V}(Y|V)/Q_Y(Y))] - \mathbb{E}[\log(P_Y(Y)/Q_Y(Y))]$; the second term here is a KL divergence, and is therefore non-negative. We obtain (16.26) from (16.24) by noting that Q_Y can be chosen to be any of the $P_Y(\cdot|v')$, and the remaining inequalities (16.25) and (16.27) are trivial.

The upper bounds in (16.24)–(16.27) are closely related, and often essentially equivalent in the sense that they lead to very similar converse bounds. In the authors' experience, it is usually slightly simpler to choose a suitable auxiliary distribution Q_Y and apply (16.25), rather than bounding the pair-wise divergences as in (16.27). Examples will be given in Sections 16.4 and 16.6.

REMARK 16.5 We have used the generic notation Y in Lemma 16.4, but in applications this may represent either the entire vector $\mathbf{Y}$, or a single one of its entries Y_i. Hence, the lemma may be used to bound $I(V;\mathbf{Y})$ directly, or one may first apply tensorization and then use the lemma to bound each $I(V;Y_i)$.

REMARK 16.6 Lemma 16.4 can also be used to bound *conditional* mutual information terms such as $I(V;Y|X)$. Conditioned on any $X = x$, we can upper-bound $I(V;Y|X = x)$ using Lemma 16.4, with an auxiliary distribution $Q_{Y|X=x}$ that may depend on x. For instance, doing this for (16.25) and then averaging over X, we obtain for any $Q_{Y|X}$ that

$$I(V;Y|X) \le \max_{v} D\Big(P_{Y|X,V}(\cdot\,|\,\cdot,v)\,\Big\|\,Q_{Y|X}\Big|P_X\Big) \tag{16.28}$$

$$\le \max_{x,v} D\Big(P_{Y|X,V}(\cdot\,|\,x,v)\,\Big\|\,Q_{Y|X}(\cdot\,|\,x)\Big). \tag{16.29}$$

The bound (16.25) in Lemma 16.4 is useful when there exists a single auxiliary distribution Q_Y that is "close" to each $P_{Y|V}(\cdot|v)$ in KL divergence, i.e., $D(P_{Y|V}(\cdot\,|\,v)\,\|\,Q_Y)$ is small. It is natural to extend this idea by introducing multiple auxiliary distributions, and requiring only that any one of them is close to a given $P_{Y|V}(\cdot|v)$. This can be viewed as "covering" the conditional distributions $\{P_{Y|V}(\cdot|v)\}_{v\in\mathcal{V}}$ with "KL-divergence balls," and we will return to this viewpoint in Section 16.5.3.

LEMMA 16.5 (Mutual information bound via covering) *Under the setup of Lemma 16.4, suppose there exist N distributions $Q_1(y),\dots,Q_N(y)$ such that, for all v and some $\epsilon > 0$, it holds that*

$$\min_{j=1,\dots,N} D\Big(P_{Y|V}(\cdot\,|\,v)\,\Big\|\,Q_j\Big) \le \epsilon. \tag{16.30}$$

Then we have

$$I(V;Y) \le \log N + \epsilon. \tag{16.31}$$

The proof is based on applying (16.24) with $Q_Y(y) = (1/N)\sum_{j=1}^{N} Q_j(y)$, and then lower-bounding this summation over j by the value $j^*(v)$ achieving the minimum in (16.30). We observe that setting $N = 1$ in Lemma 16.5 simply yields (16.25).

16.3.4 Relations between KL Divergence and Other Measures

As evidenced above, the KL divergence plays a crucial role in applications of Fano's inequality. In some cases, directly characterizing the KL divergence can still be difficult, and it is more convenient to bound it in terms of other divergences or distances. The following lemma gives a few simple examples of such relations; the reader is referred to [36] for a more thorough treatment.

LEMMA 16.6 (Relations between divergence measures) *Fix two distributions P and Q, and consider the KL divergence $D(P\|Q) = \mathbb{E}_P[\log(P(Y)/Q(Y))]$, total variation (TV) $d_{TV}(P,Q) = \frac{1}{2}\mathbb{E}_Q[|P(Y)/Q(Y) - 1|]$, squared Hellinger distance $H^2(P,Q) = \mathbb{E}_Q[(\sqrt{P(Y)/Q(Y)} - 1)^2]$, and χ^2-divergence $\chi^2(P\|Q) = \mathbb{E}_Q[(P(Y)/Q(Y) - 1)^2]$. We have*

- *(KL versus TV) $D(P\|Q) \ge 2d_{TV}(P,Q)^2$, whereas if P and Q are probability mass functions and each entry of Q is at least $\eta > 0$, then $D(P\|Q) \le (2/\eta)d_{TV}(P,Q)^2$;*
- *(Hellinger versus TV) $\frac{1}{2}H^2(P,Q) \le d_{TV}(P,Q) \le H(P,Q)\sqrt{1 - H^2(P,Q)/4}$;*
- *(KL versus χ^2-divergence) $D(P\|Q) \le \log(1 + \chi^2(P\|Q)) \le \chi^2(P\|Q)$.*

16.4 Applications – Discrete Settings

In this section, we provide two examples of statistical estimation problems in which the quantity being estimated is discrete: group testing and graphical model selection. Our goal is not to treat these problems comprehensively, but rather to study particular instances that permit a simple analysis while still illustrating the key ideas and tools introduced in the previous sections. We consider the *high-dimensional* setting, in which the underlying number of parameters being estimated is much higher than the number of measurements. To simplify the final results, we will often write them using the asymptotic notation $o(1)$ for asymptotically vanishing terms, but non-asymptotic variants are easily inferred from the proofs.

16.4.1 Group Testing

The group-testing problem consists of determining a small subset of "defective" items within a larger set of items on the basis of a number of pooled tests. A given test contains some subset of the items, and the binary test outcome indicates, possibly in a noisy manner, whether or not *at least one* defective item was included in the test. This problem has a history in medical testing [37], and has regained significant attention following applications in communication protocols, pattern matching, database systems, and more.

In more detail, the setup is described as follows.

- In a population of p items, there are k unknown *defective items*. This defective set is denoted by $S \subseteq \{1,\dots,p\}$, and is assumed to be uniform on the set of $\binom{p}{k}$ subsets having cardinality k. Hence, in this example, we are in the Bayesian setting with a uniform prior. We focus on the sparse setting, in which $k \ll p$, i.e., defective items are rare.
- There are n tests specified by a *test matrix* $\mathbf{X} \in \{0,1\}^{n\times p}$. The (i, j)th entry of $\mathbf{X}$, denoted by X_{ij}, indicates whether item j is included in test i. We initially consider the *non-adaptive* setting, where $\mathbf{X}$ is chosen in advance. We allow this choice to be random; for instance, a common choice of random design is to let the entries of $\mathbf{X}$ be i.i.d. Bernoulli random variables.
- To account for possible noise, we consider the following observation model:

$$Y_i = \left(\bigvee_{j\in S} X_{ij}\right) \oplus Z_i, \tag{16.32}$$

  where $Z_i \sim \text{Bernoulli}(\epsilon)$ for some $\epsilon \in [0, \frac{1}{2})$, $\oplus$ denotes modulo-2 addition, and $\vee$ is the "OR" operation. In the channel coding terminology, this corresponds to passing the noiseless test outcome $\bigvee_{j\in S} X_{ij}$ through a binary symmetric channel. We assume that the noise variables Z_i are independent of each other and of $\mathbf{X}$, and we define the vector of test outcomes $\mathbf{Y} = (Y_1, \dots, Y_n)$.
- Given $\mathbf{X}$ and $\mathbf{Y}$, a decoder forms an estimate $\widehat{S}$ of S. We initially consider the exact recovery criterion, in which the error probability is given by

$$P_e = \mathbb{P}[\widehat{S} \neq S], \tag{16.33}$$

  where the probability with respect to S, $\mathbf{X}$, and $\mathbf{Y}$.

In the following sections, we present several results and analysis techniques that are primarily drawn from [2, 3].

Exact Recovery with Non-Adaptive Testing

Under the exact recovery criterion (16.33), we have the following lower bound on the required number of tests. Recall that $H_2(\alpha) = \alpha\log(1/\alpha) + (1-\alpha)\log(1/(1-\alpha))$ denotes the binary entropy function.

THEOREM 16.4 (Group testing with exact recovery) *Under the preceding noisy group-testing setup, in order to achieve $P_e \le \delta$, it is necessary that*

$$n \ge \frac{k\log(p/k)}{\log 2 - H_2(\epsilon)}(1-\delta-o(1)) \tag{16.34}$$

as $p \to \infty$, possibly with $k \to \infty$ simultaneously.

Proof Since S is discrete-valued, we can use the trivial reduction to multiple hypothesis testing with $V = S$. Applying Fano's inequality (Theorem 16.1) with conditioning on $\mathbf{X}$ (Section 16.2.3), we obtain

$$I(S;\mathbf{Y}|\mathbf{X}) \ge (1-\delta)\log\binom{p}{k} - \log 2, \tag{16.35}$$

where we have also upper-bounded $I(S;\widehat{S}|\mathbf{X}) \le I(S;\mathbf{Y}|\mathbf{X})$ using the data-processing inequality (from the second part of Lemma 16.1), which in turn uses the fact that $S \to \mathbf{Y} \to \widehat{S}$ conditioned on $\mathbf{X}$.

Let $U_i = \bigvee_{j\in S} X_{ij}$ denote the hypothetical noiseless outcome. Since the noise variables $\{Z_i\}_{i=1}^n$ are independent and Y_i depends on $(S,\mathbf{X})$ only through U_i (see (16.32)), we can apply tensorization (from the third part of Lemma 16.2) to obtain

$$I(S;\mathbf{Y}|\mathbf{X}) \le \sum_{i=1}^{n} I(U_i;Y_i) \tag{16.36}$$

$$\le n(\log 2 - H_2(\epsilon)), \tag{16.37}$$

where (16.37) follows since Y_i is generated from U_i according to a binary symmetric channel, which has capacity $\log 2 - H_2(\epsilon)$. By substituting (16.37) and $\binom{p}{k} \ge (p/k)^k$ into (16.35) and rearranging, we obtain (16.34).

Theorem 16.4 is known to be tight in terms of scaling laws whenever $\delta \in (0,1)$ is fixed and $k = o(p)$, and, perhaps more interestingly, tight including constant factors as $\delta \to 0$ under the scaling $k = O(p^\theta)$ for sufficiently small $\theta > 0$. The matching achievability result in this regime can be proved using maximum-likelihood decoding [38]. However, achieving such a result using a computationally efficient decoder remains a challenging problem.

Approximate Recovery with Non-Adaptive Testing

We now move to an approximate recovery criterion. The decoder outputs a list $\mathcal{L} \subseteq \{1,\dots,p\}$ of cardinality $L \ge k$, and we require that at least a fraction $(1-\alpha)k$ of the defective items appear in the list, for some $\alpha \in (0,1)$. It follows that the error probability can be written as

$$P_e(t) = \mathbb{P}[d(S,\mathcal{L}) > t], \tag{16.38}$$

where $d(S,\mathcal{L}) = |S \setminus \mathcal{L}|$, and $t = \alpha k$. Notice that a higher value of L means that more non-defective items may be included in the list, whereas a higher value of α means that more defective items may be absent.

THEOREM 16.5 (Group testing with approximate recovery) *Under the preceding noisy group-testing setup with list size $L \geq k$, in order to achieve $P_e(\alpha k) \leq \delta$ for some $\alpha \in (0,1)$ (not depending on p), it is necessary that*

$$n \geq \frac{(1-\alpha)k\log(p/L)}{\log 2 - H_2(\epsilon)}(1-\delta-o(1)) \tag{16.39}$$

as $p \to \infty$, $k \to \infty$, and $L \to \infty$ simultaneously with $L = o(p)$.

Proof We apply the approximate recovery version of Fano's inequality (Theorem 16.2) with $d(S,\mathcal{L}) = |S \setminus \mathcal{L}|$ and $t = \alpha k$ as above. For any $\mathcal{L}$ with cardinality L, the number of S with $d(S,\mathcal{L}) \leq \alpha k$ is given by $N_{\max}(t) = \sum_{j=0}^{\lfloor \alpha k \rfloor} \binom{p-L}{j}\binom{L}{k-j}$, which follows by counting the number of ways to place $k-j$ defective items in $\mathcal{L}$, and the remaining j defective items in the other $p-L$ entries. Hence, using Theorem 16.2 with conditioning on $\mathbf{X}$ (see Section 16.2.3), and applying the data-processing inequality (from the second part of Lemma 16.1), we obtain

$$I(S;\mathbf{Y}|\mathbf{X}) \geq (1-\delta)\log\left(\frac{\binom{p}{k}}{\sum_{j=0}^{\lfloor \alpha k \rfloor}\binom{p-L}{j}\binom{L}{k-j}}\right) - \log 2. \tag{16.40}$$

By upper-bounding the summation by $\lfloor \alpha k \rfloor + 1$ times the maximum value, and performing some asymptotic simplifications via the assumption $L = o(p)$, we can simplify the logarithm to $(k\log(p/L))(1+o(1))$ [39]. The theorem is then established by upper-bounding the conditional mutual information using (16.37).

Theorem 16.5 matches Theorem 16.4 up to the factor of $1-\alpha$ and the replacement of $\log(p/k)$ by $\log(p/L)$, suggesting that approximate recovery provides a minimal reduction in the number of tests even for moderate values of α and L. However, under approximate recovery, a near-matching achievability bound is known under the scaling $k = O(p^\theta)$ *for all* $\theta \in (0,1)$, rather than only for sufficiently small θ [38].

Adaptive Testing

Next, we discuss the adaptive-testing setting, in which a given input vector $X_i \in \{0,1\}^p$, corresponding to a single row of $\mathbf{X}$, is allowed to depend on the previous inputs and outcomes, i.e., $X_1^{i-1} = (X_1,\dots,X_{i-1})$ and $Y_1^{i-1} = (Y_1,\dots,Y_{i-1})$. In fact, it turns out that Theorems 16.4 and 16.5 still apply in this setting. Establishing this simply requires making the following modifications to the above analysis.

- Apply the data-processing inequality in the form of the *third* part of Lemma 16.1, yielding (16.35) and (16.40) with $I(S;\mathbf{X},\mathbf{Y})$ in place of $I(S;\mathbf{Y}|\mathbf{X})$.
- Apply tensorization via Lemma 16.3 to deduce (16.36) and (16.37) with $I(S;\mathbf{X},\mathbf{Y})$ in place of $I(S;\mathbf{Y}|\mathbf{X})$.

In the regimes where Theorems 16.4 and/or 16.5 are known to have matching upper bounds with non-adaptive designs, we can clearly deduce that adaptivity provides no asymptotic gain. However, as with approximate recovery, adaptivity can significantly broaden the conditions under which matching achievability bounds are known, at least in the noiseless setting [40].

Discussion: General Noise Models

The preceding analysis can easily be extended to more general group-testing models in which the observations $(Y_1, \ldots, Y_n)$ are conditionally independent given $\mathbf{X}$. A broad class of such models can be written in the form $(Y_i|N_i) \sim P_{Y|N}$, where $N_i = \sum_{j \in S} \mathbb{1}\{X_{ij} = 1\}$ denotes the number of defective items in the ith test. In such cases, the preceding results hold true more generally when $\log 2 - H_2(\epsilon)$ is replaced by the capacity $\max_{P_N} I(N;Y)$ of the "channel" $P_{Y|N}$.

For certain models, we can obtain a better lower bound by applying a *genie argument*, along with the conditional form of Fano's inequality in Theorem 16.3. Fix $\ell \in \{1, \ldots, k\}$, and suppose that a uniformly random subset $S^{(1)} \subseteq S$ of cardinality $k - \ell$ is revealed to the decoder. This extra information can only make the group-testing problem easier, so any converse bound for this modified setting remains valid for the original setting. Perhaps counter-intuitively, this idea can lead to a better final bound.

We only briefly outline the details of this more general analysis, and refer the interested reader to [3, 41]. Using Theorem 16.3 with $A = S^{(1)}$, and applying the data-processing inequality and tensorization, one can obtain

$$P_e \geq 1 - \frac{\sum_{i=1}^{n} I(N_i^{(0)}; Y_i | N_i^{(1)}) - \log 2}{\log \binom{p-k+\ell}{\ell}}, \tag{16.41}$$

where $N_i^{(1)} = \sum_{j \in S^{(1)}} \mathbb{1}\{X_{ij} = 1\}$, and $N_i^{(0)} = N_i - N_i^{(1)}$. The intuition is that we condition on $N_i^{(1)}$ since it is known via the genie, while the remaining information about Y_i is determined by $N_i^{(0)}$. Once (16.41) has been established, it remains only to simplify the mutual information terms; see [3, 41] for further details.

16.4.2 Graphical Model Selection

Graphical models provide compact representations of the conditional independence relations between random variables, and frequently arise in areas such as image processing, statistical physics, computational biology, and natural-language processing. The fundamental problem of *graphical model selection* consists of recovering the graph structure given a number of independent samples from the underlying distribution.

Graphical model selection has been studied under several different families of joint distributions, and also several different graph classes. We focus our attention on the commonly used *Ising model* with binary observations, and on a simple graph class known as *forests*, defined to contain the graphs having no cycles.

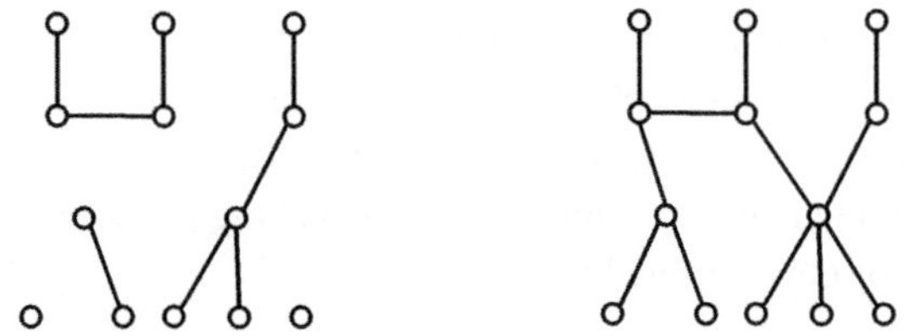

Figure 16.2 Two examples of graphs that are forests (i.e., acyclic graphs); the graph on the right is also a tree (i.e., a connected acyclic graph).

Formally, the setup is described as follows.

- We are given n independent samples $Y_1,\dots,Y_n$ from a p-dimensional joint distribution: $Y_i = (Y_{i1},\dots,Y_{ip})$ for $i = 1,\dots,n$. This joint distribution is encoded by a graph $G = (V,E)$, where $V = \{1,\dots,p\}$ is the *vertex set*, and $E \subseteq V \times V$ is the *edge set*. We use the terminology *vertex* and *node* interchangeably. We assume that there are no edges from a vertex to itself, and that the edges are *undirected*: $(i,j) \in E$ and $(j,i) \in E$ are equivalent, and count as only one edge.
- We focus on the *Ising model*, in which the observations are binary-valued, and the joint distribution of a given sample, say $Y_1 = (Y_{11},\dots,Y_{1p}) \in \{-1,1\}^p$, is

$$P_G(y_1) = \frac{1}{Z}\exp\left(\lambda \sum_{(i,j)\in E} y_{1i}y_{1j}\right), \tag{16.42}$$

where Z is a normalizing constant. Here $\lambda > 0$ is a parameter to the distribution dictating the edge strength; a higher value means it is more likely that $Y_{1i} = Y_{1j}$ for any given edge $(i,j) \in E$.
- We restrict the graph $G = (V,E)$ to be the set of all *forests*:

$$\mathcal{G}_{\text{forest}} = \{G : G \text{ has no cycles}\}, \tag{16.43}$$

where a *cycle* is defined to be a path of distinct edges leading back to the start node, e.g., $(1,4),(4,2),(2,1)$. A special case of a forest is a *tree*, which is an acyclic graph for which a path exists between any two nodes. One can view any forest as being a disjoint union of trees, each defined on some subset of V. See Fig. 16.2 for an illustration.
- Let $\mathbf{Y} \in \{-1,1\}^{n\times p}$ be the matrix whose ith row contains the p entries of the ith sample. Given $\mathbf{Y}$, a decoder forms an estimate $\widehat{G}$ of G, or equivalently, an estimate $\widehat{E}$ of E. We initially focus on the exact-recovery criterion, in which the minimax error probability is given by

$$\mathcal{M}_n(\mathcal{G}_{\text{forest}},\lambda) = \inf_{\widehat{G}} \sup_{G\in\mathcal{G}_{\text{forest}}} \mathbb{P}_G[\widehat{G} \neq G], \tag{16.44}$$

where $\mathbb{P}_G$ denotes the probability when the true graph is G, and the infimum is over all estimators.

To the best of our knowledge, Fano's inequality has not been applied previously in this exact setup; we do so using the general tools for Ising models given in [26, 27, 42, 43].

Exact Recovery

Under the exact-recovery criterion, we have the following.

THEOREM 16.6 (Exact recovery of forest graphical models) *Under the preceding Ising graphical model selection setup with a given edge parameter $\lambda > 0$, in order to achieve $\mathcal{M}_n(\mathcal{G}_{\text{forest}}, \lambda) \le \delta$, it is necessary that*

$$n \ge \max\left\{\frac{\log p}{\log 2}, \frac{2\log p}{\lambda \tanh \lambda}\right\}(1 - \delta - o(1)) \tag{16.45}$$

as $p \to \infty$.

Proof Recall from Section 16.1.1 that we can lower-bound the worst-case error probability over $\mathcal{G}_{\text{forest}}$ by the average error probability over any subset of $\mathcal{G}_{\text{forest}}$. This gives us an important degree of freedom in the reduction to multiple hypothesis testing, and corresponds to selecting a hard subset $\theta_1, \ldots, \theta_M$ as described in Section 16.1.1. We refer to a given subset $\mathcal{G} \subseteq \mathcal{G}_{\text{forest}}$ as a *graph ensemble*, and provide two choices that lead to the two terms in (16.45).

For any choice of $\mathcal{G} \subseteq \mathcal{G}_{\text{forest}}$, Fano's inequality (Theorem 16.1) gives

$$n \ge \frac{(1-\delta)\log|\mathcal{G}| - \log 2}{I(G;Y_1)}, \tag{16.46}$$

for G uniform on $\mathcal{G}$, where we used $I(G;\widehat{G}) \le I(G;\mathbf{Y}) \le nI(G;Y_1)$ by the data-processing inequality and tensorization (from the first parts of Lemmas 16.1 and 16.2).

Restricted Ensemble 1. Let $\mathcal{G}_1$ be the set of all trees. It is well known from graph theory that the number of trees on p nodes is $|\mathcal{G}_1| = p^{p-2}$ [44]. Moreover, since Y_1 is a length-p binary sequence, we have $I(G;Y_1) \le H(Y_1) \le p\log 2$. Hence, (16.46) yields $n \ge ((1-\delta)(p-2)\log p - \log 2)/(p\log 2)$, implying the first bound in (16.45).

Restricted Ensemble 2. Let $\mathcal{G}_2$ be the set of graphs containing a single edge, so that $|\mathcal{G}_2| = \binom{p}{2}$. We will upper-bound the mutual information using (16.25) in Lemma 16.4, choosing the auxiliary distribution Q_Y to be $P_{\overline{G}}$, with $\overline{G}$ being the empty graph. Thus, we need to bound $D(P_G \| P_{\overline{G}})$ for each $G \in \mathcal{G}_2$.

We first give an upper bound on $D(P_G \| P_{\overline{G}})$ for *any* two graphs $(G, \overline{G})$. We start with the trivial bound

$$D(P_G \| P_{\overline{G}}) \le D(P_G \| P_{\overline{G}}) + D(P_{\overline{G}} \| P_G). \tag{16.47}$$

Recall the definition $D(P\|Q) = \mathbb{E}_P[\log(P(Y)/Q(Y))]$, and consider the substitution of P_G and $P_{\overline{G}}$ according to (16.42), with different normalizing constants Z_G and $Z_{\overline{G}}$. We see that when we sum the two terms in (16.47), the normalizing constants inside the logarithms cancel out, and we are left with

$$\begin{aligned} D(P_G \| P_{\overline{G}}) \le & \sum_{(i,j)\in E\setminus \overline{E}} \lambda(\mathbb{E}_G[Y_{1i}Y_{1j}] - \mathbb{E}_{\overline{G}}[Y_{1i}Y_{1j}]) \\ & + \sum_{(i,j)\in \overline{E}\setminus E} \lambda(\mathbb{E}_{\overline{G}}[Y_{1i}Y_{1j}] - \mathbb{E}_G[Y_{1i}Y_{1j}]) \end{aligned} \tag{16.48}$$

for $G = (V, E)$ and $\overline{G} = (V, \overline{E})$.

In the case that G has a single edge (i.e., $G \in \mathcal{G}_2$) and $\overline{G}$ is the empty graph, we can easily compute $\mathbb{E}_{\overline{G}}[Y_{1i}Y_{1j}] = 0$, and (16.48) simplifies to

$$D(P_G \| P_{\overline{G}}) \leq \lambda \mathbb{E}_G[Y_{1i}Y_{1j}], \tag{16.49}$$

where (i, j) is the unique edge in G. Since Y_{1i} and Y_{1j} only take values in $\{-1, 1\}$, we have $\mathbb{E}_G[Y_{1i}Y_{1j}] = (+1)\mathbb{P}[Y_{1i} = Y_{1j}] + (-1)\mathbb{P}[Y_{1i} \neq Y_{1j}] = 2\mathbb{P}[Y_{1i} = Y_{1j}] - 1$, and letting E have a single edge in (16.42) yields $\mathbb{P}_G[(Y_{1i}, Y_{1j}) = (y_i, y_j)] = e^{\lambda y_i y_j}/(2e^{\lambda} + 2e^{-\lambda})$, and hence $\mathbb{P}_G[Y_{1i} = Y_{1j}] = e^{\lambda}/(e^{\lambda} + e^{-\lambda})$. Combining this with $\mathbb{E}_G[Y_{1i}Y_{1j}] = 2\mathbb{P}[Y_{1i} = Y_{1j}] - 1$ yields $\mathbb{E}_G[Y_{1i}Y_{1j}] = 2e^{\lambda}/(e^{\lambda} + e^{-\lambda}) - 1 = \tanh\lambda$. Hence, using (16.49) along with (16.25) in Lemma 16.4, we obtain $I(G; Y_1) \leq \lambda \tanh\lambda$. Substitution into (16.46) (with $\log|\mathcal{G}| = (2\log p)(1 + o(1)))$ yields the second bound in (16.45).

Theorem 16.6 is known to be tight up to constant factors whenever $\lambda = O(1)$ [44, 45]. When λ is constant, the lower bound becomes $n = \Omega(\log p)$, whereas for asymptotically vanishing λ it simplifies to $n = \Omega((1/\lambda^2)\log p)$.

Approximate Recovery

We consider the approximate recovery of $G = (V, E)$ with respect to the *edit distance* $d(G, \widehat{G}) = |E \backslash \widehat{E}| + |\widehat{E} \backslash E|$, which is the number of edge additions and removals needed to transform G into $\widehat{G}$ or vice versa. Since any forest can have at most $p - 1$ edges, it is natural to consider the case in which an edit distance of up to αp is permitted, for some $\alpha > 0$. Hence, the minimax risk is given by

$$\mathcal{M}_n(\mathcal{G}_{\text{forest}}, \lambda, \alpha) = \inf_{\widehat{G}} \sup_{G \in \mathcal{G}_{\text{forest}}} \mathbb{P}_G[d(G, \widehat{G}) > \alpha p]. \tag{16.50}$$

In this setting, we have the following.

THEOREM 16.7 (Approximate recovery of forest graphical models) *Under the preceding Ising graphical model selection setup with a given edge parameter $\lambda > 0$ and approximate recovery parameter $\alpha \in (0, \frac{1}{2})$ (with the latter not depending on p), in order to achieve $\mathcal{M}_n(\mathcal{G}_{forest}, \lambda, \alpha) \leq \delta$, it is necessary that*

$$n \geq \max\left\{\frac{(1-\alpha)\log p}{\log 2}, \frac{2(1-\alpha)\log p}{\lambda \tanh\lambda}\right\}(1 - \delta - o(1)) \tag{16.51}$$

as $p \to \infty$.

Proof For any $\mathcal{G} \subseteq \mathcal{G}_{\text{forest}}$, Theorem 16.2 provides the following analog of (16.46):

$$n \geq \frac{(1-\delta)\log(|\mathcal{G}|/N_{\max}(\alpha p)) - \log 2}{I(G; Y_1)} \tag{16.52}$$

for G uniform on $\mathcal{G}$, where $N_{\max}(t) = \max_{\widehat{G}} \sum_{G \in \mathcal{G}} \mathbb{1}\{d(G, \widehat{G}) \leq t\}$ implicitly depends on $\mathcal{G}$. We again consider two restricted ensembles; the first is identical to the exact recovery setting, whereas the second is modified due to the fact that learning single-edge graphs with approximate recovery is trivial.

Restricted Ensemble 1. Once again, let $\mathcal{G}_1$ be the set of all trees. We have already established $|\mathcal{G}_1| = (p - 2)\log p$ and $I(G; Y_1) \leq n\log 2$ for this ensemble, so it remains only to characterize $N_{\max}(\alpha p)$.

While the decoder may output a graph $\widehat{G}$ not lying in $\mathcal{G}_1$, we can assume without loss of generality that $\widehat{G}$ is always selected such that $d(\widehat{G}, G^*) \le \alpha p$ for some $G^* \in \mathcal{G}_1$; otherwise, an error would be guaranteed. As a result, for any $\widehat{G}$, and any $G \in \mathcal{G}_1$ such that $d(G, \widehat{G}) \le \alpha p$, we have from the triangle inequality that $d(G, G^*) \le d(G, \widehat{G}) + d(\widehat{G}, G^*) \le 2\alpha p$, which implies that

$$N_{\max}(\alpha p) \le \sum_{G \in \mathcal{G}_1} \mathbb{1}\{d(G, G^*) \le 2\alpha p\}. \tag{16.53}$$

Now observe that, since all graphs in $\mathcal{G}_1$ have exactly $p-1$ edges, transforming G to G^* requires removing j edges and adding j different edges, for some $j \le \alpha p$. Hence, we have

$$N_{\max}(\alpha p) \le \sum_{j=0}^{\lfloor \alpha p \rfloor} \binom{p-1}{j} \binom{\binom{p}{2} - p + 1}{j}. \tag{16.54}$$

By upper-bounding the summation by $\lfloor \alpha p \rfloor + 1$ times the maximum, and performing some asymptotic simplifications, we can show that $\log N_{\max}(\alpha p) \le (\alpha p \log p)(1 + o(1))$. By substituting into (16.52) and recalling that $|\mathcal{G}_1| = (p-2)\log p$ and $I(G; Y_1) \le p \log 2$, we obtain the first bound in (16.51).

Restricted Ensemble 2a. Let $\mathcal{G}_{2a}$ be the set of all graphs on p nodes containing exactly $p/2$ isolated edges; if p is an odd number, the same analysis applies with an arbitrary single node ignored. We proceed by characterizing $|\mathcal{G}_{2a}|$, $I(G; Y_1)$, and $N_{\max}(\alpha p)$. The number of graphs in the ensemble is $|\mathcal{G}_{2a}| = \binom{p}{2}\binom{p-2}{2} \cdots \binom{4}{2}\binom{2}{2} = p!/2^{p/2}$, and Stirling's approximation yields $\log |\mathcal{G}_{2a}| \ge (p \log p)(1 + o(1))$.

Since the KL divergence is additive for product distributions, and we established in the exact-recovery case that the KL divergence between the distributions of a single-edge graph and an empty graph is at most $\lambda \tanh \lambda$, we deduce that $D(P_G \| P_{\overline{G}}) \le (p/2)\lambda \tanh \lambda$ for any $G \in \mathcal{G}_{2a}$, where $\overline{G}$ is the empty graph. We therefore obtain from Lemma 16.4 that $I(G; Y_1) \le (p/2)\lambda \tanh \lambda$.

A similar argument to that of Ensemble 1 yields $N_{\max}(\alpha p) \le \sum_{j=0}^{\lfloor \alpha p \rfloor} \binom{p/2}{j}\binom{\binom{p}{2} - p/2}{j}$, in analogy with (16.54). This again simplifies to $N_{\max}(\alpha p) \le (\alpha p \log p)(1 + o(1))$, and, having established $\log |\mathcal{G}_{2a}| \ge (p \log p)(1 + o(1))$ and $I(G; Y_1) \le (p/2)\lambda \tanh \lambda$, substitution into (16.52) yields the second bound in (16.51).

The bound in Theorem 16.7 matches that of Theorem 16.6 up to a multiplicative factor of $1 - \alpha$, thus suggesting that approximate recovery does not significantly help in reducing the required number of samples, at least in the minimax sense, for the Ising model and forest graph class.

Adaptive Sampling

We now return to the exact-recovery setting, and consider a modification in which we have an added degree of freedom in the form of *adaptive sampling*.

- The algorithm proceeds in rounds; in round i, the algorithm queries a subset of the p nodes indexed by $X_i \in \{0, 1\}^p$, and the corresponding sample Y_i is generated as follows.

- The joint distribution of the entries of Y_i, corresponding to the entries where X_i is one, coincides with the corresponding marginal distribution of P_G, with independence between rounds.
- The values of the entries of Y_i, corresponding to the entries where X_i is zero, are given by $*$, a symbol indicating that the node was not observed.

We allow X_i to be selected on the basis of past queries and samples, namely, $X_1^{i-1} = (X_1, \ldots, X_{i-1})$ and $Y_1^{i-1} = (Y_1, \ldots, Y_{i-1})$.

- Let $n(X_i)$ denote the number of ones in X_i, i.e., the number of nodes observed in round i. While we allow the total number of rounds to vary, we restrict the algorithm to output an estimate $\widehat{G}$ after observing at most n_{node} nodes. This quantity is related to n in the non-adaptive setting according to $n_{\text{node}} = np$, since in the non-adaptive setting we always observe all p nodes in each sample.
- The minimax risk is given by

$$\mathcal{M}_{n_{\text{node}}}(\mathcal{G}_{\text{forest}}, \lambda) = \inf_{\widehat{G}} \sup_{G \in \mathcal{G}_{\text{forest}}} \mathbb{P}_G[\widehat{G} \neq G], \tag{16.55}$$

where the infimum is over all adaptive algorithms that observe at most n_{node} nodes in total.

THEOREM 16.8 (Adaptive sampling for forest graphical models) *Under the preceding Ising graphical model selection problem with adaptive sampling and a given parameter $\lambda > 0$, in order to achieve $\mathcal{M}_{n_{node}}(\mathcal{G}_{forest}, \lambda) \leq \delta$, it is necessary that*

$$n_{node} \geq \max\left\{\frac{p \log p}{\log 2}, \frac{2p \log p}{\lambda \tanh \lambda}\right\}(1 - \delta - o(1)) \tag{16.56}$$

as $p \to \infty$.

Proof We prove the result using Ensemble 1 and Ensemble 2a above. We let N denote the number of rounds; while this quantity is allowed to vary, we can assume without loss of generality that $N = n_{\text{node}}$ by adding or removing rounds where no nodes are queried. For any subset $\mathcal{G} \subseteq \mathcal{G}_{\text{forest}}$, applying Fano's inequality (Theorem 16.1) and tensorization (from the first part of Theorem 16.3) yields

$$\sum_{i=1}^{N} I(G; Y_i | X_i) \geq (1 - \delta) \log |\mathcal{G}| - \log 2, \tag{16.57}$$

where G is uniform on $\mathcal{G}$.

Restricted Ensemble 1. We again let $\mathcal{G}_1$ be the set of all trees, for which we know that $|\mathcal{G}| = p^{p-2}$. Since the $n(X_i)$ entries of Y_i differing from $*$ are binary, and those equaling $*$ are deterministic given X_i, we have $I(G; Y_i | X_i = x_i) \leq n(x_i) \log 2$. Averaging over X_i and summing over i yields $\sum_{i=1}^{N} I(G; Y_i | X_i) \leq \sum_{i=1}^{N} \mathbb{E}[n(X_i)] \log 2 \leq n_{\text{node}} \log 2$, and substitution into (16.57) yields the first bound in (16.56).

Restricted Ensemble 2. We again use the above-defined ensemble $\mathcal{G}_{2a}$ of graphs with $p/2$ isolated edges, for which we know that $|\mathcal{G}_{2a}| \geq (p \log p)(1 + o(1))$. In this case, when we observe $n(X_i)$ nodes, the subgraph corresponding to these observed nodes has at most $n(X_i)/2$ edges, all of which are isolated. Hence, using Lemma 16.4, the above-established fact that the KL divergence from a single-edge graph to the empty graph is at most $\lambda \tanh \lambda$, and the additivity of KL divergence for product distributions, we deduce that $I(G; Y_i | X_i = x_i) \le (n(x_i)/2)\lambda \tanh \lambda$. Averaging over X_i and summing over i yields $\sum_{i=1}^{N} I(G; Y_i | X_i) \le \frac{1}{2} n_{\text{node}} \lambda \tanh \lambda$, and substitution into (16.57) yields the second bound in (16.56).

The threshold in Theorem 16.8 matches that of Theorem 16.6, and, in fact, a similar analysis under approximate recovery also recovers the threshold in Theorem 16.7. This suggests that adaptivity is of limited help in the minimax sense for the Ising model and forest graph class. There are, however, other instances of graphical model selection where adaptivity provably helps [43, 46].

Discussion: Other Graph Classes

Degree and edge constraints. While the class $\mathcal{G}_{\text{forest}}$ is a relatively easy class to handle, similar techniques have also been used for more difficult classes, notably including those that place restrictions on the maximal degree d and/or the number of edges k. Ensembles 2 and 2a above can again be used, and the resulting bounds are tight in certain scaling regimes where $\lambda \to 0$, but loose in other regimes due to their lack of dependence on d and k. To obtain bounds with such a dependence, alternative ensembles consisting of subgraphs with highly correlated nodes have been proposed [26, 27, 42].

For instance, suppose that a group of $d+1$ nodes has all possible edges connected except one. Unless d or the edge strength λ is small, the high connectivity makes the nodes very highly correlated, and the subgraph is difficult to distinguish from a fully connected subgraph. This is in contrast with Ensembles 2 and 2a above, whose graphs are difficult to distinguish from the empty graph.

Bayesian setting. Beyond minimax estimation, it is also of interest to understand the fundamental limits of random graphs. A particularly prominent example is the Erdős–Rényi random graph, in which each edge is independently included with some probability $q \in (0, 1)$. This is a case where the conditional form of Fano's inequality has proved useful; specifically, one can apply Theorem 16.3 with $A = G$, and $\mathcal{A}$ equal to the following *typical set* of graphs:

$$\mathcal{T} = \left\{ G : (1-\epsilon) q \binom{p}{2} \le |E| \le (1+\epsilon) q \binom{p}{2} \right\}, \tag{16.58}$$

where $\epsilon > 0$ is a constant. Standard properties of typical sets [34] yield that $\mathbb{P}[G_{\text{ER}} \in \mathcal{T}] \to 1$, $|\mathcal{T}| = e^{(H_2(q)\binom{p}{2})(1+O(\epsilon))}$, and $H(V | V \in \mathcal{T}) = (H_2(q)\binom{p}{2})(1 + O(\epsilon))$ whenever $q\binom{p}{2} \to \infty$, and once these facts have been established, Theorem 16.3 yields the following necessary condition for $P_{\text{e}} \le \delta$:

$$n \ge \frac{p H_2(q)}{2 \log 2} (1 - \delta - o(1)). \tag{16.59}$$

For instance, in the case that $q = O(1/p)$ (i.e., there are $O(p)$ edges on average), we have $H_2(q) = \Theta((\log p)/p)$, and we find that $n = \Omega(\log p)$ samples are necessary. This scaling is tight when λ is constant [45], whereas improved bounds for other scalings can be found in [27].

16.5 From Discrete to Continuous

Thus far, we have focused on using Fano's inequality to provide converse bounds for the estimation of discrete quantities. In many, if not most, statistical applications, one is instead interested in estimating continuous quantities; examples include linear regression, covariance estimation, density estimation, and so on. It turns out that the discrete form of Fano's inequality is still broadly applicable in such settings. The idea, as outlined in Section 16.1, is to choose a finite subset that still captures the inherent difficulty in the problem. In this section, we present several tools used for this purpose.

16.5.1 Minimax Estimation Setup

Recall the setup described in Section 16.1.1: a parameter θ is known to lie in some subset Θ of a continuous domain (e.g., $\mathbb{R}^p$), the samples $\mathbf{Y} = (Y_1, \ldots, Y_n)$ are drawn from a joint distribution $P_\theta^n(\mathbf{y})$, an estimate $\widehat{\theta}$ is formed, and the loss incurred is $\ell(\theta, \widehat{\theta})$. For clarity of exposition, we focus primarily on the case in which there is no input, i.e., $\mathbf{X}$ in Fig. 16.1 is absent or deterministic. However, the main results (Theorems 16.9 and 16.10 below) extend to settings with inputs as described in Section 16.1.1; the mutual information $I(V;\mathbf{Y})$ is replaced by $I(V;\mathbf{Y}|\mathbf{X})$ in the non-adaptive setting, or $I(V;\mathbf{X},\mathbf{Y})$ in the adaptive setting.

In continuous settings, the reduction to multiple hypothesis testing (see Fig. 16.1) requires that the loss function is sufficiently well behaved. We focus here on a widely considered class of functions that can be written as

$$\ell(\theta, \widehat{\theta}) = \Phi(\rho(\theta, \widehat{\theta})), \tag{16.60}$$

where $\rho(\theta, \theta')$ is a metric, and $\Phi(\cdot)$ is an increasing function from $\mathbb{R}_+$ to $\mathbb{R}_+$. For instance, the squared-ℓ_2 loss $\ell(\theta, \theta') = \|\theta - \theta'\|_2^2$ clearly takes this form.

We focus on the minimax setting, defining the *minimax risk* as follows:

$$\mathcal{M}_n(\Theta, \ell) = \inf_{\widehat{\theta}} \sup_{\theta \in \Theta} \mathbb{E}_\theta[\ell(\theta, \widehat{\theta})], \tag{16.61}$$

where the infimum is over all estimators $\widehat{\theta} = \widehat{\theta}(\mathbf{Y})$, and $\mathbb{E}_\theta$ denotes expectation when the underlying parameter is θ. We subsequently define $\mathbb{P}_\theta$ analogously.

16.5.2 Reduction to the Discrete Case

We present two related approaches to reducing the continuous estimation problem to a discrete one. The first, which is based on the standard form of Fano's inequality in Theorem 16.1, was discovered much earlier [12], and, accordingly, it has been used in a much wider range of applications. However, the second approach, which is based on the approximate recovery version of Fano's inequality in Theorem 16.2, has recently been shown to provide added flexibility in the reduction [35].

Reduction with Exact Recovery

As we discussed in Section 16.1, we seek to reduce the continuous problem to multiple hypothesis testing in such a way that successful minimax estimation implies success in the hypothesis test with high probability. To this end, we choose a *hard subset* $\theta_1,\dots,\theta_M$, for which the elements are sufficiently well separated that the index $v \in \{1,\dots,M\}$ can be identified from the estimate $\widehat{\theta}$ (see Fig. 16.1). This is formalized in the proof of the following result.

THEOREM 16.9 (Minimax bound via reduction to exact recovery) *Under the preceding minimax estimation setup, fix* $\epsilon > 0$, *and let* $\{\theta_1,\dots,\theta_M\}$ *be a finite subset of* Θ *such that*

$$\rho(\theta_v,\theta_{v'}) \geq \epsilon, \quad \forall v,v' \in \{1,\dots,M\}, v \neq v'. \tag{16.62}$$

Then, we have

$$\mathcal{M}_n(\Theta,\ell) \geq \Phi\left(\frac{\epsilon}{2}\right)\left(1 - \frac{I(V;\mathbf{Y})+\log 2}{\log M}\right), \tag{16.63}$$

where V *is uniform on* $\{1,\dots,M\}$, *and the mutual information is with respect to* $V \to \theta_V \to \mathbf{Y}$. *Moreover, in the special case* $M = 2$, *we have*

$$\mathcal{M}_n(\Theta,\ell) \geq \Phi\left(\frac{\epsilon}{2}\right) H_2^{-1}(\log 2 - I(V;\mathbf{Y})), \tag{16.64}$$

where $H_2^{-1}(\cdot) \in [0,0.5]$ *is the inverse binary entropy function.*

Proof As illustrated in Fig. 16.1, the idea is to reduce the estimation problem to a multiple-hypothesis-testing problem. As an initial step, we note from Markov's inequality that, for any $\epsilon_0 > 0$,

$$\sup_{\theta\in\Theta} \mathbb{E}_\theta[\ell(\theta,\widehat{\theta})] \geq \sup_{\theta\in\Theta} \Phi(\epsilon_0)\mathbb{P}_\theta[\ell(\theta,\widehat{\theta}) \geq \Phi(\epsilon_0)] \tag{16.65}$$

$$= \Phi(\epsilon_0)\sup_{\theta\in\Theta} \mathbb{P}_\theta[\rho(\theta,\widehat{\theta}) \geq \epsilon_0], \tag{16.66}$$

where (16.66) uses (16.60) and the assumption that $\Phi(\cdot)$ is increasing.

Suppose that a random index V is drawn uniformly from $\{1,\dots,M\}$, the samples $\mathbf{Y}$ are drawn from the distribution P_θ^n corresponding to $\theta = \theta_V$, and the estimator is applied to produce $\widehat{\theta}$. Let $\widehat{V}$ correspond to the closest θ_j according to the metric ρ, i.e., $\widehat{V} = \arg\min_{v=1,\dots,M} \rho(\theta_v,\widehat{\theta})$. Using the triangle inequality and the assumption (16.62), if $\rho(\theta_v,\widehat{\theta}) < \epsilon/2$ then we must have $\widehat{V} = v$; hence,

$$\mathbb{P}_v\left[\rho(\theta_v,\widehat{\theta}) \geq \frac{\epsilon}{2}\right] \geq \mathbb{P}_v[\widehat{V} \neq v], \tag{16.67}$$

where $\mathbb{P}_v$ is a shorthand for $\mathbb{P}_{\theta_v}$.

With the above tools in place, we proceed as follows:

$$\sup_{\theta\in\Theta} \mathbb{P}_\theta\left[\rho(\theta,\widehat{\theta}) \geq \frac{\epsilon}{2}\right] \geq \max_{v=1,\dots,M} \mathbb{P}_v\left[\rho(\theta_v,\widehat{\theta}) \geq \frac{\epsilon}{2}\right] \tag{16.68}$$

$$\geq \max_{v=1,\dots,M} \mathbb{P}_v[\widehat{V} \neq v] \tag{16.69}$$

$$\geq \frac{1}{M} \sum_{v=1,\dots,M} \mathbb{P}_v[\widehat{V} \neq v] \tag{16.70}$$

$$\geq 1 - \frac{I(V;\mathbf{Y}) + \log 2}{\log M}, \tag{16.71}$$

where (16.68) follows upon maximizing over a smaller set, (16.69) follows from (16.67), (16.70) lower-bounds the maximum by the average, and (16.71) follows from Fano's inequality (Theorem 16.1) and the fact that $I(V;\widehat{V}) \leq I(V;\mathbf{Y})$ by the data-processing inequality (Lemma 16.1).

The proof of (16.63) is concluded by substituting (16.71) into (16.66) with $\epsilon_0 = \epsilon/2$, and taking the infimum over all estimators $\widehat{\theta}$. For $M = 2$, we obtain (16.64) in the same way upon replacing (16.71) by the version of Fano's inequality for $M = 2$ given in Remark 16.1.

We return to this result in Section 16.5.3, where we introduce and compare some of the most widely used approaches to choosing the set $\{\theta_1,\dots,\theta_M\}$ and bounding the mutual information.

Reduction with Approximate Recovery

The following generalization of Theorem 16.9, which is based on Fano's inequality with approximate recovery (Theorem 16.2), provides added flexibility in the reduction. An example comparing the two approaches will be given in Section 16.6 for the sparse linear regression problem.

THEOREM 16.10 (Minimax bound via reduction to approximate recovery) *Under the preceding minimax estimation setup, fix $\epsilon > 0$, $t \in \mathbb{R}$, a finite set $\mathcal{V}$ of cardinality M, and an arbitrary real-valued function $d(v,v')$ on $\mathcal{V} \times \mathcal{V}$, and let $\{\theta_v\}_{v\in\mathcal{V}}$ be a finite subset of Θ such that*

$$d(v,v') > t \Rightarrow \rho(\theta_v, \theta_{v'}) \geq \epsilon, \quad \forall v,v' \in \mathcal{V}. \tag{16.72}$$

Then we have for any $\epsilon \geq 0$ that

$$\mathcal{M}_n(\Theta,\ell) \geq \Phi\Big(\frac{\epsilon}{2}\Big)\bigg(1 - \frac{I(V;\mathbf{Y}) + \log 2}{\log(M/N_{\max}(t))}\bigg), \tag{16.73}$$

where V is uniform on $\{1,\dots,M\}$, the mutual information is with respect to $V \to \theta_V \to \mathbf{Y}$, and $N_{\max}(t) = \max_{v'\in\mathcal{V}} \sum_{v\in\mathcal{V}} \mathbb{1}\{d(v,v') \leq t\}$.

The proof is analogous to that of Theorem 16.9, and can be found in [35].

16.5.3 Local versus Global Approaches

Here we highlight two distinct approaches to applying the reduction to exact recovery as per Theorem 16.9, termed the *local* and *global* approaches. We do not make such a distinction for the approximate-recovery variant in Theorem 16.10, since we are not aware of a global approach having been used previously for this variant.

Local approach. The most common approach to applying Theorem 16.9 is to construct a set $\{\theta_1,\ldots,\theta_M\}$ of elements that are "close" in KL divergence. Specifically, upper-bounding the mutual information via Lemma 16.4 (with the vector $\mathbf{Y}$ playing the role of Y therein), one can weaken (16.63) as follows.

COROLLARY 16.1 (Local approach to minimax estimation) *Under the setup of Theorem 16.9 with a given set* $\{\theta_1,\ldots,\theta_M\}$ *satisfying* (16.62), *it holds for any auxiliary distribution* $Q^n(\mathbf{y})$ *that*

$$\mathcal{M}_n(\Theta,\ell) \geq \Phi\left(\frac{\epsilon}{2}\right)\left(1 - \frac{\min_{v=1,\ldots,M} D(P^n_{\theta_v}\|Q^n) + \log 2}{\log M}\right). \tag{16.74}$$

Moreover, the same bound holds true when $\min_v D(P^n_{\theta_v}\|Q^n)$ *is replaced by any one of* $(1/M)\sum_v D(P^n_{\theta_v}\|Q)$, $(1/M^2)\sum_{v,v'} D(P^n_{\theta_v}\|P^n_{\theta_{v'}})$, *or* $\max_{v,v'} D(P^n_{\theta_v}\|P^n_{\theta_{v'}})$.

Attaining a good bound in (16.74) requires choosing $\{\theta_1,\ldots,\theta_M\}$ to trade off two competing objectives: (i) a larger value of M means that more hypotheses need to be distinguished; and (ii) a smaller value of $\min_v D(P^n_{\theta_v}\|Q^n)$ means that the hypotheses are more similar. Generally speaking, there is no single best approach to optimizing this trade-off, and the size and structure of the set can vary significantly from problem to problem. Moreover, the construction need not be explicit; one can instead use probabilistic arguments to prove the existence of a set satisfying the desired properties. Examples are given in Section 16.6. Naturally, an analog of Corollary 16.1 holds for $M = 2$ as per Theorem 16.9, and a counterpart for approximate recovery holds as per Theorem 16.10.

We briefly mention that Corollary 16.1 has interesting connections with the popular Assouad method from the statistics literature, as detailed in [47]. In addition, the counterpart of Corollary 16.1 with $M = 2$ (using (16.10) in its proof) is similarly related to an analogous technique known as Le Cam's method.

Global approach. An alternative approach to applying Theorem 16.9 is the *global* approach, which performs the following: (i) construct a subset of Θ with as many elements as possible subject to the assumption (16.62); and (ii) construct a set that *covers* Θ, in the sense of Lemma 16.5, with as few elements as possible. The following definitions formalize the notions of forming "as many" and "as few" elements as possible. We write these in terms of a general real-valued function $\rho_0(\theta,\theta')$ that need not be a metric.

DEFINITION 16.1 *A set* $\{\theta_1,\ldots,\theta_M\} \subseteq \Theta$ *is said to be an* $\epsilon_{\rm p}$*-packing set of* Θ *with respect to a measure* $\rho_0 : \Theta\times\Theta \to \mathbb{R}$ *if* $\rho_0(\theta_v,\theta_{v'}) \geq \epsilon_{\rm p}$ *for all* $v,v' \in \{1,\ldots,M\}$ *with* $v' \neq v$. *The* $\epsilon_{\rm p}$*-packing number* $M^*_{\rho_0}(\Theta,\epsilon_{\rm p})$ *is defined to be the maximum cardinality of any* $\epsilon_{\rm p}$*-packing.*

DEFINITION 16.2 *A set* $\{\theta_1,\ldots,\theta_N\} \subseteq \Theta$ *is said to be an* $\epsilon_{\rm c}$*-covering set of* Θ *with respect to* $\rho_0 : \Theta\times\Theta \to \mathbb{R}$ *if, for any* $\theta \in \Theta$, *there exists some* $v \in \{1,\ldots,N\}$ *such that* $\rho_0(\theta,\theta_v) \leq \epsilon_{\rm c}$. *The* $\epsilon_{\rm c}$*-covering number* $N^*_{\rho_0}(\Theta,\epsilon_{\rm c})$ *is defined to be the minimum cardinality of any* $\epsilon_{\rm c}$*-covering.*

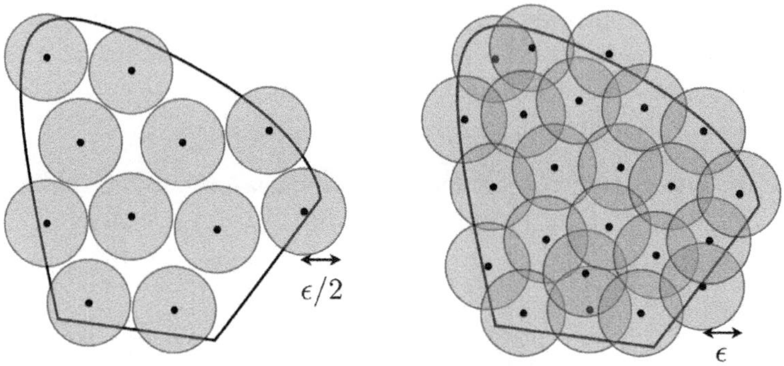

Figure 16.3 Examples of ϵ-packing (left) and ϵ-covering (right) sets in the case in which ρ_0 is the Euclidean distance in $\mathbb{R}^2$. Since ρ_0 is a metric, a set of points is an ϵ-packing if and only if their corresponding $\epsilon/2$-balls do not intersect.

Observe that assumption (16.62) of Theorem 16.9 precisely states that $\{\theta_1,\ldots,\theta_M\}$ is an ϵ-packing set, though the result is often applied with M far smaller than the ϵ-packing number. The logarithm of the covering number is often referred to as the *metric entropy*.

The notions of packing and covering are illustrated in Fig. 16.3. We do not explore the properties of packing and covering numbers in detail in this chapter; the interested reader is referred to [48, 49] for a more detailed treatment. We briefly state the following useful property, showing that the two definitions are closely related in the case in which ρ_0 is a metric.

LEMMA 16.7 (Packing versus covering numbers) *If ρ_0 is a metric, then $M^*_{\rho_0}(\Theta, 2\epsilon) \le N^*_{\rho_0}(\Theta,\epsilon) \le M^*_{\rho_0}(\Theta,\epsilon)$.*

We now show how to use Theorem 16.9 to construct a lower bound on the minimax risk in terms of certain packing and covering numbers. For the packing number, we will directly consider the metric ρ used in Theorem 16.9. On the other hand, for the covering number, we consider the density $P^n_{\theta_v}(\mathbf{y})$ associated with each $\theta \in \Theta$, and use the associated KL divergence measure:

$$N^*_{\mathrm{KL},n}(\Theta,\epsilon) = N^*_{\rho^n_{\mathrm{KL}}}(\Theta,\epsilon), \quad \rho^n_{\mathrm{KL}}(\theta,\theta') = D(P^n_\theta \| P^n_{\theta'}). \tag{16.75}$$

COROLLARY 16.2 (Global approach to minimax estimation) *Under the minimax estimation setup of Section 16.5.1, we have for any $\epsilon_{\mathrm{p}} > 0$ and $\epsilon_{c,n} > 0$ that*

$$\mathcal{M}_n(\Theta,\ell) \ge \Phi\left(\frac{\epsilon_{\mathrm{p}}}{2}\right)\left(1 - \frac{\log N^*_{KL,n}(\Theta,\epsilon_{c,n}) + \epsilon_{c,n} + \log 2}{\log M^*_\rho(\Theta,\epsilon_{\mathrm{p}})}\right). \tag{16.76}$$

In particular, if $P^n_\theta(\mathbf{y})$ is the n-fold product of some single-measurement distribution $P_\theta(y)$ for each $\theta \in \Theta$, then we have for any $\epsilon_{\mathrm{p}} > 0$ and $\epsilon_{\mathrm{c}} > 0$ that

$$\mathcal{M}_n(\Theta,\ell) \ge \Phi\left(\frac{\epsilon_{\mathrm{p}}}{2}\right)\left(1 - \frac{\log N^*_{KL}(\Theta,\epsilon_{\mathrm{c}}) + n\epsilon_{\mathrm{c}} + \log 2}{\log M^*_\rho(\Theta,\epsilon_{\mathrm{p}})}\right), \tag{16.77}$$

where $N^*_{KL}(\Theta,\epsilon) = N^*_{\rho_{KL}}(\Theta,\epsilon)$ with $\rho_{KL}(\theta,\theta') = D(P_\theta \| P_{\theta'})$.

Proof Since Theorem 16.9 holds for any packing set, it holds for the maximal packing set. Moreover, using Lemma 16.5, we have $I(V;\mathbf{Y}) \le \log N^*_{\mathrm{KL},n}(\Theta,\epsilon_{\mathrm{c},n}) + \epsilon_{\mathrm{c},n}$ in (16.63), since covering the entire space Θ is certainly enough to cover the elements in the packing set. By combining these, we obtain the first part of the corollary. The second part follows directly from the first part on choosing $\epsilon_{\mathrm{c},n} = n\epsilon_{\mathrm{c}}$ and noting that the KL divergence is additive for product distributions.

Corollary 16.2 has been used as the starting point to derive minimax lower bounds for a wide range of problems [13]; see Section 16.6 for an example. It has been observed that the global approach is mainly useful for infinite-dimensional problems such as density estimation and non-parametric regression, with the local approach typically being superior for finite-dimensional problems such as vector or matrix estimation.

16.5.4 Beyond Estimation – Fano's Inequality for Optimization

While the minimax estimation framework captures a diverse range of problems of interest, there are also interesting problems that it does not capture. A notable example, which we consider in this section, is *stochastic optimization*. We provide a brief treatment, and refer the reader to [20] for further details and results.

We consider the following setup.

- We seek to minimize an unknown function $f : \mathcal{X} \to \mathbb{R}$ on some input domain $\mathcal{X}$, i.e., to find a point $x \in \mathcal{X}$ such that $f(x)$ is as low as possible.
- The algorithm proceeds in iterations. At the ith iteration, a point $x_i \in \mathcal{X}$ is queried, and an *oracle* returns a sample y_i depending on the function, e.g., a noisy function value, a noisy gradient, or a tuple containing both. The selected point x_i can depend on the past queries and samples.
- After iteratively sampling n points, the optimization algorithm returns a final point $\widehat{x}$, and the *loss* incurred is $\ell_f(\widehat{x}) = f(\widehat{x}) - \min_{x\in\mathcal{X}} f(x)$, i.e., the gap to the optimal function value.
- For a given class of functions $\mathcal{F}$, the *minimax risk* is given by

$$\mathcal{M}_n(\mathcal{F}) = \inf_{\widehat{X}} \sup_{f\in\mathcal{F}} \mathbb{E}_f[\ell_f(\widehat{X})], \qquad (16.78)$$

  where the infimum is over all optimization algorithms that iteratively query the function n times and return a final point $\widehat{x}$ as above, and $\mathbb{E}_f$ denotes the expectation when the underlying function is f.

In the following, we let $\mathbf{X} = (X_1,\dots,X_n)$ and $\mathbf{Y} = (Y_1,\dots,Y_n)$ denote the queried locations and samples across the n rounds.

THEOREM 16.11 (Minimax bound for noisy optimization) *Fix $\epsilon > 0$, and let $\{f_1,\dots,f_M\} \subseteq \mathcal{F}$ be a finite subset of $\mathcal{F}$ such that for each $x \in \mathcal{X}$, we have $\ell_{f_v}(x) \le \epsilon$ for at most one value of $v \in \{1,\dots,M\}$. Then we have*

$$\mathcal{M}_n(\mathcal{F}) \ge \epsilon \cdot \left(1 - \frac{I(V;\mathbf{X},\mathbf{Y}) + \log 2}{\log M}\right), \qquad (16.79)$$

where V is uniform on $\{1,\dots,M\}$, and the mutual information is with respect to $V \to f_V \to (\mathbf{X},\mathbf{Y})$. Moreover, in the special case $M = 2$, we have

$$\mathcal{M}_n(\mathcal{F}) \geq \epsilon \cdot H_2^{-1}(\log 2 - I(V;\mathbf{X},\mathbf{Y})), \tag{16.80}$$

where $H_2^{-1}(\cdot) \in [0,0.5]$ is the inverse binary entropy function.

Proof By Markov's inequality, we have

$$\sup_{f\in\mathcal{F}} \mathbb{E}_f[\ell_f(\widehat{X})] \geq \sup_{f\in\mathcal{F}} \epsilon \cdot \mathbb{P}_f[\ell_f(\widehat{X}) \geq \epsilon]. \tag{16.81}$$

Suppose that a random index V is drawn uniformly from $\{1,\dots,M\}$, and the triplet $(\mathbf{X},\mathbf{Y},\widehat{X})$ is generated by running the optimization algorithm on f_V. Given $\widehat{X} = \widehat{x}$, let $\widehat{V}$ index the function among $\{f_1,\dots,f_M\}$ with the lowest corresponding value: $\widehat{V} = \arg\min_{v=1,\dots,M} f_v(\widehat{x})$.

By the assumption that any x satisfies $\ell_{f_v}(x) \leq \epsilon$ for at most one of the M functions, we find that the condition $\ell_{f_v}(\widehat{x}) \leq \epsilon$ implies $\widehat{V} = v$. Hence, we have

$$\mathbb{P}_v[\ell_{f_v}(\widehat{X}) > \epsilon] \geq \mathbb{P}_{f_v}[\widehat{V} \neq v]. \tag{16.82}$$

The remainder of the proof follows (16.68)–(16.71) in the proof of Theorem 16.9. We lower-bound the minimax risk $\sup_{f\in\mathcal{F}} \mathbb{P}_f[\ell_f(\widehat{X}) \geq \epsilon]$ by the average over V, and apply Fano's inequality (Theorem 16.1 and Remark 16.1) and the data-processing inequality (from the third part of Lemma 16.3).

REMARK 16.7 Theorem 16.10 is based on reducing the optimization problem to a multiple-hypothesis-testing problem with exact recovery. One can derive an analogous result reducing to approximate recovery, but we are unaware of any works making use of such a result for optimization.

16.6 Applications – Continuous Settings

In this section, we present three applications of the tools introduced in Section 16.5: sparse linear regression, density estimation, and convex optimization. Similarly to the discrete case, our examples are chosen to permit a relatively simple analysis, while still effectively exemplifying the key concepts and tools.

16.6.1 Sparse Linear Regression

In this example, we extend the 1-sparse linear regression example of Section 16.1.1 to the more general scenario of k-sparsity. The setup is described as follows.

- We wish to estimate a high-dimensional vector $\theta \in \mathbb{R}^p$ that is *k-sparse*: $\|\theta\|_0 \leq k$, where $\|\theta\|_0$ is the number of non-zero entries in θ.
- The vector of n measurements is given by $\mathbf{Y} = \mathbf{X}\theta + \mathbf{Z}$, where $\mathbf{X} \in \mathbb{R}^{n\times p}$ is a known deterministic matrix, and $\mathbf{Z} \sim \mathcal{N}(\mathbf{0},\sigma^2\mathbf{I}_n)$ is additive Gaussian noise.

- Given knowledge of **X** and **Y**, an estimate $\widehat{\theta}$ is formed, and the loss is given by the squared ℓ_2-error, $\ell(\theta,\widehat{\theta}) = \|\theta-\widehat{\theta}\|_2^2$, corresponding to (16.60) with $\rho(\theta,\widehat{\theta}) = \|\theta-\widehat{\theta}\|_2$ and $\Phi(\cdot) = (\cdot)^2$. Overloading the general notation $\mathcal{M}_n(\Theta,\ell)$, we write the minimax risk as

$$\mathcal{M}_n(k,\mathbf{X}) = \inf_{\widehat{\theta}} \sup_{\theta\in\mathbb{R}^p : \|\theta\|_0\le k} \mathbb{E}_\theta[\|\theta-\widehat{\theta}\|_2^2], \tag{16.83}$$

where $\mathbb{E}_\theta$ denotes the expectation when the underlying vector is θ.

Minimax Bound

The lower bound on the minimax risk is formally stated as follows. To simplify the analysis slightly, we state the result in an asymptotic form for the sparse regime $k = o(p)$; with only minor changes, one can attain a non-asymptotic variant attaining the same scaling laws for more general choices of k [35].

THEOREM 16.12 (Sparse linear regression) *Under the preceding sparse linear regression problem with $k = o(p)$ and a fixed regression matrix* **X**, *we have*

$$\mathcal{M}_n(k,\mathbf{X}) \ge \frac{\sigma^2 kp\log(p/k)}{32\|\mathbf{X}\|_F^2}(1+o(1)) \tag{16.84}$$

as $p\to\infty$. In particular, under the constraint $\|\mathbf{X}\|_F^2 \le np\Gamma$ for some $\Gamma > 0$, achieving $\mathcal{M}_n(k,\mathbf{X}) \le \delta$ requires $n \ge (\sigma^2 k\log(p/k)/(32\delta\Gamma)(1+o(1))$.

Proof We present a simple proof based on a reduction to approximate recovery (Theorem 16.10). In Section 16.6.1, we discuss an alternative proof based on a reduction to exact recovery (Theorem 16.9).

We define the set

$$\mathcal{V} = \{v \in \{-1,0,1\}^p : \|v\|_0 = k\}, \tag{16.85}$$

and with each $v\in\mathcal{V}$ we associate a vector $\theta_v = \epsilon' v$ for some $\epsilon' > 0$. Letting $d(v,v')$ denote the Hamming distance, we have the following properties.

- For $v,v'\in\mathcal{V}$, if $d(v,v') > t$, then $\|\theta_v - \theta_{v'}\|_2 > \epsilon'\sqrt{t}$.
- The cardinality of $\mathcal{V}$ is $|\mathcal{V}| = 2^k\binom{p}{k}$, yielding $\log|\mathcal{V}| \ge \log\binom{p}{k} \ge k\log(p/k)$.
- The quantity $N_{\max}(t)$ in Theorem 16.10 is the maximum possible number of $v'\in\mathcal{V}$ such that $d(v,v')\le t$ for a fixed v. Setting $t = k/2$, a simple counting argument gives $N_{\max}(t) \le \sum_{j=0}^{\lceil k/2\rceil} 2^j\binom{p}{j} \le (\lceil k/2\rceil+1)\cdot 2^{\lceil k/2\rceil}\cdot\binom{p}{\lceil k/2\rceil}$, which simplifies to $\log N_{\max}(t) \le ((k/2)\log(p/k))(1+o(1))$ due to the assumption $k = o(p)$.

From these observations, applying Theorem 16.10 with $t = k/2$ and $\epsilon = \epsilon'\sqrt{k/2}$ yields

$$\mathcal{M}_n(k,\mathbf{X}) \ge \frac{k\cdot(\epsilon')^2}{8}\left(1 - \frac{I(V;\mathbf{Y}) + \log 2}{((k/2)\log(p/k))(1+o(1))}\right). \tag{16.86}$$

Note that we do not condition on **X** in the mutual information, since we have assumed that **X** is deterministic.

To bound the mutual information, we first apply tensorization (from the first part of Lemma 16.2) to obtain $I(V;\mathbf{Y}) \le \sum_{i=1}^n I(V;Y_i)$, and then bound each $I(V;Y_i)$ using equation (16.24) in Lemma 16.4. We let Q_Y be the $\mathcal{N}(0,\sigma^2)$ density function, and we let

$P_{v,i}$ denote the density function of $\mathcal{N}(X_i^{\mathrm{T}}\theta_v, \sigma^2)$, where X_i^{T} is the transpose of the ith row of $\mathbf{X}$. Since the KL divergence between the $\mathcal{N}(\mu_0, \sigma^2)$ and $\mathcal{N}(\mu_1, \sigma^2)$ density functions is $(\mu_1 - \mu_0)^2/(2\sigma^2)$, we have $D(P_{v,i} \| Q_Y) = |X_i^{\mathrm{T}}\theta_v|^2/(2\sigma^2)$. As a result, Lemma 16.4 yields $I(V; Y_i) \leq (1/|\mathcal{V}|)\sum_v D(P_{v,i}\|Q_Y) = (1/(2\sigma^2))\mathbb{E}[|X_i^{\mathrm{T}}\theta_V|^2]$ for uniform V. Summing over i and recalling that $\theta_v = \epsilon' v$, we deduce that

$$I(V; \mathbf{Y}) \leq \frac{(\epsilon')^2}{2\sigma^2}\mathbb{E}[\|\mathbf{X}V\|_2^2]. \tag{16.87}$$

From the choice of $\mathcal{V}$ in (16.85), we can easily compute $\mathrm{Cov}[V] = (k/p)\mathbf{I}_p$, which implies that $\mathbb{E}[\|\mathbf{X}V\|_2^2] = (k/p)\|\mathbf{X}\|_{\mathrm{F}}^2$. Substitution into (16.87) yields $I(V; \mathbf{Y}) \leq (\epsilon')^2/(2\sigma^2) \cdot (k/p)\|\mathbf{X}\|_{\mathrm{F}}^2$, and we conclude from (16.86) that

$$\mathcal{M}_n(k, \mathbf{X}) \geq \frac{k \cdot (\epsilon')^2}{8}\left(1 - \frac{(\epsilon')^2/2\sigma^2 \cdot (k/p)\|\mathbf{X}\|_{\mathrm{F}}^2 + \log 2}{((k/2)\log(p/k))(1 + o(1))}\right). \tag{16.88}$$

The proof is concluded by setting $(\epsilon')^2 = \sigma^2 p \log(p/k)/(2\|\mathbf{X}\|_{\mathrm{F}}^2)$, which is chosen to make the bracketed term tend to $\frac{1}{2}$.

Up to constant factors, the lower bound in Theorem 16.12 cannot be improved without additional knowledge of $\mathbf{X}$ beyond its Frobenius norm [5]. For instance, in the case in which $\mathbf{X}$ has i.i.d. Gaussian entries, a matching upper bound holds with high probability under maximum-likelihood decoding.

Alternative Proof: Reduction with Exact Recovery

In contrast to the proof given above (which has been adapted from [35]), the first known proof of Theorem 16.12 was based on packing with *exact* recovery (Theorem 16.9) [5]. For the sake of comparison, we briefly outline this alternative approach, which turns out to be more complicated.

The main step is to prove the existence of a set $\{\theta_1, \ldots, \theta_M\}$ satisfying the following properties.

- The number of elements satisfies $M = \Omega(k \log(p/k))$.
- Each element is k-sparse with non-zero entries equal to ± 1.
- The elements are well separated in the sense that $\|\theta_v - \theta_{v'}\|_2^2 = \Omega(k)$ for $v \neq v'$.
- The empirical covariance matrix is close to a scaled identity matrix in the following sense: $\left\|(1/M)\sum_{v=1}^M \theta_v\theta_v^{\mathrm{T}} - (k/p) \cdot \mathbf{I}_p\right\|_{2\to 2} = o(k/p)$, where $\|\cdot\|_{2\to 2}$ denotes the ℓ_2/ℓ_2-operator norm, i.e., the largest singular value.

Once this has been established, the proof proceeds along the same lines as the proof we gave above, scaling the vectors down by some $\epsilon' > 0$ and using Theorem 16.9 in place of Theorem 16.10.

The existence of the packing set is proved via a probabilistic argument. If one generates $\Omega(k \log(p/k))$ uniformly random k-sparse sequences with non-zero entries equaling ± 1, then these will satisfy the remaining two properties with positive probability. While it is straightforward to establish the condition of being well separated, the proof of the condition on the empirical covariance matrix requires a careful application of the non-elementary matrix Bernstein inequality.

Overall, while the two approaches yield the same result up to constant factors in this example, the approach based on approximate recovery is entirely elementary and avoids the preceding difficulties.

16.6.2 Density Estimation

In this section, we consider the problem of estimating an entire probability density function given samples from its distribution, which is commonly known as *density estimation*. We consider a non-parametric view, meaning that the density does not take any specific parametric form. As a result, the problem is inherently infinite-dimensional, and lends itself to the global packing and covering approach introduced in Section 16.5.3.

While many classes of density functions have been considered in the literature [13], we focus our attention on a specific setting for clarity of exposition.

- The density function f that we seek to estimate is defined on the domain $[0,1]$, i.e., $f(y) \geq 0$ for all $y \in [0,1]$, and $\int_0^1 f(y)dy = 1$.
- We assume that f satisfies the following conditions:

$$f(y) \geq \eta, \forall y \in [0,1], \qquad \|f\|_{\mathrm{TV}} \leq \Gamma \tag{16.89}$$

for some $\eta \in (0,1)$ and $\Gamma > 0$, where the *total variation* (TV) norm is defined as $\|f\|_{\mathrm{TV}} = \sup_L \sup_{0 \leq x_1 \leq \ldots \leq x_L \leq 1} \sum_{l=2}^{L} (f(x_l) - f(x_{l-1}))$. The set of all density functions satisfying these constraints is denoted by $\mathcal{F}_{\eta,\Gamma}$.
- Given n independent samples $\mathbf{Y} = (Y_1, \ldots, Y_n)$ from f, an estimate $\widehat{f}$ is formed, and the loss is given by $\ell(f,\widehat{f}) = \|f - \widehat{f}\|_2^2 = \int_0^1 (f(x) - \widehat{f}(x))^2 dx$. Hence, the minimax risk is given by

$$\mathcal{M}_n(\eta,\Gamma) = \inf_{\widehat{f}} \sup_{f \in \mathcal{F}_{\eta,\Gamma}} \mathbb{E}_f[\|f - \widehat{f}\|_2^2], \tag{16.90}$$

where $\mathbb{E}_f$ denotes the expectation when the underlying density is f.

Minimax Bound

The minimax lower bound is given as follows.

THEOREM 16.13 (Density estimation) *Consider the preceding density estimation setup with some $\eta \in (0,1)$ and $\Gamma > 0$ not depending on n. There exists a constant $c > 0$ (depending on η and Γ) such that, in order to achieve $\mathcal{M}_n(\eta,\Gamma) \leq \delta$, it is necessary that*

$$n \geq c \cdot \left(\frac{1}{\delta}\right)^{3/2} \tag{16.91}$$

when δ is sufficiently small. In other words, $\mathcal{M}_n(\eta,\Gamma) = \Omega(n^{-2/3})$.

Proof We specialize the general analysis of [13] to the class $\mathcal{F}_{\eta,\Gamma}$. Recalling the packing and covering numbers from Definitions 16.1 and 16.2, we adopt the shorthand notation $M_2^*(\epsilon_{\mathrm{p}}) = M_\rho^*(\mathcal{F}_{\eta,\Gamma}, \epsilon_{\mathrm{p}})$ with $\rho(f,f') = \|f - f'\|_2$, and similarly

$N_2^*(\epsilon_c) = N_\rho^*(\mathcal{F}_{\eta,\Gamma}, \epsilon_c)$. We first show that N_{KL}^* (Corollary 16.2) can be upper-bounded in terms of M_2^*, which will lead to a minimax lower bound that depends only on the packing number M_2^*. For $f_1, f_2 \in \mathcal{F}_{\eta,\Gamma}$, we have

$$D(f_1 \| f_2) \le \int_0^1 \frac{(f_1(x) - f_2(x))^2}{f_2(x)}\,dx \tag{16.92}$$

$$\le \frac{1}{\eta}\int_0^1 (f_1(x) - f_2(x))^2\,dx \tag{16.93}$$

$$= \frac{1}{\eta}\|f_1 - f_2\|_2^2, \tag{16.94}$$

where (16.92) follows since the KL divergence is upper-bounded by the χ^2-divergence (Lemma 16.6), and (16.93) follows from the assumption that the density is lower-bounded by η. From the definition of N_{KL}^* in Corollary 16.2, we deduce the following for any $\epsilon_c > 0$:

$$N_{\mathrm{KL}}^*(\epsilon_c) \le N_2^*(\sqrt{\eta\epsilon_c}) \le M_2^*(\sqrt{\eta\epsilon_c}), \tag{16.95}$$

where the first inequality holds because any $\sqrt{\eta\epsilon_c}$-covering in the ℓ_2-norm is also a ϵ_c-covering in the KL divergence due to (16.94), and the second inequality follows from Lemma 16.7.

Combining (16.95) with Corollary 16.2 and the choice $\Phi(\cdot) = (\cdot)^2$ gives

$$\mathcal{M}_n(\eta,\Gamma) \ge \left(\frac{\epsilon_p}{2}\right)^2\left(1 - \frac{\log M_2^*(\sqrt{\eta\epsilon_c}) + n\epsilon_c + \log 2}{\log M_2^*(\epsilon_p)}\right). \tag{16.96}$$

We now apply the following bounds on the packing number of $\mathcal{F}_{\eta,\Gamma}$, which we state from [13] without proof:

$$\underline{c}\cdot\epsilon^{-1} \le \log M_2^*(\epsilon) \le \overline{c}\cdot\epsilon^{-1}, \tag{16.97}$$

for some constants $\underline{c}, \overline{c} > 0$ and sufficiently small $\epsilon > 0$. It follows that

$$\mathcal{M}_n(\eta,\Gamma) \ge \left(\frac{\epsilon_p}{2}\right)^2\left(1 - \frac{\overline{c}\cdot(\eta\epsilon_c)^{-1/2} + n\epsilon_c + \log 2}{\underline{c}\cdot\epsilon_p^{-1}}\right). \tag{16.98}$$

The remainder of the proof amounts to choosing ϵ_p and ϵ_c to balance the terms appearing in this expression.

First, choosing ϵ_c to equate the terms $\overline{c}\cdot(\eta\epsilon_c)^{-1/2}$ and $n\epsilon_c$ leads to $\epsilon_c = (c'/n)^{2/3}$ with $c' = \overline{c}\eta^{-1/2}$, yielding $(\overline{c}\cdot(\eta\epsilon_c)^{-1/2} + n\epsilon_c + \log 2)/(\underline{c}\cdot\epsilon_p^{-1}) = (2n(c'/n)^{2/3} + \log 2)/(\overline{c}\cdot\epsilon_p^{-1})$. Next, choosing ϵ_p to make this fraction equal to $\frac{1}{2}$ yields $\epsilon_p^{-1} = (2/\overline{c})(2(c')^{2/3}n^{1/3} + \log 2)$, which means that $\epsilon_p \ge c'' \cdot n^{-1/3}$ for suitable $c'' > 0$ and sufficiently large n. Finally, since we made the fraction equal to $\frac{1}{2}$, (16.98) yields $\mathcal{M}_n(\eta,\Gamma) \ge \epsilon_p^2/8 \ge (c'')^2 n^{-2/3}/8$. Setting $\mathcal{M}_n(\eta,\Gamma) = \delta$ and solving for n yields the desired result.

The scaling given in Theorem 16.13 cannot be improved; a matching upper bound is given in [13], and can be achieved even when $\eta = 0$.

16.6.3 Convex Optimization

In our final example, we consider the optimization setting introduced in Section 16.5.4. We provide an example that is rather simple, yet has interesting features not present in the previous examples: (i) an example departing from estimation; (ii) a continuous example with adaptivity; and (iii) a case where Fano's inequality with $|\mathcal{V}| = 2$ is used.

We consider the following special case of the general setup of Section 16.5.4.

- We let $\mathcal{F}$ be the set of differentiable and *strongly convex* functions on $\mathcal{X} = [0,1]$, with strong convexity parameter equal to one:

$$\mathcal{F}_{\mathrm{scv}} = \left\{ f : f \text{ is differentiable} \cap f(x) - \frac{1}{2}x^2 \text{ is convex} \right\}. \tag{16.99}$$

  The analysis that we present can easily be extended to functions on an arbitrary closed interval with an arbitary strong convexity parameter.
- When we query a point $x \in \mathcal{X}$, we observe a noisy sample of the function value and its gradient:

$$Y = (f(x) + Z, f'(x) + Z'), \tag{16.100}$$

  where Z and Z' are independent $\mathcal{N}(0, \sigma^2)$ random variables, for some $\sigma^2 > 0$. This is commonly referred to as the *noisy first-order oracle*.

Minimax Bound

The following theorem lower-bounds the number of queries required to achieve δ-optimality. The proof is taken from [20] with only minor modifications.

THEOREM 16.14 (Stochastic optimization of strongly convex functions) *Under the preceding convex optimization setting with noisy first-order oracle information, in order to achieve $\mathcal{M}_n(\mathcal{F}_{scv}) \le \delta$, it is necessary that*

$$n \ge \frac{\sigma^2 \log 2}{40\delta} \tag{16.101}$$

when δ is sufficiently small.

Proof We construct a set of two functions satisfying the assumptions of Theorem 16.11. Specifically, we fix (ϵ, ϵ') such that $0 < \epsilon < \epsilon' < \frac{1}{8}$, define $x_1^* = \frac{1}{2} - \sqrt{2\epsilon'}$ and $x_2^* = \frac{1}{2} + \sqrt{2\epsilon'}$, and set

$$f_v(x) = \frac{1}{2}(x - x_v^*)^2, \quad v = 1, 2. \tag{16.102}$$

These functions are illustrated in Fig. 16.4.

Since $\epsilon' \in (0, \frac{1}{8})$, both x_1^* and x_2^* lie in $(0,1)$, and hence $\min_{x\in[0,1]} f_1(x) = \min_{x\in[0,1]} f_2(x) = 0$. Moreover, a direct evaluation reveals that $f_1(x) + f_2(x) = (x - \frac{1}{2})^2 + 2\epsilon' > 2\epsilon$, which implies that any ϵ-optimal point for one function cannot be

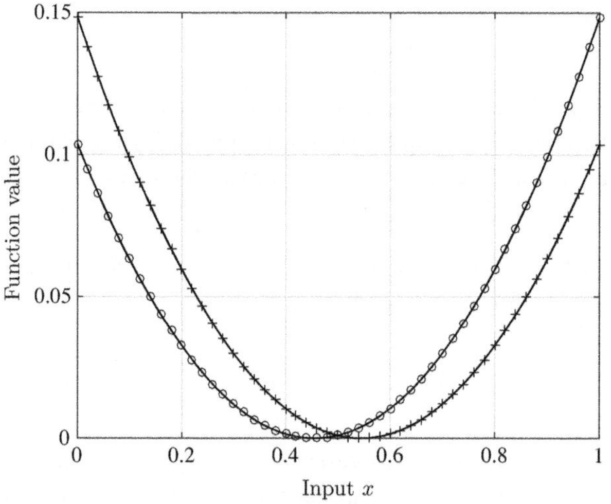

Figure 16.4 Construction of two functions in $\mathcal{F}_{\text{scv}}$ that are difficult to distinguish, and such that any point $x \in [0,1]$ can be ϵ-optimal for only one of the two functions.

ϵ-optimal for the other function. This is the condition needed in order for us to apply Theorem 16.11, yielding from (16.80) that

$$\mathcal{M}_n(\mathcal{F}_{\text{scv}}) \geq \epsilon \cdot H_2^{-1}(\log 2 - I(V;\mathbf{X},\mathbf{Y})). \tag{16.103}$$

To bound the mutual information, we first apply tensorization (from the first part of Lemma 16.3) to obtain $I(V;\mathbf{X},\mathbf{Y}) \leq \sum_{i=1}^{n} I(V;Y_i|X_i)$. We proceed by bounding $I(V;Y_i|X_i)$ for any given i. Fix $x \in [0,1]$, let P_{Y_x} and $P_{Y'_x}$ be the density functions of the noisy samples of $f_1(x)$ and $f_1'(x)$, and let Q_{Y_x} and $Q_{Y'_x}$ be defined similarly for $f_0(x) = \frac{1}{2}(x-\frac{1}{2})^2$. We have

$$D(P_{Y_x} \times P_{Y'_x} \| Q_{Y_x} \times Q_{Y'_x}) = D(P_{Y_x} \| Q_{Y_x}) + D(P_{Y'_x} \| Q_{Y'_x}) \tag{16.104}$$

$$= \frac{(f_1(x) - f_0(x))^2}{2\sigma^2} + \frac{(f_1'(x) - f_0'(x))^2}{2\sigma^2}, \tag{16.105}$$

where (16.104) holds since the KL divergence is additive for product distributions, and (16.105) uses the fact that the divergence between the $\mathcal{N}(\mu_0,\sigma^2)$ and $\mathcal{N}(\mu_1,\sigma^2)$ density functions is $(\mu_1-\mu_0)^2/(2\sigma^2)$.

Recalling that $f_1(x) = \frac{1}{2}(x-\frac{1}{2}+\sqrt{2\epsilon'})^2$ and $f_0(x) = \frac{1}{2}(x-\frac{1}{2})^2$, we have

$$(f_1(x) - f_0(x))^2 = \frac{1}{4}\left(2\epsilon' + 2\left(x - \frac{1}{2}\right)\sqrt{2\epsilon'}\right)^2 \leq \left(\epsilon' + \sqrt{\frac{\epsilon'}{2}}\right)^2 \leq 2\epsilon', \tag{16.106}$$

where the first inequality uses the fact that $x \in [0,1]$, and the second inequality follows since $\epsilon' < \frac{1}{8}$ and hence $\epsilon' = \sqrt{\epsilon'} \cdot \sqrt{\epsilon'} \leq \sqrt{\epsilon'/8}$ (note that $\left(1/\sqrt{8}+1/\sqrt{2}\right)^2 \leq 2$). Moreover, taking the derivatives of f_0 and f_1 gives $(f_1'(x) - f_0'(x))^2 = 2\epsilon'$, and substitution into (16.105) yields $D(P_{Y_x} \times P_{Y'_x} \| Q_{Y_x} \times Q_{Y'_x}) \leq 2\epsilon'/\sigma^2$.

The preceding analysis applies in a near-identical manner when f_2 is used in place of f_1, and yields the same KL divergence bound when $(P_{Y_x}, P_{Y'_x})$ is defined with respect to f_2. As a result, for any $x \in [0,1]$, we obtain from (16.25) in Lemma 16.4 that $I(V;Y_i|X_i = x) \le 2\epsilon'/\sigma^2$. Averaging over X, we obtain $I(V;Y_i|X_i) \le 2\epsilon'/\sigma^2$, and substitution into the above-established bound $I(V;\mathbf{X},\mathbf{Y}) \le \sum_{i=1}^n I(V;Y_i|X_i)$ yields $I(V;\mathbf{X},\mathbf{Y}) \le 2n\epsilon'/\sigma^2$. Hence, (16.103) yields

$$\mathcal{M}_n(\mathcal{F}_{\mathrm{scv}}) \ge \epsilon \cdot H_2^{-1}\left(\log 2 - \frac{2n\epsilon'}{\sigma^2}\right). \tag{16.107}$$

Now observe that if $n \le \sigma^2(\log 2)/4\epsilon'$ then the argument to $H_2^{-1}(\cdot)$ is at least $(\log 2)/2$. It is easy to verify that $H_2^{-1}((\log 2)/2) > \frac{1}{10}$, from which it follows that $\mathcal{M}_n(\mathcal{F}_{\mathrm{scv}}) > \epsilon/10$. Setting $\epsilon = 10\delta$ and noting that ϵ' can be chosen arbitrarily close to ϵ, we conclude that the required number of samples $\sigma^2(\log 2)/4\epsilon'$ recovers (16.101).

Theorem 16.14 provides tight scaling laws, since stochastic gradient descent is known to achieve δ-optimality for strongly convex functions using $O(\sigma^2/\delta)$ queries. Analogous results for the multidimensional setting can be found in [20].

16.7 Discussion

16.7.1 Limitations of Fano's Inequality

While Fano's inequality is a highly versatile method with successes in a wide range of statistical applications (see Table 16.1), it is worth pointing out some of its main limitations. We briefly mention some alternative methods below, as well as discussing some suitable generalizations of Fano's inequality in Section 16.7.2.

Non-asymptotic weakness. Even in scenarios where Fano's inequality provides converse bounds with the correct asymptotics including constants, these bounds can be inferior to alternative methods in the non-asymptotic sense [50, 51]. Related to this issue is the distinction between the *weak converse* and *strong converse*: we have seen that Fano's inequality typically provides necessary conditions of the form $n \ge n^*(1-\delta-o(1))$ for achieving $P_e \le \delta$, in contrast with strong converse results of the form $n \ge n^*(1-o(1))$ for *any* $\delta \in (0,1)$. Alternative techniques addressing these limitations are discussed in the context of communication in [50], and in the context of statistical estimation in [52, 53].

Difficulties in adaptive settings. While we have provided examples where Fano's inequality provides tight bounds in adaptive settings, there are several applications where alternative methods have proved to be more suitable. One reason for this is that the conditional mutual information terms $I(V;Y_i|X_i)$ (see Lemma 16.3) often involve complicated conditional distributions that are difficult to analyze. We refer the reader to [54–56] for examples in which alternative techniques proved to be more suitable for adaptive settings.

Restriction to KL divergence. When applying Fano's inequality, one invariably needs to bound a mutual information term, which is an instance of the KL divergence. While the KL divergence satisfies a number of convenient properties that can help in this process, it is sometimes the case that other divergence measures are more convenient to work with, or can be used to derive tighter results. Generalizations of Fano's inequality have been proposed specifically for this purpose, as we discuss in the following section.

16.7.2 Generalizations of Fano's Inequality

Several variations and generalizations of Fano's inequality have been proposed in the literature [57–62]. Most of these are not derived from the most well-known proof of Theorem 16.1, but are instead based on an alternative proof via the data-processing inequality for KL divergence: For any event E, one has

$$I(V;\widehat{V}) = D(P_{V\widehat{V}} \| P_V \times P_{\widehat{V}}) \geq D_2(P_{V\widehat{V}}[E] \,\|\, (P_V \times P_{\widehat{V}})[E]), \tag{16.108}$$

where $D_2(p\|q) = p\log(p/q) + (1-p)\log((1-p)/(1-q))$ is the binary KL divergence function. Observe that, if V is uniform and E is the event that $V \neq \widehat{V}$, then we have $P_{V\widehat{V}}[E] = P_{\rm e}$ and $(P_V \times P_{\widehat{V}})[E] = 1 - 1/|\mathcal{V}|$, and Fano's inequality (Theorem 16.1) follows on substituting the definition of $D_2(\cdot\|\cdot)$ into (16.108) and rearranging. This proof lends itself to interesting generalizations, including the following.

Continuum version. Consider a continuous random variable V taking values on $\mathcal{V} \subseteq \mathbb{R}^p$ for some $p \geq 1$, and an error probability of the form $P_{\rm e}(t) = \mathbb{P}[d(V,\widehat{V}) > t]$ for some real-valued function d on $\mathbb{R}^p \times \mathbb{R}^p$. This is the same formula as (16.11), which we previously introduced for the discrete setting. Defining the "ball" $\mathbb{B}_d(\widehat{v},t) = \{v \in \mathbb{R}^p : d(v,\widehat{v}) \leq t\}$ centered at $\widehat{v}$, (16.108) leads to the following for V uniform on $\mathcal{V}$:

$$P_{\rm e}(t) \geq 1 - \frac{I(V;\widehat{V}) + \log 2}{\log\left(\mathrm{Vol}(\mathcal{V}) / \sup_{\widehat{v}\in\mathbb{R}^p} \mathrm{Vol}(\mathcal{V} \cap \mathbb{B}_d(\widehat{v},t))\right)}, \tag{16.109}$$

where $\mathrm{Vol}(\cdot)$ denotes the volume of a set. This result provides a continuous counterpart to the final part of Theorem 16.2, in which the cardinality ratio is replaced by a volume ratio. We refer the reader to [35] for example applications, and to [62] for the simple proof outlined above.

Beyond KL divergence. The key step (16.108) extends immediately to other measures that satisfy the data-processing inequality. A useful class of such measures is the class of f*-divergences*: $D_f(P\|Q) = \mathbb{E}_Q[f(P(\mathbf{Y})/Q(\mathbf{Y}))]$ for some convex function f satisfying $f(1) = 0$. Special cases include KL divergence ($f(z) = z\log z$), total variation ($f(z) = \frac{1}{2}|z-1|$), squared Hellinger distance ($f(z) = (\sqrt{z}-1)^2$), and χ^2-divergence ($f(z) = (z-1)^2$). It was shown in [60] that alternative choices beyond the KL divergence can provide improved bounds in some cases. Generalizations of Fano's inequality beyond f-divergences can be found in [61].

Non-uniform priors. The first form of Fano's inequality in Theorem 16.1 does not require V to be uniform. However, in highly non-uniform cases where $H(V) \ll \log|\mathcal{V}|$, the term $P_{\rm e}\log(|\mathcal{V}|-1)$ may be too large for the bound to be useful. In such cases, it is often useful to use different Fano-like bounds that are based on the alternative proof above. In particular, the step (16.108) makes no use of uniformity, and continues to hold even in the non-uniform case. In [57], this bound was further weakened to provide simpler lower bounds for non-uniform settings with discrete alphabets. Fano-type lower bounds in *continuous* Bayesian settings with non-uniform priors arose more recently, and are typically more technically challenging; the interested reader is referred to [18, 63].

Acknowledgments

J. Scarlett was supported by an NUS startup grant. V. Cevher was supported by the European Research Council (ERC) under the European Union's Horizon 2020 research and innovation programme (grant agreement 725594 – time-data).

References

[1] R. M. Fano, "Class notes for MIT course 6.574: Transmission of information," 1952.
[2] M. B. Malyutov, "The separating property of random matrices," *Math. Notes Academy Sci. USSR*, vol. 23, no. 1, pp. 84–91, 1978.
[3] G. Atia and V. Saligrama, "Boolean compressed sensing and noisy group testing," *IEEE Trans. Information Theory*, vol. 58, no. 3, pp. 1880–1901, 2012.
[4] M. J. Wainwright, "Information-theoretic limits on sparsity recovery in the high-dimensional and noisy setting," *IEEE Trans. Information Theory*, vol. 55, no. 12, pp. 5728–5741, 2009.
[5] E. J. Candès and M. A. Davenport, "How well can we estimate a sparse vector?" *Appl. Comput. Harmonic Analysis*, vol. 34, no. 2, pp. 317–323, 2013.
[6] H. Hassanieh, P. Indyk, D. Katabi, and E. Price, "Nearly optimal sparse Fourier transform," in *Proc. 44th Annual ACM Symposium on Theory of Computation*, 2012, pp. 563–578.
[7] V. Cevher, M. Kapralov, J. Scarlett, and A. Zandieh, "An adaptive sublinear-time block sparse Fourier transform," in *Proc. 49th Annual ACM Symposium on Theory of Computation*, 2017, pp. 702–715.
[8] A. A. Amini and M. J. Wainwright, "High-dimensional analysis of semidefinite relaxations for sparse principal components," *Annals Statist.*, vol. 37, no. 5B, pp. 2877–2921, 2009.
[9] V. Q. Vu and J. Lei, "Minimax rates of estimation for sparse PCA in high dimensions," in *Proc. 15th International Conference on Artificial Intelligence and Statistics*, 2012, pp. 1278–1286.
[10] S. Negahban and M. J. Wainwright, "Restricted strong convexity and weighted matrix completion: Optimal bounds with noise," *J. Machine Learning Res.*, vol. 13, no. 5, pp. 1665–1697, 2012.
[11] M. A. Davenport, Y. Plan, E. Van Den Berg, and M. Wootters, "1-bit matrix completion," *Information and Inference*, vol. 3, no. 3, pp. 189–223, 2014.
[12] I. Ibragimov and R. Khasminskii, "Estimation of infinite-dimensional parameter in Gaussian white noise," *Soviet Math. Doklady*, vol. 236, no. 5, pp. 1053–1055, 1977.
[13] Y. Yang and A. Barron, "Information-theoretic determination of minimax rates of convergence," *Annals Statist.*, vol. 27, no. 5, pp. 1564–1599, 1999.
[14] L. Birgé, "Approximation dans les espaces métriques et théorie de l'estimation," *Probability Theory and Related Fields*, vol. 65, no. 2, pp. 181–237, 1983.

[15] G. Raskutti, M. J. Wainwright, and B. Yu, "Minimax-optimal rates for sparse additive models over kernel classes via convex programming," *J. Machine Learning Res.*, vol. 13, no. 2, pp. 389–427, 2012.

[16] Y. Yang, M. Pilanci, and M. J. Wainwright, "Randomized sketches for kernels: Fast and optimal nonparametric regression," *Annals Statist.*, vol. 45, no. 3, pp. 991–1023, 2017.

[17] Y. Zhang, J. Duchi, M. I. Jordan, and M. J. Wainwright, "Information-theoretic lower bounds for distributed statistical estimation with communication constraints," in *Advances in Neural Information Processing Systems*, 2013, pp. 2328–2336.

[18] A. Xu and M. Raginsky, "Information-theoretic lower bounds on Bayes risk in decentralized estimation," *IEEE Trans. Information Theory*, vol. 63, no. 3, pp. 1580–1600, 2017.

[19] J. C. Duchi, M. I. Jordan, and M. J. Wainwright, "Local privacy and statistical minimax rates," in *Proc. 54th Annual IEEE Symposium on Foundations of Computer Science*, 2013, pp. 429–438.

[20] M. Raginsky and A. Rakhlin, "Information-based complexity, feedback and dynamics in convex programming," *IEEE Trans. Information Theory*, vol. 57, no. 10, pp. 7036–7056, 2011.

[21] A. Agarwal, P. L. Bartlett, P. Ravikumar, and M. J. Wainwright, "Information-theoretic lower bounds on the oracle complexity of stochastic convex optimization," *IEEE Trans. Information Theory*, vol. 58, no. 5, pp. 3235–3249, 2012.

[22] M. Raginsky and A. Rakhlin, "Lower bounds for passive and active learning," in *Advances in Neural Information Processing Systems*, 2011, pp. 1026–1034.

[23] A. Agarwal, S. Agarwal, S. Assadi, and S. Khanna, "Learning with limited rounds of adaptivity: Coin tossing, multi-armed bandits, and ranking from pairwise comparisons," in *Proc. Conference on Learning Theory*, 2017, pp. 39–75.

[24] J. Scarlett, "Tight regret bounds for Bayesian optimization in one dimension," in *Proc. International Conference on Machine Learning*, 2018, pp. 4507–4515.

[25] Z. Bar-Yossef, T. S. Jayram, R. Kumar, and D. Sivakumar, "Information theory methods in communication complexity," in *Proc. 17th IEEE Annual Conference on Computational Complexity*, 2002, pp. 93–102.

[26] N. Santhanam and M. Wainwright, "Information-theoretic limits of selecting binary graphical models in high dimensions," *IEEE Trans. Information Theory*, vol. 58, no. 7, pp. 4117–4134, 2012.

[27] K. Shanmugam, R. Tandon, A. Dimakis, and P. Ravikumar, "On the information theoretic limits of learning Ising models," in *Advances in Neural Information Processing Systems*, 2014, pp. 2303–2311.

[28] N. B. Shah and M. J. Wainwright, "Simple, robust and optimal ranking from pairwise comparisons," *J. Machine Learning Res.*, vol. 18, no. 199, pp. 1–38, 2018.

[29] A. Pananjady, C. Mao, V. Muthukumar, M. J. Wainwright, and T. A. Courtade, "Worst-case vs average-case design for estimation from fixed pairwise comparisons," http://arxiv.org/abs/1707.06217.

[30] Y. Yang, "Minimax nonparametric classification. i. rates of convergence," *IEEE Trans. Information Theory*, vol. 45, no. 7, pp. 2271–2284, 1999.

[31] M. Nokleby, M. Rodrigues, and R. Calderbank, "Discrimination on the Grassmann manifold: Fundamental limits of subspace classifiers," *IEEE Trans. Information Theory*, vol. 61, no. 4, pp. 2133–2147, 2015.

[32] A. Mazumdar and B. Saha, "Query complexity of clustering with side information," in *Advances in Neural Information Processing Systems*, 2017, pp. 4682–4693.

[33] E. Mossel, "Phase transitions in phylogeny," *Trans. Amer. Math. Soc.*, vol. 356, no. 6, pp. 2379–2404, 2004.

[34] T. M. Cover and J. A. Thomas, *Elements of information theory*. John Wiley & Sons, 2006.

[35] J. C. Duchi and M. J. Wainwright, "Distance-based and continuum Fano inequalities with applications to statistical estimation," http://arxiv.org/abs/1311.2669.

[36] I. Sason and S. Verdú, "f-divergence inequalities," *IEEE Trans. Information Theory*, vol. 62, no. 11, pp. 5973–6006, 2016.

[37] R. Dorfman, "The detection of defective members of large populations," *Annals Math. Statist.*, vol. 14, no. 4, pp. 436–440, 1943.

[38] J. Scarlett and V. Cevher, "Phase transitions in group testing," in *Proc. ACM-SIAM Symposium on Discrete Algorithms*, 2016, pp. 40–53.

[39] J. Scarlett and V. Cevher, "How little does non-exact recovery help in group testing?" in *Proc. IEEE International Conference on Acoustics, Speech and Signal Processing*, 2017, pp. 6090–6094.

[40] L. Baldassini, O. Johnson, and M. Aldridge, "The capacity of adaptive group testing," in *Proc. IEEE Int. Symp. Inform. Theory*, 2013, pp. 2676–2680.

[41] J. Scarlett and V. Cevher, "Converse bounds for noisy group testing with arbitrary measurement matrices," in *Proc. IEEE International Symposium on Information Theory*, 2016, pp. 2868–2872.

[42] J. Scarlett and V. Cevher, "On the difficulty of selecting Ising models with approximate recovery," *IEEE Trans. Signal Information Processing over Networks*, vol. 2, no. 4, pp. 625–638, 2016.

[43] J. Scarlett and V. Cevher, "Lower bounds on active learning for graphical model selection," in *Proc. 20th International Conference on Artificial Intelligence and Statistics*, 2017.

[44] V. Y. F. Tan, A. Anandkumar, and A. S. Willsky, "Learning high-dimensional Markov forest distributions: Analysis of error rates," *J. Machine Learning Res.*, vol. 12, no. 5, pp. 1617–1653, 2011.

[45] A. Anandkumar, V. Y. F. Tan, F. Huang, and A. S. Willsky, "High-dimensional structure estimation in Ising models: Local separation criterion," *Annals Statist.*, vol. 40, no. 3, pp. 1346–1375, 2012.

[46] G. Dasarathy, A. Singh, M.-F. Balcan, and J. H. Park, "Active learning algorithms for graphical model selection," in *Proc. 19th International Conference on Artificial Intelligence and Statistics*, 2016, pp. 1356–1364.

[47] B. Yu, "Assouad, Fano, and Le Cam," in *Festschrift for Lucien Le Cam*. Springer, 1997, pp. 423–435.

[48] J. Duchi, "Lecture notes for statistics 311/electrical engineering 377 (MIT)," http://stanford.edu/class/stats311/.

[49] Y. Wu, "Lecture notes for ECE598YW: Information-theoretic methods for high-dimensional statistics," www.stat.yale.edu/~yw562/ln.html.

[50] Y. Polyanskiy, V. Poor, and S. Verdú, "Channel coding rate in the finite blocklength regime," *IEEE Trans. Information Theory*, vol. 56, no. 5, pp. 2307–2359, 2010.

[51] O. Johnson, "Strong converses for group testing from finite blocklength results," *IEEE Trans. Information Theory*, vol. 63, no. 9, pp. 5923–5933, 2017.

[52] R. Venkataramanan and O. Johnson, "A strong converse bound for multiple hypothesis testing, with applications to high-dimensional estimation," *Electron. J. Statistics*, vol. 12, no. 1, pp. 1126–1149, 2018.

[53] P.-L. Loh, "On lower bounds for statistical learning theory," *Entropy*, vol. 19, no. 11, p. 617, 2017.

[54] T. L. Lai and H. Robbins, "Asymptotically efficient adaptive allocation rules," *Advances Appl. Math.*, vol. 6, no. 1, pp. 4–22, 1985.

[55] P. Auer, N. Cesa-Bianchi, Y. Freund, and R. E. Schapire, "Gambling in a rigged casino: The adversarial multi-armed bandit problem," in *Proc. 26th Annual IEEE Conference on Foundations of Computer Science*, 1995, pp. 322–331.

[56] E. Arias-Castro, E. J. Candès, and M. A. Davenport, "On the fundamental limits of adaptive sensing," *IEEE Trans. Information Theory*, vol. 59, no. 1, pp. 472–481, 2013.

[57] T. S. Han and S. Verdú, "Generalizing the Fano inequality," *IEEE Trans. Information Theory*, vol. 40, no. 4, pp. 1247–1251, 1994.

[58] L. Birgé, "A new lower bound for multiple hypothesis testing," *IEEE Trans. Information Theory*, vol. 51, no. 4, pp. 1611–1615, 2005.

[59] A. A. Gushchin, "On Fano's lemma and similar inequalities for the minimax risk," *Probability Theory and Math. Statistics*, vol. 2003, no. 67, pp. 26–37, 2004.

[60] A. Guntuboyina, "Lower bounds for the minimax risk using f-divergences, and applications," *IEEE Trans. Information Theory*, vol. 57, no. 4, pp. 2386–2399, 2011.

[61] Y. Polyanskiy and S. Verdú, "Arimoto channel coding converse and Rényi divergence," in *Proc. 48th Annual Allerton Conference on Communication, Control, and Compution*, 2010, pp. 1327–1333.

[62] G. Braun and S. Pokutta, "An information diffusion Fano inequality," http://arxiv.org/abs/1504.05492.

[63] X. Chen, A. Guntuboyina, and Y. Zhang, "On Bayes risk lower bounds," *J. Machine Learning Res.*, vol. 17, no. 219, pp. 1–58, 2016.

Index